Phage Display

A Laboratory Manual

Front cover: Structure of the p3 and p8 filamentous phage-display systems. The native fd phage is shown in red. Each p8 protein subunit (PDB entry 1IFJ) in the array is represented as a space-filling coil of about 1 nm diameter following the α-helical protein backbone. The array of p8 subunits in the figure extends from the pointed C-terminal end (thought to be the site of the minor proteins p3 and p6, about five copies each) for about 30 nm, or about 3% of the total phage length. Near the C-terminal end of the array, the D1/D2 region (PDB entry 1G3P) of a single p3 protein is shown, also in red. The complex between the p6 minor protein and the D3 domain of p3 is shown as an oval, roughly to scale (both of these proteins bind to the p8 array near the C-terminus, but their structure and the nature of their binding to p8 are unknown). The D3 domain is joined to D2 by a flexible glycine-rich linker. (See Marvin, *Curr. Opin. Struct. Biol.* **8:** 150 [1998] for further description of the structure.) A Fab fragment, heavy chain in yellow and light chain in blue (PDB entry 1IFH), replaces the D1/D2 region of a second p3. GPGRAF epitopes inserted near the N-terminus of random p8 proteins are shown in purple, with a bound Fab fragment in green (after Malik et al., *J. Mol. Biol.* **260:** 9 [1996]); This is shown in a larger scale on the gray phage body in the background. The figure was prepared using the molecular graphics packages BOBSCRIPT (Kraulis, *J. Appl. Cryst.* **24:** 946 [1991]; Esnouf, *Mol. Graph.* **15:** 132 [1997]) and Raster3D (Merritt and Bacon, *Methods. Enzymol.* **277:** 505 [1997]).

Back cover: Human antibody Fab fragments expressed on the surface of filamentous phage using the pComb3 vector system. The ball-like structures at the end of the phage are particles of hepatitis B surface antigen that are bound specifically by the antibody fragment displayed on the phage surface. (Reprinted, with permission, from Zebedee et al., 1992 *Proc. Natl. Acad. Sci.* **89:** 3175–3179.)

Phage Display

A Laboratory Manual

Carlos F. Barbas III
The Scripps Research Institute

Dennis R. Burton
The Scripps Research Institute

Jamie K. Scott
Simon Fraser University

Gregg J. Silverman
University of California, San Diego

COLD SPRING HARBOR LABORATORY PRESS
Cold Spring Harbor, New York

Phage Display: A Laboratory Manual

Printed in the United States of America

Developmental Editor	*Tracy Kuhlman*	**Desktop Editor**	*Danny deBruin*
Project Coordinator	*Mary Cozza*	**Interior Designer**	*Denise Weiss*
Production Editor	*Patricia Barker*	**Cover Designer**	*Tony Urgo*

Library of Congress Cataloging-in-Publication Data

Phage display: a laboratory manual / by Carlos F. Barbas III [et al.].
p.cm.
Includes bibliographical references and index.
ISBN 0-87969-546-3 (alk. paper)
1. Bacteriophages--Laboratory manuals. 2. Microbial biotechnology--Laboratory manuals. 3. Viral proteins--Laboratory manuals. 4. Affinity chromatography--Laboratory manuals. 5. Peptides--Biotechnology--Laboratory manuals. 6. Proteins--Biotechnology--Laboratory manuals. I. Barbas, Carlos F.

QR342.P454 2000
579.2′6′078—dc21 00-030834

10 9 8 7 6 5 4 3 2 1

Students and researchers using the procedures in this manual do so at their own risk. Cold Spring Harbor Laboratory makes no representations or warranties with respect to the material set forth in this manual and has no liability in connection with the use of these materials.

Procedures for the humane treatment of animals must be observed at all times. Check with the local animal facility for guidelines. Certain experimental procedures in this manual may be the subject of national or local legislation or agency restrictions. Users of this manual are responsible for obtaining the relevant permissions, certificates, or licenses in these cases. Neither the authors of this manual nor Cold Spring Harbor Laboratory assumes any responsibility for failure of a user to do so.

Several techniques in this manual may be or are covered by certain patent and proprietary rights. Users of this manual are responsible for obtaining any licenses necessary to practice these techniques or to commercialize the results of such use. COLD SPRING HARBOR LABORATORY MAKES NO REPRESENTATION THAT USE OF THE INFORMATION IN THIS MANUAL WILL NOT INFRINGE ANY PATENT OR OTHER PROPRIETARY RIGHT.

All Cold Spring Harbor Laboratory Press publications may be ordered directly from Cold Spring Harbor Laboratory Press, 10 Skyline Drive, Plainview, New York 11803-2500. Phone: 1-800-843-4388 in Continental U.S. and Canada. All other locations: (516) 349-1930. FAX: (516) 349-1946. E-mail: cshpress@cshl.org. For a complete catalog of all Cold Spring Harbor Laboratory Press publications, visit our World Wide Web Site http://www.cshl.org/

For orders from Europe, the Middle East, and Africa, British pound pricing is provided. Orders are fufilled and shipped from Cold Spring Harbor Laboratory Press - Europe c/o Lavis Marketing, 73 Lime Walk, Headington, Oxford OX3 7AD, U.K. Phone: +44(0) 1865 741541. FAX: +44 (0) 1865 750079. E-mail: cshlpress.europe@cshl.org. WWW Site: WWW:cshl.press.co.uk

Contents

Section 5 • Appendices

Preface

MOLECULAR RECOGNITION IS CENTRAL TO BIOLOGY, and the discovery and characterization of interacting partners are major endeavors of biological scientists. Phage display, largely developed in the 1990s, has begun to make critical contributions to these endeavors. The approach is based on two pivotal concepts. The first is that phage, viruses that infect bacteria, can be used to link protein recognition and DNA replication. The protein (or peptide) is displayed on the surface of the phage particle and the genes encoding it are contained within the particle. The second concept is that large libraries of the DNA sequences encoding these molecules can be cloned into phage. Individual phage can then be rescued from libraries by virtue of interaction of the displayed protein with the cognate ligand, and the phage can be amplified by infection of bacteria.

The broad strategy is one that was adopted long ago by nature in the immune system. There, vast immune repertoires or libraries of molecules (antibodies, T-cell receptors) permit recognition of virtually any foreign entity. Protein recognition and replication are then linked; for example, when specific antibody-producing cells are stimulated to divide by interaction of antigen and antibody cell-surface receptors for antigen. The result is a system for efficiently generating molecular species capable of specifically recognizing almost any molecular shape.

In 1985, George Smith first showed that the linkage between phenotype and genotype could be established in filamentous bacteriophage and gave birth to the new technology of phage display. Smith showed that foreign DNA fragments could be inserted into filamentous phage *gene III*, which codes for the phage coat protein pIII, to create a fusion protein with the foreign sequence in the amino-terminal domain. The fusion protein was incorporated into the virion, which retained infectivity and displayed the foreign peptide in a form accessible to specific antibody to the peptide. This "fusion phage" could be greatly enriched relative to ordinary phage by affinity selection on immobilized antibody (a process usually termed "panning"). Subsequently, in 1990, Scott and Smith, Dower and colleagues, and Devlin and colleagues independently cloned libraries of peptides and showed that peptides of specific activity could be retrieved from these libraries by panning. Concurrent with these developments, in 1989, Richard Lerner and colleagues reported that libraries of randomly recombined antibody heavy and light chains could provide an alternative route to monoclonal antibodies of defined specificity. These studies were performed by cloning into phage lambda and involved plaque screening rather than phage display. The expression of proteins such as antibodies as phage-displayed libraries followed shortly thereafter. These two types of libraries created an explosion of activity in the area.

Despite rapid growth in the field, there has been a relative dearth of publications dealing with practical aspects of phage display. The technology has the reputation of requiring some considerable technical expertise. The aim of this manual is to provide comprehensive instruction in theoretical and applied aspects of phage-display technologies, so that any scientist with even modest molecular biology experience can effectively employ them.

This manual is the direct descendant of materials prepared for the Cold Spring Harbor Laboratory course on "Phage Display of Combinatorial Antibody Libraries." Following a conversation between Jim Watson of Cold Spring Harbor Laboratory and Richard Lerner of The Scripps Research Institute, the first course was presented by two of us (C.F.B. and D.R.B.) in the fall of 1992. Thanks to the outstanding support of the CSHL staff, the course was a success and has been modified every year since then to take account of the experience and comments of the students, and to reflect developments in the field. Much of this manual is thus the result of nearly a decade of experience with students of greatly varying technical expertise and experience from all over the world. All of these students made the writing of this manual possible. In addition to antibody libraries, the content of this manual has been expanded to include other types of libraries displayed on phage. We have included our most up-to-date laboratory protocols, and the accompanying didactic material provides all of the essential information and references needed by both the novice and the experienced practitioner of expression-cloning techniques to design experiments of their own.

This manual is divided into five sections. The first gives an overview of some of the key aspects of phage display. This manual was not intended to reproduce all the information contained in the many excellent reviews that are currently available. Rather, we felt it appropriate to present the more important concepts in phage display in one text as a background for understanding the practical approaches described. Thus, this section reviews phage structure, genetics, and physiology and, within this context, presents the phage vectors. The crucial features of antibody, peptide, protein fragment, and cDNA libraries are also summarized, and the section ends with an overview of emerging technologies.

The second section deals with the construction, screening, and analysis of antibody libraries on the surface of phage. These libraries are expressed as fusions with the pIII coat protein of filamentous phage. This section includes chapters on the production and purification of recombinant antibody fragments, as well as antibody engineering and the construction of specialty libraries.

The third section deals with the construction, screening, and analysis of phage-displayed peptide libraries. The focus is principally on the use of peptide expressed as fusions with the pVIII phage coat protein. A chapter is also included that considers peptide libraries constructed by fusion with the pIII phage coat protein.

The fourth section covers the construction, screening, and analysis of gene-fragment and cDNA-expression libraries. It also deals with affinity selection of libraries with more complex targets (namely, cells) as well as in vivo selection techniques. The final section includes a number of appendices that summarize commonly used experimental procedures, data, recipes, suppliers, and important precautions.

We gratefully acknowledge the excellent contributions of our friends and collaborators who took the time to share their practical experience. They have made important contributions to the CSHL Antibody Library course and have provided outstanding chapters to this manual.

We acknowledge Dr. George Smith, from whose work all phage-display methods derive. He demonstrated the concept of phage display using expressed cDNA fragments on phage, developed affinity-selection methods for phage, and invented the concept of peptide libraries on filamentous phage. With outstanding collegiate spirit, he has freely shared all of his inventions, vectors, and libraries with the scientific community.

We are indebted to Dr. Richard Lerner for his contributions to and support of this work, and especially for his longstanding vision of antibody libraries. He fostered the group at The Scripps Research Institute that designed and constructed the first Fab libraries from immune antibody responses. That work, along with the pComb3 phage-display vector that emerged from it, has made possible the CSHL Antibody Display course and this manual. We acknowledge the help and support of John Inglis, Mary Cozza, Tracy Kuhlman, Danny deBruin, and Pat Barker, who have nurtured this project to completion.

This work is dedicated to our teachers, colleagues, collaborators, and students for their many contributions over the years; and to our families for their patience, encouragement, and support.

Carlos F. Barbas III
Dennis R. Burton
Jamie K. Scott
Gregg J. Silverman

Abbreviations

(+)	viral strand of DNA
(–)	complementary strand of DNA
ABTS	2,2′-azino-di-[3-ethylbenzthiazoline sulfonate (6)]
ANP	atrial naturetic peptide
APS	ammonium persulfate
AU_{260}	absorbance unit at 260 nm
B	biotinylated
bp	base pair
BPTI	bovine pancreatic trypsin inhibitor
BSA	bovine serum albumin
C	complement
carb	carbenicillin
ccc	covalently closed circular
CD	circular dichroism
cDNA	complementary DNA
CDR	complementarity determining region
CFA	complete Freund's adjuvant
cfu	colony-forming unit
C_H	constant region, heavy chain
C_κ	constant region, kappa light chain
C_L	constant region, light chain
CNTF	ciliary neurotrophic factor
CT	carboxy-terminal (domain of pIII)
CWS	cell wall skeleton
D	diversity (gene segment)
D1, D2, D3	domain 1, 2, 3
DIRE	direct interaction rescue
DMEM	Dulbecco's modified Eagle's medium
DMP	dimethyl pimelimidate
DNase	deoxyribonuclease
dNTP	deoxynucleotide triphosphate
DO	dissolved oxygen

dsDNA	double-stranded DNA
DTT	dithiothreitol
ELISA	enzyme-linked immunosorbent assay
ERD	ERF repressor domain
Fab	fragment antigen binding
Fc	fragment crystalline
FCS	fetal calf serum
FITC	fluorescein isothiocyanate
FR	framework region
gIII	gene III
gVIII	gene VIII
GAP	glyceraldehyde-3-phosphate dehydrogenase promoter
GH	growth hormone
GST	glutathione *S*-transferase
H	heavy (chain)
HA	hemagglutinin
HABA	4′-hydroxyazobenzene-2-carboxylic acid
HCDR3	heavy chain CDR3
hGH	human growth hormone
HNE	human neutrophil elastase
HPLC	high performance liquid chromatography
HRP	horseradish peroxidase
IFA	incomplete Freund's adjuvant
Ig	immunoglobulin
IgG	immunoglobulin G
IL	interleukin
IMAC	immobilized metal affinity chromatography
IPTG	isopropyl-β-D-thiogalactopyranoside
J	joining (gene segment)
J_H	heavy-chain joining
J_L	light-chain joining
Kb	kilobase
Kbp	kilobase pair
K_d	dissociation constant
KRAB	Krüppel-associated box
L	light (chain)
LB	Luria broth
LCDR3	light chain CDR3
LES	lipid emulsion system
LL	long linker
mA	milliampere
mAb	monoclonal antibody
MBP	maltose-binding protein
moi	multiplicity of infection
MPL	monophosphoryl lipid A

mRNA	messenger RNA
MW	molecular weight
MWM	molecular weight marker
NEM	N-ethylmaleimide
NMR	nuclear magnetic resonance
NPR	naturetic peptide receptor
NTA	nitrilotriacetic acid
oc	open-circular
OD	optical density
OD_{260}	optical density at 260 nm
OMP	outer membrane protein
OPD	*o*-phenylenediamine
ORF	open reading frame
pIII	protein III
PA	plasminogen activator
PAGE	polyacrylamide gel electrophoresis
PBL	peripheral blood lymphocyte
PBS	phosphate-buffered saline
PCR	polymerase chain reaction
PEG	polyethylene glycol
pfu	plaque-forming unit
PI	protease inhibitors
PMSF	phenylmethylsulfonyl fluoride
PPI	peptidylprolyl isomerase
PrP	prion protein
PS	packaging signal
PSM	prostate-specific membrane
PSTI	pancreatic secretory trypsin inhibitor
RBC	red blood cell
RF	replicative form
Rh	Rhesus
RNase	ribonuclease
Sa	streptavidin
SAP	selective amplification of phages
SAS	saturated ammonium sulfate
SB	super broth
scFv	single-chain Fv
SD	Shine-Dalgarno
SDS-PAGE	sodium dodecyl sulfate polyacrylamide gel electrophoresis
SID	mSIN3 interaction domain
SIP	selectively infective phages
SL	short linker
SLE	systemic lupus erythematosus
SM	screening molecule
SpA	staphylococcal protein A

SRP	signal recognition particle
ss	single-stranded
STM	*Salmonella typhimurium* mitogen
$t_{1/2}$	half-life
TDM	trehalose dicorynomycolate
TE	Tris/EDTA buffer
TEMED	N,N,N´,N´-tetramethylenediamine
tet	tetracycline
TPO	thrombopoietin
TU	transducing unit
V	variable gene segment
V_H	variable domain, heavy chain
V_L	variable domain, light chain
(v/v)	volume/volume
(w/v)	weight/volume
X	variable or randomized residue/amino acid

1 Filamentous Phage Biology

ROBERT WEBSTER

Duke University Medical Center, Department of Biochemistry, Durham, North Carolina 27710

THE FILAMENTOUS BACTERIOPHAGES (GENUS *INOVIRUS*) are a group of viruses that contain a circular single-stranded DNA genome encased in a long protein capsid cylinder. Many use some type of bacterial pilus to facilitate the infection process. The Ff class of the filamentous phages (f1, fd, and M13) have been the most extensively studied. As the name implies, these bacteriophage use the tip of the F conjugative pilus as a receptor and thus are specific for *Escherichia coli* containing the F plasmid. The DNA sequence of these three phages shows them to be 98% homologous; consequently, the protein sequences of the gene products are practically the same.

CONTENTS

The Ff phages do not kill their host during productive infection. The single-stranded viral DNA is replicated via a double-stranded intermediate by a mixture of bacterial and phage-encoded components. The result of this replicative process is a newly synthesized viral single-stranded DNA in a complex with many copies of a phage-encoded single-stranded DNA-binding protein. The capsid proteins are all synthesized as integral membrane proteins that remain in the membrane until they are assembled around the DNA. Assembly occurs at specific sites in the bacterial envelope where the cytoplasmic and outer membranes are in close contact. During the assembly process, the viral DNA is extruded through the membrane-associated assembly site, where the phage DNA-binding proteins are removed and the capsid proteins are packaged around the DNA. This process continues until the end of the DNA is reached, so there is little if any constraint on the size of the DNA packaged. The bacteria tolerate this process quite well and continue to grow and divide with a generation time approximately 50% longer than that of uninfected bacteria. There is a burst of about 1000 phage particles produced in the first generation after infection, and then the bacteria produce about 100–200 particles per generation. This continues for many generations, resulting in titers of 10^{11} to 10^{12} particles per ml. The plaques are turbid and of varying size and contain about 10^{8} infective phage particles.

The phage structure and its mode of replication have made it a valuable tool for biological research. Phage can be used as cloning vehicles, because insertion of DNA into a nonessential region of the phage genome results in a longer phage that contains a single-stranded copy of the inserted DNA. The ability to isolate the single-stranded viral DNA and its double-stranded replication intermediate makes it possible to easily create substrates for studying recombination and repair of mismatches in DNA. The membrane-associated assembly process has made it possible to display foreign peptides or proteins on the surface of the phage particle, as described in this manual. To aid in understanding the techniques involved in "phage display," this chapter describes aspects of the biology of the phage and bacteria. In the first section, the phage particle and its life cycle are described. The next section relates the phage life cycle to some of the basic principles involved in displaying proteins on the phage surface. Because the replication of phage is governed to a great extent by the physiology of bacteria, the last section briefly discusses some aspects of bacterial biology that can have a direct relation to the phage-display technique.

This chapter is intended to give the reader only an overview of the biology of the Ff bacteriophage. Therefore, it is brief and does not fully discuss all aspects of the subject or the many papers that have contributed to the study of this organism. In some

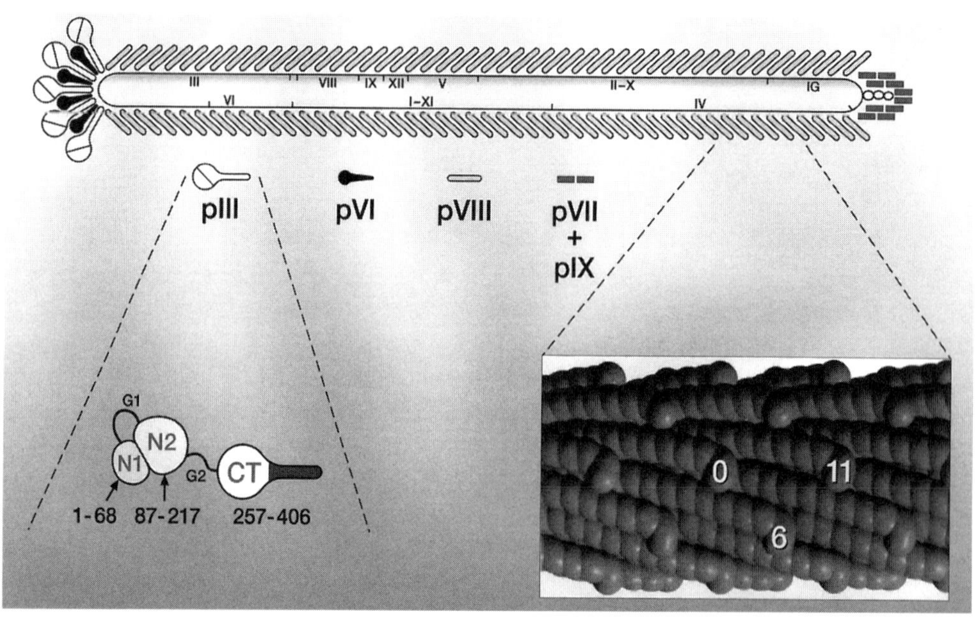

Figure 1.1. The Ff bacteriophage particle. Schematic representation of the phage particle showing the location of the capsid proteins and the orientation of the DNA. The lower left is a schematic of the structure of pIII. N1, N2, and CT refer to domains, and G1 and G2 refer to glycine-rich regions. The lower right is a representation of the orientation of the pVIII molecules along the cylinder part of the phage. (Adapted, with permission, from Marvin 1998 © Elsevier Science.) The three nearest neighbors indexed as 0, 6, and 11 are indicated. Because the amino-terminal regions face to the right, this depiction of the phage would have the cone end at the pVII–pIX end of the particle.

cases, conclusions may be stated that probably are correct but are not absolutely proven by the present experimental data. For readers wanting to explore a particular area more deeply, each section mentions a number of recent reviews or papers with good introductions related to the various topics discussed.

THE Ff BACTERIOPHAGE

Structure of the Bacteriophage

The Ff phage particle is approximately 6.5 nm in diameter and 930 nm in length (Fig. 1.1). The mass of the particle is approximately 16.3 MD, of which 87% is contributed by protein. The genome is a single-stranded, covalently closed DNA molecule of about 6400 nucleotides that is encased in a somewhat flexible protein cylinder. The length of the cylinder consists of approximately 2700 molecules of the 50-amino-acid major coat protein, also called *gene VIII* protein (pVIII). At one end of the particle, there are about 5 molecules each of the 33-residue *gene VII* protein (pVII) and the 32-residue *gene IX* protein (pIX). The other end contains approximately 5 molecules each of the 406-residue *gene III* and 112-residue *gene VI* proteins (pIII and pVI). The DNA is oriented within the virion such that a 78-nucleotide hairpin region called the packaging signal (PS) is always located at the end of the particle containing the pVII and pIX proteins.

There now exists a fairly complete description of the pVIII cylinder portion of the virion (Marvin et al. 1994; Overman and Thomas 1995; Williams et al. 1995; Marvin 1998). The pVIII monomers are present in the particle as an uninterrupted α-helix except for the amino-terminal 5 residues. The proteins are arranged in an overlapping shingle-type array with a symmetry defined by a fivefold rotational axis with a twofold screw axis of pitch 3.2 nm (Fig. 1.1, lower right). The axis of the helical pVIII monomer is tilted approximately 20° to the long axis of the particle, gently wrapping around the long axis of the virus in a right-handed way. The pVIII molecules are packed quite tightly, as only the outside 3 residues are accessible to digestion by proteases (Terry et al. 1997). The carboxy-terminal 10–13 residues of pVIII form the inside wall of the cylinder. This region contains 4 positively charged lysine residues that reside on one face of an amphiphilic helix. These positive charges interact with the sugar phosphate backbone of the DNA that is present in the particle with the bases pointed inward (Greenwood et al. 1991a; Marvin et al. 1994). The amino-terminal portion of pVIII is present on the outside of the particle. The residues connecting the amino and carboxyl regions of pVIII interact with the same region of other pVIII molecules to form the stable inner core of the protein cylinder. Most of this middle portion of pVIII spans the cytoplasmic membrane before being assembled into phage particles.

One end of the particle has approximately 5 molecules each of the small hydrophobic pVII and pIX proteins. This end contains the PS and is the first part of the phage to be assembled. It is not known how these two proteins are arranged at the end of the phage or how they interact with the pVIII cylinder. Attempts to model the ends of the particle suggest that one of these proteins must be buried close to the DNA, whereas the other is exposed at the surface (Makowski 1992). The observation that antibodies to pIX but not pVII are able to interact with one end of the phage particle suggests that only pIX is exposed and pVII is buried (Endemann and Model 1995). Immunoprecipitation experiments on detergent-disrupted phage suggest an interaction between pVII and the pVIII major capsid protein in the intact particle.

The other end of the particle contains approximately 5 molecules each of pIII and pVI, accounting for about 10–16 nm of the phage length (Specthrie et al. 1992). pIII is made up of three domains separated by glycine-rich regions (Stengle et al. 1990). These domains have been designated N1 or D1, N2 or D2, and CT or D3 by different groups; the N1 designation will be used here (Fig. 1.1, lower left). The first domain, N1, contains the amino-terminal 68 amino acids and is required during infection for the translocation of the DNA into the cytoplasm and the insertion of the coat proteins into the membrane. The second domain, N2, is made up of residues 87–217 and is responsible for binding to the F pilus (Deng et al. 1999). Both domains contain cysteine molecules that are involved in intramolecular disulfide bonds within each domain (Kremser and Rasched 1994). The structure of these two domains (residues 1–217) has been determined by NMR spectroscopy and X-ray crystallography (Riechmann and Holliger 1997; Lubkowski et al. 1998; Holliger et al. 1999). Domains N1 and N2 specifically interact with each other to form a horseshoe shape. Under certain conditions, this portion of pIII can be visualized by electron microscopy as a knob-like structure at one end of the phage (Gray et al. 1981). N1 and N2 are exposed on the surface of the phage particles; removal of these domains by protease treatment produces noninfectious phage (Gray et al. 1981; Armstrong et al. 1981).

The carboxy-terminal 150 residues make up the third domain of pIII (CT) and are essential for forming a stable phage particle (Crissman and Smith 1984; Kremser and Rasched 1994). It has been proposed that part or all of the CT domain, together with pVI, interacts with pVIII to form one end of the particle (Rakonjac et al. 1999). There is also evidence that pVI and pIII interact with each other in the phage particle, as they remain associated after disruption of the phage particle with certain detergents (Gailus and Rasched 1994; Endemann and Model 1995). Nothing is known about the structure of pVI in the virion, but it has been proposed that the hydrophobic amino terminus of pVI is buried within the particle (Makowski 1992). The carboxyl terminus of pVI may be near the surface of the phage particle, as virions can be produced containing foreign proteins fused to this end of pVI (Jespers et al. 1995). However, pVI is not accessible to anti-pVI antiserum in the phage (Endemann and Model 1995).

At which end of the virion the minor proteins are located, relative to the configuration of pVIII in the protein tube, is still unknown (Marvin 1998). Because the pVIII monomers overlap each other in the particle with the amino terminus on the outside, the end of the tube containing the amino-terminal end of pVIII has a cone-shaped structure (Fig. 1.1). The other end, containing the carboxyl terminus, has a pointed shape. Based on the orientation of pVIII in the membrane, it is generally assumed that the amino-terminal end of pVIII leaves the membrane first during assembly (see p. 1.13, The assembly process). This would place the pVII and pIX proteins at the cone end of the particle, shown in Figure 1.1, although such an orientation of pVII and pIX relative to pVIII has not been experimentally demonstrated.

For more information about bacteriophage structure, see Marvin et al. (1994) and Marvin (1998).

Table 1.1. Genes and gene products of the f1 bacteriophage

Gene	Function	No. of amino acids	Protein MW
II	DNA replication	410	46,137
X	DNA replication	111	12,672
V	binding ssDNA	87	9,682
VIII	major capsid protein	50	5,235
III	minor capsid protein	406	42,522
VI	minor capsid protein	112	12,342
VII	minor capsid protein	33	3,599
IX	minor capsid protein	32	3,650
I	assembly	348	39,502
IV	assembly	405	43,476
XI	assembly	108	12,424

The number of amino acids and the molecular weight are for the mature proteins. The initiating methionine is included in proteins that do not contain an amino-terminal signal sequence.

The Phage Genome

The genomes of the Ff phage (M13, f1, and fd) have been completely sequenced (Van Wezenbeek et al. 1980; Beck and Zink 1981; Hill and Petersen 1982). Each genome encodes 11 genes, whose products are listed in Table 1.1. Two of the gene products, pX

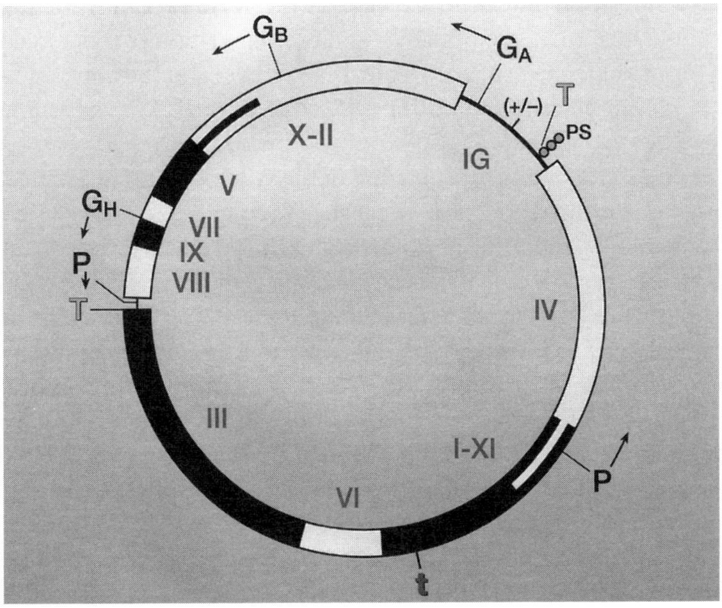

Figure 1.2. Genome of the Ff bacteriophage. This diagram shows the relative positions of the genes and the important terminators and promoters. IG refers to the intergenic region; T, the two strong terminators; t, the weak terminator in pI; G_A, G_B, and G_H, the promoters of the frequently transcribed region; P, the promoters for the infrequently transcribed regions; PS, the packaging signal; and (+/–), the relative position of the origins of replication for the viral (+) and complementary (–) DNA strands. (Adapted, with permission, from Model and Russel 1988.)

and pXI, are the result of a translational start at an internal methionine codon in genes II and I, respectively (Model and Russel 1988; Guy-Caffey et al. 1992; Rapoza and Webster 1995). These internal methionine codons are in-frame, so the smaller proteins have the same sequence as the carboxy-terminal portions of their larger counterparts. The genes are grouped in the genome according to their function in the life cycle of the phage (Fig. 1.2). One group (*genes II, V*, and *X*) encodes the proteins required for the replication of the phage genome. Another encodes the capsid proteins (pVII, pIX, pVIII, pIII, and pVI), and the third group encodes three proteins (pI, pXI, and pIV) that are involved in the membrane-associated assembly of the bacteriophage. There also is a short sequence called the intergenic region that does not code for protein. It contains the sites of origin for the synthesis of viral (+) and complementary (–) DNA. The PS is also in the intergenic region near the end of *gene IV*.

Transcription uses the (–) strand of the double-stranded intermediate in viral DNA replication. Therefore, the messages produced have the same sequence as the (+) strand. The direction of transcription is counterclockwise in Figure 1.2, going from *gene II* through *gene IV*. There are two strong terminators of transcription (T), a rho-independent one just after *gene VIII* and a rho-dependent one in the intergenic region. These terminators divide the genome into two transcription regions: a frequently transcribed region containing *genes II* through *VIII* and an infrequently transcribed region containing *genes III* through *IV*.

Transcription in the frequently transcribed region initiates from three strong promotors, G_A, G_B, and G_H, and leads to three transcripts terminated after *gene VIII* (Fig. 1.2). The G_H transcript is not processed and is the smallest and most stable of these RNA molecules. The 5′ portions of the G_A and G_B transcripts are processed rapidly into five smaller, more stable messenger RNA (mRNA) molecules (Stump and Steege 1996). Because of the common terminator, all six mRNAs encode pVIII. The four largest of the processed mRNAs also encode pV. Both of these proteins are needed and produced in large amounts; pVIII forms the tube portion of the virion and pV interacts with the single-stranded DNA during replication (see p. 1.7, The phage life cycle). In contrast, only small amounts of pVII and pIX are produced, even though their genes are located on the mRNA between *genes V* and *VIII*. The translation of *genes VII* and *IX* is coupled to the translation of *gene V* (Ivey-Hoyle and Steege 1989; 1992), because the initiation site for translation of *gene VII* is inherently inactive, and the initiation site for *gene IX* is masked in the secondary structure of the RNA. The ribosomes are only able to continue through the pV–pVII junction with low frequency, resulting in the low amounts of pVII and pIX that are needed for phage production. There is also translational regulation for the synthesis of pII and pX (Model and Russel 1988). The pV single-stranded binding protein can specifically bind to the *gene II* and *X* mRNA and thus inhibit translation. Binding only occurs at high concentrations of pV, when there is more pV than needed to bind the newly synthesized viral DNA.

The major promoters (P) for the infrequently transcribed region are located just before genes III and IV. There is a weak rho-dependent terminator (t) in the beginning of *gene I*, such that transcription from the promoter upstream of *gene III* gives two classes of mRNA. The most abundant class encodes only *genes III* and *VI*, whereas there is less mRNA encoding *genes III, VI*, and *I/XI*. The down-regulation of pI is important, because the bacteria can tolerate only small amounts of this protein because of pI's ability to form channels. Substantial amounts of pIV mRNA are produced from the promoter at the end of pI/pXI.

There are very few regions in the phage genome that do not code for protein. Cassettes encoding antibiotic resistance are generally inserted in the intergenic region or in the space between the end of *gene VIII* and the beginning of *gene III*. In the latter case, some alterations of the positions of the terminator and promoter in this region must be made, and care must be taken not to interfere with the origins of replication or other control areas. Similarly, one should not disturb normal terminators or promoters. There appears to be a delicate balance in the synthesis of the phage proteins that allows phage production without seriously affecting bacterial cell growth.

For more information about the phage genome, see Model and Russel (1988) and Webster (1996).

THE PHAGE LIFE CYCLE

The Infection Process

Infection is a multistep process requiring interactions with the F conjugative pilus and the bacterial TolQ, R, and A cytoplasmic membrane proteins. The F pilus is a protein tube assembled from pilin subunits in the cytoplasmic membrane and extends from the membrane out into the medium. Proteins required for its structure, assembly, and disassembly are encoded by genes in the *tra* operon on the F conjugative plasmid (Frost et al. 1994). The F pilus is required for the conjugal transfer of the F plasmid DNA from a donor cell into a recipient bacterium lacking the plasmid (Firth et al. 1996). This transfer of DNA, called conjugation, is initiated by interaction of the tip of the F pilus with the envelope of the recipient bacterium. The pilus is thought to retract, drawing the donor and recipient cells together to facilitate the processes required for replication and transfer of a single strand of the F plasmid DNA. The result of this process is the presence of the plasmid in both the donor and recipient bacteria.

The TolQ, R, and A proteins are bacterial proteins that appear necessary for maintaining the integrity of the bacterial outer membrane (Lazzaroni et al. 1999). Bacteria containing mutations in *tolQ*, *R*, or *A* become hypersensitive to various detergents and drugs, release periplasmic proteins into the medium, and form outer membrane vesicles (Webster 1991; Bernadac et al. 1998). These three Tol proteins are required during phage infection for translocation of the filamentous phage DNA into the cytoplasm and translocation of the phage coat proteins into the cytoplasmic membrane (Russel et al. 1988; Click and Webster 1998). The TolQRA proteins also are involved in translocation of the group A colicins, called bacteriocins, to their respective sites of action (Lazdunski et al. 1998). Each of these colicins is a plasmid-encoded protein toxin that can enter and kill bacteria lacking the plasmid encoding that colicin. In *tol* mutant bacteria, the filamentous phage can interact with the F pilus but not infect the bacteria, and the colicins can bind to their respective outer membrane receptors but not kill the bacteria. Therefore, these mutant bacteria have been designated as "tolerant" (*tol*) to filamentous phage and the group A colicins.

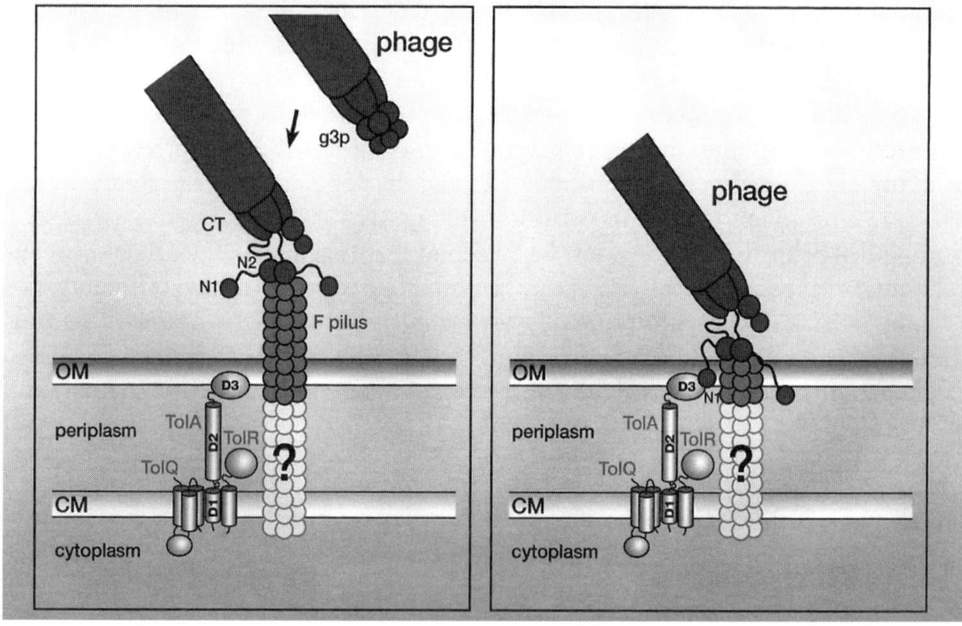

Figure 1.3. Infection process of the Ff bacteriophage. The membrane topology of the Tol proteins is shown. D1, D2, and D3 refer to the three domains of TolA. The question mark on the pilus means that it is unknown how far it may retract into the envelope. (Reprinted, with permission, from Lubkowski et al. 1999 © Elsevier Science.)

All three of these Tol proteins are located in the cytoplasmic membrane (Fig. 1.3) and appear to form a complex via interactions among their transmembrane regions (Lazzaroni et al. 1995; Germon et al. 1998). TolQ spans the membrane three times, with most of its residues located in the cytoplasm (Kampfenkel and Braun 1993; Vianney et al. 1994). TolA and TolR are each anchored in the membrane by one transmembrane region located near the amino-terminal portion of each protein, with the bulk of their residues exposed in the periplasm (Levengood et al. 1991; Muller et al. 1993). TolA has a three-domain structure, with each domain separated by a glycine-rich region (Levengood et al. 1991). Domain 1 (D1) comprises the amino-terminal 43 residues, which anchor TolA to the cytoplasmic membrane. Domain 3 (D3) is a carboxy-terminal domain of approximately 108 residues that appears to interact with the outer membrane (Levengood-Freyermuth et al. 1993). D1 and D3 are connected by the central domain (D2), which appears to contain an α-helical structure long enough to span the periplasm.

Infection is initiated by the binding of the tip of the F pilus to the N2 domain of the phage pIII protein (Fig. 1.3) (Stengle et al. 1990; Deng et al. 1999). The exact binding site on N2 is not known. It probably is not the same binding pocket as that recognized by N1, because purified pIII fragments pIII-N2 and pIII-N1–N2 appear to compete equally well with wild-type phage for binding the F pilus (Krebber et al. 1997; Deng et al. 1999). After the phage binds to the pilus, the pilus retracts, bringing the pIII end of the phage particle to the periplasm. It is not known whether this retraction is the result of normal assembly–retraction cycles inherent to the pilus or whether phage attachment triggers the retraction process (Firth et al. 1996).

Binding of N2 to the pilus releases N1 from N2, and allows N1 to interact with domain 3 of TolA (TolA-D3) (Riechmann and Holliger 1997). The tip of the pilus is the receptor for phage attachment, and TolA-D3 is referred to as the coreceptor. Structural analysis shows that both pIII-N2 and TolA-D3 bind the same region of the pIII-N1, even though there is no topological similarity between pIII-N2 and TolA-D3 (Lubkowski et al. 1999). Therefore, the binding of the pilus tip to the N2 domain of pIII must displace the N1 domain, making it available for interacting with TolA-D3. This hypothesis is consistent with the observation that purified TolA-D3 does not inhibit infection when added to free phage but will inhibit infection if soluble TolA-D3 is present in the periplasm, where it can interact with the pilus–phage complex (Click and Webster 1997). The primary role of the pilus appears to be to bring the phage to a position where pIII-N1 can interact with TolA-D3. This can be bypassed by treating the bacteria with 50 mM Ca^{++}. Infection of F^- bacteria with wild-type phage or phage missing pIII-N2 can be enhanced two to four orders of magnitude after Ca^{++} treatment (Russel et al. 1988; Krebber et al. 1997). The rate of infection of these treated F^- bacteria with wild-type phage, however, is still four to five orders of magnitude less than the rate of infection of pilus-bearing F^+ bacteria. Presumably, the Ca^{++} alters the outer membrane enough to allow pIII-N1 to find TolA-D3.

The subsequent steps involved in phage infection are unclear. The major capsid protein, pVIII, and probably the pVII and pIX minor capsid proteins, disassemble into the cytoplasmic membrane as the phage DNA is translocated into the cytoplasm (Webster and Lopez 1985; Model and Russel 1988). Once in the membrane, the pVIII protein of the infecting phage joins the pool of newly synthesized pVIII, and both are assembled into newly formed particles. It is possible that the disassembly of pVIII into the membrane may be part of the driving force for translocating the phage DNA into the cytoplasm. Even if this is the case, TolQ, TolR, and TolA are absolutely required for entrance of the DNA into the cytoplasm and of the coat protein into the membrane (Sun and Webster 1987; Russel et al. 1988; Click and Webster 1997). The role that TolQ and TolR play in the process is unknown. Because the transmembrane regions of these proteins interact with each other (Lazzaroni et al. 1995; Germon et al. 1998), it may be that TolQ and TolR are involved in formation of some channel or protein complex required for the DNA to traverse the membrane. It even has been suggested that the CT domain of pIII may also be involved in forming an entrance pore (Glaser-Wuttke et al. 1989). Further research is required to better understand the role that the Tol proteins play in the mechanism of infection.

For more information about the infection process, see Webster (1996) and Lubkowski et al. (1999).

Replication and Protein Synthesis

Once the viral (+) strand DNA enters the cytoplasm, the complementary (–) strand is synthesized by bacterial enzymes. The resulting double-stranded DNA is acted on by

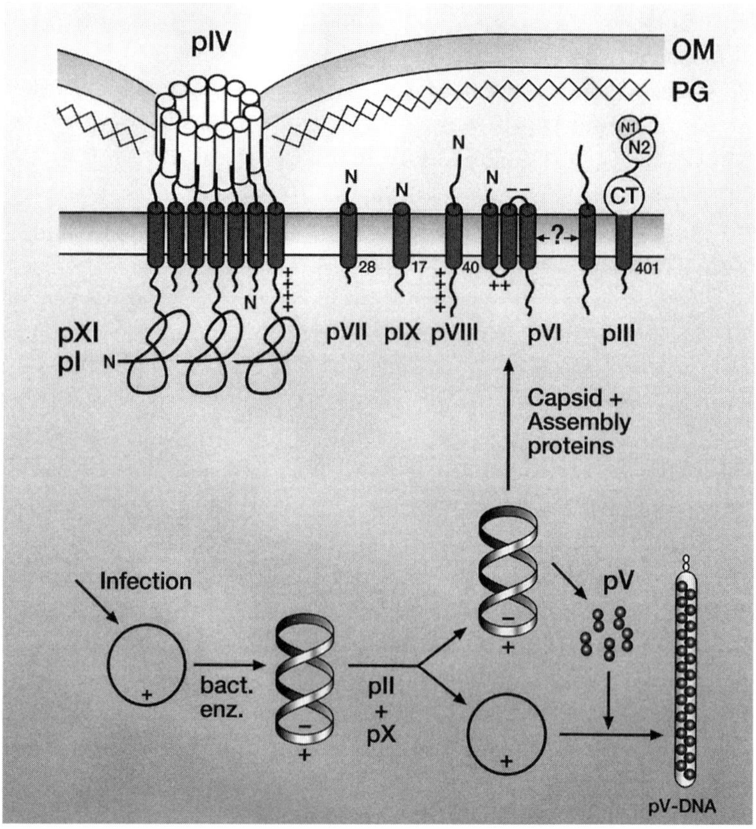

Figure 1.4. DNA replication, protein synthesis, and protein location in the Ff filamentous bacteriophage. The bottom of the figure shows the replication of the DNA. (+) refers to the viral strand and (–) the complementary stand. The top shows the position of the proteins in the bacterial periplasm, cytoplasmic membrane, and outer membrane (OM). PG refers to the peptidoglycan, or murein, layer. The + symbols on the proteins represent the positive side chains of the amphiphilic helices of pI and pVIII adjacent to the cytoplasmic membrane. The numbers on the cytoplasmic side of the membrane refer to the residue at the end of the membrane-spanning region for that protein.

gyrase, a type II topoisomerase that catalyzes the formation of negative supercoils in double-stranded DNA. The final product is a covalently closed, supercoiled, double-stranded DNA called the parental replicative form (RF) DNA (Fig. 1.4). The (–) strand of this RF is the template for transcription, and the resulting mRNAs are translated into all of the phage proteins. One of these phage proteins, pII, nicks the (+) strand in the RF at a specific place in the intergenic region. The resulting 3′-hydroxyl acts as a primer for synthesis of a new viral strand via a "rolling-circle" mode of replication that uses bacterial enzymes. After one round, pII circularizes the displaced viral (+) strand DNA, which then is converted to a covalently closed, supercoiled, double-stranded progeny RF molecule by bacterial enzymes. In this way, a pool of progeny double-stranded RF molecules are produced that can direct the synthesis of the phage proteins.

RF DNA synthesis continues until the amount of pV reaches a critical concentration. Dimers of pV are able to cooperatively bind to newly synthesized viral single-stranded DNA and prevent its conversion to RF DNA. Therefore, the presence of adequate amounts of pV essentially switches most of the DNA replication to the synthesis of single-stranded (+) viral DNA. The result of this switch is the formation of pV–DNA structures about 800 nm long and 8 nm in diameter, containing approximately 800 pV dimers and one single-stranded viral DNA molecule (Gray 1989; Skinner et al. 1994; Guan et al. 1995; Konings et al. 1995; Olah et al. 1995). The overall structure has a left-handed helical appearance. Each monomer of the pV dimer binds a strand of the single-stranded DNA genome, such that the two bound portions of the circular DNA run in opposite directions along the length of the pV–DNA structure. The DNA is wrapped inside the protein but is not completely enclosed by it. The 78-nucleotide hairpin packaging signal (PS) is at one end of the pV–DNA particle and is probably not bound to any pV protein (Bauer and Smith 1988). The pV–DNA complex is the substrate for the phage assembly reaction, which initiates at the end containing the PS.

pX also is required for proper replication of phage DNA. It is created by an in-frame translational start within the pII messenger RNA and thus has the same sequence as the carboxy-terminal third of pII. Its exact role in replication is unclear, but it appears to function as an inhibitor of pII function, and thus helps to regulate the amount of viral DNA produced (Fulford and Model 1984).

The only phage proteins residing in the cytoplasm are pII, pX, and pV, the three proteins engaged in the replication of DNA. All of the other phage proteins are synthesized and inserted into the cytoplasmic or outer membranes (Fig. 1.4). The most studied phage protein has been the major capsid protein, pVIII. It is synthesized with a 23-amino acid amino-terminal signal peptide, which directs its spontaneous insertion into the cytoplasmic membrane (see p. 1.21, Protein translocation across the cytoplasmic membrane). After removal of the signal sequence, the mature 50-residue pVIII spans the membrane one time with its amino-terminal 20 amino acids in the periplasm and its carboxy-terminal 10 amino acids exposed in the cytoplasm. Evidence suggests that this membrane-associated pVIII is composed of two α-helical regions, an amino-terminal amphiphilic helix and a carboxy-terminal helix. It is proposed that the amino-terminal amphiphilic helix lies parallel to the periplasmic surface of the membrane, whereas the carboxy-terminal 30 residues are oriented perpendicular to the surface of the membrane (McDonnell et al. 1993; Williams et al. 1996; Papavoine et al. 1997). The carboxy-terminal 10 residues of the pVIII helix also have an amphiphilic nature because all four lysine residues are aligned on one side of the helix. The pVIII protein is able to form dimers in the membrane by packing along the hydrophobic face of this amphiphilic helix and extending through the membrane-spanning region presumably perpendicular to the plane of the membrane (Haigh and Webster 1998). It may well be the interaction of the positively charged lysines at the carboxy-terminal end of pVIII with the acidic phospholipids in the membrane that is the driving force behind dimer formation (Woolford et al. 1974). However, not all of the pVIII proteins are in dimeric form (Haigh and Webster 1998). Another hypothesis has been proposed recently, which suggests that pVIII traverses the membrane in a tilted orientation with the apolar faces of both amphiphilic ends buried in the membrane (Marvin 1998).

All of the minor capsid proteins have been shown to reside in the cytoplasmic membrane before being assembled into phage (Endemann and Model 1995). pIII is synthesized with an 18-residue amino-terminal signal peptide, which is removed after membrane insertion. It spans the membrane one time with the carboxy-terminal residues in the cytoplasm (Davis et al. 1985). The region spanning the membrane is part of the carboxy-terminal portion of the 150-residue CT domain, which is involved in anchoring pIII to the end of the phage particle (Fig. 1.4). Thus, the N1, N2, and most of the CT domains are exposed to the periplasm.

The remaining three minor capsid proteins, pVI, pVII, and pIX, are synthesized without signal peptides, and their mechanism of insertion into the membrane is unknown. It is thought that both pVII and pIX span the membrane once, with their amino termini facing the periplasm. This topology is based only on the observation that these molecules, when overproduced from a plasmid, retain their amino-terminal formyl group after membrane insertion (Simons et al. 1981). In addition, the sequence of each of these molecules predicts one membrane-spanning region. Recent data suggest that pIX has a helical conformation in the membrane (Houbiers et al. 1999). There are no data regarding the topology of pVI in the membrane. It is predicted to have three potential membrane-spanning regions and may adopt the topology shown in Figure 1.4, based on the hypothesis that the positively charged region remains in the cytoplasm (von Heijne 1992). Other predictions have suggested that pVI has only one membrane-spanning region, so it is shown with both configurations in Figure 1.4.

The three assembly proteins, pIV, pI, and pXI, also are integral membrane proteins. pIV is synthesized with a 21-residue amino-terminal signal peptide and is translocated into the periplasm, probably using the Sec system of the host bacterium (Raposa and Webster 1993; Russel and Kazmierczac 1993) (see p. 1.21, Protein translocation across the cytoplasmic membrane). Twelve to fourteen pIV molecules then form an oligomer that integrates into the outer membrane with the amino-terminal portion of the subunits exposed in the periplasm (Kazmierczak et al. 1994). These multimers have been isolated and shown to be large cylindrical structures by scanning electron microscopy (Linderoth et al. 1997), and when placed in planar lipid bilayers, they form highly conductive channels (Maricino et al. 1999). These properties are consistent with the pIV complex forming a channel across the outer membrane with a diameter of 6–8 nm, large enough to accommodate an extruding phage particle. This channel must be gated in some way, as it does not let large molecules in or out of the periplasm in the absence of other phage proteins (Maricino et al. 1999). There is a great deal of homology between the pIV complex and the outer membrane components, or "secretins," common to the type II bacterial secretion systems (Russel 1998).

The other two assembly proteins, pI and pXI, are synthesized without an amino-terminal signal peptide and are inserted into the membrane in a manner that requires SecA (Rapoza and Webster 1993). They both span the membrane once, and their carboxy-terminal 75 amino acids are in the periplasm (Guy-Caffey et al. 1992). Production of moderate amounts of pI from a plasmid results in loss of membrane potential and cessation of growth, suggesting that these molecules can interact to form a channel through the cytoplasmic membrane (Horabin and Webster 1988; Guy-Caffey and Webster 1993). Interestingly, production of moderate amounts of pXI, which has the same sequence as the carboxy-terminal third of pI, does not result in cessation of growth (Rapoza and Webster 1995). Because both proteins are required for phage assembly and have the same topology in the membrane, it is assumed that they interact with each other to form some type of channel through the cytoplasmic membrane. Such a channel could then interact with the outer membrane pIV complex to form a channel through both membranes during phage assembly. A temperature-sensitive mutation in the amino-terminal portion of pIV has been isolated that can be rescued by a periplasmic mutation in pI/XI, consistent with the existence of such a channel (Russel, 1993). However, it has not been possible to isolate a complex containing all three assembly proteins.

For more information about replication and protein synthesis, see Model and Russel (1998) and Webster (1996).

The Assembly Process

The assembly of the filamentous phage is a membrane-associated event requiring the five capsid proteins, the three assembly proteins, ATP, a proton motive force, and at least one bacterial protein, thioredoxin (Model and Russel 1988; Feng et al. 1997). Assembly occurs at sites resembling bacterial adhesion zones, where the inner and outer membranes are in close contact (Lopez and Webster 1985). Presumably it is the interaction of the outer membrane pIV oligomer with cytoplasmic pI and pXI that forms these sites (see above). The substrate for the assembly process is the pV–DNA complex. The overall reaction is a process in which the pV dimers are removed and the capsid proteins are assembled around the DNA as it is extruded through the assembly site. Because the bacteria continue growing, the integrity of the membrane must be maintained during this extrusion process so as not to destroy the membrane potential. To reflect the different reactions involved in forming the ends and the tube portion of the phage particle, the assembly process has been divided into initiation, elongation, and termination.

During the process of initiation, pVII and pIX, together with the first set of pVIII molecules, must interact with the DNA packaging signal (PS), because it is this end of the phage that emerges first from the bacterium (Lopez and Webster 1983). Genetic evidence suggests that the PS interacts with the cytoplasmic portion of pI and membrane-associated pVII and pIX (Russel and Model 1989). The presence of the PS is not absolutely required for assembly, as very low quantities of phage can be produced when this site is altered. This may be the reason that low levels of phagemid single-stranded DNA lacking a PS can be assembled in the presence of a helper phage that supplies coat proteins. However, a PS is required for efficient assembly, and the presence of a PS on any single-stranded circular DNA allows it to be encapsulated into phage (Model and Russel 1988; Russel and Model 1989).

The order in which pVII, pIX, and pVIII associate with the PS during initiation is unknown. A logical hypothesis is that pI is able to direct an ordered assembly of these three capsid proteins around the PS to form the tip of the particle. This assembly might induce a conformational change in pI, allowing its periplasmic region to interact with the amino-terminal portion of pIV and open the gate in the pIV exit pore

(Russel et al. 1997; Marciano et al. 1999). The presence of the already formed tip would plug any channel formed by this interaction. This model is consistent with the observation that infection of bacteria with phage bearing amber mutations in pVII or pIX results in the formation of assembly sites accompanied by cessation of bacterial growth (Lopez and Webster 1985), suggesting that pI/XI and pIV have interacted to form an open channel that is not blocked by the phage tip. Alternatively, pVII, pIX, and pVIII may first interact with each other before recognizing the PS in a pI-directed interaction. Such a hypothesis is supported by immunoprecipitation data that suggest an association of pVII with pVIII in the membrane (Endemann and Model 1995).

After initiation, the phage is elongated in a processive set of reactions that remove the pV dimers and replace them with pVIII as the DNA is extruded through the membrane (Fig. 1.5). The positively charged carboxy-terminal portions of the pVIII molecules must interact with the DNA while the transmembrane portions interact with each other to form the capsid tube. Presumably, these reactions require both ATP hydrolysis and thioredoxin and are catalyzed by the cytoplasmic and transmembrane portions of pI. The redox activity of thioredoxin is not necessary, as changing the active-site cysteines to serines has no effect on phage production (Russel and Model 1986). Thus, thioredoxin is only required in its reduced conformation. It has been suggested that it is the concerted action of thioredoxin and pI that confers processivity on the elongation reaction. ATP hydrolysis is probably involved in at least this step of assembly, because there is a Walker nucleotide-binding motif near the amino-terminal portion of pI that is required for phage assembly.

The structure of the DNA in the pV–DNA complex is different from its structure in the phage particle, suggesting that some conformational change in the DNA is required to allow interaction with the carboxy-terminal portion of pVIII. The 10

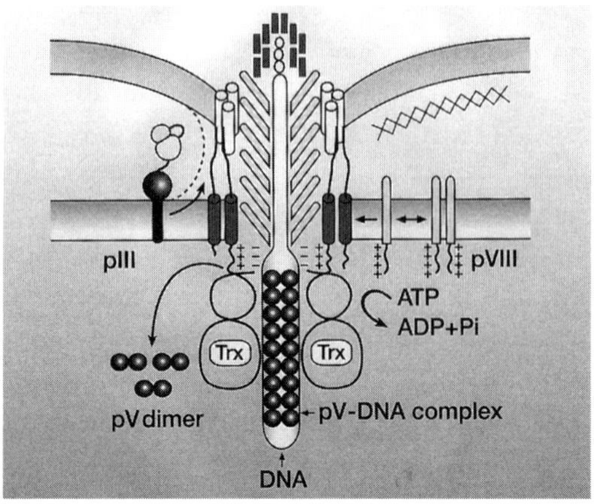

Figure 1.5. Schematic representation of the elongation process on the assembly of the Ff bacteriophages. Trx refers to thioredoxin. The symbols for the pVII–pIX and pVIII on the phage are the same as in Fig. 1.1. This figure only depicts what may happen during assembly and does not represent any structure of the DNA or protein. The arrow next to pIII indicates that its orientation must change during the termination of assembly. (Adapted, with permission, from Webster 1996.)

residues of pI and pXI adjacent to the cytoplasmic face of the cytoplasmic membrane have an amphiphilic character extremely similar to the carboxy-terminal 10 residues of pVIII that interact with the DNA in the phage particle (Rapoza and Webster 1995). These regions of pI or pXI may interact with the DNA that has just been stripped of pV, to facilitate the DNA's adopting the proper configuration for interaction with pVIII.

Conformational changes occur in the pVIII protein before it can assemble around the DNA. The structure of the protein is somewhat different in the phage than in the membrane, where it tends to form dimers that are not suitable for packaging (Haigh and Webster 1998; Marvin 1998). The packaging reaction dissociates dimers and delivers the positively charged carboxy-terminal region of pVIII from the environment of the acidic phospholipid headgroups in the membrane to the sugar phosphate backbone of the phage DNA. Again, this transition may be accomplished by an interaction of the cytoplasmic and transmembrane regions of pVIII with its counterpart regions in pI and/or pXI. There is genetic evidence for an interaction between the transmembrane regions of pVIII and pI and/or pXI (Russel 1993).

It is generally assumed that the major coat protein interacts with the DNA during assembly in the same orientation as in the membrane. If this is the case, the amino-terminal portion of pVIII should emerge first, and the cone-shaped portion of the phage pVIII tube should be at the end containing the pVII and pIX protein (Fig. 1.1). If the carboxy-terminal portion of the pVIII tube emerges first, there should exist some type of membrane connection between the cytoplasmic and outer membranes that allows the carboxyl terminus of pVIII to flip from the cytoplasmic side to the periplasmic side of the membrane (Marvin 1998). Knowledge of which end of pVIII emerges first would be useful in determining the mechanism of elongation.

As elongation continues, the phage particle emerges through the pIV exit pore. At this point in the elongation process, it is not known whether the interactions between the periplasmic portions of the pIV and pI/XI assembly proteins are still needed. When the end of the DNA is reached, assembly is terminated by the addition of pVI and pIII. In the absence of pIII, elongation continues, with pVIII encapsulating another phage DNA (Model and Russel 1988). In this way, polyphage containing multiple copies of the genome are produced, which remain attached to the bacteria (Rakonjac and Model 1998). Addition of pIII in *trans* releases these phage, suggesting that the CT domain of pIII is required for release as well as for terminating assembly.

The mechanism by which pVI and pIII are added to the end of the particle is unknown. The 150-residue CT domain appears to be composed of two subdomains involved in the termination reaction (Rakonjac et al. 1999). Phage particles can be completed and released from the bacteria when only the carboxy-terminal 93 residues of the CT domain are present. However, these phage are unstable in the presence of some detergents that normally do not affect wild-type phage. An additional 30-residue upstream "subdomain" (121 total carboxy-terminal residues of CT) is required for making stable phage particles. On the basis of these observations, it has been proposed that termination is a two-step process (Rakonjac et al. 1999). First, pVI and pIII interact with the pVIII molecules assembled around the end of the DNA to form a pretermination complex. Then a structural change occurs in the CT region that allows pIII to break free of the membrane. It is the carboxyl terminus of pIII that is anchored in the membrane, and it is also this region that is anchored in the end of the phage particle. Therefore, this region must flip around to interact with the phage and be the first part of pIII to exit (see Fig. 1.5). It is possible that this is a property of the structural change in the pVI–pIII–pVIII pretermination complex. An alternative explanation is that the cytoplasmic and outer membranes may fuse to allow pIII to flip. Such an interaction may cause a breakdown of the inner membrane portion of the assembly site. In this case, only one phage would be made per site and a new site would be formed for each phage produced.

For more information about the assembly process, see Russel (1995), Webster (1996), and Rakonjac et al. (1999).

DISPLAY OF PEPTIDES AND PROTEINS ON PHAGE PARTICLES

The major interactions that occur among the capsid proteins during the assembly process involve the hydrophobic regions of these proteins, which initially span the membrane. These membrane-associated assembly properties of the capsid proteins allow packaging of chimeric proteins into the phage particle. Any protein fused to the periplasmic portion of these capsid proteins theoretically should have a good chance of being packaged into a phage particle, provided it can be translocated efficiently across the inner membrane and not interfere with the processes that occur at the assembly site. This concept was proven by George Smith (1985), who showed that fragments of the *Eco*RI endonuclease could be fused to the amino-terminal portion of pIII, to produce a chimeric protein that was packaged into phage particles. These phage were somewhat defective with regard to infectivity but could be propagated quite well. Phage containing the *Eco*RI fragment could be affinity-purified from a mixture of phage in which the majority did not contain the *Eco*RI insert. *Eco*RI antibody was adsorbed to a petri dish, the phage mixture was added, the dish was washed to remove nonbinding phage, and the *Eco*RI-containing phage were released from the antibody by acidic buffer. These released phage were amplified by growth in bacteria, and the resulting phage were subjected to another round of affinity purification. Multiple rounds of this procedure led to a large enrichment in the phage containing the *Eco*RI fragment. This type of purification, using several rounds of specific binding between the displayed peptide and its immobilized binding partner, is sometimes referred to as "panning." Many variations of this procedure have been devised to reduce the nonspecific binding of unwanted species as well as to increase the specificity of binding for the desired species.

Large libraries of peptides and proteins have been made using pIII as the display vehicle (Smith and Scott 1993; Winter et al. 1994), leading to the development of a number of techniques for selecting the molecule(s) desired from such libraries (Clackson and Wells 1994; Hoogenboom 1997). Peptides and proteins have also been fused to the amino-terminal portion of the major capsid protein pVIII (Iannolo et al. 1995; Malik et al. 1996). There has even been a report of phage particles displaying proteins that are fused to the carboxy-terminal portion of pVI, although the efficiency of display appears to be lower in this case (Jespers et al. 1995). Recently, antibody heavy- and light-chain variable regions have been fused to the amino terminus of pVII and pIX and displayed on phage, showing that these two minor coat proteins can also be used for display (Gao et al. 1999). Generally, proteins are displayed on filamentous phage fused to either pVIII or pIII and selected by some type of panning, affinity columns, or two-hybrid technology, as briefly described below.

Proteins also can be displayed on smaller filamentous particles called phagemids (Bass et al. 1990; Breitling et al. 1991). The phagemid genome contains the filamentous phage intergenic region with its origin of replication for viral and complementary strand synthesis as well as the hairpin packaging signal. The genome also contains a plasmid origin of replication and a gene encoding resistance to a specific antibiotic. Chimeric genes encoding peptide–phage protein fusions can be placed under control of a specific promoter in these phagemid genomes. The phagemid can maintain itself as a plasmid, directing the expression of the protein in bacteria if desired. Infection of the bacteria with a filamentous helper phage activates the phage origin of replication, resulting in single-stranded phagemid DNA being encapsulated into filamentous phage-like particles using helper phage proteins. A helper phage containing a defective packaging signal can be used so that the majority of particles produced contain the phagemid single-stranded DNA (Russel et al. 1986). Bacteria can be infected with the phagemid–phage mixtures and colonies selected that are resistant to the antibiotic. The resistant colonies will contain only the phagemid DNA, which can then be propagated again by infection with helper phage. Because the phagemid particles can transmit antibiotic resistance, they are sometimes referred to as "transducing particles."

Display of Peptides and Proteins as pVIII and pIII Fusions

The DNA sequence encoding the peptide to be displayed on pVIII is generally inserted between the sequence for the pVIII signal peptide and the amino-terminal coding region for the mature capsid protein. This insertion leads to the production of phage particles with the peptide exposed along the surface of the phage and therefore potentially immunogenic and susceptible to proteases (Fig. 1.6) (Terry et al. 1997). This type of display on every copy of pVIII is limited to peptides 6 or 8 amino acids long (Iannolo et al. 1995; Malik 1996; Petrenko et al. 1996). Sometimes pVIII containing inserts up to 12 residues can be less efficiently packaged into phage, but the number of particles produced is not enough to form plaques. If the phage genome contains an antibiotic resistance marker, infection will confer resistance to the bacteria. Thus, if any of these larger chimeric pVIII molecules can lead to the production of low levels of phage, they can be detected as antibiotic-resistant colonies resulting from infection. Like the phagemids described above, phage producing such antibiotic-resistant colonies are generally referred to as "transducing particles."

pVIII containing large peptides or even proteins must be displayed in hybrid virions, in which 80% or more of the pVIII molecules are wild type (Fig. 1.6) (Greenwood et al. 1991b; Kang et al. 1991). The DNA sequence for the protein is inserted in the same position in the coat protein gene as for the smaller peptides, and the resulting gene is placed into a plasmid, preferably under control of an inducible promoter. When bacteria producing this chimeric pVIII–fusion protein are infected with wild-

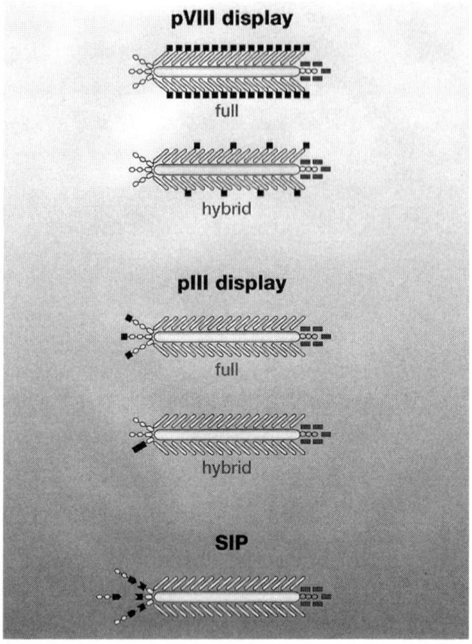

Figure 1.6. Display of proteins on the Ff filamentous bacteriophage. The dark solid symbols attached to pVIII, pIII, or parts of pIII represent the foreign peptides or proteins attached as described in the text.

type phage, the chimeric pVIII protein joins the pool of wild-type pVIII, and phage are produced that have the modified pVIII interspersed with wild-type pVIII. Proteins displayed this way appear to be quite accessible, as they are immunogenic (Minenkova et al. 1993), can interact with specific ligands (Hart et al. 1994), and are susceptible to cleavage by proteases (Terry et al. 1997).

It is not clear why inserts into pVIII of more than 6 to 8 amino acids are not tolerated very well. One reason may be that the inserts interfere with the rate of successful targeting and translocation of the chimeric pVIII into the membrane. In some cases, increasing the amount of leader peptidase I, which removes the signal peptide, can increase the number of chimeric pVIII molecules that are incorporated into hybrid phage particles (Malik et al. 1996; 1998). Phage assembly may also be affected by the presence of a large insert at the amino-terminal end of pVIII. It has been suggested that the insert may interfere with the interaction between pVIII and pVII during the initiation of assembly (Endemann and Model 1995). A third, more plausible, explanation may be that a phage containing pVIII with a large peptide may be too large in diameter to pass though the 7-nm pIV exit pore in the outer membrane (Linderoth et al. 1997; Marciano et al. 1999).

pIII is the capsid protein most used for display of proteins. It has the disadvantage that only about 5 molecules can be displayed per phage particle. However, its advantage is that chimeric pIII molecules containing large inserts appear to package into phage reasonably well. In most cases, the proteins are inserted between the signal sequence and the beginning of the first domain (N1) of pIII (Fig. 1.6). Once in the membrane, the foreign protein presumably does not interfere with assembly, a process that requires the CT domain of pIII. Also, because this insertion would place the foreign protein at the very end of the packaged particle, it may create less steric hindrance when passing through the pIV exit pore. The problem is that large inserts tend to lower phage infectivity or even make the phage noninfective, thus limiting the ability to select a particular displayed protein (Smith 1985). This problem can be overcome by making hybrid phage, in which only one of the pIII molecules contains the displayed protein. This is generally done by expressing the chimeric pIII from a phagemid so that the majority of the pIII in the cell is wild type, encoded by a helper phage. Proteins fused to the carboxy-terminal portion of pIII (missing domains N1 and N2) can be displayed in hybrid phage that retain their infectivity because of the presence of wild-type pIII (Barbas et al. 1991). The ability to display an active protein depends on whether the fusion protein can be translocated properly into the membrane, fold properly, escape degradation in the periplasm, and be acceptable for packaging into phage.

Foreign proteins also can be inserted between the N1 and N2 domains as well as between the N2 and CT domains of pIII (Krebber et al. 1997). Such inserts can retain infectivity, albeit at a lower level, provided that the N1 and N2 domains can interact to generate a pilus-binding site. It has been suggested that by inserting the protein between the N1 and N2 domains, one can select for protease-resistant proteins, because cleavage would result in the loss of N1 and, consequently, infectivity (Krebber et al. 1997).

For more information about the display of peptides and proteins as pVIII and pIII fusions, see Dunn (1996) and Kay et al. (1996).

Selectively Infective Phage

The realization that pIII is composed of three domains led to the creation of a type of phage "two-hybrid system" in which infective phage would depend on the interaction between two different peptides. In these systems, one of the interacting peptides is fused to the carboxyl terminus of the portion of pIII containing the amino-terminal penetration (N1) and adsorption (N2) domains. The other interacting peptide is fused to the amino-terminal side of the pIII CT domain. The presence of the chimeric CT domain of pIII results in the production of noninfectious phage particles containing one of the interacting peptides present at one end of the phage (Crissman and Smith 1984; Krebber et al. 1997). These particles can bind via the interacting peptides to the chimeric amino-terminal portion of pIII containing the N1–N2 domains to give an infectious phage particle. Some of the systems depend on an antibody–antigen interaction with the interacting antibody species, such as Fab or scFv molecules, attached to the CT region of pIII. Successful systems of this type have been called SAP, for *s*election and *a*mplification of *p*hages (Duenas and Borrebaeck 1994), and SIP, for *s*electively *i*nfective *p*hages (Krebber et al. 1995). Another system uses the Jun–Fos leucine zipper interactions to join the pIII domains together and was called DIRE, for *d*irect *i*nteraction *re*scue (Gramatikoff et al. 1994).

The most studied of these systems is SIP, and phage display requiring such protein–ligand interactions between chimeric pIII domains for infection is generally called SIP technology. It has been found that the interacting molecules can be present either between pIII domains N1 and N2 or between domains N2 and CT (Krebber et al. 1997). Highest titers of infective phage were obtained if the antigen was attached to the carboxy-terminal end of N1 and the scFv antibody to the amino-terminal portion of the N2–CT region of pIII (Fig. 1.6). Even in this study, however, only one in 10,000 of the phage particles was infective. The inability to get high titers of infective phage when the antigen was at the carboxy-terminal end of the N1–N2 domains of pIII appears to be partially because of the inhibitory effect of free N1–N2–antigen molecules on infection, presumably caused by binding to the pilus. SIP phage can also be made in which the N2 domain is missing, but in this case the bacteria must be treated with calcium, presumably to allow N1 to interact with domain 3 of TolA in the absence of N2 interaction with the pilus.

One of the interesting aspects of SIP technology is that the interaction of the chimeric pIII N1–N2 domains with the noninfectious chimeric CT phage particles can occur either in vivo or in vitro. For example, the chimeric N1–N2–antigen protein and noninfectious phage containing the scFv–CT can be produced in the same bacterium. In this case, the interaction between the antigen and antibody portions of the chimeric proteins occurs in the periplasm, resulting in the release of infective phage. Alternatively, the noninfectious phage containing the pIII CT–antibody chimera can be isolated and added to isolated pIII N1–N2–antigen chimeric proteins in vitro to obtain infective phage. Although the in vitro method suffers from the additional steps required for isolation of the components, it does have the advantage that the relative concentrations of the reactants can be better controlled.

One of the advantages of SIP technology over conventional phage display is that selection of good interactions does not require panning, which is always subject to background resulting from nonspecific adsorption of unwanted phage (Adey et al. 1995). The major drawback to the SIP technology, however, is the relatively low level of infectivity of the phage particles produced. This low infectivity could be related to the binding affinities of the interacting molecules responsible for joining the pIII domains. The binding affinities required also depend on the number of intact pIII molecules that are required for infection. Experiments using different Fab fragments have indicated that tighter binding between interacting species results in phage with higher infectivity (Duenas et al. 1996). A better understanding of the infection process will help in developing this technique further.

For more information about SIP, see Spada et al. (1997).

CONSIDERATIONS FOR SUCCESSFUL PROTEIN DISPLAY

The wide range of peptides and proteins that can be expressed on the surface of the phage particle is truly remarkable. This great success in protein display has led to a variety of clever methods designed to select the protein of choice, as briefly described above. It must be recognized, however, that not every protein or peptide can be packaged into or attached to the phage particle. The chimeric protein must be acceptable for packaging into the phage capsid. Assembly of the phage particle, as well as production of chimeric proteins, is dependent on the bacterial physiology. The proteins must be synthesized in the controlled and reducing environment of the cytoplasm and be translocated, probably in an unfolded state, across the cytoplasmic membrane into the periplasm. The periplasm is a gel-like compartment, with diffusion coefficients 100-fold lower than those in the cytoplasm. It has an oxidizing environment and is subject to changes in growth conditions such as pH and osmolarity (Oliver 1996). Perhaps because of these variable conditions, many of the periplasmic or secreted proteins contain disulfide bonds that aid in their folding and stability. The periplasm does not contain ATP, so chaperones such as the Hsp60 and Hsp70 classes are not present. Protein folding appears to depend more on folding catalysts such as the thiol-disulfide oxidoreductases and peptidyl-prolyl *cis/trans* isomerases, although there is some evidence for the presence of one possible chaperone-type molecule (Missiakas and Raina 1997a). The periplasm appears to have at least two systems that can respond to stress caused by conditions that might lead to misfolding of proteins.

It is rather amazing how versatile and flexible the phage-display system is, considering the possible constraints that the processes of protein translocation, folding, and phage assembly might place on it. To provide a better understanding of the designs of the various phage-display systems, a brief description is given below of translocation into and across the cytoplasmic membrane, together with an update about what is known about protein folding in the periplasm and systems that may respond to the possible stess induced by the presence of misfolded proteins in the periplasm.

Protein Translocation across the Cytoplasmic Membrane

At the present time, there are four identified mechanisms in *E. coli* by which proteins are able to translocate across the membrane. Three of these, the Sec, SRP, and Tat systems, require the presence of proteins that appear to be specific for the translocation process. The fourth mechanism does not require other proteins and is sometimes termed spontaneous insertion. All mechanisms usually require that proteins to be transported have an additional sequence of approximately 15–25 amino acids at the amino-terminal end, generally called the signal or leader sequence. Nascent proteins containing such a sequence are called pre- or proproteins. There is no sequence homology among different signal peptides, but they all share similar characteristics. They contain a positively charged amino-terminal region, a central hydrophobic region, and a polar carboxyl region. After translocation, the signal peptide is usually removed by a specific signal peptidase to yield the mature protein.

In general, proteins must be in an unfolded state during the translocation process. In *E. coli*, most, if not all, proteins are completely synthesized and then translocated. It is thought that there is something inherent in the sequence of the signal peptide and full-length protein that helps the proteins remain unfolded. There also are specific chaperone proteins that bind to the proteins to keep them unfolded while delivering them to the translocation machinery. The best studied of these chaperones is SecB, a homotetramer that appears to interact with large regions of the full-length protein. Other cytoplasmic chaperones that appear to act as chaperones for some translocated proteins are DnaK, DnaJ, and GrpE (Wild et al. 1996).

The **Sec** system is responsible for translocation of most proteins to the periplasm and therefore probably is responsible for the display of most proteins on the phage.

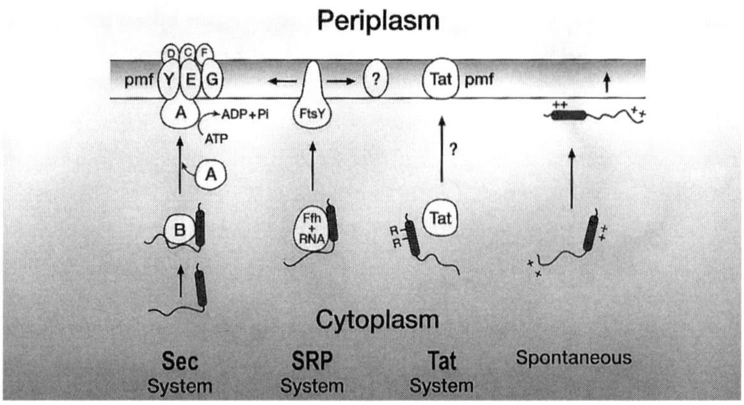

Figure 1.7. The different mechanisms for translocation of proteins into or across the bacterial cytoplasmic membrane. The letters in the Sec system refer to the proteins involved. pmf refers to the proton motive force. The filled-in cylinders refer to the signal peptides. R refers to the conserved arginines in the signal sequence required for targeting to the Tat translocation system. The arrows on either side of FtsY indicate that it is unknown whether the bacterial SRP system has a special (?) transport channel or uses the Sec system (see text). The ? on the Tat system indicates the lack of knowledge of how this system translocates folded proteins across the membrane.

This translocation system is composed of the Sec A, B, D, E, F, G, Y, and probably yajC proteins (Fig. 1.7). Newly synthesized proteins are delivered in an unfolded state to SecA via a chaperone such as SecB. The SecA–proprotein complex then interacts with the SecYEG translocation channel, and the protein is inserted through the membrane, approximately 20–25 residues at a time, in a series of ATP-driven events catalyzed by Sec A (Duong and Wickner 1997; van der Wolk et al. 1997). In addition to ATP hydrolysis, a proton motive force across the membrane is required for translocation. The SecDF–yajC complex is thought to help maintain the proton motive force as well as stabilize the membrane-inserted form of SecA. Following translocation into the periplasm, the signal peptide is removed by the action of the membrane-bound signal peptidase and subsequently degraded by a specific protease. In the case of an integral membrane protein, the insertion of the hydrophobic transmembrane region into the hydrocarbon layer of the membrane appears to stop the transfer. The pIII capsid protein and the pI, pIV, and pXI assembly proteins all require SecA for insertion and presumably are inserted by the Sec system (Rapoza and Webster 1993).

The **Tat** translocation system has recently been shown to translocate proteins that contain metal or other cofactors into or across the cytoplasmic membrane, suggesting that phage display of factor-containing proteins may require this system (Dalbey and Roberson 1999). These proteins are synthesized with a long signal peptide containing two adjacent arginines at a conserved site near the amino terminus of the signal peptide (Berks 1996). After synthesis, the proproteins appear to first bind their respective cofactors before they are translocated, implying that they are translocated in a folded state (Santini et al. 1998). Four genetic loci have been identified whose products must be present for these cofactor-containing proteins to reach the periplasm (Chanal et al. 1998). These loci have been identified as *tat* (for *t*win *a*rginine *t*ranslocation) and define the Tat system (Fig. 1.7). Four of the genes, *tatA*, *B*, *C*, and *D* are present in one operon, whereas *tatE* is independently transcribed. In *E. coli* missing the TatA, B, C, and E proteins, cofactor-containing proteins accumulate in the cytoplasm (Bogsch et al. 1998; Sargent et al. 1998; Weiner et al. 1998).

The twin arginine signal sequence appears to be the element mainly responsible for targeting proteins (Cristobal et al. 1999) and even multiprotein complexes to the Tat translocation system (Rodrigue et al. 1999). An example of the latter is the bacterial hydrogenase 2 complex, which consists of a small subunit containing a twin arginine signal sequence and a nickel-containing large subunit. In the absence of either subunit or the Tat system, the components accumulate in the cytoplasm and do not take up nickel. The mechanism by which the Tat proteins are able to translocate these folded proteins into or across the cytoplasmic membrane is not known but is thought to require a proton motive force or ΔpH gradient across the membrane. This theory is based on the homology between the bacterial TatA and B proteins and the Hcf106 protein in plants, which is an essential part of a ΔpH-driven translocation system across the thylakoid membrane (Settles et al. 1997).

The bacterial **SRP** translocation system is similar to the signal recognition particle (SRP) system required for translocation of proteins across the endoplasmic reticulum in mammalian cells. Mammalian SRP is composed of six proteins and one RNA molecule, and it arrests translation by binding to the ribosome–proprotein complex. This SRP–ribosome complex then binds to the SRP receptor in the endoplasmic membrane, where translation resumes and the protein is cotranslationally translocated into the lumen through a protein translocation channel.

The *E. coli* bacterial SRP consists of the Ffh protein bound to a 4.5S RNA. Ffh and the 4.5S RNA are homologous to the 54-kD protein and 7S RNA components of mammalian SRP. The Ffh–4.5S RNA particle (bacterial SRP) recognizes the nascent protein and delivers it to the membrane-bound FtsY protein, a protein that is homologous to the mammalian SRP receptor (Fig. 1.7). There is some specificity to the proteins recognized by bacterial SRP, as it seems to be required for insertion of membrane proteins that have multiple membrane-spanning regions (Seluanov and Bibi 1997; Ulbrandt et al. 1997). Some of the specificity for bacterial SRP resides in the nature of the signal peptide: The more hydrophobic the amino acids in the core of the signal peptide are, the greater the affinity for the Ffh–4.5S RNA SRP (De Gier et al. 1997).

It is not clear whether bacterial SRP binds cotranslationally or posttranslationally to the nascent protein or whether there is a special SRP transport channel at the membrane. Cotranslational targeting to the membrane may be possible, as the Ffh–4.5S RNA particle and FtsY protein can target nascent secretory proteins to microsomal membranes in vitro (Powers and Walter 1997). In this case translocation is cotranslational, although the bacterial Ffh–4.5S RNA particle does not arrest translation. If cotranslational translocation does occur in bacteria, there may be a special transport channel. Independent of any such special translocation system, there is strong evidence that the bacterial SRP–nascent protein can be delivered to the Sec translocation system (Valent et al. 1998). There is even recent evidence that SecA may be required for insertion of proteins targeted to the membrane by the bacterial SRP (Oi and Bernstein 1999).

The **spontaneous insertion** pathway is responsible for membrane insertion of certain small proteins into the cytoplasmic membrane (Fig. 1.7). The phage pVIII major capsid protein is the one most studied for this pathway of insertion. The 50-residue mature protein is synthesized with a typical 23-residue signal peptide. After synthesis, it is targeted to the membrane via an interaction between the positive charges located at both ends of the proprotein and the acidic phospholipids of the membrane. No other bacterial proteins are required, but a proton motive force appears to help drive the acidic loops across the membrane. After translocation, the signal peptide is removed by signal peptidase I.

The protein translocation systems in *E. coli* are quite versatile. In general, addition of a known signal peptide to a protein and signal peptidase cleavage will suffice to allow translocation of the protein into the periplasm. This is almost always true for molecules such as antibody scFv fragments that normally are translocated across a membrane. Creation of such chimeric proteins with sequences that do not normally cross the membrane can lead to difficulties. For example, fusing the signal sequence portion of the maltose-binding protein (a periplasmic protein) to the cytoplasmic β-galactosidase results in a chimera that is targeted to the membrane, but the β-galactosidase gets stuck crossing the membrane (Danese and Silhavy 1998a). This appears to be because of the formation of an aberrant disulfide bond in the portion of the protein that reaches the periplasm (Rietsch and Beckwith 1998). Chimeric proteins that include a long stretch of helical hydrophobic residues might be prevented from crossing the membrane if the hydrophobic stretch acts as a transmembrane stop transfer region, resulting in the inability of the protein to be displayed on phage.

Another potential insertion problem is that the addition of a foreign protein to the signal peptide may alter the targeting pathway used. Placement of a large peptide at the amino terminus of the mature pVIII capsid protein (after the signal sequence) appears to lower the rate of insertion, perhaps by driving the chimera to use the Sec system rather than spontaneously insert into the membrane. It has been observed that chimeras of pVIII containing relatively large peptides are inserted into the membrane and processed at different rates (Malik et al. 1996). In some cases, part of the problem may be release of the signal peptide, as increasing the amount of signal peptidase I will increase the rate of proper insertion of these proteins into the membrane (Malik et al. 1998).

For more information about the Sec system, see Danese and Silhavy (1998a) and Driessen et al. (1998); about the Tat system, see Dalbey and Robinson (1999); and about the SRP system, see De Gier et al. (1997). For more information about spontaneous insertion, see Kuhn and Troschel (1992).

Protein Folding and Degradation in the Periplasm

Many proteins secreted into the periplasm form specific disulfide bonds that aid in both the folding and stabilization of the mature protein. Therefore, the periplasm has an oxidizing environment to allow formation of these structural disulfide bonds. In contrast, the cytoplasm has a reducing environment to maintain the active-site cysteines in many of the enzymes in this compartment. This difference appears to be the result of the particular enzymatic systems present in these compartments, which are responsible for the oxidation, reduction, and isomerization of disulfide bonds in proteins. The cytoplasm has two systems that catalyze the NADP-dependent reduction of disulfide bridges in target proteins: the thioredoxin/thioredoxin reductase system and the glutaredoxin/glutathione system (Holmgren 1989). The proper formation of the disulfide bonds in the periplasm is catalyzed by specific thiol-disulfide oxidoreductases. In both the periplasmic and cytoplasmic systems, the proteins responsible for catalyzing oxidation or reduction have a common thioredoxin active-site motif, CxxC, where "C" refers to cysteine and "x" refers to any amino acid. Proteins containing this sequence are members of the thioredoxin family because they all contain a similar three-dimensional structure called the thioredoxin fold (Eklund et al. 1991; Martin et al. 1993). The cysteines in the CxxC motif engage in a thiol-disulfide exchange during

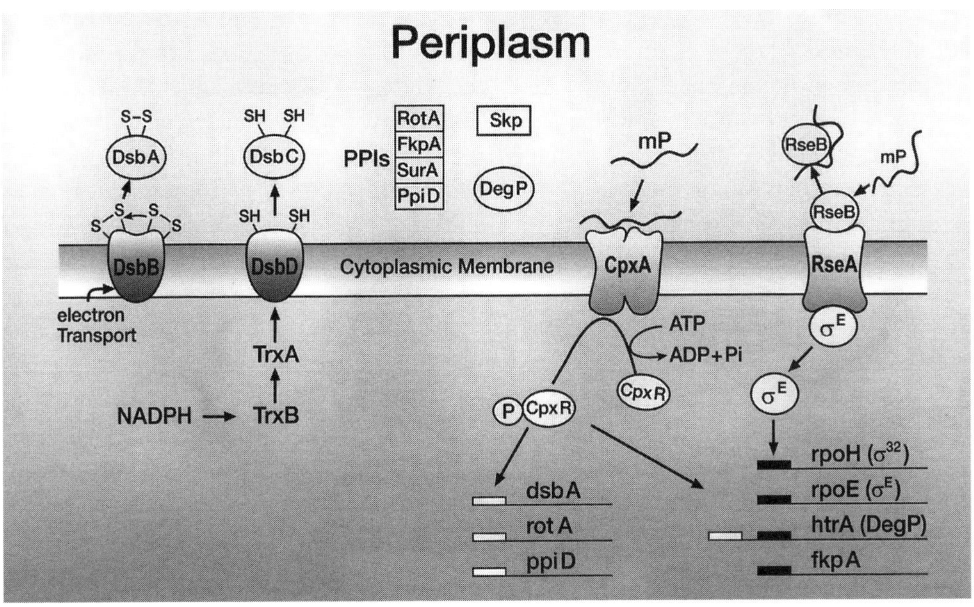

Figure 1.8. Schematic diagram of the thiol-disulfide oxidoreductases, peptidyl-prolyl isomerases (PPIs), and other molecules involved in periplasmic protein folding and degradation. The left side depicts the thiol-disulfide oxidoreductase systems involved in forming proper disulfide bonds in periplasmic proteins. The reducing power of DsbD is acquired from cytoplasmic thioredoxin (TrxA), which in turn is reduced by thioredoxin reductase (TrxB) at the expense of NADPH. DsbB maintains its ability to oxidize DsbA by being coupled to the respiratory chain (see text). The center part of the figure depicts the various peptidylprolyl isomerases (PPIs), possible chaperone (Skp), and proteolytic enzymes (DegP). The right side of the figure depicts the σ^E and Cpx periplasmic stress response systems. mP refers to misfolded protein. Interaction of a mP with CpxA is thought to activate it to phosphorylate CpxR, which then activates the promoters for the genes as shown by the arrows pointing to the white rectangles. It is also proposed that interaction of mP with RseB leads to its release from RseA, which then releases σ^E, allowing it to induce the synthesis of a number of genes (indicated by the arrows pointing to the black rectangles). See text for more complete explanation.

reduction or oxidation of cysteines in specific proteins. The ability to oxidize or reduce cysteines partially resides in the redox potential of the disulfide at the active site. It is the sequence of amino acids located between the cysteines that greatly influences the redox potential of a particular active-site motif (Chivers et al. 1997).

There are a number of membrane and soluble thiol-disulfide oxidoreductases that contribute to proper oxidation of structural disulfide bonds in periplasmic proteins (Fig. 1.8). DsbA is one such oxidoreductase that contains a thioredoxin-type motif whose active-site disulfide bond has a high redox potential (Bardwell et al. 1991; Kamitami et al. 1992; Wunderlich and Glockshuler 1993; Zapun et al. 1993). Thus, it actively catalyzes the formation of disulfide bonds in target proteins, leaving its active site in a reduced state. The reduced form of DsbA is reoxidized by the integral cytoplasmic membrane protein DsbB (Bardwell et al. 1993; Missiakas et al. 1993). DsbB contains two periplasmic disulfide bonds, one in a thioredoxin-type motif (CxxC) and the other separated by 25 residues. The latter disulfide reoxidizes the active site in DsbA and in turn is oxidized back to a disulfide bond by interacting with the CxxC domain in DsbB (Kishigami and Ito 1996). The maintenance of the disulfide bond in the CxxC motif is coupled to the respiratory chain (Kobayashi and Ito 1999). Recently, another periplasmic protein, DsbG, has been described which can act as an effective catalyst for the formation of disulfide bonds in periplasmic proteins (van Straaten et al. 1998).

Although DsbA and DsbG can catalyze the formation of disulfide bonds, they are inefficient in catalyzing the rearrangement or isomerization of incorrectly formed disulfide bonds. This function is carried out by the periplasmic DsbC protein, which contains a highly reactive CxxC active-site motif (Rietsch et al. 1996; Sone et al. 1997). This active site is usually found in the reduced state, making it competent for disulfide rearrangement. An integral membrane protein, DsbD, maintains DsbC in the reduced state via an interaction with its own reduced CxxC active-site motif (Missiakas et al. 1995; Rietsch et al 1996). Evidence suggests that the reducing power of DsbD is acquired by electron transfer from cytoplasmic thioredoxin (Rietsch et al. 1997).

The peptidyl-prolyl *cis/trans* isomerases (PPI) are another class of periplasmic folding catalysts (Fig. 1.8). These enzymes catalyze the interconversion between the *cis* and *trans* form of the peptide bond X-Pro, where X is any amino acid residue. Four PPIs (RotA, FkpA, SurA, and PpiD) have been identified in the periplasm and belong to one of the three classes of peptidyl-prolyl *cis/trans* isomerases (Liu and Walsh 1990; Missiakas et al. 1996; Dartigalongue and Raina 1998). RotA is in the cyclophilin class because of its homology with the eukaryotic PPI cyclophilin. FkpA is a member of the FKBP class because of its homology with PPIs that are targets for the immunosuppressive drug FK506. Both SurA and PpiD are members of the parvulin class because of their homology with the 10-kD *E. coli* PPI parvulin.

The specificities and exact role that these proteins play in promoting periplasmic protein folding are not entirely clear. Individually, each is nonessential because mutation of each of their corresponding genes has little effect on growth. However, a double mutant in both *surA* and *ppiD* genes is lethal, suggesting that these PPIs have overlapping functions. This idea is substantiated by the observation that both SurA and PpiD are involved in proper folding of outer membrane proteins (OMPs). RotA has good isomerase activity in vitro, but no specific function can be assigned to it in the periplasm. Probably the best evidence that these proteins are involved in protein folding is that some have been shown to be under control of extracytoplasmic stress response systems (see below). These systems respond to the presence of misfolded proteins in the periplasm, and in many cases, these stress response systems are activated in bacteria that are missing some of these PPIs or oxidoreductases.

One other periplasmic protein, Skp, has been isolated that may be involved in protein folding (Missiakas and Raina 1997a). This protein was originally isolated because of its ability to selectively bind to denatured OmpF outer membrane protein, suggesting that it acts as a chaperone for the assembly of OmpF (Chen and Henning 1996). Recent work has shown that periplasm containing an increased amount of Skp resulted in better incorporation of a poorly folding scFv–pIII chimeric protein into phage particles (Bothmann and Plückthun 1998). This would argue that Skp does, in some way, act as a molecular chaperone.

The fate of misfolded proteins in the periplasm presumably is the formation of inclusion bodies or degradation by periplasmic proteases, depending on the kinetics of folding, degradation, and aggregation (Betton et al. 1998). The periplasm contains many proteases, some of which have been shown to be involved in the degradation of abnormal proteins (Miller 1996; Oliver 1996). DegP, the product of the *htrA* locus, is a serine endopeptidase that recognizes proteins with abnormal structures. It is essential for bacterial survival above 42°C and is under heat shock control (see below). Protease III, the product of the *ptr* locus, is a zinc-containing endopeptidase that appears to degrade unusually folded proteins in the periplasm. Prc is an endoprotease that degrades nonpolar carboxy-terminal regions of proteins. OmpT, an outer membrane protein, also can degrade various fusion proteins present in the periplasm. Strains have been constructed that are missing one or more of these proteases and have been shown to have a decreased rate of degradation of fusion proteins in the periplasm (Baneyx and Georgio 1991). Unfortunately, decreasing the number of proteases in the periplasm also decreases the growth rate of the bacteria, thus negating any advantage these strains might have for phage display.

For more information about thiol-disulfide oxidoreductases, see Rietsch and Beckwith (1998); peptidyl-prolyl isomerases and chaperones, see Danese and Silhavy (1998a); periplasmic proteases, see Oliver (1996).

Periplasmic Stress Response Systems

All cells appear to have systems that respond to stresses, such as heat, which may cause deleterious effects such as promoting misfolding of proteins. The classic heat shock response of *E. coli* is to induce the synthesis of σ^{32}, the *rpoH* gene product. This transcription factor is responsible for inducing the synthesis of a number of stress-inducible loci encoding such proteins as the GroEL/ES and DnaK chaperones and the Lon protease (Gross 1996). However, σ^{32} appears to have only some control over the production of PpiD folding catalyst in the periplasm. All of the other proteins involved in periplasmic protein folding are under control of two separate extracytoplasmic stress response systems, the σ^E pathway and the two-component Cpx pathway (Fig. 1.8).

The σ^E protein is the product of the *rpoE* gene and appears to control the synthesis of at least 11 proteins, some of which are involved in protein folding or degradation in the periplasm. σ^E regulates the synthesis of DegP protease and FkpA peptidyl-prolyl isomerase, as well as its own synthesis. At extremely high temperatures it also regulates the synthesis of σ^{32} from the *rpoH* gene. The σ^E pathway is induced by conditions that lead to misfolding of periplasmic proteins such as heat shock, overproduction of outer membrane proteins, or inactivation of the SurA isomerase or Skp protein. Because the transcription of the *rpoE* locus is controlled by its own product, σ^E, other factors are responsible for modulating the activity of σ^E in response to stress signals in the periplasm. These factors are the RseA and B proteins, which appear to be negative regulators of σ^E (de la Penas et al. 1997).

RseA is an integral cytoplasmic membrane protein that spans the membrane with its amino terminus in the cytoplasm and its carboxyl terminus in the periplasm. The periplasmic RseB protein binds to the periplasmic portion of RseA and enhances RseA's ability to inhibit σ^E activity, presumably by promoting the binding of σ^E to the cytoplasmic portion of RseA. It has been proposed that when periplasmic protein folding or degradation is impaired, misfolded proteins may bind RseB, lowering the binding affinity of RseA for σ^E in the cytoplasm. This would result in the release of free σ^E to activate transcription of its own gene and of the genes for σ^{32}, DegP protease, and FkpA isomerase (Missiakas and Raina 1997b).

The second system that monitors stress conditions in the periplasm is the two-component Cpx pathway (Fig. 1.8). This pathway appears to regulate directly the expression of genes encoding the DsbA thiol-disulfide oxidoreductase, the RotA and PpiD peptidyl-prolyl *cis/trans* isomerases, and the DegP protease (Danese and Silhavy 1997; Pogliano et al. 1997; Dartigalongue and Raina 1998). CpxP, another periplasmic protein that genetically has been shown to respond to unfolded or misprocessed proteins, is also under control of the Cpx regulon (Danese and Silhavy 1998b).

The Cpx pathway is composed of two proteins, CpxA and CpxR. CpxA is an integral membrane protein that senses alterations in the periplasmic components. CpxR is a cytoplasmic DNA-binding protein that alters the expression of the genes encoding the proteases and folding catalysts under control of the Cpx pathway. CpxA has been shown to be activated by overproduction of the outer membrane lipoprotein, NlpE, or by conditions that might alter protein folding such as alkaline pH or null mutations in *dsbB*, a part of the periplasmic disulfide isomerase system. The activated CpxA then phosphorylates CpxR, which activates the transcription of *degP*, *ppiD*, *dsbA*, and *rotA* by binding to their promoters.

These two stress response systems appear to regulate three classes of genes involved in protein folding or degradation in the periplasm. One class of genes, such as *dsbA* and *rotA*, appears to be only under the control of Cpx regulon. The second class of genes is under the control of just the σ^E regulon. This class includes the genes encoding σ^{32}, FkpA, and σ^E. The third class of genes is controlled by two stress response regulons. For example, the synthesis of DegP can be controlled by both the σ^E and Cpx regulons, whereas PpiD is under control of the σ^{32} and Cpx regulons.

There is another operon that appears to be induced under certain stress conditions, but the role that its products play in the physiology of the bacterial cell is not clear. Infection by filamentous phage or production of pIV from a plasmid results in the production of PspA, a 25.5-kD peripheral cytoplasmic membrane protein (Brissette et al. 1990). PspA is encoded by the first gene in the pspA-E operon, transcription of which is controlled by σ^{54} and an activator protein. PspA expression can be induced by stress such as heat shock, hyperosmotic shock, or the expression of outer membrane proteins defective in folding (Model et al. 1997). The effect that PspA or other proteins encoded by this operon have on bacteria is not known. The presence of PspA allows *E. coli* to survive in stationary phase at alkaline pH (Weiner and Model, 1994), as well as maintain the proton motive force under stress conditions (Kleerebezem et al. 1996). Whether PspA has a direct or indirect role in these events is unknown.

For more information about the Cpx and σ^E response systems, see Danese and Silhavy (1998a) and Missiakas and Raina (1997b); the psp operon, see Model et al. (1997).

REFERENCES

Adey N.B., Mataragnon A.H., Rider J.E., Carter J.M., and Kay B.K. 1995. Characterization of phage that bind plastic from phage-displayed random peptide libraries. *Gene* **156:** 27–31.

Armstrong J., Perham R.N., and Walker J.E. 1981. Domain structure of the bacteriophage fd adsorption protein. *FEBS Lett.* **135:** 167–172.

Baneyx F. and Georgiou G. 1991. Construction and characterization of *Escherichia coli* strains deficient in multiple secreted proteases: Protease III degrades high molecular weight substrates in vivo. *J. Bacteriol.* **173:** 2696–2703.

Barbas C.F., Kang A.S., Lerner R.A., and Benkovic S.J. 1991. Assembly of combinatorial antibody libraries on phage surfaces: The gene III site. *Proc. Natl. Acad. Sci.* **88:** 7978–7982.

Bardwell J.C., McGovern K. and Beckwith J. 1991. Identification of a protein required for disulfide bond formation *in vivo*. *Cell* **67:** 581–589.

Bardwell J.C.A., Lee J.O., Jander G., Martin N., Belin D., and Beckwith J. 1993. A pathway for disulfide bond formation in vivo. *Proc. Natl. Acad. Sci.* **90:** 1038–1042.

Bass S., Greene R., and Wells J.A. 1990. Hormone phage: An enrichment method for variant proteins with altered binding properties. *Proteins* **8:** 309–314.

Bauer E. and Smith G.P. 1988. Filamentous phage morphogenetic signal sequence and orientation of DNA in the virion and gene V protein complex. *Virology* **167:** 166–175.

Beck E. and Zink B. 1981. Nucleotide sequence and gene organization of filamentous bacteriophages f1 and fd. *Gene* **16:** 35–58.

Bernadac A., Gavioli M., Lazzaroni J.C., Raina S., and Lloubes R. 1998. *Escherichia coli tol/pal* mutants form outer membrane vesicles. *J. Bacteriol.* **180:** 4872–4878.

Berks B.C. 1996. A common export pathway for proteins binding complex redox factors? *Mol. Microbiol.* **22:** 393–404.

Betton J-M., Sassoon N., Hofnung M., and Laurent M. 1998. Degredation versus aggregation of misfolded maltose-binding protein in the periplasm of *Escherichia coli*. *J. Biol. Chem.* **273:** 8897–8902.

Bogsch E.G., Sargent F., Stanley N.R., Berks B.C., Robinson C., and Palmer T. 1998. An essential component of a novel bacterial protein export system with homologues in plastids and mitochondria. *J. Biol. Chem.* **273:** 18003–18006.

Bothmann H. and Plückthun A. 1998. Selection for a periplasmic factor improving phage dis-

play and functional periplasmic expression. *Nat. Biotechnol.* **16:** 376–380.
Breitling F., Dübel S., Seehaus T., Klewinghaus, and Little M. 1991. A surface infection vector for antibody screening. *Gene* **104:** 147–153.
Brissette J.L., Russel M., Weiner L., and Model P. 1990. Phage shock protein, a stress protein of *Escherichia coli. Proc. Natl. Acad. Sci.* **87:** 862–866.
Chanal A., Santni C., and Wu L. 1998. Potential receptor function of three homologous components, TatA, TatB, and TatE, of the twin-arginine signal sequence-dependent metalloenzyme translocation pathway in *Escherichia coli. Mol. Microbiol.* **30:** 674–676.
Chen R. and Henning U. 1996. A periplasmic protein (Skp) of *Escherichia coli* selectively binds a class of outer membrane proteins. *Mol. Microbiol.* **19:** 1287–1294.
Chivers P.T., Prehoda K.E, and Raines R.T. 1997. The CXXC motif: A rheostat in the active site. *Biochemistry* **36:** 4061–4066.
Clackson T. and Wells J. A. 1994. *In vitro* selection fom protein and peptide libraries. *Trends Biotechnol.* **12:** 173–184.
Click E.M. and Webster R.E. 1997. Filamentous phage infection: Required interactions with the TolA protein. *J. Bacteriol.* **179:** 6464–6471.
———. 1998. The TolQRA proteins are required for membrane insertion of the major capsid protein of the filamentous phage f1 during infection. *J. Bacteriol.* **180:** 1723–1728.
Crissman J.W. and Smith G.P. 1984. Gene-III protein of filamentous phages: evidence for a carboxy-terminal domain with a role in morphogenesis. *Virology* **132:** 445–455.
Cristobal S., de Gier J.W., Nielsen H., and von Heijne G. 1999. Competition between Sec- and Tat-dependent protein translocation in *Escherichia coli. EMBO J.* **18:** 2982–2990.
Dalbey R.E. and Robinson C. 1999. Protein translocation into and across the bacterial plasma membrane and the plant thylakoid membrane. *Trends Biochem. Sci.* **24:** 17–22.
Danese P.N. and Silhavy T.J. 1997. The σ^E and the Cpx signal transduction systems control the synthesis of periplasmic folding proteins in *Escherichia coli. Genes Dev.* **11:** 1183–1193.
———. 1998a. Targeting and assembly of periplasmic and outer membrane proteins in *Escherichia coli. Annu. Rev. Genet.* **32:** 59–94.
———. 1998b. CpxP, a stress combative member of the Cpx regulon. *J. Bacteriol.* **180:** 831–839.
Dartigalongue C. and Raina S. 1998. A new heat-shock protein gene ppiD encoded a peptidyl-prolyl isomerase required for folding outer membrane proteins in *Escherichia coli. EMBO J.* **17:** 3968–3980.
Davis N.G., Boeke J.D., and Model P. 1985. Fine structure of a membrane anchor domain. *J. Mol. Biol.* **181:** 111–121.
De Gier J.-W.L., Scotti P.A., Sääf A., Valent Q.A., Kuhn A., Liurink J., and von Heijne G. 1998. Differential use of the signal recognition particle translocase targeting pathway for inner membrane protein assembly in *Escherichia coli. Proc. Natl. Acad. Sci.* **95:** 14646–14651.
De Gier J.-W.L., Valent, Q.A., von Heijne, G. and Liurink J. 1997. The *E. coli* SRP: Preferences of a targeting factor. *FEBS Lett.* **408:** 1–4.
De las Penas A., Connolly L., and Gross C. 1997. The σ^E-mediated response to extracytoplasmic stress in *Escherichia coli* is transduced by RseA and RseB, two negative regulators of σ^E. *Mol. Mocrobiol.* **24:** 373–385.
Deng L.W., Malik P., and Perham R.N. 1999. Interaction of the globular domains of pIII protein of filamentous bacteriophage fd with the F-pilus of *Escherichia coli. Virology* **253:** 271–277.
Driessen A.J., Fekkes P., and van der Wolk J.P. 1998. The Sec system. *Curr. Opin. Microbiol.* **1:** 216–222.
Duenas M. and Borrebaeck C.A.K. 1994. Clonal selection and amplification of phage displayed antibodies by linking antigen recognition and phage replication. *Biotechnology* **12:**

999–1002.
Duenas M., Malmborg A.-C., Casavilla R., Ohlin M, and Borrebaeck C.A.K. 1996. Selection of phage displayed antibodies based on kinetic constants. *Mol. Immunol.* **33:** 279–285.
Dunn I.S. 1996. Phage display of proteins. *Curr. Opin. Bio/Technol.* **7:** 547–553.
Duong F. and Wickner W. 1997. Distinct catalytic roles of the SecYE, SecG, and SecDFyajC subunits of preprotein translocase holoenzyme. *EMBO J.* **16:** 2756–2768.
Eklund H., Gleason F.K., and Holmgren A. 1991. Structural and functional relations among thioredoxins of different species. *Proteins* **11:**13–28.
Endemann H. and Model P. 1995. Location of filamentous phage minor coat proteins in phage and in infected bacteria. *J. Mol. Biol.* **250:** 496–506.
Feng J.N., Russel M., and Model P. 1997. A permeabilized system that assembles filamentous bacteriophage. *Proc. Natl. Acad. Sci.* **94:** 4068–4073.
Firth N., Ippen-Ihler K., and Skurray R. 1996. Stucture and function of the F factor and mechanism of conjugation. In Escherichia coli *and* Salmonella *cellular and molecular biology* (Neidhardt F.C. ed.), pp. 2377–2401. American Society of Microbiology, Washington, D.C.
Frost L.F., Ippen-Ihler K., and Skurry R.A. 1994. Analysis of the sequence and gene products of the transfer region of the F factor. *Microbiol. Rev.* **58:** 162–210.
Fulford W. and Model P. 1984. Gene X of bacteriophage f1 is required for phage DNA synthesis: Mutagenesis of in-frame overlapping genes. *J. Mol. Biol.* **178:** 137–153.
Gao C., Mao S., Low C-H. L., Wirsching P., Lerner R.A., and Janda K.D. 1999. Making artificial antibodies: A format for phage display of combinatorial heterodynamic arrays. *Proc. Natl. Acad. Sci.* **96:** 6025–6030.
Germon P., Clavel T., Vianney A., Portalier R., and Lazzaroni J.C. 1998. Mutational analysis of the *Escherichia coli* K-12 N-terminal region and characterization of its TolQ-interacting domain by genetic suppression. *J. Bacteriol.* **180:** 6433–6439.
Glaser-Wuttke G., Keppner J., and Rasched I. 1989. Pore forming properties of the adsorption protein of filamentous phage fd. *Biochim. Biophys. Acta* **985:** 239–247.
Gramatikoff K., Georgiev O., and Schaffner W. 1994. Direct interaction rescue, a novel filamentous phage technique to study protein-protein interactions. *Nucleic Acids Res.* **22:** 5761–5762.
Gray C.W. 1989. Three-dimentional structure of complexes of single-stranded DNA-binding proteins with DNA. Ike and fd gene 5 proteins form left-handed helices with single-stranded DNA. *J. Mol. Biol.* **208:** 57–64.
Gray C.W., Brown R.S., and Marvn D.A. 1981. Adsorption complex of filamentous fd virus. *J. Mol. Biol.* **146:** 621–627.
Greenwood J., Hunter G.J., and Perham R.N. 1991a. Regulation of bacteriophage length by modification of electrostatic interactions between coat protein and DNA. *J. Mol. Biol.* **217:** 223–227.
Greenwood J., Willis A.E., and Perham R.N. 1991b. Multiple display if foreign peptides on a filamentous bacteriophage. Peptides from *Plasmodium falciparum* circumsporozoite protein as antigens. *J. Mol. Biol.* **220:** 821–827.
Gross C.A. 1996. Function and regulation of the heat shock proteins. In Escherichia coli *and* Salmonella *cellular and molecular biology* (Neidhardt F.C. ed.), pp. 1383–1399. American Society of Microbiology Washington, D.C.
Guan Y., Zhang H., and Wang A.H.J. 1995. Electrostatic potential distribution of the gene V protein from Ff phage facilitates cooperative DNA binding: A model of the GVP-ssDNA complex. *Protein Sci.* **4:** 187–197.
Guy-Caffey J.K. and Webster R.E. 1993. The membrane domain of a bacteriophage assembly protein: Membrane insertion and growth inhibition. *J. Biol. Chem.* **268:** 5496–5503.

Guy-Caffey J.K., Rapoza M.P., Jolley K.A., and Webster R.E. 1992. Membrane localization and topology of a viral assembly protein. *J. Bacteriol.* **174:** 2460–2465.

Haigh N.G. and Webster R.E. 1998. The major coat protein of filamentous bacteriophage f1 specifically pairs in the bacterial cytoplasmic membrane. *J. Mol. Biol.* **279:** 19–29.

Hart S.L., Knight A.M., Harbottle R.P., Mistry A., Hunger H.-D., Cutler D.F., Williamson R., and Coutelle C. 1994. Cell binding and internalization by filamentous bacteriophage displaying a cyclic Arg-Gly-Asp containing peptide. *J. Biol. Chem.* **269:** 12468–12474.

Hill D.F. and Petersen G.B. 1982. Nucleotide sequence of bacteriophage fl DNA. *J. Virol.* **44:** 32–46.

Holliger P., Riechmann L., and Williams R.L. 1999. Crystal structure of the two N-terminal domains of g3p from filamentous phage fd at 1.9 Å: Evidence for conformational liability. *J. Mol. Biol.* **288:** 649–657.

Holmgren A. 1989. Thioredoxin and glutaredoxin systems. *J. Biol. Chem.* **264:** 13963–13966.

Hoogenboom H.R. 1997. Designing and optimizing selection strategies for generating high affinity antibodies. *Trends Biotechnol.* **15:** 62–70.

Horabin J.L. and Webster R.E. 1988. An amino acid sequence which directs membrane insertion causes loss of membrane potential. *J. Biol. Chem.* **263:** 11575–11583.

Houbiers M.C., Spruijt R.B., Wolfs C.J.A.M., and Hemminga M.A. 1999. Coformational and aggregational properties of the gene 9 minor coat protein of bacteriophage M13 in membrane-mimicking systems. *Biochemistry* **38:** 1128–1135.

Iannolo G., Minenkova O., Petruzzelli R., and Cesareni G. 1995. Modifying filamentous phage capsid: Limits in the size of the major capsid protein. *J. Mol. Biol.* **248:** 835–844.

Ivey-Hoyle M. and Steege D.A. 1989. Translation of phage f1 gene VII occurs fron an inherently defective initiation site made functional by coupling. *J. Mol. Biol.* **208:** 233–244.

———. 1992. Mutational analysis of an inherently defective translation initiation site. *J. Mol. Biol.* **224:** 1039–1054.

Jespers L.S., Messens J.H., Keyser A.D., Eeckhout D., Brande I.V.D., Gansemans Y.G., Lauwereys M.J., Viasuk G.P., and Stassens P.E. 1995. Surface expression and ligand-based selection of cDNA's fused to filamentous phage gene VI. *Bio/Technology* **13:** 378–382.

Kamitami S., Akiyama Y., and Ito K. 1992. Identification and characterization of an *Eschericia coli* gene required for correctly folded alkaline phosphatase, a periplasmic enzyme. *EMBO J.* **11:** 57–62.

Kang A.S., Barbas C.F., Janda K.D., Benkovic S.J., and Lerner R.A. 1991. Linkage of recognition and replication functions by assembling combinatorial antibody Fab libraries along phage surfaces. *Proc. Natl. Acad. Sci.* **88:** 4363–4366.

Kay B.K., Winter J., and MaCafferty J. 1996. *Phage display of peptides and proteins.* Academic Press, San Diego.

Kampfenkel K. and Braun V. 1993. Membane toplogies of the TolQ and TolR proteins of *Escherichia coli:* Inactivation of TolQ by a missense mutation in the proposed first transmembrane segment. *J. Bacteriol.* **175:** 4485–4491.

Kazmierczak B., Mielke D.L., Russel L., and Model P. 1994. Filamentous phage pIV forms a multimer that mediates phage export across the bacterial cell envelope. *J. Mol. Biol.* **238:** 187–198.

Kishigami S. and Ito K. 1996. Roles of cysteine residues of DsbB in its activity to reoxidize DsbA, the protein disulfide bond catalyst of *Escherichia coli. Genes Cells* **1:** 201–208.

Kleerebezem M., Crielaard W., and Tommassen J. 1996. Involvement of stress protein PspA (phage shock protein A) of *Escherichia coli* in maintenance of the proton motive force under stress conditions. *EMBO J.* **15:** 162–171.

Kobayashi T. and Ito I. 1999. Respiratory chain strongly oxidizes the CXXC motif of DsbB in the *Escherichia coli* disulfide bond formation pathway. *EMBO J.* **18:** 1192–1198.

Konings R.N.H., Folmer R.H.A., Folkers P.J.M., Niglis M., and Hilbers C.W. 1995. Three-dimensional structure of the single-stranded binding protein encoded by gene V of the filamentous bacteriophage M13 and a model of its complex with single-stranded DNA. *FEMS Microbiol. Rev.* **17:** 57.

Krebber C., Spada S., Desplancq D., and Plückthun A. 1995. Co-selection of cognate antibody-antigen pairs by selectively-infective phages. *FEBS Lett.* **377:** 227–231.

Krebber C., Spada S., Desplancq D., Krebber A., Ge L., and Plückthun A. 1997. Selectively infective phage (SIP): A mechanistic dissection of a novel *in vivo* selection for protein-ligand interactions. *J. Mol. Biol.* **268:** 607–618.

Kremser A. and Rasched I. 1994. The adsorption protein of the filamentous phage fd: Assignment of its disulfide bridges and identification of the domain incorporated in the coat. *Biochemistry* **33:** 13954–13958.

Kuhn A. and Troschel D. 1992. Distinct steps in the insertion pathway of the bacteriophage coat proteins. In *Membrane biogenesis and protein targeting* (Newport W. and Lill R. ed.), pp. 33–47. Elsevier, New York.

Lazdunski C.J., Bouveret E. Rigal A., Journet L., Lloubes R., and Benedetti H. 1998. Colicin import into *Escherichia coli* cells. *J. Bacteriol.* **180:** 4993–5002.

Lazzaroni J.C., Germon P., Ray M.C., and Vianney A. 1999. The Tol proteins of *Escherichia coli* and their involvement in the uptake of biomolecules and outer membrane stability. *FEBS Microbiol. Lett.* **177:** 191–197.

Lazzaroni J.C., Vianney A., Popot J.C., Benedetti H., Samatey F., Lazdunski C., Portalier R., and Geli V. 1995. Transmembrane α-helix interactions are required for the funtional assembly of the *Escherichia coli* Tol complex. *J. Mol. Biol.* **246:** 1–7.

Levengood S.K., Beyer W.F., and Webster R.E. 1991. TolA: A membrane protein involved in colicin uptake contains and extended helical region. *Proc. Natl. Acad. Sci.* **88:** 5939–5943.

Levengood-Freyermuth S.K., Click E.M., and Webster R.E. 1993. Role of the carboxyl-terminal domain of TolA in protein import and integrity of the outer membrane. *J. Bacteriol.* **175:** 222–228.

Linderoth N.A., Simon M.N., and Russel M. 1997. The filamentous phage pIV multimer visualized by scanning transmission electron microscopy. *Science* **278:** 1635–1638.

Liu J. and Walsh C.T. 1990. Peptidyl-prolyl cis-trans isomerase from *Escherichia coli*: A periplasmic homologue of cyclophilin that is not inhibited by cyclosporin A. *Proc. Natl. Acad. Sci.* **87:** 4028–4032.

Lopez J. and Webster R.E. 1983. Morphogenesis of filamentous bacteriophage f1: Orientation of extrusion and production of polyphage. *Virology* **127:** 177–193.

———. 1985. Assembly site of bacteriophage f1 corresponds to adhesion zones between the inner and outer membranes of the host cell. *J. Bacteriol.* **163:** 1270–1274.

Lubkowski J., Hennecke F., Plückthun A., and Wlodawer A. 1998. The structural basis of phage display elucidated by the crystal structure of the N-terminal domains of g3p. *Nat. Struct. Biol.* **5:** 140–147.

———. 1999. Filamentous-phage infection: crystal structure of g3p in complex with its coreceptor, the C-terminal domain of TolA. *Structure* **7:** 711–722.

Makowski L. 1992. Terminating a macromolecular helix. Structural model for the minor proteins of bacteriophage M13. *J. Mol. Biol.* **228:** 885–892.

Malik P., Terry T.D., Bellintani F., and Perham, R.N. 1998. Factors limiting display of foreign peptides on the major coat protein of filamentous bacteriophage capsids and a potential role for leader peptidase. *FEBS Lett.* **436:** 263–266.

Malik P., Terry T.D., Gouda L.R., Langara A., Petukov S.A., Symmons M.F., Welch L.C., Marvin

D.A., and Perham R.N. 1996. Role of capsid structure and membrane protein processing in determining the size and copy number of peptides displayed on the major coat protein of filamentous bacteriophage. *J. Mol. Biol.* **260:** 9–21.

Marciano D.K., Russel M., and Simon S.M. 1999. An aqueous channel for filamentous phage export. *Science* **284:** 1516–1519.

Martin J.L., Bardwell J.C., and Kuriyan J. 1993. Crystal structure of the DsbA protein required for disulfide bond formation *in vivo. Nature* **365:** 464–468.

Marvin D.A. 1998. Filamentous phage structure, infection and assembly. *Curr. Opin. Sruct. Biol.* **8:** 150–158.

Marvin D.A., Hale R.D., Nave C., and Helmer Citterich M. 1994. Molecular models and structural comparisons of native and mutant class I filamentous bacteriophages. *J. Mol. Biol.* **235:** 260–286.

McDonnell P.A., Shon K., Kim Y., and Opella S.J. 1993. fd coat protein structure in membrane environments. *J. Mol. Biol.* **233:** 447–463.

Miller C.G. 1996. Protein degredation and proteolytic modification. In Escherichia coli *and* Salmonella *cellular and molecular biology* (Neidhardt F.C., ed.), pp. 938–954. American Society of Microbiology, Washington, D.C.

Minekova O.O., Il'ichev A.A., Kishchenko G.P., and Petrenko V.A. 1993. Design of specific immunogens using filamentous bacteriophage as the carrier. *Gene* **128:** 85–88.

Missiakas D. and Raina S. 1997a. Protein folding in the bacterial periplasm. *J. Bacterial.* **179:** 2465–2471.

———. 1997b. Protein misfolding in the cell envelope of *Escherichia coli*: New signalling pathways. *Trends Biochem. Sci.* **22:** 59–63.

Missiakas D., Betton J.-M., and Raina S. 1996. New components of protein folding in extracytoplasmic compartments of *Escherichia coli*: SurA, FkpA and Skp/OmpH. *Mol. Microbiol.* **21:** 871–874.

Missiakas D., Georgopoulos C., and Raina S. 1993. Identification and characterization of the *Escherichia coli* gene *dsbB*, whose product is involved in the formation of disulfide bonds *in vivo. Proc. Natl. Acad. Sci.* **90:** 7084–7088.

Missiakas D., Schwager F., and Raina S. 1995. Identification and characterization of a new disulfide isomerase-like protein (DsbD) in *Escherichia coli. EMBO J.* **14:** 3415–3424.

Model P. and Russel M. 1988. Filamentous bacteriophage. In *The bacteriophages* (Calendar, R. ed.), vol 2, pp. 375–456. Plenum Publishing, New York.

Model P., Jovanovic G., and Dworkin J. 1997. The *Escherichia coli* phage shock protein (psp) operon. *Mol. Microbiol.* **24:** 255–261.

Muller M.M., Vianney A., Lazzaroni J.C., Webster R.E., and Portalier R. 1993. Membrane topology of the *Escherichia coli* TolR protein required for cell envelope integrity. *J. Bacteriol.* **175:** 6059–6061.

Oi H.-Y. and Bernstein H.D. 1999. SecA is required for the insertion of inner membrane proteins targeted by the *Escherichia coli* signal recognition particle. *J. Biol. Chem.* **274:** 8993–8997.

Olah G.A., Gray D.M., Gray C.W., Kergil D.L., Sosnick T.R., Mark B.L., Vaughan M.R., and Trewhella J. 1995. Structures of fd gene 5 protein-nucleic acid complexes: A combined solution, scattering and electronmicroscopy study. *J. Mol. Biol.* **249:** 576–594.

Oliver D. 1996. Periplasm. In Escherichia coli *and* Salmonella *cellular and molecular biology* (Neidhardt F. C., ed.), pp. 88-103. American Society of Microbiology, Washington, D.C.

Overman S.A. and Thomas G.J., Jr. 1995. Raman spectroscopy of the filamentous virus Ff (fd, f1, M13): Structural interpretations for coat protein aromatics. *Biochemistry* **34:** 5440–5451.

Papavoine C.H.M., Remerowski M.L., Horstink K.L.M., Konings R.N.H., Hilbers C.W., and van de Ven F.J.M. 1997. Backbone dynamics of the major coat protein of bacteriophage M13 in detergent micelles by N^{15}-nuclear magnetic resonance relaxation measurements using the model-free approach and reduced spectral density mapping. *Biochemistry* **36:** 4015–4026.

Petrenko V.A., Smith G.P., Gong X., and Quinn T. 1996. A library of organic landscapes on filamentous bacteriophage. *Protein Eng.* **9:** 797–801.

Pogliano J., Lynch A.S., Belin D., Lin E.C.C., and Beckwith J. 1997. Regulation of *Escherichia coli* cell envelope proteins involved in protein folding and degredation by the Cpx two-component system. *Genes Dev.* **11:** 1169–1182.

Powers T. and Walter P. 1997. Co-translational protein targeting catalyzed by the *Escherichia coli* signal recognition particle and its receptor. *EMBO J.* **16:** 4880–4886.

Rakonjac J. and Model P. 1998. Roles of pIII in filamentous phage assembly. *J. Mol. Biol.* **282:** 25–41.

Rakonjac J., Feng J.N., and Model P. 1999. Filamentous phage are released from the bacterial membrane by a two-step mechanism involving a short C-terminal fragment of pIII. *J. Mol. Biol.* **289:** 1253–1265.

Rapoza M.P. and Webster R.E. 1993. The filamentous phage assembly proteins require the bacterial SecA protein for correct localization to the membrane. *J. Bacteriol.* **175:** 1856–1859.

———. The products of gene I and the overlapping in-frame gene XI are required for filamentous phage assembly. *J. Mol. Biol.* **248:** 627–638.

Rietsch A. and Beckwith J. 1998. The genetics of disulfide bond metabolism. *Annu. Rev. Genet.* **32:** 163–184.

Rietsch A., Belin D., Martin N., and Beckwith J. 1996. An in vivo pathway for disulfide bond isomerization in *Escherichia coli. Proc. Natl. Acad. Sci.* **93:** 13048–13053.

Rietsch A., Bessette P., Georgiou G., and Beckwith J. 1997. Reduction of the periplasmic disulfide bond isomerase, DsbC, occurs by passage of electrons from cytoplasmic thioredoxin. *J. Bacteriol.* **179:** 6602–6608.

Riechmann L. and Hollinger P. 1997. The C-terminal domain of TolA is the coreceptor for filamentous infection of *E coli. Cell* **90:** 351–360.

Rodrigue A., Chanal A., Beck K., Müller M., and Wu. L.-F. 1999. Co-translocation of a periplasmic enzyme complex through the bacterial Tat pathway. *J. Biol. Chem.* **274:** 13223–13228.

Russel M. 1993. Protein-protein interactions during filamentous phage assembly. *J. Mol. Biol.* **231:** 689–697.

———. 1995. Moving through the membrane with filamentous phages. *Trends Microbiol.* **3:** 223–228.

———. 1998. Macromolecular assembly and secretion across the bacterial envelope: type II protein secretion systems. *J. Mol. Biol.* **279:** 485–299.

Russel M. and Kazmierczak B. 1993. Analysis of the structure and subcellular location of filamentous phage pIV. *J. Bacteriol.* **175:** 3998–4007.

Russel M. and Model P. 1986. The role of thioredoxin in filamentous phage assembly. Construction, isolation, and characterization of mutant thioredoxins. *J. Biol. Chem.* **261:** 14997–15005.

Russel M. and Model P. 1989. Genetic analysis of the filamentous bacteriophage packaging signal and the proteins that interact with it. *J. Virol.* **63:** 3284–3295.

Russel M., Kidd S., and Kelley M.R. 1986. An improved filamentous helper phage for generating single-stranded plasmid DNA. *Gene* **45:** 333–338.

Russel M., Linderoth N.A., and Sali A. 1997. Filamentous phage assembly: Variation on a protein export theme. *Gene* **192:** 23–32.

Russel M., Whirlow H., Sun T.P., and Webster R.E. 1988. Low-frequency infection of F-bacteria by transducing particles of filamentous bacteriophages. *J. Bacteriol.* **170:** 5312–5316.

Santini C.-L., Ize B., Chanal A., Müller M., Giordano G., and Wu L.-F. 1998. A novel Sec-independent periplasmic translocation pathway in *Escherichia coli*. *EMBO J.* **17:** 101–112.

Sargent F., Bogsch E.G., Stanly N.R., Wexler M., Robinson C., Berks B.C., and Palmer T. 1998. Overlapping functions of components of a bacterial Sec-independent protein export pathway. *EMBO J.* **17:** 3640–3650.

Seluanov A. and Bibi E. 1997. FtsY, the procaryotic signal recognition particle receptor homalogue is essential for biogenesis of membrane proteins. *J. Biol. Chem.* **272:** 2053–2055.

Settles A.M., Yonitani A., Baron A., Bush D.R., Cline K., and Martienssen R. 1997. Sec-independent protein translocation by the maize Hcf106 protein. *Science* **278:** 1467–1470.

Simons G.F., Konings R.N., and Schoenmakers J.G. 1981. Genes VI, VII, and IX of phage M13 code for minor capsid proteins of the virion. *Proc. Natl. Acad. Sci.* **78:** 4194–4198.

Skinner M.M., Zhang H., Leschnitzer D.H., Guan Y., Bellamy H., Sweet R.M., Gray C.W., Konings R.N.H., Wang A.H.J., and Terwilliger T.C. 1994. Structure of the gene V protein of bacteriophage f1 determined by multiwavelength x-ray diffraction on the selinomethionyl protein. *Proc. Natl. Acad. Sci.* **91:** 2071–2075.

Smith G.P. 1985. Filamentous fusion phage: Novel expression vectors that display cloned antigens on the virion surface. *Science* **228:** 1315–1317.

Smith G.P. and Scott J.K. 1993. Libraries of peptides and proteins displayed on filamentous phage. *Methods Enzymol.* 217: 228–257.

Sone M., Akiyama Y., and Ito K. 1997. Differential roles played by DsbA and DsbC in the formation of protein disulfide bonds. *J. Biol. Chem.* **272:** 10349–10352.

Spada S., Krebber C., and Plückthun A. 1997. Selectively infective phages (SIP). *Biol. Chem.* **378:** 445–456.

Specthrie L., Bullitt E., Horiuchi K., Model P., Russel M., and Makowski L. 1992. Construction of a microphage variant of filamentous bacteriophage. *J. Mol. Biol.* **228:** 720–724.

Stengle I., Bross P., Garces X., Giray J., and Rasched I. 1990. Disection of functional domains on phage fd adsorption protein-descrimination between attachment and penetration sites. *J. Mol. Biol.* **212:** 143–149.

Stump M.D. and Steege, D.A. 1996. Functional analysis of filamentous phage f1 processing sites. *RNA* **2:** 1286–1294.

Sun T.P. and Webster R.E. 1987. Nucleotide sequence of a gene cluster involved in entry of E colicins and single-stranded DNA of infecting filamentous bacteriophages into *Escherichia coli*. *J. Bacteriol.* **169:** 2667–2674.

Terry T.D., Malik P., and Perham R.N. 1997. Accessibility of peptides displayed on filamentous virions: Susceptibility to proteinases. *Biol. Chem.* **378:** 523–530.

Ulbrandt N.D., Newitt J.A., and Bernstein H.D. 1997. The *Escherichia coli* signal recognition particle is required for the insertion of a subset of inner membrane proteins. *Cell* **88:** 187–196.

Valent O.A., Scotti P.A., High S.L., de Gier J.-W., vonHeijne G., Lenzen G., Wintermeyer W., Oudega B., and Luirink J. 1998. The *Escherichia coli* SRP and SecB targeting pathways converge at the translocan. *EMBO J.* **17:** 2504–2512.

Van der Wolk J.P.W., De Wit J.G., and Driessen A.J.M. 1997. The catalytic cycle of the *Escherichia coli* SecA ATPase comprises two distinct preprotein translocation events. *EMBO J.* **16:** 7297–7304.

Van Straaten M., Missiakas D., Raina S., and Darby N.J. 1998. The functional properties of DsbG, a thiol-disulfide oxidoreductase from the periplasm of *Escherichia coli*. *FEBS Lett.* **428:** 255–258.

Van Wezebbeek P.M.G.F., Hulsbos T.J.M., and Schoenmakers, J.G.G. 1980. Nucleotide sequence of the filamentous bacteriophage M13 DNA genome: Comparison with phage fd. *Gene* **11:** 129–148.

Vianney A., Lewin T.M., Beyer W.F., Lazzaroni J.C., Portalier R., and Webster R.E. 1994. Membrane topology and mutational analysis of the TolQ protein of *Escherichia coli* required for the uptake of macromolecules and cell envelope integrity. *J. Bacteriol.* **176:** 822–829.

von Heijne G. 1992. Membrane protein structure prediction hydrophobicity analysis and the positive-inside rule. *J. Mol. Biol.* **225:** 487–404.

Webster R.E. 1991. The *tol* gene products and the import of macromolecules into *Escherichia coli. Mol. Microbiol.* **5:** 1005–1011.

———. 1996. Biology of the filamentous bacteriophage In *Phage display of peptides and proteins* (Kay B.K. et al., ed.), Academic Press, San Diego.

Webster R.E. and Lopez J. 1985. Structure and assembly of the class 1 filamentous bacteriophage. In *Virus structure and assembly* (Casjens S., ed.), pp. 235–268, Jones and Bartlett, Boston.

Weiner J.H., Bilous P.T., Shaw G.M., Lubitz S.P., Frost L., Thomas G.H., Cole J.A. and Turner R.J. 1998. A novel and ubiquitous system for membrane targeting and secretion of cofactor containing proteins. *Cell* **93:** 93–101.

Weiner L. and Model P. 1994. Role of an *Escherichia coli* stress-response operon in stationary-phase survival. *Proc. Natl. Acad. Sci.* **91:** 2191–2195.

Wild J., Rossmeissl P., Walter W.A., and Gross C.A. 1996. Involvement of DnaK-DnaJ-GrpE chaperone team in protein secretion in *Escherichia coli. J. Bacteriol.* **178:** 3608–3613.

Williams K.A., Glibowicka M., Li H., Kahn A.R., Chen Y.M.Y., Wang J., Marvin D.A. and Deber, C.M. 1995. Packing of coat protein amphipathic and transmembrane helices in filamentous bacteriophage M13: Role of small residues in protein oligomerization. *J. Mol. Biol.* **252:** 6–14.

Williams K.A., Farrow N.A., Deber C.M., and Kay L.E. 1996. Structure and dynamics of bacteriophage Ike major coat protein in MPG micelles by solution NMR. *Biochemistry* **35:** 5145–5157.

Winter G., Griffiths A.D., Hawkins R.E., and Hoogenboom H.R. 1994. Making antibodies by phage display technology. *Annu. Rev. Immunol.* **12:** 433–455.

Woolford J., Cashman J., and Webster R.E. 1974. f1 coat protein synthesis and altered phospholipid metabolism in f1 infected *Escherichia coli.* Virology **58:** 544–560.

Wunderlich M. and Glockshuler R. 1993. Redox properties of protein disulfide isomerase (DsbA) from *Escherichia coli. Protein Sci.* **2:** 717–726.

Zapun A., Bardwell J.C.A., and Creighton T.E. 1993. The reactive abd destabilizing disulfide bond of DsbA, a protein required for protein disulfide bond formation *in vivo. Biochemistry* **32:** 5083–5092.

2 Phage-display Vectors

[1]JAMIE K. SCOTT AND [2]CARLOS F. BARBAS III
[1] *Department of Molecular Biology and Biochemistry and Department of Biological Sciences, Simon Fraser University, Burnaby, B.C., Canada V5A 1S6*
[2] *Department of Molecular Biology, The Scripps Research Institute, La Jolla, California 92037*

ALL OF THE COAT PROTEINS OF THE FILAMENTOUS PHAGE can be fused to foreign proteins and peptides with varying degrees of success; however, most currently used vectors use the minor coat protein, pIII, or the major coat protein, pVIII, for such fusions. In George Smith's (1993) classification of the main types of phage-display vectors, each type of vector is differentiated on the basis of:

CONTENTS

- the coat protein used for display (i.e., pIII or pVIII)
- whether the protein to be displayed can be fused to all copies of pIII or pVIII, or to only a fraction of them
- whether the recombinant fusion is encoded on the phage genome or on a separate genome (e.g., a phagemid).

Table 2.1 uses this classification scheme to summarize the different types of vectors. Examples of vectors that are used in this manual are shown, and examples of other well-known phage-display systems are referenced. For a more complete catalog of the phage vectors, see Smith and Petrenko (1997).

Phage-display vectors also vary according to the "background" genome used for expressing the coat protein fusion. The biology of these vectors ranges from a mainly "wild-type" phage genome, whose coat protein gene has been altered for the insertion of foreign DNA (e.g., M13KE), to a "crippled" phage genome that can be propagated as a plasmid under antibiotic selection (f88-4), to a phagemid — a plasmid that carries the recombinant coat protein gene as well as a phage origin of replication (pComb3H). All three types of vectors are described in this chapter in some detail. These descriptions are meant to give the reader a practical understanding of the biology of each vector, its most appropriate uses, and its limitations.

Table 2.1. Classification of phage-display vectors

Vector type	Coat protein used for display	Display on all or some copies of coat protein	# of coat protein genes	Fusion encoded on phage or phagemid genome	Examples
Type 3	pIII	all	1	phage	M13KE[a,b,c]
Type 8	pVIII	all	1	phage	[d]
Type 33	pIII	some	2	phage	[e]
Type 88	pVIII	some	2	phage	f88-4[f,g]
Type 3+3	pIII	some	2	phagemid	pComb3[h]
Type 8+8	pVIII	some	2	phagemid	[i,j]

[a]Scott and Smith (1990).
[b]Cwirla et al. (1990).
[c]Kay et al. (1993).
[d]Petrenko et al. (1996).
[e]Unknown.
[f]McLafferty et al. (1993).
[g]Haaparanta and Huse (1995).
[h]Barbas et al. (1991).
[i]Kang et al. (1991).
[j]Wrighton et al. (1996).

WILD-TYPE PHAGE VECTORS: M13KE

The simplest phage-display vectors are those whose basic functions have not been altered; the gene for the coat protein that will be used as the scaffold for display is modified to carry restriction sites for cloning the protein or peptide to be displayed. These sites are introduced into the regions encoding the carboxyl terminus of the coat protein's leader sequence and the amino terminus of the mature coat protein. Thus, the sites flank the signal peptidase cleavage site, and after processing at the bacterial inner membrane, the displayed protein or peptide is located at the amino terminus of the mature coat protein. Wild-type phage vectors behave as wild-type filamentous phage in that they produce plaques with wild-type morphology in soft agar, and, in liquid culture, produce wild-type levels of phage and replicative form (RF) DNA.

Such vectors are almost all Type 3, and they are almost exclusively used to display peptide libraries. Although the amino terminus of mature pIII can tolerate fairly long peptide fusions (~30 amino acids; Kay et al. 1993), the display of proteins such as scFv or Fabs can be problematic on three levels. First, it can disturb the efficiency with which pIII is produced and affect phage assembly, for example, by impairing pIII insertion into the bacterial inner membrane. Although the expression of wild-type pIII may be independent of the bacterial secretory apparatus, the expression of complex pIII fusions may not be (Peters et al. 1994). Second, the presence of a large polypeptide domain at the amino terminus of pIII may reduce infectivity (Parmley and Smith 1988). Third, the larger the protein fusion, the more likely that it will be misfolded and/or cleaved by bacterial proteases, and thus it will not be displayed in its intended form. Consequently, as described below in the discussion of the pComb3H vector and its derivatives, larger proteins are displayed by phagemid (Type 3+3) systems.

There are few Type 8 vectors, because the pVIII protein is highly restricted in the length of polypeptide and amino acid sequences to which it can be fused without compromising phage assembly (Il'ichev et al. 1989; Greenwood et al. 1991; Petrenko et al.

1996). Thus, Type 8 vectors display only short peptides (of five residues or less) that comprise a biased group of amino acid sequences (Iannolo et al. 1995; Petrenko et al. 1996). Most of the more commonly used pVIII display vectors produce hybrid phage, with only a portion of the pVIII molecules bearing fusions; these include phagemid (Type 8+8) and phage (Type 88) vectors. One of the Type 88 vectors (Haaparanta and Huse 1995) is wild type, in that a second copy of *gene VIII* is inserted into a site in the intergenic region that does not greatly affect phage replication, and hence the phage produce wild-type plaques. As described below, the M13KE vector also bears an insert, in this instance of a *lacZ* gene, in a similar intergenic site that allows it to behave essentially as a "wild-type" phage.

The M13KE vector is a Type 3 vector that was derived from the M13mp19 phage vector (Messing et al. 1977; Messing 1983). It is essentially a wild-type M13 vector except for:

1. A *lacZ* gene containing a polylinker cloning site that was inserted into or near the plus-strand origin of replication to create M13mp19 (Messing et al. 1977; Messing 1983)
2. Unique *Kpn*I and *Eag*I restriction sites that were engineered into *gene III* for construction of pIII display libraries (See Fig. 2.1, A and B, for the map of M13KE and the DNA sequence of the cloning region, respectively.)

Foreign DNA encoding pIII-displayed proteins and peptides is inserted into *gene III* via the *Kpn*I and *Eag*I sites, and libraries can be made using double-stranded RF DNA or single-stranded phage DNA (see Chapters 19 and 16, respectively). Clones are grown in medium without antibiotic, and phage yields from liquid cultures are typically between 10^{13} and 10^{14} particles/ml; RF DNA yields are excellent, too. Such yields make it easy to prepare phage for DNA sequencing (Chapters 15 and 19) and for binding studies such as ELISAs (Chapters 18 and 19); these yields also simplify the preparation of high-quality RF DNA for cloning (Chapter 19).

M13KE phage behave almost identically to wild-type phage; plated clones are visualized as plaques and infectious units are counted as plaque-forming units (pfu). The amplification of phage after panning is simple, and the efficiency of M13KE infection is excellent, with an infectivity of about 1–2 particles/pfu. Libraries made with the M13KE vector can be contaminated by trace amounts of wild-type filamentous phage, which may become dominant in pannings in which tight-binding clones are not selected or when a library is repeatedly amplified. The presence of these contaminants can be detected by their inability to produce blue plaques during growth on plates containing X-gal and isopropyl-β-D-thiogalactopyranoside (IPTG); sequencing of *gene III* will also reveal a wild-type sequence with neither the *Kpn*I nor the *Eag*I site. Thus, it is prudent to use X-gal/IPTG detection when titering M13KE phage during panning or when evaluating an amplified library.

Because of the excellent infectivity of M13KE, almost all of the phage that are selected during a given round of panning will be amplified during phage propagation in the bacterial growth step. This is especially important in the first round of panning, in which each clone in the library is represented by a small number of phage. For example, if there are 10^9 clones in a library and 10^{12} phage particles are panned, each clone will be represented by approximately 1000 phage particles. Phage clones having high yields (i.e., > 1% of the particles of a given clone are retained by the screening molecule during the panning) will yield approximately 10 phage particles per clone after the first round of panning. The high infectivity of M13KE ensures that those phage will be amplified. See Chapter 17 for a more detailed discussion of phage panning.

fd-tet-DERIVED PHAGE VECTORS: f88-4

The first paper demonstrating the concept of phage display used the wild-type f1 phage, which bears a unique *BamHI* restriction endonuclease site in the region of *gene III* that encodes the middle portion of pIII (Smith 1985). Thus, f1 was the first Type 3 vector. The first instance of phage display involved fusion with a 57-amino-acid polypeptide fragment derived from the *Eco*RI restriction endonuclease. Smith showed that phage bearing this fragment could be affinity selected by a polyclonal antibody against the *Eco*RI enzyme with an enrichment factor of more than 1000 over M13mp8 phage.

In this work, however, Smith also mentioned that the presence of the insert greatly reduced the phage's plaque size (an indication of poor phage production) and decreased the phage titer from growth in liquid culture by more than 100-fold. Whole-virion electrophoresis and electron microscopy of the f1 fusions revealed multimeric polyphage, whose morphology consisted of very long phage bodies containing multiple single-stranded DNA genomes. Smith had shown earlier that phage that cannot produce pIII form polyphage (Crissman and Smith 1984), so the morphology of the f1 fusions was taken as another indication that production of the pIII fusion protein was impaired. Moreover, phage infectivity was reduced from two phage particles/pfu for f1 to more than 50 particles/pfu for f1 bearing the fusion; this indicated that expression of the *Eco*RI fragment hindered infection via the F pilus, a function mediated by the amino-terminal domains of pIII. Because of the multiple defects in phage production and infectivity caused by display of the *Eco*RI fragment by f1, Smith suggested the use of his fd-tet cloning vector (Zacher et al. 1980) for phage display. He had shown that it could be propagated as a tetracycline-resistant plasmid, independent of *gene III* function (Crissman and Smith 1984).

The fd-tet vector was originally developed as a "crippled" DNA cloning vector (Zacher et al. 1980) that could not survive on its own without antibiotic selection. Its biology is quite different from that of wild-type phage. Its ability to confer tetracycline resistance comes from the presence of a fragment bearing genes from the Tn10 transposon that is inserted into the phage minus-strand origin of replication (see Fig. 2.1C). As a result, fd-tet has defective minus-strand replication, which is expressed after infection, when the viral plus strand is converted to RF by synthesis of the minus strand (see Chapter 1 for details). The RF molecules serve as templates for rolling-circle production of more plus strands, which, in turn, can serve as templates for making more RF. The cycle is broken by the production of pV (single-strand binding protein), which binds viral DNA and prevents it from being converted into RF. Instead, pV-coated viral strands are used in the assembly of new phage, which occurs at junctions between the inner and outer membranes of the bacterial cell.

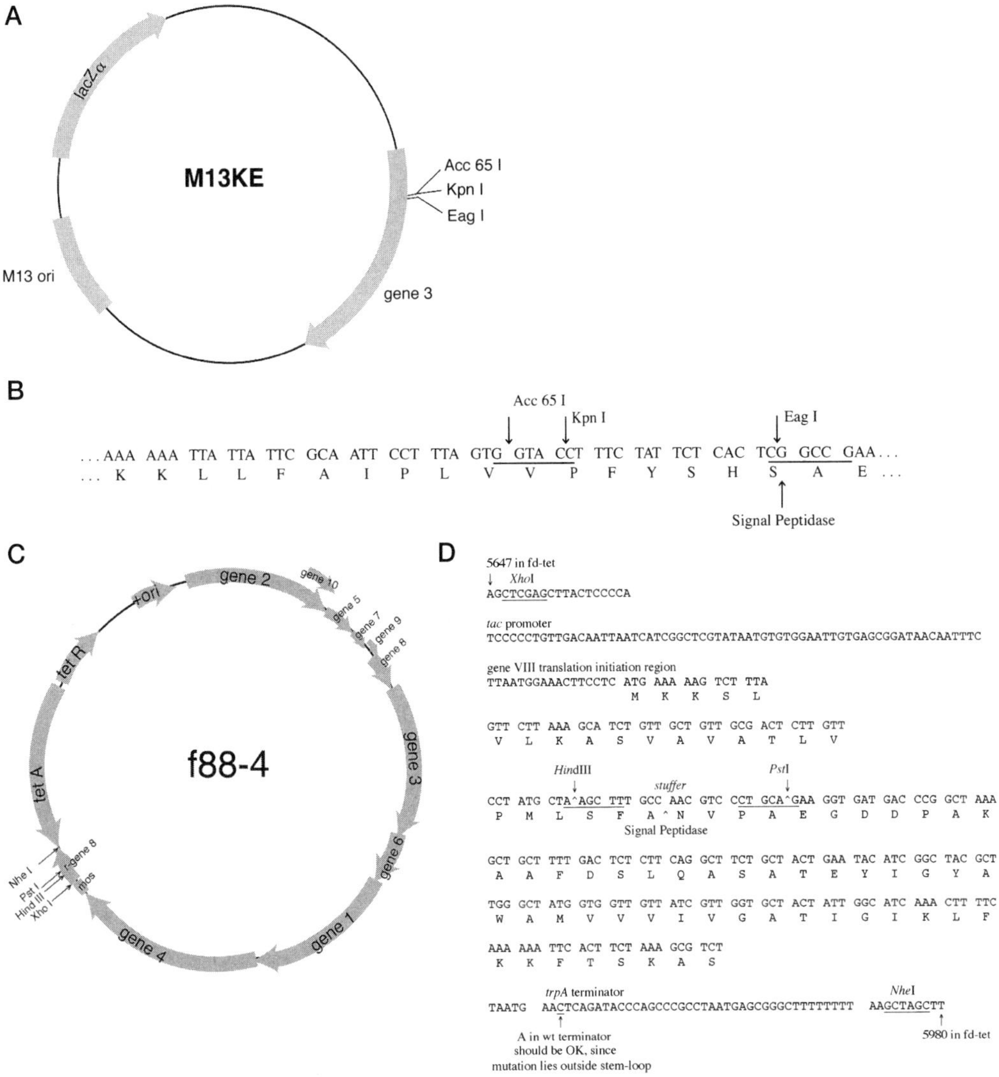

Figure 2.1. Ph.D. (M13KE) and f88-4 vectors. (*A*) Map of the Ph.D. (M13KE) vector. (*B*) Sequence of the 5′ end of the pIII gene of Ph.D. phage. (*C*) Map of the f88-4 genome. The mos site initiates packaging. The plus-strand origin is shown (+ ori); the minus-strand origin is disrupted by insertion of Tn10, which encodes the tetracycline resistance genes. (*D*) Sequence of the recombinant *gene VIII* cassette.

The defect in minus-strand replication decreases the infectivity of fd-tet phage and its derivatives. Presumably, the slowed conversion of the genome from viral to RF DNA makes the genome more vulnerable to destruction. It takes approximately 20 viral particles to produce one tetracycline-resistant transducing unit (TU), indicating that the incoming viral genome is often lost before being converted to RF and replicated. Moreover, the RF copy number is low (less than one copy per cell), and, because of the low level of RF from which new viral strands are made, the level of phage production is also reduced by about 10- to 20-fold. In the absence of tetracycline selection, F^- cells transformed with fd-tet can cure themselves of the phage; this is most likely due to the very low RF copy number of fd-tet per cell. While under tetracycline selection, some fd-tet-infected cells very probably also cure themselves of the phage, and grow as long as the tetracycline-resistance apparatus is in place in the cell. Thus, in contrast to the growth of F^+ cells infected by wild-type phage, in which an infected cell retains the same phage clone through repeated cell divisions, cells infected by fd-tet phage may start out being infected by one phage clone, become cured of it, and then be reinfected by an entirely different phage clone that allows it to continue to grow under tetracycline selection.

One of the advantages of fd-tet is that it does not kill the infected cell when pIII production is defective, whereas cells infected by phage that are otherwise wild type will die if pIII is not produced. This property indicates that "toxic" effects of pIII fusions, and perhaps fusions to other coat proteins, may be better tolerated when carried by fd-tet than by wild-type phage. Presumably, this protective effect results from the reduced phage copy number in fd-tet-infected cells. As described above, however, the downside of using fd-tet-based vectors for the production of phage libraries involves the poor infectivity of fd-tet and its low production of phage and RF DNA.

Parmley and Smith (1988) developed fd-tet into the more effective Type 3 display vectors fUSE 1 and 2, which can display peptide and protein-fragment sequences. With these vectors, the site for pIII display was moved near the amino terminus of the mature protein. This improvement reduced the impairment in infectivity caused by foreign-protein fusions. In another example, McCafferty et al. (1990) engineered this site in fd-tet (renamed fd-CAT1) to display a single-chain antibody that was successfully affinity-selected by antigen. Thus, fusions expressed by these vectors seem to be better tolerated from the standpoint of infectivity, propagation of the genome, and phage production. Based on the design of the fUSE vectors introduced by Parmley and Smith (1988), two out of the first three random peptide libraries reported used fd-tet-derived vectors for displaying peptides fused near or at the amino terminus of pIII (Scott and Smith 1990; Cwirla et al. 1990, respectively).

The f88-4 vector (Zhong et al. 1994), whose use is described in this manual, is a Type 88 vector derived from fd-tet. A gene cassette encoding recombinant pVIII was inserted into the already-disrupted minus-strand origin of fd-tet, next to the tetracycline-resistance genes (see Fig. 2.1C). The recombinant *gene VIII* cassette includes a *tac* promoter linked to a synthetic *gene VIII*, whose DNA sequence is quite different from that of wild-type *gene VIII*, but which encodes practically the same protein sequence (see Fig. 2.1D). This low homology prevents recombination between the wild-type *gene VIII* and the synthetic one. The number of pVIII-fusion proteins per phage can be modulated to some extent by IPTG induction of the *tac* promoter; however, the fusion protein copy number on any given phage will also depend greatly on the sequence displayed. For f88-4-displayed peptides, the fusion copy number per phage typically varies from 1% to 10% of the total pVIII. A number of pVIII-displayed peptide libraries have been constructed using f88-4, and they have been successfully screened with a variety of different antibodies (Zhong et al. 1994; Bonnycastle et al. 1996) and other molecules (e.g., Blancafort et al. 1999).

A contaminant wild-type phage or a mutant with wild-type characteristics can overgrow the f88-4 phage population during panning experiments in which tight-binding phage are not selected, or during phage-library amplification. Our unpublished experiments have shown that f88-4 phage, when repeatedly amplified in liquid culture, can evolve a mutant that forms wild-type-sized plaques, has high infectivity (1–5 particles/TU), has lost the tetracycline-resistance genes, and whose genome size is approximately 6 kb (similar to wild-type fd) rather than 9.3 kb (the length of f88-4). Moreover, this wild-type-like mutant arises more quickly in high-nutrient medium; it appeared after 2–3 passages when f88-4 was propagated in high-nutrient medium (e.g., superbroth; Sambrook et al. 1989), 3–4 passages in Luria broth, and 5–6 passages in NZY (K. Brown et al., unpubl.). Because phage production is not greatly different under these conditions, we now propagate f88-4 only in NZY and rarely see this problem.

It is curious that an apparently wild-type phage can propagate under tetracycline selection. We believe its ability to do so lies in the unusual biology of fd-tet and its derivatives. In contrast to wild-type phage (and even slightly defective phage like M13KE), which, once established in a cell line, remain there, fd-tet-infected cells are constantly curing themselves of fd-tet phage, even when under tetracycline selection. Thus, during the time that a given cell still has the tetracycline-resistance apparatus in place, but is no longer making phage, it can be infected by the mutant. Once a cell is infected by the mutant, it can make more phage, but will eventually stop dividing once the tetracycline-resistance apparatus is gone. The mutant phage have the advantages of higher infectivity and phage production, which allow it to increase in the culture as long as cells infected with fd-tet become available for takeover.

PHAGEMID VECTORS

Phagemid systems offer an alternative to cloning directly into the phage genome and provide several distinct advantages that are particularly relevant to the display of large proteins. A phagemid is simply a plasmid that, in addition to its plasmid origin of replication, bears a phage-derived origin of replication (also called the major intergenic region). An antibiotic-resistance marker is provided to allow selection and propagation as with typical plasmids. Unlike plasmids, however, the phagemid genomes can be packaged in the phage coat. Propagation of phagemids in cells superinfected with a helper phage or wild-type phage results in the packaging of the phagemid DNA as phage particles in a fashion identical to that of the phage DNA itself. The helper phage provides all of the phage-derived proteins and enzymes required for phage replication. These proteins act in *trans* on the phage origins of replication carried on both the helper phage and the phagemid genomes. The helper phage also provides the structural proteins that encapsulate both the helper-phage and phagemid genomes. Thus, the helper phage is acting to "help" replicate and package the phagemid genome. M13K07 (Amersham Pharmacia Biotech) and its derivative VCSM13 (Stratagene) are commonly used helper phage. They each bear an independent selection marker, the kanamycin-resistance gene, which, along with the ampicillin-resistance gene carried by compatible phagemids (e.g., pComb3), aids in the selection of cells that contain the genomes of both the helper phage and phagemid.

For the purpose of phage display, the phagemid genome carries an expression cassette that encodes the fusion-coat protein to be displayed; thus, the phagemid encodes the library to be displayed. Two types of infectious particles are produced from cells carrying both phagemid and helper-phage genomes: those containing the phagemid genome and those containing the helper-phage genome. The coat-protein fusion is displayed on particles encapsulating both the phagemid and the helper-phage genomes.

Typically, a helper phage is used that bears a defective origin of replication or packaging signal, which allows the preferential packaging of the phagemid genome over the helper-phage genome, and a greater output of phagemid phage over helper phage. This is important during selection experiments, for, although both helper phage and phagemid phage bearing the same fusion protein will be affinity selected in the panning step, only the phagemid phage will be able to drive the production of the displayed fusion protein in the amplification step that follows the selection. Thus, the efficiency of phagemid-phage production has direct bearing on the yields of phagemid phage obtained in selection experiments.

With respect to cloning, phagemid systems offer several distinct advantages. First, high yields of double-stranded DNA are easily obtained by simple plasmid preparation. Second, large DNA inserts are more readily maintained by phagemid genomes than by phage genomes. Third, two-gene display systems (Type 3+3 and 8+8 phagemid and Type 33 and 88 phage systems) allow modulation of the valency (i. e., the number of copies per phage particle) of the displayed fusion protein.

Phagemid Display on pIII

Display of proteins fused to pIII is greatly affected by pIII function. Early studies of pIII revealed two functional properties of this 406-residue protein (Gray et al. 1981; Crissman and Smith 1984). An infectivity function was shown to map to roughly the 5′ half of the gene, whereas normal (non-polyphage) morphogenesis mapped to the 3′ half of the gene. The amino-terminal domain of pIII is required for phage infectivity; it mediates the attachment of the phage to the tip of the F′ pilus of *Escherichia coli*. The carboxy-terminal portion of pIII acts in combination with pVI to cap the trailing end of the filament during phage assembly, allowing its release from the cell (see Chapter 1 for details). Crissman and Smith (1984) mapped this function to residues 198–406 of pIII and showed that in its absence, polyphage are produced. Another important property of the amino-terminal domain of pIII is its role in providing immunity to a cell already infected with a filamentous phage. This immunity function prevents superinfection by other Ff phage.

These three features of pIII function have important consequences for the design of phagemid vectors that express pIII fusions. The construction of phagemid libraries requires the introduction of phagemid DNA into F′ *E. coli* host cells (usually by electroporation to facilitate the preparation of large libraries). This step is followed by infection with helper phage, which is required for phage production (including packaging of phagemid genomes) and display of the fusion proteins constituting the library. To allow infection with helper phage, the amino-terminal domain of pIII in the fusion expressed by the phagemid should be deleted. Otherwise, the pIII produced by the phagemid will make the cell immune to superinfection by the helper phage. A truncated pIII, comprising the carboxy-terminal residues of pIII, such as residues P198–S406, can serve as the fusion partner for the displayed protein, because it will mediate incorporation of the fusion into the phage coat without producing immunity to helper-phage infection. This feature was incorporated in the first vector designed for Fab display, pComb3 (Barbas et al. 1991; Barbas and Lerner 1991) and its derivatives pComb3H and pComb3X (Yang et al. 1995; Rader and Barbas 1997), as well as in other vectors (e.g., Garrard et al. 1991). Several Type 3+3 phagemid systems, including pHEN (Hoogenboom et al. 1991) and pCANTAB (Amersham Pharmacia Biotech), use the entire pIII as the fusion partner. In these systems, the phagemid-driven expression of the pIII fusion must first be shut down to allow superinfection by the helper phage, after which expression is induced to allow production of the fusion for phage display.

The phage produced by Type 3+3 phagemid systems can potentially display from 0 to 4 or 5 copies of wild-type pIII on their surface. This is because the two types of pIII (wild type produced by the helper phage and fusion protein produced by the phagemid) compete for incorporation into the virion during phage assembly in the bacterial inner membrane. Consequently, the average valency of the fusion protein on the phage particle can be modulated by controlling the expression of the phagemid cassette. This feature has important implications for the choice of the selection experiment. For example, as described below in the section on panning methods, low valency can be important for the selection of phage bearing high-affinity fusion proteins when using solid-phase panning methods.

Note also that the display of single-chain antibodies (scFvs), rather than Fab fragments, on pIII can be accompanied by secondary effects on valency, because scFvs can spontaneously dimerize. Phage-displayed dimers are believed to form by association of phage-displayed scFv with soluble scFv that has been released from its pIII fusion partner by proteolysis in the periplasmic space. For this reason, affinity-based selections of scFvs using solid-phase capture are generally less efficient for the selection of high-affinity monovalent interactions. Avidity effects often result in the isolation of multimeric scFvs.

pComb3H and pComb3X Vectors

A map of the cloning regions of the pComb3, pComb3H, and pComb3X vectors is given in Figure 2.2. Let us first consider the features of pComb3H (Fig. 2.2B). This vector is a second-generation vector designed to overcome some of the stability problems sometimes encountered with the original pComb3 vector. Phagemid pComb3H does not contain the repeated *lac* promoter and *pelB* leader sequences of pComb3, but uses a single *lac* promoter and a combination of *ompA* and *pelB* leader sequences to direct expression of antibody light chains and heavy-chain Fd–pIII fusion proteins, respectively. A slightly smaller *gene III* fragment is used in the pComb3H vectors, amino acid residues 230–406, as compared with residues 198–406 in pComb3. Directional cloning with the single, rare-cutting restriction enzyme *Sfi*I was incorporated in the vector design. This enzyme recognizes the 8-bp sequence GGCCNNNN^NGGCC and cuts within the degenerate region of its interrupted palindromic recognition site. Use of

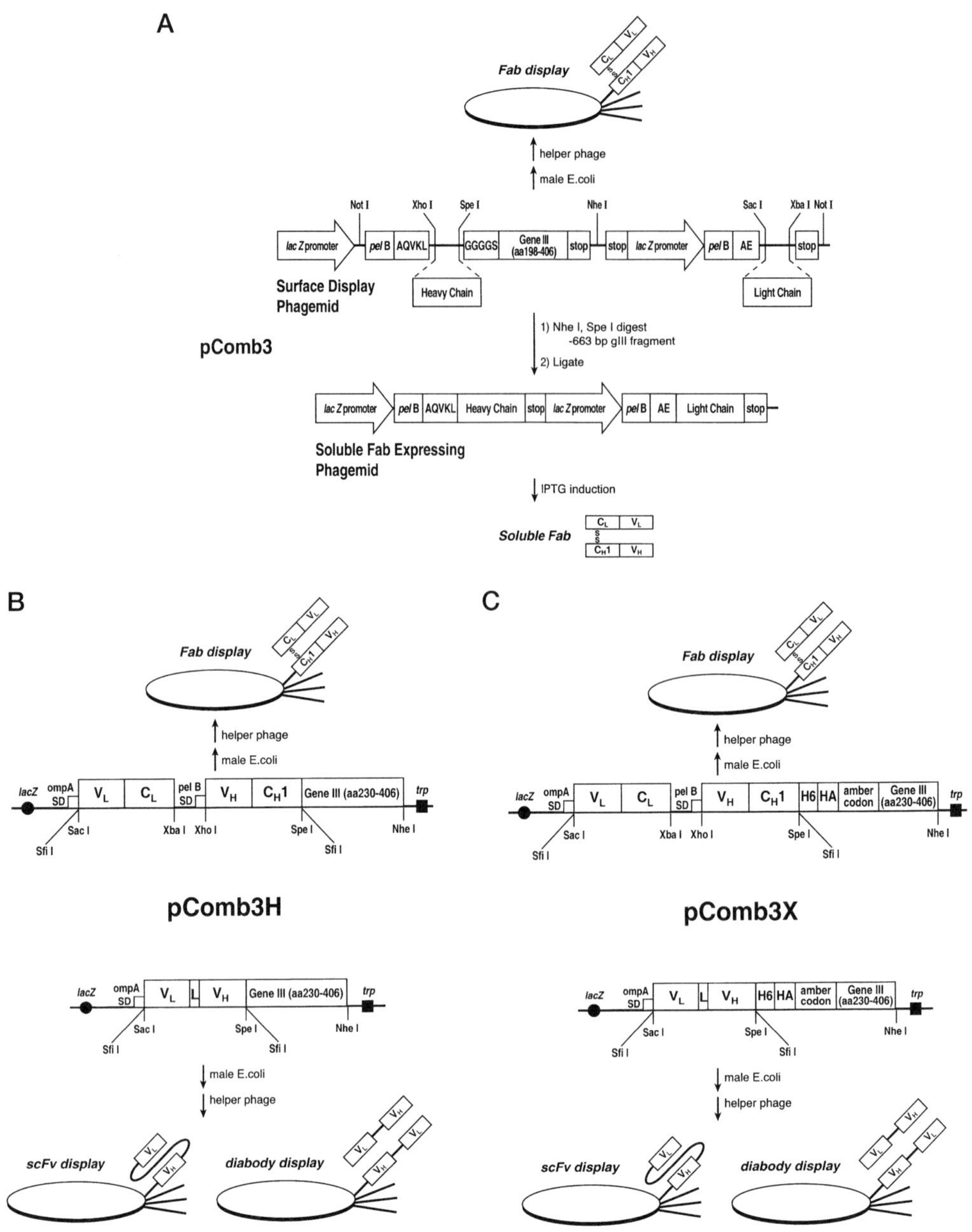

Figure 2.2. *(See facing page for legend)*

unique 5′(GGCCCAGG^CGGCC) and 3′ (GGCCAGGC^CGGCC) *Sfi*I sites allows cleaved fragments to be ligated in the correct orientation and facilitates construction of large libraries with minimal biases (see Fig. 2.2B). Sites for this enzyme are virtually never found in immunoglobulin sequences and are very uncommon in most genes, making it a versatile cloning enzyme. Thus, antibody fragments are readily cloned into pComb3H.

Single-chain antibodies and other monomeric proteins, for example, zinc-finger proteins (Segal et al. 1999), are cloned as single gene fragments using *Sfi*I cloning, and expressed as fusions to the *omp*A leader sequence. Thus, cloning excises the *pel*B leader that is used in Fab expression. Fab library construction can be accomplished using a single *Sfi*I cloning step or, alternatively, by sequential cloning of the light chain using *Sac*I and *Xba*I followed by Fd fragment cloning with *Xho*I and *Spe*I (see Fig. 2.2).

After the selection of a phage clone displaying a binding protein, one often wants to study the protein in a purified, soluble form (i.e., detached from the pIII protein). This can be done by simply purifying the soluble protein released by proteolysis of the labile fusion protein, or by excision of the *gene III* portion of the fusion fragment from the phagemid genome. Excision of the *gene III* fragment is accomplished via the *Spe*I and *Nhe*I restriction sites flanking it (Fig. 2.2). Cleavage with these enzymes generates compatible, sticky ends, and after re-ligation, a stop codon located 3′ of the *Nhe*I site directs translation termination. Although expression from *gene III*-excised phagemids yields soluble protein for study, its yield in many cases is not substantially greater than that achieved by proteolysis.

The pComb3X vector has all the features of pComb3H, along with several additional ones that facilitate the isolation and study of soluble proteins following selec-

Figure 2.2. pComb3 phagemid vectors: A series of type 3 + 3 vectors for phage display. The pComb3 phagemid vectors are designed to express antibody fragments and other proteins on the surface of filamentous phage or to express them as soluble proteins. The antibody fragments are fused to the carboxy-terminal domain of the minor coat protein (coat protein III). (*A*) Original pComb3 vector (Barbas III et al. 1991). The Fd fragment of the heavy chain is cloned via the *XhoI* and *SpeI* restriction sites, whereas the light chain is cloned using *SacI* and *XbaI*. Transcription of heavy-chain Fd/*gene III* and light chain is driven by separate *lacZ* promoters. Two ribosome-binding sites (Shine-Dalgarno, SD) give rise to separate polypeptides that are directed by the *pelB* signal peptide to the periplasm, where they are assembled and displayed on the surface of the phage particle as a Fab fragment fused to coat protein III. Expression of soluble protein that is not dependent on proteolysis requires the excision of *gene III* by restriction digest with *SpeI* and *NheI*, followed by self-ligation of the remaining vector (shown). This feature is maintained in all the pComb3 vector variants. Cloning and display of scFv proteins or other single gene proteins can be performed using *XhoI* and *SpeI*. (*B*) pComb3H vector. The pComb3H phagemid vector is a modified version of the original pComb3. The order of the genes has been reversed and a single *lacZ* promoter gives rise to a dicistronic message containing the light chain and the heavy-chain Fd. However, the presence of two SD sequences still gives rise to separate polypeptide chains. Transcription termination occurs via the *trp* transcription terminator. The light chain is directed to the periplasm by the *ompA* signal peptide, whereas the heavy-chain Fd/ coat protein III is trafficked by the *pelB* signal peptide. Two asymmetric *SfiI* restriction enzyme sites allow single-step directional cloning of Fab, scFv, diabody, or other genes. Alternatively, for Fab display the two gene fragments can be cloned separately as with the pComb3 vector. Expression of soluble protein that is not dependent on proteolysis requires excision of the *gene III* fragment and self-ligation as with pComb3. The phage display of Fab, scFv, and diabody is shown. (*C*) pComb3X vector. The pComb3X vector is identical to the pComb3H vector with three sequence insertions. The amber codon has been inserted between the 3′ *SfiI* restriction site and the 5′ end of *gene III*. This allows for soluble protein expression in nonsuppressor strains of bacteria without excising the *gene III* fragment. The 6x histidine (HIS) tag has been inserted carboxy-terminal to the Fd fragment for universal protein purification. The hemagglutinin (HA) decapeptide tag has been inserted at the 3′ end of the HIS tag for universal detection using an anti-HA antibody. This vector allows the expression of Fab, scFv, diabody, or other proteins.

tion (see Fig. 2.2C). The first feature is the addition of two peptide tags at the carboxyl terminus of the displayed protein. These are (1) a hexa-histidine (His_6) tail that facilitates purification of proteins with immobilized metal affinity chromatography (IMAC) and (2) the influenza hemagglutinin (HA) epitope tag, YPYDVPDYAS, that facilitates detection of the protein using commercially available anti-HA antibody–enzyme conjugates. At the junction of the regions encoding the peptide tags and pIII is an amber stop codon, TAG. Propagation of phage in suppressor strains such as XL1-Blue and ER2537 (*supE* or *supF* is acceptable) allows production of *gene III* fusion proteins for phage display. However, if a male nonsuppressor strain is infected with the phagemid phage, the stop codon will be read and soluble protein will be produced. Thus, pComb3X facilitates the expression, detection, and purification of soluble protein following phage selection. Details of the pComb3X cloning region are given in Figure 2.3.

Phagemid approaches can suffer from stability problems of both vector and DNA inserts. The packaging of phagemid into the phage coat is in itself a biological selection that works against large vectors and their inserts. Small phagemid genomes are more rapidly packaged than large ones, so deleted phagemids will have an advantage following passage through the phage-packaged form. This becomes particularly obvious (i.e.,

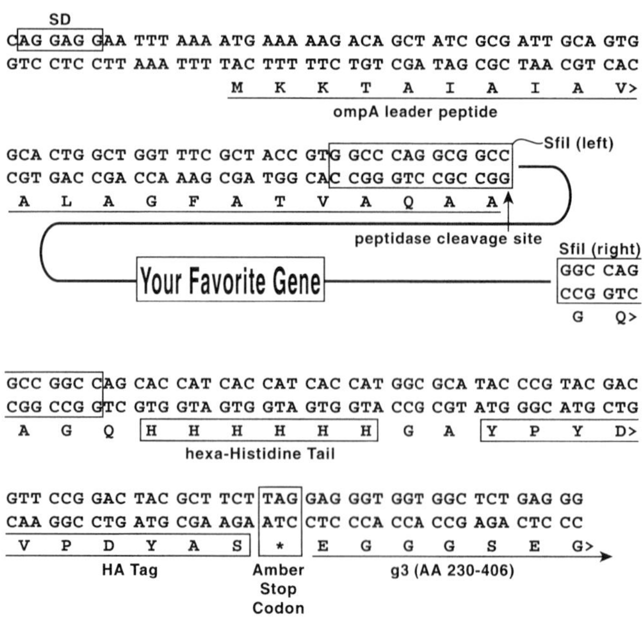

Figure 2.3. Detail of the cloning region of pComb3X. Note the differences in the sequences of the required *SfiI* sites. The second or right *SfiI* site overlaps an *FseI* site (GGCCGGCC) that can also be used in cloning. Proteolysis of the *ompA* signal peptide in the periplasm generates the native amino terminus as determined by sequencing studies of expressed Fab proteins. The cloning region of pComb3H is identical to pComb3X but the vector lacks the hexahistidine sequences, the HA epitope tag, and the amber codon. A wide variety of proteins including gene fragments and peptide libraries have been successfully displayed using these vector systems.

deleted phagemid forms are seen after multiple rounds of phage-based propagation) when no functional selection for the displayed protein is applied. Antibody fragments themselves have been shown to exhibit various degrees of toxicity when expressed in *E. coli* hosts. This type of selection may result in more rapid growth of *E. coli* bearing deleted phagemids and, in the absence of selection for the antibody, results in the degradation of the library after multiple passages. This was particularly evident with the original pComb3 vector, in which repeated promoter and leader sequences facilitated recombination events that led to insert deletion. Stability was not problematic with this vector, however, in the presence of positive selection of the phage; for example, for binding to a particular antigen via a displayed Fab fragment. To increase phagemid stability, the pComb3H and pComb3X vectors should be passaged in *E. coli* strains that overexpress the *lac* repressor protein (i.e., *lacI*^{q} mutant strains). By minimizing expression of the fusion protein, the negative-selective effect of pIII toxicity can be minimized. The limited transcriptional leakage from the uninduced *lac* promoter in these phagemids vectors is sufficient, and indeed optimal, for monovalent phage display.

The *E. coli* strains recommended for use with these phagemids are: XL1-Blue (Stratagene), ER2537 (New England Biolabs), and TOP10F′ (Invitrogen, a nonsuppressor strain). The *recA* mutation found in XL1-Blue cells also enhances stability. Strains such as ER2537 (genotype *tonA*) are resistant to environmental lytic phage of the T1 type, an infrequent, but devastating, lytic phage contamination that can plague a laboratory once it is established.

In comparison to phage display, working with phagemid-display systems is somewhat more arduous, because it requires the additional step of helper-phage superinfection. In all panning experiments, noninfected F′ *E. coli* serve as hosts for the amplification of phage recovered from the panning protocol. A typical problem encountered during phagemid panning is the lack of ampicillin-resistant colonies on output plates. This is almost always due to preinfection of the *E. coli* strain with wild-type or helper phage. Because the *E. coli* then become immune to superinfection, the introduction of the phagemid genome and its ampicillin-resistance marker is prevented. Care must be taken to avoid contamination of the *E. coli* used for propagation of libraries through the panning process. This may require the use of "phage-free" shaker-incubator and inoculation areas.

Phagemid Display on pVIII

The major coat protein pVIII has also been exploited for protein and peptide display using phagemid-based systems. The 8+8 phagemid system pComb8 has been used for the display and selection of large proteins such as antibody Fab fragments (Kang et al. 1991). This system, however, is much more prone to deletion than the 3+3 vectors when large proteins are displayed. This may be the result of size-exclusion effects during phage assembly imposed by the phage-encoded channel protein, pIV, which may select against phage bearing large pVIII fusion proteins (Marciano et al. 1999). In contrast, the orientation of pIII fusion proteins with respect to the pIV channel appears to allow passage of large fusion proteins with minimal size exclusion. The 8+8 system has been useful, however, for the selection of anti-carbohydrate antibodies (Dinh et al.

1996; Wang et al. 1997). Anti-carbohydrate antibodies are typically of very low affinity, and the enhanced multivalency of the 8+8 system makes their isolation possible (Wang et al. 1997). The pComb8 vector and other phagemid vectors of this type have a tremendous range of accessible display valencies. Electron microscopy of phage produced using this system has shown phage bearing from 1 to more than 20 copies of displayed Fab fusion proteins on their surface. Increased valency can be used advantageously to isolate peptides that interact with their targets with low affinity (Gardsvoll et al. 1998; Hart et al. 1999). For the cloning of antibody fragments with affinities greater than 10^6 M^{-1}, however, the pComb3 vector systems are suggested. The pComb3 vectors and their derivatives have been used in more than 100 published studies (Burton and Barbas 1994; Rader and Barbas 1997).

THE CHOICE OF PHAGE VECTOR AND ITS CONSEQUENCE IN AFFINITY-SELECTION EXPERIMENTS

The choice of vector depends mainly on the type of library to be screened, whereas the optimal method of screening appears to be more a function of the vector. For peptide libraries, either a pIII vector (Type 3, like M13KE), or a pVIII vector (Type 8+8 or 88, like f88-4) is most appropriate. Low-valency phagemid-display systems, like pComb3, are rarely used for peptide display, mainly because they are more difficult to amplify (because of the need for superinfection with helper phage) and to quantitate (because of the presence of helper phage in phagemid-phage preparations). Thus, peptide display is multivalent, comprising either a low number of closely apposed peptides (5 copies grouped together at one tip of the phage by M13KE), or a higher number of well-separated copies (~100 copies dispersed among ~3000 copies of pVIII by f88-4). There does not appear to be a "best" means of peptide display; both types of libraries have been used to identify peptide ligands for a variety of antibodies, receptors, enzymes, and more. In our studies, in which a panel of 16 different f88-4-displayed peptide libraries were screened side by side with two different M13KE-displayed peptide libaries, we noted that when some of the f88-4 libraries yielded enriched pools of binding phage, the M13KE libraries very often also yielded such pools. It is probable that the M13KE libraries were sometimes negative when the f88-4 libraries were positive not because of the vectors themselves, but because a certain type of peptide library was present among the f88-4 libraries and not among the M13KE libraries (A. Menendez et al., unpubl.).

Type 3+3 phagemid systems are excellent for the display of proteins such as Fabs, enzymes, and receptor domains. Such proteins, when fused to every copy of pIII (as in a Type 3 system), can produce poorly infective phage. Moreover, the overproduction of poorly folded or poorly secreted fusions has a toxic effect on cells. This more than justifies the extra effort required for screening Type 3+3 phagemid libraries, in which the fusion proteins are mixed with wild-type pIII. Smaller proteins, like zinc fingers, have been expressed with success both on pIII (using any of the pComb3 vectors; Segal et al. 1999) and on pVIII (using f88-4; Blancafort et al. 1999); however, phage production by the latter display vector was decreased, indicating difficulty with phage assembly caused by the pVIII-displayed zinc fingers.

"Gene-fragment" libraries are expressed from fragments of cDNA that are linked to the pIII or pVIII coat-protein genes; such libraries have been successfully used to isolate binding subdomains with receptors and antibodies (Jacobsson and Frykberg 1995, 1996; van Zonneveld et al. 1995). As originally described by Smith (1985), gene-fragment fusions were toxic as fusions to the pIII molecule of wild-type phage. Even with some phagemid systems, however, there appears to be strong selection against expressed fusions. In two such cases (Jacobsson and Frykberg 1995, 1996), the cDNAs encoding the selected fusions were out of frame with the coat protein genes (pIII in the former case and pVIII in the latter), indicating that there was strong selection against direct fusions. Unexpected stop codons and frameshifts were also observed in peptides selected from a pIII display library (Carcamo et al. 1998). Crameri and Suter (1993) modified pComb3 to express cDNA fragments by a novel design based on the leucine-zippered coiled coil produced by interaction of amphipathic, helical portions of the fos and jun transcription factors. To the 5′end of the cDNA fragments to be displayed, they linked DNA encoding a leader sequence, followed by the helix from the fos transcription factor; this helix was flanked on either side by a cysteine residue. Between the leader sequence and the 5′end of the truncated *gene III*, they linked a DNA fragment encoding the helix from the jun transcription factor, also flanked by cysteine residues. Thus, this "fos–jun" display system secretes expressed cDNA fragments, which are then tethered to truncated pIII via the interaction between fos and jun. The interaction becomes covalent when the cysteine residues flanking each helix of the coiled coil are oxidized in the periplasmic space of the cell. A similar system has also been used to identify tight-binding mutants of an enzyme (Pederson et al. 1998).

Affinity Selection

Phage display was initially invented to allow the affinity selection of protein fragments encoded by a corresponding cDNA fragment. Smith's first selection experiment (1985) was performed by coating a polystyrene dish with a polyclonal antibody against the *Eco*RI endonuclease, and then selectively binding phage that displayed a fragment of the protein via pIII from a pool containing a large excess of phage bearing no insert. Phage expressing functional insert were retained on the plate during extensive washing. They were eluted in acidic conditions (pH 2.2) that denatured the immobilized antibody (allowing release of the phage) but did not affect the integrity of the phage. After neutralization, the eluates were used to infect *E. coli* cells, which amplified the enriched pool of antibody-binding phage.

In their paper introducing the concept of peptide libraries, Parmley and Smith (1988) presented an improved selection procedure, which they called "biopanning." This procedure entailed biotinylating the selecting molecule, mixing it with phage displaying a targeted peptide or protein fragment, and then capturing the complexes of phage and biotinylated screening molecule on a plate coated with streptavidin (which binds biotin with a 10^{-14} to 10^{-15} K_d). Panning entailed capture of a biotinylated screening molecule; hence the name biopanning. Phage library screening entailed several rounds of selection, with each round followed by a phage-amplification step. Multiple, consecutive rounds of selection without amplification were found to be less effective, because background binding by phage increased significantly after the buffered-acid elution step.

The above two examples nicely describe what we now term "solid-phase capture" and "capture out of solution" (also called "solution-phase capture"; see Fig. 17.2), respectively; most affinity-selection procedures use one or the other approach. The theory and practice of both methods are described in detail in Chapter 17 with regard to the behavior of background phage versus weak- and tight-binding phage during the selection experiment. However, the parts of the discussion concerning the discrimination between weak- and tight-binding phage are limited to peptides displayed multivalently by the f88-4 vector. This section, therefore, will deal with in-solution versus solid-phase capture methods in terms of the different types of vectors.

When choosing a panning experiment, one should decide whether one wants to maximize phage capture or affinity discrimination. The highest yields of a binding phage are obtained by multivalent display coupled with a high density of screening molecule adsorbed to a solid surface. In contrast, discrimination between tight- and weak-binding phage (affinity discrimination) is greatest when the screening molecule binds the phage-borne fusion monovalently. Affinity discrimination can be accomplished in two ways. First, the phage itself can have only a single copy of fusion molecule and be captured by a screening molecule on a solid phase. The density of the screening molecule should not greatly affect the efficiency of capture; however, the phage concentration will. Given a low concentration of phage (i.e., below the K_d of the screening molecule–ligand interaction), more tight-binding phage will be captured on a high density of screening molecule than weak-binding phage; hence, tight-binding phage will be favored in the screening over weak-binding phage. Direct capture on solid phase is the best means of screening large proteins displayed by Type 3+3 phagemid libraries (as these will be displayed in very low copy number). Affinity discrimination between peptides displayed by Type 3 vectors can be achieved by solid-phase capture on low-density screening molecule. Because proteins adsorb to polystyrene in "islands," they can be in very high density at focal points of adsorption. Thus, to ensure low density, the screening molecule can, for instance, be biotinylated and bound in non-saturating amounts to plate-immobilized streptavidin.

The second way to achieve affinity discrimination involves capture out of solution. This method works best for peptides, because they are displayed in high copy number. In this approach, the phage are incubated with biotinylated screening molecule until the binding reaction reaches equilibrium. At this point, given a fairly constant number of peptides per phage, there will be more screening molecules bound per phage for the clones displaying the tighter-binding peptides. Only the phage bearing multiple screening molecules will be effectively captured by the streptavidin, because even for a nanomolar binder, the off-rate is seconds, whereas the washing step usually takes minutes. Thus, the avidity boost produced by the binding of multiple screening molecules to the phage allows the capture of phage bearing the tighter-binding peptides to be more effective. (Importantly, the capture reaction must be allowed to proceed for only a short time, approximately 10 minutes, so that the phage won't have a chance to bind to screening molecules that have been captured on the plate.) Type 8+8 or 88 phage display vectors are best suited for this type of capture. Type 3 vectors are also well suited for in-solution panning, as revealed by successful biopanning experiments with the fUSE5 vector (Scott and Smith 1990; Scott et al. 1992). For in-solution capture, affinity discrimination is achieved by using concentrations of the screening molecule at or below the K_d (e.g., see Table 17.2).

In general, direct capture is best used in the first round of panning for all phage panning experiments (to ensure effective capture of all binding phage), whereas in subsequent rounds, in-solution capture will produce better affinity discrimination for multivalent-display phage. Optimized screening conditions can be designed for virtually any phage-display system, as long as one is aware of the type of yield required from a given round of panning and the level of affinity discrimination desired.

REFERENCES

Barbas C.F., III and Lerner R.A. 1991. Combinatorial immunoglobulin libraries on the surface of phage (Phabs): Rapid selection of antigen specific Fabs. *Methods* **2:** 119–124.

Barbas C.F., III, Kang A.S., Lerner R.A., and Benkovic S.J. 1991. Assembly of combinatorial antibody libraries on phage surfaces: The gene III site. *Proc. Natl. Acad. Sci.* **88:** 7978–7982.

Blancafort P., Steinberg S.V., Paquin B., Klinck R., Scott J.K., and Cedergren R. 1999. The recognition of a noncanonical RNA base pair by a zinc finger protein. *Chem. Biol.* **6:** 585–597.

Bonnycastle L.L., Mehroke J.S., Rashed M., Gong X., and Scott J.K. 1996. Probing the basis of antibody reactivity with a panel of constrained peptide libraries displayed by filamentous phage. *J. Mol. Biol.* **258:** 747–762.

Burton D.R. and Barbas C.F., III. 1994. Human antibodies from combinatorial libraries. *Adv. Immunol.* **57:** 191–280.

Carcamo J., Ravera M.W., Brissette R., Dedova O., Beasley J.R., Alam-Moghe A., Wan C., Blume A., and Mandecki W. 1998. Unexpected frameshifts from gene to expressed protein in a phage-displayed peptide library. *Proc. Natl. Acad. Sci.* **95:** 11146–11151.

Crameri R. and Suter M. 1993. Display of biologically active proteins on the surface of filamentous phages: A cDNA cloning system for selection of functional gene products linked to the genetic information responsible for their production. *Gene* **137:** 69–75.

Crissman J.W. and Smith G.P. 1984. Gene-III protein of filamentous phages: Evidence for a carboxyl-terminal domain with a role inmorphogenesis. *Virology* **132:** 445–455.

Cwirla S.E., Peters E.A., Barrett R.W., Dower W.J. 1990. Peptides on phage: A vast library of peptides for identifying ligands. *Proc. Natl. Acad. Sci.* **87:** 6378–6382.

Dinh Q., Weng N.P., Kiso M., Ishida H., Hasegawa A., and Marcus D.M. 1996. High affinity antibodies against Lex and sialyl Lex from a phage display library. *J. Immunol.* **157:** 732–738.

Gardsvoll H., van Zonneveld A.J., Holm A., Eldering E., van Meijer M., Dano K., and Pannekoek H. 1998. Selection of peptides that bind to plasminogen activator inhibitor 1 (PAI-1) using random peptide phage-display libraries. *FEBS Lett.* **431:** 170–174.

Garrard L.J., Yang M., O'Connell M.P., Kelley R.F., and Henner D.J. 1991. Fab assembly and enrichment in a monovalent phage display system. *Bio/Technology* **9:** 1373-1377.

Gray C.W., Brown R.S., and Marvin D.A. 1981. Adsorption complex of filamentous fd viruses. *J. Mol. Biol.* **146:** 621–627.

Greenwood J., Willis A.E., and Perham R.N. 1991. Multiple display of foreign peptides on filamentous bacteriophage. Peptides from *Plasmodium falciparum* sporozoite protein as antigens. *J. Mol. Biol.* **220:** 821–827.

Haaparanta T. and Huse W.D. 1995. A combinatorial method for constructing libraries of long peptides displayed by filamentous phage. *Mol. Divers.* **1:** 39–52.

Hart C.P., Martin J.E., Reed M.A., Keval A.A., Pustelnik M.J., Northrop J.P., Patel D.V., and Grove J.R. 1999. Potent inhibitory ligands of the GRB2 SH2 domain from recombinant peptide libraries. *Cell. Signaling* **11:** 453–464.

Hoogenboom H.R., Griffiths A.D., Johnson K.S., Chiswell D.J., Hudson P., and Winter G. 1991. Multi-subunit proteins on the surface of filamentous phage: Methodologies for displaying antibody (Fab) heavy and light chains. *Nucleic Acids Res.* **19:** 4133–4137.

Iannolo G., Minenkova O., Petruzzelli R., and Cesareni G. 1995. Modifying filamentous phage capsid: Limits in the size of the major capsid protein. *J. Mol. Biol.* **248:** 835–844.

Il'ichev A.A., Minenkova O.O., Tat'kov S.I., Karpyshev N.N., Eroshkin A.M., Petrenko V.A., and Sandakhchiev L.S. 1989. Production of a viable variant of the M13 phage with a foreign peptide inserted into the basic coat protein. *Dokl. Akad. Nauk. S.S.S.R.* **307:** 481–483.

Jacobsson K. and Frykberg L. 1995. Cloning of ligand-binding domains of bacterial receptors by phage display. *BioTechniques* **18:** 878–885.

———. 1996. Phage display shot-gun cloning of ligand-binding domains of prokaryotic receptors approaches 100% correct clones. *BioTechniques* **20:** 1070–1081.

Kang A.S., Barbas C.F., III, Janda K.D., Benkovic S.J., and Lerner R.A. 1991. Linkage of recognition and replication functions by assembling combinatorial antibody Fab libraries along phage surfaces. *Proc. Natl. Acad. Sci.* **88:** 4363–4366.

Kay B.K., Adey N.B., He Y.-S., Manfredi J.P., Mataragnon A.H., and Fowlkes D.M. 1993. An M13 phage library displaying random 38-amino-acid peptides as a source of novel sequences with affinity to selected targets. *Gene* **128:** 59–65.

Marciano D.K., Russel M., and Simon S.M. 1999. An aqueous channel for filamentous phage export. *Science* **284:** 1516–1519.

McCafferty J., Griffiths A.D., Winter G., and Chiswell D.J. 1990. Phage antibodies: Filamentous phage displaying antibody variable domains. *Nature* **348:** 552–554.

McLafferty M.A., Kent R.B., Ladner R.C., and Markland W. 1993. M13 bacteriophage displaying disulfide-constrained microproteins. *Gene* **128:** 29–36.

Messing J. 1983. New M13 vectors for cloning. *Methods Enzymol.* **101:** 20–78.

Messing J., Gronenborn B., Muller-Hill B., and Hans Hopschneider P. 1977. Filamentous coliphage M13 as a cloning vehicle: Insertion of a *Hin*dII fragment of the lac regulatory region in M13 replicative form in vitro. *Proc. Natl. Acad. Sci.* **74:** 3642–3646.

Parmley S.F. and Smith G.P. 1988. Antibody-selectable filamentous fd phage vectors: Affinity purification of target genes. *Gene* **73:** 305–318.

Pedersen H., Holder S., Sutherlin D.P., Schwitter U., King D.S., and Schultz P.G. 1998. A method for directed evolution and functional cloning of enzymes. *Proc. Natl. Acad. Sci.* **95:** 10523–10528.

Peters E.A., Schatz P.J., Johnson S.S., and Dower W.J. 1994. Membrane insertion defects caused by positive charges in the early mature region of protein pIII of filamentous phage fd can be corrected by prlA suppressors. *J. Bacteriol.* **176:** 4296–4305.

Petrenko V.A., Smith G.P., Gong X., and Quinn T. 1996. A library of organic landscapes on filamentous phage. *Protein Eng.* **9:** 797–801.

Rader C. and Barbas C.F., III. 1997. Phage display of combinatorial antibody libraries. *Curr. Op. Biotechnol.* **8:** 503–508.

Sambrook J., Fritsch E.F., and Maniatis T. 1989. *Molecular cloning: A laboratory manual,* 2nd. edition. Cold Spring Harbor Laboratory Press, Cold Spring Harbor, New York.

Scott J.K. and Smith G.P. 1990 Searching for peptide ligands with an epitope library. *Science* **249:** 386–390.

Scott J.K., Loganathan D., Easley R.B., Gong X., and Goldstein I.J. 1992. A family of concanavalin A-binding peptides from a hexapeptide epitope library. *Proc. Natl. Acad. Sci.* **89:** 5398–5402.

Segal D.J., Dreier B., Beerli R.R., Ghiara J.B., and Barbas C.F., III. 1999. Towards controlling gene expression at will: Selection and design of zinc finger domains recognizing each of the 5′ - GNN-3′ DNA target sequences. *Proc. Natl. Acad. Sci.* **96:** 2758–2763.

Smith G.P. 1985. Filamentous fusion phage: novel expression vectors that display cloned antigens on the virion surface. *Science* **228:** 1315-1317.

———. 1993. Surface display and peptide libraries. *Gene* **128:** 1–2.

Smith G.P. and Petrenko V.A. 1997. Phage display. *Chem. Rev.* **97:** 391–410.

van Zonneveld A.-J., van den Berg B.M.M., van Meijer M., and Pannekoek H. 1995. Identification of functional interaction sites on proteins using bacteriophage-displayed random epitope libraries. *Gene* **167:** 49–52.

Wang L., Radic M.Z., Siegel D., Chang T., Bracy J., and Galili U. 1997. Cloning of anti-Gal Fabs from combinatorial phage display libraries: Structural analysis and comparison of Fab expression in pComb3H and pComb8 phage. *Mol. Immunol.* **34:** 609-618.

Wrighton N.C., Farrell F.X., Chang R., Kashyap A.K., Barbone F.P., Mulcahy L.S., Johnson D.L., Barrett R.W., Jolliffe L.K., and Dower W.J. 1996. Small peptides as potent mimetics of the protein hormone erythropoietin. *Science* **273:** 458–464.

Yang W.-P., Green K., Pinz-Sweeney S., Briones A.T., Burton D.R., and Barbas C.F., III. 1995. CDR walking mutagenesis for the affinity maturation of a potent human anti-HIV-1 antibody into the picomolar range. *J. Mol. Biol.* **254:** 392–403.

Zacher A.N. III, Stock C.A., Golden J.W., II, and Smith G.P. 1980. A new filamentous phage cloning vector: fd-tet. *Gene* **9:** 127–140.

Zhong G.-M. and Brunham R.C. 1994. Conformational mimicry of a chlamydial neutralization epitope on filamentous phage. *J. Biol. Chem.* **269:** 24183–24188.

3 Antibody Libraries

Dennis R. Burton

Departments of Immunology and Molecular Biology, The Scripps Research Institute, La Jolla, California 92037

PRINCIPLES OF ANTIBODY STRUCTURE AND GENETICS

Antibody structure, function, and genetics are comprehensively covered in a number of textbooks. Here we summarize the principal features with particular emphasis on those of relevance to recombinant antibodies.

CONTENTS

Principles of Antibody Structure

The antibody molecule is based on a 4-chain structure organized into three structural units as represented in Figure 3.1. Two of the units are identical and enable binding to antigen; these units are called the Fab (fragment antigen binding) arms of the molecule. The third unit, Fc (fragment crystalline), is generally involved in interaction with effector systems such as complement. The two identical heavy (H) chains of the molecule span the Fab and Fc regions, and the two identical light (L) chains are associated with Fab alone. There are five classes of antibodies or immunoglobulins, termed immunoglobulin G (IgG), IgM, IgA, IgD, and IgE. These classes differ in their H chains, termed γ, μ, α, δ, and ε, respectively. The most pronounced differences among these classes are in the Fab–Fc joining and Fc regions, which lead to differences in the triggering of effector function and in polymerization state.

Structure of IgG and the Other Classes of Ig

In IgG, the Fab arms are linked to the Fc via a region of polypeptide chain known as the hinge. Each of the Fab and Fc units is organized into domains as represented in Figure 3.2b and c. The L chains (Fig. 3.2a and b) exist in two forms known as kappa (κ) and lambda (λ). In humans, the H chain exists in four forms called $\gamma 1$, $\gamma 2$, $\gamma 3$, and $\gamma 4$, and these forms give rise to the four human subclasses IgG1, IgG2, IgG3, and IgG4. In mouse there are also four subclasses, denoted IgG1, IgG2a, IgG2b, and IgG3. The subclasses, particularly in humans, have very similar primary sequences, with the greatest difference being observed in the hinge region. The subclasses are an important feature, as they show marked differences in their ability to activate effector functions.

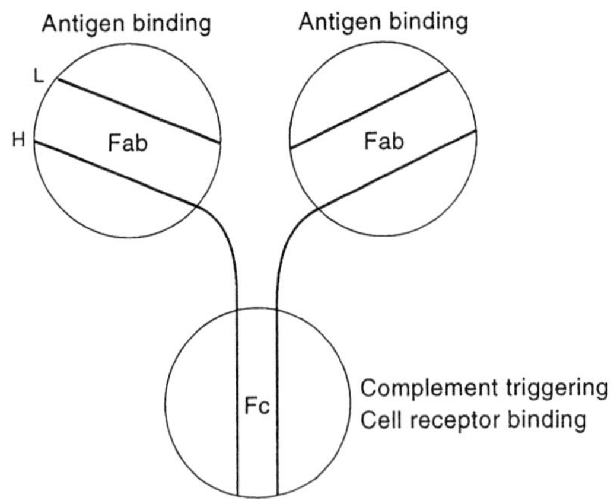

Figure 3.1. Schematic representation of the basic antibody structure emphasizing the relationship between structure and function. L: light chain, H: heavy chain.

Disulfide bonds are an important component of the IgG structure. Intrachain disulfide bonds stabilize domain folding, and interchain disulfide bonds stabilize the interaction of H and L chains and the interaction of H chains (Fig. 3.2a). The number of disulfide bonds between heavy chains in the hinge region varies from two in human IgG1 to eleven in human IgG3. Models proposed for the human IgG subclasses in solution, shown in Figure 3.3, illustrate how the hinge differences between subclasses are believed to affect average IgG conformation and, in particular, the relative disposition of Fab and Fc units. Figure 3.4 shows schematic representations of the structures of the five antibody classes, including structures for IgG, IgM, IgA, IgD, and IgE.

The Single-chain Fv Molecule

Cleavage of the IgG molecule with papain yields Fab and Fc fragments (Fig. 3.2a). These fragments are well-behaved, stable, small molecular entities that exhibit most of

Figure 3.2. Representations of the four-chain structure of IgG. (*a*) Linear representation. Disulfide bridges link the two H chains, and also the L and H chains. A regular arrangement of intrachain disulfide bonds is also found. Fragments generated by proteolytic cleavage at the indicated sites are represented. This representation should be interpreted in terms of Fig. 3.3b and c for a fuller understanding. (*b*) Domain representation. Each heavy chain (darkly shaded) is folded into two domains in the Fab arms, forms a region of extended polypeptide chain in the hinge, and is then folded into two domains in the Fc region. The light chain forms two domains associated only with a Fab arm. Domain pairing leads to close interaction of heavy and light chains in the Fab arms supplemented by a disulfide bridge. The two heavy chains are disulfide-bridged in the hinge (the number of bridges depends on the IgG subclass; shown here is human IgG1) and are in close domain-paired interaction at their carboxyl termini. Y shapes in the Fc region are N-linked carbohydrates. (*c*) Domain nomenclature and modes of flexibility. The heavy chain is composed of V_H, C_H1, C_H2, and C_H3 domains. The light chain is composed of V_L and C_L domains. All the domains are paired except for the C_H2 domains, which have two branched N-linked carbohydrate chains interposed between them. Each domain has a molecular weight of approximately 12,000, leading to a molecular weight of 50,000 for Fc and Fab and 150,000 for the whole IgG molecule. Antigen recognition involves residues from the V_H and V_L domains, and complement triggering involves the C_H2 domain (Burton 1987).

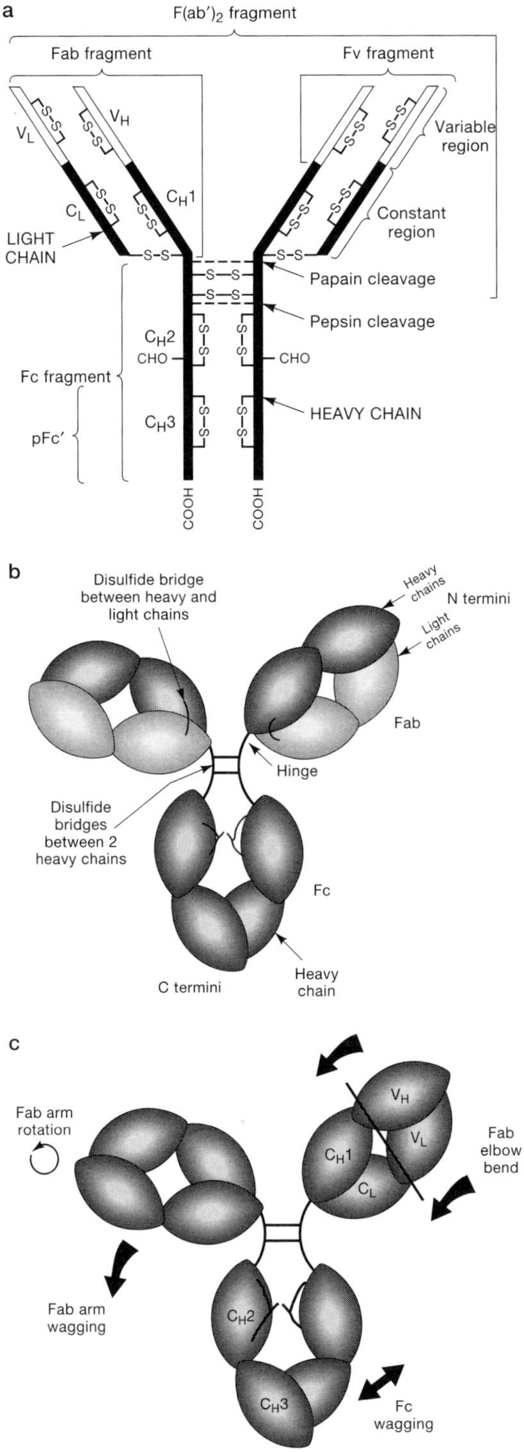

Figure 3.2. *(See facing page for legend.)*

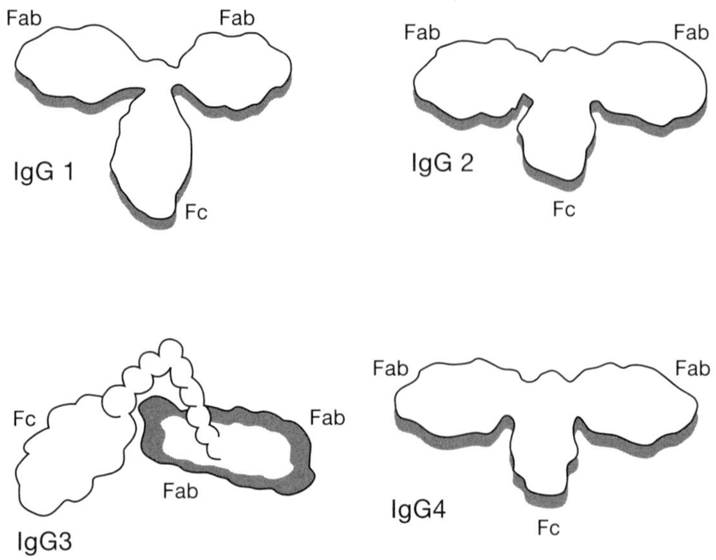

Figure 3.3. Models proposed for the conformations of the human IgG subclasses. IgG2 and IgG4 are somewhat more compact than IgG1. IgG3 has a long hinge and a relatively extended structure.

the same properties as when they are part of the whole antibody molecule. The smaller Fv fragment is considerably more difficult to produce by enzymatic digestion, and it is also much less stable. For in vitro bacterial expression, it is common to introduce a peptide linker between the V_H and V_L domains to produce single-chain Fv (scFv) molecules with improved antibody folding and stability (Fig. 3.5A). Such linkers can be incorporated in either a V_H-linker-V_L or V_L-linker-V_H orientation. The length of the linker appears to be crucial for the association state of the scFv molecule (Fig. 3.5B). Long linkers (about 20 amino acids) appear to favor monomeric molecules, whereas short linkers (e.g., 5 amino acids) favor dimers, in which the V_H of one scFv molecule is paired with the V_L of the other and vice versa. Intermediate linkers give mixed populations depending on variable region sequence. In the complete absence of a linker, trimers have been described.

Antibody Recognition of Antigen

Antigen recognition is mediated by a combining site in the Fab fragment that requires contributions from both the variable heavy (V_H) and variable light (V_L) domains. Within each variable region there are three noncontiguous linear intervals of greatest variability, which have been termed hypervariable regions or complementarity determining regions (CDRs). Separating these CDRs are intervals termed framework regions (FR) that are more highly conserved. In crystallographic analyses that have elucidated the β-barrel structure of antibodies, the CDRs were found to represent loops that are juxtaposed to form the classic antigen-binding site. In contrast, the FR subdomains fold into relatively rigid β strands that maintain the overall Ig structure. To create the antigen combining site, the CDRs from the V_H and V_L regions are juxtaposed at one end of the antibody to form a composite surface (see Fig. 3.6). The CDR

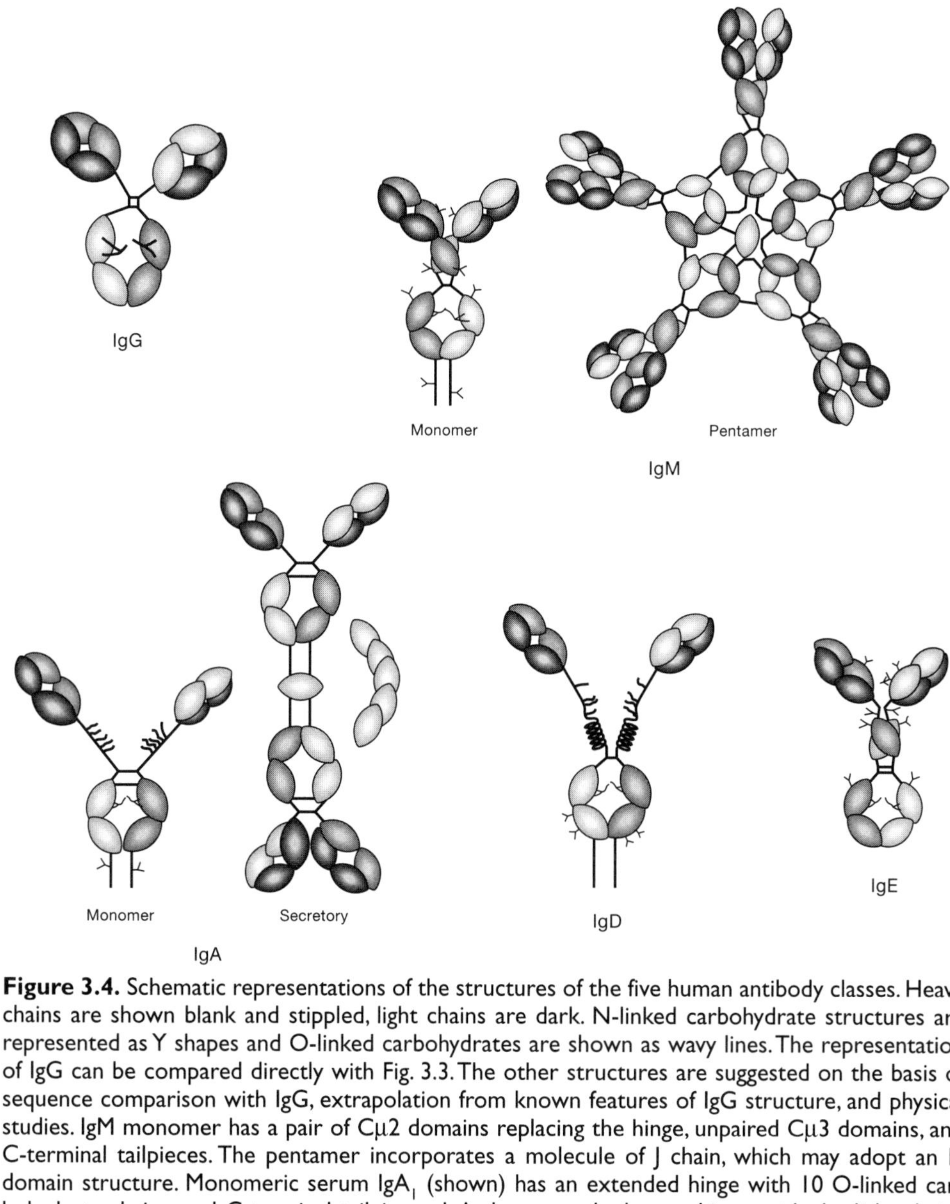

Figure 3.4. Schematic representations of the structures of the five human antibody classes. Heavy chains are shown blank and stippled, light chains are dark. N-linked carbohydrate structures are represented as Y shapes and O-linked carbohydrates are shown as wavy lines. The representation of IgG can be compared directly with Fig. 3.3. The other structures are suggested on the basis of sequence comparison with IgG, extrapolation from known features of IgG structure, and physical studies. IgM monomer has a pair of Cμ2 domains replacing the hinge, unpaired Cμ3 domains, and C-terminal tailpieces. The pentamer incorporates a molecule of J chain, which may adopt an Ig domain structure. Monomeric serum IgA_1 (shown) has an extended hinge with 10 O-linked carbohydrate chains, and C-terminal tailpieces. IgA_2 has a much shorter hinge, with the light chains generally disulfide-linked to one another rather than to the heavy chain. Secretory IgA is a dimer in which monomers are disulfide-linked via J chain. This arrangement is stabilized by secretory component, a structure of five Ig-like domains. IgD has a very extended hinge divided into a region rich in O-linked carbohydrate and a highly charged region, possibly in helical conformation, and short tailpieces. IgE closely resembles IgM monomer (Burton 1987).

that shows the greatest variation in terms of length and sequence is the heavy-chain CDR3 (HCDR3). This region appears to be an even greater source of diversity in human than in mouse antibodies. Comparison of the crystal structures of a number of antibodies suggests that the CDRs, with the exception of HCDR3, adopt a limited

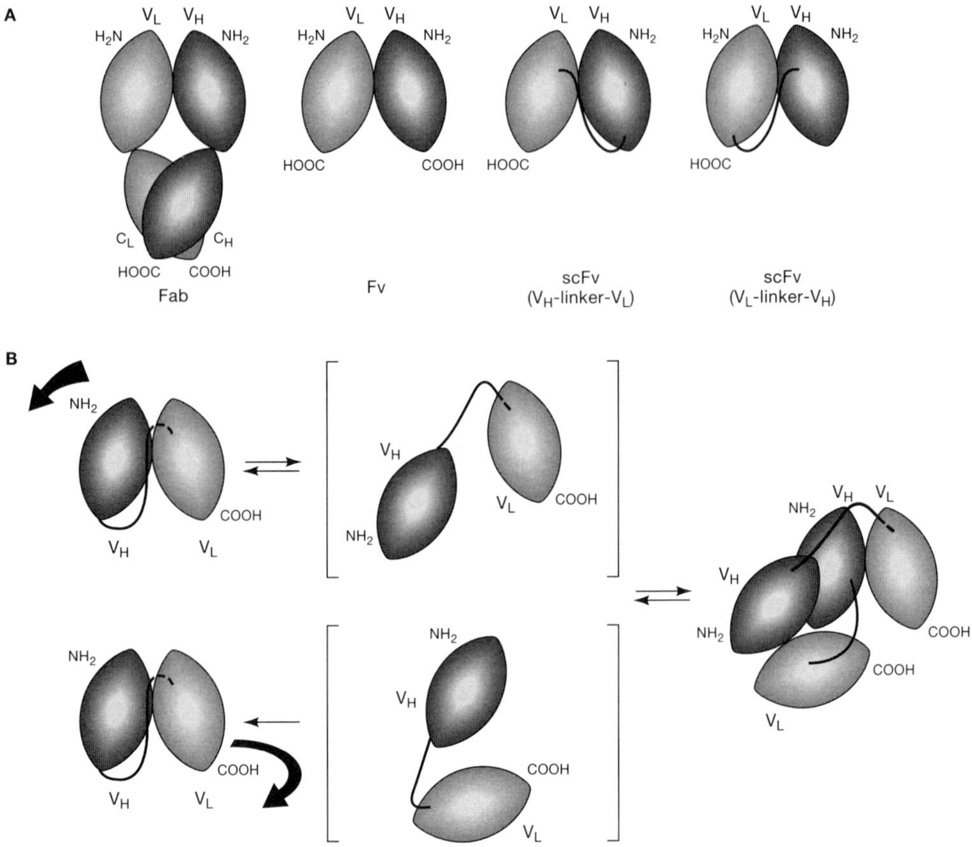

Figure 3.5. (*A*) Monovalent fragments of the antibody molecule capable of binding antigen. (*B*) Relationship between monomeric and dimeric scFv molecules. It is proposed that an equilibrium exists which is shifted to the left for longer linkers and to the right for shorter linkers. Both binding sites in the dimeric molecule can in principle bind antigen (Plückthun and Pack 1997).

number of conformations or "canonical structures." Analysis of the structures of the many antigen–antibody co-complexes solved to date indicates that in each antibody, residues in 3–5 of the CDR loops actually contribute contact sites to the interaction. The HCDR3 has been shown to make important contacts in every case.

Antibody Genetics

Primarily because of conserved framework sequences, inherited (or germ-line) gene segments encoding V_H regions have been organized into multigene families based on their sharing of more than 80% DNA sequence homology. Moreover, from early sequence analyses of human and murine antibody genes, sets of V_H gene families were shown to share highly conserved features, primarily in the FR1 and FR3 domains, enabling their clustering into three major subgroups or clans. Investigations of the antibody systems in diverse species have led to the hypothesis that the progenitors of these V_H clans diverged even prior to the emergence of the mammalian radiation, as genes representative of the clans first appeared more than 200 million years ago.

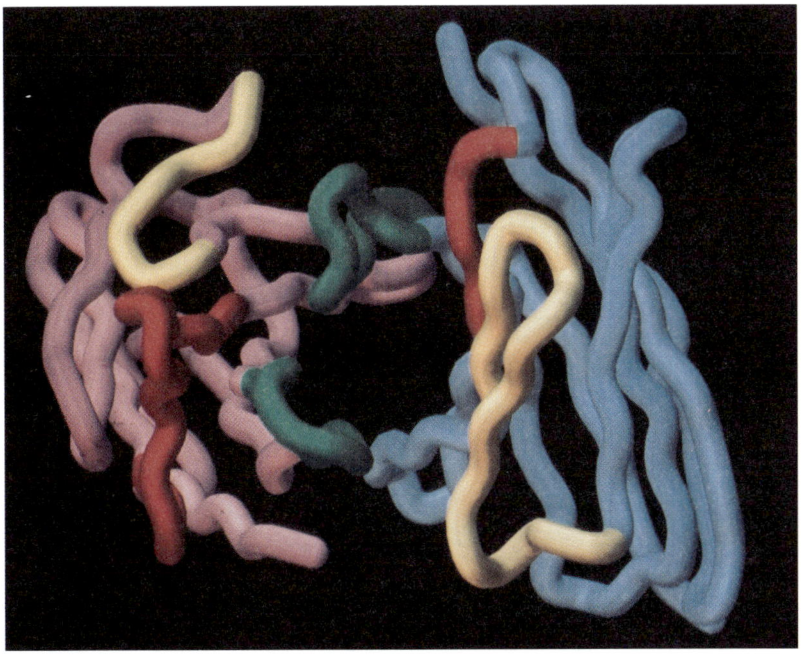

Figure 3.6. View of the binding site of an antibody. This view highlights the arrangement of the six complementarity determining regions (CDRs). The heavy-chain framework is shown in blue, the light chain in purple. CDRs 1, 2, and 3 of both chains are shown in red, yellow, and green, respectively. It is principally the CDRs that make contact with antigen and therefore are the targets for mutagenesis in optimization protocols. Molecular models were constructed by Mike Pique using AVS software (Barbas 1995).

In humans, the antibody gene loci have been well characterized. There are approximately 50 potentially functional heavy-chain germ-line sequences (V_H gene segments) that have been organized into seven families. The major family is V_H3, with 22 genes, followed by V_H1 and V_H4. V_H5 has only two members and V_H6 only one. The relative expression of these families is generally $V_H3 > V_H1, V_H4 > V_H6 > V_H5 > V_H2$. There are about 40 V_κ and about 30 V_λ human light-chain germ-line genes, which have been organized into six κ and ten λ families. There is a single human C_κ gene encoding the constant region of the κ light chains but several C_λ genes. In human expressed antibodies, the κ:λ ratio is typically of the order of 3:2. The characterization of the genetic origins of a human antibody has been greatly simplified by the accessibility of databases and computational resources at websites like V BASE (www.mrc-cpe.cam.ac.uk/imt-doc/public/INTRO.html).

Diversity in human antibodies is generated by somatic recombination in B lymphocytes, whereby V gene segments are linked to other gene segments. For the heavy chain, one of 27 diversity, or D, gene segments is linked to one of 6 joining, or J_H, gene segments. In turn, this D–J_H fusion is linked to a V_H gene segment to produce a V_H rearrangement V_H–D–J_H segment that contains the complete coding region for the heavy-chain variable domain. Later, the constant domain coding segment is connected by RNA splicing that dictates the isotype of the expressed antibody. For the κ light chain, a V_κ gene segment is joined to one of 5 J_κ segments; for the λ light chain, a V_λ

segment is joined to one of 4 J_λ segments. Additional diversity is generated during recombination by imprecise joining at the junctions of some of the gene segments. In particular, extra P-nucleotides (so denoted because they tend to be palindromic) can be added at the junctions for both heavy and light chains, and N-nucleotides (non-template encoded) can be added at the V_H–D and D–J_H junctions. The greatest diversity is therefore typically seen at and between these two junctions, which structurally correspond to the heavy-chain CDR3 region. In the characterization of the antibody sequences from libraries, determination of the heavy-chain CDR3 sequence often provides highly diagnostic information about the clonal origins and clonal relatedness of recovered antibodies. Finally, during an in vivo immune response, the interaction of B cells with antigen often results in additional diversity in antibody genes. These variations are introduced by the mechanism of somatic hypermutation that occurs in peripheral lymphoid tissue. The cellular progeny with improved antigen binding affinity because of expression of these mutated membrane-associated immunoglobulins are then selected prior to differentiation into antibody-producing cells. The net result of this mutation and selection process is the affinity maturation of the antibody response.

In mice, the organization and diversity of the antibody gene loci have not been as well characterized. Based on several different analyses, it is estimated that there are more than 100 heavy-chain germ-line sequences organized into 15 or more families, and a similar number of κ light-chain sequences. The expression of λ chains in mice is far less than in humans, with only two potentially functional genes in most strains. In Balb/c mice, light chains are expressed with a typical κ:λ ratio of about 20:1. Somatic diversification of antibody genes in the mouse appears to rely on similar mechanisms to those used in humans.

In rabbits, there are more than 20 functional heavy-chain germ-line sequences, but most heavy chains result from rearrangement of a single V_H gene. Somatic diversification by gene conversion accounts for much of the heavy-chain diversity observed in specific antibodies. The κ:λ ratio in expressed antibodies is of the order of 10–20:1. There are two C_κ genes, but one is used almost exclusively. The germ line contains multiple V_κ genes; relative expression levels are not known.

In chicken, single germ-line V_H and V_λ genes are expressed, and diversity is generated by gene conversion involving recombination with a pool of pseudogene elements, and also by somatic hypermutation. More limited diversity comes from combinatorial combinations and insertion of nontemplated nucleotides at joining sites.

ANTIBODY LIBRARIES

This section considers some general points with regard to antibody libraries, with particular emphasis on the appropriate choice of immune or nonimmune library, library format, species, and tissue source.

Immune Versus Nonimmune Libraries

In principle, very large naive or semi-synthetic established libraries offer the possibility of selection of high-affinity antibodies of any desired specificity without the need for immunization or for the investigator to be involved in library construction. In practice, there are now many reports that many moderate and some high-affinity antibodies have been selected by this route. Antibodies with great specificity have been selected that would be adequate for many applications. At present, therefore, this is an attractive starting point if one simply desires an antibody as a tool to track a specific molecule. However, if an antibody of useful affinity is not readily isolated from such a library, or if one desires a range of antibodies against a particular molecule, immunization may be the best option. Of course, this approach is also required if one is interested in studying biologic antibody responses. It is sometimes stated that a moderate affinity antibody selected from a large nonimmune library can readily be improved by a variety of strategies usually based on mutagenesis and selection. In fact, this is not a trivial task (see Chapter 13), and immunization and construction of a library may be easier than a facile selection directly from a large library followed by more time-consuming affinity improvement procedures.

Very large, semi-synthetic, nonimmune human antibody libraries have been generated by Cambridge Antibody Technology (CAT) and by MorphoSys AG (Munich, Germany).

Species for Immune Libraries

If antibodies are required for therapeutic purposes, human libraries are the obvious choice. Primates like chimpanzees, and to a lesser extent macaques, are an alternative because the immunoglobulin sequences of these primates are genetically very close to those of humans. Human antibodies have also been generated from severe combined immunodeficiency mice populated with human peripheral blood lymphocytes (hu-PBL-SCID mice) and could be obtained from transgenic mice with human antibody repertoires. Mice have become very popular for immunization because of the ease of generation of murine monoclonal antibodies using hybridoma technology. An extensive range of PCR primers have been reported for mice (particularly for the Balb/c strain), and methods for construction of mouse libraries have also been widely reported. The advantages of mice include the availability of a wide range of secondary reagents for the detection of mouse antibodies. Sequence variation in the amino terminus of the heavy chain poses a disadvantage, however, as it necessitates the use of many PCR primers to capture a good representation of the repertoire. Rabbits present a number of advantages for the library approach. First, very few antibody gene segments are rearranged and accessed during immune responses in rabbits compared to mice, which greatly reduces the number of PCR primers required for cloning of a repertoire. Second, the larger size of rabbits makes drawing blood easier and provides more tissue material for library construction. Libraries created from immune chickens may also provide advantages. Chickens, like rabbits, use very few germ-line antibody genes. Moreover, avians are phylogenetically more distant from humans than are other mammalian laboratory animals, which can provide advantages in terms of raising antibodies to molecules that have been well conserved during evolution. In an alternative approach, antibodies to highly conserved molecules can be raised in transgenic "knockout" mice, in which the target protein of interest has been knocked out.

Tissue Source for Immune Libraries

For immune libraries, it is generally preferable to use a donor or animal with a high serum antibody titer to the antigen of interest. A high serum titer is presumed to reflect relatively high levels of antibody production and, therefore, higher levels of specific mRNA should be obtainable for creation of the library. In practice, we have prepared a human Fab to measles virus nucleoprotein using bone marrow from a donor vaccinated some 20 years previously, who had a serum titer (defined as the last serum dilution significantly above background) to measles antigens of approximately 1:100. However, from these efforts only a single specific Fab was isolated, in contrast to other instances where arrays of structurally diverse Fabs were derived from donors with higher serum titers. Therefore, before library construction begins, it is crucial that serologic studies be carried out on the potential donor. In general, it is wise to preserve a generous aliquot of serum at the time of tissue collection to enable later studies that may be needed to address questions arising from the library studies.

Choice of tissue source is critical in the library approach. Ideally one begins with the source richest in plasma cells that secrete antibodies against the antigen of interest, because this source will contain the highest levels of specific mRNA. "Richest" can refer to numbers relative to irrelevant specificities and to diversity of clones that secrete antibody to the given antigen. In practice, the two probably go together.

From studies of human antibody responses to tetanus toxoid, it appears that shortly after re-exposure to the antigen (i.e., 3–10 days), there is a great increase in the number of specific antibody-secreting cells in the peripheral blood. Indeed, human Fabs to tetanus toxoid have been derived from peripheral blood lymphocyte (PBL) libraries from boosted donors. However, at later time points following the antigen boost, the number of antibody-secreting cells in the periphery declines very rapidly to a low baseline level. This decline appears to reflect the in vivo trafficking of antigen-specific B cells to anatomic sites responsible for long-term antibody generation, and such trafficking likely explains our failure to detect anti-tetanus toxoid antibodies in a library prepared from an individual with a high serum titer to toxoid but without recent antigen re-exposure. In humans, bone marrow has been shown to be a major repository of these antibody-producing cells, and for this reason bone marrow has become the most frequently used source for libraries. We have shown that antibodies to many different pathogens can be derived from bone marrow libraries constructed from a single donor, even without recent in vivo re-exposure. Libraries have also been prepared from other tissue sources, including spleen, lymph node, and tonsil. To investigate the tissue-infiltrating B-cell repertoire associated with certain diseases, libraries have also been constructed from thyroid (Graves and Hashimoto's diseases) and thymus (myasthenia gravis) of autoimmune donors. Notably, to study the autoimmune responses of patients with systemic lupus erythematosus (SLE), PBL libraries have been constructed and high-affinity binders to DNA have been isolated. In this case, the nature of the autoimmune disease leads to chronic antigen stimulation, and antibody-secreting cells continuously circulate in the periphery. Libraries have also been constructed from the lymph nodes that drain tumors, enabling the generation of tumor-specific antibodies.

For mice, libraries have been most frequently constructed from spleen, largely due to extensive experimental experience using splenocytes in cell fusions in applications of conventional hybridoma technology. More recently, both spleen and bone marrow have been used in the creation of libraries. It should be appreciated that recent studies suggest that spleen is generally a good source of post-immunization antibody-producing cells at about 4–6 days following antigen boost, but by 11 days following the boost, antigen-specific antibody-producing cells are more prevalent in the bone marrow.

Single-chain Fv Compared to Fab Fragment Libraries

A major choice facing the investigator is whether to construct a library in the single-chain Fv (scFv) or Fab antibody fragment format. The advantages of the scFv format are several. First, library construction can be simplified by overlap extension PCR that effectively reduces the number of steps involved. Second, the ability of scFv to multimerize can enhance avidity for antigen and facilitate selection against certain antigens such as cell-surface molecules. Third, yields of the smaller scFv molecule in *E. coli* tend to be better than for Fabs. The disadvantages are mostly related to the propensity of the molecule to form oligomeric structures. Thus, if the intention is to eventually generate a whole antibody molecule, a scFv that derives much of its affinity from multivalent interactions may not be a good starting point. This is because later attempts to improve the affinity of a scFv by in vitro techniques may succeed via changes in oligomerization rather than improvements in intrinsic affinity of the antibody combining site. However, if the final application for the sought antibody is best served by a scFv, then a strategy using a scFv library is a good option.

The advantages of working with a Fab are mostly related to the fact that this is a stable, well-characterized protein fragment. Although the Fab was originally part of a larger molecule, in practice it functions well as an isolated unit. In addition, Fabs do not generally show a tendency to multimerize. Improvements in the affinity of a Fab are likely to result directly from improvements in intrinsic affinity of the antibody combining site. A Fab can be converted to a whole antibody with a predictable maintenance of, or increase in, antigen affinity. The major disadvantage of Fabs relative to scFvs is the generally lower expression levels in *E. coli.*

REFERENCES

Antibody Structure, Function, and Genetics: General

Burton D.R. 1987. Structure and function of antibodies. In *Molecular genetics of immunoglobulin.* (ed. F. Calabi and M.S. Neuberger) pp. 1–50, Elsevier Science, Amsterdam.

———. 1990a. The conformation of antibodies in "Fc receptors and the action of antibodies." In *American society of microbiology series* (ed. H. Metzger) pp. 31–54, ASM Press, Washington, D.C.

———.1990b. Antibody: The flexible adaptor molecule. *Trends Biochem. Sci.* **15:** 64–69.

Harlow E. and Lane D. 1998. *Using antibodies: A laboratory manual.* Cold Spring Harbor Laboratory Press, Cold Spring Harbor, New York.

Harris L.J., Larson S.B., McPherson A. 1999. Comparison of intact antibody structures and the implications for effector function. *Adv. Immunol.* **72:** 191–208.
Janeway C.A., Jr. and Travers P. 1994. *Immunobiology: The immune system in health and disease.* Garland Publishing, New York.
Padlan E.A. 1993. Anatomy of the antibody molecule. *Mol. Immunol.* **31:** 169–217.
Roitt I.M., Brostoff J., and Male D.K. 1989. *Immunology.* Gower Medical, London.
———. 1993. *Immunology.* Mosby-Year Book Europe, London.
Roux K.H. 1999. Immunoglobulin structure and function as revealed by electron microscopy. *Int. Arch. Allergy Immunol.* **120:** 85–99.

Antibody–Antigen Interaction

Braden B.C., Goldman E.R., Mariuzza R.A., and Poljak R.J. 1998. Anatomy of an antibody molecule: Structure, kinetics, thermodynamics and mutational studies of the antilysozyme antibody D1.3. *Immunol. Rev.* **163:** 45–57.
Chothia C. and Lesk A.M. 1987. Canonical structures for the hypervariable regions of immunoglobulins. *J. Mol. Biol.* **196:** 901–917.
Chothia C., Lesk A.M., Tramontano A., Levitt M., Smith-Gill S.J., Air G., Sheriff S., Padlan E.A., Davies D., Tulip W.R., et al. 1989. Conformations of immunoglobulin hypervariable regions. *Nature* **342:** 877–883.
Colman P.M. 1988. Structure of antibody-antigen complexes: Implications for immune recognition. *Adv. Immunol.* **43:** 99–132.
Davies D.R., Padlan E.A., and Sheriff S. 1990. Antibody-antigen complexes. *Annu. Rev. Biochem.* **59:** 439–473.
Graille M., Stura E.A., Corper A.L., Sutton B.J., Taussig M.J., Charbonnier J.-B., and Silverman G.J. 2000. Crystal structure of a *Staphylococcus aureus* protein A domain complexed with the Fab fragment of a human IgM antibody: Structural basis for recognition of B-cell receptors and superantigen activity. *Proc. Natl. Acad. Sci.* **97:** 5399–5404.
Mariuzza R.A., Poljak R.J., and Schwarz F.P. 1994. The energetics of antigen-antibody binding. *Res. Immunol.* **145:** 70–72.
Poljak R.J. 1991. Structure of antibodies and their complexes with antigens. *Mol. Immunol.* **28:** 1341–1345.
Wilson I.A. and Stanfield R.L. 1993. Antibody-antigen interactions. *Curr. Opin. Struct. Biol.* **3:** 113–118.
———. 1994. Antibody-antigen interactions: New structures and new conformational changes. *Curr. Opin. Struct. Biol.* **4:** 857–867.

Antibody Effector Function

Brekke O.H., Michaelsen T.E., Aase A., and Sandlie I. 1994. Structure-function relationships of human IgG. *Immunologist* **2:** 125.
Burton D.R. and Woof J.M. 1992. Human antibody effector function. *Adv. Immunol.* **51:** 118–131.
Hulett M.D. and Hogarth P.M. 1994. Molecular basis of Fc receptor function. *Adv. Immunol.* **57:** 1–127.
Jefferis R., Lund J., and Pound J.D. 1998. IgG-Fc-mediated effector functions: Molecular definition of interaction sites for effector ligands and the role of glycosylation. *Immunol. Rev.* **163:** 59–76.

Ravetch J.V. and Clynes R.A. 1998. Divergent roles for Fc receptors and complement in vivo. *Annu. Rev. Immunol.* **16:** 421–432.
Sim R.B. 1993. *Activators and inhibitors of the complement system.* Kluwer, Amsterdam.
van de Winkel J.G.J. and Capel P.J.A. 1996. *Human IgG Fc receptors.* R.G. Landes, Austin.
Ward E.S. and Ghetie V. 1996. The effector functions of immunoglobulins: Implications for therapy. *Ther. Immunol.* **2:** 77–94.

Single-chain Antibodies

Atwell J.L., Pearce L.A., Lah M., Gruen L.C., Kortt A.A., and Hudson P.J. 1997. Design and expression of a stable bispecific scFv dimer with affinity for both glycophorin and N9 neuraminidase. *Mol. Immunol.* **33:** 1301–1312.
Bird R.E., Hardman K.D., Jacobson J.W., Kaufman J.S., Lee S.M., Lee T., Pope S.H., Riordan G.S., and Whitlow M. 1988. Single-chain antigen-binding proteins. *Science* **242:** 423–426.
Cary S., Lee J., Wagenknecht R., and Silverman G.J. 2000. Characterization of superantigen induced clonal deletion with a novel clan III-restricted avian monoclonal antibody: Exploiting evolutionary distance to create antibodies specific for a conserved VH region surface. *J. Immunol.* **164:** 4730–4741.
Desplancq D., King D.J., Lawson A.D., and Mountain A. 1994. Multimerization behaviour of single chain Fv variants for the tumour-binding antibody B72.3. *Protein Eng.* **7:** 1027–1033.
Essig N.Z., Wood J.F., Howard A.J., Raag R., and Whitlow M. 1993. Crystallization of single-chain Fv proteins. *J. Mol. Biol.* **234:** 897–901.
Glockshuber R., Malia M., Pfitzinger I., and Plückthun A. 1990. A comparison of strategies to stabilize immunoglobulin Fv-fragments. *Biochemistry* **29:** 1362–1367.
Griffiths A.D., Malmqvist M., Marks J.D., Bye J.M., Embleton M.J., McCafferty J., Baier M., Holliger K.P., Gorick B.D., and Hughes-Jones N.C. 1993. Human anti-self antibodies with high specificity from phage display libraries. *EMBO J.* **12:** 725–734.
Holliger P., Prospero T., and Winter G. 1993. "Diabodies": Small bivalent and bispecific antibody fragments. *Proc. Natl. Acad. Sci.* **90:** 6444–6448.
Huston J.S., Levinson D., Mudgett-Hunter M., Tai M.S., Novotny J., Margolies M.N., Ridge R.J., Bruccoleri R.E., Haber E., Crea R., and Oppermann H. 1988. Protein engineering of antibody binding sites: Recovery of specific activity in an anti-digoxin single-chain Fv analogue produced in *Escherichia coli. Proc. Natl. Acad. Sci.* **85:** 5879–5883.
Iliades P., Kortt A.A., and Hudson P.J. 1997. Triabodies: Single chain Fv fragments without a linker form trivalent trimers. *FEBS Lett.* **409:** 437–441.
Kortt A.A., Malby R.L., Caldwell J.B., Gruen L.C., Ivancic N., Lawrence M.C., Howlett G.J., Webster R.G., Hudson P.J., and Colman P.M. 1994. Recombinant anti-sialidase single-chain variable fragment antibody. Characterization, formation of dimer and higher-molecular-mass multimers and the solution of the crystal structure of the single-chain variable. *Eur. J. Biochem.* **221:** 151–157.
Kortt A.A., Lah M., Oddie G.W., Gruen L.C., Burns J.E., Pearce L.A., Atwell J.A., McCoy A.J., Howlett G.J., Metzger D.W., Webster R.G., and Hudson P.J. 1997. Single chain Fv fragments of anti-neuraminidase antibody NC10 containing five and ten residue linkers form dimers and with zero residue linker a trimer. *Protein Eng.* **10:** 423–433.
Krebber A., Bornhauser S., Burmester J., Honegger A., Willuda J., Bosshard H.R., and Plückthun A. 1997. Reliable cloning of functional antibody variable domains from hybridomas and spleen cell repertoires employing a reengineered phage display system. *J. Immunol. Methods* **201:** 35–55.

Malby R.L., McCoy A.J., Kortt A.A., Hudson P.J., and Colman P.M. 1998. Three-dimensional structures of single-chain Fv-neuraminidase complexes. *J. Mol. Biol.* **279:** 901–910.

Plückthun A. and Pack P. 1997. New protein engineering approaches to multivalent and bispecific antibody fragments. *Immunotechnology* **3:** 83–105.

Plückthun A. and Skerra A. 1989. Expression of functional antibody Fv and Fab fragments in *Escherichia coli. Methods Enzymol.* **178:** 497–515.

Schodin B.A. and Kranz D.M. 1993. Binding affinity and inhibitory properties of a single-chain anti-T cell receptor antibody. *J. Biol. Chem.* **268:** 25722–25727.

Whitlow M., Filpula D., Rollence M.L., Feng S.L., and Wood J.F. 1994. Multivalent Fvs: Characterization of single-chain Fv oligomers and preparation of a bispecific Fv. *Protein Eng.* **7:** 1017–1026.

Whitlow M., Bell B.A., Feng S.L., Filpula D., Hardman K.D., Hubert S.L., Rollence M.L., Wood J.F., Schott M.E., and Milenic D.E. 1993. An improved linker for single-chain Fv with reduced aggregation and enhanced proteolytic stability. *Protein Eng.* **6:** 989–995.

Antibody Genetics

Berek C. and Milstein C. 1988. The dynamic nature of the antibody repertoire. *Immunol. Rev.* **105:** 5–26.

Cook G.P. and Tomlinson I.M. 1995. The human immunoglobulin V_H repertoire. *Immunol. Today* **16:** 237–242.

Coutinho A., Freitas A.A., Holmberg D., and Grandien A. 1992. Expression and selection of murine antibody repertoires. *Int. Rev. Immunol.* **8:** 173–187.

Cox J.P., Tomlinson I.M., and Winter G. 1994. A directory of human germ-line V kappa segments reveals a strong bias in their usage. *Eur. J. Immunol.* **24:** 827–836.

Kelsoe G. 1995. The germinal center reaction. *Immunol. Today* **16:** 324–326.

Knight K.L. and Crane M.A. 1994. Generating the antibody repertoire in rabbit. *Adv. Immunol.* **56:** 179–218.

———. 1995. Development of the antibody repertoire in rabbits. *Ann. N.Y. Acad. Sci.* **764:** 198–206.

Knight K.L. and Winstead C.R. 1997. Generation of antibody diversity in rabbits. *Curr. Opin. Immunol.* **9:** 228–232.

McCormack W.T., Tjoelker L.W., and Thomspson C.B. 1991. Avian B-cell development: Generation of an immunoglobulin repertoire by gene conversion. *Annu. Rev. Immunol.* **9:** 219–241.

Milner E.C., Hufnagle W.O., Glas A.M., Suzuki I., and Alexander C. 1995. Polymorphism and utilization of human VH genes. *Ann. N.Y. Acad. Sci.* **764:** 50–61.

Rajewsky K. 1996. Clonal selection and learning in the antibody system. *Nature* **381:** 751–758.

Reynaud C.A., Bertocci B., Dahan A., and Weill J.C. 1994. Formation of the chicken B-cell repertoire: Ontogenesis, regulation of Ig gene rearrangement, and diversification by gene conversion. *Adv. Immunol.* **57:** 353–378.

Sanz I. 1991. Multiple mechanisms participate in the generation of diversity of human H chain CDR3 regions. *J. Immunol.* **147:** 1720–1729.

Schable K.F. and Zachau H.G. 1993. The variable genes of the human immunoglobulin kappa locus. *Biol. Chem. Hoppe. Seyler* **374:** 1001–1022.

Schroeder H.W., Jr., Mortari F., Shiokawa S., Kirkham P.M., Elgavish R.A., and Bertrand F.E., III. 1995. Developmental regulation of the human antibody repertoire. *Ann. N.Y. Acad. Sci.* **764:** 242–260.

Stollar B.D. 1995. The expressed heavy chain V gene repertoire of circulating B cells in normal adults. *Ann. N.Y. Acad. Sci.* **764:** 265–274.

Storb U. 1996. The molecular basis of somatic hypermutation of immunoglobulin genes. *Curr. Opin. Immunol.* **8:** 206–214.

Tomlinson I.M., Cox J.P., Gherardi E., Lesk A.M., and Chothia C. 1995a. The structural repertoire of the human V kappa domain. *EMBO J.* **14:** 4628–4638.

Tomlinson I.M., Cook G.P., Walter G., Carter N.P., Riethman H., Buluwela L., Rabbitts T.H., and Winter G. 1995b. A complete map of the human immunoglobulin VH locus. *Ann. N.Y. Acad. Sci.* **764:** 43–46.

Tomlinson I.M., Cook G.P., Carter N.P., Elaswarapu R., Smith S., Walter G., Buluwela L., Rabbitts T.H., and Winter G. 1994. Human immunoglobulin V_H and D segments on chromosomes 15q11.2 and 16p11.2. *Hum. Mol. Genet.* **3:** 853–860.

Williams S.C., Frippiat J.P., Tomlinson I.M., Ignatovich O., Lefranc M.P., and Winter G. 1996. Sequence and evolution of the human germline V lambda repertoire. *J. Mol. Biol.* **264:** 220–232.

Libraries: General

Barbas C.F. 1993. Recent advances in phage display. *Curr. Opin. Biotechnol.* **4:** 526–530.

Barbas C.F., III, Kang A.S., Lerner R.A., and Benkovic S.J. 1991. Assembly of combinatorial antibody libraries on phage surfaces: The gene III site. *Proc. Natl. Acad. Sci.* **88:** 7978–7982.

Better M., Chang C.P., Robinson R.R., and Horwitz A.H. 1988. *Escherichia coli* secretion of an active chimeric antibody fragment. *Science* **240:** 1041–1043.

Burton D.R. 1992. Monoclonal antibodies from combinatorial libraries. *Accounts Chem. Res.* **26:** 405–411.

Burton D.R. and Barbas C.F. 1994. Human antibodies from combinatorial libraries. *Adv. Immunol.* **57:** 191–280.

De Kruif J., van der Vuurst de Vries A.-R., Cilenti L., Boel E., van Ewijk W., and Logtenberg T. 1996. New perspectives on recombinant human antibodies. *Immunol. Today* **17:** 453–455.

Hudson P. 1998. Recombinant antibody fragments. *Curr. Opin. Biotechnol.* **9:** 395–402.

———. 1999. Recombinant antibody constructs in cancer therapy. *Curr. Opin. Immunol.* **11:** 548–557.

Huse W.D., Sastry L., Iverson S.A., Kang A.S., Alting-Mees M., Burton D.R., Benkovic S.J., and Lerner R.A. 1989. Generation of a large combinatorial library of the immunoglobulin repertoire in phage lambda. *Science* **246:** 1275–1281.

Kang A.S., Barbas C.F., Janda K.D., Benkovic S.J., and Lerner R.A. 1991. Linkage of recognition and replication functions by assembling combinatorial antibody Fab libraries along phage surfaces. *Proc. Natl. Acad. Sci.* **88:** 4363–4366.

Larrick J.W., Danielsson L., Brenner C.A., Abrahamson M., Fry K.E., and Borrebaeck C.A.K. 1989. Rapid cloning of rearranged immunoglobulin genes from human hybridoma cells using mixed primers and the polymerase chain reaction. *Biochem. Biophys. Res. Commun.* **160:** 1250–1256.

Marks J.D., Tristem M., Karpas A., and Winter G. 1991. Oligonucleotide primers for polymerase chain reaction amplification of human immunoglobulin variable genes and design of family-specific oligonucleotide probes. *Eur. J. Immunol.* **21:** 985–991.

McCafferty J., Griffiths A.D., Winter G., and Chiswell D.J. 1990. Phage antibodies: Filamentous phage displaying antibody variable domains. *Nature* **348:** 552–554.

Rader C. and Barbas C.F., III 1997. Phage display of combinatorial antibody libraries. *Curr. Opin. Biotechnol.* **8:** 503–508.

Rader C., Cheresh D.A., Barbas C.F., III. 1998. A phage display approach for rapid antibody humanization: Designed combinatorial V gene libraries. *Proc. Natl. Acad. Sci.* **95:** 8910–8915.

Skerra A. and Plückthun A. 1988. Assembly of a functional immunoglobulin Fv fragment in *Escherichia coli. Science* **240:** 1038–1041.

Söderlind E., Simonsson A.C., and Borrebaeck C.A.K. 1992. Phage display technology in antibody engineering: Design of phagemid vectors and *in vitro* maturation systems. *Immunol. Rev.* **130:** 109–124.

Winter G. 1998. Synthetic human antibodies and a strategy for protein engineering. *FEBS Lett.* **430:** 92–94.

Winter G., Griffiths A.D., Hawkins R.E., and Hoogenboom H.R. 1994. Making antibodies by phage display technology. *Annu. Rev. Immunol.* **12:** 433–455.

Nonimmune Libraries

Barbas C.F. 1995. Synthetic human antibodies. *Nat. Med.* **1:** 837–839.

Barbas C.F., Languino L.R., and Smith J.W. 1993a. High-affinity self-reactive human antibodies by design and selection: Targeting the integrin ligand binding site. *Proc. Natl. Acad. Sci.* **90:** 10003–10007.

Barbas C.F., Rosenblum J.S., and Lerner R.A. 1993b. Direct selection of antibodies that coordinate metals from semisynthetic combinatorial libraries. *Proc. Natl. Acad. Sci.* **90:** 6385–6389.

Barbas C.F., Amberg W., Simoncsits A., Jones T.M., and Lerner R.A. 1993c. Selection of human anti-hapten antibodies from semisynthetic libraries. *Gene* **137:** 57–62.

Barbas C.F., Bain J.D., Hoekstra D.M., and Lerner R.A. 1992. Semisynthetic combinatorial antibody libraries: A chemical solution to the diversity problem. *Proc. Natl. Acad. Sci.* **89:** 4457–4461.

Clackson T., Hoogenboom H.R., Griffiths A.D., and Winter G. 1991. Making antibody fragments using phage display libraries. *Nature* **352:** 624–628.

———. 1994. Making antibody fragments using phage display libraries. *Nature* **352:** 624–648.

Glaser S.M., Yelton D.E., and Huse W.D. 1992. Antibody engineering by codon-based mutagenesis in a filamentous phage vector system. *J. Immunol.* **149:** 3903–3913.

Gram H., Marconi L.A., Barbas C.F., Collet T.A., Lerner R.A., and Kang A.S. 1992. In vitro selection and affinity maturation of antibodies from a. *Proc. Natl. Acad. Sci.* **89:** 3576–3580.

Griffiths A.D., Williams S.C., Hartley O., Tomlinson I.M., Waterhouse P., Crosby W.L., Kontermann R.E., Jones P.T., Low N.M., Allison T.J., Prospero T.D., Hoogenboom H.R., Nissim, A., Cox, J.P.L., Harrison, J.L., Zaccolo, M., Gherardi, E., and Winter, G. 1994. Isolation of high affinity human antibodies directly from large synthetic repertoires. *EMBO J.* **13:** 3245–3260.

Hoogenboom H.R. and Winter G. 1992. By-passing immunisation. Human antibodies from synthetic repertoires of germline V_H gene segments rearranged in vitro. *J. Mol. Biol.* **227:** 381–388.

Lerner R.A., Kang A.S., Bain J.D., Burton D.R., and Barbas C.F. 1992. Antibodies without immunization. *Science* **258:** 1313–1314.

Marks J.D., Griffiths D., Malmqvist M., Clackson T., Bye J.M., and Winter G. 1992. By-passing immunization: Building high affinity human antibodies by chain shuffling. *Bio/Technology* **10:** 779–783.

Marks J.D., Hoogenboom H.R., Bonnert T.P., McCafferty J., Griffiths D., and Winter G. 1991. By-passing immunization. Human antibodies from V-gene libraries displayed on phage. *J. Mol. Biol.* **222:** 581–597.

McCafferty J., Griffiths A.D., Winter G., and Chiswell D.J. 1990. Phage antibodies: Filamentous phage displaying antibody variable domains. *Nature* **348:** 552–554.

Schier R., McCall A., Adams G.P., Marshall K.W., Merritt H., Yim M., Crawford R.S., Weiner L.M., Marks C., and Marks J.D. 1996. Isolation of picomolar affinity anti-c-erbB-2 single-chain Fv by molecular evolution of the complementarity determining regions in the center of the antibody binding site. *J. Mol. Biol.* **263:** 551–567.

Winter G., Griffiths A.D., Hawkins R.E., and Hoogenboom H.R. 1994. Making antibodies by phage display technology. *Annu. Rev. Immunol.* **12:** 433–455.

Yang W.-P., Green K., Pinz-Sweeney S., Briones A.T., Burton D.R., and Barbas C.F. 1995. CDR walking mutagenesis for the affinity maturation of a potent human anti-HIV-1 antibody into the picomolar range. *J. Mol. Biol.* **254:** 392–403.

Immune Libraries

Barbas C.F., Collet T.A., Amberg W., Roben P., Binley J.M., Hoekstra D., Cababa D., Jones T.M., Williamson R.A., Pilkington G.R., Haigwood N.L., Cabezas E., Satterthwait A.C., Sanz I., and Burton D.R. 1993. Molecular profile of an antibody response to HIV-1 as probed by combinatorial libraries. *J. Mol. Biol.* **230:** 812–823.

Bender E., Pilkington G.J., and Burton D.R. 1994. Monoclonal antibodies from a library from a donor with a low serum titer to a virus. *Hum. Antib. Hybrid.* **5:** 3–8.

Burton D.R. 1995. Human antibodies to viral pathogens from phage display libraries. In *Vaccines 95: Molecular approaches to the control of infectious diseases* (ed. R.M. Chanock et al.), pp. 1–11. Cold Spring Harbor Laboratory Press, Cold Spring Harbor, New York.

Burton D.R., Barbas C.F., Persson M.A., Koenig S., Chanock R.M., and Lerner R.A. 1991. A large array of human monoclonal antibodies to type 1 human immunodeficiency virus from combinatorial libraries of asymptomatic seropositive individuals. *Proc. Natl. Acad. Sci.* **88:** 10134–10137.

Burton D.R., Pyati J., Koduri R., Sharp S.J., Thornton G.B., Parren P.W.H.I., Sawyer L.S.W., Hendry R.M., Dunlop N., Nara P.L., Lamacchia M., Garratty E., Stiehm E.R., Bryson Y.J., Cao Y., Moore J.P., Ho D.D., and Barbas C.F. 1994. Efficient neutralization of primary isolates of HIV-1 by a recombinant human monoclonal antibody. *Science* **266:** 1024–1027.

Chanock R.M., Crowe J.E. Jr., Murphy B.R., and Burton D.R. 1993. Human monoclonal antibody Fab fragments cloned from combinatorial libraries: Potential usefulness in prevention and/or treatment of major human viral diseases. *Infect. Agents Dis.* **2:** 118–131.

Clackson T., Hoogenboom H.R., Griffiths A.D., and Winter G. 1991. Making antibody fragments using phage display libraries. *Nature* **352:** 624–628.

Crowe J.E., Jr., Murphy B.R., Chanock R.M., Williamson R.A., Barbas C.F., and Burton D.R. 1994. Recombinant human respiratory syncytial virus (RSV) monoclonal antibody Fab is effective therapeutically when introduced directly into the lungs of RSV-infected mice. *Proc. Natl. Acad. Sci.* **91:** 1386–1390.

Ohlin M. and Borrebaeck C.A.K. 1996. Characteristics of human antibody repertoires following active immune responses in vivo. *Mol. Immunol.* **33:** 583–592.

Persson M.A., Caothien R.H., and Burton D.R. 1991. Generation of diverse high-affinity human monoclonal antibodies by repertoire cloning. *Proc. Natl. Acad. Sci.* **88:** 2432–2436.

Rapoport B., Portolano S., and McLachlan S.M. 1995. Combinatorial libraries: New insights into human organ-specific autoantibodies. *Immunol. Today* **16:** 43–49.

Reason D.C., Wagner T.C., and Lucas A.H. 1997. Human Fab fragments specific for the *Haemophilus influenzae* b polysaccharide isolated from a bacteriophage combinatorial library use variable region gene combinations and express an idiotype that mirrors in vivo expression. *Infect. Immun.* **65:** 261–266.

Roben P., Barbas S.M., Sandoval L., Lecerf J.-M., Stollar B.D., Solomon A., and Silverman G.J. 1996. Repertoire cloning of lupus anti-DNA autoantibodies. *J. Clin. Invest.* **98:** 2827–2837.

Williamson R.A., Burioni R., Sanna P.P., Partridge L.J., Barbas C.F., and Burton D.R. 1993. Human monoclonal antibodies against a plethora of viral pathogens from single combinatorial libraries. *Proc. Natl. Acad. Sci.* **90:** 4141–4145.

Location and Kinetics of Antibody-producing Cells

Barbas S.M., Ditzel H.J., Salonen E.M., Wei-Ping Y., Silverman G.J., and Burton D.R. 1995. Human autoantibody recognition of DNA. *Proc. Natl. Acad. Sci.* **92:** 2529–2533.

Ershler W.B., Moore A.L., and Hacker M.P. 1982. Specific in vivo and in vitro antibody response to tetanus toxoid immunization. *Clin. Exp. Immunol.* **49:** 552–558.

Graus Y.F., de Baets M.H., Parren P.W.H.I., Berrih-Aknin S., Wokke J., van Breda Vriesman P.J., and Burton D.R. 1997. Human anti-nicotinic acetylcholine receptor recombinant Fab fragments isolated from thymus-derived phage display libraries from myasthenia gravis patients reflect predominant specificities in serum and block the action of pathogenic serum antibodies. *J. Immunol.* **158:** 1919–1929.

Lum L.G., Burns E., and Janson M.M. 1990. IgG anti-tetanus toxoid antibody synthesis by human bone marrow. I. Two distinct populations of marrow B cells and functional differences between marrow and peripheral blood B cells. *J. Clin. Immunol.* **10:** 255–264.

Manz R.A., Thiel A., and Radbruch A. 1997. Lifetime of plasma cells in the bone marrow. *Nature* **388:** 133–134.

Mullinax R.L., Gross E.A., Amberg J.F., Hay B.N., Hogrefe H.H., Kubitz M.M., Greener A., Alting-Mees M., Ardourel D., Short J.M., Sorge J.A., and Shopes B. 1990. Identification of human antibody fragment clones specific for tetanus toxoid in a bacteriophage lambda immunoexpression library. *Proc. Natl. Acad. Sci.* **87:** 8095–8099.

Persson M.A., Caothien R.H., and Burton D.R. 1991. Generation of diverse high-affinity human monoclonal antibodies by repertoire cloning. *Proc. Natl. Acad. Sci.* **88:** 2432–2436.

Rapoport B., Portolano S., and McLachlan S.M. 1995. Combinatorial libraries: New insights into human organ-specific autoantibodies. *Immunol. Today* **16:** 43–49.

Slifka M.K., Antia R., Witmire J.K., and Ahmed R. 1998. Humoral immunity due to long-lived plasma cells. *Immunity* **8:** 363–372.

Stevens R.H., Macy E., Morrow C., and Saxon A. 1979. Characterization of a circulating subpopulation of spontaneous antitetanus toxoid antibody producing B cells following in vivo booster immunization. *J. Immunol.* **122:** 2498–1504.

Thiele C.J., Morrow C.D., and Stevens R.H. 1981. Multiple subsets of anti-tetanus toxoid antibody-producing cells in human peripheral blood differ by size, expression of membrane receptors, and mitogen reactivity. *J. Immunol.* **126:** 1146–1153.

Volkman D.J., Allyn S.P., and Fauci A.S. 1982. Antigen-induced in vitro antibody production in humans: Tetanus toxoid specific antibody synthesis. *J. Immunol.* **129:** 107–112.

4 Peptide Libraries

Jamie K. Scott

Department of Molecular Biology and Biochemistry and Department of Biological Sciences, Simon Fraser University, Burnaby, B.C., Canada V5A 1S6

Peptide libraries comprise vast numbers of peptides of a given length, whose sequences have been randomly generated to vary the amino acid residues at each position. The peptides in some libraries also have fixed residues that support predetermined constraints and secondary structures. In phage-displayed peptide libraries, these include disulfide-bridged loops (O'Neil et al. 1992; Felici et al. 1993; McLafferty et al. 1993) and α helices (Rebar and Pabo 1994; Bianchi et al. 1995; Wu et al. 1995). More recently, peptide libraries have been constructed as variegated patches on the surface of folded proteins, such as on cytochrome b_{562} (Ku and Schultz 1995), and in the binding site of a Fab (Barbas et al. 1993). Discontinuous epitope libraries have been constructed as two adjacent loops on the "minibody" (Martin et al. 1994) and on tendamistat (McConnell and Hoess 1995) and as adjacent helices on a zinc-finger protein (Nord et al. 1995). The purpose of each of these libraries is to display a variable "epitope" (and sometimes "paratope," i.e., the site that binds an epitope) that comprises a minimal number of structurally contiguous residues. The review by Smith and Petrenko (1997) contains a comprehensive list of phage-displayed peptide libraries.

CONTENTS

The short, variegated regions in peptide libraries are searched for sequences that bind to a given screening molecule. Such ligands can be found in peptide libraries because of the limited number of critical residues that are required to bind the screening molecule. In the interactions between two large protein surfaces such as in the complex between the D1.3 monoclonal antibody and hen egg white lysozyme (Braden et al. 1998), each protein might contribute over 20 residues at the binding interface, but only 4 to 7 residues are responsible for most of the binding energy. The remaining residues in the interface mainly contribute to binding by forming complementary surfaces that act as scaffolds in promoting the interactions between the critical binding residues; these regions of the epitope and paratope also contribute to binding through multiple, lower-energy contacts. Thus, peptides may bind to a given binding site by recapitulating some or all of the critical binding residues that are found at the epitope–paratope interface.

Peptides can bind to a variety of antibodies, enzymes, receptors (Smith and Petrenko 1997; Zwick et al. 1998a), and even other peptides (Bremnes et al. 1998). Usually, they bind at or near the "natural" binding site of the molecule and thus act as functional mimics of the native ligand. There has been some question about the structural basis of this functional mimicry; are the same types of binding interactions between critical residues in the native ligand–receptor complex recapitulated in the peptide–receptor interaction? If so, a binding peptide would also be a structural mimic of the epitope on the native ligand. Whether the peptide mimic binds by the same mechanism as the native ligand can be important in the design of bioactive inhibitors or agonists and in the design of immunogenic mimics (i.e., peptides that, when used as hapten-immunogens, stimulate the production of antibodies against a targeted epitope). Peptides have been found that functionally mimic the epitopes on proteins, carbohydrates, nucleic acids, and other nonproteinaceous substances. Such peptides have a wide range of potential uses, especially in the development of drugs, diagnostics, vaccines, and affinity matrices.

In this chapter, two general types of peptide libraries are discussed: recombinant-displayed and synthetic peptide libraries, with phage display being the main subject in the discussion of the former type of library. For the scientist interested in using peptide libraries, much of the choice of library technology will depend on the methodology with which one is most comfortable, the relative costs, and one's access to the different types of libraries. (See Scott [1994] for a more detailed comparison of the different types of recombinant-display and synthetic peptide libraries and their methods of ligand identification.)

PHAGE-DISPLAYED PEPTIDE LIBRARIES

The idea of displaying random peptides on the surface of filamentous phage began with George P. Smith, who showed that fragments of a target protein could be fused to the pIII coat protein and captured by "panning" with immobilized antibodies against the whole target protein (Smith 1985). He suggested that vast peptide libraries could be constructed by fusing randomly generated peptides to phage coat proteins, and that they could be screened by panning to identify the epitope for any antibody (Parmley and Smith 1988). The first peptide libraries were soon made, and the concept was proven (Cwirla et al. 1990; Scott and Smith 1990). Moreover, Devlin et al. (1990) showed that targets other than antibodies could select phage-borne peptides. Since that time, many investigators have identified ligand-peptides by screening peptide libraries displayed on the minor coat protein, pIII, or the major coat protein, pVIII, with antibodies, enzymes, receptors, and other protein and nonprotein molecules (for review, see Smith and Petrenko 1997; Zwick et al. 1998a). Peptide libraries have also been made using non-phage-based recombinant-expression systems such as the *Escherichia coli* phage lambda receptor (Brown 1992), fimbriae (Schembri et al. 1999), and a flagellar-thioredoxin fusion protein (Lu et al. 1995) for bacterial surface display;

T7 and lambda bacteriophage (Stolz et al. 1998; Houshmand et al. 1999); the lac repressor for display linked to a plasmid (Cull et al. 1992); and in vitro translation systems in which a newly made peptide is complexed, via a ribosome, to mRNA (Mattheakis et al. 1994; Roberts and Szostak 1997).

In all of these "recombinant-display" systems, peptide libraries are encoded by a degenerate oligonucleotide that has been introduced into a phage or phagemid genome, a receptor- or repressor-encoding plasmid, or an RNA. This physically links each peptide in the library to a "read-out" nucleic acid, which can be amplified and decoded by sequencing. Each variable residue (X) in the library's peptide sequence is usually encoded by an NNK or NNS degenerate codon (in which N stands for A, G, C, and T; K for G and T; and S for G and C) that represents 32 codons and encodes all 20 amino acids and one amber stop codon. Although NNK and NNS represent different codon sets, they encode the same distribution of amino acids. There are three codons for arginine, leucine, and serine; two codons for valine, proline, threonine, alanine, and glycine; and one codon for each of the remaining 12 amino acids.

Ideally, one would prefer to synthesize degenerate oligonucleotides for production of a "primary" or fully randomized library based on codon mixtures instead of nucleotide mixtures. This is because, in the former case, a single codon could be used to encode each amino acid (for a total of 20 codons), thereby eliminating the amino acid bias inherent in the 32 codons that are encoded by NNK or NNS, including the amber stop codon. Virnekaes et al. (1994) encode fully randomized (X) residues simply by synthesizing oligonucleotides on a single column with mixed trinucleotide phosphoramidites (codons). At any step in the synthesis, the desired codons can be added as a mixture to the growing oligonucleotide chain. This type of degenerate oligonucleotide can be obtained (at least by industrial contract) through MorphoSys (Munich, Germany). In an alternative approach, Ixsys Corporation (San Diego, California) synthesizes oligonucleotides with X residues by following the resin-dividing procedure of Glaser et al. (1992; see "doped" libraries below), but using 10, instead of 2, columns. A 5′ fixed sequence is synthesized on a single column, and where X residues are to be encoded, the resin is split among 10 columns, with two codons being synthesized on each. For example, on one of the columns, codons for serine and arginine are synthesized together by adding dA in the first position, dG in the second, and a 1-to-1 mixture of dC and dA in the third position, respectively. After synthesis of the 20 trinucleotides, the resin from the 10 columns is thoroughly mixed and split again among the 10 columns for the addition of the nucleotides that will encode the next X residue in the sequence. This process is repeated until the 3′ fixed sequence is reached; that sequence is synthesized, as before, on a single column (W.D. Huse, Ixsys Corp., pers. comm.). Column syntheses of codon-based degenerate oligonucleotides that add a single codon to each of 20 columns are limited by the number of resin beads that each column can hold (10^8); hence, a library's size is limited to 2×10^9 sequences. However, by synthesizing 4 codons together on each column (by using mixed nucleotides, such as T.T/C.T/A, encoding Phe/Ser/Leu), the number of codon sequences for each bead increases. This allows the total number of sequences on the resin to surpass 10^9. Phage libraries typically cannot exceed 10^{10} clones, so there really is little need to synthesize more than 10^{11} DNA sequences.

Binding peptides are identified from most recombinant-display libraries by affinity selection experiments in which a phage (or other entity bearing a peptide-encoding molecule) is captured, via its displayed peptide, by a target molecule (see Chapters 17 and 19); such experiments are also called screening or panning. The target–peptide–phage complex is immobilized (e.g., on a magnetic bead or microtiter well) to allow the removal of nonbinding phage by washing. The target-bound phage clones are eluted, usually by denaturation of the target, and then amplified by infection of and growth in *E. coli*. The resulting phage pool is thus enriched for target-binding phage, and this is reflected in an increase in phage yield that is seen in subsequent rounds of selection. Clones are chosen from affinity-enriched pools having high phage yields (relative to a control phage) and tested for binding by ELISA (or other methods, see Chapters 18 and 19). The amino acid sequence of the peptide displayed by a given clone is easily obtained by DNA sequencing (Chapters 15 and 19). Consensus sequences (shared amino acid sequences among different peptides) can often be deduced by comparing the peptide sequences from a number of selected, non-sibling clones. Consensus sequences are important because they are often composed of amino acids that confer binding affinity to a peptide.

SYNTHETIC PEPTIDE LIBRARIES

Although the term "combinatorial chemistry" can include recombinant-display libraries, it mostly refers to libraries that have been synthesized directly from monomeric components. These include synthetic peptide libraries made from natural and unnatural amino acids, and libraries made from non-amino-acid-based, organic monomers (for a review of this field, see Ellman and Gallop 1998). There are a variety of synthetic peptide libraries, including libraries synthesized on plastic pins (Geysen et al. 1986), on paper (Kramer et al. 1994), on beads (Lam et al. 1991; Needels et al. 1993), and as free peptides in solution (Houghten et al. 1991).

There are two general means by which binding peptides are identified from a library: affinity selection and deduction. In the affinity-selection procedures, peptides are synthesized on beads with one peptide sequence per bead (these may be encoded by an accompanying "decoding" molecule, like an oligonucleotide; Needels et al. 1993). Target molecule is reacted with the library, and its presence on beads bearing ligand peptides is detected with a secondary reagent, such as an antibody linked to alkaline phosphatase; this allows the beads to be picked or sorted. In the libraries pioneered by Lam et al. (1991), the peptide sequence of a given bead is simply obtained from direct amino acid sequencing, whereas oligonucleotide-encoded peptides (Needels et al. 1993) can be deciphered by PCR sequencing. With some synthetic peptide libraries, the sequence of affinity-selected peptides can be determined by mass spectrometric methods.

Mario Geysen originally coined the term "combinatorial chemistry" to refer to libraries with which a deductive method is used to determine the sequences of peptide ligands. Deductive methods depend on the synthesis of multiple pools of peptides from which optimal peptide sequences are deduced after one round of synthesis and assay (viz., the positional scanning method of Pinilla et al. 1992) or after iterative rounds of synthesis and assay (Geysen et al. 1986, and its derivatives; Blake and Litzi-Davis 1992). The peptides in a library can be synthesized on plastic pins in a 96-pin format, using mixed amino acids in each addition step, and then tested for binding by a direct ELISA-like approach (Geysen et al. 1986). Alternatively, peptide libraries can be synthesized as peptides free in solution; activated resin is placed in "teabags," with each bag being placed in a different amino acid addition reaction. After reaction completion, the resin is removed from all the teabags, mixed, and divided among a new set of teabags before beginning the synthesis of the next randomized residue in the library sequence. Free peptides are released from the beads and tested for binding, usually in a competition ELISA format; however, they can be used in other assays, such as cell-based assays.

Positional scanning libraries for, say, a 6-mer peptide length, comprise six sets of peptide libraries; each library is made of a single, fixed residue in one position, with the other five residues being fully randomized (X residues). Thus, 19 libraries are made for every position in the hexapeptide (Cys is excluded), and, for a hexapeptide, there would be 19 libraries for each of six positions, or 114 libraries. All the libraries are assayed side by side for binding activity, and the binding residues at specific positions in the hexapeptide sequence are deduced from the libraries giving positive signals, because binding activity by a given library implies that its fixed residue is involved. Thus, the sequences of binding peptides are deduced from the pattern of reactivity with the panel of fixed-residue libraries, synthesized, and tested for their binding activity.

In the iterative screening strategies, multiple peptide pools are synthesized and tested for binding to the target molecule; then, on the basis of those outcomes, a new set of pools with reduced complexity is synthesized and tested for binding. The complexity of the pools synthesized is sequentially decreased after each assay until each pool contains a single ligand. Thus, using the example of a hexapeptide again, two positions of a hexapeptide are fixed in all possible amino acid combinations with the remaining four residues being fully randomized (with Cys excluded, this equals 19 x 19 or 361 pools). These libraries are screened for binding activity, and based on the best-binding pool, a new set of 19 libraries having three fixed residues is synthesized, with the first two residues being those of the best-binding pool and the third residue being one of the 19 amino acids; the remaining three positions are X residues. These libraries are screened to determine the best amino acid to fix for the third position in the sequence. With this accomplished, a new set of libraries is made in which the first three positions are fixed with optimally binding residues, the fourth position is fixed with each of the 19 residues, and the last two positions are X residues. These are screened to define the optimal fourth residue, and the procedure is repeated to define the fifth and sixth positions. Of course, in the final screening, each "library" comprises only a single peptide, and from testing these 19 peptides, the best binders are observed.

COMPARISON OF RECOMBINANT-DISPLAY AND SYNTHETIC PEPTIDE LIBRARIES

Ultimately, the utility of any library is best determined by its success in yielding specific, high-affinity ligands for a variety of target molecules. The success rate of a given library is often difficult to determine from published work, because usually only positive results are reported. This lack of published success rates for the different libraries is compounded by the fact that some target molecules do not cross-react well with peptides. Bonnycastle et al. (1996) determined the relative success rates for a panel of 11 different disulfide-constrained and unconstrained peptide libraries in finding ligand-peptides for antibodies that were raised against a number of different immunogens. They screened the libraries with 17 antibodies, including 15 monoclonal antibodies directed against peptides, linear and discontinuous epitopes on proteins, or carbohydrates, as well as two polyclonal antibodies against proteins of known structure. All of the antibodies directed against peptides, linear epitopes on folded proteins, and, surprisingly, carbohydrates, isolated binding peptides from several libraries. The peptides isolated by most of the monoclonal antibodies bore consensus sequences; notable exceptions included polyspecific monoclonal antibodies known to bind to a host of seemingly unrelated peptide sequences. In contrast to these successes, only 2 of the 4 monoclonal antibodies directed against discontinuous protein epitopes isolated binding peptides. Moreover, the polyclonal antibodies that were directed against folded proteins exclusively isolated peptides whose sequences matched short, linear stretches (linear epitopes) on the surface of the protein immunogen. We have since shown that antibodies against discontinuous protein epitopes will often isolate only a few weak-binding peptides from the libraries (Craig 1998; J.K. Scott, unpubl.). The affinities of several of these peptides have been significantly improved by the construction and screening of second-generation "sublibraries" in which the peptide sequences included X residues and a limited number of fixed consensus residues (J.K. Scott et al., unpubl.). It should also be noted that although peptides appear to cross-react with most anticarbohydrate antibodies, binding peptides have not been found for most plant and animal lectins (Harris et al. 1997; J.K. Scott et al., unpubl.). In addition, little has been published on the ability of peptides to cross-react with small molecules (e.g., haptens).

Only one side-by-side comparative study has been performed to assess a synthetic versus recombinant-display peptide library (Linn et al. 1997), and in this work, both types of libraries converged on the same set of selected residues (consensus sequence). From reports in the literature in which similar targets are screened by both means, it appears that libraries composed of similar peptides (e.g., synthetic or phage-displayed 6-mer libraries) probably isolate the same types of sequences, regardless of the mode of construction or screening method used. (For example, compare the data of Cwirla et al. 1990 and Barrett et al. 1992 with those of Houghten et al. 1991 and of Lam et al. 1991.) This comparison, however, cannot be considered a stringent test, because antibodies raised against peptides were used as the screening molecule, thus biasing the different library screenings toward similar outcomes. Such a test should be performed with more difficult screening molecules like enzymes.

Both synthetic and recombinant-display approaches have obvious advantages. On one hand, synthetic peptide libraries can incorporate a wide variety of non-amino acid entities to add both constraint and chemical variation to the libraries. Additionally, for most purposes, one is working with the peptides in something closer to the final form intended for the ligand-peptide. The deductive screening strategies also offer a more complete set of related peptides as screening outcome than do affinity-selection strategies. Screening can be performed with peptides in solution, which offers optimal affinity data. Moreover, some synthetic libraries (such as the positional scanning ones; Pinilla et al. 1992) can be screened in as little as a day. On the other hand, the recombinant-display libraries can encompass longer peptides, and they can be constrained by cysteine bridges or by a fixed framework. Multiple consensus sequences can often be identified by screening with polyclonal targets, such as polyclonal IgG (Bonnycastle et al. 1996), or polyspecific targets, such as a chaperone (Blond-Elguindi et al. 1993) or monoclonal antibody (Bonnycastle et al. 1996; De Ciechi et al. 1996); this is because binding phage are drawn independently from a library. (Note that this advantage is also true of the peptide-on-bead libraries of Lam et al. [1991] and their derivatives.)

Both types of libraries have unique biases. Synthetic peptide libraries are biased against peptides that are insoluble. The "window" size can be a problem in choosing which library to screen. For example, screening results may be confounded if the critical binding residues are in a 4-residue sequence, but a 6-mer library was screened. This problem can be minimized by screening a panel of starting libraries of different lengths (either positional libraries or the first-step libraries of an iterative screening). Positional scanning offers quick results from a single round of assays; however, the preferences for particular residues at each position in the sequence are usually assumed to be part of a single binding sequence, and thus may be confounded by the presence of multiple binding sequences for a given screening molecule. Thus, the final composition of optimally binding peptides must be determined after making a set of peptides bearing different combinations of the preferred residues, and in some cases, multiple binding motifs have been discovered (Pinilla et al. 1995). A problem with mixed-pool screenings is that a small family of high-affinity peptides that exists at a low concentration might not be detected over a larger family of mediocre binders that exists at a higher concentration. Another problem with iterative screening is that a set of related binding sequences is obtained, but the outcome can be changed dramatically by changing the positions of the fixed residues in the beginning of the screening. Again, it is useful to know the minimal window size of the binding sequence before embarking on iterative screenings. The recombinant-display libraries have the drawback of poor display of some peptides, and/or poor production of phage clones displaying certain peptides (these can be related phenomena, see Chapters 1 and 2). This poor display may have little to do with a given peptide's solubility, but rather with its ability to impede the processing of the coat protein fusion at the inner membrane of the bacterial cell. Thus, in some cases there may be distinct peptides that are present in one type of library, but not in another type. Moreover, there are several published examples describing phage-borne peptides that lost their ability to bind the target molecule when synthesized chemically (e.g., see Felici et al. 1993); peptides that are soluble on the phage may be insoluble when free in solution. As compared to its synthetic form, attachment to a fusion protein may allow a peptide to adopt a stable structure (Jelinek et al. 1997) or a preferred binding structure (J.K. Scott et al., unpubl.). There are also significant drawbacks to phage-library screening; for example, binding peptides may be missed due to analysis of too few clones or to overpanning. When a library has been overpanned, a limited number of clones come to predominate in the population to the exclusion of others having similar or better affinities (see Chapter 17 for a discussion of the dynamics of phage panning experiments). To avoid this effect, Folgori et al. (1994) and others use affinity purification only in the first one or two rounds of screening. Binding clones are identified by secondary screens in which the phage are plated and transferred from plaques to membranes; binding of target molecule to the membrane-bound phage is detected.

STRATEGIES FOR OBTAINING AND OPTIMIZING PEPTIDE LIGANDS FROM PHAGE-DISPLAY LIBRARIES

The investigator wishing to use phage-displayed peptide libraries and, ultimately, to obtain a synthetic peptide ligand may consider the following strategy. First, the libraries are screened with the target molecule. There are two approaches to screening: capturing the phage on an immobilized target, or binding the target to the phage in solution and then capturing the phage–target complexes on a surface (see Chapters 2 and 17). It is best to screen a panel of libraries, as this maximizes information about the specificity of the target's binding site, both with regard to the variation in consensus sequence groups (i.e., target polyspecificity) and the preferred constraints within a consensus sequence group (Bonnycastle et al. 1996).

Second, the phage pools are assessed for binding by comparing the phage yields from the different pools screened by a given molecule (Chapter 17) and by comparing ELISA signals produced by each pool (Chapter 18). Clones are isolated from the best pools, tested for direct binding to target by ELISA (Chapter 18), and sequenced (Chapter 15). Specificity of phage binding can be tested by competition ELISAs, in which a preexisting ligand for the target molecule should be able to inhibit binding between the phage clones and the target. Consensus sequences may be deduced by comparing the peptide sequences of clones having dissimilar DNA sequences (i.e., non-sibling clones). At this point, if the binding strength of the best-binding clone is weak, one could consider making a "focused" sublibrary (Wrighton et al. 1996; Cwirla et al. 1997). If a consensus sequence is apparent, then a degenerate oligonucleotide bearing fixed codons encoding the consensus residues can be made, with the remaining residues in the peptide randomized via NNK codons. More complex libraries can be made following the approaches of W.D. Huse (pers. comm.) or Virnekaes et al. (1994), using oligonucleotides bearing specified codon mixtures.

If a consensus sequence is not apparent and a weak-binding clone has been identified, one can prepare a "doped" library to identify tighter binders; that is, a library in which a fixed peptide sequence is supplemented (doped) at some positions with randomized amino acids. Such libraries usually entail synthesizing the DNA encoding the binding clone's peptide sequence, but doping each residue with an equimolar mixture of the four nucleotides (this DNA can be obtained from any oligonucleotide-synthesis facility). The drawback in using such libraries is that extra analysis is required to decide whether a given residue that is favored in the screening was truly affinity-selected or was simply overrepresented in the NNN doping. For example, for a methionine residue that is varied by a 1:1 doping with an N-mixture at each nucleotide in the methionine codon, the synthesis mixture will contain each nucleotide in the methionine codon at 62.5% versus 12.5% for each of the other three nucleotides; hence, there will be intrinsic bias toward codons (like isoleucine, valine, and threonine) whose sequences are most closely related to the methionine codon. This bias should be absent in doped libraries that are made following the approach of Virnekaes et al. (1994), in which trinucleotide phosphoramidites for codons can be mixed at any desired ratio, or that of Glaser et al. (1992), in which the library oligonucleotide is made using two columns to synthesize in parallel the codons in the peptide sequence and NNK codons, with mixing of the resins and their redistribution between the two columns after each codon addition (as described in the preceding section).

If the best-binding phage clone from a screening yields a strong signal in the binding assay, the sequence of that clone can be synthesized and tested, independent of the phage, for binding in a competitive binding assay against a preexisting ligand for the target molecule. If the peptide maintains its binding properties after transformation to the soluble form, it can be synthetically mutated to determine (or confirm) the critical residues for binding. There may be prior information about critical residues from the consensus sequence on the phage, and this can be used to make peptides bearing replacements that allow binding. Thus, a phage-borne peptide sequence may be analyzed and optimized using synthetic peptides. It is probably most cost-effective to carry out such studies with peptides synthesized on pins (Chiron Technologies, San Diego, California) or on cellulose (Jerini Bio Tools GmbH, Berlin, Germany). The pin method allows one to produce the peptides oneself, on demand, at a cost of about $20 a peptide; the cellulose-bound peptides, although cheaper on a per-peptide basis, must be ordered in advance as a standard mutational analysis. Chiron Technologies also offers a system devised by H.M. Geysen (Maeji et al. 1990) in which peptides are synthesized on pins in 1–2-mg quantities and then cleaved from the pins. This allows the peptides to be purified (by HPLC) and analyzed for the correct molecular weight (by mass spectroscopy). Moreover, if disulfide constraints are planned for the peptide, one can optimize intramolecular cyclization in dilute, oxidizing conditions, and then analyze the reaction products for the correct form (by HPLC and mass spectrometry). This approach provides free peptide that can be tested in competitive binding assays, which are usually better suited for affinity discrimination than the direct-binding assays that must be used with pin- or cellulose-bound peptides (this is especially so for assay with multivalent binding molecules like antibodies).

Alternatively, binding peptides can be transferred by recombinant means to a suitable expression protein, such as alkaline phosphatase (Grihalde et al. 1995; Yamabhai and Kay 1997), the maltose-binding protein (MBP; Zwick et al. 1998b), or glutathione *S*-transferase (GST; Tang et al. 1999). This approach allows one to transfer selected peptides either from enriched phage pools en masse or from single clones to a monovalent system (in the case of MBP) or divalent system (in the case of alkaline phosphatase or GST) that is easy to use. This is especially helpful for long peptides (which are difficult to synthesize in useful amounts), peptides whose synthetic analog does not have binding activity, peptides with multiple disulfide bridges (with active structures that may be difficult to analyze or prepare as synthetic analogs), or epitopes formed of discontinuous elements on the surface of a folded protein. Binding studies, mutational analyses, and optimization can be carried out with these high-expression systems; however, the production and purification of such constructs is far more labor-intensive than for synthetic peptides. Transfer of a peptide to a nonphage milieu offers the chance of removing specific effects the phage might have on peptide structure, while retaining nonspecific properties, like improved solubility, that result from being fused to a larger protein. Peptide-transfer to maltose-binding protein also overcomes the problem of determining the affinities of multiply displayed peptides, as occur on the phage surface.

During the past decade, a large number of bioactive peptides have been isolated from peptide libraries, and phage-displayed peptide libraries continue to be popular, available, and relatively inexpensive sources of peptide ligands for many investigators. The next decade will likely hold new advances in this technology and its application, especially in the areas of proteinomics and high-throughput screening.

REFERENCES

Barbas C.F., III, Amberg W., Simoncsits A., Jones T.M. and Lerner R.A. 1993. Selection of human anti-hapten antibodies from semisynthetic libraries. *Gene* **137:** 57–62.

Barrett R.W., Cwirla S.E., Ackerman M.S., Olson A.M., Peters E.A., and Dower W.J. 1992. Selective enrichment and characterization of high affinity ligands from collections of random peptides on filamentous phage. *Anal. Biochem.* **204:** 357–364.

Bianchi E., Folgori A., Wallace A., Nicotra M., Acali S., Phalipon A., Barbato G., Bazzo R., Cortese R., Felici F. et al. 1995. A conformationally homogeneous combinatorial peptide library. *J. Mol. Biol.* **247:** 154-160.

Blake J. and Litzi-Davis L. 1992. Evaluation of peptide libraries: An iterative strategy to analyze the reactivity of peptide mixtures with antibodies. *Bioconjug. Chem.* **3:** 510–513.

Blond-Elguindi S., Cwirla S.E., Dower W.J., Lipshutz R.J., Sprang S.R., Sambrook J.F., and Gething M.J. 1993. Affinity panning of a library of peptides displayed on bacteriophages reveals the binding specificity of BiP. *Cell* **75:** 717–728.

Bonnycastle L.L., Mehroke J.S., Rashed M., Gong X. and Scott J.K. 1996. Probing the basis of antibody reactivity with a panel of constrained peptide libraries displayed by filamentous phage. *J. Mol. Biol.* **258:** 747–762.

Braden B.C., Goldman E.R., Mariuzza R.A., and Poljak R.J. 1998. Anatomy of an antibody molecule: Structure, kinetics, thermodynamics and mutational studies of the antilysozyme antibody D1.3. *Immunol. Rev.* **163:** 45–57.

Bremnes T., Lauvrak V., Lindqvist B., and Bakke O. 1998. Selection of phage displayed peptides from a random 10-mer library recognising a peptide target. *Immunotechnology* **4:** 21–28.

Brown S. 1992. Engineered iron oxide-adhesion mutants of the *Escherichia coli* phage lambda receptor. *Proc. Natl. Acad. Sci.* **89:** 8651–8655.

Craig, L. 1998. "Exploring the basis of antibody-mediated protein-peptide cross-reactivity, pp. 65–87." PhD thesis. Simon Fraser University, Canada.

Cull M.G., Miller J.F., and Schatz P.J. 1992. Screening for receptor ligands using large libraries of peptides linked to the C terminus of the lac repressor. *Proc. Natl. Acad. Sci.* **89:** 1865–1869.

Cwirla S.E., Peters E.A., Barrett R.W., and Dower W.J. 1990. Peptides on phage: A vast library of peptides for identifying ligands. *Proc. Natl. Acad. Sci.* **87:** 6378–6382.

Cwirla S.E., Balasubramanian P., Duffin D.J., Wagstrom C.R., Gates C.M., Singer S.C., Davis A.M., Tansik R.L., Mattheakis L.C., Boytos C.M., Schatz P.J., Baccanari D.P., Wrighton N.C., Barrett R.W., and Dower W.J. 1997. Peptide agonist of the thrombopoietin receptor as potent as the natural cytokine. *Science* **276:** 1696–1699.

De Ciechi P.A., Devine C.S., Lee S.C., Howard S.C., Olins P.O., and Caparon M.H. 1996. Utilization of multiple phage display libraries for the identification of dissimilar peptide motifs that bind to a B7-1 monoclonal antibody. *Mol. Divers.* **1:** 79–86.

Devlin J.J., Panganiban L.C., and Devlin P.E. 1990. Random peptide libraries: A source of specific protein binding molecules. *Science* **249:** 404–406.

Duenas M. and Borrebaeck C.A. 1994. Clonal selection and amplification of phage displayed antibodies by linking antigen recognition and phage replication. *Bio/Technology* **12:** 999–1002.

Ellman J.A. and Gallop M.A. 1998. Combinatorial chemistry. *Curr. Opin. Chem. Biol.* **2:** 317–319.

Felici F., Luzzago A., Folgori A., and Cortese R. 1993. Mimicking of discontinuous epitopes by phage-displayed peptides., II. Selection of clones recognized by a protective monoclonal antibody against the *Bordetella pertussis* toxin from phage peptide libraries. *Gene* **128:** 21–27.

Folgori A., Tafi R., Meola A., Felici F., Galfre G., Cortese R., Monaci P., and Nicosia A. 1994. A general strategy to identify mimotopes of pathological antigens using only random peptide libraries and human sera. *EMBO J.* **13:** 2236–2243.

Geysen H.M., Rodda S.J., and T.J. Mason. 1986. A priori delineation of a peptide which mimics a discontinuous antigenic determinant. *Mol. Immunol.* **23:** 709–715.

Glaser S.M., Yelton D.E., and Huse W.D. 1992. Antibody engineering by codon-based mutagenesis in a filamentous phage vector system. *J. Immunol.* **149:** 3903–3913.

Grihalde N.D., Chen Y.C., Golden A., Gubbins E., and Mandecki W. 1995. Epitope mapping of anti-HIV and anti-HCV monoclonal antibodies and characterization of epitope mimics using a filamentous phage peptide library. *Gene* **166:** 187–195.

Harris S.L., Craig L., Mehroke J.S., Rashed M., Zwick M.B., Kenar K., Toone E.J., Auzanneau F.-I., Marino-Albernas J.-R., Pinto B.M., and Scott J.K. 1997. Exploring the basis of peptide-carbohydrate cross-reactivity: Evidence for discrimination by peptides between closely related, anti-carbohydrate antibodies. *Proc. Natl. Acad. Sci.* **94:** 2454–2459.

Houghten R.A., Pinilla C., Blondelle S.E., Appel J.R., Dooley C.T., and Cuervo J.H. 1991. Generation of synthetic-peptide, combinatorial libraries for basic research and drug discovery. *Nature* **354:** 84–86.

Houshmand H., Froman G., and Magnusson G. 1999. Use of bacteriophage T7 displayed peptides for determination of monoclonal antibody specificity and biosensor analysis of the binding reaction. *Anal. Biochem.* **268:** 363–370.

Jelinek R., Terry T.D., Gesell J.J., Malik P., Perham R.N., and Opella S.J. 1997. NMR structure of the principal neutralizing determinant of HIV-1 displayed in filamentous bacteriophage coat protein. *J. Mol. Biol.* **266:** 649–655.

Kramer A., Schuster A., Reineke U., Malin R., Volkmer-Engert R., Landgraf C., and Schneider-Mergener J. 1994. Combinatorial cellulose-bound peptide libraries: Screening tool for the identification of peptides that bind ligands with predefined specificity. *Methods* **6:** 388–395.

Ku J. and Schultz P.G. 1995. Alternate protein frameworks for molecular recognition. *Proc. Natl. Acad. Sci.* **92:** 6552–6656.

Lam K.S., Salmon S.E., Hersh E.M., Hruby V.J., Kazmierski W.M., and Knapp R.J. 1991. A new type of synthetic peptide library for identifying ligand-binding activity. *Nature* **354:** 82–84. [Published errata appear in *Nature* 1992, **358:** 434 and 1992 **360:** 768.]

Linn H., Ermekova K.S., Rentschler S., Sparks A.B., Kay B.K., Sudol M. 1997. Using molecular repertoires to identify high-affinity peptide ligands of the WW domain of human and mouse YAP. *Biol. Chem.* **378:** 531–537.

Lu Z., Murray K.S., Van Cleave V., LaVallie E.R., Stahl M.L., and McCoy J.M. 1995. Expression of thioredoxin random peptide libraries on the *Escherichia coli* cell surface as functional fusions to flagellin: A system designed for exploring protein-protein interactions. *Bio/Technology* **13:** 366–372.

Maeji N.J., Bray A.M., and Geysen H.M. 1990. Multi-pin peptide synthesis strategy for T cell determinant analysis. *J. Immunol. Methods* **134:** 23–33.

Martin F., Toniatti C., Salvati A.L., Venturini S., Ciliberto G., Cortese R., and Sollazzo M., 1994. The affinity-selection of a minibody polypeptide inhibitor of human interleukin-6. *EMBO J.* **13:** 5303–5309.

Mattheakis L.C., Bhatt R.R., and Dower W.J. 1994. An in vitro polysome display system for identifying ligands from very large peptide libraries. *Proc. Natl. Acad. Sci.* **91:** 9022–9026.

McConnell S.J. and Hoess R.H. 1995. Tendamistat as a scaffold for conformationally constrained phage peptide libraries. *J. Mol. Biol.* **250:** 460–470.

McLafferty M.A., Kent R.B., Ladner R.C., and Markland W. 1993. M13 bacteriophage displaying disulfide-constrained microproteins. *Gene* **128:** 29–36.

Needels M.C., Jones D.G., Tate E.H., Heinkel G.L., Kochersperger L.M., Dower W.J., Barrett R.W., and Gallop M.A. 1993. Generation and screening of an oligonucleotide-encoded synthetic peptide library. *Proc. Natl. Acad. Sci.* **90:** 10700–10704.

Nord K., Nilsson J., Nilsson B., Uhlen M., and Nygren P.A. 1995. A combinatorial library of an alpha-helical bacterial receptor domain. *Protein Eng.* **8:** 601–608.

O'Neil K.T., Hoess R.H., Jackson S.A., Ramachandran N.S., Mousa S.A., and DeGrado W.F. 1992. Identification of novel peptide antagonists for GPIIb/IIIa from a conformationally constrained phage peptide library. *Proteins* **14:** 509–515.

Parmley S.F. and Smith G.P. 1988. Antibody-selectable filamentous fd phage vectors: Affinity purification of target genes. *Gene* **73:** 305–318.

Pinilla C., Appel J.R., Blanc P., and Houghten R.A. 1992. Rapid identification of high affinity peptide ligands using positional scanning synthetic peptide combinatorial libraries. *BioTechniques* **13:** 901–905.

Pinilla C., Chendra S., Appel J.R., and Houghten R.A. 1995. Elucidation of monoclonal antibody polyspecificity using a synthetic combinatorial library. *Pept. Res.* **8:** 250–257.

Rebar E.J. and Pabo C.O. 1994. Zinc finger phage: Affinity selection of fingers with new DNA-binding specificities. *Science* **263:** 671–673.

Roberts R.W. and Szostak J.W. 1997. RNA-peptide fusions for the in vitro selection of peptides and proteins. *Proc. Natl. Acad. Sci.* **94:** 12297–122302.

Schembri M.A., Kjaergaard K., and Klemm P. 1999. Bioaccumulation of heavy metals by fimbrial designer adhesins. *FEMS Microbiol. Lett.* **170:** 363–371.

Scott J.K. 1994. Identifying lead peptides from epitope libraries. In *Biological approaches to rational drug design* (ed. Weiner D.B. and Williams W.V.), pp. 1–27. CRC Press, Boca Raton, Florida.

Scott J.K. and Smith G.P. 1990. Searching for peptide ligands with an epitope library. *Science* **249:** 386–390.

Scott J.K., Loganathan D., Easley R.B., Gong X., and Goldstein I.J. 1992. A family of concanavalin A-binding peptides from a hexapeptide epitope library. *Proc. Natl. Acad. Sci.* **89:** 5398–5402.

Smith G.P. 1985. Filamentous fusion phage: Novel expression vectors that display cloned antigens on the virion surface. *Science* **228:** 1315–1317.

Smith G.P. and Petrenko V.A. 1997. Phage display. *Chem. Rev.* **97:** 391–410.

Stolz J., Ludwig A., and Sauer N. 1998. Bacteriophage lambda surface display of a bacterial biotin acceptor domain reveals the minimal peptide size required for biotinylation. *FEBS Lett.* **440:** 213–217.

Tang Y., Beuerlein G., Pecht G., Chilton T., Huse W.D., and Watkins J.D. 1999. Use of a peptide mimotope to guide the humanization of MRK-16, an anti-P-glycoprotein monoclonal antibody. *J. Biol. Chem.* **274:** 27371–27378.

Virnekaes B., Ge L., Plueckthun A., Schneider K.C., Wellnhofer G., and Moroney S.E. 1994. Trinucleotide phosphoramidites: Ideal reagents for the synthesis of mixed oligonucleotides for random mutagenesis. *Nucleic Acids Res.* **22:** 5600–5607.

Wrighton N.C., Farrell F.X., Chang R., Kashyap A.K., Barbone F.P., Mulcahy L.S., Johnson D.L., Barrett R.W., Jolliffe L.K., and Dower W.J. 1996. Small peptides as potent mimetics of the protein hormone erythropoietin. *Science* **273:** 458–464.

Wu H., Yang W.P., and Barbas C.F., III. 1995. Building zinc fingers by selection: Toward a therapeutic application. *Proc. Natl. Acad. Sci.* **92:** 344–348.

Wu H,. Nie Y., Huse W.D., and Watkins J.D. 1999. Humanization of a murine monoclonal antibody by simultaneous optimization of framework and CDR residues. *J. Mol. Biol.* **294:** 151–162.

Yamabhai M. and Kay B.K. 1997. Examining the specificity of Src homology 3 domain—ligand interactions with alkaline phosphatase fusion proteins. *Anal. Biochem.* **247:** 143–151.

Zwick M.B., Shen J., and Scott J.K. 1998a. Phage-displayed peptide libraries. *Curr. Opin. Biotechnol.* **9:** 427–436.

Zwick M.B., Bonnycastle L.L.C., Noren K.A., Venturini S., Leong E., Barbas C.F., III, Noren C.J., and Scott J.K. 1998b. The maltose binding protein as a scaffold for monovalent display of peptides derived from phage libraries. *Anal. Biochem.* **264:** 87–97.

5 Functional Domains and Scaffolds

GREGG J. SILVERMAN
Department of Medicine, University of California, San Diego, La Jolla, California 92093-0663

SINCE THE FIRST PHAGE-DISPLAY REPORT in 1985 (Smith 1985), this technology has rapidly evolved into an efficient tool used by structural biologists for the discovery and characterization of diverse ligand–receptor-binding interactions. By starting with the phage display of an entire protein or functional domain, altered and novel properties can be engineered, even in cases in which a detailed understanding of structure–function relationships is not available (for review, see O'Neil and Hoess 1995; Katz 1997). Novel structure-based designer ligands potentially offer promising new treatments for infectious diseases, immunologic diseases, and malignancies. Phage-display technology also offers a means to harness the explosion of DNA sequences from host genomes and the insights from advanced structural analyses, to propel the translational studies required for the development of new therapeutic agents. Whereas other chapters in this manual focus on the discovery of small peptide ligands or the recovery or refinement of antibodies, this chapter describes other types of applications that involve small non-antibody proteins.

CONTENTS

FUNCTIONAL DOMAINS: DISPLAY, VARIEGATION, AND OPTIMIZATION

Protein domains are independent folding units that represent the minimal polypeptide sequences required to create three-dimensional structures conveying defined functional activities. Ideally, by altering the composition, and at times the size, of the polypeptide, the functional capacity of a domain can be improved. However, despite several decades of investigation and many impressive successes for in vitro protein expression, all too often a functional eukaryotic protein has proven to be incompatible with bacterial expression systems. These limitations commonly derive from the inherent inability of bacteria to perform many posttranslational modifications. In other cases, unpredictable pitfalls are attributed to problems in protein folding, as the folding of mammalian proteins into a functional conformation depends on chaperonin proteins that perform essential but poorly understood functions. Because local structural context can also affect protein folding, the fusion of a gene sequence to a foreign gene can interfere with the native functional activity. From this perspective, it was almost unexpected that mammalian proteins could be displayed fused to the coat protein of a filamentous bacteriophage and still retain native functional capacity. The list of reports of successfully displayed proteins is now quite extensive (for review, see Lowman et al. 1991; Chiswell and McCafferty 1992; Wells and Lowman 1992; O'Neil and Hoess 1995).

Phage display of a functional protein has now become a standard first step of proof of principles prior to the application of combinatorial strategies using the cloned DNA template to evaluate or remold functional activity (Bass et al. 1990; Roberts et al. 1992b). Even though a natural functional domain can represent the end product of a highly directed evolutionary process, phage-display approaches can create variations of the domain with altered binding affinity or fine specificity, or with structural refinements that greatly enhance stability. Hence, by rational design and the application of combinatorial approaches, novel small proteins can be created with properties that are greatly enhanced compared to naturally occurring proteins.

The most efficient applications of phage-display technology require an appreciation of several practical limitations (for discussion, see Clackson and Wells 1994). Most importantly, because the isolation of a protein variant with improved properties requires the creation of phage(mid) libraries with sequence variations, it is important to understand the size limits of libraries that can be created. With the most commonly applied cloning and transfection methods, libraries of up to about 10^8–10^9 independent members represent the practical size limit. Oligonucleotide-based methods can completely randomize (variegate) the codons for all possible amino acids expressed at critical positions, but this requires 64 different nucleic acid sequences for a trinucleotide sequence (i.e., 4 x 4 x 4). Hence, for the variegation of N different codon positions in a gene, a library of up to 64^N members is required. Therefore, in each library it is possible to completely randomize only a limited number of codons in a DNA sequence of interest. Because of these practical considerations, at most, 6–8 amino acid positions can be completely randomized in a library (representing 20^6 to 20^8 distinct members)(Clackson and Wells 1994). However, the targeted mutagenesis of this small number of positions may greatly affect binding affinity and specificity, as in many cases the binding of large and discontinuous surfaces may be mediated by only a small subset of critical contact side chains (Li et al. 1995). Methods that enable the generation of larger libraries have been reported, including the use of lambda packaging systems to boost transformation efficiencies (Alting-Mees and Short 1993). A combinatorial approach has also been described that uses Cre Lox combination site to assemble a very large combinatorial repertoire ($>10^{11}$ members) from two different libraries of exons (Fisch et al. 1996). However, neither of these approaches has yet been widely adopted.

In practice, the redundancy of the genetic code enables the representation of all natural amino acids with fewer than 64 codon variations, which is desirable because there is a strict numerical relationship between the size of a library that can be created (i.e., the number of independent transformants in a library) and the maximum number of nucleotide positions that can be completely randomized. To reduce the codon diversity required to provide sufficient amino acid diversity, potentially allowing an increase in the number of randomized codons at the greatest number of positions, several nucleotide doping strategies have been used. One example of a common codon doping strategy is the use of the degenerate codon, NNK, in which N is any nucleotide, and K is either G or T. This strategy allows for all 20 amino acid combinations within 32 codons. Alternatively, NNS has been used, in which S is either a G or C. In another approach, the degenerate triplet VNS has been used, in which V represents A, C, or G, and S represents G or C; the 24 resultant codons code for diverse amino acids except for Tyr, Phe, Trp, and Cys, and do not include any stop codons. When used to generate libraries of a defined size, each of these doping strategies enables the inclusion of members that are representative of a greater diversity of protein sequences.

A domain engineering project is more likely to be successful if the strategies are based on an understanding of the structural basis of the functional activity (Fig. 5.1) (see Kast and Hilvert 1997). Of special value is a specific knowledge of which amino acid residues are involved directly in binding interactions or the functional properties of the domain, as is identification of the positions that must be conserved to stabilize the overall structure of the protein (also termed the framework residues). For these reasons, crystallographic analysis or NMR studies of the domain of interest are highly useful, as they help to define the structural basis of the functional activity. In addition, methods like alanine scanning, which enable the sequential removal of the influence of a single residue, can provide an in-depth understanding of structure–function relationships. By coordinating these approaches, it is possible to identify the amino acid positions that are most critical to determining the activities of a protein, and to predict which substitutions may be altered to improve these properties.

ENGINEERING OF PROTEIN HORMONES

Phage-display technology has been very successfully applied in several systems directed at the development of functional domains with enhanced properties. In pioneer studies using a model hormone system, Lowman et al. (1991) demonstrated that high-affinity nonantibody domains could be displayed on monovalent phage and successfully selected in vitro on the basis of ligand-binding activities. Growth hormone (GH) is a peptide product of the pituitary gland involved in the metabolic regulation of a wide range of cell types in the body. Human GH (hGH) represents an α-helical bundle, and extensive structural data from crystallographic data of growth hormone (Abdel-Meguid et al. 1987) and from mutagenesis studies (Cunningham and Wells 1989; 1991) guided the design of the strategy. Using an NNS doping regimen, three

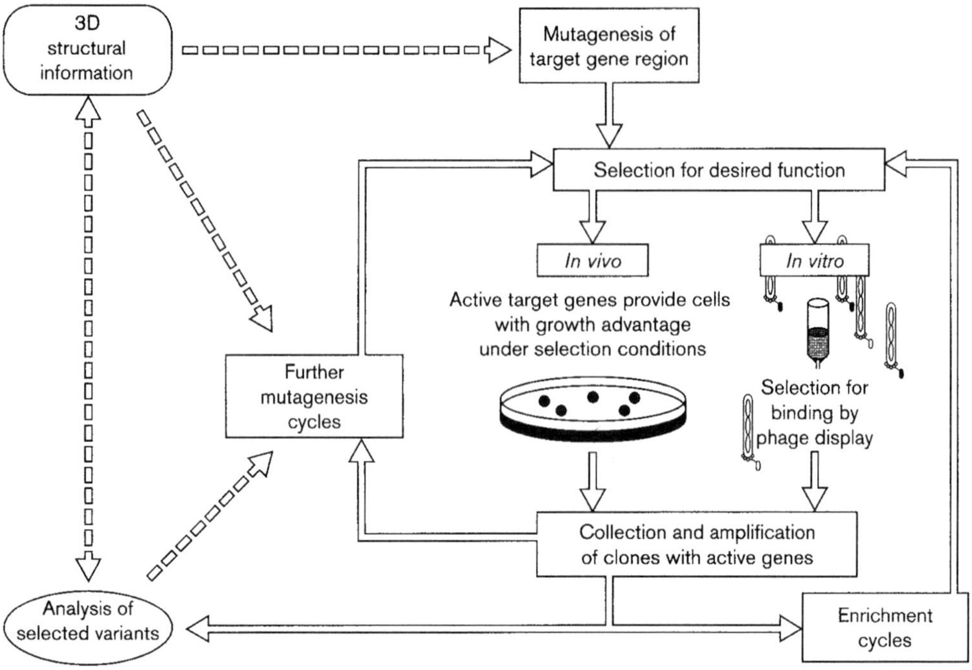

Figure 5.1. Evolutionary approaches to protein engineering based on structural information. 3D structural information is an invaluable guide for targeting randomizing mutagenesis to regions of proteins presumed to be critical for function. Structural data can also be used to interpret the results of protein evolution experiments. The outcome of such experiments in turn yields new information on the structure and interactions of proteins. The evolutionary scheme depicts two examples for successful strategies for in vivo or in vitro selection of desired variants from large combinatorial libraries, with the dashed arrows indicating the flow of information. (Reprinted, with permission, from Kast and Hilvert 1997.)

separate random mutant growth hormone libraries with ~10^6 random mutations were created, which encompassed 12 sites previously shown to modulate binding to hGH receptor or human prolactin receptor. With these libraries, variants of hGH with increased affinity and specificity for hGH receptor were isolated from an engineered phage-display system. Selections were performed in vitro using hGH receptors on beads, and the output libraries were examined after 3–6 cycles of selection. Residues previously identified as important to binding by alanine scanning mutagenesis were more highly conserved in the selected libraries. Significantly, several of these mutations were found to be additive for improved affinity (Wells 1990). When several of these mutations were introduced together, the variant hormone exhibited a cumulative 400-fold increase in affinity (Lowman and Wells 1993). Unexpectedly, these studies also identified protein variations with enhanced functional properties that had residue replacements at positions that were not at the ligand–receptor interface (Ultsch et al. 1994). In addition, mutations associated with improved binding affinity were identified that had no detectable conformational or surface changes. Hence, these findings illustrate the capacity of phage-display methods to select beneficial mutations independent of involvement in a known active site, and in this case the functional improvements may have derived from altered flexibility of the protein that improved functional activity.

In another set of studies, phage-display approaches were used to localize a functional activity to specific domains in thrombopoietin (TPO), a peptide hormone responsible for the induction of megakaryocyte proliferation and production of blood platelets that play a central role in control of bleeding. To characterize the sites in the 153-amino-acid amino-terminal domain responsible for binding and stimulation of the TPO receptor, the domain was fused to the gIII coat protein of M13 bacteriophage (Pearce et al. 1997). Based on the high sequence homology with IL-4, which has a known structure involving a four-α-helical bundle, 40 residues were identified that were predicted to be solvent exposed. Each of these positions was separately mutated to alanine and displayed as a single copy on the phage surface. The functional properties of the mutants were then evaluated in a phage-binding immunoassay. By this approach, the epitopes recognized by a panel of monoclonal antibodies were characterized, and it was demonstrated that helix-4 was involved in receptor binding and bioactivity, whereas determinants in helix-1 and helix-3 blocked bioactivity but not receptor binding (Pearce et al. 1997). These studies contributed to a model in which a TPO molecule has two receptor-binding sites that bind and activate the TPO receptor in a sequential dimerization mechanism similar to the mechanism of human growth hormone.

From an integrated approach combining phage display, alanine-scanning mutagenesis, and structural analysis, small peptide analogs of atrial naturetic peptide (ANP) were created with minimized size and altered fine specificity (Cunningham et al. 1994; Fairbrother et al. 1994; Li et al. 1995). ANP is a 28-residue polypeptide hormone with regulatory roles in blood pressure, electrolyte, and fluid homeostasis. Three distinct cellular naturetic peptide receptor types (NPR-A, NPR-B, and NPR-C) have been described, which are associated with different signaling pathways and, presumably, different physiologic activities. In one approach, libraries were created containing variant ANP molecules with randomized patches of three or four contiguous residues. To obtain variants with preferential fine specificity for one type of receptor, positive selection was performed using the NPR-A immobilized to a bead, with concurrent negative selection using the NPR-C type in solution phase (Cunningham et al. 1994). This strategy succeeded in isolating ANP variants that had both enhanced receptor fine specificity and enhanced expression levels in *E. coli.* Structural analyses later indicated that the enhanced ANP had less flexibility in solution phase and greater order in secondary structure as determined by NMR spectroscopy, which may contribute to the observed differences in functional activity (Fairbrother et al. 1994).

Although smaller ANPs had been previously isolated following screening of panels of synthetic analogs, these peptides demonstrated at least 500-fold weaker receptor binding affinity. To improve upon this approach, structural observations were used to identify the underlying scaffold in ANP responsible for preserving the presentation of critical binding determinants. Alanine-scanning mutagenesis was then used to identify the residues important for binding to the receptor. Next, libraries of phage displaying variant ANP molecules with randomized codons at nonessential sites were selected to obtain optimized binders. Derivatives were then isolated by selection from libraries based on the recovered ANP variants using the conserved scaffolds in which nonessential, non-scaffolding residues were removed, and the remaining adjacent codons were randomized. Following these sequential steps, a 15-amino-acid peptide was identified that represented a reduction of nearly 50% in size compared to the parental ANP, but which had only ~8-fold lower affinity for the receptor (Li et al. 1995).

Other types of peptide hormones have also been functionally displayed on the surface of filamentous phage. In studies by Gram et al. (1993), human cytokine IL-3 was successfully displayed on M13 phage in active form. After injection into mice, the induced response to the recombinant phage also provoked an immune response to human IL-3. These mice were shown to contain B cells at high frequency that produced antibodies efficient at neutralizing native human IL-3.

In other work, Saggio et al. demonstrated the display of functional human ciliary neurotropic factor (CNTF), a 23-kD neurocytokine that promotes the survival and differentiation of a variety of neuronal and glial cells. Subsequently, by creation of CNTF-based libraries with randomized codons in a putative D helix portion, they were able to select receptor-specific superagonists (Saggio et al. 1995). In addition to enhanced receptor binding affinity, these analogs also exhibited improved 4-fold to 8-fold enhanced bioactivity for promoting the in vitro survival of ciliary ganglion neurons.

PHAGE DISPLAY AND HUMAN COMPLEMENT PROTEINS

Complement (C) proteins are critical components in innate immunity that are involved in the regulation of immune responsiveness and leukocyte migration. In studies of the human anaphylotoxin hC5a, several adaptations from standard approaches were required for the phage display of a functional form of the protein (Hennecke et al. 1997). Functional activity of hC5a was known to require a free carboxyl terminus, so the gene was cloned downstream of the Fos gene in the special phagemid vector for gene product III (pIII) array, pJuFo (described in Chapters 2 and 21). Unexpectedly, when arrayed as a fusion protein with pIII, the native hC5a was not functionally active, presumably because folding was adversely affected by the local protein environment. Functional activity was restored when a nonconserved cysteine residue at position 27 in hC5a was replaced with an alanine, enabling the induction of enzyme release from differentiated U937 cells. Further evidence of the functional activity of the altered hC5a was demonstrated with its selection from dilute libraries by biopanning on differentiated U937 cells.

In subsequent studies, libraries based on a modified hC5a form were used to evaluate the effect of simultaneous substitutions within the C5a effector domain (Hennecke et al. 1998). Codons at four positions involved in binding were randomized, then the activity of each of the selected clones was assessed to determine the tolerance for amino acid replacement. Surprisingly, most selected clones included a phenylalanine residue at a position that was originally a histidine residue, and the associated binding affinity and signaling activity were not affected. Selection studies also yielded clones with novel sequence variations that provided 4- to 10-fold higher binding affinity, compared to the parental form. These studies also characterized one mutant with a high binding affinity that was devoid of signaling activity. Thus, random mutagenesis of phage-displayed C5a enabled identification of receptor-binding contacts and also provided a means to select active C5aR antagonists based on the structure of the natural ligand.

In studies exploiting a different type of recognition unit, Swimmer et al. (1992) reported the engineering of the B chain of ricin toxin, a domain with lectin activity. In initial studies, they first demonstrated that this domain could be displayed as a fusion protein on the surface of filamentous phage with preservation of the lectin activity for specific binding of a carbohydrate. Subsequently, variant phage were created that displayed targeted point mutations, introduced at sites to alter key residues implicated in hydrogen bonding with substrate. The isolated phage were shown to have altered carbohydrate-binding fine specificities, confirming the localization of critical carbohydrate-binding contact sites in the ricin molecule.

ENGINEERED ENZYME INHIBITORS

In studies aimed at engineering proteins that control coagulation cascades, enzymes that regulate tissue plasminogen activator (PA) have become attractive targets for the development of new therapeutic agents. In studies aimed at adapting a bacterial enzyme inhibitor to have a specificity for human urokinase-type PA, mutagenesis experiments were performed with ecotrin, a serine protease inhibitor found in the periplasm of *E. coli.* Ecotrin is representative of a set of inhibitors with a dyad-related secondary binding site that permits recognition of a wide range of proteases. After array on phage via the *gene III* phage coat protein, the 142-amino-acid domain of ecotrin was functionally active and was also recognized by an ecotrin-specific antibody (Wang et al. 1995). Libraries were created with variegated codons for positions 84 and 85 that are at or near the reactive site of this example of the "substrate-like" class of inhibitors (Laskowski and Kato 1980). Following sequential rounds of selection, variants with increased affinity for urokinase-type PA were recovered (Wang et al. 1995), and these demonstrated nanomolar inhibitory activity. Compared to wild-type ecotrin, the selected mutant displayed a 2800-fold increase in binding affinity for human PA. Ultimately, these types of studies may yield domains with clinically useful activities that are compatible with high-yield and low-cost bacterial expression systems.

Human neutrophil elastase (HNE) is an abundant serine protease involved in the elimination of pathogens. However, during connective tissue remodeling, uncontrolled elastolytic activity of neurophil elastase can have significant pathologic implications (e.g., HNE contributes to smoker's emphysema and hereditary forms of emphysema associated with reduction of circulation of α-1 antiprotease inhibitor). In studies designed for the development of novel inhibitors of HNE, Roberts et al. (1992a, b) reported a successful strategy for engineering variants of bovine pancreatic trypsin inhibitor (BPTI) that displayed altered and clinically useful activity profiles. Before library construction, four site-specific mutations were introduced that were shown to enhance the affinity of a BPTI variant for HNE. Using a single-stranded 76-mer oligonucleotide with limited randomization, this BPTI gene variant served as a template to create a library of fusion phage displaying 1000 variant BPTI-type domains. This mutagenic oligonucleotide was designed to randomize the codon representation at specific sites involved in binding. In the selected BPTI variants, codons for both of the possible variants at position 16, glycine and alanine, were found, even though glycine does not naturally occur at this position. At positions 18 and 19, codons for several amino acid variations that do not naturally occur were also selected. Cumulatively, this study documented the selection of a restricted set of variant domains with K_d in the picomolar range, which exceeded by more than 50-fold the highest affinity previously reported for any reversible human neutrophil elastase inhibitor.

In an alternative approach, Rottgen and Collins (1995) reported the construction of a library based on the gene for the human pancreatic secretory trypsin inhibitor (PSTI). In these libraries, randomized seven- and eight-codon sequences were inserted into the native PSTI gene sequence, which was fused to pIII. In the design of these constructs, the inserts were meant to code for an extended peptide held between two disulfide bridges, at the exposed tip of the PSTI domains. The final library contained an estimated 31 million individual clones, and sequential rounds of selection against chymotrypsin were performed. In contrast to the parental clone that encoded a trypsin inhibitor, the selected engineered domains displayed altered sequences and were specific for chymotrypsin.

PEPTIDE LIBRARIES AND IDENTIFICATION OF ACTIVE SITES

Phage-display systems have also been used to provide an understanding of the structural basis of an interaction in systems in which insight is limited. By screening libraries that displayed random cyclic and non-cyclic octopeptides, Wright et al. (1995) selected for binders to monoclonal antibodies specific for human somatostatin, a protein hormone involved in glucose metabolism and cellular homeostasis. These authors demonstrated the selection of clones coding for a recurring tetrapeptide consensus amino acid sequence, FWKT. One peptide in particular, CRFWKTWC, exhibited nanomolar affinities to the selecting monoclonal antibody, and also bound to human somatostatin receptors with submicromolar binding constants. Significantly, this primary amino acid sequence is also shared by somatostatin, suggesting that the monoclonal antibody specific for somatostatin could select its binding surface even from a library of randomized peptides.

In other studies performed with large incomplete libraries containing sequences for 37 and 43 randomized codons, similar selection studies were performed with a monoclonal antibody that recognizes the prostate-specific membrane (PSM) antigen (McConnell et al. 1996). Here, isolated phage also encoded a consensus sequence that is homologous to a portion of the native PSM molecule. Cumulatively, these studies provided impressive evidence that random peptide libraries can reveal structural information about the section in the primary sequence of a ligand that is responsible for binding interaction contacts.

ANTIBODY DOMAIN-BASED ARTIFICIAL SCAFFOLDS

From a protein engineer's perspective, the vertebrate antibody system represents an efficient workhorse for the creation of binding sites of virtually limitless versatility. In these domains, loops between β strands are primarily responsible for antigen contact, and they have been termed hypervariable regions (or complementarity determining regions, CDR; for review, see Kirkham and Schroeder 1994). The framework region forms a highly conserved β-barrel structure with intra- and interchain disulfide bonds. In most mammalian species, the diversity of the antibody repertoire derives in part from the inheritance of families of related variable region genes that share conserved framework regions. In a family, the greatest variability between members is associated with the CDR subdomains. Further structural diversification is created in immunoglobulin (Ig) repertoires with the superimposition of somatic mechanisms that provide an enormous range of potential ligand recognition units expressed by B-lymphocyte clones that compete in secondary lymphoid tissue for binding to an antigen.

For an antibody, the β-barrel structure of the Ig fold distributes functional capacities to different variable region subdomains, based on an organization that is the exact opposite of the common structural pattern in enzymes of related functional activity (see Kirkham and Schroeder 1994). Analyses of the same enzyme expressed in different species often reveal a strong sequence conservation at the site responsible for functional activity. In these enzymes, the peptide stretches that maintain overall structure and conformational alignment (i.e., the frameworks) often display much greater sequence variations. In contrast, natural antibody systems have evolved with highly conserved framework subdomains that maintain the Ig fold. Therefore, the greatest differences in antibody sequences occur primarily in the hypervariable subdomains that convey functional activity (i.e., the diverse binding activities).

The tolerance for variation in the loops of the Ig fold can be exploited by the protein engineer for the creation of libraries of artificial stable proteins with potential binding surfaces composed of diverse amino acid sequences. These antibody-based domains have highly malleable binding surfaces that are amenable to randomization without harming secondary structure. However, the native six-loop structure of most mammalian Igs can be associated with expression problems in prokaryotic systems (Jung and Pluckthun 1997). Therefore, the Ig fold has been used to develop simpler and better-adapted scaffolding systems that employ a binding surface created by contiguous residues in only one or two loops between β strands (Tramontano et al. 1994). Notably, using this approach, functional proteins can be refined on a scaffold support that may have no relationship to the structures of proteins that naturally have these functional activities (for review, see Nygren and Uhlen 1997; Sollazzo et al. 1997).

The utility of an artificial scaffold based on the Ig fold was first demonstrated in a 61-residue β protein, termed the "minibody" (see Fig. 5.2) (Pessi et al. 1993). This designed structure is based on the two β sheets of the McPC603 antibody that includes three β strands and the CDR1 and CDR2 loops. Circular dichroism (CD) spectrum analysis confirmed retention of the parental β secondary structure. The engineered functional specificity for metal binding was associated with an estimated dissociation constant (K_d) of 10^{-6} M. On the basis of this scaffold, minibody phage-display libraries

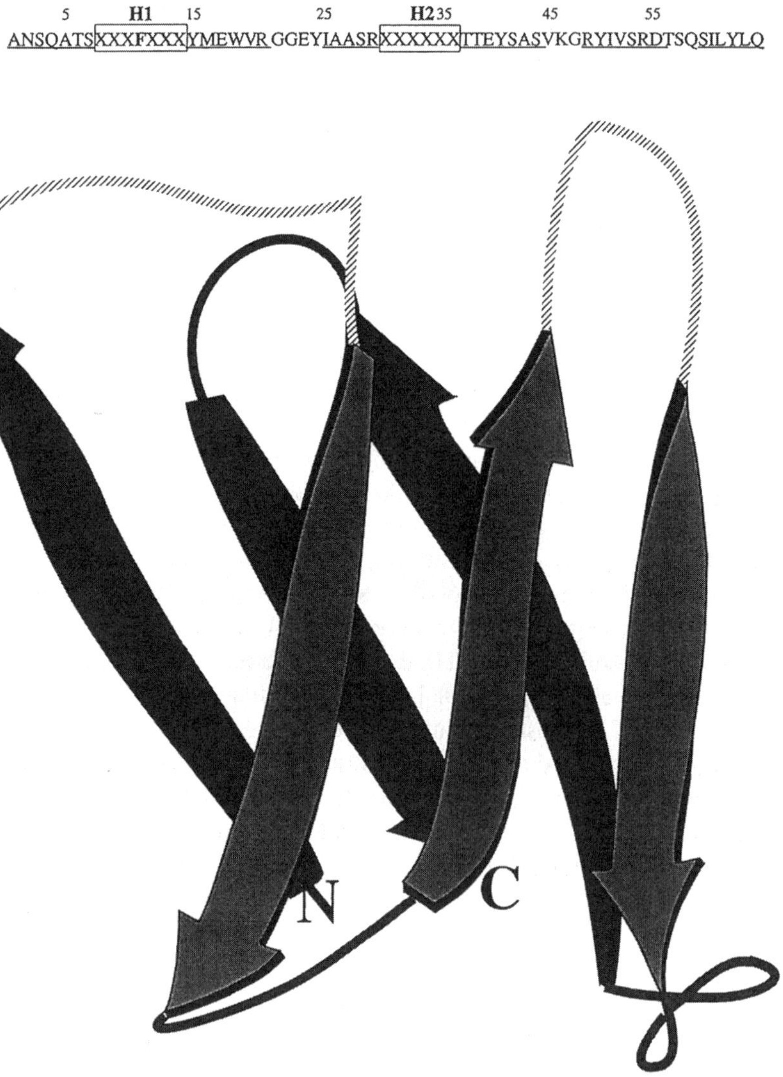

Figure 5.2. Amino acid sequences and hypothetical structural model of minibody. Positions corresponding to the H1 and H2 hypervariable loops are shown within boxes. β strands are underlined. In the model, β strands are depicted as arrows, whereas the two hypervariable subdomains are represented as dashed loops. (Reprinted, with permission, from Sollazzo et al. 1997.)

with randomized amino acid sequences in the CDR loops were subsequently created, and high-affinity minibodies were selected that were inhibitors for the hepatitis C virus NS3, a serine protease required for viral life cycle (Martin et al. 1997). In other studies, a functional minibody inhibitor to human cytokine IL-6 has also been recovered (Martin et al. 1994).

The design of minimal antibody-based scaffolds has been advanced by refinements modeled on the camel antibody system (Davies and Riechmann 1994; Riechmann 1996), one of the few known natural systems in which specific antibodies can be composed of a single chain (Hamers-Casterman et al. 1993). In these antibodies, only three hypervariable loops are employed in antigen contact. There are also unusual variations of the conserved hydrophobic amino acids that in other species are involved in V_H–V_L chain-pairing interactions. In several reports, single V_H region "camelized" antibodies have been shown to be highly effective small recognition units, enabling the creation of high-affinity binders using encoding genes from either synthetic or natural sources (Dimasi et al. 1997; Lauwereys et al. 1998).

PHAGE DISPLAY AND NON-ANTIBODY SCAFFOLDS

In studies that evaluated whether proteins with structures other than the Ig fold are useful to mimic the ligand-binding properties of antibodies, Ku et al. generated libraries based on the structure of the four-helix-bundle protein, cytochrome b_{562} (Ku and Schultz 1995). In these libraries, two loops were randomized, and the library was screened for binders to a hapten conjugated to bovine serum albumin (BSA). The selected mutants folded into a native-like structure and were shown to bind selectively to the BSA conjugate with micromolar K_d values, in comparison to a monoclonal antibody that binds selectively to the same antigenic determinant with a K_d of 290 nM.

As another alternate protein framework, libraries have been generated based on tendamistat, a 74-amino-acid β-sheet protein that is an α-amylase inhibitor isolated from *Streptomyces tendae* (McConnell and Hoess 1995). After the demonstration that phage-displayed tendamistat was functional, because it inhibited the hydrolysis of starch by α-amylase, libraries were created using tendamistat as a molecular scaffold for the presentation of constrained random peptides at two loops. These two loops were randomized using oligonucleotides with degenerate codons (Fig. 5.3). By this approach, libraries of approximately 10^8 different mutant molecules were assembled. The libraries were then subjected to selection based on binding to a monoclonal antibody that recognizes endothelin, to which native tendamistat does not bind. After several cycles of biopanning, phage that specifically bound this antibody were selected, and subsequent analyses demonstrated that only one of the loops was responsible for binding. Hence, this work documented the utility of the tendamistat scaffold to create novel, small proteins with new functional activities.

In more recent studies, Smith et al. (1998) reported the creation of libraries based on scaffolds from the knottins, a group of structurally related disulfide-bonded proteins, generally less than 40 amino acids, that bind with high specificity to their target molecules. The proteins share a common scaffold comprising a triple-stranded antiparallel β sheet and disulfide-bond framework, and they appear to use different faces for interactions with different targets. Libraries were created based on the cellulose-binding domain of a fungal knottin enzyme protein, cellobiohydrolase I, and sequence variations were introduced at seven codons responsible for the face of the protein that binds cellulose (Fig. 5.4). The resulting library was selected for binding to cellulose or to one of three different enzymes, and variant knottins were successfully isolated that were specific for cellulose or for bovine alkaline phosphatase. These recovered knottins also exhibited differences in sequence from the parent knottin. Taken together, these non-antibody-based scaffolds appear to provide promising architectures for the

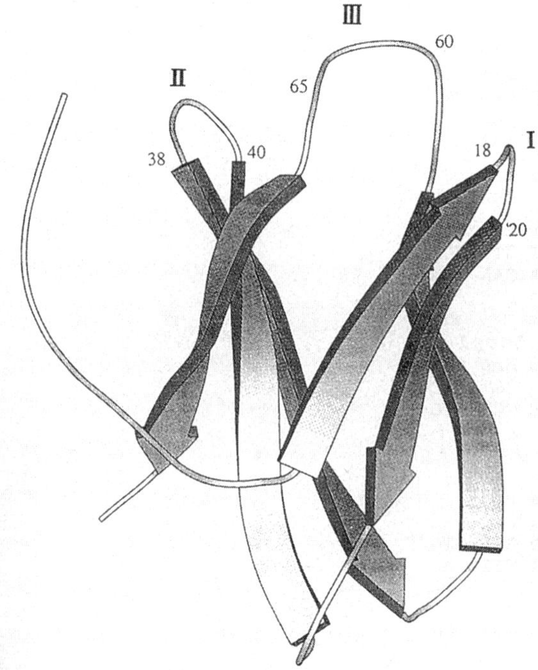

Figure 5.3. Ribbon diagram of tendamistat. The three surface loops are indicated with roman numerals, and the amino acid residue numbers are indicated. Loop I contains the conserved residues necessary for binding α-amylase. Loops II and III are those residues that are randomized in the peptide libraries. (Reprinted, with permission, from McConnell and Hoess 1995.)

creation of novel, small, folded proteins with engineered binding activities. These domains have the potential for improvement of binding affinities by mutation, and may possibly enable the creation of different activities in a single domain.

ZINC-FINGER PROTEINS AND ENGINEERED DNA-BINDING PROTEINS

The central role of DNA-binding proteins in the regulation of gene transcription and translation has resulted in a great interest in the possibility of creating assembled binding proteins specific for defined DNA sequences (for review, see Choo and Klug 1995; Segal and Barbas 2000). Zinc-finger proteins have proven to be an ideal system for studying the structural basis for sequence-specific DNA recognition. The zinc-finger motif was first identified in the TFIIIA transcription factor (Brown et al. 1985; Miller et al. 1985), and these Cys_2His_2 zinc fingers represent the most frequently utilized nucleic acid-binding motif in eukaryotes. Each finger domain is composed of about 30 amino acids that create a simple structure of two antiparallel β strands followed by an α helix. The structural stability of this fold is achieved by the coordination of a single zinc atom with two cysteines and two histidines. The sequence-specific binding of DNA is mediated by amino acids in the amino-terminal portion of the α helix that

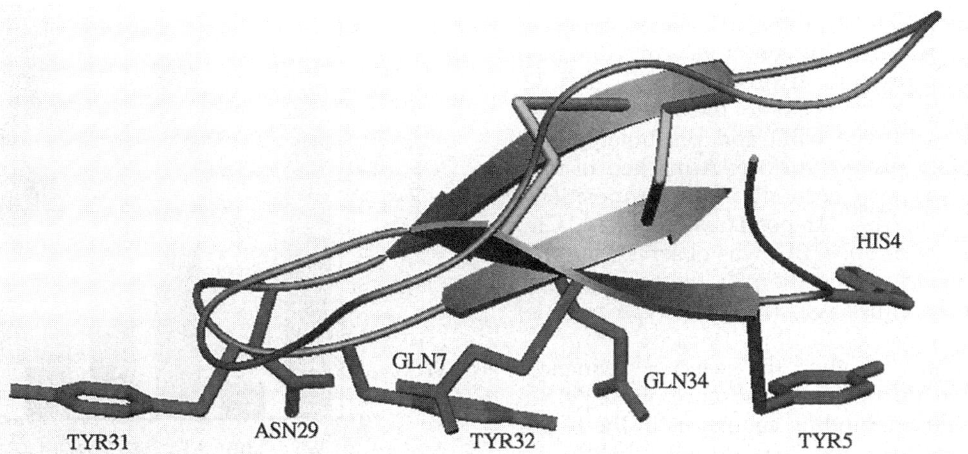

Figure 5.4. Structure of the cellulose binding domain of a knottin from the fungus *Trichoderma reesei*. Depicted are residues 4, 5, and 7, and residues 29, 31, 32, and 34, which together compose a face involved in cellulose binding, and which were randomized in a scaffolded library. The polypeptide backbone is shown as a gray tube, β strands are drawn as arrows, and disulfide bonds are yellow. (Reprinted, with permission, from Smith et al. 1998.)

typically recognizes three base pairs of DNA sequence (see Fig. 5.5) (Pavletich and Pabo 1991; Elrod-Erickson et al. 1996).

The zinc finger protein was one of the first non-antibody domains to be proposed for phage display (Barbas and Lerner 1991). At least four different research groups have reported the phage display of zinc finger proteins for the development of new nucleic acid sequence specificities (Choo and Klug 1994; Jamieson et al. 1994; Rebar and Pabo 1994; Wu et al. 1995). Each of these studies has been guided by information obtained from the crystal structure of the murine transcription factor Zif268 in complex with DNA. By targeting specific residues shown to be involved in DNA-binding specificity, libraries of zinc finger proteins were displayed on phage. Since each zinc finger domain recognizes three contiguous base pairs of DNA, selection of the phage libraries using specific DNA sequences provides for the isolation of zinc finger domains that recognize novel 3-bp sequences of DNA. In a recent report, an elegant systematic approach was presented for the creation of modular zinc fingers specific for each of the different codons that encompass the degenerate 5′-GNN-3′ motif. These studies employed two large ($>10^9$) independent member pIII-linked phage-display libraries that included nearly exhaustive representation of variegation of six positions in the second finger. The selection strategies for sequence-specific zinc finger variants employed both positive and negative selection components (e.g., blocking with DNA of unwanted sequences), and up to seven rounds of biopanning (Fig. 5.6). The specificity for selecting the DNAs and the binding activities of these zinc finger variants were confirmed by ELISA and gel retardation studies, documenting the highly specific activities of these engineered domains (Segal et al. 1999). By this approach, zinc fingers capable of distinguishing operator sequences that differ by a single base change

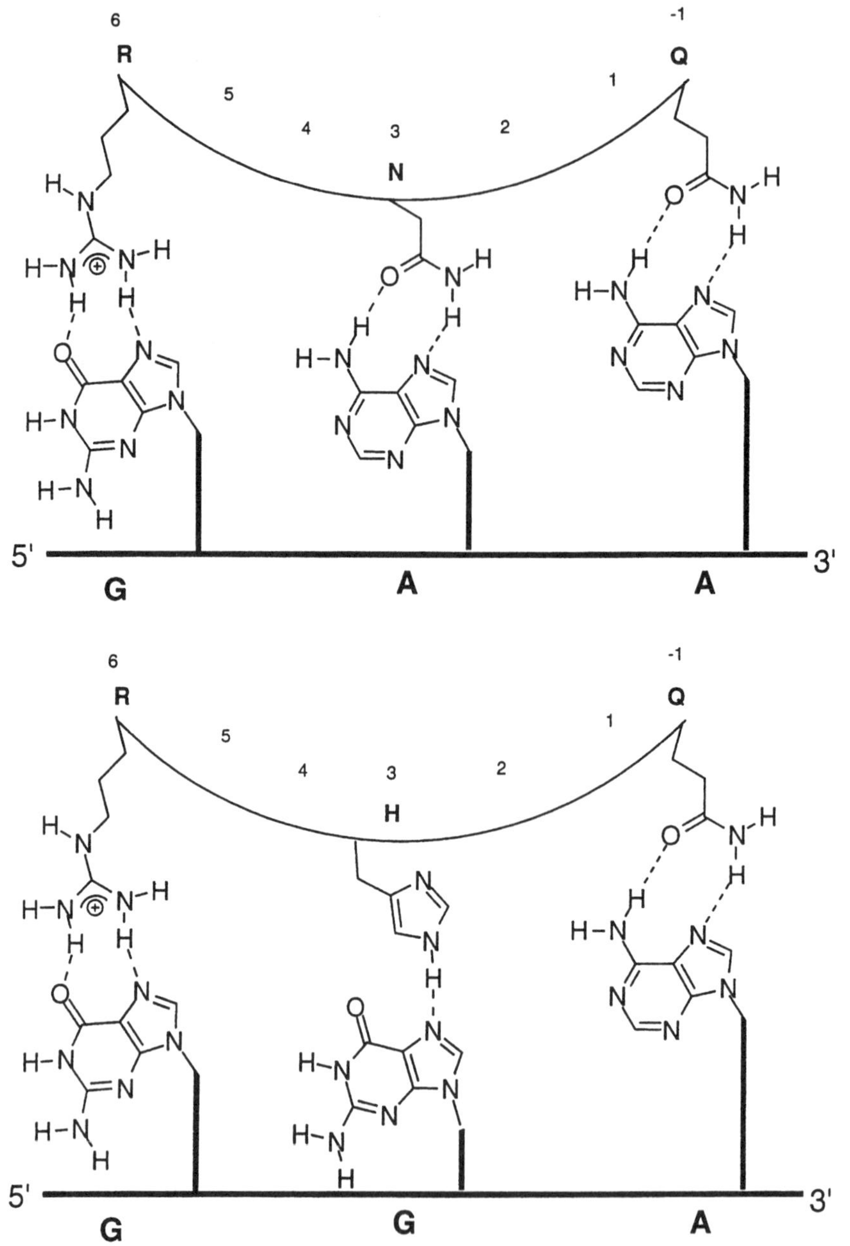

Figure 5.5. Molecular recognition of GAA and GGA DNA triplets by zinc finger domains. (Courtesy of Dr. C.F. Barbas, III.)

were isolated (see Fig. 5.6). These reported successes represent significant progress toward an idealized goal of creating a complete panel of zinc finger proteins specific for each of the 64 possible codon sequences.

Ideally, the creation of zinc fingers that recognize all possible codon sequences would represent the building blocks for creation of proteins that can recognize any

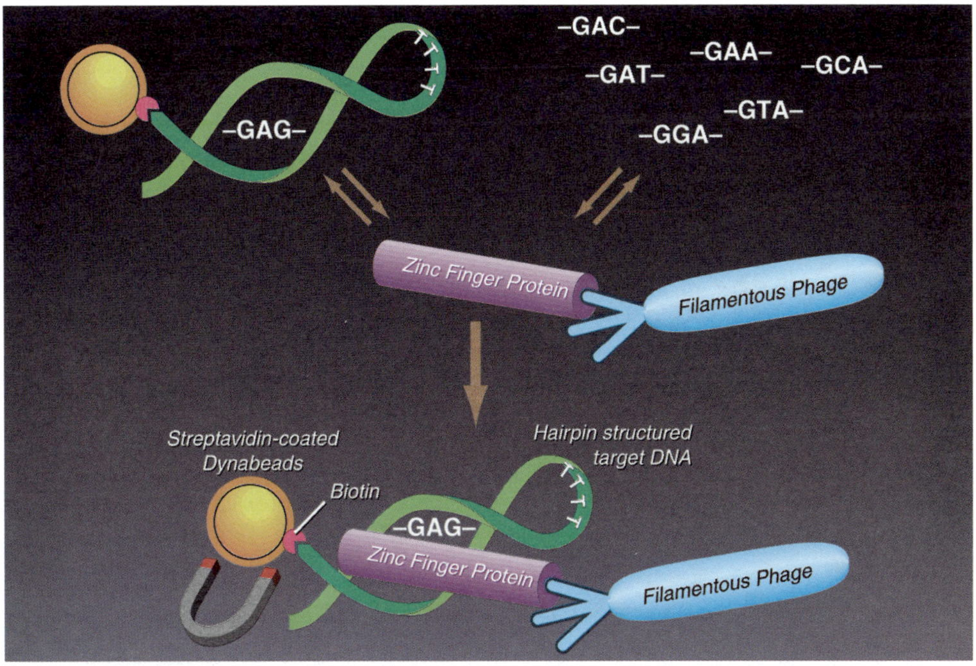

Figure 5.6. Components of a phage-display system used for the selection of zinc-finger motifs. Zinc fingers displayed on the surface of filamentous phage are bound to 5′ biotinylated DNA, which is captured on streptavidin-coated paramagnetic beads. These beads can be harvested using a magnet. Stringent negative selction is provided by adding non-biotinylated competitor DNAs that are closely related to the selecting DNA. In this way, only highly specific zinc-finger domains are selected. (See Segal et al. 1999.) (Reprinted, with permission, from Choo and Klug 1995.)

DNA sequence. In natural zinc finger proteins, simple covalent tandem repeats of the zinc finger domain allows for the recognition of longer asymmetric sequences of DNA (Pavletich and Pabo 1991; Swirnoff and Milbrandt 1995). The modularity of the zinc finger domain suggested that fusing individually selceted domains might enable the creation of molecules capable of recognizing any DNA sequence for the selective and controlled activation or repression of a defined gene. For example, if each zinc finger identifies a nucleotide triplet, then a six-finger construct should be capable of identifying a a unique 18-bp site in the human genome (see Fig. 5.7) (Liu et al. 1997).

Although no natural zinc finger proteins have been found that bind specifically to 18 contiguous nucleotides, phage-selected domains have been used to construct them. Polydactyl proteins specifically recognizing an 18-nucleotide sequence in the 5′-untranslated region of the human *erbB-2* and *erbB-3* genes have been constructed. These proteins bind their target sequences with subnanomolar affinity. When expressed as a fusion protein with repressor or activator domains, the transcription factor was able to specifically regulate the *erbB-2* promoter (Beerli et al. 1998). Extension of this work to study of the regulation of the endogenous *erbB-2* gene has resulted in the first demonstration of the modulation of endogenous genes with designed transcription factors (Beerli et al. 2000). These proteins were shown to be effective in cells derived from three mammalian species that conserve the *erbB-2*-bind-

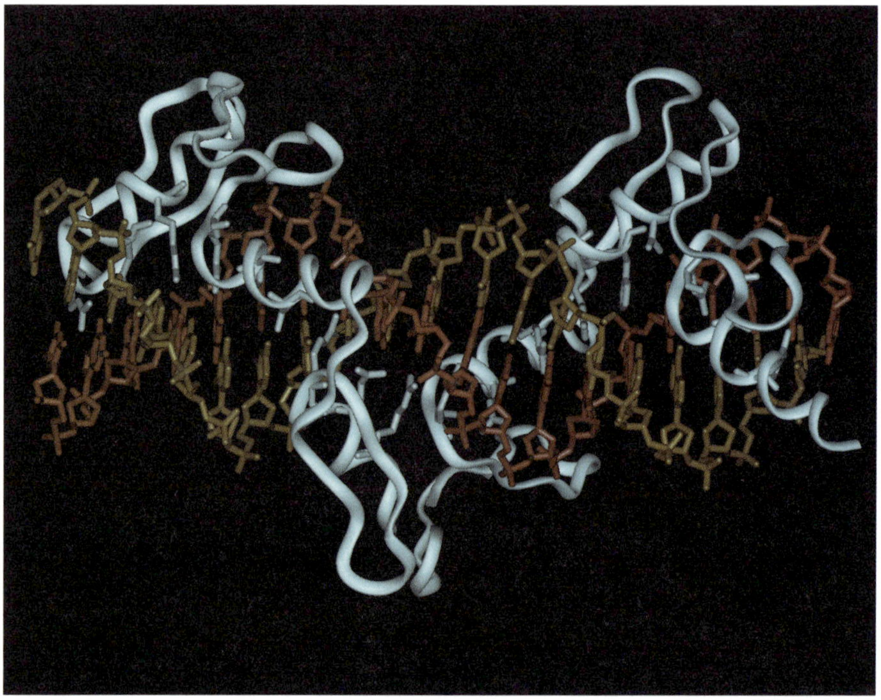

Figure 5.7. Model of a six zinc-finger–DNA complex. The zinc fingers (numbered F1–F3 and F4–F6) are based on the structure of the three zinc-finger protein Zif268. The six-finger protein was constructed by connecting F3 to F4 by building in the conserved linker sequence TGEKP, represented in magenta, with INSIGHTII. The guanine-rich strand, with which most of the base-specific contacts are made, is represented in red (3—TGCGGGTGCGGCGGGTGCGA-5′). The zinc atoms are represented in yellow. The structure of the modeled linker is similar to that of the natural linker peptides between F1 and F2 (TGQKP) and between F2 and F3 (TGEKP), respectively (with an rmsd of 0.3 Å for backbone atoms). The modeled linker also retains the general position and hydrogen bond characteristics observed for the natural linkers. (Reprinted, with permission, from Liu et al. 1997 © National Academy of Sciences U.S.A.)

ing site. Amazingly, the *erbB-3* gene is not regulated despite a 15 of 18 nucleotide match to the *erbB-2* target site at the same relative position. Furthermore, the *erbB-3* targeting transcription factor exclusively regulated *erbB-3* and not *erbB-2*. Thus, phage display coupled with transcription factor design has now provided the first proteins that allow for genome-specific transcriptional regulation.

The zinc-finger domain has also been used as a scaffold for novel protein–protein interaction unrelated to natural activities. In these studies, the codons for five discontinuous positions in the α-helical domain of the parental Cys_2His_2 zinc finger were variegated (Bianchi et al. 1995). The randomization strategy avoided codons for hydrophobic residues that were predicted to have adverse effects on structural stability. Library selection against a monoclonal antibody specific for *Shigella flexneri*, a common enteric bacterial pathogen, yielded a consensus epitope, His^1 Phe^2 Val^3 Gln^4 His/Arg^5 at the five targeted positions. Subsequent ELISA, circular dichroism, and NMR studies confirmed that the synthetic ligand and the natural target have closely related structures in solution (Bianchi et al. 1995). These studies documented the utility of scaffolds with defined secondary structure for the characterization of discontinuous epitopes of biologic significance.

The zinc-finger domain has also been used as a scaffold for novel protein–protein interaction unrelated to natural activities. In these studies, the codons for five discontinuous positions in the α-helical domain of the parental Cys_2His_2 zinc finger were variegated (Bianchi et al. 1995). The randomization strategy avoided codons for hydrophobic residues that were predicted to have adverse effects on structural stability. Library selection against a monoclonal antibody specific for *Shigella flexneri*, a common enteric bacterial pathogen, yielded a consensus epitope, His^1 Phe^2 Val^3 Gln^4 His/Arg^5 at the five targeted positions. Subsequent ELISA, circular dichroism, and NMR studies confirmed that the synthetic ligand and the natural target have closely related structures in solution (Bianchi et al. 1995). These studies documented the utility of scaffolds with defined secondary structure for the characterization of discontinuous epitopes of biologic significance.

STAPHYLOCOCCAL PROTEIN A AND A MINIMIZED IgG Fc-BINDING DOMAIN

Because of their highly ordered secondary structures, bacterial Ig-binding proteins have become attractive model scaffold systems. These proteins have great structural stability without the involvement of internal disulfide bonds, and they are often produced at high expression levels in bacterial systems. The prototypic Ig-binding protein, staphylococcal protein A (SpA), is composed of five highly homologous Ig-binding

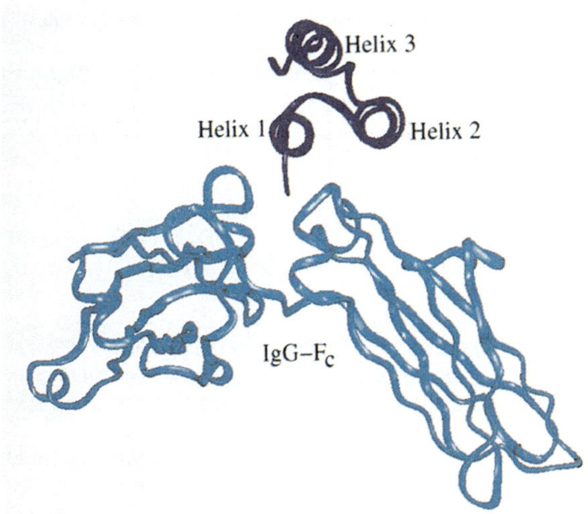

Figure 5.8. Ribbon diagram of the B domain of staphylococcal protein A (blue) in complex with the CH2, CH3 fragment of an IgG_1 (*gray*) taken from x-ray coordinates. Helix 3 appears disordered in the crystal structure, although NMR experiments indicate that it is in fact helical. In this figure, helix 3 has been modeled as a helix. (Reprinted, with permission, from Braisted and Wells 1996 © National Academy of Sciences, U.S.A.)

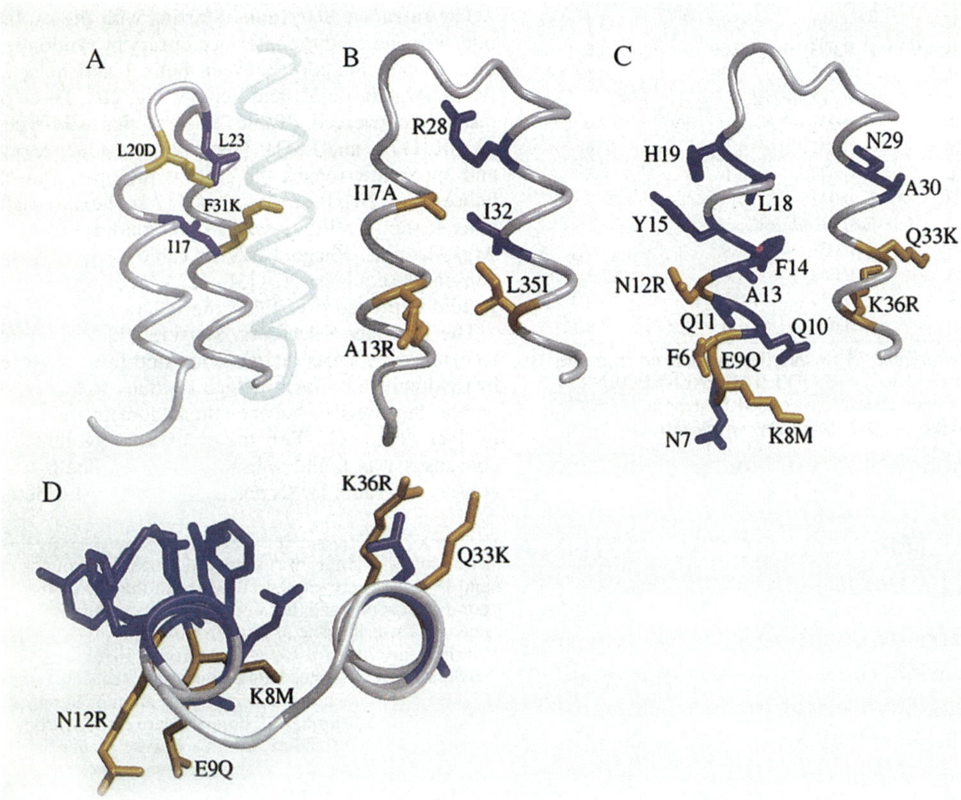

Figure 5.9. Ribbon diagrams showing a model of the helix 1 and helix 2 of the truncated Z domain of staphylococcal protein A. (*A*) Residues that were randomly mutated and selected from the exoface library. Residues that were conserved as the wild type after selection are shown in blue, and those that strongly consensed to something other than the wild type are shown in yellow. The transparent helix represents where helix 3 used to be; it was modeled as in Fig. 5.7. Replacement residues were aligned on the Cα, Cβ vector of the wild-type residue. (*B*) Residues (colored as in *A*) that were randomly mutated and selected from the intraface library. (*C*) Residues (colored as in *A*) that were randomly mutated and selected from the five interface libraries. Interface library 3A covers residues that are not seen in this structure. (*D*) View of the interface libraries looking down the helical axis. (Reprinted, with permission, from Braisted and Wells 1996 © National Academy of Sciences U.S.A.)

domains (Lofdahl et al. 1983; Moks et al. 1986). Most studies have focused on the well-characterized B-domain, which is a 58-amino-acid domain that forms a highly stable triple α-helical bundle in which only helix 1 and helix 2 contribute to the IgG Fc-binding site (Deisenhofer 1981), which is primarily based on hydrophobic interactions. Single domains of SpA can be displayed as fusions at the amino terminus of the gIII coat protein with retention of Fc binding activity, enabling efficient selection from dilute libraries based on IgG binding activity (Djojonegoro et al. 1994; Kushwaha et al. 1994).

Exploiting the detailed understanding of structure of the IgG Fc–SpA complex, Braisted and Wells (1996) designed a strategy to create much smaller protein derivatives with the same functional capacity as SpA (Figs. 5.8 and 5.9). In the native molecule, only two of the three α helices in the B domain of SpA are involved in Fc binding, whereas helix 3 appears to stabilize a hydrophobic core and helix 2 and helix 3 are involved in the distinct binding activity for VH3 family-encoded Fab binding, which should provide additional domain engineering in the future (Roben et al. 1995; Silverman et al. 2000).

To optimize an Fc-binding domain, a library based on the amino-terminal 38 amino acids that comprise helix 1 and helix 2 of the homologous synthetic Z domain was created. Following several rounds of randomization of codons followed by library in vitro selection, a variant was isolated in which the hydrophobic interaction of helix 1 and helix 2 was stabilized, and Fc binding activity was retained. Hence, starting from a domain of 59 amino acids that creates a triple α-helical bundle that binds the Fc fragment of IgG with a K_d of 10 nM, a 38-residue form (termed Z38) that binds with a K_d of 43 nM was isolated and characterized (Braisted and Wells 1996). NMR analysis later demonstrated that the Z38 protein is composed of two antiparallel α helices in a tertiary structure (Starovasnik et al. 1997) that is remarkably similar to the first two helices of the B domain in the B domain/Fc complex (Deisenhofer 1981). Subsequently, a 34-residue analog (Z34C) was designed in which an interhelical disulfide bond enhanced stability and was associated with a 9-fold higher affinity IgG binding activity. Z34C, like Z38, is structurally almost identical to the equivalent region in native SpA domains. Thus, the stabilized, minimized, two-helix peptide, about half the size of the native domain and with one-third of its residues altered from the original sequence, accurately mimics both the structure and function of the native protein (Braisted and Wells 1996; Starovasnik et al. 1997). The small engineered protein analog, Z34C, currently represents one of the smallest functional peptides with features typical of a folded protein.

The well-ordered triple α-helical-bundle structure of SpA has also been exploited to create stable domains with novel binding activities. Using a solid-state DNA synthesis strategy, the gene for a synthetic 58-amino-acid SpA domain was created with variegation of the codons for 13 noncontiguous surface-exposed residues of helix 1 and helix 2, enabling the creation of incomplete libraries of less than 10^8 members (Nord et al. 1995; 1997). Seven of these randomized residues are directly involved in the native Fc binding activity. Residues involved in stabilizing the hydrophobic core were conserved. In this type of scaffold, because of the prominence of the α helix in the structure, functional sites are created by an amalgamation of residues that are dispersed along the primary sequence. From phage-display libraries, "affibodies" of diverse specificity were subsequently recovered after 3–5 rounds of selection, including binders to *Taq* DNA polymerase, human insulin, and human apolipoprotein. Recombinant clones expressed in *E.coli* were evaluated by SDS-PAGE, biosensor studies, and CD spectroscopy, demonstrating the conservation of the parental α-helical structure, and highly specific micromolar binding activities. These studies demonstrate the use of the stable secondary scaffolding of the SpA domain to build novel binding activities. In these studies, the exceptional stability and high solubility of SpA was a significant asset, and the possible utility of expressing these affibodies as intracellular proteins with therapeutic activities was also suggested.

CONCLUSION

As described in the preceding sections, phage-display methods have now been successfully applied to a variety of domain types for the molding of functional sites with enhanced or altered activities. Using nucleotide-doping strategies that decrease the representation of nonfunctional and defective variants, libraries with effectively greater sizes can be more readily generated. Successful strategies often integrate into the mutagenesis approach insights into the structural and functional features that are associated with different parts of the domain. In certain cases, however, improved properties have been associated with amino acid variations that were not predicted to be at a site directly involved in the interaction. In general, these reports present the basic principles that should be applicable to the engineering of any domain, and the potential now exists to introduce novel activities that are completely distinct from those of the parental molecule.

REFERENCES

Abdel-Meguid S.S., Shieh H.S., Smith W.W., Dayringer, H.E., Violand, B.N., and Bentle, L.A. 1987. Three-dimensional structure of a genetically engineered variant of porcine growth hormone. *Proc. Natl. Acad. Sci.* **84:** 6434–6437.

Alting-Mees M.A. and Short J.M. 1993. Polycos vectors: A system for packaging filamentous phage and phagemid vectors using lambda phage packaging extracts. *Gene* **137:** 93–100.

Barbas C.F., III and Lerner R.A. 1991. Combinatorial immunoglobulin libraries on the surface of phage (Phabs): Rapid selection of antigen specific Fabs. *Methods: Companion Methods Enzymol.* **2:** 119–124.

Bass S., Greene R., and Wells J.A. 1990. Hormone phage: An enrichment method for variant proteins with altered binding properties. *Proteins* **8:** 309–314.

Beerli R.R., Dreier B., and Barbas C.F.,III. 2000. Selective positive and negative regulation of endogenous genes by designed transcription factors. *Proc. Natl. Acad. Sci.* **97:** 1495–1500.

Beerli R.R., Segal D.J., Dreier B., and Barbas C.F., III. 1998. Toward controlling gene expression at will: Specific regulation of the erbB-2/HER-2 promoter by using polydactyl zinc finger proteins constructed from modular building blocks. *Proc. Natl. Acad. Sci.* **25:** 14628–14633.

Bianchi E., Folgori A., Wallace A., Nicotra M., Acali S., Phalipon A., Barbato G., Bazzo R., Cortese R., and Felici F. 1995. A conformationally homogeneous combinatorial peptide library. *J. Mol. Biol.* **247:** 154–160.

Braisted A.C. and Wells J.A. 1996. Minimizing a binding domain from protein A. *Proc. Natl. Acad. Sci.* **93:** 5688–5692.

Brown R.S., Sander C., and Argos P. 1985. The primary structure of transcription factor IIIA has 12 consecutive repeats. *FEBS Lett.* **186:** 271–274.

Chiswell D.J. and McCafferty J. 1992. Phage antibodies: Will new 'coliclonal' antibodies replace monoclonal antibodies? *Trends Biotechnol.* **10:** 80–84.

Choo Y. and Klug A. 1994. Selection of DNA binding sites for zinc fingers using rationally randomized DNA reveals coded interactions. *Proc. Natl. Acad. Sci.* **91:** 11168–11172.

Clackson T. and Wells J.A. 1994. In vitro selection from protein and peptide libraries. *Trends Biotechnol.* **12:** 173–184.

Cunningham B.C. and Wells J.A. 1989. High-resolution epitope mapping of hGH-receptor interactions by alanine-scanning mutagenesis. *Science* **244:** 1081–1085.

———. 1991. Rational design of receptor-specific variants of human growth hormone. *Proc.*

Natl. Acad. Sci. **88:** 3407–3411.

Cunningham B.C., Lowe D.G., Li B., Bennett B.D., and Wells J.A. 1994. Production of an atrial natriuretic peptide variant that is specific for type A receptor. *EMBO J.* **13:** 2508–2515.

Davies J. and Riechmann L. 1994. 'Camelising' human antibody fragments: NMR studies on VH domains. *FEBS Lett.* **339:** 285–290.

Deisenhofer J. 1981. Crystallographic refinement and atomic models of a human Fc fragment and its complex with fragment B of protein A from *Staphylococcus aureus* at 2.9- and 2.8-Å resolution. *Biochemistry* **20:** 2361–2370.

Dimasi N., Martin F., Volpari C., Brunetti M., Biasiol G., Altamura S., Cortese R., De Francesco R., Steinkuhler C., and Sollazzo M. 1997. Characterization of engineered hepatitis C virus NS3 protease inhibitors affinity selected from human pancreatic secretory trypsin inhibitor and minibody repertoires. *J. Virol.* **71:** 7461–7469.

Djojonegoro B.M., Benedik M.J., and Willson R.C. 1994. Bacteriophage surface display for optimization of separations ligands: Display of immunoglobulin-binding protein A of *Staphylococcus aureus. Bio/Technology* **12:** 169–72.

Elrod-Erickson M., Rould M.A., Nekludova L., and Pabo C.O. 1996. Zif268 protein-DNA complex refined at 1.6 Å: A model system for understanding zinc finger-DNA interactions. *Structure* **4:** 1171–1180.

Fairbrother W.J., McDowell R.S., and Cunningham B.C. 1994. Solution conformation of an atrial natriuretic peptide variant selective for the type A receptor. *Biochemistry* **33:** 8897–8904.

Fisch I., Kontermann R.E., Finnern R., Hartley O., Soler-Gonzalez A.S., Griffiths A.D., and Winter G. 1996. A strategy of exon shuffling for making large peptide repertoires displayed on filamentous bacteriophage. *Proc. Natl. Acad. Sci.* **93:** 7761–7766.

Graille M., Stura E.A., Corper A.L., Sutton B.I., Taussig M.J., Charbonnier J.-B., and Silverman G.J. 2000. Crystal structure of a *Staphylococcus aureus* protein A domain complexed with the Fab fragment of a human Ig M antibody: Structural basis for recognition of B-cell receptors and superantigen activity. *Proc. Natl. Acad. Sci.* **97:** 5399–5404.

Gram H., Strittmatter U., Lorenz M., Gluck D., and Zenke G. 1993. Phage display as a rapid gene expression system: Production of bioactive cytokine-phage and generation of neutralizing monoclonal antibodies. *J. Immunol. Methods* **161:** 169–176.

Hamers-Casterman C., Atarhouch T., Muyldermans S., Robinson G., Hamers C., Songa E.B., Bendahman N., and Hamers R. 1993. Naturally occurring antibodies devoid of light chains. *Nature* **363:** 446–448.

Hennecke M., Kola A., Baensch M., Wrede A., Klos A., Bautsch W., and Kohl J. 1997. A selection system to study C5a-C5a-receptor interactions: phage display of a novel C5a anaphylatoxin, Fos-C5aAla27. *Gene* **184:** 263–272.

Hennecke M., Otto A., Baensch M., Kola A., Bautsch W., Klos A., and Kohl J. 1998. A detailed analysis of the C5a anaphylatoxin effector domain: Selection of C5a phage libraries on differentiated U937 cells. *Eur. J. Biochem.* **252:** 36–44.

Jamieson A.C., Kim S.H., and Wells J.A. 1994. In vitro selection of zinc fingers with altered DNA-binding specificity. *Biochemistry* **33:** 5689–5695.

Jung S. and Pluckthun A. 1997. Improving in vivo folding and stability of a single-chain Fv antibody fragment by loop grafting. *Protein. Eng.* **10:** 959–966.

Kast P. and Hilvert D. 1997. 3D structural information as a guide to protein engineering using genetic selection. *Curr. Opin. Struct. Biol.* **7:** 470–479.

Katz B.A. 1997. Structural and mechanistic determinants of affinity and specificity of ligands discovered or engineered by phage display. *Annu. Rev. Biophys. Biomol. Struct.* **26:** 27–45.

Kirkham P.M. and Schroeder H.W.J. 1994. Antibody structure and the evolution of immunoglobulin V gene segments. *Semin. Immunol.* **6:** 347–360.

Ku J. and Schultz P.G. 1995. Alternate protein frameworks for molecular recognition. *Proc. Natl. Acad. Sci.* **92:** 6552–6556.

Kushwaha A., Chowdhury P.S., Arora K., Abrol S., and Chaudhary V.K. 1994. Construction and characterization of M13 bacteriophages displaying functional IgG-binding domains of staphylococcal protein A. *Gene* **151:** 45–51.

Laskowski M.J. and Kato I. 1980. Protein inhibitors of proteinases. *Annu. Rev. Biochem.* **49:** 593–626.

Lauwereys M., Arbabi Ghahroudi M., Desmyter A., Kinne J., Holzer W., De Genst E., Wyns L. and Muyldermans S. 1998. Potent enzyme inhibitors derived from dromedary heavy-chain antibodies. *EMBO J.* **17:** 3512–3520.

Li B., Tom J.Y., Oare D., Yen R., Fairbrother W.J., Wells J.A., and Cunningham B.C. 1995. Minimization of a polypeptide hormone. *Science* **270:** 1657–1560.

Liu Q., Segal D.J., Ghiara J.B., and Barbas C.F. 1997. Design of polydactyl zinc-finger proteins for unique addressing within complex genomes. *Proc. Natl. Acad. Sci.* **94:** 5525–5530.

Lofdahl S., Guss B., Uhlen M., Philipson L. and Lindberg M. 1983. Gene for staphylococcal protein A. *Proc. Natl. Acad. Sci.* **80:** 697–701.

Lowman H.B. and Wells J.A. 1993. Affinity maturation of human growth hormone by monovalent phage display. *J. Mol. Biol.* **234:** 564–578.

Lowman H.B., Bass S.H., Simpson N., and Wells J.A. 1991. Selecting high-affinity binding proteins by monovalent phage display. *Biochemistry* **30:** 10832–10838.

Martin F., Toniatti C., Salvati A.L., Venturini S., Ciliberto G., Cortese R., and Sollazzo M. 1994. The affinity-selection of a minibody polypeptide inhibitor of human interleukin-6. *EMBO J.* **13:** 5303–5309.

Martin F., Volpari C., Steinkuhler C., Dimasi N., Brunetti M., Biasiol G., Altamura S., Cortese R., De Francesco R., and Sollazzo M. 1997. Affinity selection of a camelized V(H) domain antibody inhibitor of hepatitis C virus NS3 protease. *Protein Eng.* **10:** 607–614.

McConnell S.J. and Hoess R.H. 1995. Tendamistat as a scaffold for conformationally constrained phage peptide libraries. *J. Mol. Biol.* **250:** 460–470.

McConnell S.J., Uveges A.J., Fowlkes D.M., and Spinella D.G. 1996. Construction and screening of M13 phage libraries displaying long random peptides. *Mol. Divers.* **1:** *165–176.*

Miller J., McLachlan A.D., and Klug A. 1985. Repetitive zinc-binding domains in the protein transcription factor IIIA from *Xenopus* oocytes. *EMBO J.* **4:** 1609–1614.

Moks T., Abrahmsen L., Nilsson B., Hellman U., Sjoquist J., and Uhlen M. 1986. Staphylococcal protein A consists of five IgG-binding domains. *Eur. J. Biochem.* **156:** 637–643.

Nord K., Nilsson J., Nilsson B., Uhlen M., and Nygren P.A. 1995. A combinatorial library of an alpha-helical bacterial receptor domain. *Protein Eng.* **8:** 601–608.

Nord K., Gunneriusson E., Ringdahl J., Stahl S., Uhlen M., and Nygren P.A. 1997. Binding proteins selected from combinatorial libraries of an alpha-helical bacterial receptor domain. *Nat. Biotechnol.* **15:** 772–777.

Nygren P.A. and Uhlen M. 1997. Scaffolds for engineering novel binding sites in proteins. *Curr. Opin. Struct. Biol.* **7:** 463–469.

O'Neil K.T. and Hoess R.H. 1995. Phage display: Protein engineering by directed evolution. *Curr. Opin. Struct. Biol.* **5:** 443–449.

Pavletich N.P. and Pabo C.O. 1991. Zinc finger-DNA recognition: Crystal structure of a Zif268-DNA complex at 2.1 Å. *Science* **252:** 809–817.

Pearce K.H.J., Potts B.J., Presta L.G., Bald L.N., Fendly B.M., and Wells J.A. 1997. Mutational analysis of thrombopoietin for identification of receptor and neutralizing antibody sites. *J. Biol. Chem.* **272:** 20595–20602.

Pessi A., Bianchi E., Crameri A., Venturini S., Tramontano A., and Sollazzo M. 1993. A designed metal-binding protein with a novel fold. *Nature* **362:** 367–369.

Rebar E.J. and Pabo C.O. 1994. Zinc finger phage: Affinity selection of fingers with new DNA-binding specificities. *Science* **263:** 671–673.

Riechmann L. 1996. Rearrangement of the former VL interface in the solution structure of a camelised, single antibody VH domain. *J. Mol. Biol.* **259:** 957–969.

Roben P., Salem A., and Silverman G.J. 1995. VH3 antibodies bind domain D of staphylococcal protein A. *J. Immunol.* **154:** 6437–6446.

Roberts B.L., Markland W., Siranosian K., Saxena M.J., Guterman S.K., and Ladner R.C. 1992a. Protease inhibitor display M13 phage: Selection of high-affinity neutrophil elastase inhibitors. *Gene* **121:** 9–15.

Roberts B.L., Markland W., Ley A.C., Kent R.B., White D.W., Guterman S.K., and Ladner R.C. 1992b. Directed evolution of a protein: Selection of potent neutrophil elastase inhibitors displayed on M13 fusion phage. *Proc. Natl. Acad. Sci.* **89:** 2429–2433.

Rottgen P. and Collins J. 1995. A human pancreatic secretory trypsin inhibitor presenting a hypervariable highly constrained epitope via monovalent phagemid display. *Gene* **164:** 243–250.

Saggio I., Gloaguen I., Poiana G., and Laufer R. 1995. CNTF variants with increased biological potency and receptor selectivity define a functional site of receptor interaction. *EMBO J.* **14:** 3045–3054.

Segal D. and Barbas C.F., III. 2000. Design of novel sequence specific DNA-binding proteins. *Curr. Opin. Chem. Biol.* **4:** 34–39.

Segal D.J., Dreier B., and Barbas C.F. 1999. Toward controlling gene expression at will: Selection and design of zinc finger domains recognizing each of the 5′-GNN-3′ DNA target sequences. *Proc. Natl. Acad. Sci.* **96:** 2758–2763.

Silverman G.J., Cary S., Graille M., Curtiss V.E., Wagenknecht R., Luo L., Dwyer D., Goodyear C., Corper A.L., Stura E.A., and Charbonnier J.B. 2000. A B-cell superantigen that targets B-1 lymphocytes. *Curr. Topics in Microbiol. Immunol.* (in press).

Smith G.P. 1985. Filamentous fusion phage: Novel expression vectors that display cloned antigens on the virion surface. *Science* **228:** 1315–1317.

Smith G.P., Patel S.U., Windass J.D., Thornton J.M., Winter G., and Griffiths A.D. 1998. Small binding proteins selected from a combinatorial repertoire of knottins displayed on phage. *J. Mol. Biol.* **277:** 317–332.

Sollazzo M., Venturini S., Lorenzetti S., Pinola M., and Martin F. 1997. Engineering minibody-like ligands by design and selection. *Chem. Immunol.* **65:** 1–17.

Starovasnik M.A., Braisted A.C., and Wells J.A. 1997. Structural mimicry of a native protein by a minimized binding domain. *Proc. Natl. Acad. Sci.* **94:** 10080–10085.

Swimmer C., Lehar S.M., McCafferty J., Chiswell D.J., Blattler W.A., and Guild B.C. 1992. Phage display of ricin B chain and its single binding domains: System for screening galactose-binding mutants. *Proc. Natl. Acad. Sci.* **89:** 3756–3760.

Swirnoff A.H. and Milbrandt J. 1995. DNA-binding specificity of NGFI-A and related zinc finger transcription factors. *Mol. Cell Biol.* **15:** 2275–2287.

Tramontano A., Bianchi E., Venturini S., Martin F., Pessi A., and Sollazzo M. 1994. The making of the minibody: An engineered beta-protein for the display of conformationally constrained peptides. *J. Mol. Recognit.* **7:** 9–24.

Ultsch M.H., Somers W., Kossiakoff A.A., and de Vos A.M. 1994. The crystal structure of affinity-matured human growth hormone at 2 Å resolution. *J. Mol. Biol.* **236:** 286–299.

Wang C.I., Yang Q., and Craik C.S. 1995. Isolation of a high affinity inhibitor of urokinase-type plasminogen activator by phage display of ecotin. *J. Biol. Chem.* **270:** 12250–12256.

Wells J.A. 1990. Additivity of mutational effects in proteins. *Biochemistry* **29:** 8509–8517.

Wells J.A. and Lowman H.B. 1992. Rapid evolution of peptide and protein binding properties in vitro. *Curr. Opin. Biotechnol.* **3:** 355–362.

Wright R.M., Gram H., Vattay A., Byme S., Lake P., and Dottavio D. 1995. Binding epitope of somatostatin defined by phage-displayed peptide libraries. *Bio/Technology* **13:** 165–169.

Wu H., Yang W.P., and Barbas C.F. 1995. Building zinc fingers by selection: Toward a therapeutic application. *Proc. Natl. Acad. Sci.* **92:** 344–348.

6 Gene Fragment Libraries and Genomic and cDNA Expression Cloning

R. ANTHONY WILLIAMSON[1] AND GREGG J. SILVERMAN[2]

[1]*Department of Immunology, Scripps Research Institute, La Jolla, California 92037*
[2]*Department of Medicine, University of California, San Diego, La Jolla, California 92093-0663*

EXPRESSION CLONING METHODS THAT USE the display of proteins on the surface of phage particles offer special advantages for the selection and characterization of specific protein–ligand interactions. Several chapters in this manual present different applications of phage-display technology for the isolation of monoclonal antibodies and recovery of small oligopeptide ligands. This chapter, however, reviews the application of phage-display methods for the characterization of the structural features of known domains and the discovery of novel, naturally occurring functional domains.

CONTENTS

cDNA AND GENOMIC EXPRESSION LIBRARIES ON THE SURFACE OF PHAGE

During the past two decades, many different strategies for expression cloning have been described for the isolation of genes of unknown sequence and function (Helfman et al. 1983; Young and Davis 1983; Seed and Aruffo 1987; Sikela and Hahn 1987; Singh et al. 1988; Fields and Song 1989; Germino et al. 1993). In many of these approaches, the gene encoding a protein of interest is identified from complex, cloned libraries that are expressed in prokaryotic hosts, and specific binders are isolated by screening with an appropriate ligand. These expression vector systems, however, suffer from a variety of inherent technical limitations. Notably, many of these approaches require the immobilization of the expressed library of protein translation products onto a solid support, such as a transfer membrane, which may result in altered protein conformations that compromise the structural and ligand-binding properties of these recombinant molecules. Moreover, in these approaches only a limited number of individual clones can be easily examined by available screening methods, and it is difficult to scale up these methods to increase the size of the library that can be thoroughly screened.

Hence, the identification of genes represented at low frequencies in these libraries is technically demanding and labor-intensive.

Techniques incorporating phage-display libraries offer attractive alternatives that can circumvent these potential limitations, and possess certain advantages over established methodologies for the selection of specific cDNA or genomic clones from large and diverse expression libraries. First, by incorporating the protein and genetic components into a single phage particle, the phage-display approach creates a direct physical link between the expressed protein and the encoding gene. This linkage of protein and genetic information enables successive rounds of selection and amplification. As a consequence, during library panning, specific phage clones are enriched on the basis of their affinity for ligand. Thus, relatively rare ligand-binding clones can be rapidly and efficiently rescued from large libraries. As outlined above, the conformational and functional integrity of the expressed proteins can be a very important consideration when pursuing selection strategies from cDNA or genomic expression libraries. A wide variety of molecules of divergent size and character, including enzymes (McCafferty et al. 1991), antibodies (Burton and Barbas 1994; Winter et al. 1994), cytokines (Gram et al. 1993), protein fragments (Petersen et al. 1995; Williamson et al. 1998), and peptides (Smith 1985; Cwirla et al. 1990; Scott and Smith 1990), have been successfully displayed on the surface of filamentous phage. Together, these studies suggest that when recombinant proteins are directed to the periplasmic space of *E. coli*, they often refold to adopt conformational and functional properties on the phage surface that are similar to those associated with their native environment.

STRATEGIES FOR THE DESIGN OF PHAGE-DISPLAY VECTORS

The design of phage vectors used for the surface display of proteins encoded by cDNA or genomic fragments integrates a number of important considerations. One of the most important is the choice of the phage coat protein with which the recombinant protein of interest will be fused (for a more detailed discussion, see Chapters 1 and 2). To date, the most commonly adopted approach has been to effect fusion to the amino termini of mature or truncated M13 phage coat proteins III or VIII. There are an estimated 2700 copies of pVIII per individual virion, which can afford high-multivalent display and enable selection based on lower-affinity interactions that might not be recovered in *gene III*-based systems (Folgori et al. 1994). However, perhaps due to defective phage assembly, the fusion of larger polypeptides to pVIII is not as well tolerated (Malik et al. 1996). In contrast, there are only five copies of pIII, and these are localized to one tip of the phage particle. Hence, vectors that create fusions to pIII may endow more oligovalent (or even monovalent) interactions (discussed below). In addition, pIII has been shown to readily accommodate fusion with larger peptides and proteins and has been more often used for displaying polypeptides encoded by cDNA and genomic sequences. It is noteworthy that foreign proteins have also been displayed via fusion with either M13 pVI (Jespers et al. 1995; Fransen et al. 1999) or pVII and pIX (Gao et al. 1999), and these systems also may be modified to construct cDNA or genomic expression libraries. Moreover, several other display systems have been developed using nonfilamentous lytic phage. These include fusions with the bacteriophage lambda capsid protein D, which has been shown to accept large fusions at the amino or carboxyl termini (Sternberg and Hoess 1995; Santini et al. 1998), and with the tail protein V of the same virus (Maruyama et al. 1994; Dunn 1995; Kuwabara et al. 1997). In addition, T4 and T7 bacteriophage have also been exploited for displaying peptides and larger protein domains (Efimov et al. 1995; Ren et al. 1996; Houshmand et al. 1999; Zozulya et al. 1999). In some instances, these latter approaches may be advantageous because they do not rely on the cloned protein being properly processed through the bacterial secretory machinery.

Due to the versatility of filamentous phage systems, fusions of gene products with M13 phage coat proteins, as well as other components of the phage particle, can be encoded within a single phage vector. Phagemid systems are another commonly used variation in which sequences encoding the coat protein fusion are carried within a plasmid (phagemid) that contains an M13 filamentous phage origin of replication. The phagemid sequences can be packaged into phage particles by rescue with helper phage that carries sequence encoding all of the native phage proteins. Because the helper phage contains a defective origin of replication, its genome is relatively inefficiently packaged in comparison to the phagemid sequence. In some applications, phagemid vector systems may hold advantages over phage vectors. For example, in bacterial expression of native pIII and pIII fusions, an average of less than one copy of the fusion protein is incorporated into each extruding nascent phage particle. Monovalent display can be an advantage for the selection of phage from large libraries on the basis of affinity, because multivalent display may cause avidity effects in the binding interaction that result in the recovery of phage containing proteins with relatively modest affinity for ligand and relatively inefficient interclonal competition.

Phage libraries prepared from DNAs with well defined 5′ and 3′ sequences, such as peptides or antibodies, are readily inserted into display vectors in the correct orientation, with maintenance of the correct reading frame in its position between the carboxyl terminus of a bacterial export signal sequence and the amino terminus of either pIII or pVIII filamentous coat protein. However, in most cases in which diverse libraries are created for expression cloning, genomic DNA fragments and cDNAs possess undefined 5′ and 3′ ends. Hence, the cloning of these types of genes can be relatively inefficient, because potentially only one out of every 18 clones created (2 x 3 x 3) will contain a fusion that is both in the correct orientation (i.e., twofold variation) and in the correct reading frame with both the signal sequence (i.e., threefold variation) and the phage coat protein (i.e., threefold variation). An additional complication stems from the presence of translational stop codons at the 3′ end of the mRNA of cloned eukaryotic genes, as the insertion of these stop signals before the amino terminus of the phage coat proteins truncates the fusion product, effectively preventing the display of their expressed polypeptides. If it were feasible, the fusion of the library to the carboxyl terminus of the coat protein would avoid this problem, but in practice the resulting fusion protein with pIII or pVIII would not be incorporated into the phage particle.

To overcome this potential bottleneck, Crameri and Suter (1993) developed a highly innovative strategy. Starting with the pComb3 vector (see Chapter 2), these investigators devised a vector system, pJuFo, in which the cloned cDNA translation product is linked covalently to pIII via the interaction of the Fos–Jun leucine zipper heterodimeric protein. In this approach, a leader sequence for secretion to the bacterial periplasmic space is linked to the amino terminus of a Fos moiety, which at its carboxyl terminus has a cloning site for the cDNA sequence. Also within the vector is the gene for the other component of the zipper, Jun, which is fused to the amino terminus of pIII. A bacterial leader sequence also directs this translation product to the periplasm where it is anchored to the inner membrane. In the periplasmic space, the Fos and Jun moieties assemble together, and their interaction is stabilized by two intermolecular covalent bonds formed by flanking cysteine residues introduced at both ends of the Fos and Jun peptide sequences. Hence, production of the phage form of the vector results in display of the cDNA gene product linked via the zipper domains to the virion surface. Moreover, because recombinant sequences are inserted in a directional fashion between the pIII and leader sequences, the pJuFo vector enables cloning in the correct reading frame and correct orientation for 1 in 6 DNA inserts (rather than 1 in 18, using amino-terminal pIII fusion and nondirectional cloning).

As an alternative approach, Jespers et al. (1995) avoided the limitations of the pIII-based systems by creating a vector for cloning cDNAs or genomic fragments fused to the carboxyl terminus of coat protein VI. Cloning at this site does not block integration of the pVI-chimeric protein into the viral capsid. Moreover, in this system the fusion protein is believed to be monovalently displayed on the surface of filamentous phage. Consequently, it was anticipated that clonal selection strategies would be an efficient means to identify variants conveying even minor improvements in relative binding affinity. This approach, however, has not been extensively used, so its relative merit is uncertain.

PRACTICAL APPLICATION OF PHAGE-DISPLAY EXPRESSION LIBRARIES

As described above, phage-display systems offer considerable promise as a robust and efficient way to recover functional ligand/receptors from genomic or cDNA expression libraries. Although these systems may come to be considered as a viable complement to existing methodologies, relatively few reports involving the use of these technologies have been published since the first descriptions of this approach (Crameri and Suter 1993; Crameri et al. 1994). Indeed, by comparison to applications of antibody and peptide display libraries, there have been relatively few reports of using phage-display methods for the efficient recovery of biologically relevant functional domains. A brief review of the literature is provided below.

In an early report of phage-display expression cloning from a diverse library, a genomic library of randomly fragmented (100–700 bp in size) chromosomal DNA from a strain of *Staphylococcus aureus* (Jacobson and Frykberg 1995) was cloned into the pHEN1 phagemid vector (Hoogenboom et al. 1991). By use of specific selection ligands in a solid-phase-based selection strategy, previously known bacterial IgG- and fibronectin-binding genes were recovered, and a previously unknown IgG-binding protein was also discovered. However, the authors also noted that in most experiments ligand-specific clones represented only 1–10% of the recovered clones. Moreover, they also determined that in studies using a pIII-based system, the genes encoding specific binders were selected despite the fact that they had been cloned out of frame. To explain the expression cloning of genes that were apparently out of frame, the authors attributed their success to ribosomal slippage that enabled expression of the insert as a fusion protein with coat proteins. The authors also speculated that the relative inefficiencies during phage selection that were observed in their experiments might have been partly a consequence of the carryover of phage that displayed peptides with "sticky properties."

In subsequent reports, these authors demonstrated a pVIII-based phagemid vector system that enabled more efficient selection of ligand-binding domains of prokaryotic receptors directly from chromosomal DNA (Jacobsson and Frykberg 1996; 1998). Following several rounds of ligand selection, up to 100% of the recovered phage contained the specific ligand-encoding gene, and undesirable frameshifts in the DNA sequence were identified in only 1/41 of the clones with the correct sequence. In these studies, the relatively higher effective level of oligo- or multivalency of display of the fusion protein was indicated to contribute to enhanced efficiency of selection and stability of the relevant clones in the library. These issues remain to be more completely defined for pVIII systems, as an earlier report indicated that whereas virtually all pVIII molecules can be displayed as fusions with up to six codons, only about one-third of the pVIII molecules on a single phage particle can represent fusion proteins with 12 introduced amino acids. Much lower frequencies of incorporation have been reported for larger polypeptide fusions with pVIII (for review, see Perham et al. 1995).

In a separate study, a genomic library prepared from chromosomal DNA of *Staphylococcus epidermidis* and cloned into a phagemid vector via a pVIII fusion was used to isolate a fibrinogen-binding protein (Nilsson et al. 1998). Similarly, Palzkill et al. (1998) described the application of a modified pJuFo vector to express a genomic library from *E. coli* for the identification of dominant antibody-binding sites (i.e., epitopes) recognized by an anti-RecA protein polyclonal antibody sera.

Expression cloning from a complex mammalian cDNA library was reported by Hottiger et al., who created a filamentous phage cDNA expression library to select for ligands that bind to HIV-1 reverse transcriptase (Hottiger et al. 1995). In these studies, a previously generated library from Epstein-Barr virus-transformed human lymphocytes was transferred into the pJuFo vector. To establish the integrity of the library, and to define the appropriateness of selection conditions, studies were first performed in which the library was biopanned against staphylococcal protein A (SpA), a well-characterized immunoglobulin-binding protein. These experiments succeeded in isolating gene fragments encoding the Fc portion of human IgG constant region that interacts with SpA. Subsequently, when the same library was selected against bacterially expressed monomeric large subunit of reverse transcriptase (p66), phage encoding host β-actin were recovered. Furthermore, the authors showed that eukaryotic β-actin binds to either the large subunit of reverse transcriptase or to the Pol precursor polyprotein in vitro. These observations suggested that an interaction between HIV-1 reverse transcriptase and human β-actin may be important for the secretion of HIV-1 virions.

In studies directed toward understanding the pathophysiology of common environmental allergies, Crameri and coworkers displayed cDNA libraries of *Aspergillus fumigatus* on the phage surface using the pJuFo vector. Solid-phase immobilized serum IgE obtained from patients allergic to *A. fumigatus* were successfully used to select the fungal antigenic targets of the IgE antibodies. Using this approach, the investigators could dissect allergen-specific IgE responses by identifying the individual sensitizing proteins from this organism that are responsible for clinical allergy. Specifically, these studies provided evidence that humoral reactivity against acidic ribosomal phosphoprotein type 2 is an important immunologic feature of clinical allergies triggered by *A. fumigatus* antigens (Crameri 1998; Mayer et al. 1999).

GENE FRAGMENT LIBRARIES AND CHARACTERIZATION OF ANTIGENIC DETERMINANTS AND ACTIVE SITES

Phage display has been used to localize the epitopes of monoclonal (Balass et al. 1993; Bottger et al. 1995; Peterson et al. 1995; van Zonneveld et al. 1995; Fack et al. 1997; Williamson et al. 1998) and polyclonal (Dybawd et al. 1995; Blüthner et al. 1996) antibodies and many other binding proteins (Katz 1997). For the identification of antibody-binding epitopes or for the identification of minimal linear ligand-binding sequences, phage display of random peptide libraries, often in the context of small, constrained loops, has frequently yielded highly specific sequences. In certain cases, these consensus features are homologous or even identical to actual sequences in the native binding partner. Recovered sequences have also been shown to mimic or inhibit interactions with the native protein (Balass et al. 1993; Wright et al. 1995).

To investigate the essential structural features that convey epitopes or functional activities, methods have been described in which the gene (cDNA or genomic) encoding the ligand of interest is reduced to smaller gene fragments by DNase I-mediated random digestion (Petersen et al. 1995; van Zonneveld et al. 1995). These libraries of limited diversity have proven to be an efficient means for the characterization of linear epitopes and binding sites, and in some cases can identify motifs with conformational character (Williamson et al. 1998). As discussed in other chapters, libraries of random peptides have also been used to attain the same goal of identification and localization of an active site. To evaluate the relative accuracy and utility of these approaches, Fack et al. (1997) reported parallel studies using both random peptide libraries and gene fragment libraries, which were selected against the same specific murine monoclonal antibodies. The recovered protein sequences were highly similar, documenting the validity of both approaches, but also demonstrating the relative simplicity of the gene fragment approach (methods are discussed in Chapter 20).

Phage-display expression libraries have also been recently used to investigate the molecular basis for the development of prion diseases, which are slowly developing neurologic diseases caused by the accumulation of pathologic proteins, termed prions, in deposits of insoluble proteinaceous polymers. Prions have been speculated to represent abnormal posttranslational products of cellular protein(s) in an aberrant conformation. To characterize the diverse prion protein (PrP) epitopes recognized by a series of specific monoclonal antibodies, a PrP fragment library was constructed in pFRAG, a vector derived from pComb3 (Williamson et al. 1998). Within these studies, independent efforts were made to characterize these same binding epitopes using synthetic peptide-based ELISA studies. Importantly, the data collected by these two divergent approaches proved highly consistent, as several of these monoclonal antibodies were shown by both methods to interact with very similar primary amino acid determinants. In addition, by use of the fragment libraries, discontinuous epitopes on PrP were also characterized that are recognized by two other monoclonal antibodies.

Recently reported investigations of the poorly understood autoimmune disease primary biliary cirrhosis have also benefited from the use of gene fragment libraries. To identify the determinants recognized by antibodies in the serum of patients suffering from primary biliary cirrhosis, Blüthner et al. (1999) used a protein fragment library derived from the gene encoding the sp100 autoantigen. By this approach, phage were selected from the library using sera obtained from several different patients. These efforts identified two epitope primary sequences of 16 and 20 amino acids that are recognized by the disease-associated antibodies in more than one-third of these patients. Further characterization of the molecular requirements for autoantigen binding was subsequently attained by use of a panel of synthetic peptides based on the sp100 autoantigen, which were immobilized on a solid phase for binding assays.

SUMMARY

The power of phage display has been applied successfully in a number of systems designed for the recovery of antibodies, oligopeptide ligand analogs, and variant domains with limited portions of randomized codons. Much more limited experience has been reported for the recovery of novel functional domains from natural DNA/RNA sources. For the cloning of bacterial genes, the current systems have been shown to be well suited. However, for the cloning of genes from more complex species, there are many possible factors that may inherently limit the applicability of these approaches. Size limitations of the cloned DNA are not likely to be the sole explanation, as proteins of 70 kD or greater have been reported to be functionally expressed on the surface of filamentous phage. More likely, there are problems inherent to expression of eukaryotic proteins, in that bacterial systems lack many modes of post-translational modification that can affect functional activity. Representation of critical, functionally active clones in the library may also be adversely affected by the absence of compatible bacterial chaperonin-type molecules that assist folding of proteins into functionally active conformations. This may result in lack of expression of the functional form on the phage surface and prevent selection. Moreover, the accumulation of incorrectly folded aggregates may convey toxicity to a bacterial clone, resulting in dropout of these clones from expanded libraries. Following the recognition of this vexing problem, approaches to the augmentation of chaperonin activity in bacteria have been reported (Bothmann and Pluckthun 1998; Hayhurst and Harris 1999), but a practical way to remedy this intrinsic limitation has not yet been validated. This problem, therefore, likely poses a substantial obstacle to the routine use of phage- display expression libraries that are composed of genetic inserts which are highly heterogeneous in their expressed protein sizes, in their peptide compositions, in their functional capacities, and presumably in their folding capabilities. Hence, the biologic limitations of bacterial systems may be a greater problem for cDNA/genomic libraries that should contain highly diverse genetic/protein species, than for peptide and antibody libraries in which the diversity is comparatively much less heterogeneous.

From this perspective, despite the great potential advantages associated with phage-display methods, at the present time there remain a number of significant technical obstacles for the creation of and selection from libraries of diverse molecular composition. It is, therefore, advisable to employ special care in the creation of libraries to ensure the highest possible representation of every clonal type. In addition, every effort should be made to limit or avoid any round of library expansion that is conducted without concurrent positive selection of desirable clones, as overgrowth of irrelevant and functionally impaired clones will continually be working to adversely bias the composition of the library. For the future, it is hoped that new expression systems with enhanced performance will become available (Crameri and Walter 1999), perhaps by incorporating methods that improve the bacterial capacity for expression and appropriate folding of mammalian proteins in the periplasmic space (Bothmann and Pluckthun 1998; Hayhurst and Harris 1999).

REFERENCES

Balass M., Heldman Y., Cabilly S., Givol D., Katchalski-Katzir E., and Fuchs S. 1993. Identification of a hexapeptide that mimics a conformation-dependent binding site of acetylcholine receptor by use of a phage-epitope library. *Proc. Natl. Acad. Sci.* **90:** 10638–10642.

Blüthner M., Bautz E.K.F., and Bautz F.A. 1996. Mapping of epitopes recognized by PM/Scl autoantibodies with gene-fragments phage display libraries. *J. Immunol. Methods* **198:** 187–198.

Blüthner M., Schafer C., Schneider C., and Bautz F.A. 1999. Identification of major linear epitopes on the sp100 nuclear PBC autoantigen by the gene-fragment phage-display technology. *Autoimmunity* **29:** 33–42.

Bothmann H. and Pluckthun A. 1998. Selection for a periplasmic factor improving phage display and functional periplasmic expression. *Nat. Biotechnol.* **16:** 376–80.

Bottger V., Bottger A., Lane E.B., and Spruce B.A. 1995. Comprehensive epitope analysis of monoclonal anti-proenkephalin antibodies using phage display libraries and synthetic peptides: Revelation of antibody fine specificities caused by somatic mutations in the variable region genes. *J. Mol. Biol.* **247:** 932–946.

Burton D.R. and Barbas C.F. 1994. Human antibodies from combinatorial libraries. *Adv. Immunol.* **57:** 191–280.

Crameri R. 1998. Recombinant *Aspergillus fumigatus* allergens: From the nucleotide sequences to clinical applications. *Allerg. Immunol.* **115:** 99–114.

Crameri R. and Suter M. 1993. Display of biologically active proteins on the surface of filamentous phages: A cDNA cloning system for selection of functional gene products linked to the genetic information responsible for their production. *Gene* **137:** 69–75.

Crameri R. and Walter G. 1999. Selective enrichment and high-throughput screening of phage surface-displayed cDNA libraries from complex allergenic systems. *Comb. Chem. High Throughput Screen* **2:** 63–72.

Crameri R., Jaussi R., Menz G., and Blaser K. 1994. Display of expression products of cDNA libraries on phage surfaces. A versatile screening system for selective isolation of genes by specific gene-product/ligand interaction. *Eur. J. Biochem.* **226:** 53–58.

Cwirla S.E., Peters E.A., Barrett R.W., and Dower W.J. 1990. Peptides on phage: A vast library of peptides for identifying ligands. *Proc. Natl. Acad. Sci.* **87:** 6378–6382.

Dunn I.S. 1995. Assembly of functional bacteriophage lambda virions incorporating C-terminal peptide or protein fusions with the major tail protein. *J. Mol. Biol.* **248:** 497–506.

Dybwad A., Forre O., Natvig J.B., and Sioud M. 1995. Structural characterization of peptides that bind synovial fluid antibodies from RA patients: A novel strategy for identification of disease-related epitopes using a random peptide library. *Clin. Immunol. Immunopathol.* **75:** 45–50.

Efimov V.P., Nepluev I.V., and Mesyanzhinov V.V. 1995. Bacteriophage T4 as a surface display vector. *Virus Genes* **10:** 173–177.

Fack F., Hugle-Dorr B., Song D., Queitsch I., Peterson G., and Bautz E.K.F. 1997. Epitope mapping by phage display: Random versus gene-fragment libraries. *J. Immunol. Methods* **206:** 43–52.

Fields S. and Song O.K. 1989. A novel genetic system to detect protein-protein interactions. *Nature* **340:** 245–246.

Folgori A., Tafi R., Meola A., Felici F., Galfré G., Cortese R., Monaci P. and Nicosia A. 1994. A general strategy to identify mimotopes of pathological antigens using only random peptide libraries and human sera. *EMBO J.* **13:** 2236–2243.

Fransen M., Van Veldhoven P.P., and Subramani S. 1999. Identification of peroxisomal proteins by using M13 phage protein VI phage display: Molecular evidence that mammalian peroxisomes contain a 2,4-dienoul-CoA reductase. *Biochem. J.* **340:** 561–568.

Gao C., Mao S., Lo C.H., Wursching P., Lerner R.A., and Janda K.D. 1999. Making artificial antibodies: A format for phage display of combinatorial heterodimeric arrays. *Proc. Natl. Acad. Sci.* **96:** 6025–6030.

Germino F.J., Wang Z.X., and Weissman S.M. 1993. Screening for in vivo protein-protein interactions. *Proc. Natl. Acad. Sci.* **90:** 933–937.

Gram H., Strittmatter U., Lornzo M., and Glück Z.G. 1993. Phage display as a rapid gene expression system: Production of bioactive cytokine-phage and generation of neutralizing monoclonal antibodies. *J. Immunol.* **16:** 169–176.

Hayhurst A. and Harris W.J. 1999. *Escherichia coli* skp chaperone coexpression improves solubility and phage display of single-chain antibody fragments. *Protein Expr. Purif.* **15:** 336–43.

Helfman D.M., Fiddes J.R., Thomas G.P., and Hughes S. 1983. Identification of clones that encode chicken tropomyosin. *Proc. Natl. Acad. Sci.* **80:** 31–35.

Hoogenboom H.R., Griffiths A.D., Johnson K.S., Chiswell D.J., Hudson P., and Winter G. 1991. Multi-subunit proteins on the surface of filamentous phage: Methodologies for displaying antibody (Fab) heavy and light chains. *Nucleic Acids Res.* **19:** 4133–4137.

Hottiger M., Gramatikoff K., Georgiev O., Chaponnier C., Schaffner W., and Hubscher U. 1995. The large subunit of HIV-1 reverse transcriptase interacts with β-actin. *Nucleic Acids Res.* **23:** 736–741.

Houshmand H., Froman G., and Magnusson G. 1999. Use of bacteriophage T7 displayed peptides for determination of monoclonal antibody specificity and biosensor analysis of the binding reaction. *Anal. Biochem.* **268:** 363–70.

Jacobsson K. and Frykberg L. 1995. Cloning of ligand-binding domains of bacterial receptors by phage display. *BioTechniques* **18:** 878–885.

———. 1996. Phage display shot-gun cloning of ligand-binding domains of prokaryotic receptors approaches 100% correct clones. *BioTechniques* **20:** 1070–1081.

———. 1998. Gene VIII-based, phage-display vectors for selection against complex mixtures of ligands. *BioTechniques* **24:** 294–301.

Jespers L.S., Messens J.H., De Keyser A., Eeckhout D., Van Den Brande I., Gansemans Y.G.,

Lauwereys M.J., Vlasuk G.P., and Stanssens P.E. 1995. Surface expression and ligand-based selection of cDNAs fused to filamentous phage gene VI. *Bio/Technology* **13:** 378–382.
Katz B.A. 1997. Structural and mechanistic determinants of affinity and specificity of ligands discovered or engineered by phage display. *Annu. Rev. Biophys. Struct.* **26:** 27–45.
Kuwabara I., Maruyama H., Mikawa Y.G., Zuberi R.I., Liu F.T., and Maruyama I.N. 1997. Efficient epitope mapping by bacteriophage λ surface display. *Nat. Biotechnol.* **15:** 74–78.
Malik P., Terry T.D., Gowda L.R., Petukhov A.L.S.A., Symmons M.F., Welsh L.C., Marvin D.A., and Perham R.N. 1996. Role of capsid structure and membrane protein processing in determining the size and copy number of peptides displayed on the major coat protein of filamentous bacteriophage. *J. Mol. Biol.* **260:** 9–21.
Maruyama I.N., Maruyama H.I., and Brenner S. 1994. λfoo: A lambda phage vector for the expression of foreign proteins. *Proc. Natl. Acad. Sci.* **91:** 8273–8277.
Mayer C., Appenzeller U., Seelbach H., Achatz G., Oberkofler H., Breitenbach M., Blaser K., and Crameri R. 1999. Humoral and cell-mediated autoimmune reactions to human acidic ribosomal P2 protein in individuals sensitized to *Aspergillus fumigatus* P2 protein. *J. Exp. Med.* **189:** 1507–1512.
McCafferty J., Jackson R.H., and Chiswell D.J. 1991. Phage-enzymes: Expression and affinity chromatography of functional alkaline phosphatase on the surface of bacteriophage. *Protein Eng.* **4:** 955–961.
Nilsson M., Frykberg L., Flock J.I., Pei L., Kindberg M., and Guss B. 1998. A fibrinogen-binding protein of *Staphylococcus epidermidis. Infect. Immun.* **66:** 2666–2673.
Perham R.N., Terry T.D., Willis A.E., Greenwood J., di Marzo Veronese F., and Appella E. 1995. Engineering a peptide epitope display system on filamentous bacteriophage. *FEMS Microbiol. Rev.* **17:** 25–31.
Petersen G., Song D., Hugle-Dorr B., Oldenburg I., and Bautz E.K. 1995. Mapping of linear epitopes recognized by monoclonal antibodies with gene-fragment phage display libraries. *Mol. Gen. Genet.* **249:** 425–431.
Plazkill T., Huang W., and Weinstock G.M. 1998. Mapping protein-ligand interactions using whole genome phage display libraries. *Gene* **221:** 79–83.
Ren Z.J., Lewis G.K., Wingfield P.T., Locke E.G., Steven A.C., and Black L.W. 1996. Phage display of intact domains at high copy number: A system based on SOC, the small outer capsid protein of bacteriophage T4. *Protein Sci.* **5:** 1833–1843.
Santini C., Brennan D., Mennuni C., Hoess R.H., Nicosia A., Cortese R. and Luzzago A. 1998. Efficient display of an HCV cDNA expression library as C-terminal fusion to the capsid protein D of bacteriophage lambda. *J. Mol. Biol.* **282:** 125–135.
Scott J.K. and Smith G.P. 1990. Searching for peptide ligands with an epitope library. *Science* **249:** 386–390.
Seed B. and Aruffo A. 1987. Molecular cloning of the CD2 antigen, the T-cell erythrocyte receptor, by a rapid immunoselection procedure. *Proc. Natl. Acad. Sci.* **84:** 3365–3369.
Sikela J.M. and Hahn W. 1987. Screening an expression library with a ligand probe: Isolation and sequence of a cDNA corresponding to a brain calmodulin binding protein. *Proc. Natl. Acad. Sci.* **84:** 3038–3042.
Singh S.J.H., LeBowitz A.S., Baldwin J., and Sharp P.A. 1988. Molecular cloning of an enhancer binding protein: Isolation by screening of an expression library with a recognition site DNA. *Cell* **52:** 415–423.
Smith G.P. 1985. Filamentous fusion phage: Novel expression vectors that display cloned antigens on the virion surface. *Science* **228:** 1315–1317.
Sternberg N. and Hoess R.H. 1995. Display of peptides and proteins on the surface of bacteriophage λ. *Proc. Natl. Acad. Sci.* **92:** 1609–1613.

van Zonneveld A.J., van den Berg B.M., van Meijer M., and Pannekoek H. 1995. Identification of functional interaction sites on proteins using bacteriophage-displayed random epitope libraries. *Gene* **167:** 49–52.

Williamson R.A., Peretz D., Pinilla C., Ball H., Bastidas R.B., Rozenshteyn R., Houghten R.A., Prusiner S.B., and Burton D.R. 1998. Mapping the prion protein using recombinant antibodies. *J. Virol.* **72:** 9413–9418.

Winter G., Griffiths A.D., Hawkins R.E., and Hoogenboom H.R. 1994. Making antibodies by phage display technology. *Annu. Rev. Immunol.* **12:** 433–455.

Wright R.M., Gram H., Vattay A., Byrne S., Lake P., and Dottavio D. 1995. Binding epitope of somatostatin defined by phage-displayed peptide libraries. *Bio/Technology* **13:** 165–169.

Young R.A. and Davis R.W. 1983. Efficient isolation of genes using antibody probes. *Proc. Natl. Acad. Sci.* **80:** 1194–1198.

Zozulya S., Lioubin M., Hill R.J., Abram C., and Gishizky M.L. 1999. Mapping signal transduction pathways by phage display. *Nat. Biotechnol.* **17:** 1193–1198.

7 Overview: Amplification of Antibody Genes

DENNIS R. BURTON
Departments of Immunology and Molecular Biology, The Scripps Research Institute, La Jolla, California 92037

THE AMPLIFICATION OF ANTIBODY GENES by the polymerase chain reaction (PCR) is a key part of library construction. Primer design determines the nature of the libraries generated. The 5′ or forward primers are generally designed to correspond to the sense strand at the 5′ end of the heavy- or light-chain genes in the first framework region (FR1, Fig. 7.1). In human heavy and κ light chains there is a great deal of conserved sequence in this region, and a relatively limited set of primers can be expected to amplify a major part of the antibody repertoire. A standard practice is to design "family-based" primers, although this is a loose term because within these particular short sequences there are great similarities between different families that routinely lead to considerable cross-priming of diverse genes. As an example, in past reports we have generally used a single 5′ oligonucleotide primer for the amplification of human genes from the V_H1, V_H3, and V_H5 families; one primer for the V_H2 family; one primer for the V_H6 family; and two primers for the V_H4 family. Clearly, somatic mutation can introduce changes from germ-line sequence at any position in the variable region gene, and germ lines can show variation between individuals, so that there is no guarantee that all antibody genes in the in vivo repertoire will be amplified. This problem is likely to be most apparent if the primer set is used to amplify the expressed genes from a monoclonal population of cells, e.g., a hybridoma or Epstein-Barr virus-transformed cell line. In some cases, no PCR amplification will be found with commonly used primers, and it may be necessary to carry out amino-terminal protein sequencing of the antibody of interest to design a compatible amplifying primer.

Sequence variation in FR1 is more pronounced in human λ chains and in mouse heavy and light chains than in human heavy and κ light chains. This means that more primers are required to amplify the corresponding repertoire. In contrast, because there is less germ-line diversity of inherited genes in rabbit and chicken, fewer primers are required.

The restriction site for cloning of PCR inserts is generally placed in the 5′ primer at a position corresponding to encoded amino acid sequence in FR1 and will usually lead to some differences between the recombinant antibody fragment and the native protein. Primers are designed to minimize such differences. The success in cloning many specific high-affinity antibodies from libraries documents that the differences

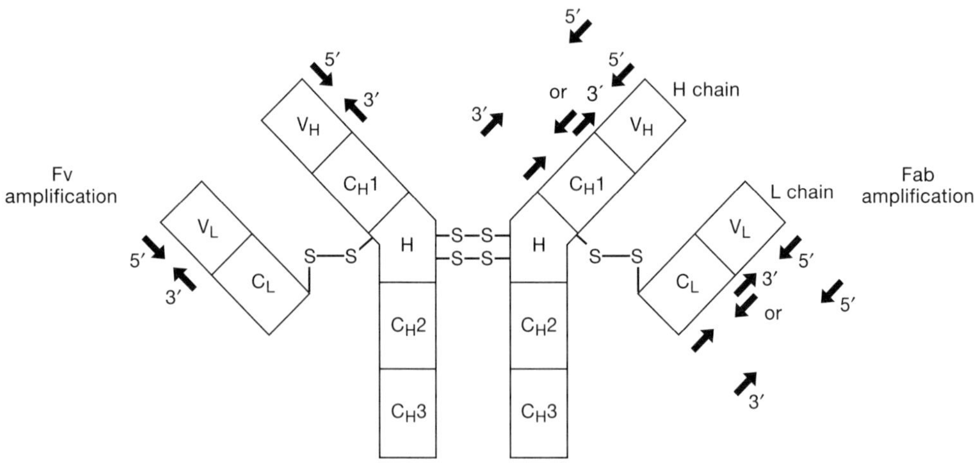

Figure 7.1. Correspondence of PCR primers to features of IgG structure. 5′ or forward primers correspond to sequence at the amino termini of heavy and light chains. For Fv amplification, 3′ or back primers correspond to the carboxyl termini of the variable domains, which are encoded by J gene segments. For Fab amplification there are two formats. In the first, 3′ primers are placed at the carboxyl terminus of the light chain and a position in the heavy chain to include the cysteine involved in heavy–light chain disulfide formation. For human and mouse IgG1 this residue is part of the hinge exon. For other isotypes, it is encoded in the C_H1 domain. Nevertheless, a 3′ primer at the amino terminus of the exon adjacent to C_H1 (hinge or C_H2, depending on antibody isotype, see Fig. 3.2) may be advisable, as the folded C_H1 domain probably extends beyond the sequence defined by the C_H1 exon. In the second format (Chapter 9), 3′ primers are placed at the carboxyl termini of the variable domains, and PCR products are linked to human IgG1 C_H1 or C_κ gene segments by PCR overlap extension.

are often tolerated. However, for any given antibody molecule, the differences could be important, and again, this might be critical in the rescue of antibody genes from a hybridoma that expresses a single pair of antibody genes.

The restriction site in the phage-display vector is placed in a position adjacent to the bacterial leader sequence with or without a short antibody coding sequence. In pComb3H and pComb3X, a human consensus sequence of four residues is encoded in the vector and contributes the first four residues of the FR1 of the recombinant heavy chain. The amino terminus of the recombinant light chain is formed of residues encoded within the restriction site. Amino acid changes resulting from the cloning process may be undesirable in antibodies for therapeutic purposes. They are generally reengineered in the construction of whole antibody molecules from phage-derived antibody fragments.

For the expression of single-chain Fv molecules, the 3′ or back primers are designed to correspond to the antisense strand in the J region. Several 3′ primers are used in combination with each 5′ primer.

For the expression of Fab molecules, two design formats have been used. In the original format, 3′ primers correspond to sequence of the antisense strand from the constant domains. Each heavy-chain isotype generally requires a unique 3′ primer. The positioning of the heavy-chain 3′ primer can be crucial in terms of the amplification of a heavy-chain fragment capable of forming a heavy–light chain disulfide bond. For human γ1 and mouse γ1, the cysteine forming the heavy–light chain bond is encoded in the hinge exon, and the 3′ primer is placed accordingly. For other isotypes, this cysteine is encoded in the C_H1 domain exon. However, the likelihood that sequence at the beginning of the next exon (hinge or C_H2, depending on isotype) is involved in the folded Fab structure has led us to position 3′ primers for other isotypes equivalently to human γ1, i.e., at the 5′ end of the hinge or C_H2 domain exons. For the light chain, 3′ primers correspond to sequence at the 3′ terminus of the chain.

In the newer format described in detail in Chapter 9, the variable domain is amplified by means of a 3′ primer corresponding to the J region, and then the C_H1 region of human IgG1 or the C_κ region of human Ig is linked by PCR overlap extension. This yields human IgG1κ Fab constant domains irrespective of the species or isoptype source of the variable domains.

SUGGESTED READINGS

Cary S.P., Lee J., Wagenknecht R., Silverman G.J. 2000. Characterization of superantigen-induced clonal deletion with a novel clan III-restricted avian monoclonal antibody: Exploiting evolutionary distance to create antibodies specific for a conserved VH region surface. *J. Immunol.* **164:** 4730–4741.

Barbas C.F., III, Kang A.S., Lerner R.A., and Benkovic S.J. 1991. Assembly of combinatorial antibody libraries on phage surfaces: The gene III site. *Proc. Natl. Acad. Sci.* **88:** 7978–7982.

de Boer M., Chang S.-Y., Eichinger G., and Wong H.C. 1994. Design and analysis of PCR primers for the amplification and cloning of human immunoglobulin Fab fragments. *Hum. Antib. Hybrid.* **5:** 57–64.

Huse W.D., Sastry L., Iverson S.A., Kang A.S., Alting-Mees M., Burton D.R., Benkovic S.J., and Lerner R.A. 1989. Generation of a large combinatorial library of the immunoglobulin repertoire in phage lambda. *Science* **246:** 1275–1281.

Kang A.S., Burton D.R., and Lerner R.A. 1991. Combinatorial immunoglobulin libraries in phage. *Methods* **2:** 111–118.

Knight K.L. and Winstead C.R. 1997. Generation of antibody diversity in rabbits. *Curr. Opin. Immunol.* **9:** 228–232.

Larrick J.W., Danielsson L., Brenner C.A., Abrahamson M., Fry K.E., and Borrebaeck C.A.K. 1989. Rapid cloning of rearranged immunoglobulin genes from human hybridoma cells using mixed primers and the polymerase chain reaction. *Biochem. Biophys. Res. Commun.* **160:** 1250–1256.

Marks J.D., Tristem M., Karpas A., and Winter G. 1991. Oligonucleotide primers for polymerase chain reaction amplification of human immunoglobulin variable genes and design of family-specific oligonucleotide probes. *Eur. J. Immunol.* **21:** 985–991.

McCormack W.T., Tjoelker L.W., and Thompson C.B. 1991. Avian B-cell development: Generation of an immunoglobulin repertoire by gene conversion. *Annu. Rev. Immunol.* **9:** 219–241.

Rader C., Cheresh D.A., and Barbas C.F., III. 1998. A phage display approach for rapid antibody humanization: Designed combinatorial V gene libraries. *Proc. Natl. Acad. Sci.* **95:** 8910–8915.

Rader C., Ritter G., Nathan S., Elia M., Gout I., Jungbluth A.A., Cohen L.S., Welt S., Old L.J., and Barbas C.F., III. 2000. The rabbit antibody repertoire as a novel source for the generation of therapeutic human antibodies. *J. Biol. Chem.* (in press).

Williamson R.A., Burioni R., Sanna P., Partridge L., Barbas C.F., III, and Burton D.R. 1993. Human monoclonal antibodies against a plethora of viral pathogens from single combinatorial libraries. *Proc. Natl. Acad. Sci.* **90:** 4141–4145.

8 Generation of Antibody Libraries: Immunization, RNA Preparation, and cDNA Synthesis

JENNIFER ANDRIS-WIDHOPF, CHRISTOPH RADER, AND CARLOS F. BARBAS III
Department of Molecular Biology, The Scripps Research Institute, La Jolla, California 92037

BACTERIOPHAGE DISPLAY OF COMBINATORIAL ANTIBODY libraries is one means by which monoclonal antibodies of a desired specificity can be selected without the use of conventional hybridoma technology. The isolation of specific antibodies from a cloned immunological repertoire requires a large, diverse library, as well as an efficient selection procedure. The key to achieving this goal is the generation of a good immune response and preparation of quality RNA and cDNA from which the library is constructed. The first step in the construction of an "immune" library is a successful course of vaccination with an immunogen of choice. This requires planning and preparation of enough antigen to complete the immunization protocol. Detection of a strong serum antibody titer will theoretically result in a pool of RNA and cDNA that is enriched for antigen-binding immunoglobulin genes that make up the building blocks of a combinatorial antibody library. This chapter presents the protocols currently being used to immunize mice, rabbits, and chickens; extract RNA from spleen and/or bone marrow; and synthesize cDNA for library construction.

CONTENTS

PROTOCOL 8.1

Immunization Schedules (Mouse, Chicken, and Rabbit)

Materials

immunizing antigen (Use a storage buffer acceptable for use in animals, such as PBS. See Harlow and Lane [1999] for more details regarding immunizing animals.)
syringes
needles
Freund's Complete adjuvant (Sigma, Cat. # F 5881)
RIBI's adjuvant (RIBI ImmunoChem Research)
MPL + TDM Emulsion (monophosphoryl lipid A + synthetic trehalose dicorynomycolate, adjuvant for mice, Cat. # R700)
MPL + TDM + CWS (monophosphoryl lipid A + synthetic trehalose dicorynomycolate + cell wall skeleton, adjuvant for rabbits, Cat. # R730)
LES + STM (lipid emulsion system + *Salmonella typhimurium* mitogen, adjuvant for chickens, Cat. # R550)
Freund's Incomplete adjuvant (alternative to RIBI's, Sigma, Cat. # F 5506)

To generate a good "immune" antibody library, each animal should have an immunization schedule whereby injections are given at designated intervals in appropriate doses. Each species is different with regard to the amount of injection than can be tolerated (both in quantity of antigen and total volume of injection). It is also important to use animals of an appropriate age, as animals that are too young become tolerant more easily than adult animals. The quantity of immunogen should also be controlled, as too much can lead to tolerance and too little may result in a lesser immune response.

Table 8.1 lists guidelines for the immunization of mice, chickens, and rabbits. Although these are standard guidelines developed for library construction, the conditions can be varied depending on individual needs. For example, if the immunogen is

Table 8.1. Standard immunization schedule

Species	Breed	Age/ Weight	Sex	Immunogen quantity	Injection interval	Bleed interval
Mouse	Balb/c	6 weeks	either	50 μg	2–3 wk	1 wk post injection
Rabbit	New Zealand White	2.5 kg	either	200 μg	2–3 wk	1 wk post injection
Chicken	White Leghorn	3–6 mos.	either (female)	200 μg	2–3 wk	1 wk post injection

available in small quantities, the later injections can usually be decreased without a noticeable difference in the antibody response. In most instances, the sex of the animal does not make a difference (unless the immunizations are related to sex-specific antigens), but, for example, if chickens are the species of choice, females are sometimes preferred because large of amounts of antibody can be extracted from the eggs of immunized animals (see Support Protocol 1). A regular bleeding schedule is important for following the progress of the antibody titer (see Protocol 8.2). Knowing whether the titer is improving over time may help determine whether alterations in the immunization protocol are needed. These alterations might include increasing or decreasing the quantity of antigen or changing the frequency of injections.

Restrictions on the use of human material for library construction are usually more limiting than those for the use of laboratory animals. Human bone marrow and other tissues are available commercially and can usually be obtained from local hospitals with consent and approval of the clinical review committee. Blood donations are less difficult to obtain. However, more than 80% of blood B cells are resting and/or naive IgM^+/IgD^+ B cells, which are not the best source of an immune library. If immunized libraries are needed, one is usually limited to blood or bone marrow from volunteer donors who have been exposed to the desired antigen by vaccination or illness. In some instances it is possible to obtain spleen, lymph nodes, thyroid tissue, thymus tissue, and brain tissue for selection of disease-specific antibodies (for review, see Burton and Barbas 1994; Graus et al. 1997; and Burgoon et al. 1999).

There are several adjuvants available for immunization with different antigens. Depending on the adjuvant, different formulations are used for different species. Freund's complete adjuvant is the adjuvant of choice, but in most circumstancs animal protocols allow only a single injection with Freund's complete, because it can promote tumor formation. We recommend the use of Freund's complete for the first injection and RIBI's adjuvant for subsequent injections. Different formulations are available for immunizing different species. The adjuvant used for mice is MPL + TDM Emulsion (monophosphoryl lipid A + synthetic trehalose dicorynomycolate). The adjuvant used for rabbits is composed of MPL + TDM + CWS (monophosphoryl lipid A + synthetic trehalose dicorynomycolate + cell wall skeleton). The formulation recommended for chicken immunization contains LES + STM (lipid emulsion system + *S. typhimurium* mitogen). Freund's incomplete adjuvant can also be used for subsequent immunizations; however, the RIBI's is sold with emulsifier and adjuvant in a single vial, whereas the Freund's incomplete is not. The manufacturer provides specific instructions for use with each species. The use of cells as immunogens does not require adjuvants, because they express a large number of different proteins that serve as self-adjuvants.

PROTOCOL 8.2

Titering Immune Serum by ELISA

As part of every immunization protocol, it is important to determine the titer of serum antibody at regular intervals during the course of injections. Knowing the antibody titer provides a basis for making changes in the injection schedule or quantity of immunogen if necessary. Once the titer reaches an appropriate level, the spleen and/or bone marrow is ready for harvest and antibody libraries can be prepared. It is also useful to compare titers with those of previous bleeds. We often find that eventually the serum titers reach a plateau, at which time the immunizations are stopped, the animals are euthanized, and the organs are harvested. The following is a standard protocol for titering any animal sera.

Materials

serum sample from immunized animal
antigen (same as or similar to immunizing antigen)
ELISA plates (Corning Costar, Cat. # 3690)
1x PBS, pH 7.4 (phosphate-buffered saline, see Appendix 2)
5% reconstituted dry milk in 1x PBS (BioRad, Cat. # 170-6404)
3% BSA (bovine serum albumin) reconstituted in 1x PBS (Sigma, Cat. # A 7906)
species-specific, enzyme-conjugated, anti-Ig antibody (Jackson Immunoresearch)
  goat anti-mouse IgG H+L, Cat. # 115036-003 or 115-056-003
  goat anti-rabbit IgG H+L, Cat. # 111-036-003 or 111-056-003
  rabbit anti-chicken IgG H+L, Cat. # 303-036-003 or 303-056-003
  rabbit anti-human IgG H+L, Cat. # 309-035-003 or 309-055-003
PNPP tablets (*p*-nitrophenyl phosphate, Sigma, Cat. # N9389)
PNPP buffer: 10% **diethanolamine,** 0.01% **$MgCl_2$**, 3 mM **sodium azide;** adjust to pH 9.8 with 12 N **HCl**
alkaline phosphatase developing buffer: One 5-mg PNPP tablet, 5 ml of PNPP buffer; 5 ml of developing buffer will develop one 96-well plate at 50 µl/well.
ABTS (**2,2′-azino-di-[3-ethyl-benzthiazoline-sulfonate (6)**], Roche Molecular Biochemicals, Cat # 102946)
10x citrate buffer: 0.316 M citric acid monohydrate, 0.185 M sodium citrate dihydrate, adjust to pH 4.0
ABTS developing buffer (for horseradish peroxidase-conjugated antibodies): 120 µl of 50x **ABTS** (20 mg/ml), 1.8 µl of 30% **H_2O_2**, 0.6 ml of 10x citrate buffer; add H_2O to a final volume of 6 ml. 6 ml of developing buffer will develop one 96-well plate at 50 µl/well.
ELISA reader with 405-nm filter

CAUTIONS: sodium azide, HCl, ABTS, hydrogen peroxide, PNPP, diethanolamine, $MgCl_2$ ***(see Appendix 4)***

Procedure

1. Immunize the animal of choice according to the immunization schedule described in Protocol 8.1, Table 8.1.
2. Bleed the animal approximately 5–7 days postinjection, according to facility regulations. A typical amount of blood to obtain from a single bleed for a given species is as follows:

   Mouse: 100–300 µl

   Rabbit: 1–2 ml

   Chicken: 1–2 ml
3. Scrape the inner surface of the tube containing the blood with an applicator stick to dislodge the clotted blood from the sides. Refrigerate overnight at 4°C to ensure complete clotting.
4. Centrifuge the sample at 3500 rpm for 15 minutes in a tabletop centrifuge to separate the serum from debris and clotted cells. For smaller samples, centrifuge at full speed in a microcentrifuge for 5 minutes.
5. Transfer the serum to a second tube and store at 4°C (6–12 months) or –20°C (for several years, if not repeatedly thawing and freezing). Discard the clot in an appropriate biohazard waste container.
6. Prepare ELISA plates by incubating an appropriate number of wells with 25 µl of the designated antigen(s) at a concentration between 0.1 and 1 µg/well in 1x PBS. Use the higher concentrations for weaker antigens. Incubate overnight at 4°C. We prefer to use Costar EIA/RIA 96-well half-area, flat bottom, high binding plates.
7. Wash the antigen from the wells by filling them with distilled water, and blot dry by firmly slapping the plate on clean paper towels. Block the plate by filling the wells with 150 µl of 5% reconstituted dry milk in 1x PBS. Alternatively, the plate can be blocked with 3% BSA. Incubate for 1 hour at 37°C.
8. Prepare twofold serial dilutions of the serum samples in 5% milk, beginning with a 1:100 dilution and ending with a 1:51,200 dilution. If you have reason to believe that the titer is extraordinarily high, the dilutions can be carried further or prepared as fivefold or higher dilutions. Likewise, if you have reason to believe that the immune response is exceptionally weak, the diutions can be prepared beginning at a point lower than 1:100.
9. Rinse the blocking reagent (Step 7) from the wells with water and blot dry. Apply 25 µl of each serum dilution to the designated wells and incubate for 1 hour at 37°C.

10. Wash the plate 10 times with water and blot dry. Dilute the conjugated secondary antibody in 5% milk and apply 25 µl to each well. Incubate for 1 hour at 37°C. If high background is a problem, the plates can also be washed with 0.05–0.5% Tween 20 in 1X PBS.

    Note: The best choice for a secondary reagent is one that is species-specific for Heavy chain + Light chain, as these reagents generally recognize whole IgG, Fab, and $F(ab')_2$.

11. Prepare enough alkaline phosphatase developing buffer (for alkaline-phosphatase-conjugated antibodies) or ABTS developing buffer (for horseradish-peroxidase-conjugated antibodies) for the number of plates used. Wash the plate 10 times with water and blot dry. Apply 50 µl of developing reagent to each well.

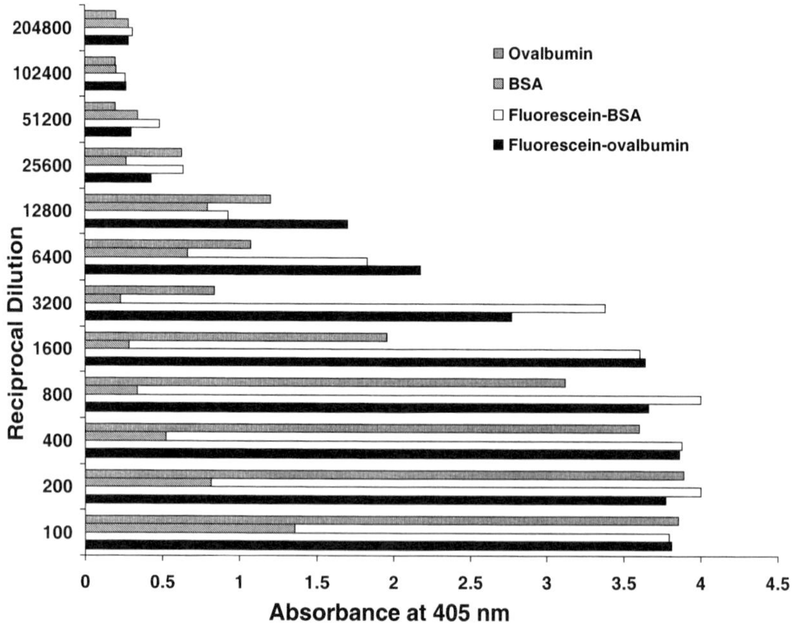

Figure 8.1. Titration of serum antibodies. In this example, an adult female chicken was immunized with a fluorescein–ovalbumin conjugate over a period of 2 months, receiving injections of 200 µg every 2–3 weeks. The serum was titered against the original immunizing conjugate (Fluorescein–ovalbumin), a second fluorescein–BSA conjugate (Fluorescein–BSA), a fluorescein–biotin conjugate, ovalbumin, and bovine serum albumin (BSA) according to Protocol 8.2. The secondary antibody was a rabbit anti-chicken IgG (Heavy chain + Light chain) (Jackson Immunoresearch, Cat. # 303-046-003) conjugated to horseradish peroxidase. The animal in this example has a fluorescein antibody titer between 1:25,600 and 1:51,200, determined by comparing flourescein–BSA binding with BSA binding. The ovalbumin antibody titer is approximately 1:25,600, determined by comparing ovalbumin binding with BSA binding.

Read the absorbance at the designated wavelength at 2–3 time points. The first reading is usually 5–15 minutes after addition of developer, but the time points may vary depending on how much antibody is present and how quickly color development first appears. If the immunizing antigen is a peptide conjugated to a carrier protein, it is important to note that the carrier protein often elicits a greater response earlier in the immunization than the peptide itself. If high background is a problem, the plates can also be washed with 0.05–0.5% Tween 20 in 1x PBS.

12. Graph the results of the titration to compare the response to different antigens and/or consecutive bleeds (see Fig. 8.1). The "titer" refers to the highest dilution at which antigen-specific binding is detectable above background binding to an irrelevant antigen. For the purposes of phage display, determining whether the titer is relatively high or low is more important than obtaining a precise titer (1:1000, for example, is considered relatively low).

PROTOCOL 8.3

Isolation of Total RNA for the Generation of Mouse, Rabbit, Chicken, and Human Antibody Libraries

This section consists of protocols for the preparation of total RNA from mouse spleen, rabbit spleen and bone marrow, chicken spleen and bone marrow, and human bone marrow. The tissue source of choice is one rich in plasma cells that secrete antibodies against the antigen of interest, because it will contain the highest levels of specific mRNA. Bone marrow and spleen have been shown to be a major repository of plasma cells and, thus, are most frequently used for the generation of antibody libraries. In contrast, the number of specific antibody-secreting cells in the peripheral blood is only high for a few days after an antigen boost and declines very rapidly (Burton and Barbas 1994). Human peripheral blood lymphocytes (PBLs), however, are a ready source of mRNA for the generation of naive human antibody libraries.

Special Handling Precautions

Wear gloves (change frequently) and safety glasses.
Use RNase-free plasticware, e.g.: Stripette pipets (Corning Costar), ART pipet tips (Molecular Bio-Products), 50-ml polypropylene tubes (Corning), 15-ml conical tubes, 50-ml NALGENE polypropylene copolymer centrifuge tubes (Nalge), 1.5-ml BIOPUR microcentrifuge tubes (Eppendorf).
Keep all reagents RNase-free; hands and dust are major sources of RNase contamination (cf. Sambrook et al. 1989).

ISOLATION OF TOTAL RNA FROM MOUSE SPLEEN:

Materials

immunized mouse
70% **ethanol** in a bucket
TRI Reagent (contains **phenol** and **guanidine thiocyanate**) and **1-bromo-3-chloropropane** (BCP) from Molecular Research Center, store protected from light at 4°C.
isopropanol
75% **ethanol** (made with RNase-free water)
RNase-free water
RNase-free 3 M sodium acetate, pH 5.2
ethanol

CAUTIONS: animal treatment, 1-bromo-3-chloropropane, ethanol, phenol, guanidine thiocyanate, isopropanol, METOFANE *(see Appendix 4)*

Procedure

1. Anesthetize the mouse with a suitable inhalation anesthetic, e.g., methoxyflurane (**METOFANE**; Schering-Plough Animal Health Corporation, Madison, NJ), and sacrifice by cervical dislocation. Rinse the mouse in a bucket filled with 70% ethanol and lay it on a dissection tray with the face down and the left side up. Open the abdominal cavity with scissors, pull out the spleen (~1.5-cm long, dark red organ in the upper left quadrant of the abdomen) with forceps and remove the attached connective tissue with scissors. Transfer the spleen into 10 ml of TRI Reagent in a 50-ml tube.

2. Homogenize the sample in TRI Reagent using a homogenizer, e.g., Tissumizer (Tekmar-Dohrmann) at 50% output for 1 minute, and store for 5 minutes at room temperature.

3. Centrifuge at 2500*g* (3500 rpm in a Beckman GPR tabletop centrifuge) for 10 minutes at 4°C.

4. Transfer the supernatant to a 50-ml centrifuge tube and discard the pellet. Add 1 ml of BCP to the supernatant, vortex for 15 seconds, and incubate for 15 minutes at room temperature.

5. Centrifuge at 17,500*g* (12,000 rpm in a Beckman JA-20 rotor) for 15 minutes at 4°C.

6. Transfer the upper, colorless aqueous phase to a fresh 50-ml centrifuge tube. Add 5 ml of isopropanol, vortex for 15 seconds, and incubate for 10 minutes at room temperature.

   Note: The interphase and lower red organic phase can be saved for subsequent DNA isolation (cf. TRI Reagent protocol).

7. Centrifuge at 17,500*g* for 15 minutes at 4°C.

8. Remove the supernatant carefully and discard. Add 10 ml of 75% ethanol to the pellet, but do not resuspend. Centrifuge at 17,500*g* for 10 minutes at 4°C.

9. Remove the supernatant carefully and discard. Air-dry the pellet briefly at room temperature (do not dry under vacuum), dissolve in 250 μl of RNase-free water, and transfer to a microcentrifuge tube. Remove a 5-μl aliquot and store the rest of the isolated total RNA at –80°C. For storage periods exceeding a few weeks, precipitate the isolated total RNA by adding 0.1 volume of RNase-free 3 M sodium acetate, pH 5.2, and 2 volumes of ethanol.

10. Use the aliquot to prepare a 1:250 dilution of the total RNA in water. Determine the optical density (OD) at 260 and 280 nm. Calculate the ratio of OD_{260}/OD_{280} to determine RNA purity (typically in the range of 1.6 to 1.9), and determine the RNA concentration (40 ng/μl RNA gives an $OD_{260} = 1$). The expected yield of total RNA isolated from the spleen of an immune mouse of the Balb/c strain is in the range of 1–2 mg.

ISOLATION OF TOTAL RNA FROM RABBIT OR CHICKEN BONE MARROW AND SPLEEN

Materials

immunized rabbit or chicken

TRI Reagent (contains **phenol** and **guanidine thiocyanate**) and 1-bromo-3-chloropropane (BCP) from Molecular Research Center, store protected from light at 4°C

isopropanol

75% **ethanol** (made with RNase-free water)

RNase-free water

RNase-free 3 M sodium acetate, pH 5.2

ethanol

CAUTIONS: **animal treatment, phenol, guanidine thiocyanate, isopropanol, ethanol** *(see Appendix 4)*

Procedure

1. The bone marrow and spleen from rabbits and chickens should be harvested only by trained personnel. After removal, immediately transfer the bone marrow from one leg and the spleen to separate 50-ml tubes, each containing 10 ml of TRI Reagent.
2. Homogenize the samples in TRI Reagent using a homogenizer, e.g., Tissumizer (Tekmar-Dohrmann) at 50% output for 1 minute. Incubate for 5 minutes at room temperature.
3. Add 20 ml of TRI Reagent to each sample and centrifuge at 2,500*g* (3,500 rpm in a Beckman GPR tabletop centrifuge) for 10 minutes at 4°C.
4. Transfer the supernatants to 50-ml centrifuge tubes and discard the pellets. Add 3 ml of BCP to each supernatant, vortex for 15 seconds, and incubate for 15 minutes at room temperature.
5. Centrifuge at 17,500*g* (12,000 rpm in a Beckman JA-20 rotor) for 15 minutes at 4°C.
6. Transfer the upper, colorless aqueous phase to fresh 50-ml centrifuge tubes. Add 15 ml of isopropanol, vortex for 15 seconds, and incubate for 10 minutes at room temperature.

   Note: The interphase and lower red organic phase may be saved for subsequent DNA isolation (cf. TRI Reagent protocol).
7. Centrifuge at 17,500*g* for 10 minutes at 4°C.
8. Remove the supernatants carefully and discard. Add 30 ml of 75% ethanol to each pellet, but do not resuspend. Centrifuge at 17,500*g* for 10 minutes at 4°C.

9. Remove the supernatants carefully and discard. Air-dry the pellets briefly at room temperature (do not dry under vacuum), dissolve each in 500 μl of RNase-free water, and combine the isolated total RNA derived from bone marrow and spleen in a microcentrifuge tube. Remove a 5-μl aliquot and store the rest at –80°C. For storage periods exceeding a few weeks, precipitate the isolated total RNA by adding 0.1 volume of RNase-free 3 M sodium acetate, pH 5.2, and 2 volumes of ethanol.

10. Use the aliquot to prepare a 1:250 dilution of the total RNA in water. Determine the optical density at 260 and 280 nm. Calculate the ratio of OD_{260}/OD_{280} to determine RNA purity (typically in the range of 1.6 to 1.9), and determine the RNA concentration (40 ng/μl RNA gives an $OD_{260} = 1$). The expected yield of total RNA isolated from the bone marrow of one leg and the spleen of a New Zealand White rabbit or a White Leghorn chicken is in the range of 6–7 mg.

ISOLATION OF TOTAL RNA FROM HUMAN BONE MARROW

The isolation of total RNA from human PBLs, which can be prepared by Ficoll-Paque density gradient centrifugation (Amersham Pharmacia Biotech), also follows the protocol described below. Start with 100–500 ml of human blood.

Materials

human bone marrow or PBLs
TRI Reagent (contains **phenol** and **guanidine thiocyanate**) and 1-bromo-3-chloropropane (BCP) from Molecular Research Center, store protected from light at 4°C
isopropanol
75% **ethanol** (made with RNase-free water)
RNase-free water (e.g., Promega, Cat. # P119C)
RNase-free 3 M sodium acetate, pH 5.2
ethanol

CAUTIONS: blood products, phenol, guanidine thiocyanate, isopropanol, ethanol *(see Appendix 4)*

Special Handling Precautions

Consider hepatitis B vaccination.
Inform your lab colleagues.
Wear double gloves (change frequently) and safety glasses.
Treat waste that is derived from or has been in contact with human blood and tissue as a biohazard.

Procedure

1. Start with 25 ml of freshly aspirated human bone marrow. If less material is available, scale the procedure down appropriately. 5 ml of bone marrow is sufficient for the isolation of a reasonable amount of total RNA.

2. Transfer the bone marrow to two 15-ml conical tubes and centrifuge at 500*g* (1500 rpm in a Beckman GPR tabletop centrifuge) for 10 minutes at 4°C.
3. Remove the supernatant (~12.5 ml). Like serum, the supernatant may be used to analyze the immune response to a particular antigen; store at –80°C. Resuspend the pellets (~ 12.5 ml) in a total of 120 ml of TRI Reagent distributed evenly among four 50-ml tubes.
4. Homogenize the samples in TRI Reagent using a homogenizer, e.g., a Tissumizer (Tekmar-Dohrmann) at 50% output for 1 minute, and incubate for 5 minutes at room temperature.
5. Centrifuge at 2,500*g* (3,500 rpm in a Beckman GPR tabletop centrifuge) for 10 minutes at 4°C.
6. Transfer the supernatants to 50-ml centrifuge tubes and discard the pellets. Add 3 ml of BCP to each supernatant, vortex for 15 seconds, and incubate for 15 minutes at room temperature.
7. Centrifuge at 17,500*g* (12,000 rpm in a Beckman JA-20 rotor) for 15 minutes at 4°C.
8. Transfer the upper colorless aqueous phase to fresh 50-ml centrifuge tubes. Add 15 ml of isopropanol, vortex for 15 seconds, and incubate for 10 minutes at room temperature.

   Note: The interphase and lower red organic phase may be saved for subsequent DNA isolation (cf. TRI Reagent protocol).
9. Centrifuge at 17,500*g* for 10 minutes at 4°C.
10. Remove the supernatants carefully and discard. Add 30 ml of 75% ethanol to each pellet; do not resuspend. Centrifuge at 17,500*g* for 10 minutes at 4°C.
11. Remove the supernatants carefully and discard. Air-dry the pellets briefly at room temperature (do not dry under vacuum), dissolve each in 250 µl of RNase-free water, and combine the isolated total RNA in a microcentrifuge tube. Remove a 5-µl aliquot and store the rest at –80°C. For storage periods exceeding a few weeks, precipitate the isolated total RNA by adding 0.1 volume of RNase-free 3 M sodium acetate, pH 5.2, and 2 volumes of ethanol.
12. Use the aliquot to prepare a 1:250 dilution of the total RNA in water. Determine the optical density at 260 and 280 nm. Calculate the ratio of OD_{260}/OD_{280} to determine RNA purity (typically in the range of 1.6–1.9), and determine the RNA concentration (40 ng/µl RNA gives an $OD_{260} = 1$). The expected yield of total RNA isolated from 25 ml of human bone marrow is in the range of 1–2 mg.

LITHIUM CHLORIDE PRECIPITATION TO PURIFY TOTAL RNA

If the DNA derived from total RNA yields no or only weak amplification of light- and heavy-chain-encoding sequences, a dramatic improvement can be achieved by first purifying the total RNA by lithium chloride precipitation.

Materials

RNase-free 7.5 M **LiCl**
RNase-free 3 M sodium acetate, pH 5.2
RNase-free water
ethanol (store at –20°C)

CAUTIONS: **LiCl, ethanol** ***(see Appendix 4)***

Procedure

1. Transfer 300 μl of the isolated total RNA to a 1.5-ml microcentrifuge tube. Add 32 μl of RNase-free 7.5 M LiCl, vortex, and incubate for 2 hours on ice.
2. Centrifuge at full speed in a microcentrifuge for 30 minutes at 4°C.
3. Remove the supernatant carefully and discard. Dissolve the pellet in 300 μl of RNase-free water, add 32 μl of RNase-free 7.5 M LiCl, vortex, and incubate for 2 hours on ice.
4. Centrifuge at full speed in a microcentrifuge for 30 minutes at 4°C.
5. Remove the supernatant carefully and discard. Dissolve the pellet in 300 μl of RNase-free water. Add 30 μl of RNase-free 3 M sodium acetate, pH 5.2, and 600 μl of ethanol (-20°C), vortex, and incubate for 30 minutes at –20°C.
6. Centrifuge at full speed in a microcentrifuge for 30 minutes at 4°C. Remove the supernatant carefully and discard. Add 500 μl of 70% ethanol (–20°C) and centrifuge at full speed in a microcentrifuge for 10 minutes at 4°C.
7. Remove the supernatant carefully and discard; air-dry the pellet briefly at room temperature and dissolve in 100 μl of RNase-free water. Remove a 5-μl aliquot and store the rest at –80°C. For storage exceeding a couple of weeks, the purified total RNA should be precipitated by adding 0.1 volume of RNase-free 3 M sodium acetate, pH 5.2, and 2 volumes of ethanol.
8. Determine the concentration of the purified total RNA as described above (Isolation of total RNA from human bone marrow, Step 12).

PROTOCOL 8.4

First-strand cDNA Synthesis from Total RNA

In this protocol, first-strand cDNA is synthesized from total RNA using an oligo (dT) primer and reverse transcriptase in a standard procedure (Sambrook et al. 1989; see Fig. 8.2).

Materials

purified total RNA (see Protocol 8.3)

SUPERSCRIPT Preamplification System for First Strand cDNA Synthesis (Life Technologies, Cat. # 18089-011), store at –20°C. Contains all components needed for cDNA synthesis.

CAUTIONS: **magnesium chloride, dithiothreitol *(see Appendix 4)***

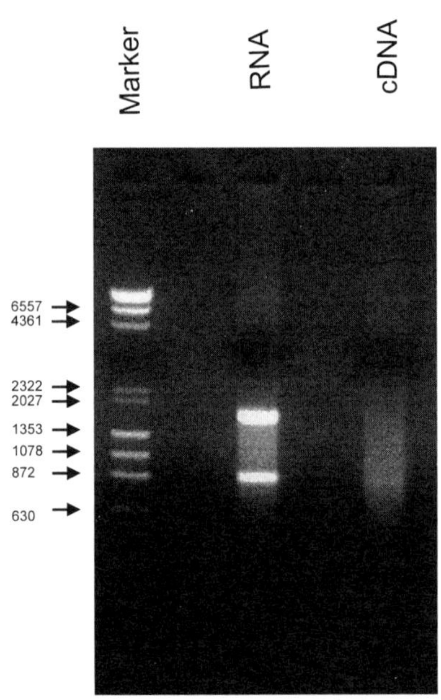

Figure 8.2. Preparation of cDNA by reverse transcription of mRNA from human peripheral blood mononuclear cells. RNA was extracted from human mononuclear cells with TRI-Reagent as described in Protocol 8.3. Using an oligo-(dT) primer and a SUPERSCRIPT II reverse transcriptase kit, mRNA was then reverse transcribed into cDNA (Protocol 8.4). Samples of RNA and cDNA were electrophoretically separated on a 1% agarose gel. The major ribosomal RNA species at 2 kb and 850 bp were not reverse transcribed because of the absence of polyadenylation sequences. Molecular weight DNA mass markers of ϕX174 *Hae*III digests and λDNA *Hind*III digests are included (New England Biolabs). (Provided by Dr. Peter Shaw, UCSD, La Jolla, CA.)

Procedure

1. Mix 20 µg of the isolated total RNA with 8 µl of 0.5 µg/µl oligo(dT)$_{12\text{-}18}$ and add DEPC-treated water to 96 µl. Incubate the sample for 10 minutes at 70°C and store for at least 1 minute on ice. Centrifuge briefly (~3-sec pulse in a microcentrifuge).
2. Prepare the following reaction mixture:

16 µl	10x PCR buffer (200 mM Tris-HCl, pH 8.4; 500 mM KCl)
16 µl	25 mM $MgCl_2$
8 µl	10 mM dNTP mix (10 mM each dATP, dCTP, dGTP, dTTP)
16 µl	100 mM DTT

3. Add the reaction mixture (56 µl) to the RNA sample and incubate for 5 minutes at 42°C.
4. Add 8 µl of 200 Units/µl SUPERSCRIPT II reverse transcriptase and incubate for 1 hour at 42°C.
5. Incubate the sample for 15 minutes at 70°C and store for at least 1 minute on ice. Centrifuge briefly (~3 sec pulse in a microcentrifuge).
6. Add 8 µl of 2 Units/µl *E. coli* RNase H and incubate for 20 minutes at 37°C. The first-strand cDNA is stable when stored at –20°C.

The synthesized first-strand cDNA is now ready for use in library construction (Chapter 9). Use 1–2 µl in a 100-µl PCR with 60 pmole each of sense and antisense primer.

SUPPORT PROTOCOL 1

Purification of IgY (IgG) from Chicken Eggs

If the serum titer of an immunized chicken is high, it is often desirable to collect a large amount of polyclonal antibody. This antibody preparation can serve as control antibodies in any assay in which phage-display-selected clones are tested. Because egg yolks are a good source of relatively large amounts of polyclonal antibody, it is important to immunize female chickens (Song et al. 1985). The IgG homolog is referred to as IgY and does not bind to protein A or protein G. There are a few published protocols for the isolation of chicken IgY, but they are time-consuming (Jensenius et al. 1981; Akita and Nakai 1992, 1993a,b). Alternatively, Promega Corporation markets a kit, EGGstract IgY purification system, that can yield 55–80 mg of 75% pure IgY per egg. An additional precipitation can result in approximately 90% pure IgY without significant loss of protein. We recommend the use of this product for purification of large amounts of polyclonal chicken antibody.

Materials

egg(s) from immunized chickens
EGGstract IgY purification system (Promega Corporation, Cat. # G2610)

Procedure

The manufacturer provides a detailed protocol with the purchase of the EGGstract IgY purification system. Below is a summary of the procedure.

1. Separate the yolks from the egg white and precipitate the lipids from the yolks with the designated precipitation solution.
2. Remove the lipids by centrifugation.
3. Precipitate the IgY from the remaining supernatant using the designated precipitation solution.
4. Recover the IgY by centrifugation and resuspension of the pellet.
5. Repeat the second precipitation step for further purification.
6. Resuspend the IgY pellet and analyze by polyacrylamide gel electrophoresis.

SUPPORT PROTOCOL 2

Purification of Rabbit IgG from Serum

Similar to the chicken, rabbits with a high titer of specific antibodies are a good source of large quantities of polyclonal IgG. These antibodies are particularly useful for phage display, because they can be used as a positive control when screening clones. Serum antibodies from the immunized rabbit can be purified from consecutive bleeds and/or the final serum obtained by exsanguination at the time of euthanization. The easiest way to obtain concentrated polyclonal antibody is by saturated ammonium sulfate (SAS) precipitation. If a purer source of IgG is needed, the SAS-precipitated IgG can be further purified using Protein G affinity chromatography.

Materials

rabbit serum (obtain from blood treated as in Steps 3–5, Protocol 8.2)
saturated **ammonium sulfate** (SAS, store at 4°C, see Appendix 2)
beakers
ice containers
stir bars
60-cc syringe with stopcock (BioRad, Cat # 7328102)
ring stand and clamps
stir plate
1x PBS, pH 7.4 (see Appendix 2)
dialysis tubing (Spectra/Por2, 12,000–14,000 MWCO, VWR, Cat. # 25218-435)

CAUTION: **ammonium sulfate** ***(see Appendix 4)***

Procedure

Perform the entire protocol at 4°C:

1. Swirl the SAS in a flask to ensure full saturation. Let the flask settle for 5 minutes.
2. Place the serum in a beaker. Place the beaker in a container of ice on a stir plate so that the serum is completely surrounded by ice. Place a stir bar in the beaker and begin stirring.
3. Set up a ring stand holding a syringe and stopcock apparatus so that the syringe is positioned over the beaker containing the serum. Determine the quantity of SAS to add to the serum for a 45% cut by multiplying the initial volume of serum by 0.833. Pipet that amount of cold SAS into the syringe with the stopcock closed (when pipetting avoid the crystals that have settled in the bottom of the flask).
4. Open the stopcock so that the SAS is added dropwise, while stirring, so that each drop is thoroughly mixed before the next drop is added. Too high a concentra-

tion of ammonium sulfate at one time may cause unwanted material to coprecipitate.

5. Once the entire volume of SAS is added, allow the serum to sit on ice for 15–30 minutes. Stirring is no longer necessary.
6. Transfer the precipitated serum into 50-ml centrifuge tubes. Centrifuge at 12,000*g* for 20 minutes at 4°C.
7. Decant the supernatant into a clean tube. This supernatant can be tested for residual antibody activity in ELISA if necessary. Add approximately the original serum volume of 1x PBS to the pellet and incubate at 37°C for 15–30 minutes to soften the pellet. Resuspend the pellet gently, with minimal frothing, to avoid denaturing the protein.
8. Transfer the resuspended material to dialysis tubing and dialyze overnight against 4–6 liters of 1x PBS, changing the buffer 2–3 times.
9. Transfer the dialyzed material to a clean tube. 1- to 1.5-ml aliquots can be stored at 4°C for 6–12 months. If frozen at –20°C without repeated thawing and freezing, aliquots can be stored for years.
10. The antibody can be titrated using the ELISA protocol described in Protocol 8.2.

REFERENCES

Akita E.M. and S. Nakai. 1993. Comparison of four purification methods for the production of immunoglobulins from eggs laid by hens immunized with an enterotoxigenic *E. coli* strain. *J. Immunol. Methods* **160:** 207–214.

———. 1993. Production and purification of Fab´ fragments from chicken egg yolk immunoglobulin Y (IgY). *J. Immunol. Methods* **162:**155–164.

Burton D.R. and Barbas C.F., III. 1994. Human antibodies from combinatorial libraries. *Adv. Immunol.* **57:** 191–280.

Burgoon M.P., Williamson R.A., Owens G.P., Ghausi O., Bastidas R.B., Burton D.R., and Gilden D.H. 1999. Cloning the antibody response in humans with inflammatory CNS disease: Isolation of measles virus-specific antibodies from phage display libraries of a subacute sclerosing panencephalitis brain. *J. Neuroimmunol.* **94:** 204–211.

Graus Y.F., de Baets M.H., Parren P.W., Berrih-Aknin A., Wokke J., Van Breda Vriesman P.J., and Burton D.R. 1997. Human anti-nicotinic acetylcholine receptor recombinant Fab fragments isolated from thymus-derived phage display libraries from myasthenia gravis patients reflect predominant specificities in serum and block the action of pathogenic serum antibodies. *J. Immunol.* **158:** 1919–1929.

Harlow E. and Lane D. 1999. *Antibodies: A laboratory manual,* 2nd edition. Cold Spring Harbor Laboratory Press, Cold Spring Harbor, New York.

Jensenius J.C., Andersen I., Hau J., Crone M., and Koch C. 1981. Eggs: Conveniently packaged antibodies. Methods for purification of yolk IgG. *J. Immunol. Methods* **46:** 63–68.

Sambrook J., Fritsch E.F., and Maniatis T. 1989. *Molecular cloning: A laboratory manual,* 2nd edition. Cold Spring Harbor Laboratory Press, Cold Spring Harbor, New York.

Song C.S., Yu J.H., Bai D.H., Hester P.Y., and Kim K.H. 1985. Antibodies to the α-subunit of insulin receptor from eggs of immunized hens. *J. Immunol.* **135:** 3354–3359.

9 Generation of Antibody Libraries: PCR Amplification and Assembly of Light- and Heavy-chain Coding Sequences

*JENNIFER ANDRIS-WIDHOPF, *PETER STEINBERGER, ROBERTA FULLER, CHRISTOPH RADER, AND CARLOS F. BARBAS III

Department of Molecular Biology, The Scripps Research Institute, La Jolla, California 92037

PHAGE-DISPLAYED ANTIBODY FRAGMENT LIBRARIES can be constructed using essentially any species that is easily immunized, as long as the immunoglobulin variable region gene sequences are known. This chapter includes protocols for the preparation of Fab and scFv antibody libraries from human, nonhuman primate, mouse, rabbit, and chicken. The general scheme for building the libraries is the same for each species, with only the oligonucleotide primer sequences being species-specific. Briefly, the individual rearranged heavy- and light-chain variable regions are amplified separately and are linked through a series of overlap PCR steps to give the final Fab or scFv products that are used for cloning. Figures 9.1 and 9.2 show a schematic overview of the steps that are used to generate the PCR products, and the flowchart in Figure 9.3 depicts the entire library construction and selection procedure.

CONTENTS

* These authors contributed equally to this work.

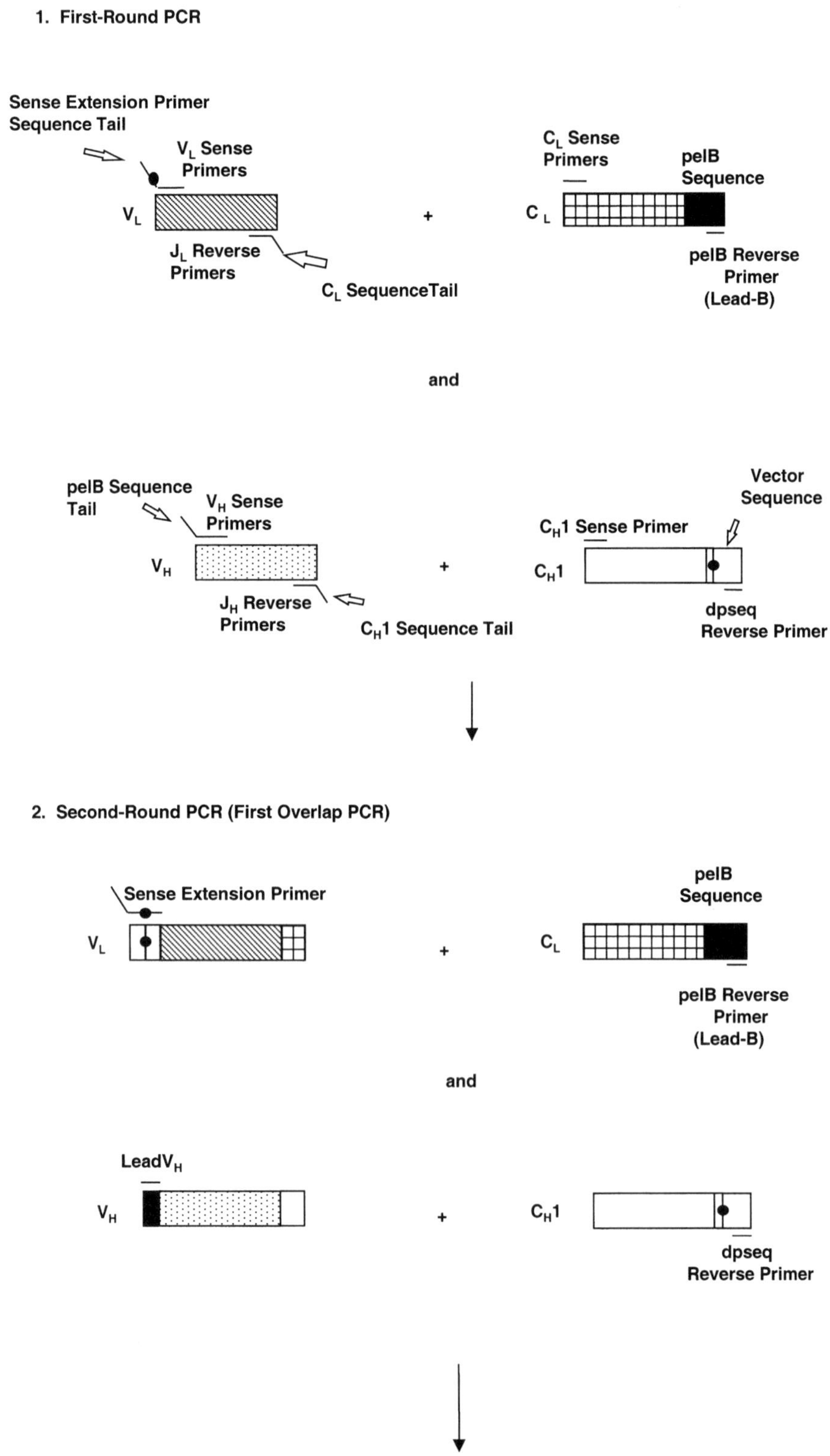
1. First-Round PCR
Sense Extension Primer Sequence Tail
V_L Sense Primers
V_L
J_L Reverse Primers
C_L SequenceTail
+
C_L Sense Primers
pelB Sequence
C_L
pelB Reverse Primer (Lead-B)
and
pelB Sequence Tail
V_H Sense Primers
V_H
J_H Reverse Primers
C_H1 Sequence Tail
+
Vector Sequence
C_H1 Sense Primer
C_H1
dpseq Reverse Primer
2. Second-Round PCR (First Overlap PCR)
Sense Extension Primer
V_L
+
pelB Sequence
C_L
pelB Reverse Primer (Lead-B)
and
LeadV_H
V_H
+
C_H1
dpseq Reverse Primer

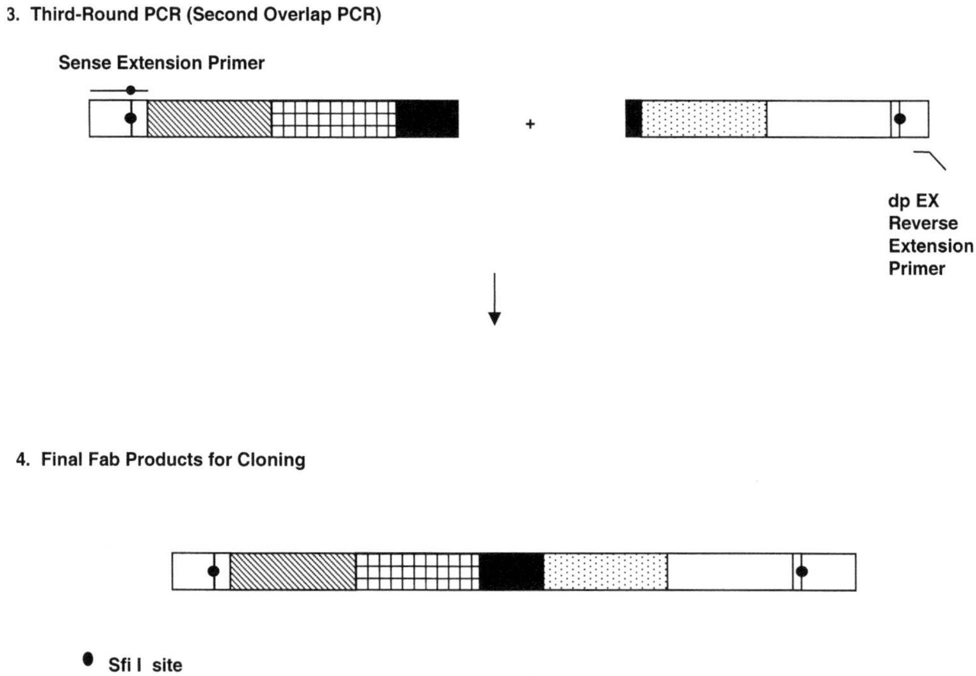

Figure 9.1. Generation of Fab antibody fragments by PCR overlap extension for cloning into the pComb3 system. In the first round of PCR, the light-chain variable regions (V_L) are amplified from cDNA by using V_L sense primers with reverse primers specific for the 5′ end of the light-chain constant region (C_L; human Fab libraries) or the light-chain joining region (J_L; chimeric Fab libraries). The V_L sense primers add a 5′ sequence tail that contains an *Sfi*I site and is recognized by the sense extension primer used in the second-round PCR. The J_L-specific reverse primers have a tail that introduces the 5′ end of a human C_L sequence into the V_L product. This tail allows the overlap extension of the V_L and C_L–pelB products in the second-round PCR. The C_L and pelB leader sequence is amplified from phagemid DNA containing a cloned human Fab by using a sense primer specific to the 5′ end of the C_L sequence and a reverse primer specific to the 3′ end of the pelB leader sequence that is located downstream of the cloned C_L fragment. The heavy-chain variable region fragments (V_H) are amplified from cDNA by using V_H sense primers and reverse primers specific for the heavy-chain joining regions (J_H). The J_H reverse primers have a sequence tail that adds the 5′ end of the human C_H1 sequence to the V_H PCR products and allows the overlap extension assembly of V_H and C_H1 fragments in the second-round PCR. The V_H sense primers have a sequence tail that corresponds to the 3′ end of the pelB leader fragment and allows the overlap extension assembly of the heavy-chain (Fd) fragment with the light-chain–pelB fragment in the final round of PCR. The heavy-chain constant region (C_H1) fragments are amplified from a template plasmid containing a cloned human Fab by using a C_H1 sense primer and a reverse primer specific for the decapeptide tag sequence that is downstream of the cloned C_H1 fragment and the 3′ *Sfi*I site in the phagemid vector. All first-round products range in size from 350 to 400 bp. In the second-round PCR, the light-chain–pelB fragments are constructed from the V_L and C_L–pelB fragments by overlap extension PCR using the V_L sense extension primer and the reverse primer specific for the 3′ end of the C_L–pelB product. The heavy-chain Fd fragment is generated by PCR overlap extension of the V_H and C_H1 fragments with the V_H sense extension primer, which is specific for the pelB sequence tail that was introduced into the V_H fragments in the first-round PCR, and the C_H1 reverse primer, which is specific for the decapeptide sequence. The second-round PCR products have a size of 750–800 bp. To generate the final Fab product, the purified light-chain–pelB products and Fd products are amplified with a V_L sense extension primer and the reverse extension primer. The 1500-bp PCR product has asymmetric *Sfi*I sites on the 5′ and 3′ ends that are used for directional cloning into the pComb3 vectors.

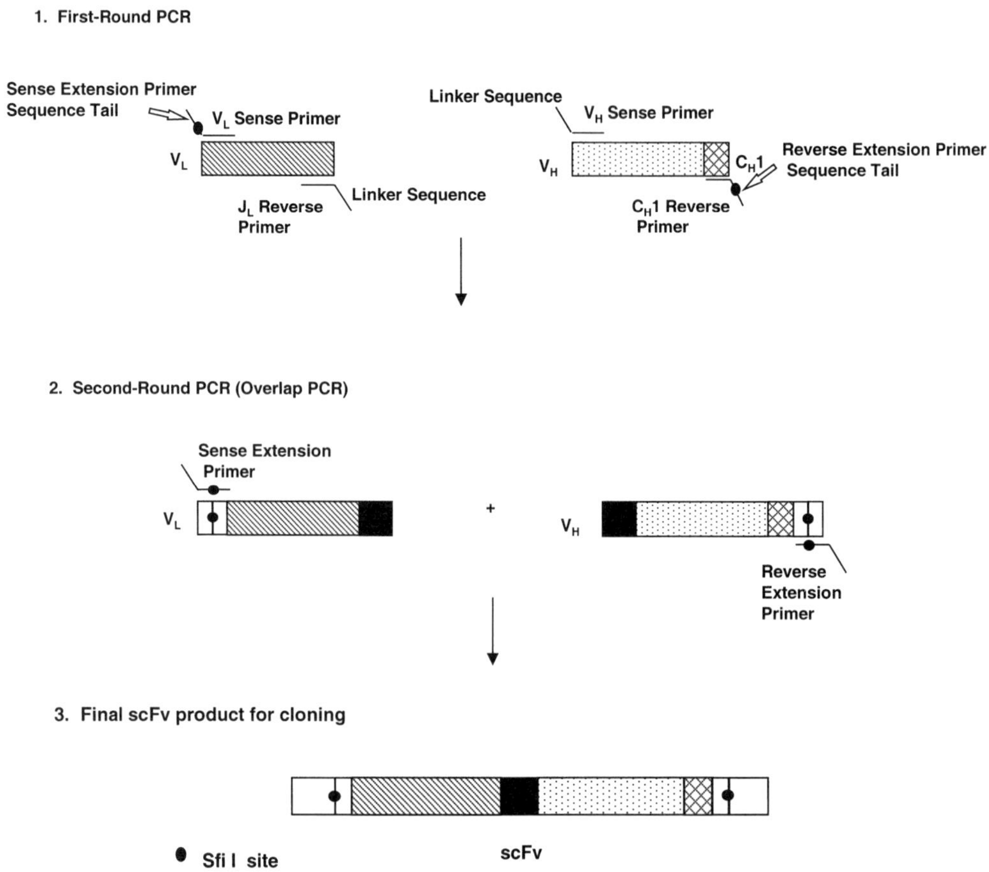

Figure 9.2. Generation of scFv fragments by PCR overlap extension for cloning into the pComb3 vector system. In the first round of PCR, the rearranged light- and heavy-chain variable regions are amplified by using V_L and V_H sense primers in conjunction with J_L and C_H1 reverse primers (or J_H reverse primers in chicken scFv libraries). The first-round products have a size of about 350–400 bp. The V_L sense primers include a 5′ sequence tail that contains an *Sfi*I site and is recognized by the sense extension primer. The C_H1 (or J_H) reverse primers introduce a 3′ sequence tail that contains an *Sfi*I site and is recognized by the reverse extension primer. The J_L reverse and V_H sense primers have overlapping sequence tails that code for the linker peptide in the final scFv PCR fragment. In the second-round PCR, the purified V_L and V_H products are fused by overlap extension PCR using the sense and reverse extension primers. The resulting product is approximately 750–800 bp in size and is referred to as the scFv PCR fragment. It has asymmetric *Sfi*I restriction sites on the 5′ and 3′ ends that are used for directional cloning into the pComb3 vectors.

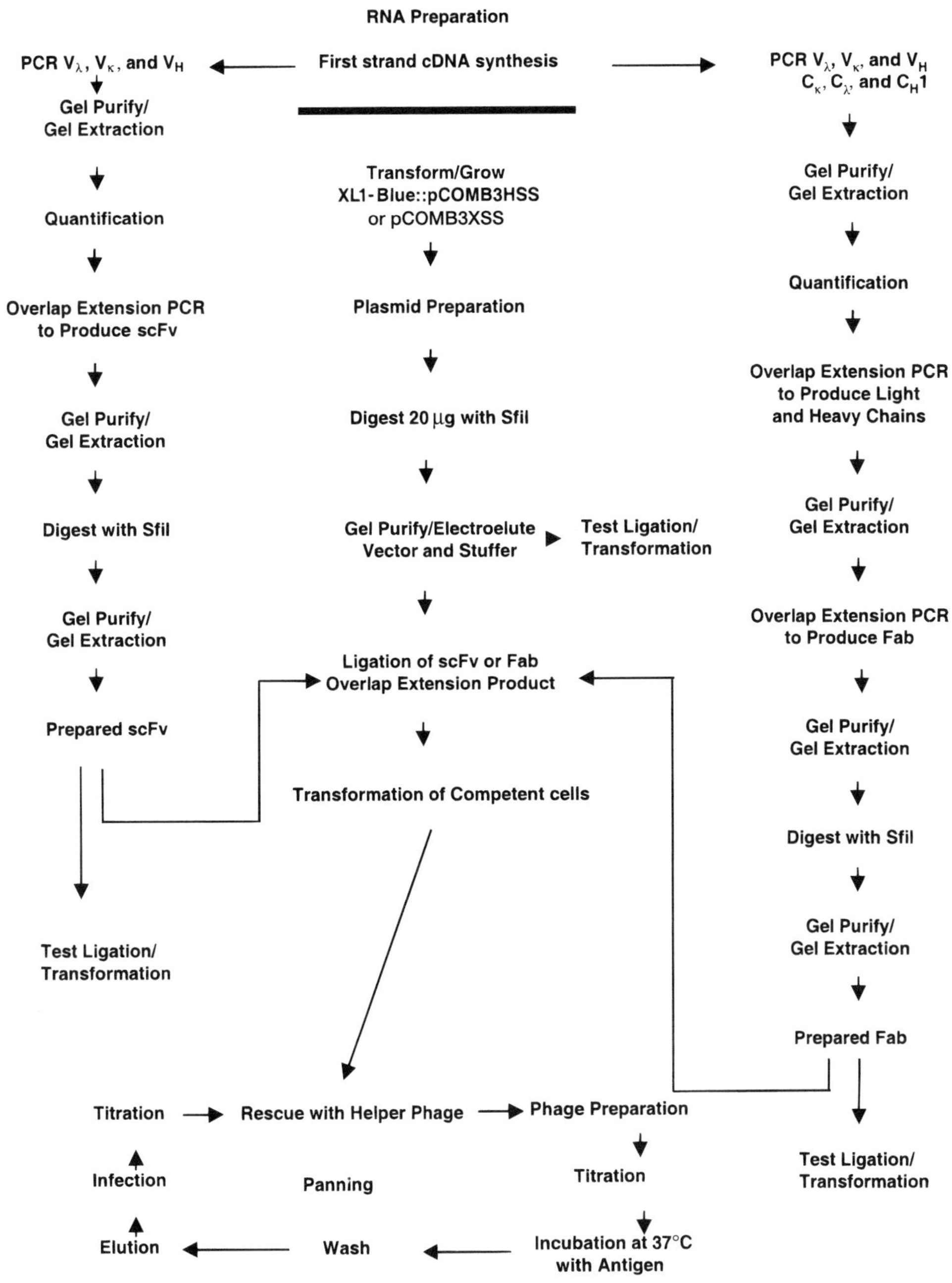

Figure 9.3. Flowchart depicting library construction and selection. Library construction begins with RNA preparation and cDNA synthesis, followed by preparation of PCR inserts for scFv (*left*) or Fab (*right*). During this preparation process, pComb3 vector is also prepared and tested for ligation efficiency. Prepared inserts are ligated into the prepared vector and transformed into competent bacterial cells. Phage are rescued by the addition of helper phage and the panning/selection process is started using the rescued phage. Panning is a cyclic procedure that is carried out in consecutive rounds over the course of several days.

PROTOCOL 9.1

Human Fab Libraries

The development of therapeutic antibodies for use in the treatment of human diseases has long been a goal for many researchers in the antibody field. Although monoclonal antibodies are easily obtained from species such as mouse or rabbit, the use of these antibodies in humans results in an immune response to the foreign immunoglobulin epitopes, severely limiting the long-term use of these reagents. It is, therefore, desirable to have human monoclonal antibodies. One way to obtain these antibodies is through phage-display libraries constructed from human lymphocytes. These immune libraries are generally constructed from the antibody genes of someone who either has been routinely immunized with an approved vaccine or has been naturally exposed to the antigen, e.g., through infection. There are numerous reports of successfully screened human libraries, some of which demonstrate that it is possible to obtain recombinant antibodies from combinatorial libraries that are representative of the natural immune response (Barbas et al. 1993; Roben et al. 1996).

To construct a representative antibody library, it is essential to possess some sequence information and understanding of the organization of human immunoglobulin genes. As with most higher vertebrates, human antibody variable regions are generated through somatic DNA recombination (Tonegawa 1983). In the heavy chain, the recombination occurs between three individual gene segments, V_H (variable), D (diversity), and J_H (joining), whereas the light chain has only two gene segments, V_L (variable) and J_L (joining). Essentially all of the heavy- and light-chain gene loci have been mapped and the gene segments have been sequenced (Anderson et al. 1984; Meindl et al. 1990; Matsuda et al. 1993; Williams and Winter 1993; Cook et al. 1994; Cox et al. 1994). In both the heavy and light chains the variable region sequences are grouped into gene families based on nucleotide sequence similarity. Each locus has a different number of families, which consist of a different number of gene segments. The similar sequence structure can be used to design oligonucleotide primer sequences to amplify essentially all members of a given family. In addition, both heavy- and light-chain loci have multiple J gene segments from which one or more oligonucleotide primers can be designed for PCR amplification. The use of all primer combinations theoretically permits the amplification of all rearranged variable regions. To amplify segments for expression of Fab, oligonucleotide primers can easily be generated that are specific for any or all constant region genes. This is particularly useful if one wishes to construct an antibody library from a particular lymphoid compartment that expresses a predominant immunoglobulin isotype.

This protocol (9.1) and the following protocol (9.2) describe the construction of human Fab and scFv libraries, respectively. There have been a number of reports of successful human library constructions with variations in the primer sequences (Burton et al. 1991; Kang et al. 1991; Persson et al. 1991). The primers listed in the following protocols are those currently being used with the present generation of pComb3 vectors (see Chapter 2 for details concerning the pComb3 vectors).

Materials

cDNA (see Protocol 8.4)
cloned human Fab template (pComb3XTT and pComb3Xλ; available from the Barbas laboratory)
oligonucleotide primers (listed below; see Appendix 1 for letter code)
2.5 mM dNTP mix (dATP/dCTP/dGTP/dTTP set, 100 mM, Amersham Pharmacia Biotech, Cat. # 27-2035-02)
AmpliTaq DNA polymerase (Perkin-Elmer, Cat. # N808-052)
10x PCR buffer for use with *Taq* polymerase (supplied with polymerase)
Expand High Fidelity Enzyme mix; 3.5 Units/μl (Roche Molecular Biochemicals, Cat. # 1 732 650)
10x Expand High Fidelity PCR buffer (supplied with Expand High Fidelity Enzyme mix)
MicroAmp PCR tubes (Perkin-Elmer, Cat. # N801-0533)
MicroAmp PCR caps (Perkin-Elmer, Cat. # N801-0534)
MicroAmp PCR trays (Perkin-Elmer, Cat. # N801-5530)
GeneAmp 9600 PCR cycler (Perkin-Elmer)
agarose (Gibco-BRL, Cat. # 15510-027)
100-bp DNA molecular weight marker (Amersham Pharmacia Biotech, Cat. #27-4001-01) or 1-kb DNA molecular weight marker (Gibco-BRL, Cat. # 15615-024)
1x TAE electrophoresis running buffer (see Appendix 2)
DNA gel loading dye (see Appendix 2)
pComb3HSS or pComb3XSS (available from the Barbas laboratory)
*Sfi*I restriction enzyme (Roche Molecular Biochemicals, 40 units/μl, Cat. # 1 288 059)
10x restriction enzyme buffer M (supplied with *Sfi*I restriction enzyme)
T4 DNA ligase (Gibco-BRL, 1 unit/μl, Cat. # 15224 090)
5x T4 DNA ligase buffer (supplied with enzyme)
glycogen, 20 mg/ml (Roche Molecular Biochemicals, Cat. # 901 393)
electroporation apparatus and materials
electrocompetent ***E. coli*** cells (see Chapter 10): XL1-Blue (Stratagene, Cat. # 200228) or ER2537 (New England Biolabs, Cat. # 801-N)
LB + 100 μg/ml carbenicillin plates (see Appendix 2)

CAUTION: ***E. coli* (see Appendix 4)**

HUMAN Fab PRIMERS

V_H5′ Sense Primers

HFabVH1-F
5′ GCT GCC CAA CCA GCC ATG GCC CAG GTG CAG CTG GTG CAG TCT GG 3′

HFabVH2-F
5′ GCT GCC CAA CCA GCC ATG GCC CAG ATC ACC TTG AAG GAG TCT GG 3′

HFabVH35-F
5′ GCT GCC CAA CCA GCC ATG GCC GAG GTG CAG CTG GTG SAG TCT GG 3′

HFabVH3a-F
5′ GCT GCC CAA CCA GCC ATG GCC GAG GTG CAG CTG KTG GAG TCT G 3′

HFabVH4-F
5′ GCT GCC CAA CCA GCC ATG GCC CAG GTG CAG CTG CAG GAG TCG GG 3′

HFabVH4a-F
5′ GCT GCC CAA CCA GCC ATG GCC CAG GTG CAG CTA CAG CAG TGG GG 3′

V_H3′ Reverse Primers

HFabVHJa-B
5′ CGA TGG GCC CTT GGT GGA GGC TGA GGA GAC GGT GAC CAG GGT TCC 3′

HFabVHJb-B
5′ CGA TGG GCC CTT GGT GGA GGC WGR GGA GAC GGT GAC CAG GGT BCC 3′

V_κ 5′ Sense Primers

HSCK1-F
5′ GGG CCC AGG CGG CCG AGC TCC AGA TGA CCC AGT CTC C 3′

HSCK24-F
5′ GGG CCC AGG CGG CCG AGC TCG TGA TGA CYC AGT CTC C 3′

HSCK3-F
5′ GGG CCC AGG CGG CCG AGC TCG TGW TGA CRC AGT CTC C 3′

HSCK5-F
5′ GGG CCC AGG CGG CCG AGC TCA CAC TCA CGC AGT CTC C 3′

V_κ 3′ Reverse Primer

HCK5-B
5′ GAA GAC AGA TGG TGC AGC CAC AGT 3′

V_λ 5′ Sense Primers

HSCLam1a
5′ GGG CCC AGG CGG CCG AGC TCG TGB TGA CGC AGC CGC CCT C 3′

HSCLam1b
5′ GGG CCC AGG CGG CCG AGC TCG TGC TGA CTC AGC CAC CCT C 3′

HSCLam2
5′ GGG CCC AGG CGG CCG AGC TCG CCC TGA CTC AGC CTC CCT CCG T 3′

HSCLam3
5′ GGG CCC AGG CGG CCG AGC TCG AGC TGA CTC AGC CAC CCT CAG TGT C 3′

HSCLam4
5′ GGG CCC AGG CGG CCG AGC TCG TGC TGA CTC AAT CGC CCT C 3′

HSCLam6
5′ GGG CCC AGG CGG CCG AGC TCA TGC TGA CTC AGC CCC ACT C 3′

HSCLam78
5′ GGG CCC AGG CGG CCG AGC TCG TGG TGA CYC AGG AGC CMT C 3′

HSCLam9
5′ GGG CCC AGG CGG CCG AGC TCG TGC TGA CTC AGC CAC CTT C 3′

HSCLam10
5′ GGG CCC AGG CGG CCG AGC TCG GGC AGA CTC AGC AGC TCT C 3′

V_λ 3′ Reverse Primer

HCL5-B
5′ CGA GGG GGC AGC CTT GGG CTG ACC 3′

C_H1 Primers

HIgGCH1-F (sense)
5′ GCC TCC ACC AAG GGC CCA TCG GTC 3′
dpseq (reverse)
5′ AGA AGC GTA GTC CGG AAC GTC 3′

C_κ Primers

HKC-F (sense)
5′ CGA ACT GTG GCT GCA CCA TCT GTC 3′

Lead-B (reverse)
5′ GGC CAT GGC TGG TTG GGC AGC 3′

C_λ Primers

HLC-F (sense)
5′ GGT CAG CCC AAG GCT GCC CCC 3′

Lead-B (reverse)
(see above)

Primers for Heavy-chain Fd Overlap Assembly

LeadVH (sense)
5′ GCT GCC CAA CCA GCC ATG GCC 3′

dpseq (reverse)
(see above)

Primers for Light-chain Overlap Assembly

RSC-F (sense)
5′ GAG GAG GAG GAG GAG GAG GCG GGG CCC AGG CGG CCG AGC TC 3′

Lead B (reverse)
(see above)

Primers for PCR Assembly of Light- and Heavy- chain (Fd) Sequences

RSC-F (sense)
(see above)

dp-EX (reverse)
5′ GAG GAG GAG GAG GAG GAG AGA AGC GTA GTC CGG AAC GTC 3′

Procedure

1. **First round of PCR.**

   Note: PCR products and other DNA samples, precipitated or in solution, can be stored for years at –20°C. The construction of antibody fragment libraries, therefore, can be interrupted at any step.

   To amplify the V gene rearrangements, perform 12 V_H amplifications, four V_κ amplifications, and nine V_λ amplifications (see Figs. 9.4–9.6).

For the V_H, V_κ, and V_λ amplifications, set up one tube for each primer combination, using tissue cDNA as template. Each reaction should contain the following:

1 µl	cDNA (0.5 µg)
60 pmole	5′ primer (primer combinations listed below)
60 pmole	3′ primer (primer combinations listed below)
10 µl	10x PCR buffer
8 µl	2.5 mM dNTPs
0.5 µl	*Taq* DNA polymerase

Add water to a final volume of 100 µl.

V_H primer combinations:

HFabVH1-F	HFabVH2-F
HFabVHJa-B	HFabVHJa-B
HFabVH35-F	HFabVH3a-F
HFabVHJa-B	HFabVHJa-B
HFabVH4-F	HFabVH4a-F
HFabVHJa-B	HFabVHJa-B
HFabVH1-F	HFabVH2-F
HFabVHJb-B	HFabVHJb-B
HFabVH35-F	HFabVH3a-F
HFabVHJb-B	HFabVHJb-B
HFabVH4-F	HFabVH4a-F
HFabVHJb-B	HFabVHJb-B

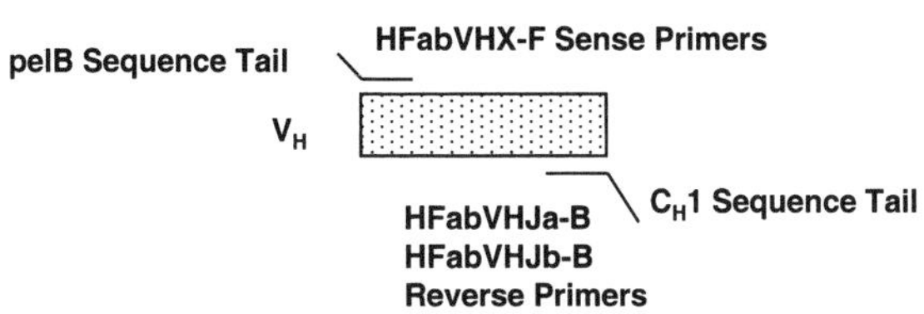

Figure 9.4. The first-round PCR amplification of V_H product. Each of the HFabVH-F sense primers is paired with each of the reverse primers, HFabVHJa-B and HFabVHJb-B, to amplify V_H gene segments from cDNA. The sense primers have a sequence tail that corresponds to the 3′ end of the pelB leader sequence and allows the overlap extension assembly of the heavy-chain (Fd) fragment with the light-chain pelB fragment in the final round of PCR. The reverse primers have a sequence tail that is specific for C_H1 that is used in the first overlap extension PCR to create the Fd fragment.

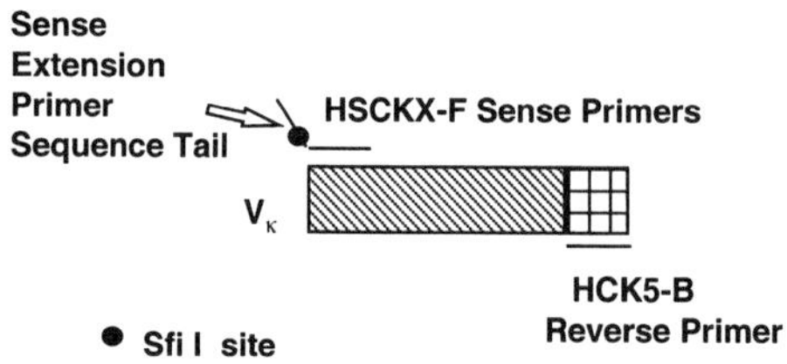

Figure 9.5. The first-round PCR amplification of V_κ products. Each of the HSCK-F sense primers is paired with the reverse primer HCK5-B to amplify V_κ gene segments from cDNA. The sense primers have a 5′ sequence tail that contains an *Sfi*I site and is recognized by the sense extension primer used in the second-round PCR. The reverse primer is specific for the 5′ end of the κ light-chain constant region.

V_κ primer combinations:

HSCK1-F
HCK5-B

HSCK24-F
HCK5-B

HSCK3-F
HCK5-B

HSCK5-F
HCK5-B

V_λ primer combinations:

HSCLam1a
HCL5-B

HSCLam1b
HCL5-B

HSCLam2
HCL5-B

HSCLam3
HCL5-B

HSCLam4
HCL5-B

HSCLam6
HCL5-B

HSCLam78
HCL5-B

HSCLam9
HCL5-B

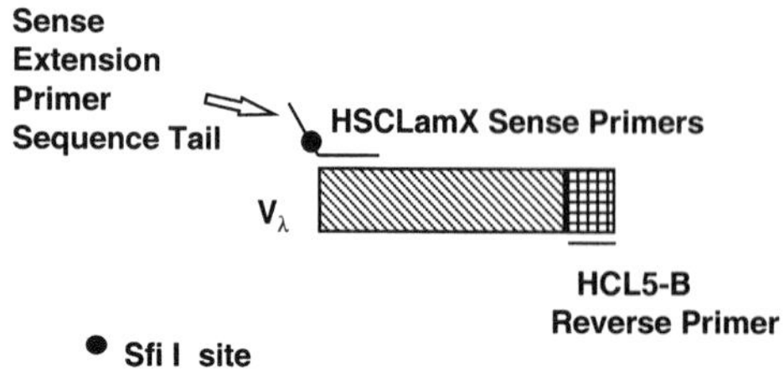

Figure 9.6. The first-round PCR amplification of V_λ products. Each of the HSCLam sense primers is paired with the reverse primer HCL5-B to amplify V_λ gene segments from cDNA. The sense primers have a 5′ sequence tail that contains an *Sfi*I site and is recognized by the sense extension primer used in the second-round PCR. The reverse primer is specific for the 5′ end of the constant region of the λ light chain.

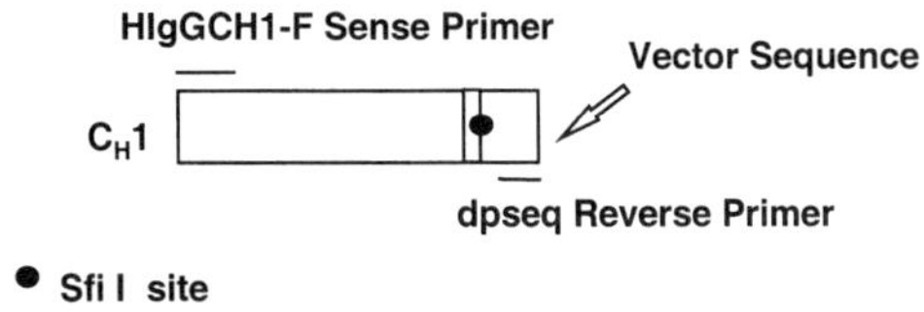

Figure 9.7. The first-round PCR amplification of C_H1 products from a cloned Fab in the vector pComb3XTT. The sense primer is specific for the 5′ region of C_H1, the region that is used in the overlap extension PCR to create the heavy-chain (Fd) fragment. The reverse primer is specific for the decapeptide sequence located downstream of the C_H1 fragment and the 3′ *SfiI* site of pComb3XTT.

HSCLam10

HCL5-B

Perform the PCR for first-round V_H, V_κ, and V_λ combinations under the following conditions:

94°C for 5 minutes

30 cycles of:

94°C for 15 seconds

56°C for 15 seconds

72°C for 90 seconds

followed by 72°C for 10 minutes

Note: Try a test amplification with the cDNA sample. If good amplification is not obtained (if a strong amplified band is not clearly visible in the agarose gel), lithium chloride precipitation of the RNA might improve the results (see Protocol 8.3).

For the amplification of human C_H1, C_κ, and C_λ fragments, a cloned human Fab template is used with the following primer combinations (see Figs. 9.7–9.9). Set up four tubes for each primer combination. Each reaction should contain the following:

20 ng	template (see template–primer combinations below)
60 pmole	5′ primer
60 pmole	3′ primer
10 µl	10x PCR buffer
8 µl	2.5 mM dNTPs
0.5 µl	*Taq* DNA polymerase

Add water to final volume of 100 µl

C_H1 template–primer combination:

pComb3XTT

HIgGCH1-F

dpseq

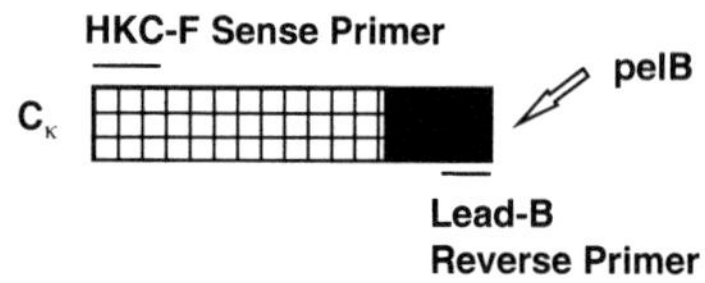

Figure 9.8. The first-round PCR amplification of C_κ products and the pelB leader sequence from the vector pComb3XTT. The sense primer is specific for the 5′ region of C_κ, the region used in the overlap extension PCR to create the kappa light chain. The reverse primer is specific for the 3′ end of the pelB leader sequence.

C_κ-pelB template–primer combination:

pComb3XTT

HKC-F

Lead-B

C_λ-pelB template–primer combination:

pComb3Xλ

HLC-F

Lead-B

Perform the PCR for first-round C_H1, C_κ-pelB, and C_λ-pelB combinations under the following conditions:

94°C for 5 minutes

20 cycles of:

94°C for 15 seconds

56°C for 15 seconds

72°C for 90 seconds

followed by 72°C for 10 minutes

Note: A "hot start" PCR protocol can improve specificity, sensitivity, and yield. In hot start PCR, either an essential reaction component is not added until the first denaturing step, or a reversible inhibitor of the polymerase is used. This protocol prevents low-stringency primer extension, which can generate nonspecific products.

Evaluate 5–10 μl of each PCR on a 2% agarose gel using DNA gel loading dye and an appropriate molecular weight marker (MWM). From these amplifications expect a 350–400 bp product for each type of reaction.

2. **Isolate the PCR products.**

Pool the products of each type of reaction (for example, pool all of the V_H reactions in one pool, all of the V_κ reactions in another pool, etc.), ethanol-precipitate, and wash as described in Appendix 3. Run the products on a 2% agarose gel, cut out the correct-sized bands, and purify the DNA by freeze-squeeze, electroelution with an Elutrap (see Appendix 3), or resin binding (e.g., QIAEX II Gel Extraction Kit, QIAGEN). Quantitate yields by reading the optical density (O.D.) at 260 nm (1 O.D. unit = 50 μg/ml). If yields are too low, repeat the first round of

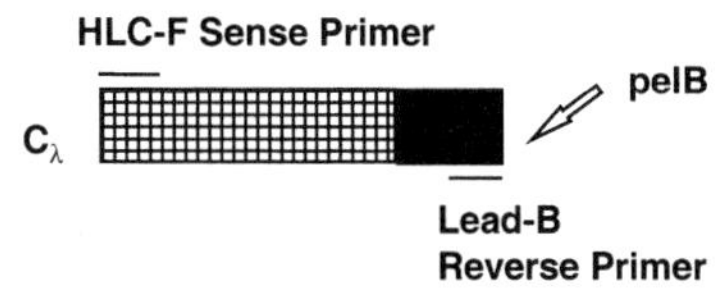

Figure 9.9. The first-round PCR amplification of C_λ products and the pelB leader sequence from the vector pComb3Xλ. The sense primer is specific for the 5′ region of C_λ, the region used in the overlap extension PCR to create the lambda light chain. The reverse primer is specific for the 3′ end of the pelB leader sequence.

PCR and combine the end products. Approximately 2–4 μg of each pool is required to proceed.

3. **Second round of PCR (overlap extension).**

In the second round of PCR, the appropriate first-round products are mixed in equal ratios to generate the heavy-chain Fd, κ light-chain, and λ light-chain overlap products (see Figs. 9.10–9.12). The primers in the first round of PCR create identical sequences in the upstream region of the light-chain constant regions and the downstream regions of the light-chain variable region. These identical sequences serve as the overlap for the second-round extension of the full-length κ or λ light-chain products. Similarly, the primers in the first round of PCR create identical sequences in the upstream region of the heavy-chain constant region and the downstream region of the heavy-chain variable region that serve as the overlap for the second-round extension of the full-length heavy-chain Fd product.

Assemble ten 100-μl reactions for each overlap. Each reaction consists of the following:

100 ng	purified V product (see template–primer combinations below)
100 ng	purified C product
60 pmole	5′ primer
60 pmole	3′ primer
10 μl	10x PCR buffer
8 μl	2.5 mM dNTPs
0.5 μl	*Taq* DNA polymerase

Add water to a final volume of 100 μl.

Heavy-chain Fd template–primer combination:

100 ng	heavy-chain variable region product
100 ng	heavy-chain C_H1 product
60 pmole	LeadVH primer
60 pmole	dpseq primer

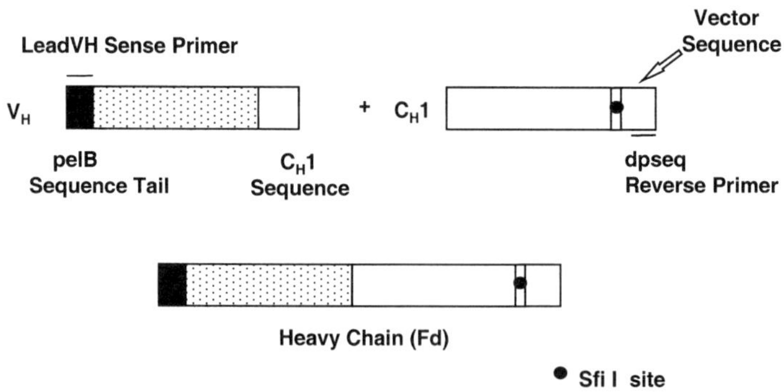

Figure 9.10. The overlap extension PCR amplification of the heavy-chain (Fd) fragment. Equimolar quantities of the V_H and C_H1 PCR products are used in the overlap PCR to create the Fd fragment. The sense primer recognizes the sequence tail corresponding to the 3′ end of the pelB leader sequence that was introduced through the V_H primers in the first-round amplification of the V_H sequences. The same reverse primer that was used for the generation of the C_H1 fragment is used in the overlap extension PCR of the heavy-chain (Fd) fragment.

κ light-chain template–primer combination:

100 ng	κ variable region product
100 ng	C_κ-pelB product
60 pmole	RSC-F primer
60 pmole	Lead-B primer

λ light-chain template–primer combination:

100 ng	λ variable region product
100 ng	C_λ-pelB product
60 pmole	RSC-F primer
60 pmole	Lead-B primer

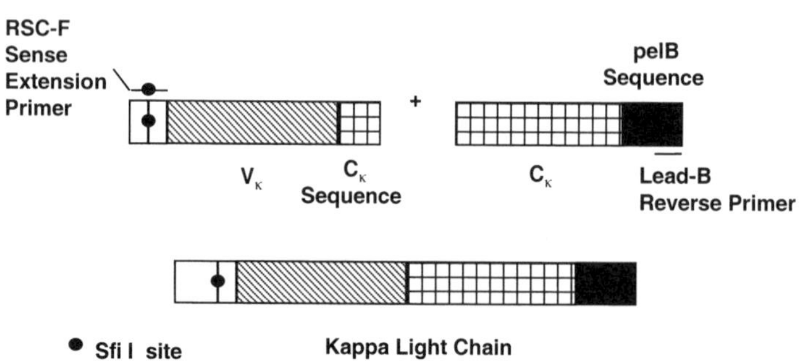

Figure 9.11. The overlap extension PCR for amplification of κ light chain and the pelB leader sequence. Equimolar quantities of the V_κ and C_κ–pelB PCR products are used in the overlap PCR to create the kappa light-chain–pelB fragment. The sense extension primer recognizes the sequence tail created by the V_κ primers in the first-round amplification of the V_κ sequences. The same reverse primer that was used for the amplification of the C_κ–pelB fragment in the first-round PCR is used in the overlap extension PCR.

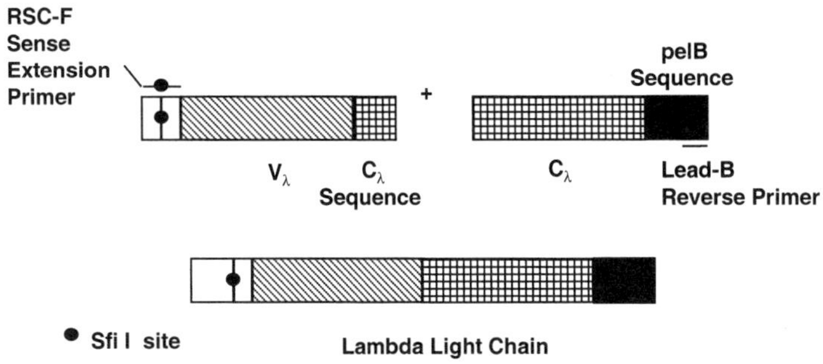

Figure 9.12. The overlap extension PCR for amplification of the λ light chain and the pelB leader sequence. Equimolar quantities of V_λ and C_λ–pelB PCR products are used in the overlap PCR to create the λ light-chain–pelB fragment. The sense extension primer recognizes the sequence tail created by the V_λ primers in the first-round amplification of the V_λ sequences. The same reverse primer that was used for the amplification of the C_λ–pelB fragment in the first-round PCR is used in the overlap extension PCR.

Perform the second-round PCR under the following conditions:

94°C for 5 minutes

15 cycles of:

  94°C for 15 seconds

  56°C for 15 seconds

  72°C for 2 minutes

followed by 72°C for 10 minutes

Evaluate 5–10 µl of each reaction on a 2% agarose gel using DNA gel loading dye and an appropriate MWM. For each type of reaction expect a ~750- to 800-bp product.

> *Note:* The long PCR extension time favors full-length product. If a strong band of the correct size is not clearly visible in the gel, the number of PCR cycles can be increased, or more overlap PCRs can be performed. To retain diversity of the library, it is better to increase the number of PCRs.

4. **Isolate the PCR products.**

Pool the products of each type of reaction, ethanol-precipitate, and wash as described in Appendix 3. Purify the DNA on a 2% agarose gel as in Step 2. Quantitate yields by reading the O.D. at 260 nm. If yields are too low, repeat the overlap PCR and combine the end products. Approximately 2–4 µg of each pool is needed to proceed.

5. **Third round of PCR (second overlap extension).**

In the third round of PCR, the appropriate second-round products are mixed in equal ratios to generate the overlap product. The primers used in the second-round PCR create identical sequences in the downstream region of the light chains and the upstream region of the heavy-chain Fd that serve as the overlap for the third-round extension of the full-length Fab products (see Fig. 9.13).

Note: This protocol describes the construction of κ and λ human Fab libraries as two separate entities. Depending on the needs of the user, the κ and λ light chains can be combined before the third round of PCR such that only one final overlap extension PCR is needed (see Protocol 9.4). Alternatively, the κ and λ Fab products can be combined at a later step, before or after *Sfi*I digestion.

Assemble ten 100-μl reactions for each overlap. Each reaction should contain the following:

100 ng	purified heavy-chain product (see template–primer combinations below)
100 ng	purified light-chain product
60 pmole	RSC-F sense primer
60 pmole	dp-EX reverse primer
10 μl	10x PCR buffer
8 μl	2.5 mM dNTPs
0.75 μl	Expand HF Polymerase mix

Add water to a final volume of 100 μl.

Heavy-chain/κ light-chain Fab template combination:

100 ng	heavy-chain Fd product
100 ng	κ light-chain product

Heavy-chain/λ light-chain Fab template combination:

100 ng	heavy-chain Fd product
100 ng	λ light-chain product

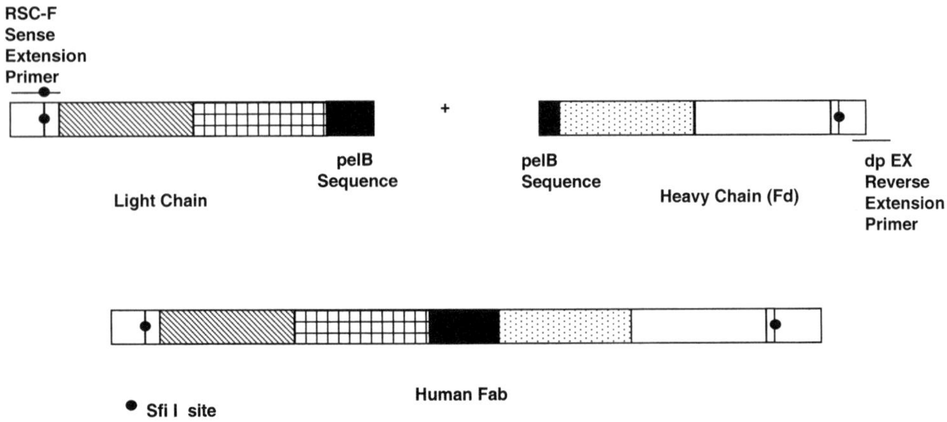

Figure 9.13. The final overlap amplification of the human Fab PCR fragment. Equimolar quantities of the heavy chain (Fd) products and light-chain–pelB products are used in the final overlap extension PCR to create the full-length Fab PCR fragment. The sense extension primer was used in the overlap extension amplification of the light-chain–pelB fragment. The reverse extension primer recognizes the decapeptide sequence downstream of the C_H1 region.

Perform the third-round PCR under the following conditions:

10 cycles of:

- 94°C for 15 seconds
- 56°C for 30 seconds
- 72°C for 3 minutes

followed by 72°C for 10 minutes

Evaluate 5–10 μl of each reaction on a 2% agarose gel using DNA gel loading dye and an appropriate MWM. For each type of reaction expect a ~1500-bp product.

Note: Efficiency in generating the 1.5-kb overlap PCR product is often improved by using a long extension time and the Expand High Fidelity PCR System. This system has a unique enzyme mix that contains thermostable *Taq* and *Pwo* DNA polymerases. It is designed to give high yield, high fidelity, and high specificity PCR products.

The overlap products can also be generated with *AmpliTaq* polymerase. To amplify with *AmpliTaq* polymerase, use 0.5 μl of enzyme in the 100-μl amplification reaction. Perform the PCR under the following conditions:

94°C for 5 minutes

25 cycles of:

- 94°C for 15 seconds
- 56°C for 15 seconds
- 72°C for 3 minutes

followed by 72°C for 10 minutes

6. Isolate the PCR products.

Pool the PCR products, ethanol-precipitate, and wash as described in Appendix 3. Purify the DNA on a 2% agarose gel as in Step 2. Quantitate yields by reading the O.D. at 260 nm. If yields are too low, repeat the second overlap PCR and combine the end products. 10–15 μg should be sufficient to continue.

7. Restriction-digest the overlap Fab PCR product and the pComb3HSS or pComb3XSS vector.

Prepare the combined PCR products and the vector for cloning by performing restriction digests with *Sfi*I. We prefer the high concentration (40 units/μl) *Sfi*I and the 10x buffer M from Roche Molecular Biochemicals.

The digest of the PCR products should contain:

- 10 μg of purified Fab product
- 160 Units of *Sfi*I (16 Units per μg of DNA)
- 20 μl of 10x buffer M
- Add water to a volume of 200 μl.

The digest of the vector should contain:

20 μg of pCombHSS or pComb3XSS (contains stuffer fragment between the two *Sfi*I cloning sites)

120 Units of *Sfi*I (6 Units per μg of DNA)

20 μl of 10x buffer M

Add water to a volume of 200 μl.

Note: If the loss of digested PCR product in the final purification step is high or if several large-size ligations are needed to obtain a library of the desired size, digest more PCR product with *Sfi*I. The amount of added enzyme should not exceed 10% of the reaction volume. If desired, larger quantities of *Sfi*I-cut and purified vector DNA can be prepared and tested in advance.

Incubate both digests for 5 hours at 50°C. Ethanol-precipitate and wash as described in Appendix 3. Purify the digested Fab product on a 2% agarose gel, and purify the vector and stuffer fragment on a 1% agarose gel. We recommend electroelution with an Elutrap for extraction of the vector from the gel (see Appendix 3). Suboptimal digestion or purification will yield vector with low transformation efficiencies and high background after ligation. Use the stuffer fragment in a test ligation to determine the quality of the vector prior to use in library ligations (see Step 8). The stuffer fragment is slightly larger than an Fab fragment, approximately 1600 bp. Quantify the purified, digested PCR products, vector, and stuffer fragment by measuring the O.D. at 260 nm.

Note: In some cases *Sfi*I-digested DNA is partly insoluble after ethanol precipitation. To avoid this problem, the DNA can be loaded directly after digestion onto the agarose gel. When purifying the *Sfi*I-digested vector DNA, let the DNA run long enough to separate linearized vector DNA (size ~ 5000 bp; cut only once) and uncut vector DNA from the desired double-cut product (size ~ 3400 bp).

8. **Ligate the digested overlap PCR product with the vector DNA.**

Perform small-scale ligations to assess the suitability of the vector and inserts for high-efficiency ligation and transformation. The ligation efficiency of the vector DNA can be tested by ligating it with the gel-purified stuffer fragment that is generated during the *Sfi*I digestion of the vector DNA. The vector preparation should contain little uncut vector DNA or DNA that is cut only once. The amount of uncut or partly cut vector DNA can be estimated by setting up a ligation reaction that contains only vector DNA. These two types of control ligations should be done in parallel with a small-scale test ligation and should both include the same amount of vector DNA.

Test Ligations

Assemble the following small-scale ligation reactions:

small-scale test ligation (one reaction for each PCR insert):

140 ng	pComb3HSS or pComb3XSS, *Sfi*I-digested and purified

140 ng	Fab overlap PCR product, *Sfi*I-digested and purified
4 μl	5x ligase buffer
1 μl	ligase

Add water to a volume of 20 μl.

control ligation 1 (control insert):

140 ng	pComb3HSS or pComb3XSS, *Sfi*I-digested and purified
140 ng	stuffer fragment, *Sfi*I-digested and purified
4 μl	5x ligase buffer
1 μl	ligase

Add water to a volume of 20 μl.

control ligation 2 (test for vector self-ligation):

140 ng	pComb3HSS or pComb3XSS, *Sfi*I-digested and purified
4 μl	5x ligase buffer
1 μl	ligase

Add water to a volume of 20 μl.

Incubate between 4 hours and overnight at room temperature. Transform 0.5–1 μl of each reaction by electroporation into 50 μl of XL1-Blue or ER2537 (see Support Protocol). Dilute the transformed cultures 10-fold and 100-fold with prewarmed (37°C) SOC, and plate 100 μl of each dilution on LB+carb plates. Incubate the plates overnight at 37°C. Count the colonies on the vector + Fab insert plates and calculate the number of transformants per μg of vector DNA. If this number does not exceed 1×10^7, do not proceed with the large-scale library ligation. Ideally, the final library size should be several times 10^8 but at least 5×10^7 total transformants. Determine the number of scaled-up library ligations needed to achieve this size. Count the colonies obtained from control ligation 1 for an indication of vector quality and ligation efficiency. A good vector DNA preparation should yield at least 10^8 colony-forming units (cfu) per μg of vector DNA and should have less than 10% (ideally less than 5%) background ligation (calculated cfu per μg of vector DNA in control ligation 2). If the background is greater than 10% or the ligation efficiency is low, digest new vector and repeat the ligations. If the ligations are reasonably good, but there is not enough vector or insert to make a library of the desired size, perform additional digests.

Library Ligation

Assemble enough reactions to produce at least 5×10^7 transformants. Combine the following:

1.4 μg	pComb3HSS or pComb3XSS, *Sfi*I-digested and purified

1.4 μg	Fab overlap product, *Sfi*I digested and purified
40 μl	5x ligase buffer
10 μl	ligase

Add water to a volume of 200 μl.

Incubate overnight at room temperature. Ethanol-precipitate and wash the pellet as described in Appendix 3. The ethanol-precipitated ligation reaction can be stored for months at –20°C. Transform the ligation reaction by resuspending in 15 μl of water and electroporating into 300 μl of XL1-Blue or ER2537 electrocompetent cells, as described in Chapter 10.

Proceed with the preparation and panning of the primary library (see Chapter 10).

PROTOCOL 9.2

Human scFv Libraries

The following is a protocol for the generation of human scFv libraries with a short linker (GGSSRSS) or a long linker (GGSSRSSSSGGGGSGGGG). Linkers of different compositions may result in different levels of oligomerization and may also alter the interface of the V_H and V_L regions. Single-chain fragments in which the light- and heavy-chain variable regions are connected with a short peptide linker tend to form dimers, called bivalent diabodies, whereas scFvs with long linkers tend to be monomers (Holliger et al. 1993; McGuinness et al. 1996; Zhu et al. 1996; see also Chapter 3). Bivalent diabodies can have the advantage of binding with higher avidity to their antigen, but the use of a short linker can also lead to selection of unwanted low-affinity binders. Alternative linkers that enhance scFv phage-binding activity have also been reported (Tang 1996; Turner et al. 1997). The primers listed in this protocol are those currently being used with the pComb3 vectors. For a brief introduction to the use and construction of human immune libraries, see Protocol 9.1. For details concerning the pComb3 vectors, see Chapter 2.

Materials

cDNA (see Protocol 8.4)
oligonucleotide primers (listed below; see Appendix 1 for letter code)
2.5 mM dNTP mix (dATP/dCTP/dGTP/dTTP set, 100 mM, Amersham Pharmacia Biotech, Cat. # 27-2035-02)
AmpliTaq DNA polymerase (Perkin-Elmer, Cat. # N808-0152)
10x PCR buffer (Perkin-Elmer, supplied with *Taq* DNA polymerase)
MicroAmp PCR tubes (Perkin-Elmer, Cat. # N801-0533)
MicroAmp PCR caps (Perkin-Elmer, Cat. # N801-0534)
MicroAmp PCR trays (Perkin-Elmer, Cat. # N801-5530)
GeneAmp 9600 PCR cycler (Perkin-Elmer)
agarose (Gibco-BRL, Cat. # 15510-027)
100-bp DNA molecular weight marker (Amersham Pharmacia Biotech, Cat. # 27-4001-01) or 1-kb DNA molecular weight marker (Gibco-BRL, Cat. # 15615-024)
1x TAE electrophoresis running buffer (see Appendix 2)
DNA gel loading dye (see Appendix 2)
pComb3HSS or pComb3XSS (available from the Barbas laboratory)
*Sfi*I restriction enzyme (Roche Molecular Biochemicals, 40 Units/μl, Cat. # 1 288 059)
10x restriction enzyme buffer M (supplied with *Sfi*I restriction enzyme)
T4 DNA ligase (Gibco-BRL, 1 Unit/μl, Cat. # 15224 090)
5x T4 DNA ligase buffer (supplied with enzyme)
electroporation apparatus and materials
electrocompetent ***E. coli*** cells (see Chapter 10): XL1-Blue (Stratagene, Cat # 200228) or ER2537 (New England Biolabs, Cat. # 801-N)
LB + 100 μg/ml carbenicillin plates (see Appendix 2)

CAUTION: *E. coli (see Appendix 4)*

HUMAN scFv PRIMERS, SHORT AND LONG LINKER

V_H Primers, 5′ Sense, Short Linker

HSCVH1-F
5′ GGT GGT TCC TCT AGA TCT TCC CAG GTG CAG CTG GTG CAG TCT GG 3′

HSCVH2-F
5′ GGT GGT TCC TCT AGA TCT TCC CAG ATC ACC TTG AAG GAG TCT GG 3′

HSCVH35-F
5′ GGT GGT TCC TCT AGA TCT TCC GAG GTG CAG CTG GTG SAG TCT GG 3′

HSCVH3a-F
5′ GGT GGT TCC TCT AGA TCT TCC GAG GTG CAG CTG KTG GAG TCT G 3′

HSCVH4-F
5′ GGT GGT TCC TCT AGA TCT TCC CAG GTG CAG CTG CAG GAG TCG GG 3′

HSCVH4a-F
5′ GGT GGT TCC TCT AGA TCT TCC CAG GTG CAG CTA CAG CAG TGG GG 3′

V_H Primers, 5′ Sense, Long Linker

HSCVH1-FL
5′ GGT GGT TCC TCT AGA TCT TCC TCC TCT GGT GGC GGT GGC TCG GGC GGT GGT GGG CAG GTG CAG CTG GTG CAG TCT GG 3′

HSCVH2-FL
5′ GGT GGT TCC TCT AGA TCT TCC TCC TCT GGT GGC GGT GGC TCG GGC GGT GGT GGG CAG ATC ACC TTG AAG GAG TCT GG 3′

HSCVH35-FL
5′ GGT GGT TCC TCT AGA TCT TCC TCC TCT GGT GGC GGT GGC TCG GGC GGT GGT GGG GAG GTG CAG CTG GTG SAG TCT GG 3′

HSCVH3a-FL
5′ GGT GGT TCC TCT AGA TCT TCC TCC TCT GGT GGC GGT GGC TCG GGC GGT GGT GGG GAG GTG CAG CTG KTG GAG TCT G 3′

HSCVH4-FL
5′ GGT GGT TCC TCT AGA TCT TCC TCC TCT GGT GGC GGT GGC TCG GGC GGT GGT GGG CAG GTG CAG CTG CAG GAG TCG GG 3′

HSCVH4a-FL
5′ GGT GGT TCC TCT AGA TCT TCC TCC TCT GGT GGC GGT GGC TCG GGC GGT GGT GGG CAG GTG CAG CTA CAG CAG TGG GG 3′

V_H Primers, 3′ Reverse, Short and Long Linker

HSCG1234-B (corresponding to human IgG isotypes 1–4)
5′ CCT GGC CGG CCT GGC CAC TAG TGA CCG ATG GGC CCT TGG TGG ARG C 3′

HSCM-B (corresponding to C_H1 domain of human IgM, see Note, Step 1)
5′ CCT GGC CGG CCT GGC CAC TAG TAA GGG TTG GGG CGG ATG CAC TCC C 3′

HSCA-B (corresponding to C_H1 domain of human IgA, see Note, Step 1)
5′ CCT GGC CGG CCT GGC CAC TAG TGA CCT TGG GGC TGG TCG GGG ATG C 3′

HSCD-B (corresponding to C_H1 domain of human IgD, see Note, Step 1)
5′ CCT GGC CGG CCT GGC CAC TAG TCA CAT CCG GAG CCT TGG TGG GTG C 3′

HSCE-B (corresponding to C_H1 domain of human IgE, see Note, Step 1)
5′ CCT GGC CGG CCT GGC CAC TAG TGA CGG ATG GGC TCT GTG TGG AGG C 3′

V_κ Primers, 5′ Sense, Short and Long Linker

HSCK1-F
5′ GGG CCC AGG CGG CCG AGC TCC AGA TGA CCC AGT CTC C 3′

HSCK24-F
5′ GGG CCC AGG CGG CCG AGC TCG TGA TGA CYC AGT CTC C 3′

HSCK3-F
5′ GGG CCC AGG CGG CCG AGC TCG TGW TGA CRC AGT CTC C 3′

HSCK5-F
5′ GGG CCC AGG CGG CCG AGC TCA CAC TCA CGC AGT CTC C 3′

V_κ Primers, 3′ Reverse, Short and Long Linker

HSCJK14o-B
5′ GGA AGA TCT AGA GGA ACC ACC TTT GAT YTC CAC CTT GGT CCC 3′

HSCJK2o-B
5′ GGA AGA TCT AGA GGA ACC ACC TTT GAT CTC CAG CTT GGT CCC 3′

HSCJK3o-B
5′ GGA AGA TCT AGA GGA ACC ACC TTT GAT ATC CAC TTT GGT CCC 3′

HSCJK5o-B
5′ GGA AGA TCT AGA GGA ACC ACC TTT AAT CTC CAG TCG TGT CCC 3′

V_λ Primers, 5′ Sense, Short and Long Linker

HSCLam1a
5′ GGG CCC AGG CGG CCG AGC TCG TGB TGA CGC AGC CGC CCT C 3′

HSCLam1b
5′ GGG CCC AGG CGG CCG AGC TCG TGC TGA CTC AGC CAC CCT C 3′

HSCLam2
5′ GGG CCC AGG CGG CCG AGC TCG CCC TGA CTC AGC CTC CCT CCG T 3′

HSCLam3
5′ GGG CCC AGG CGG CCG AGC TCG AGC TGA CTC AGC CAC CCT CAG TGT C 3′

HSCLam4
5′ GGG CCC AGG CGG CCG AGC TCG TGC TGA CTC AAT CGC CCT C 3′

HSCLam6
5′ GGG CCC AGG CGG CCG AGC TCA TGC TGA CTC AGC CCC ACT C 3′

HSCLam78
5′ GGG CCC AGG CGG CCG AGC TCG TGG TGA CYC AGG AGC CMT C 3′

HSCLam9
5′ GGG CCC AGG CGG CCG AGC TCG TGC TGA CTC AGC CAC CTT C 3′

HSCLam10
5′ GGG CCC AGG CGG CCG AGC TCG GGC AGA CTC AGC AGC TCT C 3′

V_λ Primers, 3′ Reverse, Short and Long Linker

HSCJLam1236
5′ GGA AGA TCT AGA GGA ACC ACC GCC TAG GAC GGT CAS CTT GGT SCC 3′

HSCJLam4
5′ GGA AGA TCT AGA GGA ACC ACC GCC TAA AAT GAT CAG CTG GGT TCC 3′

HSCJLam57
5′ GGA AGA TCT AGA GGA ACC ACC GCC GAG GAC GGT CAG CTS GGT SCC 3′

Overlap Extension Primers

RSC-F (sense)
5′ GAG GAG GAG GAG GAG GAG GCG GGG CCC AGG CGG CCG AGC TC 3′

RSC-B (reverse)
5′ GAG GAG GAG GAG GAG GAG CCT GGC CGG CCT GGC CAC TAG TG 3′

Procedure

1. First round of PCR.

Note: PCR products and other DNA samples, precipitated or in solution, can be stored for years at –20°C. The construction of antibody fragment libraries, therefore, can be interrupted at any step.

To amplify the V gene rearrangements for a scFv with a short or long linker, perform 12 V_H amplifications, 16 V_κ amplifications, and 27 V_λ amplifications (see Figs. 9.14–9.16). Assemble one reaction for each primer combination, using the spleen/bone marrow cDNA as template.

Each variable gene PCR consists of the following:

1 µl	cDNA (0.5 µg)
60 pmole	5′ primer (primer combinations listed below)
60 pmole	3′ primer
10 µl	10x PCR buffer
8 µl	2.5 mM dNTPs
0.5 µl	*Taq* DNA polymerase

Add water to a final volume of 100 µl.

Note: Extra V_H reverse primer sequences are provided in the primer list at the beginning of the protocol. These primers correspond to the C_H1 domain of human IgA, IgD, and IgE. Traditionally, we make human libraries primarily from the IgM and IgG circulating pools of B lymphocytes, as these are the isotypes that have the highest steady-state serum concentrations. The three additional isotypes may be prevalent in certain types of responses; e.g., IgE antibodies are elevated in

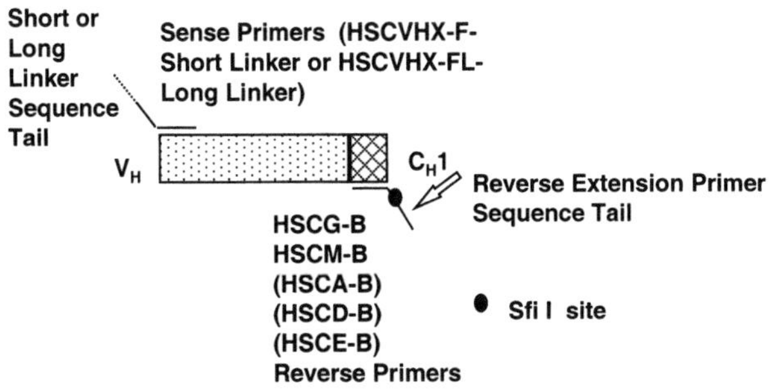

Figure 9.14. PCR amplification of human V_H regions from cDNA. Each of the HSCVH-F (short linker) or HSCVH-FL (long linker) sense primers is paired with the reverse primers specific for the 5′ end of the C_H1 regions of the different immunoglobulin isotypes. In most cases, the use of the IgG- and IgM-specific primers is sufficient. The sense primers have a sequence tail that corresponds to the linker sequence used in the overlap extension PCR. Each reverse primer has a sequence tail containing an *SfiI* site; this tail is recognized by the reverse extension primer used in the second-round PCR.

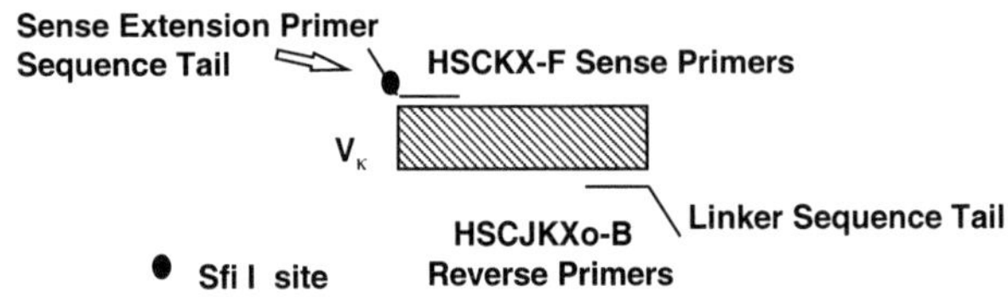

Figure 9.15. The amplification of human V_κ sequences for the construction of scFv short- and long-linker libraries. Each of the HSCK-F sense primers is paired with each of the HSCJKo-B reverse primers to amplify human V_κ gene segments from cDNA. The sense primers have a 5′ sequence tail containing an *Sfi*I site; this tail is recognized by the sense extension primer. The reverse primers have a linker sequence tail that allows the overlap extension with the V_H products in the second round of PCR.

allergic responses and IgA is most prevalent at mucosal sites. We recommend using these reverse primers to build libraries for such specific purposes. In each case, the V_H sense primers listed below should be paired with the desired reverse primer as shown with IgM and IgG. These reverse primers can be used in the amplification of V_H for both short linker and long linker scFv.

V_H primer combinations, short linker:

HSCVH1-F HSCG1234-B	HSCVH2-F HSCG1234-B
HSCVH35-F HSCG1234-B	HSCVH3a-F HSCG1234-B
HSCVH4-F HSCG1234-B	HSCVH4a-F HSCG1234-B
HSCVH1-F HSCM-B	HSCVH2-F HSCM-B
HSCVH35-F HSCM-B	HSCVH3a-F HSCM-B

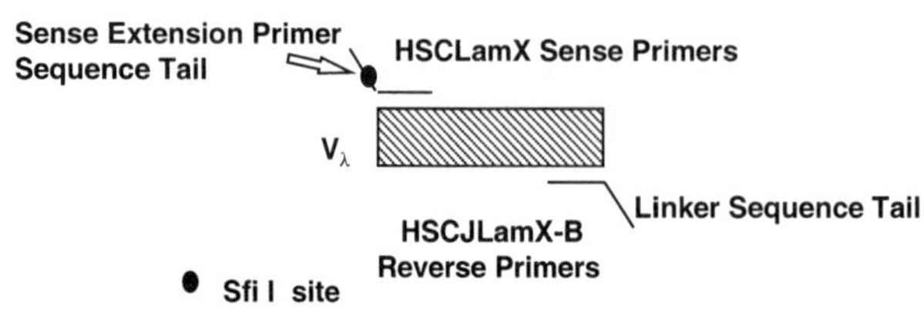

Figure 9.16. The amplification of human V_λ sequences for the construction of scFv short- and long-linker libraries. Each of the HSCLam sense primers is paired with each of the HSCJLam reverse primers to amplify human V_λ gene segments from cDNA. The sense primers have a 5′ sequence tail containing an *Sfi*I site; this tail is recognized by the sense extension primer. The reverse primers have a linker sequence tail that allows the overlap extension with the V_H products in the second round of PCR.

HSCVH4-F	HSCVH4a-F
HSCM-B	HSCM-B

V_H primer combinations, long linker:

HSCVH1-FL	HSCVH2-FL
HSCG1234-B	HSCG1234-B
HSCVH35-FL	HSCVH3a-FL
HSCG1234-B	HSCG1234-B
HSCVH4-FL	HSCVH4a-FL
HSCG1234-B	HSCG1234-B
HSCVH1-FL	HSCVH2-FL
HSCM-B	HSCM-B
HSCVH35-FL	HSCVH3a-FL
HSCM-B	HSCM-B
HSCVH4-FL	HSCVH4a-FL
HSCM-B	HSCM-B

V_κ primer combinations, short and long linker:

HSCK1-F	HSCK24-F
HSCJK14o-B	HSCJK14o-B
HSCK3-F	HSCK5-F
HSCJK14o-B	HSCJK14o-B
HSCK1-F	HSCK24-F
HSCJK2o-B	HSCJK2o-B
HSCK3-F	HSCK5-F
HSCJK2o-B	HSCJK2o-B
HSCK1-F	HSCK24-F
HSCJK3o-B	HSCJK3o-B
HSCK3-F	HSCK5-F
HSCJK3o-B	HSCJK3o-B
HSCK1-F	HSCK24-F
HSCJK5o-B	HSCJK5o-B

HSCK3-F HSCJK5o-B	HSCK5-F HSCJK5o-B

V_{λ} primer combinations, short and long linker:

HSCLam1a HSCJLam1236	HSCLam1b HSCJLam1236
HSCLam2 HSCJLam1236	HSCLam3 HSCJLam1236
HSCLam4 HSCJLam1236	HSCLam6 HSCJLam1236
HSCLam78 HSCJLam1236	HSCLam9 HSCJLam1236
HSCLam10 HSCJLam1236	
HSCLam1a HSCJLam4	HSCLam1b HSCJLam4
HSCLam2 HSCJLam4	HSCLam3 HSCJLam4
HSCLam4 HSCJLam4	HSCLam6 HSCJLam4
HSCLam78 HSCJLam4	HSCLam9 HSCJLam4
HSCLam10 HSCJLam4	
HSCLam1a HSCJLam57	HSCLam1b HSCJLam57
HSCLam2 HSCJLam57	HSCLam3 HSCJLam57
HSCLam4 HSCJLam57	HSCLam6 HSCJLam57

HSCLam78	HSCLam9
HSCJLam57	HSCJLam57

HSCLam10
HSCJLam57

Perform the PCR under the following conditions:

94°C for 5 minutes

30 cycles of:
94°C for 15 seconds
56°C for 15 seconds
72°C for 90 seconds

followed by 72°C for 10 minutes

Note: A "hot start" PCR protocol can improve specificity, sensitivity, and yield. In hot start PCR, either an essential reaction component is not added until the first denaturing step, or a reversible inhibitor of the polymerase is used. This protocol prevents low-stringency primer extension, which can generate nonspecific products.

Evaluate 5–10 µl of each reaction on a 2% agarose gel using DNA gel loading dye and an appropriate MWM. From these amplifications expect a ~400-bp product for the V_H reactions and a ~350-bp product for the V_κ and V_λ reactions.

Note: Try a test amplification with the cDNA sample. If good amplification is not obtained (if a strong amplified band is not clearly visible in the agarose gel), lithium chloride precipitation of the RNA might improve the results (see Protocol 8.3).

2. **Isolate the PCR products.**

   Pool the products of each type of reaction, ethanol-precipitate, and wash as described in Appendix 3. Run the products on a 2% agarose gel, cut out the correct-sized products, and purify the DNA by the freeze-squeeze method, electroelution with an Elutrap (see Appendix 3), or resin binding (e.g., QIAEX II Gel Extraction Kit, QIAGEN). Quantitate yields by reading the O.D. at 260 nm (1 O.D. = 50 µg/ml). If yields are too low, repeat the first round of PCR and combine the end products. Approximately 2–4 µg of each pooled product is needed to proceed.

3. **Second round of PCR (overlap extension).**

   In the second round of PCR, the appropriate first-round products are mixed in equal ratios to generate the overlap product. The primers in the first-round PCR create complementary sequences in the downstream region of the light-chain variable regions and the upstream region of the heavy-chain variable regions that serve as the overlap for the extension of the full-length product (see Fig. 9.17).

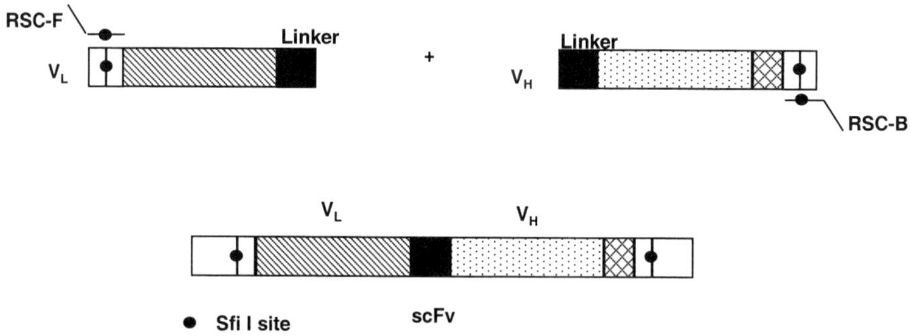

Figure 9.17. The second-round PCR to generate scFv PCR fragments. Equimolar quantities of light-chain variable and heavy-chain variable fragments are used to create the overlap extension product. The sense and reverse extension primers used in the second-round PCR recognize the sequence tails that were generated in the first-round PCR.

Assemble ten 100-μl reactions for each overlap.

Note: This protocol describes the construction of κ and λ human scFv libraries as separate entities. Depending on the needs of the user, the κ and λ light-chain products can be combined prior to the overlap extension PCR such that only one kind of reaction is needed. Alternatively, the κ and λ scFv products can be combined at a later step, before or after *Sfi*I restriction digestion.

Each second-round PCR consists of the following:

100 ng	appropriate first-round products (see template combinations below)
60 pmole	5′ primer (RSC-F)
60 pmole	3′ primer (RSC-B)
10 μl	10x PCR buffer
8 μl	2.5 mM dNTPs
0.5 μl	*Taq* DNA polymerase

Add water to a final volume of 100 μl.

template combinations for scFv with a short linker:

100 ng short linker V_H product	100 ng short linker V_H product
100 ng V_κ product	100 ng V_λ product

template combinations for scFv with a long linker:

100 ng long linker V_H product	100 ng long linker V_H product
100 ng V_κ product	100 ng V_λ product

Perform the second-round PCR under the following conditions:

94°C for 5 minutes

25 cycles of:

94°C for 15 seconds

56°C for 15 seconds

72°C for 2 minutes

followed by 72°C for 10 minutes

Evaluate 5–10 µl of each reaction on a 2% agarose gel using DNA gel loading dye and an appropriate MWM for reference. For each type of reaction expect a ~750–800-bp product.

Note: The long PCR extension time favors full-length product. If a strong band of the correct size is not clearly visible in the gel, the number of PCR cycles can be increased, or more overlap PCRs can be performed. To retain diversity of the library, it is better to increase the number of PCRs.

4. **Isolate the PCR products.**

   Pool the products of each type of reaction, ethanol-precipitate, and wash as described in Appendix 3. Purify the DNA on a 2% agarose gel as in Step 2. Quantitate yields by reading the O.D. at 260 nm. If yields are too low, repeat the overlap PCR and combine the end products. 10–15 µg should be sufficient to continue.

5. **Restriction-digest the overlap scFv PCR product and the pComb3HSS or pComb3XSS vector.**

   Prepare the PCR products and vector for cloning by performing restriction digests with *Sfi*I. We prefer the high concentration (40 units/µl) *Sfi*I and the 10x buffer M from Roche Molecular Biochemicals.

   The digest of the PCR products should contain:

   10 µg of purified scFv product

   360 Units of *Sfi*I (36 Units per µg of DNA)

   20 µl of 10x buffer M

   Add water to a volume of 200 µl.

   The digest of the vector should contain:

   20 µg of pComb3HSS or pComb3XSS (contains stuffer fragment between the two *Sfi*I cloning sites)

   120 Units of *Sfi*I (6 Units per µg of DNA)

   20 µl of 10x buffer M

   Add water to a volume of 200 µl.

   Note: If the loss of digested PCR product in the final purification step is high or if several large-size ligations are needed to obtain a library of the desired size, digest more PCR product with *Sfi*I. The amount of added enzyme should not exceed 10% of the reaction volume. If desired, larger quantities of *Sfi*I-cut and purified vector DNA can be prepared and tested in advance.

   Incubate both digests for 5 hours at 50°C. Ethanol-precipitate and wash as described in Appendix 3. Purify the digested scFv products on a 2% agarose gel, and purify the vector and stuffer fragment on a 1% agarose gel. We recommend electroelution with an Elutrap for extraction of the vector from the gel (see Appendix 3). Suboptimal digestion or purification will yield vector with low transformation efficiencies and high background after ligation. Use the stuffer fragment in a test ligation to determine the quality of the vector prior to use in library ligations (see Step 6). The stuffer fragment is approximately 1600 bp. Quantify the purified, digested PCR products, vector, and stuffer fragment by measuring the O.D. at 260 nm.

Note: In some cases, *Sfi*I-digested DNA is partly insoluble after ethanol precipitation. To avoid this problem, the DNA can be loaded directly after digestion onto the agarose gel. When purifying the *Sfi*I-digested vector DNA, let the DNA run long enough to separate linearized vector DNA (size ~ 5000 bp; cut only once) and uncut vector DNA from the desired double cut product (size ~ 3400 bp).

6. **Ligate the digested overlap PCR product with vector DNA.**

Perform small-scale ligations to assess the suitability of the vector and inserts for high-efficiency ligation and transformation. The ligation efficiency of the vector DNA can be tested by ligating it with the gel-purified stuffer fragment that is generated during the *Sfi*I digestion of the vector DNA. The vector preparation should contain little uncut vector DNA or DNA that is cut only once. The amount of uncut or partly cut vector DNA can be estimated by setting up a ligation reaction that contains only vector DNA. These two types of control ligations should be done in parallel with a small-scale test ligation of vector plus PCR insert, and both should include the same amount of vector DNA.

Test Ligations

Assemble the following small-scale ligation reactions:

small-scale test ligation (one reaction for each PCR insert):

140 ng	pComb3HSS or pComb3XSS, *Sfi*I-digested and purified
70 ng	scFv (short or long) overlap PCR product, *Sfi*I-digested and purified
4 µl	5x ligase buffer
1 µl	ligase

Add water to a volume of 20 µl.

control ligation 1 (control insert)

140 ng	pComb3HSS or pComb3XSS, *Sfi*I-digested and purified
140 ng	stuffer fragment, *Sfi*I-digested and purified
4 µl	5x ligase buffer
1 µl	ligase

Add water to a volume of 20 µl.

control ligation 2 (test for vector self-ligation)

140 ng	pComb3HSS or pComb3XSS, *Sfi*I-digested and purified
4 µl	5x ligase buffer
1 µl	ligase

Add water to a volume of 20 µl.

Incubate between 4 hours and overnight at room temperature. Transform 0.5–1 µl of each reaction by electroporation into 50 µl of XL1-Blue or ER2537 (see Support Protocol). Dilute the transformed cultures 10-fold and 100-fold with prewarmed (37°C) SOC, and plate 100 µl of each dilution on LB+carb plates. Incubate the plates overnight at 37°C. Count the colonies on the vector + Fab insert plates and calculate the number of transformants per µg pf vector DNA. If this number does not exceed 1×10^7, do not proceed with the large-scale library ligation. Ideally, the final library size should be several times 10^8 but at least 5×10^7 total transformants. Determine the number of scaled-up library ligations needed to achieve this size. Count the colonies obtained from control ligation 1 for an indication of vector quality and ligation efficiency. A good vector DNA preparation should yield at least 10^8 colony-forming units (cfu) per µg of vector DNA and should have less than 10% (ideally less than 5%) background ligation (calculated cfu per µg of vector DNA in control ligation 2). If the background is greater than 10% or the ligation efficiency is low, digest new vector and repeat the ligations. If the ligations are reasonably good, but there is not enough vector or insert to make a library of the desired size, perform additional digests.

Library Ligation

Assemble enough reactions to produce at least 5×10^7 transformants. Combine the following:

1.4 µg	pComb3HSS or pComb3XSS, *Sfi*I-digested and purified
700 ng	scFv (short or long) overlap product, *Sfi*I-digested and purified
40 µl	5x ligase buffer
10 µl	ligase

Add water to a volume of 200 µl.

Incubate overnight at room temperature. Ethanol-precipitate and wash the pellet as described in Appendix 3. The ethanol-precipitated ligation reaction can be stored for months at –20°C. Transform the ligation reaction by resuspending in 15 µl of water and electroporating into XL1-Blue or ER2537 electrocompetent cells, as described in Chapter 10.

Proceed with the preparation and panning of the primary library (see Chapter 10).

PROTOCOL 9.3

Libraries of Nonhuman Primate Antibody Fragments

Obtaining antibodies from laboratory animals is a viable alternative to obtaining antibodies from humans. Antibodies from laboratory animals can be humanized (see Chapter 13), but the humanized antibodies may lose affinity relative to the parent antibody. Another method for obtaining antibodies that are closely related to human molecules involves the use of nonhuman primates. Generally, nonhuman primates have been shown in various reports to have highly evolutionarily conserved immunoglobulin genes; in fact, the gene segments of macaques, for instance, are as closely related to human immunoglobulin genes as human genes are to each other (Kawamura et al. 1990; Meek et al. 1991; Kawamura and Ueda 1992; Andris et al. 1997). It is, therefore, possible to obtain nonhuman antibodies that do not require further manipulation to render them relatively nonimmunogenic for humans.

Although expensive to use experimentally, nonhuman primates can be a particularly useful system when working in areas of human disease research, such as HIV-1 (Samuelsson et al. 1995), for which there is no small animal model. Nonhuman primates have also been used successfully to obtain antibodies specific for other immunologically relevant targets, such as human CD4 (Newman et al. 1992). They are less commonly used for the sole purpose of obtaining monoclonal antibodies, and there are few reports regarding the use of these animals to construct combinatorial antibody libraries.

Little work has been published with regard to the genomic structure of nonhuman primate immunoglobulin loci, but the available information suggests that the variable region gene segments fall into families that closely resemble those of human. These regions are easily amplifiable with oligonucleotide primers that are specific for human variable region genes. Sequence data also indicate that D and J segments are evolutionarily conserved. Although some of the nonhuman primate constant regions differ in part of their sequence, these regions do not pertain to any of the primers used in the construction of human Fab and scFv libraries. Therefore, nonhuman primate antibody libraries can be constructed like any other animal library described in this chapter using the same oligonucleotide primers that are recommended for human libraries. For construction of such libraries we refer the user to the protocols for human library construction (Protocols 9.1 and 9.2). Differences found in the constant regions, such as in one subclass of macaque IgG (P. Calvas et al., in prep.), can be overcome by constructing macaque/human chimeric molecules similar to those described in later protocols for mouse, rabbit, and chicken.

PROTOCOL 9.4

Chimeric Mouse/Human Fab Libraries (by Overlap Extension)

The organization of the murine immunoglobulin genes is well characterized. As in humans and many other mammals, the mouse variable antibody regions are generated through somatic DNA recombination (Tonegawa 1983). The κ chain, the dominant light chain in mice, is generated by the combination of one of many V_κ genes with four J_κ genes. The variable regions of the heavy chain are formed by the combination of one of several hundred V_H genes with 1 of about 30 D segments and 1 of 6 J_H segments. The cloning of the murine antibody repertoire, therefore, requires many V-specific oligonucleotide primers.

The following protocol describes the construction of a chimeric mouse/human Fab library by overlap extension PCR. It allows the expression and selection of chimeric Fabs that have variable regions obtained from an immunized mouse and light- and heavy-chain constant regions from a cloned human Fab. This approach facilitates the humanization of selected clones with therapeutic potential using phage display. Chimeric Fabs with human constant regions can be detected with antihuman Fab reagents, and they are usually better expressed in *E. coli* than are mouse antibody fragments, which are often expressed at undetectable levels. The V-specific primers listed below are adapted for the pComb3 vectors from Krebber et al. (1997).

Materials

cDNA (prepared from immunized mice, see Protocol 8.4)
cloned human Fab template (e.g., pComb3XTT; available from the Barbas laboratory)
oligonucleotide primers (listed below; see Appendix 1 for letter code)
2.5 mM dNTP mix (dATP/dCTP/dGTP/dTTP set, 100 mM, Amersham Pharmacia Biotech, Cat. # 27-2035-02)
AmpliTaq DNA polymerase (Perkin-Elmer, Cat. # N808-052)
10x PCR buffer for use with *Taq* polymerase (supplied with polymerase)
Expand High Fidelity Enzyme mix; 3.5 Units/μl (Roche Molecular Biochemicals, Cat. # 1 732 650)
10x Expand High Fidelity PCR buffer (supplied with Expand High Fidelity Enzyme mix)
MicroAmp PCR tubes (Perkin-Elmer, Cat. # N801-0533)
MicroAmp PCR caps (Perkin-Elmer, Cat. # N801-0534)
MicroAmp PCR trays (Perkin-Elmer, Cat. # N801-5530)
GeneAmp 9600 PCR cycler (Perkin-Elmer)
agarose (Gibco-BRL, Cat. # 15510-027)
100-bp DNA molecular weight marker (Amersham Pharmacia Biotech, Cat. # 27-4001-01) or 1-kb DNA molecular weight marker (Gibco-BRL, Cat. # 15615-024)

1x TAE (see Appendix 2)
DNA gel loading dye (see Appendix 2)
pComb3HSS or pComb3XSS (available from the Barbas laboratory)
*Sfi*I restriction enzyme (Roche Molecular Biochemicals, 40 Units/μl, Cat. # 1 288 059)
10x restriction enzyme buffer M (supplied with *Sfi*I restriction enzyme)
T4 DNA ligase (Gibco-BRL, 1 Unit/μl, Cat. # 15224 090)
5x T4 DNA ligase buffer (supplied with enzyme)
glycogen, 20 mg/ml (Roche Molecular Biochemicals, Cat. # 901 393)
electroporation apparatus and materials
electrocompetent ***E. coli*** cells (see Chapter 10): XL1-Blue (Stratagene, Cat. # 200228) or ER2537 (New England Biolabs, Cat. # 801-N)
LB + 100 μg/ml carbenicillin plates (see Appendix 2)

CAUTION: ***E. coli (see Appendix 4)***

CHIMERIC MOUSE/HUMAN Fab PRIMERS

The number to the right of each primer indicates the amount (in μl, at equimolar concentrations) that should be used when setting up the primer mixes (for example, see Step 1).

V_κ 5′ Sense Primers

MSCVK-1
5′ GGG CCC AGG CGG CCG AGC TCG AYA TCC AGC TGA CTC AGC C 3′ 1

MSCVK-2
5′ GGG CCC AGG CGG CCG AGC TCG AYA TTG TTC TCW CCC AGT C 3′ 2

MSCVK-3
5′ GGG CCC AGG CGG CCG AGC TCG AYA TTG TGM TMA CTC AGT C 3′ 5

MSCVK-4
5′ GGG CCC AGG CGG CCG AGC TCG AYA TTG TGY TRA CAC AGT C 3′ 3.5

MSCVK-5
5′ GGG CCC AGG CGG CCG AGC TCG AYA TTG TRA TGA CMC AGT C 3′ 4

MSCVK-6
5′ GGG CCC AGG CGG CCG AGC TCG AYA TTM AGA TRA MCC AGT C 3′ 7

MSCVK-7
5′ GGG CCC AGG CGG CCG AGC TCG AYA TTC AGA TGA YDC AGT C 3′ 6

MSCVK-8
5′ GGG CCC AGG CGG CCG AGC TCG AYA TYC AGA TGA CAC AGA C 3′ 1.5

MSCVK-9
5′ GGG CCC AGG CGG CCG AGC TCG AYA TTG TTC TCA WCC AGT C 3′ 2

MSCVK-10
5′ GGG CCC AGG CGG CCG AGC TCG AYA TTG WGC TSA CCC AAT C 3′ 3.5

MSCVK-11
5′ GGG CCC AGG CGG CCG AGC TCG AYA TTS TRA TGA CCC ART C 3′ 8

MSCVK-12
5′ GGG CCC AGG CGG CCG AGC TCG AYR TTK TGA TGA CCC ARA C 3′ 8

MSCVK-13
5′ GGG CCC AGG CGG CCG AGC TCG AYA TTG TGA TGA CBC AGK C 3′ 6

MSCVK-14
5′ GGG CCC AGG CGG CCG AGC TCG AYA TTG TGA TAA CYC AGG A 3′ 2

MSCVK-15
5′ GGG CCC AGG CGG CCG AGC TCG AYA TTG TGA TGA CCC AGW T 3′ 2

MSCVK-16
5′ GGG CCC AGG CGG CCG AGC TCG AYA TTG TGA TGA CAC AAC C 3′ 1

MSCVK-17
5′ GGG CCC AGG CGG CCG AGC TCG AYA TTT TGC TGA CTC AGT C 3′ 1

V_κ 3′ Reverse Primers

MHybJK12-B
5′ AGA TGG TGC AGC CAC AGT TCG TTT KAT TTC CAG YTT GGT CCC 3′ 1

MHybJK4-B
5′ AGA TGG TGC AGC CAC AGT TCG TTT TAT TTC CAA CTT TGT CCC 3′ 1

MHybJK5-B
5′ AGA TGG TGC AGC CAC AGT TCG TTT CAG CTC CAG CTT GGT CCC 3′ 1

V_λ 5′ Sense Primer

MSCVL-1
5′ GGG CCC AGG CGG CCG AGC TCG ATG CTG TTG TGA CTC AGG AAT C 3′

V_λ 3′ Reverse Primer

MHybLJ-B
5′ AGA TGG TGC AGC CAC AGT TCG ACC TAG GAC AGT CAG TTT GG 3′

V_H 5′ Sense Primers

MHyVH1
5′ GCT GCC CAA CCA GCC ATG GCC CTC GAG GTR MAG CTT CAG GAG TC 3′ 4

MHyVH2
5′ GCT GCC CAA CCA GCC ATG GCC CTC GAG GTB CAG CTB CAG CAG TC 3′ 4

MHyVH3
5′ GCT GCC CAA CCA GCC ATG GCC CTC GAG GTG CAG CTG AAG SAS TC 3′ 3

MHyVH4
5′ GCT GCC CAA CCA GCC ATG GCC CTC GAG GTC CAR CTG CAA CAR TC 3′ 4

MHyVH5
5′ GCT GCC CAA CCA GCC ATG GCC CTC GAG GTY CAG CTB CAG CAR TC 3′ 7

MHyVH6
5′ GCT GCC CAA CCA GCC ATG GCC CTC GAG GTY CAR CTG CAG CAG TC 3′ 2

MHyVH7
5′ GCT GCC CAA CCA GCC ATG GCC CTC GAG GTC CAC GTG AAG CAG TC 3′ 1

MHyVH8
5′ GCT GCC CAA CCA GCC ATG GCC CTC GAG GTG AAS STG GTG GAA TC 3′ 2

MHyVH9
5′ GCT GCC CAA CCA GCC ATG GCC CTC GAG GTG AWG YTG GTG GAG TC 3′ 5

MHyVH10
5′ GCT GCC CAA CCA GCC ATG GCC CTC GAG GTG CAG SKG GTG GAG TC 3′ 2

MHyVH11
5′ GCT GCC CAA CCA GCC ATG GCC CTC GAG GTG CAM CTG GTG GAG TC 3′ 2

MHyVH12
5′ GCT GCC CAA CCA GCC ATG GCC CTC GAG GTG AAG CTG ATG GAR TC 3′ 2

MHyVH13
5′ GCT GCC CAA CCA GCC ATG GCC CTC GAG GTG CAR CTT GTT GAG TC 3′ 1

MHyVH14
5′ GCT GCC CAA CCA GCC ATG GCC CTC GAG GTR AAG CTT CTC GAG TC 3′ 2

MHyVH15
5′ GCT GCC CAA CCA GCC ATG GCC CTC GAG GTG AAR STT GAG GAG TC 3′ 2

MHyVH16
5′ GCT GCC CAA CCA GCC ATG GCC CTC GAG GTT ACT CTR AAA GWG TST G 3′ 5

MHyVH17
5′ GCT GCC CAA CCA GCC ATG GCC CTC GAG GTC CAA CTV CAG CAR CC 3′ 3.5

MHyVH18
5′ GCT GCC CAA CCA GCC ATG GCC CTC GAG GTG AAC TTG GAA GTG TC 3′ 0.7

MHyVH19
5′ GCT GCC CAA CCA GCC ATG GCC CTC GAG GTG AAG GTC ATC GAG TC 3′ 0.7

V_H 3′ Reverse Primers

MHyIgGCH1-B1
5′ CGA TGG GCC CTT GGT GGA GGC TGA GGA GAC GGT GAC CGT GGT 3′ 1

MHyIgGCH1-B2
5′ CGA TGG GCC CTT GGT GGA GGC TGA GGA GAC TGT GAG AGT GGT 3′ 1

MHyIgGCH1-B3
5′ CGA TGG GCC CTT GGT GGA GGC TGC AGA GAC AGT GAC CAG AGT 3′ 1

MHyIgGCH1-B4
5′ CGA TGG GCC CTT GGT GGA GGC TGA GGA GAC GGT GAC TGA GGT 3′ 1

Primers for Amplification of the Human C_κ Region and the pelB Leader Sequence from a Cloned Human Fab

HKC-F (sense)
5′ CGA ACT GTG GCT GCA CCA TCT GTC 3′

Lead-B (reverse)
5′ GGC CAT GGC TGG TTG GGC AGC 3′

Primers for Amplification of the Human C_H1 Chain from a Cloned Human Fab

HIgGCH1-F (sense)
5′ GCC TCC ACC AAG GGC CCA TCG GTC 3′

dpseq (reverse)
5′ AGA AGC GTA GTC CGG AAC GTC 3′

Primers for PCR Assembly of Mouse V_L Sequences with the Human C_κ PCR Product

RSC-F (sense)
5′ GAG GAG GAG GAG GAG GAG GCG GGG CCC AGG CGG CCG AGC TC 3′

Lead-B (reverse)
(see above)

Primers for PCR Assembly of Mouse V_H Sequences with the Human C_H1 PCR Product

leadVH (sense)
5′ GCT GCC CAA CCA GCC ATG GCC 3′

dpseq (reverse)
(see above)

Primers for PCR Assembly of Chimeric Light-chain Sequences with Chimeric Heavy-chain (Fd) Sequences

RSC-F (sense)
(see above)

dp-EX (reverse)
5′ GAG GAG GAG GAG GAG GAG AGA AGC GTA GTC CGG AAC GTC 3′

Procedure

1. **First round of PCR.**

   Note: PCR products and other DNA samples, precipitated or in solution, can be stored at –20°C for years. The construction of antibody fragment libraries, therefore, can be interrupted at any step.

   Amplify the mouse V_L and V_H sequences by performing nine V_κ amplifications, one V_λ amplification, and twelve V_H amplifications (see Figs. 9.18–9.20). Set up the following PCRs, using the primer combinations listed below. Make the primer mixes by combining the primers in the ratios indicated in the primer list at the beginning of the protocol.

1 µl	cDNA (~0.5 µg)
60 pmole	sense primer (primer combinations are listed below)
60 pmole	reverse primer
10 µl	10x PCR buffer

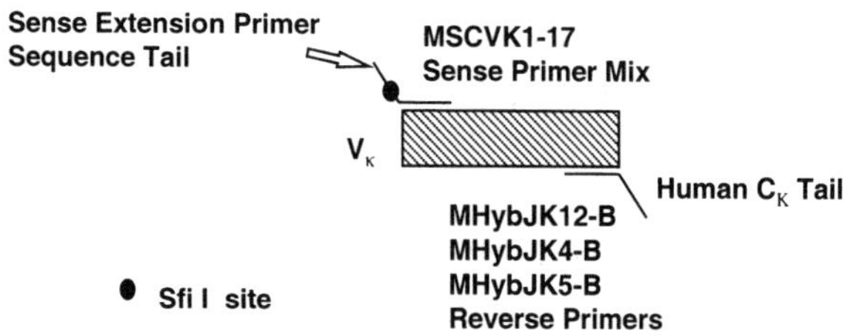

Figure 9.18. The amplification of mouse V_κ sequences for the construction of chimeric Fab libraries. A mixture of MSCVK sense primers is combined with a mixture of MHybJK reverse primers to amplify V_κ sequences from murine cDNA. The sense primers have a 5′ sequence tail that contains an *Sfi*I site and is recognized by the sense extension primer in the second- and third-round PCR. Each reverse primer has a human C_κ sequence tail that is used in the overlap extension PCR to create the chimeric κ light-chain–pelB fragment.

8 µl	2.5 mM dNTP mix
0.5 µl	*Taq* polymerase

Add water to bring the final reaction volume to 100 µl.

V_κ primer combination:

MSCVK primer mix (sense; MSCVK primers 1–17)

MHybJK primer mix (reverse; MHybJK12-B, 4-B, 5-B)

V_λ primer combination:

MSCVL-1 primer (sense)

MHybLJ-B primer (reverse)

V_H primer combination:

MHyVH primer mix (sense; MHyVH1–19)

MHyIgGCH1-B primer mix (reverse; MHyIgGCH1-B1 through B4)

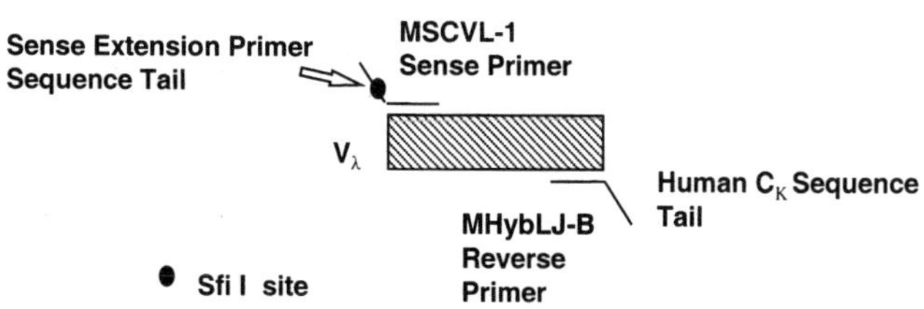

Figure 9.19. The amplification of mouse V_λ sequences for the construction of chimeric Fab libraries. The MSCVL-1 sense primer is combined with the MHybLJ-B reverse primer to amplify V_λ sequences from murine cDNA. MSCVL-1 has a 5′ sequence tail that contains an *Sfi*I site and is recognized by the sense extension primer in the second- and third-round PCR. MHybLJ-B has a human C_κ sequence tail that is used in the overlap extension PCR to create the chimeric lambda light-chain–pelB fragment.

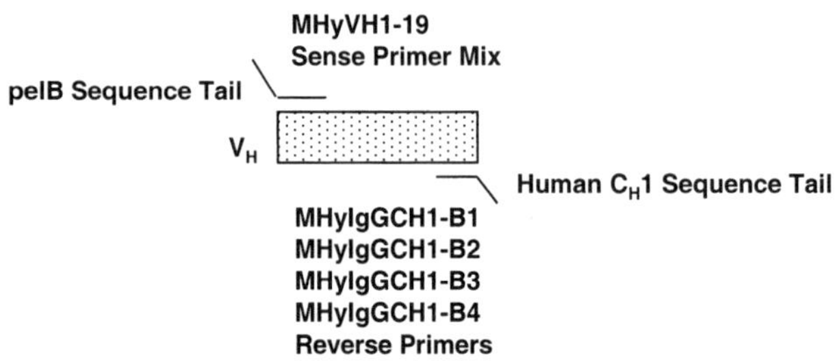

Figure 9.20. The amplification of mouse V_H sequences for the construction of chimeric Fab libraries. A mixture of MHyVH sense primers is combined with a mixture of MHyIgGCH1-B reverse primers to amplify V_H gene segments from murine cDNA. Each sense primer has a 5′ sequence tail that corresponds to the 3′ end of the pelB leader sequence and is recognized by the sense extension primer (leadVH) in the second-round PCR. Each reverse primer has a human C_H1 sequence tail that is used in the overlap extension PCR to create the chimeric heavy chain (Fd) fragment.

Perform the PCR under the following conditions:

30 cycles of:

94°C for 15 seconds

56°C for 30 seconds

72°C for 90 seconds

followed by 72°C for 10 minutes

Assemble ten PCRs to amplify the human κ constant region and the pelB leader sequence from a cloned human Fab. Assemble ten PCRs to amplify the first constant region of the human γ chain (see Figs. 9.21 and 9.22). Each reaction should contain the following:

20 ng	plasmid DNA containing cloned human Fab fragment (e.g., pComb3XTT)
60 pmole	sense primer (primer combinations are listed below)
60 pmole	reverse primer
10 µl	10x PCR buffer

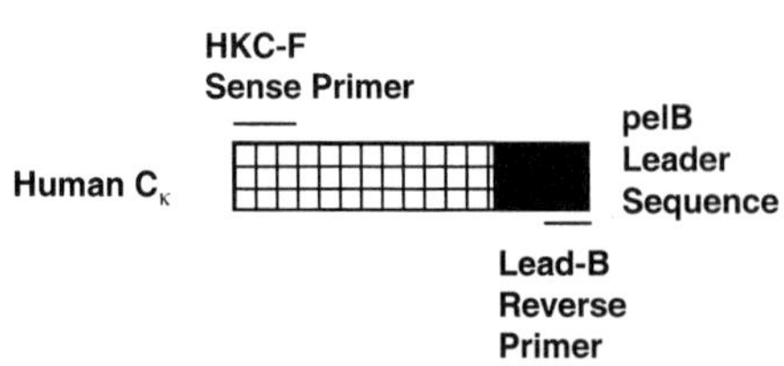

Figure 9.21. The first-round amplification of C_κ products and the pelB leader sequence from pComb3XTT. The sense primer is specific for the 5′ region of human C_κ, the region used in the overlap extension PCR to create the chimeric light chain. The reverse primer is specific for the 3′ end of the pelB leader sequence.

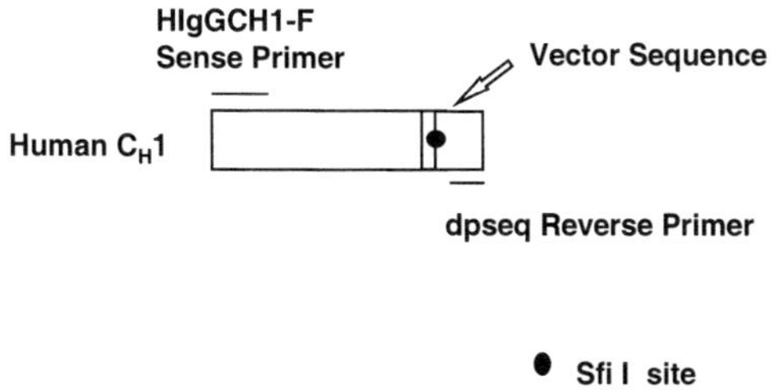

Figure 9.22. The first-round PCR amplification of C_H1 products from the pComb3XTT vector. The sense primer is specific for the 5′ region of C_H1, the region that is used in the overlap extension PCR to create the heavy-chain (Fd) fragment. The reverse primer is specific for the decapeptide sequence that is located downstream of the C_H1 fragment and the 3′ *SfiI* site of pComb3XTT.

8 µl	2.5 mM dNTP mix
0.5 µl	*Taq* polymerase

Add water to a final reaction volume of 100 µl.

Primers to amplify the constant region of the human κ chain and the pelB leader sequence:

HKC-F (sense)

Lead-B (reverse)

Primers to amplify the first constant region of the human γ chain:

HIgGCH1-F (sense)

dpseq (reverse)

Perform the PCR under the following conditions:

20 cycles of

94°C for 15 seconds

56°C for 30 seconds

72°C for 90 seconds

followed by 72°C for 10 minutes

Note: A "hot start" PCR protocol can improve specificity, sensitivity, and yield. In hot start PCR, either an essential reaction component is not added until the first denaturing step, or a reversible inhibitor of the polymerase is used. This protocol prevents low-stringency primer extension, which can generate nonspecific products.

Analyze 5–10 µl of each reaction on a 2% agarose gel using DNA gel loading dye and an appropriate MWM. The mouse variable region products (V_κ, V_λ, and V_H) should be about 350 bp, the human C_κ–pelB products should be about 400 bp, and the human C_H1 (γ chain) products should be about 350 bp.

Note: Try a test amplification with the cDNA sample. If good amplification is not obtained (if a strong amplified band is not clearly visible in the agarose gel), lithium chloride precipitation of the RNA may improve the results (see Protocol 8.3).

2. **Isolate the PCR products.**

   Pool the products of each type of reaction, ethanol-precipitate, and wash as described in Appendix 3. You should have four pools: one for the mouse light-chain products (V_κ and V_λ), one for the mouse heavy-chain products (V_H), one for the C_κ–pelB fragment, and one for the C_H1 region. Purify the pools on a 1.5% agarose gel. Cut out the correct-sized products, and purify the DNA by freeze-squeeze, electroelution with an Elutrap (see Appendix 3), or resin binding (e.g., QIAEX II Gel Extraction Kit, QIAGEN). Quantify the purified PCR products by reading the O.D. at 260 nm (1 O.D. unit = 50 µg/ml). At least 1 µg of each purified pool is required to proceed.

3. **Second round of PCR (1st step overlap extension).**

   In the second round of PCR, the first-round mouse variable region PCR products are combined with the human constant region PCR products to generate chimeric light-chain and heavy-chain fragments (see Figs. 9.23 and 9.24). Assemble at least ten 100-µl reactions for each template–primer combination.

100 ng	purified V product (template–primer combinations are listed below)
100 ng	purified C product
60 pmole	sense primer
60 pmole	reverse primer
10 µl	10x PCR buffer
8 µl	2.5 mM dNTP mix
0.5 µl	*Taq* polymerase

   Add water to a final reaction volume of 100 µl.

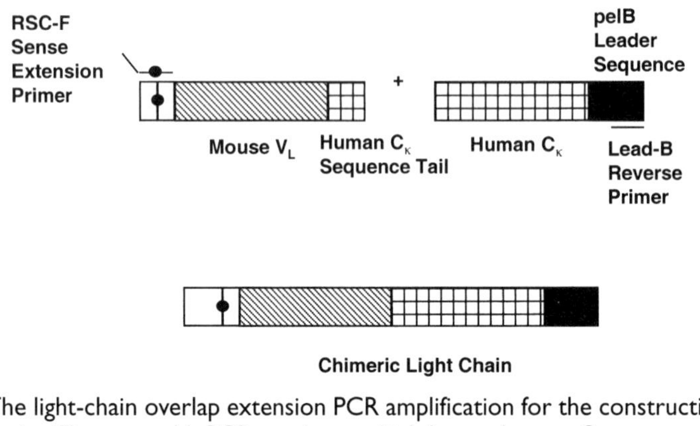

Figure 9.23. The light-chain overlap extension PCR amplification for the construction of chimeric mouse Fab libraries. The mouse V_L PCR products, which have a human C_κ sequence tail, are combined with the human C_κ–pelB products to create chimeric light-chain–pelB fragments. The sense extension primer (RSC-F) recognizes the sequence created by the first-round PCR V_L primers. The Lead-B reverse primer recognizes the 3′ end of the pelB leader sequence that was amplified together with the human C_κ region from pComb3XTT. The 3′ end of the pelB sequence serves as the overlap region in the final overlap extension PCR.

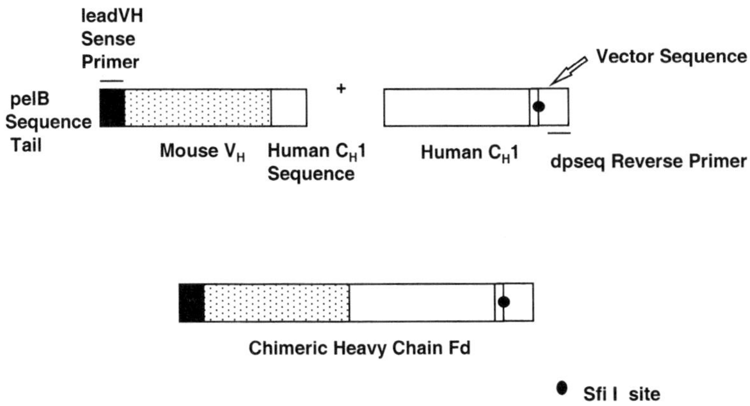

Figure 9.24. The heavy-chain (Fd) overlap extension PCR amplification for the construction of chimeric mouse Fab libraries. The mouse V_H PCR products, which have a human C_H1 sequence tail, are combined with the human C_H1 PCR products to create the chimeric Fd fragment. The sense extension primer (leadVH) recognizes the pelB leader sequence tail created by the first-round PCR V_H primers. The same reverse primer (dpseq) that was used for the generation of the C_H1 fragment is used in the overlap extension PCR of the heavy-chain fragment. The 3′ end of the pelB leader sequence serves as the overlap region in the final overlap extension PCR.

Chimeric light-chain template–primer combination:

100 ng	V_L product (V_κ plus V_λ)
100 ng	C_κ–pelB product
60 pmole	RSC-F primer (sense)
60 pmole	Lead-B primer (reverse)

Chimeric heavy-chain Fd template–primer combination:

100 ng	V_H product
100 ng	C_H1 product
60 pmole	leadVH primer (sense)
60 pmole	dpseq primer (reverse)

Perform the second-round PCR under the following conditions:

15 cycles of:

94°C for 15 seconds
56°C for 30 seconds
72°C for 2 minutes

followed by 72°C for 10 minutes

Analyze 5–10 µl of each reaction on a 1.5% agarose gel using DNA gel loading dye and an appropriate MWM. The chimeric light-chain PCR products should have a size of about 800 bp. The chimeric heavy-chain PCR products should have a size of about 750 bp.

Note: The long PCR extension time favors full-length product. If a strong band of the correct size is not clearly visible in the gel, the number of PCR cycles can be increased, or more overlap PCRs can be performed. To retain diversity of the library, it is better to increase the number of PCRs.

4. **Isolate the PCR products.**

Pool the products of each type of reaction, ethanol-precipitate, and wash as described in Appendix 3. Purify the DNA on a 1% agarose gel as in Step 2. Quantify the purified PCR products by reading the O.D. at 260 nm. At least 1 μg of each purified product is needed to continue.

5. **Third round of PCR (second overlap extension).**

In the third round of PCR, the second-round chimeric light-chain products with the pelB leader sequence and the chimeric heavy-chain Fd fragments are joined by a second overlap extension (see Fig. 9.25). Assemble at least ten 100-μl reactions for each overlap.

100 ng	purified chimeric light-chain product
100 ng	purified chimeric Fd product
60 pmole	RSC-F sense primer
60 pmole	dp-EX reverse primer
10 μl	10x PCR buffer
8 μl	2.5 mM dNTP mix
0.75 μl	Expand HF Polymerase mix

Add water to a final reaction volume of 100 μl.

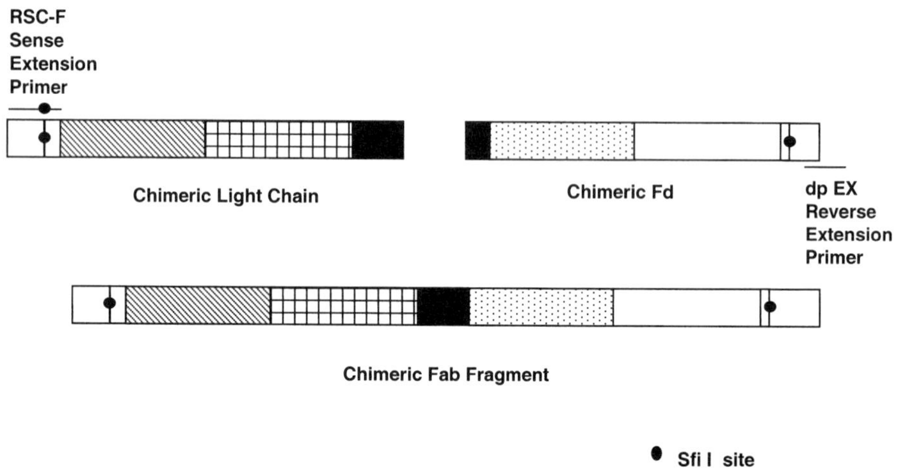

Figure 9.25. The final overlap extension PCR to combine the chimeric light chain–pelB fragment and the Fd fragment for the construction of mouse chimeric Fab libraries. The 3′ end of the pelB leader sequence serves as the overlap region for the two PCR products. The sense extension primer (RSC-F) used in this round of PCR was also used in the overlap extension amplification of the light chain. The reverse extension primer recognizes the decapeptide sequence downsteam of the C_HI region.

Perform the PCR under the following conditions:

10 cycles of:

- 94°C for 15 seconds
- 56°C for 30 seconds
- 72°C for 3 minutes

followed by 72°C for 10 minutes

> *Note:* Efficiency in generating the 1.5-kb overlap PCR product is often improved by using a long extension time and the Expand High Fidelity PCR System. This system has a unique enzyme mix that contains thermostable *Taq* and *Pwo* DNA polymerases. It is designed to give high yield, high fidelity, and high specificity PCR products.
>
> The overlap products can also be generated with *AmpliTaq* polymerase. To amplify with *AmpliTaq* polymerase, use 0.5 µl of enzyme in the 100-µl amplification reaction. Perform the PCR under the following conditions:
>
> 94°C for 5 minutes
> 25 cycles of:
> 94°C for 15 seconds
> 56°C for 15 seconds
> 72°C for 3 minutes
>
> followed by 72°C for 10 minutes.

Analyze 5–10 µl of each reaction on a 1.5% agarose gel, using DNA gel loading dye and an appropriate MWM. The size of the overlap PCR products should be around 1500 bp.

6. Isolate the PCR products.

Pool the PCR products, ethanol-precipitate, and wash as described in Appendix 3. Purify the DNA on a 1% agarose gel as in Step 2. Quantify the purified PCR products by reading the O.D. at 260 nm. 10–15 µg should be sufficient to continue.

7. Restriction-digest the purified overlap extension PCR products and vector DNA.

Prepare the PCR products and the vector for cloning by performing restriction digests with *Sfi*I. We prefer the high concentration (40 units/µl) *Sfi*I and the 10x buffer M from Roche Molecular Biochemicals.

The digest of the PCR products should contain:

- 10 µg of purified overlap PCR product
- 160 Units of *Sfi*I (16 Units per µg of DNA)
- 20 µl of 10x reaction buffer M
- Add water to a total volume of 200 µl.

The digest of the vector should contain:

20 μg of vector DNA (pComb3HSS or pComb3XSS, contains stuffer fragment between two *Sfi*I sites)

120 Units of *Sfi*I (6 Units per μg of plasmid DNA)

20 μl of 10X reaction buffer M

Add water to a total volume of 200 μl.

Note: If the loss of digested PCR product in the final purification step is high or if several large-size ligations are needed to obtain a library of the desired size, digest more PCR product with *Sfi*I. The amount of added enzyme should not exceed 10% of the reaction volume. If desired, larger quantities of *Sfi*I-cut and purified vector DNA can be prepared and tested in advance.

Incubate both digests for 5 hours at 50°C. Ethanol-precipitate and wash the DNA as described in Appendix 3 (optional, see Note). Purify the digested PCR insert (~1500 bp) on a 1% agarose gel, and purify the vector (~3400 bp) and stuffer fragment (~1600 bp) on a 0.6% agarose gel. We recommend electroelution with an Elutrap for extraction of the vector from the gel. Suboptimal digestion or purification will yield vector with low transformation efficiencies and high background after ligation. Use the stuffer fragment in a test ligation to assess vector quality (see Step 8). Quantify the purified, digested PCR products, vector, and stuffer fragment by measuring the O.D. at 260 nm.

Note: In some cases, *Sfi*I-digested DNA is partly insoluble after ethanol precipitation. To avoid this problem, the DNA can be loaded directly after digestion onto the agarose gel. When purifying the *Sfi*I-digested vector DNA, let the DNA run long enough to separate linearized vector DNA (size ~ 5000 bp; cut only once) and uncut vector DNA from the desired double cut product (size ~ 3400 bp).

8. **Ligate the digested overlap PCR product with vector DNA.**

Perform small-scale ligations to assess the suitability of the vector and inserts for high-efficiency ligation and transformation. The ligation efficiency of the vector DNA can be tested by ligating it with the gel-purified stuffer fragment that is generated during the *Sfi*I digestion of the vector DNA. The vector preparation should contain little uncut vector DNA or DNA that is cut only once. The amount of uncut or partly cut vector DNA can be estimated by setting up a ligation reaction that contains only vector DNA. These two types of control ligations should be done in parallel with a small-scale test ligation and should both include the same amount of vector DNA.

Test Ligations

Assemble the following small-scale ligation reactions:

small-scale test ligation:

140 ng	pComb3HSS or pComb3XSS, *Sfi*I-digested and purified

140 ng	overlap PCR product, *Sfi*I-digested and purified
4 μl	5x ligase buffer
1 μl	ligase

Add water to a total volume of 20 μl.

control ligation 1 (control insert):

140 ng	pComb3HSS or pComb3XSS, *Sfi*I-digested and purified
140 ng	gel-purified stuffer fragment
4 μl	5x ligase buffer
1 μl	ligase

Add water to a total volume of 20 μl.

control ligation 2 (test for vector self-ligation):

140 ng	pComb3HSS or pComb3XSS, *Sfi*I-digested and purified
4 μl	5x ligase buffer
1 μl	ligase

Add water to a total volume of 20 μl.

Incubate between 4 hours and overnight at room temperature. Transform 0.5–1 μl of each reaction by electroporation into 50 μl of XL1-Blue or ER2537 (see Support Protocol). Dilute the transformed cultures 10-fold and 100-fold with prewarmed (37°C) SOC, and plate 100 μl of each dilution on LB+carb plates. Incubate the plates overnight at 37°C. Count the colonies on the vector + Fab insert plates and calculate the number of transformants per μg of vector DNA. If this number does not exceed 1×10^7, do not proceed with the large-scale library ligation. Ideally, the final library size should be several times 10^8 but at least 5×10^7 total transformants. Determine the number of scaled-up library ligations needed to achieve this size. Count the colonies obtained from control ligation 1 for an indication of vector quality and ligation efficiency. A good vector DNA preparation should yield at least 10^8 colony-forming units (cfu) per μg of vector DNA and should have less than 10% (ideally less than 5%) background ligation (calculated cfu per μg of vector DNA in control ligation 2). If the background is greater than 10% or the ligation efficiency is low, digest new vector and repeat the ligations. If the ligations are reasonably good, but there is not enough vector or insert to make a library of the desired size, perform additional digests.

Library Ligation

Assemble one or more large-scale library ligations containing:

1.4 μg	pComb3HSS or pComb3XSS, *Sfi*I-digested and purified
1.4 μg	overlap PCR product, *Sfi*I-digested and purified

40 μl	5× ligase buffer
10 μl	ligase

Add water to a total volume of 200 μl.

Incubate overnight at room temperature, add 1 μl of glycogen solution as carrier, and ethanol-precipitate as described in Appendix 3 (incubate between 4 hours and overnight at –20°C). The ethanol-precipitated ligation reaction can be stored for months at –20°C. Transform the ligation reaction by resuspending in 15 μl water and electroporating into XL1-Blue or ER2537 electrocompetent cells, as described in Chapter 10.

Proceed with library preparation and panning as described in Chapter 10.

PROTOCOL 9.5

Mouse scFv Libraries (by Overlap Extension)

The following protocol describes the construction of mouse single-chain antibody libraries (short linker or long linker libraries) by overlap extension PCR. Depending on whether a short-linker (SL) or a long-linker (LL) library is to be constructed, different reverse primers have to be used for the amplification of the V_L regions. Sequences of two sets of V_L reverse primers are given below. The long-linker primers are used to construct a single-chain library in which the V_L polypeptide chain is linked to the V_H chain with an 18-amino-acid linker, whereas the use of the short-linker primers generates single-chain fragments with a 7-amino-acid linker. For more information regarding the use of short and long peptide linkers, see Protocol 9.2 and Chapter 3. For a brief introduction to the organization of mouse immunoglobulin genes, see Protocol 9.4.

Materials

cDNA (prepared from immunized mice, see Protocol 8.4)
oligonucleotide primers (listed below; see Appendix 1 for letter code)
2.5 mM dNTP mix (dATP/dCTP/dGTP/dTTP set, 100 mM, Amersham Pharmacia Biotech, Cat. # 27-2035-02)
AmpliTaq DNA polymerase (Perkin-Elmer, Cat. # N808-0152)
10x PCR buffer (Perkin-Elmer, supplied with *Taq* DNA polymerase)
MicroAmp PCR tubes (Perkin-Elmer, Cat. # N801-0533)
MicroAmp PCR caps (Perkin-Elmer, Cat. # N801-0534)
MicroAmp PCR trays (Perkin-Elmer, Cat. # N801-5530)
GeneAmp 9600 PCR cycler (Perkin-Elmer)
agarose (Gibco-BRL, Cat. # 15510-027)
100-bp DNA molecular weight marker (Amersham Pharmacia Biotech, Cat. # 27-4001-01) or 1-kb DNA molecular weight marker (Gibco-BRL, Cat. # 15615-024)
1x TAE electrophoresis running buffer (see Appendix 2)
DNA gel loading dye (see Appendix 2)
pComb3HSS or pComb3XSS (available from the Barbas laboratory)
*Sfi*I restriction enzyme (Roche Molecular Biochemicals, 40 Units/μl, Cat. # 1 288 059)
10x restriction enzyme buffer M (supplied with *Sfi*I restriction enzyme)
T4 DNA ligase (Gibco-BRL, 1 Unit/μl, Cat. # 15224090)
5x T4 DNA ligase buffer (supplied with enzyme)
glycogen, 20 mg/ml (Roche Molecular Biochemicals, Cat # 901 393)
electroporation apparatus and materials
electrocompetent *E. coli* cells (see Chapter 10): XL1-Blue (Stratagene, Cat. # 200228) or ER2537 (New England Biolabs, Cat. # 801-N)
LB + 100 μg/ml carbenicillin plates (see Appendix 2)

CAUTION: *E. coli (see Appendix 4)*

PRIMERS FOR MOUSE SINGLE-CHAIN LIBRARIES, SHORT LINKER (SL) AND LONG LINKER (LL)

The number to the right of each primer indicates the amount (in μl, at equimolar concentrations) that should be used when setting up the primer mixes (for example, see Step 1).

V_{κ} 5′ Sense Primers

MSCVK-1
5′ GGG CCC AGG CGG CCG AGC TCG AYA TCC AGC TGA CTC AGC C 3′ 1

MSCVK-2
5′ GGG CCC AGG CGG CCG AGC TCG AYA TTG TTC TCW CCC AGT C 3′ 2

MSCVK-3
5′ GGG CCC AGG CGG CCG AGC TCG AYA TTG TGM TMA CTC AGT C 3′ 5

MSCVK-4
5′ GGG CCC AGG CGG CCG AGC TCG AYA TTG TGY TRA CAC AGT C 3′ 3.5

MSCVK-5
5′ GGG CCC AGG CGG CCG AGC TCG AYA TTG TRA TGA CMC AGT C 3′ 4

MSCVK-6
5′ GGG CCC AGG CGG CCG AGC TCG AYA TTM AGA TRA MCC AGT C 3′ 7

MSCVK-7
5′ GGG CCC AGG CGG CCG AGC TCG AYA TTC AGA TGA YDC AGT C 3′ 6

MSCVK-8
5′ GGG CCC AGG CGG CCG AGC TCG AYA TYC AGA TGA CAC AGA C 3′ 1.5

MSCVK-9
5′ GGG CCC AGG CGG CCG AGC TCG AYA TTG TTC TCA WCC AGT C 3′ 2

MSCVK-10
5′ GGG CCC AGG CGG CCG AGC TCG AYA TTG WGC TSA CCC AAT C 3′ 3.5

MSCVK-11
5′ GGG CCC AGG CGG CCG AGC TCG AYA TTS TRA TGA CCC ART C 3′ 8

MSCVK-12
5′ GGG CCC AGG CGG CCG AGC TCG AYR TTK TGA TGA CCC ARA C 3′ 8

MSCVK-13
5′ GGG CCC AGG CGG CCG AGC TCG AYA TTG TGA TGA CBC AGK C 3′ 6

MSCVK-14
5′ GGG CCC AGG CGG CCG AGC TCG AYA TTG TGA TAA CYC AGG A 3′ 2

MSCVK-15
5′ GGG CCC AGG CGG CCG AGC TCG AYA TTG TGA TGA CCC AGW T 3′ 2

MSCVK-16
5′ GGG CCC AGG CGG CCG AGC TCG AYA TTG TGA TGA CAC AAC C 3′ 1

MSCVK-17
5′ GGG CCC AGG CGG CCG AGC TCG AYA TTT TGC TGA CTC AGT C 3′ 1

V_κ 3′ Reverse Primers, Short Linker

(linker amino acid sequence: GGSSRSS)

MSCJK12-B
5′ GGA AGA TCT AGA GGA ACC ACC TTT KAT TTC CAG YTT GGT CCC 3′ 2

MSCJK4-B
5′ GGA AGA TCT AGA GGA ACC ACC TTT TAT TTC CAA CTT TGT CCC 3′ 1

MSCJK5-B
5′ GGA AGA TCT AGA GGA ACC ACC TTT CAG CTC CAG CTT GGT CCC 3′ 1

V_κ 3′ Reverse Primers, Long Linker

(linker amino acid sequence: SSGGGGSGGGGGGSSRSS)

MSCJK12-BL
5′ GGA AGA TCT AGA GGA ACC ACC CCC ACC ACC GCC CGA GCC ACC GCC ACC
AGA GGA TTT KAT TTC CAG YTT GGT CCC 3′ 2

MSCJK4-BL
5′ GGA AGA TCT AGA GGA ACC ACC CCC ACC ACC GCC CGA GCC ACC GCC ACC
AGA GGA TTT TAT TTC CAA CTT TGT CCC 3′ 1

MSCJK5-BL
5′ GGA AGA TCT AGA GGA ACC ACC CCC ACC ACC GCC CGA GCC ACC GCC ACC
AGA GGA TTT CAG CTC CAG CTT GGT CCC 3′ 1

V_λ 5′ Sense Primer

MSCVL-1
5′ GGG CCC AGG CGG CCG AGC TCG ATG CTG TTG TGA CTC AGG AAT C 3′

V_λ 3′ Reverse Primer, Short Linker

(linker amino acid sequence: GGSSRSS):

MSCJL-B
5′ GGA AGA TCT AGA GGA ACC ACC GCC TAG GAC AGT CAG TTT GG 3′

V_λ 3′ Reverse Primer, Long Linker

(linker amino acid sequence: SSGGGGSGGGGGGSSRSS)

MSCJL-BL
5′ GGA AGA TCT AGA GGA ACC ACC CCC ACC ACC GCC CGA GCC ACC GCC ACC AGA GGA GCC TAG GAC AGT CAG TTT GG 3′

V_H 5′ Sense Primers

MSCVH1
5′ GGT GGT TCC TCT AGA TCT TCC CTC GAG GTR MAG CTT CAG GAG TC 3′ 4

MSCVH2
5′ GGT GGT TCC TCT AGA TCT TCC CTC GAG GTB CAG CTB CAG CAG TC 3′ 4

MSCVH3
5′ GGT GGT TCC TCT AGA TCT TCC CTC GAG GTG CAG CTG AAG SAS TC 3′ 3

MSCVH4
5′ GGT GGT TCC TCT AGA TCT TCC CTC GAG GTC CAR CTG CAA CAR TC 3′ 4

MSCVH5
5′ GGT GGT TCC TCT AGA TCT TCC CTC GAG GTY CAG CTB CAG CAR TC 3′ 7

MSCVH6
5′ GGT GGT TCC TCT AGA TCT TCC CTC GAG GTY CAR CTG CAG CAG TC 3′ 2

MSCVH7
5′ GGT GGT TCC TCT AGA TCT TCC CTC GAG GTC CAC GTG AAG CAG TC 3′ 1

MSCVH8
5′ GGT GGT TCC TCT AGA TCT TCC CTC GAG GTG AAS STG GTG GAA TC 3′ 2

MSCVH9
5′ GGT GGT TCC TCT AGA TCT TCC CTC GAG GTG AWG YTG GTG GAG TC 3′ 5

MSCVH10
5′ GGT GGT TCC TCT AGA TCT TCC CTC GAG GTG CAG SKG GTG GAG TC 3′ 2

MSCVH11
5′ GGT GGT TCC TCT AGA TCT TCC CTC GAG GTG CAM CTG GTG GAG TC 3′ 2

MSCVH12
5′ GGT GGT TCC TCT AGA TCT TCC CTC GAG GTG AAG CTG ATG GAR TC 3′ 2

MSCVH13
5′ GGT GGT TCC TCT AGA TCT TCC CTC GAG GTG CAR CTT GTT GAG TC 3′ 1

MSCVH14
5′ GGT GGT TCC TCT AGA TCT TCC CTC GAG GTR AAG CTT CTC GAG TC 3′ 2

MSCVH15
5′ GGT GGT TCC TCT AGA TCT TCC CTC GAG GTG AAR STT GAG GAG TC 3′ 2

MSCVH16
5′ GGT GGT TCC TCT AGA TCT TCC CTC GAG GTT ACT CTR AAA GWG TST G 3′ 5

MSCVH17
5′ GGT GGT TCC TCT AGA TCT TCC CTC GAG GTC CAA CTV CAG CAR CC 3′ 3.5

MSCVH18
5′ GGT GGT TCC TCT AGA TCT TCC CTC GAG GTG AAC TTG GAA GTG TC 3′ 0.7

MSCVH19
5′ GGT GGT TCC TCT AGA TCT TCC CTC GAG GTG AAG GTC ATC GAG TC 3′ 0.7

V_H 3′ Reverse Primers (Specific for the 5′ End of the First Constant Regions)

MSCG1ab-B
5′ CCT GGC CGG CCT GGC CAC TAG TGA CAG ATG GGG STG TYG TTT TGG C 3′ 3

MSCG3-B
5′ CCT GGC CGG CCT GGC CAC TAG TGA CAG ATG GGG CTG TTG TTG T 3′ 1

MSCM-B
5′ CCT GGC CGG CCT GGC CAC TAG TGA CAT TTG GGA AGG ACT GAC TCT C 3′ 1

Overlap Extension Primers

RSC-F (sense)
5′ GAG GAG GAG GAG GAG GAG GCG GGG CCC AGG CGG CCG AGC TC 3′

RSC-B (reverse)
5′ GAG GAG GAG GAG GAG GAG CCT GGC CGG CCT GGC CAC TAG TG 3′

Procedure

1. First round of PCR.

Note: PCR products and other DNA samples, precipitated or in solution, can be stored at –20°C for years. The construction of antibody fragment libraries, therefore, can be interrupted at any step.

Assemble the following PCRs to amplify mouse V_L sequences. Perform ten reactions to generate either short- or long-linker libraries, including nine reactions with V_κ primers and one reaction with V_λ primers (see Figs. 9.26 and 9.27). Make the primer mixes by combining the primers in the ratios indicated in the primer list at the beginning of the protocol.

1 µl	cDNA (about 0.5 µg)
60 pmole	sense primer (primer combinations are listed below)
60 pmole	reverse primer
10 µl	10x PCR buffer
8 µl	2.5 mM dNTP mix
0.5 µl	*Taq* polymerase

Add water to bring the final reaction volume to 100 µl.

V_κ short-linker primer combination:

MSCVK primer mix (sense; MSCVK1 through 17)

MSCJK-B, short primer mix (reverse; MSCJK12-B, 4-B, and 5-B)

V_λ short-linker primer combination:

MSCVL-1 primer (sense)

MSCJL-B primer (reverse)

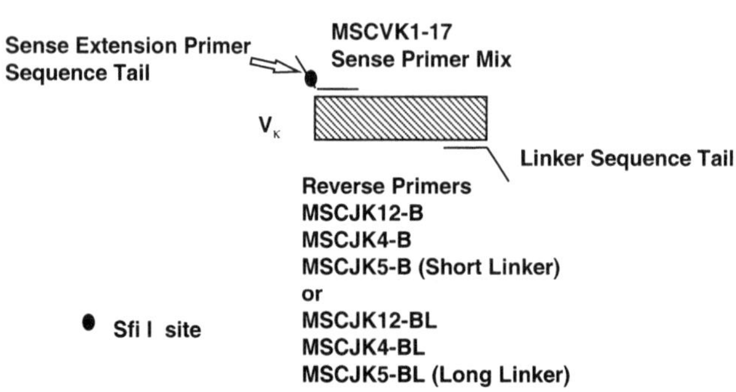

Figure 9.26. The amplification of mouse V_κ sequences for the construction of scFv short- and long-linker libraries. A mixture of MSCVK sense primers is combined with a mixture of MSCJK-B (short linker) or MSCJK-BL (long linker) reverse primers to amplify mouse V_κ gene segments from cDNA. Each sense primer has a 5′ sequence tail that contains an *Sfi*I site and is recognized by the sense extension primer in the second-round PCR. The reverse primer has a linker sequence tail that is used in the overlap extension.

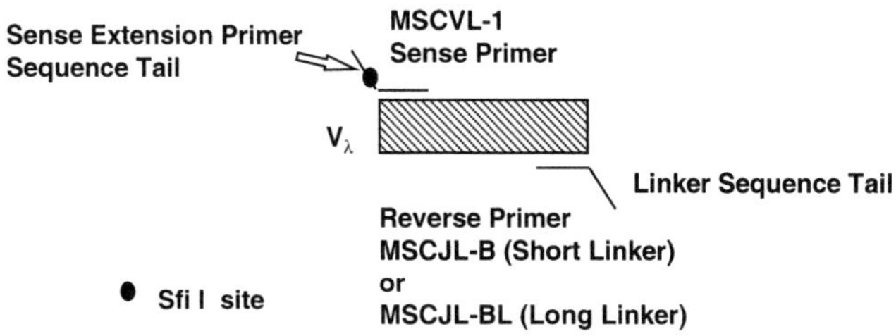

Figure 9.27. The amplification of mouse V_λ sequences for the construction of scFv short- and long-linker libraries. The MSCVL-1 sense primer is combined with the MSCJL-B (short linker) or MSCJL-BL (long linker) reverse primer to amplify mouse V_λ gene segments from cDNA. MSCVL-1 has a 5′ sequence tail that contains an *Sfi*I site and is recognized by the sense extension primer in the second-round PCR. The reverse primers have a linker sequence tail that is used in the overlap extension.

V_κ long-linker primer combination:
- MSCVK primer mix (sense; MSCVK1–17)
- MSCJK-BL, long primer mix (reverse; MSCJK12-BL, 4-BL, 5-BL)

V_λ long-linker primer combination:
- MSCVL-1 primer (sense)
- MSCJL-BL primer (reverse)

Assemble the following PCRs to amplify mouse V_H sequences (see Fig. 9.28). Twelve 100-μl amplification reactions with the V_H primer mix should yield enough PCR product to construct a large single-chain library. Because the V_H PCR products can be used for the construction of both short- and long-linker libraries, the number of V_H PCRs should be increased if both library types are to be constructed from an immune source.

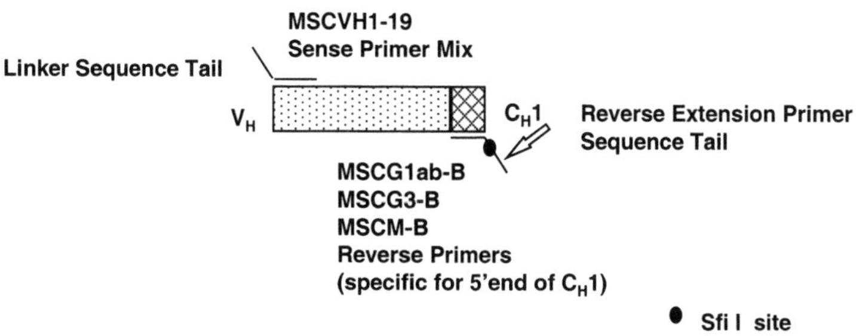

Figure 9.28. The amplification of mouse V_H sequences for the construction of scFv short- and long-linker libraries. A mixture of MSCVH sense primers is combined with reverse primers specific for the 5′ end of mouse C_H1 regions to amplify V_H gene segments from cDNA. The sense primers have a sequence tail that corresponds to the linker sequence used in the overlap extension PCR. Each reverse primer has a sequence tail containing an *Sfi*I site; this sequence tail is recognized by the reverse extension primer used in the second-round PCR.

1 µl	cDNA mix (~0.5µg)
60 pmole	MSCVH sense primer mix (MSCVH1–19)
60 pmole	MSCVH reverse primer mix (MSCG1ab-B, MSCG3-B, MSCM-B)
10 µl	10x PCR buffer
8 µl	2.5 mM dNTP mix
0.5 µl	*Taq* polymerase

Add water to bring the final reaction volume to 100 µl.

Note: The V_H reverse primers are specific for the 5′ ends of the first constant region of different heavy-chain classes and subclasses. The V_H primers in the primer list at the beginning of the protocol allow for the amplification of V_H regions of mouse IgG1, IgG2a, IgG2b, IgG3, and IgM sequences. (MSCG1ab amplifies the IgG1, IgG2a, and IgG2b V_H regions.) The use of only primer IgG1ab favors the cloning of high-affinity antibodies from a hyperimmune animal.

Perform the first-round PCR under the following conditions:

30 cycles of

- 94°C for 15 seconds
- 56°C for 30 seconds
- 72°C for 90 seconds

followed by 72°C for 10 minutes

Note: A "hot start" PCR protocol can improve specificity, sensitivity, and yield. In hot start PCR, either an essential reaction component is not added until the first denaturing step, or a reversible inhibitor of the polymerase is used. This protocol prevents low-stringency primer extension, which can generate nonspecific products.

Analyze 5–10 µl of each reaction on a 2% agarose gel, using DNA gel loading dye and appropriate MWM. The PCR products should have a size of about 350 bp.

Note: Try a test amplification with the cDNA sample. If good amplification is not obtained (if a strong amplified band is not clearly visible in the agarose gel), lithium chloride precipitation of the RNA might improve the results (see Protocol 8.3).

2. **Isolate the PCR products.**

Combine the light-chain short-linker PCR products and the light-chain long-linker PCR products into two separate pools. Combine the heavy-chain products. Ethanol-precipitate and wash as described in Appendix 3. Load and run the products on a 1.5% agarose gel, cut out the correct-sized bands, and purify the DNA by freeze-squeeze, electroelution with an Elutrap (see Appendix 3), or resin binding (e.g., QIAEX II Gel Extraction Kit, QIAGEN). Quantify the purified PCR products by reading the O.D. at 260 nm (1 O.D. = 50 µg/ml). At least 1 µg of each purified, pooled PCR product is required to proceed; if both short- and long-linker libraries are to be made, at least 2 µg of the heavy-chain PCR pool is needed.

3. Second round of PCR (overlap extension).

Assemble overlap extension PCRs to combine the V_L products with the V_H products (see Fig. 9.29). Perform at least ten reactions for either short- or long-linker single-chain fragments.

100 ng	purified light chain product (V_λ and V_κ), either short or long linker
100 ng	purified heavy-chain product
60 pmole	RSC-F sense primer
60 pmole	RSC-B reverse primer
10 µl	10x PCR buffer
8 µl	2.5 mM dNTP mix
0.5 µl	*Taq* polymerase

Add water to bring the final reaction volume to 100 µl.

Perform the second-round PCR under the following conditions:

20 cycles of
- 94°C for 15 seconds
- 56°C for 30 seconds
- 72°C for 2 minutes

followed by 72°C for 10 minutes

Analyze 5–10 µl of each reaction on a 1.5% agarose gel, using DNA gel loading dye and appropriate MWM. The size of the overlap PCR products (short and long linker) should be around 700 bp.

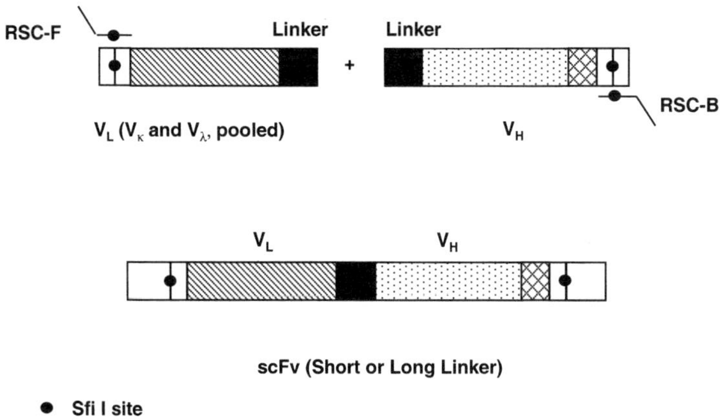

Figure 9.29. The overlap extension PCR to combine the mouse V_L and V_H fragments for the construction of scFv libraries (short or long linker). The sense and reverse extension primers used in this round of PCR (RSC-F and RSC-B) recognize the sequence tails that were generated in the first-round PCR.

Note: The long PCR extension time favors full-length product. If a strong band of the correct size is not clearly visible in the gel, the number of cycles can be increased or more overlap PCRs can be performed. To retain diversity of the library, it is better to increase the number of PCRs.

4. **Isolate the PCR products.**

Combine the short- and long-linker PCR products into two separate pools. Ethanol-precipitate and wash as described in Appendix 3. Purify the DNA on a 1% agarose gel as in Step 2. Quantify the purified PCR products by reading the O.D. at 260 nm. 10–15 μg of purified product is usually sufficient to continue.

5. **Restriction-digest the purified overlap extension product and the vector DNA.**

Prepare the PCR products and vector for cloning by performing restriction digests with *Sfi*I. We prefer high-concentration (40 units/μl) *Sfi*I and the 10x buffer M from Roche Molecular Biochemicals.

The digest of the PCR product should contain:

10 μg of purified overlap PCR product

360 Units of *Sfi*I (36 Units per μg of DNA)

20 μl of 10x reaction buffer M

Add water to a total volume of 200 μl.

The digest of the vector should contain:

20 μg of vector DNA (pComb3HSS or pComb3XSS, contains stuffer fragment between two *Sfi*I sites)

120 Units of *Sfi*I (6 Units per μg of plasmid DNA)

20 μl of 10x reaction buffer M

Add water to a total volume of 200 μl.

Note: If the loss of digested PCR product in the final purification step is high or if several large-size ligations are needed to obtain a library of the desired size, digest more PCR product with *Sfi*I. The amount of added enzyme should not exceed 10% of the reaction volume. If desired, larger quantities of *Sfi*I-cut and purified vector DNA can be prepared and tested in advance.

Incubate both digests for 5 hours at 50°C. Ethanol-precipitate and wash the DNA as described in Appendix 3 (optional, see Note). Purify the digested PCR insert (~700 bp) on a 1% agarose gel, and purify the vector (~3400 bp) and the stuffer fragment (~1600 bp) on a 0.6% agarose gel. We recommend electroelution with an Elutrap for extraction of the vector from the gel. Suboptimal digestion or purification will yield vector with low transformation efficiencies and high background after ligation. Use the stuffer fragment in a test ligation to assess vector quality (see Step 6). Quantify the purified, *Sfi*I-digested PCR products, vector, and stuffer fragment by measuring the O.D. at 260 nm.

Note: In some cases *Sfi*I-digested DNA is partly insoluble after ethanol precipitation. To avoid this problem, the DNA can be loaded directly after digestion onto the agarose gel. When purifying the *Sfi*I-digested vector DNA, let the DNA run long enough to separate linearized vector DNA (size ~ 5000 bp; cut only once) and uncut vector DNA from the desired double-cut product (size ~ 3400 bp).

6. **Ligate the digested overlap PCR product with vector DNA.**

   Perform small-scale ligations to assess the suitability of the vector and inserts for high-efficiency ligation and transformation. The ligation efficiency of the vector DNA can be tested by ligating it with the gel-purified stuffer fragment that is generated during the *Sfi*I digestion of the vector DNA. The vector preparation should contain little uncut vector DNA or DNA that is cut only once. The amount of uncut or partly cut vector DNA can be estimated by setting up a ligation reaction that contains only vector DNA. These two types of control ligations should be done in parallel with a small-scale test ligation and should both include the same amount of vector DNA.

Test Ligations

Assemble the following small-scale ligation reactions:

small-scale test ligation:

140 ng	pComb3HSS or pComb3XSS, *Sfi*I-digested and purified
70 ng	overlap PCR product, *Sfi*I-digested and purified
4 μl	5x ligase buffer
1 μl	ligase

Add water to a total volume of 20 μl.

control ligation 1 (control insert):

140 ng	pComb3HSS or pComb3XSS, *Sfi*I-digested and purified
140 ng	gel-purified stuffer fragment
4 μl	5x ligase buffer
1 μl	ligase

Add water to a total volume of 20 μl.

control ligation 2 (test for vector self-ligation):

140 ng	pComb3HSS or pComb3XSS, *Sfi*I-digested and purified
4 μl	5x ligase buffer
1 μl	ligase

Add water to a total volume of 20 μl.

Incubate between 4 hours and overnight at room temperature. Transform 0.5–1 μl of each reaction by electroporation into 50 μl of XL1-Blue or ER2537 (see Chapter 10). Dilute the transformed cultures 10-fold and 100-fold with prewarmed (37°C) SOC, and plate 100 μl of each dilution on LB+carb plates. Incubate the plates overnight at 37°C. Count the colonies on the vector + scFv insert plates and calculate the number of transformants per μg of vector DNA. If this number does not exceed 1×10^7, do not proceed with the large-scale library ligation. Ideally, the final library size should be several times 10^8 but at least 5×10^7 total transformants. Determine the number of scaled-up library ligations needed to achieve this size. Count the colonies obtained from control ligation 1 for an indication of vector quality and ligation efficiency. A good vector DNA preparation should yield at least 10^8 colony-forming units (cfu) per μg of vector DNA and should have less than 10% (ideally less than 5%) background ligation (calculated cfu per μg of vector DNA in control ligation 2). If the background is greater than 10% or the ligation efficiency is low, digest new vector and repeat the ligations. If the ligations are reasonably good, but there is not enough vector or insert to make a library of the desired size, perform additional digests.

Library Ligation

Assemble one or more large-scale library ligations containing:

1.4 μg	pComb3HSS or pComb3XSS, *Sfi*I-digested and purified
700 ng	overlap PCR product, *Sfi*I-digested and purified
40 μl	5x ligase buffer
10 μl	ligase

Add water to a total volume of 200 μl.

Incubate overnight at room temperature. Add 1 μl of glycogen solution as carrier, and ethanol-precipitate as described in Appendix 3 (incubate between 4 hours and overnight at –20°C). The ethanol-precipitated ligation reaction can be stored for months at –20°C. Transform the ligation reaction by resuspending in 15 μl of water and electroporating into XL1-Blue or ER2537 electrocompetent cells, as described in Chapter 10.

Proceed with library preparation and panning as described in Chapter 10.

PROTOCOL 9.6

Chimeric Rabbit/Human Fab Libraries (by Overlap Extension)

Rabbits are widely used for the generation of polyclonal sera to peptides and other antigens. A variety of antigens and epitopes elicit better immune responses in rabbits than in mice and rats, which are usually used for the generation of monoclonal antibodies.

Specific rabbit antibody fragments have been isolated from phage-display libraries by us and by other ivestigators (Ridder et al. 1995; Lang et al. 1996; Foti et al. 1998; Rader et al. 2000). In rabbits, unlike many mammals such as mice and primates, the generation of antibody diversity does not rely on the use of many V-segments. Most rabbit B-lymphocytes recombine the same V_H gene in the V(D)J gene rearrangements. These genes are then diversified through gene conversion and hypermutation (Currier et al. 1988; Knight and Winstead 1997). Compared to mice and primates, therefore, a relatively small number of V-region-specific primers are required for the generation of rabbit antibody fragment libraries.

This protocol describes the construction of a chimeric rabbit/human Fab library by overlap extension PCR. It allows the expression and selection of chimeric Fabs that have variable regions obtained from an immunized rabbit and light- and heavy-chain constant regions from a cloned human Fab. This approach facilitates the humanization of selected clones with therapeutic potential using phage display. Chimeric Fabs can be detected with antihuman Fab reagents, and chimeric Fabs with human constant regions are usually better expressed in *E. coli* than rabbit Fabs.

Materials

cDNA (prepared from immunized rabbits, see Protocol 8.4)
cloned human Fab template (e.g., pComb3XTT; available from the Barbas laboratory)
oligonucleotide primers (listed below; see Appendix 1 for letter code)
2.5 mM dNTP mix (dATP/dCTP/dGTP/dTTP set, 100 mM, Amersham Pharmacia Biotech, Cat. # 27-2035-02)
AmpliTaq DNA polymerase (Perkin-Elmer, Cat. # N808-052)
10x PCR buffer for use with *Taq* polymerase (supplied with polymerase)
Expand High Fidelity Enzyme mix; 3.5 Units/µl (Roche Molecular Biochemicals, Cat. # 1 732 650)
10x Expand High Fidelity PCR buffer (supplied with Expand High Fidelity Enzyme mix)
MicroAmp PCR tubes (Perkin-Elmer, Cat. # N801-0533)
MicroAmp PCR caps (Perkin-Elmer, Cat. # N801-0534)
MicroAmp PCR trays (Perkin-Elmer, Cat. # N801-5530)

GeneAmp 9600 PCR cycler (Perkin-Elmer)
agarose (Gibco-BRL, Cat. # 15510-027)
100-bp DNA molecular weight marker (Amersham Pharmacia Biotech, Cat. # 27-4001-01) or 1-kb DNA molecular weight marker (Gibco-BRL, Cat. # 15615-024)
1x TAE electrophoresis running buffer (see Appendix 2)
DNA gel loading dye (see Appendix 2)
pComb3HSS or pComb3HXX (available from the Barbas laboratory)
*Sfi*I restriction enzyme (Roche Molecular Biochemicals, 40 Units/μl, Cat. # 1 288 059)
10x restriction enzyme buffer M (supplied with *Sfi*I restriction enzyme)
T4 DNA ligase (Gibco-BRL, 1 Unit/μl, Cat. # 15224 090)
5x T4 DNA ligase buffer (supplied with T4 DNA ligase)
glycogen, 20 mg/ml (Roche Molecular Biochemicals, Cat. # 901 393)
electoporation apparatus and materials
electrocompetent ***E. coli*** cells (see Chapter 10): XL1-Blue (Stratagene, Cat. # 200228) or ER2537 (New England Biolabs, Cat. # 801-N)
LB + 100 μg/ml carbenicillin plates (see Appendix 2)

CAUTION: ***E. coli (see Appendix 4)***

PRIMERS FOR CHIMERIC RABBIT/HUMAN Fab LIBRARIES

V_κ 5′ Sense Primers

RSCVK1
5′ GGG CCC AGG CGG CCG AGC TCG TGM TGA CCC AGA CTC CA 3′

RSCVK2
5′ GGG CCC AGG CGG CCG AGC TCG ATM TGA CCC AGA CTC CA 3′

RSCVK3
5′ GGG CCC AGG CGG CCG AGC TCG TGA TGA CCC AGA CTG AA 3′

V_κ 3′ Reverse Primers

RHybK1-B
5′ AGA TGG TGC AGC CAC AGT TCG TTT GAT TTC CAC ATT GGT GCC 3′

RHybK2-B
5′ AGA TGG TGC AGC CAC AGT TCG TAG GAT CTC CAG CTC GGT CCC 3′

RHybK3-B
5′ AGA TGG TGC AGC CAC AGT TCG TTT GAC SAC CAC CTC GGT CCC 3′

V_λ 5′ Sense Primer

RSCλ1
5′ GGG CCC AGG CGG CCG AGC TCG TGC TGA CTC AGT CGC CCT C 3′

V_λ 3′ Reverse Primer

RHybL-B
5′ AGA TGG TGC AGC CAC AGT TCG GCC TGT GAC GGT CAG CTG GGT CCC 3′

V_H 5′ Sense Primers

RHyVH1
5′ GCT GCC CAA CCA GCC ATG GCC CAG TCG GTG GAG GAG TCC RGG 3′

RHyVH2
5′ GCT GCC CAA CCA GCC ATG GCC CAG TCG GTG AAG GAG TCC GAG 3′

RHyVH3
5′ GCT GCC CAA CCA GCC ATG GCC CAG TCG YTG GAG GAG TCC GGG 3′

RHyVH4
5′ GCT GCC CAA CCA GCC ATG GCC CAG SAG CAG CTG RTG GAG TCC GG 3′

V_H 3′ Reverse Primers

RHyIgGCH1-B
5′ CGA TGG GCC CTT GGT GGA GGC TGA RGA GAY GGT GAC CAG GGT GCC 3′

Primers for Amplification of the Human C_κ Region and the pelB Leader Sequence from a Cloned Human Fab

HKC-F (sense)
5′ CGA ACT GTG GCT GCA CCA TCT GTC 3′

Lead-B (reverse)
5′ GGC CAT GGC TGG TTG GGC AGC 3′

Primers for Amplification of the Human C_H1 Chain from a Cloned Human Fab

HIgGCH1-F (sense)
5′ GCC TCC ACC AAG GGC CCA TCG GTC 3′

dpseq (reverse)
5′ AGA AGC GTA GTC CGG AAC GTC 3′

Primers for PCR Assembly of Rabbit V_L Sequences with the Human C_κ PCR Product

RSC-F (sense)
5′ GAG GAG GAG GAG GAG GAG GCG GGG CCC AGG CGG CCG AGC TC 3′

Lead-B (reverse)
(see above)

Primers for PCR Assembly of Rabbit V_H Sequences with the Human C_H1 PCR Product

leadVH (sense)
5′ GCT GCC CAA CCA GCC ATG GCC 3′

dpseq (reverse)
(see above)

Primers for PCR Assembly of Chimeric Light-chain Sequences with Chimeric Heavy-chain (Fd) Sequences

RSC-F (sense)
(see above)

dp-EX (reverse)
5′ GAG GAG GAG GAG GAG GAG AGA AGC GTA GTC CGG AAC GTC 3′

Procedure

1. **First round of PCR.**

   Note: PCR products and other DNA samples, precipitated or in solution, can be stored at –20°C for years. The construction of antibody fragment libraries, therefore, can be interrupted at any step.

   For V_κ and V_λ amplifications, if spleen and bone marrow have been kept separate for RNA preparation and cDNA synthesis, one PCR of each kind should be performed with cDNA derived from spleen, and one with cDNA derived from bone marrow. For V_H amplifications, two PCRs of each kind should be performed with cDNA derived from spleen, and two with cDNA derived from bone marrow.

Assemble the following PCRs to amplify rabbit V_L and V_H sequences:

1 µl	cDNA (about 0.5 µg)
60 pmole	sense primer (primer combinations are listed below)
60 pmole	reverse primer
10 µl	10x PCR buffer
8 µl	2.5 mM dNTP mix
0.5 µl	*Taq* polymerase

Add water to bring the final reaction volume to 100 µl.

V_κ primer combinations (see Fig. 9.30): Perform two reactions for each of the nine primer combinations; 18 reactions in total.

RSCVK1 RHybK1-B	RSCVK3 RHybK2-B
RSCVK2 RHybK1-B	RSCVK1 RHybK3-B
RSCVK3 RHybK1-B	RSCVK2 RHybK3-B
RSCVK1 RHybK2-B	RSCVK3 RHybK3-B
RSCVK2 RHybK2-B	

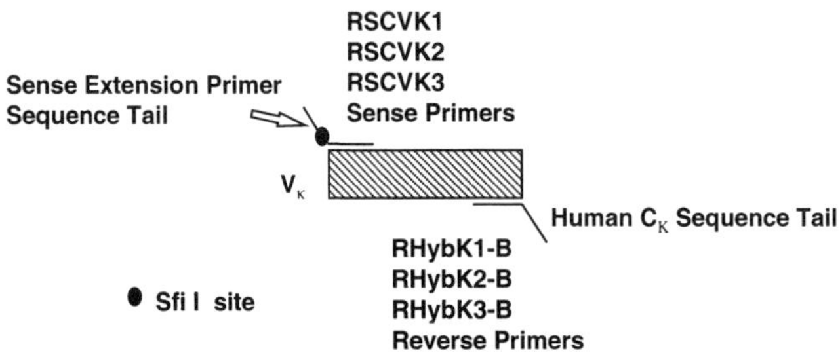

Figure 9.30. The amplification of rabbit V_κ sequences for the construction of chimeric Fab libraries. Each RSCVK sense primer is combined with each RHybK-B reverse primer to amplify V_κ gene segments from rabbit cDNA. Each sense primer has a 5′ sequence tail that contains an *SfiI* site and is recognized by the sense extension primer in the second-round PCR. Each reverse primer has a human C_κ sequence tail that is used in the overlap extension PCR to create the chimeric light-chain fragment.

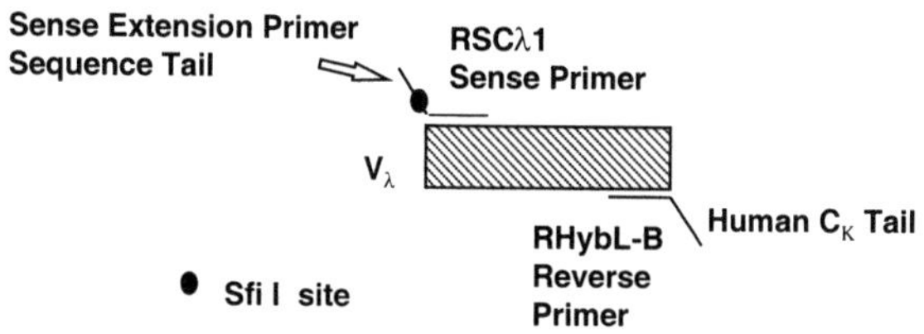

Figure 9.31. The amplification of rabbit V_λ sequences for the construction of chimeric Fab libraries. The RSCλ1 sense primer is combined with the RHybL-B reverse primer to amplify V_λ gene segments from cDNA. RSCλ1 has a 5′ sequence tail that contains an *Sfi*I site and is recognized by the sense extension primer in the second- and third-round PCR. RHybL-B has a human C_κ sequence tail that is used in the overlap extension PCR to create the chimeric light-chain fragment.

V_λ primer combination (see Fig. 9.31): Perform two reactions.

RSCλ1
RHybL-B

V_H primer combinations (see Fig. 9.32): Perform four reactions for each of the four primer combinations, 16 reactions in total.

RHyVH1	RHyVH3
RHyIgGCH1-B	RHyIgGCH1-B
RHyVH2	RHyVH4
RHyIgGCH1-B	RHyIgGCH1-B

Perform the first-round V region PCR under the following conditions:

30 cycles of:

94°C for 15 seconds
56°C for 30 seconds
72°C for 90 seconds

followed by 72°C for 10 minutes

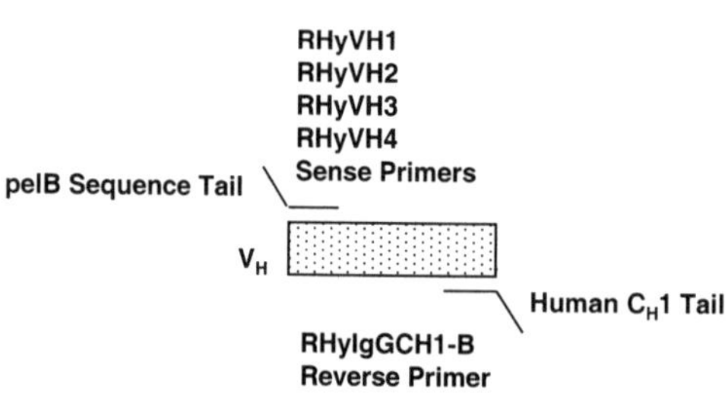

Figure 9.32. The amplification of rabbit V_H sequences for the construction of chimeric Fab libraries. Each RHyVH sense primer is combined with the RHyIgGCH1-B reverse primer to amplify rabbit V_H gene segments from cDNA. Each sense primer has a sequence tail that corresponds to the 3′ end of the pelB leader sequence and is used in the final overlap extension PCR. The reverse primer has a human C_H1 sequence tail that is used in the second-round overlap extension PCR to create the chimeric heavy-chain Fd fragment.

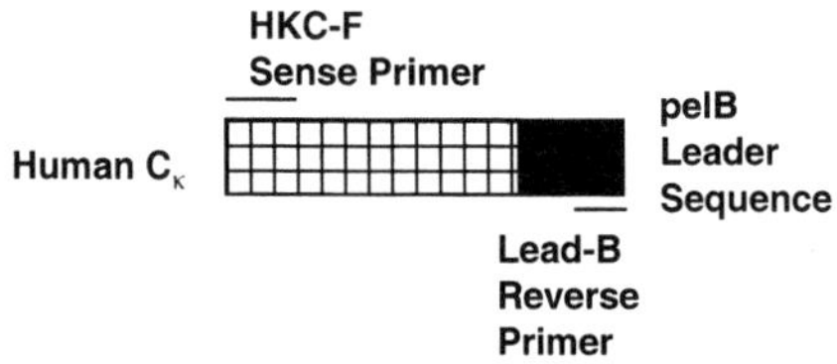

Figure 9.33. The first-round amplification of C_κ products and the pelB leader sequence from the pComb3XTT vector. The sense primer is specific for the 5′ region of human C_κ, the region used in the overlap extension PCR to create the chimeric light chain. The reverse primer is specific for the 3′ end of the pelB leader sequence.

Assemble the following PCRs to amplify the human κ constant region and the pelB leader sequence from a cloned human Fab template (see Fig. 9.33). Perform ten reactions for one library.

20 ng	plasmid template containing a cloned human Fab fragment (e. g. pComb3XTT)
60 pmole	HKC-F sense primer
60 pmole	Lead-B reverse primer
10 µl	10x PCR buffer
8 µl	2.5 mM dNTP mix
0.5 µl	*Taq* polymerase

Add water to bring the final reaction volume to 100 µl.

Assemble the following PCRs to amplify the first constant region of the γ chain from a cloned human Fab template (see Fig. 9.34). Perform four reactions for one library.

20 ng	plasmid template containing cloned human Fab fragment (e.g., pComb3XTT)
60 pmole	HIgGCH1-F sense primer
60 pmole	dpseq reverse primer
10 µl	10x PCR buffer

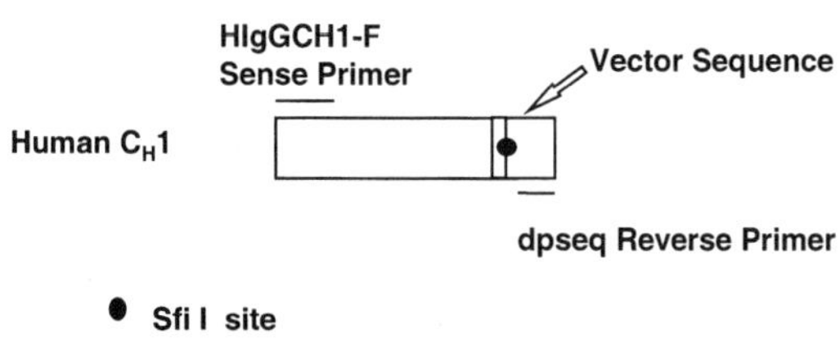

Figure 9.34. The first-round amplification of C_H1 products from the pComb3XTT vector. The sense primer is specific for the 5′ region of human C_H1, the region used in the overlap extension PCR to create the heavy-chain (Fd) fragment. The reverse primer is specific for the decapeptide sequence that is located downstream of the C_H1 fragment and the 3′ *SfiI* site of pComb3XTT.

8 µl	2.5 mM dNTP mix
0.5 µl	*Taq* polymerase

Add water to bring the final reaction volume to 100 µl.

Perform the first-round constant region PCR under the following conditions:

20 cycles of

94°C for 15 seconds

56°C for 30 seconds

72°C for 90 seconds

followed by 72°C for 10 minutes

Note: A "hot start" PCR protocol can improve specificity, sensitivity, and yield. In hot start PCR, either an essential reaction component is not added until the first denaturing step, or a reversible inhibitor of the polymerase is used. This protocol prevents low-stringency primer extension, which can generate nonspecific products.

Analyze 5–10 µl of each reaction on a 2% agarose gel, using DNA gel loading dye and appropriate MWM. The V_{κ}, V_{λ}, and V_H products should have a size of about 350 bp. The κ constant region PCR products should be about 400 bp, and the γ constant region products should be about 350 bp.

Note: Try a test amplification with the cDNA sample. If good amplification is not obtained (if a strong amplified band is not clearly visible in the agarose gel), lithium chloride precipitation of the RNA might improve the results (see Protocol 8.3).

2. **Isolate the PCR products.**

Combine the light-chain PCR products (V_{κ} and V_{λ}) into one pool and the heavy-chain products into another pool. Combine the κ constant region products and the γ constant region products into two additional pools. Ethanol-precipitate and wash as described in Appendix 3. Load the pools on a 1.5% agarose gel, cut out the correct-sized band, and purify the DNA by freeze-squeeze, electroelution with an Elutrap (see Appendix 3), or resin binding (e.g., QIAEX II Gel Extraction Kit, QIAGEN). Quantify the purified PCR products by reading the O.D. at 260 nm (1 O.D. unit = 50 µg/ml). At least 1 µg of each purified pool is needed to continue.

3. **Second round of PCR (first overlap extension).**

In the second round of PCR, heavy- and light-chain fragments are generated by an overlap extension assembly of the first-round rabbit variable region PCR products and the human constant region PCR products (see Figs. 9.35 and 9.36). Assemble at least ten or more 100-µl reactions for each template–primer combination. Each overlap reaction should contain the following:

100 ng	purified V product (template–primer combinations listed below)
100 ng	purified C product

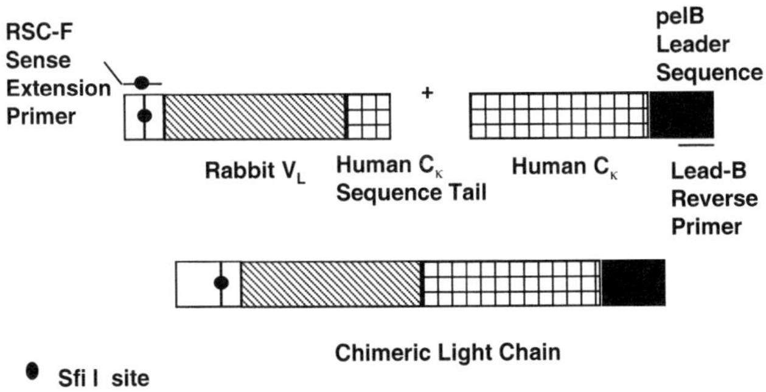

Figure 9.35. The light-chain overlap extension PCR amplification for the construction of rabbit chimeric Fab libraries. The rabbit V_L PCR products, which have a human C_κ sequence tail, are combined with the human C_κ–pelB products to create chimeric light-chain–pelB fragments. The sense extension primer (RSC-F) recognizes the sequence created by the first-round PCR V_L primers, and the Lead-B reverse primer recognizes the 3′ end of the pelB leader sequence that was amplified together with the human C_κ region from pComb3XTT. The 3′ end of the pelB sequence serves as the overlap region in the final overlap extension PCR.

60 pmole	sense primer
60 pmole	reverse primer
10 µl	10x PCR buffer (15 mM $MgCl_2$)
8 µl	2.5 mM dNTP mix
0.5 µl	*Taq* polymerase

Add water to bring the final reaction volume to 100 µl.

overlap extension PCRs to generate chimeric light-chain fragments:

100 ng	V_L product (V_κ plus V_λ)
100 ng	C_κ product

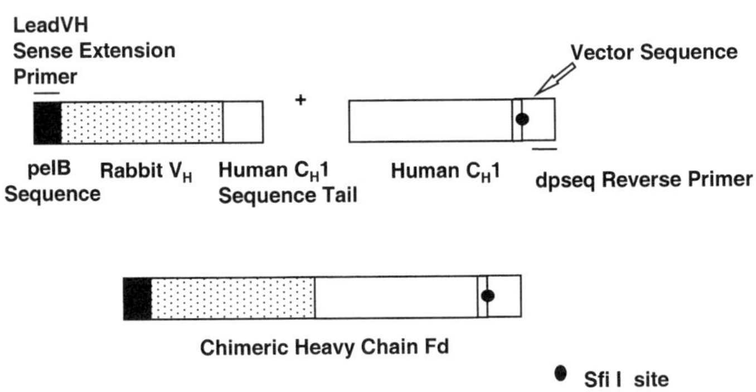

Figure 9.36. The heavy-chain (Fd) overlap extension PCR amplification for the construction of rabbit chimeric Fab libraries. The rabbit V_H PCR products, which have a human C_H1 sequence tail, are combined with the human C_H1 PCR products to create the chimeric Fd fragment. The sense extension primer (leadVH) recognizes the pelB leader sequence tail created by the first-round PCR V_H primers. The same reverse primer (dpseq) that was used for the generation of the C_H1 fragment is used in the overlap extension PCR of the heavy-chain fragment. The 3′ end of the pelB leader sequence serves as the overlap region in the final overlap extension PCR.

60 pmole	RSC-F primer (sense)
60 pmole	Lead-B primer (reverse)

overlap extension PCRs to generate chimeric heavy-chain Fd fragments:

100 ng	V_H product
100 ng	C_H1 product
60 pmole	leadVH primer (sense)
60 pmole	dpseq primer (reverse)

Perform the PCR under the following conditions:

15 cycles of

94°C for 15 seconds

56°C for 30 seconds

72°C for 2 minutes

followed by 72°C for 10 minutes

Analyze 5–10 µl of each reaction on a 1.5 % agarose gel, with DNA gel loading dye and appropriate MWM. The light-chain PCR products should have a size of about 800 bp, and the heavy-chain PCR products should have a size of about 750 bp.

Note: The long PCR extension time favors full-length product. If a strong band of the correct size is not clearly visible in the gel, the number of cycles can be increased, or more overlap PCRs can be performed. To retain diversity of the library, it is better to increase the number of PCRs.

4. **Isolate the PCR products.**

Pool the products of each type of reaction, ethanol-precipitate, and wash as described in Appendix 3. Purify the DNA on a 1% agarose gel as in Step 2. Quantify the purified PCR products by reading the O.D. at 260 nm. At least 1 µg per pool is needed to proceed.

5. **Third round of PCR (second overlap extension).**

In the third round of PCR, the chimeric light-chain products with the pelB leader sequence and the chimeric heavy-chain Fd fragments are joined by a second overlap extension step. Perform at least ten PCR assays (see Fig. 9.37).

100 ng	purified chimeric light-chain product
100 ng	purified chimeric heavy-chain product
60 pmole	RSC-F sense primer
60 pmole	dp-EX reverse primer
10 µl	10x PCR buffer
8 µl	2.5 mM dNTP mix
0.75 µl	Expand HF Polymerase mix

Add water to bring the final reaction volume to 100 µl.

Perform the PCR under the following conditions:

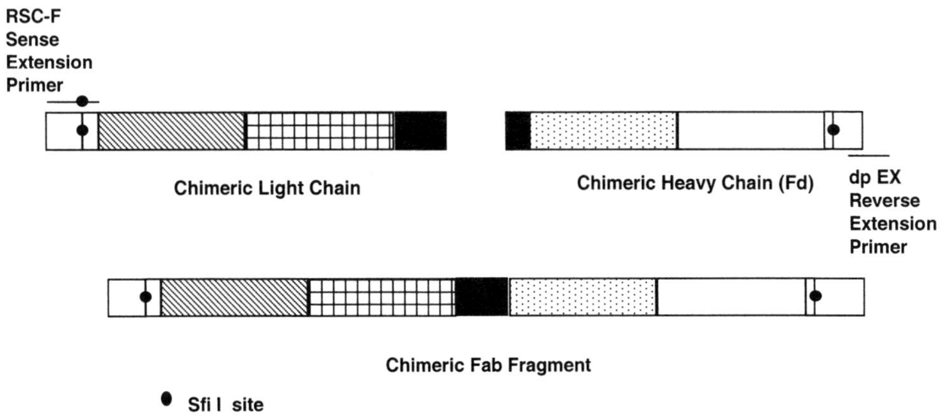

Figure 9.37. The final overlap extension PCR to combine the chimeric light chain–pelB fragment and the Fd fragment for the construction of rabbit chimeric Fab libraries. The 3′ end of the pelB leader sequence serves as the overlap region for the two PCR products. The sense extension primer (RSC-F) used in this round of PCR was also used in the overlap extension amplification of the light chain. The reverse extension primer recognizes the decapeptide sequence downstream of the C_H1 region.

10 cycles of
- 94°C for 15 seconds
- 56°C for 30 seconds
- 72°C for 3 minutes

followed by 72°C for 10 minutes

Note: Efficiency in generating the 1.5-kb overlap PCR product is often improved by using a long extension time and the Expand High Fidelity PCR System. This system has a unique enzyme mix that contains thermostable *Taq* and *Pwo* DNA polymerases. It is designed to give high yield, high fidelity, and high specificity PCR products.

The overlap products can also be generated with *AmpliTaq* polymerase. To amplify with *AmpliTaq* polymerase, use 0.5 µl of enzyme in the 100-µl amplification reaction. Perform the PCR under the following conditions:

94°C for 5 minutes

25 cycles of
- 94°C for 15 seconds
- 56°C for 15 seconds
- 72°C for 3 minutes

followed by 72°C for 10 minutes

Analyze 5–10 µl of each reaction on a 1.5% agarose gel, using DNA gel loading dye and appropriate MWM. The size of the overlap PCR product should be around 1500 bp.

6. Isolate the PCR products.

Combine the PCR products, ethanol-precipitate, and wash as described in Appendix 3. Purify the DNA on a 1% agarose gel as in Step 2. Quantify the purified PCR products by reading the O.D. at 260 nm. 10 μg of purified PCR product is usually sufficient to continue.

7. **Restriction-digest the purified overlap extension PCR products and the vector DNA.**

Prepare the PCR products and vector for cloning by performing restriction digests with *Sfi*I. We prefer high-concentration (40 Units/μl) *Sfi*I and the 10x buffer M from Roche Molecular Biochemicals.

The digest of the PCR products should contain:

10 μg	purified overlap PCR product
160 Units	*Sfi*I (16 Units per μg of DNA)
20 μl	10x reaction buffer M

Add water to a total volume of 200 μl.

The digest of the vector should contain:

20 μg	vector DNA (pComb3HSS or pComb3XSS, contains stuffer fragment between two *Sfi*I sites)
120 Units	*Sfi*I (6 Units per μg of plasmid DNA)
20 μl	10x reaction buffer M

Add water to a total volume of 200 μl.

Note: If the loss of digested PCR product in the final purification step is high or several large-size ligations are needed to obtain a library of the desired size, digest more PCR product with *Sfi*I. The amount of added enzyme should not exceed 10% of the reaction volume. If desired, larger quantities of *Sfi*I-cut and purified vector DNA can be prepared and tested in advance.

Incubate both digests for 5 hours at 50°C. Ethanol-precipitate and wash the DNA as described in Appendix 3 (optional, see Note). Purify the digested PCR insert (~1500 bp) on a 1% agarose gel, and purify the vector (~3400 bp) and the stuffer fragment (~1600 bp) on a 0.6% agarose gel. We recommend electroelution with an Elutrap for extraction of the vector from the gel. Suboptimal digestion or purification will yield vector with low transformation efficiencies and high background after ligation. Use the stuffer fragment in a test ligation to assess vector quality (see Step 8). Quantify the purified, digested PCR products, vector, and stuffer fragment by measuring the O.D. at 260 nm.

Note: In some cases *Sfi*I-digested DNA is partly insoluble after ethanol precipitation. To avoid this problem, the DNA can be loaded directly onto the agarose gel. When purifying the *Sfi*I-digested vector DNA, let the DNA run long enough to separate linearized vector DNA (size ~ 5000 bp; cut only once) and uncut vector DNA from the desired double cut product (size ~ 3400 bp).

8. **Ligate the digested overlap PCR product with the vector DNA.**

Perform small-scale ligations to assess the suitability of the vector and inserts for high-efficiency ligation and transformation. The ligation efficiency of the vector

DNA can be tested by ligating it with the gel-purified stuffer fragment that is generated during the *Sfi*I digestion of the vector DNA. The vector preparation should contain little uncut vector DNA or DNA that is cut only once. The amount of uncut or partly cut vector DNA can be estimated by setting up a ligation reaction that contains only vector DNA. These two types of control ligations should be done in parallel with a small-scale test ligation and should both include the same amount of vector DNA.

Test Ligations

Assemble the following small-scale ligation reactions:

small-scale test ligation:

140 ng	pComb3HSS or pComb3XSS, *Sfi*I-digested and purified
140 ng	overlap PCR product, *Sfi*I-digested and purified
4 µl	5x ligase buffer
1 µl	ligase

Add water to a total volume of 20 µl.

control ligation 1 (control insert):

140 ng	pComb3HSS or pComb3XSS, *Sfi*I-digested and purified
140 ng	gel-purified stuffer fragment
4 µl	5x ligase buffer
1 µl	ligase

Add water to a total volume of 20 µl.

control ligation 2 (test for vector self-ligation):

140 ng	pComb3HSS or pComb3XSS, *Sfi*I digested and purified
4 µl	5x ligase buffer
1 µl	ligase

Add water to a total volume of 20 µl.

Incubate between 4 hours and overnight at room temperature. Transform 0.5–1 µl of each reaction by electroporation into 50 µl of XL1-Blue or ER2537 (see Support Protocol). Dilute the transformed cultures 10-fold and 100-fold with prewarmed (37°C) SOC, and plate 100 µl of each dilution on LB+carb plates. Incubate the plates overnight at 37°C. Count the colonies on the vector + Fab insert plates and calculate the number of transformants per µg of vector DNA. If this number does not exceed 1×10^7, do not proceed with the large-scale library ligation. Ideally, the final library size should be several times 10^8 but at least 5×10^7 total transformants. Determine the number of scaled-up library ligations needed to achieve this size. Count the colonies obtained from control ligation 1 for an indication of vector quality and ligation efficiency. A good vector DNA preparation should yield at least 10^8 colony-forming units (cfu) per μg of vector DNA and should have less than 10% (ideally less than 5%) background ligation (calculated cfu per μg of vector DNA in control ligation 2). If the background is greater than 10% or the ligation efficiency is low, digest new vector and repeat the ligations. If the ligations are reasonably good, but there is not enough vector or insert to make a library of the desired size, perform additional digests.

Library Ligation

Set up one or more large-scale library ligations containing:

1.4 μg	pComb3HSS or pComb3XSS, *Sfi*I-digested and purified
1.4 μg	overlap PCR product, *Sfi*I-digested and purified
40 μl	5x ligase buffer
10 μl	ligase

Add nuclease-free water to a total volume of 200 μl.

Incubate overnight at room temperature. Add 1 μl of glycogen solution as carrier and ethanol-precipitate as in Appendix 3 (incubate between 4 hours and overnight at –20°C). The ethanol-precipitated ligation reaction can be stored for months at –20°C. Transform the ligation reaction by resuspending in 15 μl of water and electroporating into XL1-Blue or ER2537 electrocompetent cells, as described in Chapter 10.

Proceed with library preparation and panning as described in Chapter 10.

PROTOCOL 9.7

Rabbit scFv Libraries (by Overlap Extension)

The following protocol describes the construction of rabbit single-chain antibody libraries by overlap extension PCR. Depending on whether a short linker (SL) or a long linker (LL) library is to be constructed, different reverse primers must be used for the amplification of the V_L regions. Sequences of two sets of reverse primers are given below. The LL primers are used to construct a single-chain library in which the V_L polypeptide chain is linked to the V_H chain with an 18-amino-acid linker/spacer, whereas the use of the SL primers yields single-chain fragments with a 7-amino-acid linker. For more information regarding the use of short and long peptide linkers, see Protocol 9.2 and Chapter 3. For a brief introduction to the organization of rabbit immunoglobulin genes, see Protocol 9.6.

Materials

cDNA (prepared from immunized rabbits, see Protocol 8.4)
oligonucleotide primers (listed below; see Appendix 1 for letter code)
2.5 mM dNTP mix (dATP/dCTP/dGTP/dTTP set, 100 mM, Amersham Pharmacia Biotech, Cat. # 27-2035-02)
AmpliTaq DNA polymerase (Perkin-Elmer, Cat. # N808-052)
10x PCR buffer (Perkin-Elmer, supplied with *Taq* DNA polymerase)
MicroAmp PCR tubes (Perkin-Elmer, Cat. # N801-0533)
MicroAmp PCR caps (Perkin-Elmer, Cat. # N801-0534)
MicroAmp PCR trays (Perkin-Elmer, Cat. # N801-5530)
GeneAmp 9600 PCR cycler (Perkin-Elmer)
agarose (Gibco-BRL, Cat. # 15510-027)
100-bp DNA molecular weight marker (Amersham Pharmacia Biotech, Cat. # 27-4001-01) or 1-kb DNA molecular weight marker (Gibco-BRL, Cat. # 15615-024)
1x TAE electrophoresis running buffer (see Appendix 2)
DNA gel loading dye (see Appendix 2)
pComb3HSS or pComb3XSS (available from the Barbas laboratory)
*Sfi*I restriction enzyme (Roche Molecular Biochemicals, 40 Units/μl, Cat. # 1 288 059)
10x restriction enzyme buffer M (supplied with *Sfi*I restriction enzyme)
T4 DNA ligase (Gibco-BRL, 1 Unit/μl, Cat. # 15224 090)
5x T4 DNA ligase buffer (supplied with enzyme)
glycogen, 20 mg/ml (Roche Molecular Biochemicals, Cat. # 901 393)
electroporation apparatus and materials
electrocompetent ***E. coli*** cells (see Chapter 10): XL1-Blue (Stratagene, Cat. # 200228) or ER2537 (New England Biolabs, Cat. # 801-N)
LB + 100 μg/ml carbenicillin plates (see Appendix 2)

CAUTION: ***E. coli* (see Appendix 4)**

PRIMERS FOR RABBIT SINGLE-CHAIN LIBRARIES, SHORT LINKER (SL) AND LONG LINKER (LL)

V_κ 5′ Sense Primers

RSCVK1
5′ GGG CCC AGG CGG CCG AGC TCG TGM TGA CCC AGA CTC CA 3′

RSCVK2
5′ GGG CCC AGG CGG CCG AGC TCG ATM TGA CCC AGA CTC CA 3′

RSCVK3
5′ GGG CCC AGG CGG CCG AGC TCG TGA TGA CCC AGA CTG AA 3′

V_κ 3′ Reverse Primers, Short Linker

(linker amino acid sequence: GGSSRSS):

RKB9J1o-B
5′ GGA AGA TCT AGA GGA ACC ACC TTT GAT TTC CAC ATT GGT GCC 3′

RKB9Jo-B
5′ GGA AGA TCT AGA GGA ACC ACC TAG GAT CTC CAG CTC GGT CCC 3′

RKB42Jo-B
5′ GGA AGA TCT AGA GGA ACC ACC TTT GAC SAC CAC CTC GGT CCC 3′

V_κ 3′ Reverse Primers, Long Linker

(linker amino acid sequence: SSGGGGSGGGGGGSSRSS):

RKB9J1o-BL
5′ GGA AGA TCT AGA GGA ACC ACC CCC ACC ACC GCC CGA GCC ACC GCC ACC AGA GGA TAG GAT CTC CAG CTC GGT CCC 3′

RKB9Jo-BL
5′ GGA AGA TCT AGA GGA ACC ACC CCC ACC ACC GCC CGA GCC ACC GCC ACC AGA GGA TAG GAT CTC CAG CTC GGT CCC 3′

RKB42Jo-BL
5′ GGA AGA TCT AGA GGA ACC ACC CCC ACC ACC GCC CGA GCC ACC GCC ACC AGA GGA TTT GAC SAC CAC CTC GGT CCC 3′

V_λ 5′ Sense Primer

RSCλ1
5′ GGG CCC AGG CGG CCG AGC TCG TGC TGA CTC AGT CGC CCT C 3′

V_λ 3′ Reverse Primer, Short Linker

(linker amino acid sequence: GGSSRSS):

RJλo-B
5′ GGA AGA TCT AGA GGA ACC ACC GCC TGT GAC GGT CAG CTG GGT CCC 3′

V_λ 3′ Reverse Primer, Long Linker

(linker amino acid sequence: SSGGGGSGGGGGGSSRSS):

RJλo-BL
5′ GGA AGA TCT AGA GGA ACC ACC CCC ACC ACC GCC CGA GCC ACC GCC ACC AGA GGA GCC TGT GAC GGT CAG CTG GGT CCC 3′

V_H 5′ Sense Primers

RSCVH1
5′ GGT GGT TCC TCT AGA TCT TCC CAG TCG GTG GAG GAG TCC RGG 3′

RSCVH2
5′ GGT GGT TCC TCT AGA TCT TCC CAG TCG GTG AAG GAG TCC GAG 3′

RSCVH3
5′ GGT GGT TCC TCT AGA TCT TCC CAG TCG YTG GAG GAG TCC GGG 3′

RSCVH4
5′ GGT GGT TCC TCT AGA TCT TCC CAG SAG CAG CTG RTG GAG TCC GG 3′

V_H 3′ Reverse Primer

RSCG-B
5′ CCT GGC CGG CCT GGC CAC TAG TGA CTG AYG GAG CCT TAG GTT GCC C 3′

Overlap Extension Primers

RSC-F (sense)
5′ GAG GAG GAG GAG GAG GAG GCG GGG CCC AGG CGG CCG AGC TC 3′

RSC-B (reverse)
5′ GAG GAG GAG GAG GAG GAG CCT GGC CGG CCT GGC CAC TAG TG 3′

Procedure

1. First round of PCR.

Note: PCR products and other DNA samples, precipitated or in solution, can be stored at –20°C for years. The construction of antibody fragment libraries can therefore be interrupted at any step.

For V_κ and V_λ amplifications, if spleen and bone marrow have been kept separate for RNA preparation and cDNA synthesis, one PCR of each kind should be performed with cDNA derived from spleen, and one using cDNA derived from bone marrow. For V_H amplifications, two PCRs of each kind should be performed with cDNA derived from spleen, and two with cDNA derived from bone marrow.

Assemble the following PCRs to amplify rabbit V_L and V_H sequences:

1 µl	cDNA (~ 0.5 µg)
60 pmole	sense primer (primer combinations are listed below)
60 pmole	reverse primer
10 µl	10x PCR buffer
8 µl	2.5 mM dNTP mix
0.5 µl	*Taq* polymerase

Add water to bring the final reaction volume to 100 µl.

V_κ short linker (SL) primer combinations (see Fig. 9.38). Perform two reactions for each of the nine primer combinations; 18 reactions in total.

RSCVK1	RSCVK3
RKB9J1o-B	RKB9Jo-B

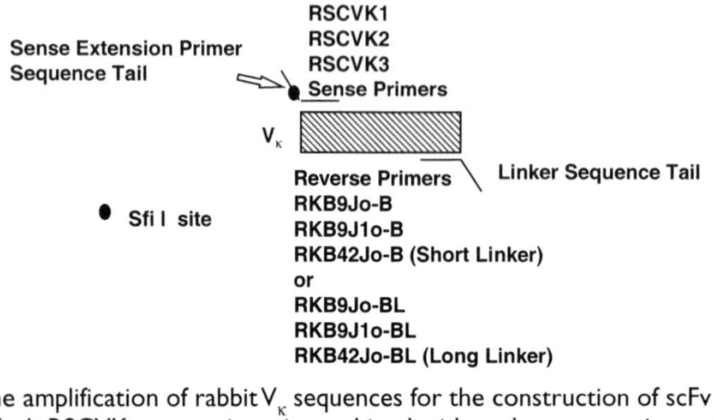

Figure 9.38. The amplification of rabbit V_κ sequences for the construction of scFv libraries (short or long linker). Each RSCVK sense primer is combined with each reverse primer to amplify rabbit V_κ gene segments from cDNA. Each sense primer has a 5′ sequence tail that contains an *Sfi*I site and is recognized by the sense extension primer used in the second-round PCR. Each reverse primer has a linker sequence tail that is used in the overlap extension.

RSCVK2 RKB9J1o-B	RSCVK1 RKB42Jo-B
RSCVK3 RKB9J1o-B	RSCVK2 RKB42Jo-B
RSCVK1 RKB9Jo-B	RSCVK3 RKB42Jo-B
RSCVK2 RKB9Jo-B	

V_λ SL primer combination, perform two reactions (see Fig. 9.39).

RSCλ1
RJλo-B

V_κ long linker (LL) primer combinations (see Fig. 9.38). Perform two reactions for each of the nine primer combinations; 18 reactions in total.

RSCVK1 RKB9J1o-BL	RSCVK3 RKB9Jo-BL
RSCVK2 RKB9J1o-BL	RSCVK1 RKB42Jo-BL
RSCVK3 RKB9J1o-BL	RSCVK2 RKB42Jo-BL
RSCVK1 RKB9Jo-BL	RSCVK3 RKB42Jo-BL

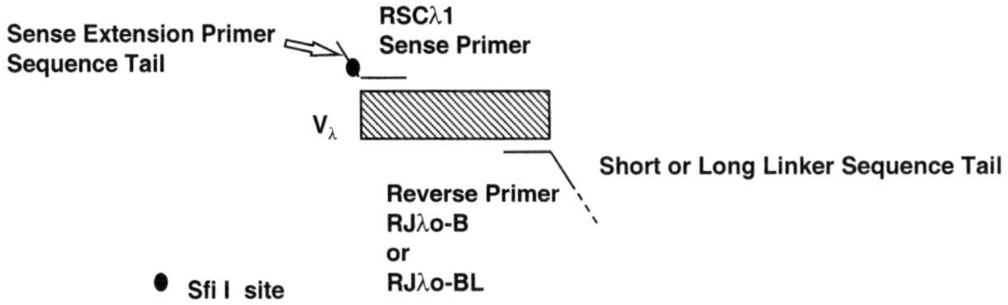

Figure 9.39. The amplification of rabbit V_λ sequences for the construction of scFv libraries (short or long linker). The RSCλ1 sense primer is combined with the RJλo-B (short linker) or RJλo-BL (long linker) reverse primer to amplify rabbit V_λ gene segments from cDNA. RSCλ1 has a 5′ sequence tail that contains an *Sfi*I site and is recognized by the sense extension primer used in the second-round PCR. Each reverse primer has a linker sequence tail that is used in the overlap extension.

RSCVK2

RKB9Jo-BL

V_λ LL primer combination (see Fig. 9.39). Perform two reactions.

RSCλ1

RJλo-BL

V_H primer combinations (see Fig. 9.40). Perform four reactions for each of the four combinations; 16 reactions in total; increase number of reactions if both SL and LL libraries will be made.

RSCVH1	RSCVH3
RSCG-B	RSCG-B
RSCVH2	RSCVH4
RSCG-B	RSCG-B

Perform the PCR under the following conditions:

30 cycles of:

94°C for 15 seconds

56°C for 30 seconds

72°C for 90 seconds

followed by 72°C for 10 minutes

Note: A "hot start" PCR protocol can improve specificity, sensitivity, and yield. In hot start PCR, either an essential reaction component is not added until the first denaturing step, or a reversible inhibitor of the polymerase is used. This protocol prevents low-stringency primer extension, which can generate nonspecific products.

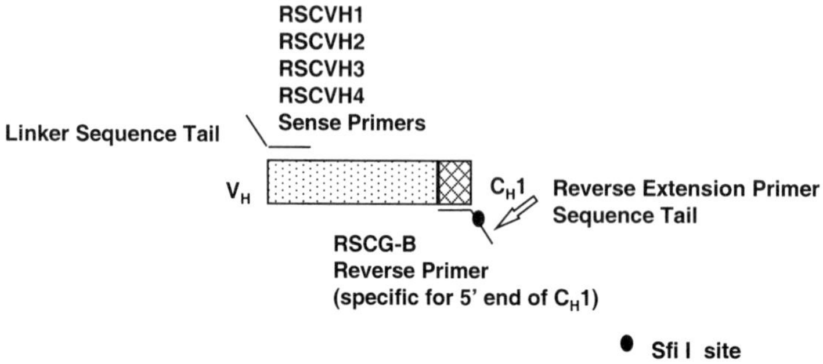

Figure 9.40. The amplification of rabbit V_H sequences for the construction of scFv libraries (short or long linker). Each of the RSCVH sense primers is combined with the RSCG-B reverse primer, which recognizes the 5′ end of the rabbit C_H1 region, to amplify V_H segments from rabbit cDNA. The sense primers have a sequence tail corresponding to the linker sequence that is used in the overlap extension PCR. The reverse primer has a sequence tail containing an *Sfi*I site; this tail is recognized by the reverse extension primer used in the second-round PCR.

Analyze 5–10 µl of each PCR on a 2% agarose gel, using DNA gel loading dye and appropriate molecular weight markers (MWM). The PCR products should have a size of about 350 bp.

Note: Try a test amplification with the cDNA sample. If good amplification is not obtained (if a strong amplified band is not clearly visible in the agarose gel), lithium chloride precipitation of the RNA may improve the results (see Protocol 8.3).

2. **Isolate the PCR products.**

Combine the light-chain short linker and light-chain long linker PCR products into two separate pools. Combine the heavy-chain products into another pool. Ethanol-precipitate and wash as described in Appendix 3. Load and run the products on a 1.5% agarose gel, cut out the bands with the correct-sized fragments, and purify the DNA by freeze-squeeze, electroelution with an Elutrap (see Appendix 3), or resin binding (e.g., QIAEX II Gel Extraction Kit, QIAGEN). Quantify the purified PCR products by reading the O.D. at 260 nm (1 O.D. unit = 50 µg/ml). At least 1 µg of each pool is needed to continue; if both short- and long-linker libraries are to be made, at least 2 µg of the heavy-chain PCR pool is needed.

3. **Second round of PCR (overlap extension).**

In the second round of PCR, the first-round V_L products are randomly joined with the first-round V_H products by overlap extension PCR (see Fig. 9.41). Perform at least ten reactions for either short linker or long linker single-chain fragments.

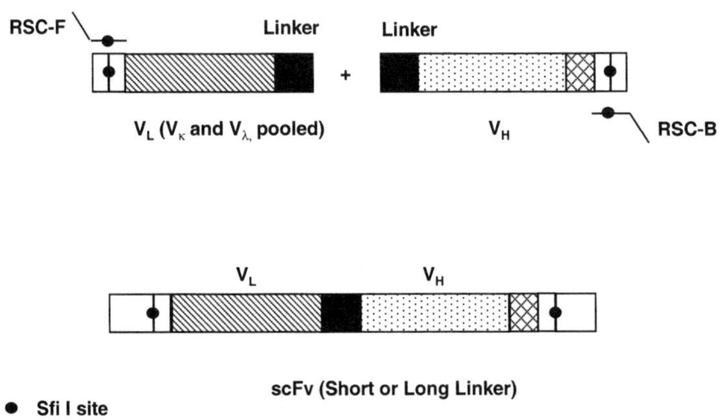

Figure 9.41. The overlap extension PCR to combine the rabbit V_L and V_H fragments for the construction of scFv libraries (short or long linker). The sense and reverse extension primers used in this second round of PCR (RSC-F and RSC-B) recognize the sequence tails that were generated in the first round of PCR.

100 ng	purified light-chain product, either short or long linker
100 ng	purified heavy-chain product
60 pmole	RSC-F sense primer
60 pmole	RSC-B reverse primer
10 µl	10x PCR buffer
8 µl	2.5 mM dNTP mix
0.5 µl	*Taq* polymerase

Add water to bring the final reaction volume to 100 µl.

Perform the PCR under the following conditions:

20 cycles of

94°C for 15 seconds

56°C for 30 seconds

72°C for 2 minutes

followed by 72°C for 10 minutes

Note: The long PCR extension time favors full-length product. If a strong band of the correct size is not clearly visible in the gel, the number of PCR cycles can be increased, or more overlap PCRs can be performed. To retain diversity of the library, it is better to increase the number of PCRs.

Analyze 5–10 µl of each reaction on a 1.5% agarose gel, using DNA gel loading dye and appropriate MWM. The PCR products should have a size of about 700 bp.

4. **Isolate the PCR products.**

Combine the short-linker and long-linker PCR products into two separate pools. Ethanol-precipitate and wash as described in Appendix 3. Purify the DNA on a 1% agarose gel as in Step 2. Quantify the purified PCR products by reading the O.D. at 260 nm. 10–15 µg should be sufficient to continue.

5. **Restriction-digest the purified overlap extension product and the vector DNA.**

Prepare the PCR products and vector for cloning by performing restriction digests with *Sfi*I. We prefer high-concentration (40 Units/µl) *Sfi*I and the 10x buffer M from Roche Molecular Biochemicals.

The digest of the PCR products should contain:

10 µg	purified overlap PCR product (SL product or LL product)
360 Units	*Sfi*I (36 Units per µg of DNA)
20 µl	10x reaction buffer M

Add water to a total volume of 200 µl.

The digest of the vector should contain:

20 μg	vector DNA (pCombt3HSS or pComb3XSS, contains stuffer fragment between two *Sfi*I sites)
120 Units	*Sfi*I (6 Units per μg of plasmid DNA)
20 μl	10x reaction buffer M

Add water to a total volume of 200 μl.

Note: If the loss of digested PCR product in the final purification step is high or if several large-size ligations are needed to obtain a library of the desired size, digest more PCR product with *Sfi*I. The amount of added enzyme should not exceed 10% of the reaction volume. If desired, larger quantities of *Sfi*I-cut and purified vector DNA can be prepared and tested in advance.

Incubate both digests for 5 hours at 50°C. Ethanol-precipitate and wash the DNA as described in Appendix 3 (optional, see Note). Purify the digested insert (~700 bp) on a 1% agarose gel, and purify the vector DNA (~3400 bp) and the stuffer fragment (~1600 bp) on a 0.6% agarose gel. We recommend electroelution with an Elutrap for extraction of the vector from the gel. Suboptimal digestion or purification will yield vector with low transformation efficiencies and high background after ligation. Use the stuffer fragment in a test ligation to assess vector quality (see Step 6). Quantify the purified, digested PCR products, vector, and stuffer fragment by measuring the O.D. at 260 nm.

Note: In some cases *Sfi*I-digested DNA is partly insoluble after ethanol precipitation. To avoid this problem, the DNA can be loaded directly after digestion onto the agarose gel. When purifying the *Sfi*I-digested vector DNA, let the DNA run long enough to separate linearized vector DNA (size ~ 5000 bp; cut only once) and uncut vector DNA from the desired double-cut product (size ~ 3400 bp).

6. **Ligate the digested overlap PCR product with the vector DNA.**

   Perform small-scale ligations to assess the suitability of the vector and inserts for high-efficiency ligation and transformation. The ligation efficiency of the vector DNA can be tested by ligating it with the gel-purified stuffer fragment that is generated during the *Sfi*I digestion of the vector DNA. The vector preparation should contain little uncut vector DNA or DNA that is cut only once. The amount of uncut or partly cut vector DNA can be estimated by setting up a ligation reaction that contains only vector DNA. These two types of control ligations should be done in parallel with a small-scale test ligation and should both include the same amount of vector DNA.

Test Ligations

Assemble the following small-scale ligation reactions:

small-scale test ligation:

140 ng	pComb3HSS or pComb3XSS, *Sfi*I-digested and purified

70 ng	overlap PCR product (SL or LL), *Sfi*I-digested and purified
4 µl	5x ligase buffer
1 µl	ligase

Add water to a total volume of 20 µl.

control ligation 1 (control insert):

140 ng	pComb3HSS or pComb3XSS, *Sfi*I-digested and purified
140 ng	gel-purified stuffer fragment
4 µl	5x ligase buffer
1 µl	ligase

Add water to a total volume of 20 µl.

control ligation 2 (test for vector self-ligation):

140 ng	pComb3HSS or pComb3XSS, *Sfi*I-digested and purified
4 µl	5x ligase buffer
1 µl	ligase

Add water to a total volume of 20 µl.

Incubate between 4 hours and overnight at room temperature. Transform 0.5–1 µl of each reaction by electroporation into 50 µl of XL1-Blue or ER2537 (see Support Protocol). Dilute the transformed cultures 10-fold and 100-fold with prewarmed (37°C) SOC, and plate 100 µl of each dilution on LB+carb plates. Incubate the plates overnight at 37°C. Count the colonies on the vector + Fab insert plates and calculate the number of transformants per µg of vector DNA. If this number does not exceed 1×10^7, do not proceed with the large-scale library ligation. Ideally, the final library size should be several times 10^8 but at least 5×10^7 total transformants. Determine the number of scaled-up library ligations needed to achieve this size. Count the colonies obtained from control ligation 1 for an indication of vector quality and ligation efficiency. A good vector DNA preparation should yield at least 10^8 colony-forming units (cfu) per µg of vector DNA and should have less than 10% (ideally less than 5%) background ligation (calculated cfu per µg of vector DNA in control ligation 2). If the background is greater than 10% or the ligation efficiency is low, digest new vector and repeat the ligations. If the ligations are reasonably good, but there is not enough vector or insert to make a library of the desired size, perform additional digests.

Library Ligation

Assemble one or more large-scale library ligations containing:

1.4 µg	pComb3HSS or pComb3XSS, *Sfi*I-digested and purified

700 ng	overlap PCR product (SL or LL), *Sfi*I-digested and purified
40 µl	5x ligase buffer
10 µl	ligase

Add water to a total volume of 200 µl.

Incubate overnight at room temperature. Add 1 µl of glycogen solution as carrier, and ethanol-precipitate as described in Appendix 3 (incubate between 4 hours and overnight at –20°C). The ethanol-precipitated ligation reaction can be stored at -20°C for months. Transform the ligation reaction by resuspending in 15 µl of water and electroporating into XL1-Blue or ER2537 electrocompetent cells, as described in Chapter 10.

Proceed with library preparation and panning as described in Chapter 10.

PROTOCOL 9.8

Chimeric Chicken/Human Fab Libraries

In many instances human monoclonal antibodies are most desirable, particularly when the development of therapeutic antibodies is the long-term goal. However, sources of human antibodies are largely limited to seropositive individuals who have been infected with or vaccinated against a pathogen. A common alternative is to obtain antibodies from immunized animals and "humanize" them (see Chapter 13). The most common route to obtaining these antibodies is through immunization of mice or rabbits, followed by hybridoma formation or phage-display selection. However, a large number of targets that possess therapeutic potential are highly conserved in mammalian evolution and, therefore, mount a limited immune response in mice and/or rabbits due to immunologic tolerance invoked during fetal development. One viable alternative is the use of chickens for immunization and selection of antibodies. Many of the highly conserved mammalian proteins are less conserved or absent in nonmammalian species, making animals like chickens a ready source of useful antibody.

To generate a large, diverse antibody repertoire, avian species use a type of DNA recombination event called gene conversion (for review, see Thompson et al. 1987; Reynaud et al. 1987, 1989; McCormack et al. 1993). Both the heavy- and light-chain loci in chickens consist of single functional V-region genes and J-region genes (and multiple D segments in the heavy chain) that are rearranged using conventional V(D)J recombination mechanisms. Individual rearrangements are further diversified by transplanting blocks of sequences from upstream pseudogenes in both the heavy- and light-chain variable regions. The majority of antibody V regions have multiple conversion events that can be traced to the pseudogenes through sequence comparison. Chickens also use somatic mutation, making it difficult at times to judge the origin of some sequences.

The use of single V_H and V_L genes makes antibody library construction relatively simple, as only one set of primers is needed for each chain. There are multiple breeds of chickens, but we have obtained the best results from White Leghorn chickens, one of the most commonly used strains. The libraries are easy to construct, and the eggs of immunized hens are a source of large quantities of polyclonal antibody that can be extracted and purified (see Chapter 8). We have found that using chickens to obtain monoclonal antibodies to mammalian proteins results in immune responses that are sometimes difficult to achieve with mice and rabbits. There are a number of documented reports regarding the application of phage-display technology to chicken immunoglobulin genes (Davies et al. 1995; Yamanaka et al. 1996). The following protocol is used to generate chimeric chicken/human Fab libraries using the pComb3 system (Andris-Widhopf et al. 2000; Cary et al. 2000).

Materials

cDNA (prepared from immunized female chickens, see Protocol 8.4)
oligonucleotide primers (listed below; see Appendix 1 for letter code)
2.5 mM dNTP mix (dATP/dCTP/dGTP/dTTP set, 100 mM, Amersham Pharmacia Biotech, Cat. # 27-2035-02)
AmpliTaq DNA polymerase (Perkin-Elmer, Cat. # N808-052)
10x PCR buffer for use with *Taq* polymerase (supplied with polymerase)
Expand High Fidelity Enzyme mix; 3.5 Units/µl (Roche Molecular Biochemicals, Cat. # 1 732 650)
10x Expand High Fidelity PCR buffer (supplied with Expand High Fidelity Enzyme mix)
MicroAmp PCR tubes (Perkin-Elmer, Cat. # N801-0533)
MicroAmp PCR caps (Perkin-Elmer, Cat. # N801-0534)
MicroAmp PCR trays (Perkin-Elmer, Cat. # N801-5530)
GeneAmp 9600 PCR cycler (Perkin-Elmer)
agarose (Gibco-BRL, Cat. # 15510-027)
100-bp DNA molecular weight marker (Amersham Pharmacia Biotech, Cat. # 27-4001-01) or 1-kb DNA molecular weight marker (Gibco-BRL, Cat. # 15615-024)
1x TAE electrophoresis running buffer (see Appendix 2)
DNA gel loading dye (see Appendix 2)
pComb3HSS or pComb3XSS (available from the Barbas laboratory)
*Sfi*I restriction enzyme (Roche Molecular Biochemicals, 40 Units/µl, Cat. # 1 288 059)
10x restriction enzyme buffer M (supplied with *Sfi*I restriction enzyme)
T4 DNA ligase (Gibco-BRL, 1 Unit/µl, Cat. # 15224 090)
5x T4 DNA ligase buffer (supplied with enzyme)
glycogen, 20 mg/ml (Roche Molecular Biochemicals, Cat. # 901 393)
electroporation apparatus and materials
electrocompetent ***E. coli*** cells (see Chapter 10): XL1-Blue (Stratagene, Cat. # 200228) or ER2537 (New England Biolabs, Cat. # 801-N)
LB + 100 µg/ml carbenicillin plates (see Appendix 2)

CAUTION: ***E. coli (see Appendix 4)***

CHICKEN/HUMAN CHIMERIC Fab PRIMERS

V_H Primers

CHybVH (sense)
5′ GCT GCC CAA CCA GCC ATG GCC GCC GTG ACG TTG GAC GAG TCC 3′

CHybIg-B (reverse)
5′ CGA TGG GCC CTT GGT GGA GGC GGA GGA GAC GAT GAC TTC GGT CCC 3′

V_λ Primers

CSCVK (sense)
5′ GTG GCC CAG GCG GCC CTG ACT CAG CCG TCC TCG GTG TC 3′

CHybL-B (reverse)
5′ AGA TGG TGC AGC CAC AGT TCG TAG GAC GGT CAG GGT TGT CCC GGC 3′

Primers for Amplification of the Human C_H1 Chain from a Cloned Human Fab

HIgGCH1-F (sense)
5′ GCC TCC ACC AAG GGC CCA TCG GTC 3′

dpseq (reverse)
5′ AGA AGC GTA GTC CGG AAC GTC 3′

Primers for Amplification of the Human C_κ Region and the pelB Leader Sequence from a Cloned Human Fab

HKC-F (sense)
5′ CGA ACT GTG GCT GCA CCA TCT GTC 3′

Lead-B (reverse)
5′ GGC CAT GGC TGG TTG GGC AGC 3′

Primers for PCR Assembly of Chicken V_H Sequences with the Human C_H1 PCR Product

LeadVH (sense)
5′ GCT GCC CAA CCA GCC ATG GCC 3′

dpseq (reverse)
(see above)

Primers for PCR Assembly of Chicken V_L Sequences with the Human C_κ PCR Product

CSC-F (sense)
5′ GAG GAG GAG GAG GAG GAG GTG GCC CAG GCG GCC CTG ACT CAG 3′

Lead-B (reverse)
(see above)

Primers for PCR Assembly of Chimeric Light-chain Sequences with Chimeric Heavy-chain Sequences

CSC-F (sense)
(see above)

dp-EX (reverse)
5′ GAG GAG GAG GAG GAG GAG AGA AGC GTA GTC CGG AAC GTC 3′

Procedure

1. **First round of PCR.**

   Note: PCR products and other DNA samples, precipitated or in solution, can be stored for years at –20°C. The construction of antibody fragment libraries, therefore, can be interrupted at any step.

   To amplify the V gene rearrangements, one chicken V_H amplification and one chicken V_λ amplification are performed (see Figs. 9.42–9.45). Set up four tubes for each type of amplification.

   Each reaction consists of the following:

   1 μl of cDNA or 100 ng of pComb3XTT template
   (template–primer combinations listed below)

60 pmole	5′ primer
60 pmole	3′ primer
10 μl	10X PCR buffer
8 μl	2.5 mM dNTPs
0.5 μl	*Taq* DNA polymerase

   Add water to final volume of 100 μl.

   V_H template–primer combination:

   1 μl of cDNA mix (0.5 μg)
   60 pmole of CHybVH primer
   60 pmole of CHybIg-B primer

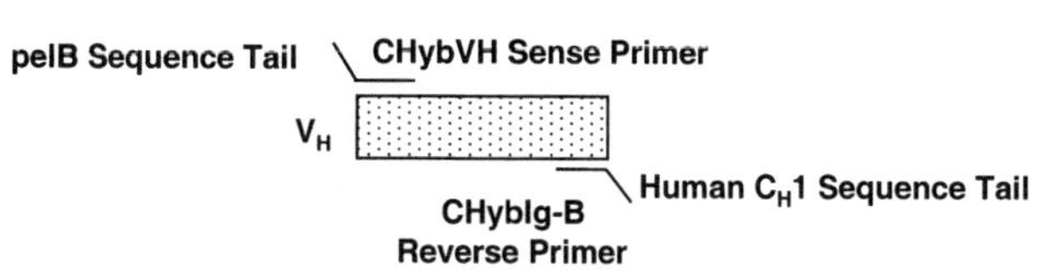

Figure 9.42. The amplification of chicken V_H sequences for the construction of chimeric Fab libraries. The CHybVH sense primer is combined with the CHybIg-B reverse primer to amplify V_H gene sequences from chicken cDNA. The sense primer has a 5′ sequence tail that corresponds to the 3′ end of the pelB leader sequence and is recognized by the sense extension primer (LeadVH) in the second-round PCR. The reverse primer has a human C_HI sequence tail that is used in the overlap extension PCR to create the chimeric heavy-chain (Fd) fragment.

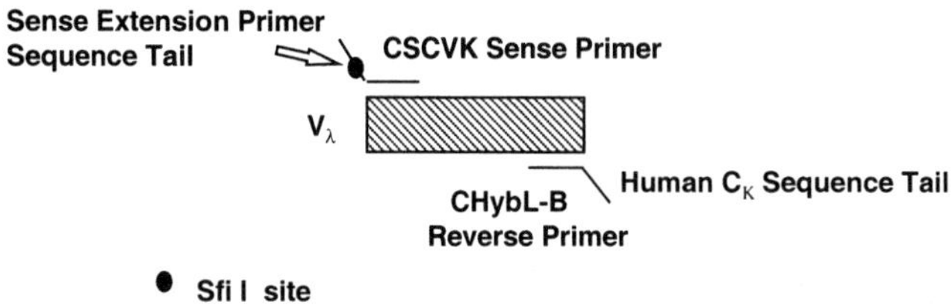

Figure 9.43. The amplification of chicken V_λ sequences for the construction of chimeric Fab libraries. The CSCVK sense primer is combined with the CHybL-B reverse primer to amplify V_λ sequences from chicken cDNA. The sense primer has a 5′ sequence tail that contains an *Sfi*I site and is recognized by the sense extension primer in the second-round PCR. The reverse primer has a human C_κ sequence tail that is used in the overlap extension PCR to create the chimeric lambda light-chain–pelB fragment.

V_λ template–primer combination:
- 1 μl of cDNA mix (0.5 μg)
- 60 pmole of CSCVK primer
- 60 pmole of CHybL-B primer

C_H1 template–primer combination:
- 20 ng of cloned pComb3XTT template
- 60 pmole of HIgGCH1-F primer
- 60 pmole of dpseq primer

C_κ–pelB template primer combination:
- 100 ng of cloned pComb3XTT template
- 60 pmole of HKC-F primer
- 60 pmole of Lead-B primer

Perform the V_H and V_λ PCR under the following conditions:

94°C for 5 minutes

30 cycles of

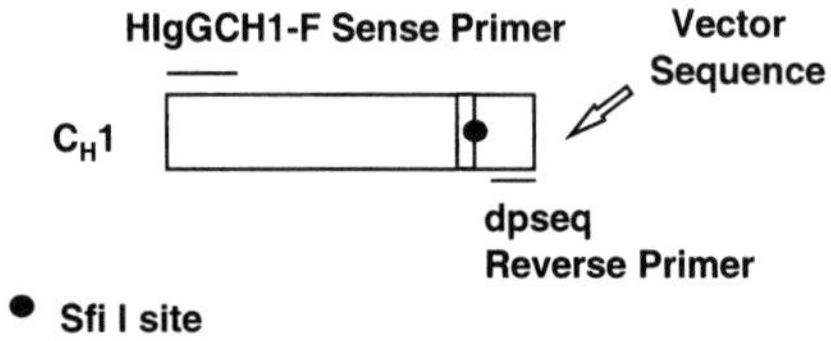

Figure 9.44. The first-round PCR amplification of human C_HI products from the pComb3XTT vector. The sense primer is specific for the 5′ region of C_HI, the region that is used in the overlap extension PCR to create the heavy-chain (Fd) fragment. The reverse primer is specific for the decapeptide sequence that is located downstream of the C_HI fragment and the 3′ *Sfi*I site of pComb3XTT.

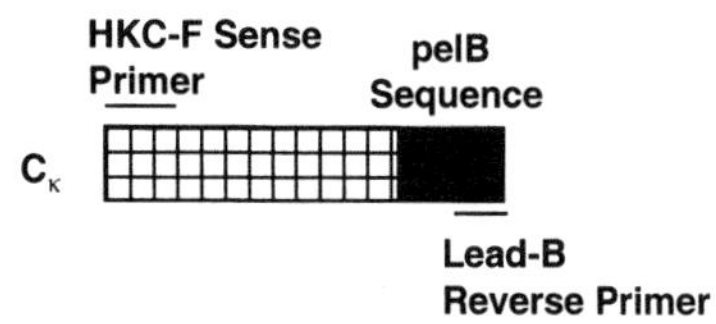

Figure 9.45. The first-round amplification of human C_κ products and the pelB leader sequence from the pComb3XTT vector. The sense primer is specific for the 5′ region of human C_κ, the region used in the overlap extension PCR to create the chimeric light chain. The reverse primer is specific for the 3′ end of the pelB leader sequence.

94°C for 15 seconds
56°C for 15 seconds
72°C for 90 seconds

followed by 72°C for 10 minutes

Perform the C_H1 and C_κ–pelB PCR under the following conditions:

94°C for 5 minutes

20 cycles of
94°C for 15 seconds
56°C for 15 seconds
72°C for 90 seconds

followed by 72°C for 10 minutes

Note: A "hot start" PCR protocol can improve specificity, sensitivity, and yield. In hot start PCR, either an essential reaction component is not added until the first denaturing step, or a reversible inhibitor of the polymerase is used. This protocol prevents low-stringency primer extension, which can generate nonspecific products.

Evaluate 5–10 µl of each reaction on a 2% agarose gel using DNA gel loading dye and appropriate molecular weight markers (MWM). From these amplifications expect a ~400 bp product for the V_H and C_κ reactions and a ~350 bp product for the V_λ and C_H1 reactions.

Note: Try a test amplification with the cDNA sample. If good amplification is not obtained (if a strong amplified band is not clearly visible in the agarose gel), lithium chloride precipitation of the RNA may improve the results (see Protocol 8.3).

2. **Isolate the PCR products.**

Pool the products of each type of reaction, ethanol-precipitate, and wash as described in Appendix 3. Gel-purify the products on a 2% agarose gel. Cut out the correct-sized products and purify the DNA by freeze-squeeze, electroelution with an Elutrap (see Appendix 3), or resin binding (e.g., QIAEX II Gel Extraction Kit, QIAGEN). Quantitate yields by reading the O.D. at 260 nm (1 O.D. unit = 50 μg/ml). If yields are too low, repeat the first round of PCR and combine the end products. Approximately 2–4 μg of each pooled product is required to proceed.

3. **Second round of PCR (overlap extension).**

In the second round of PCR, the appropriate first-round products are mixed in equal ratios to generate the overlap product. The primers in the first-round PCR create identical sequences in the downstream portion of the variable regions and the upstream portion of the constant regions that serve as the overlap for the extension of the full-length product (see Figs. 9.46 and 9.47). Assemble ten 100-μl reactions for each overlap. Each reaction should contain the following:

100 ng	purified V product (template–primer combinations listed below)
100 ng	purified C product
60 pmole	5′ primer
60 pmole	3′ primer
10 μl	10x PCR buffer
8 μl	2.5 mM dNTPs
0.5 μl	*Taq* DNA polymerase

Add water to a final volume of 100 μl.

Heavy-chain Fd template–primer combination:

100 ng of V_H product

100 ng of C_H1 product

60 pmole of LeadVH primer

60 pmole of dpseq primer

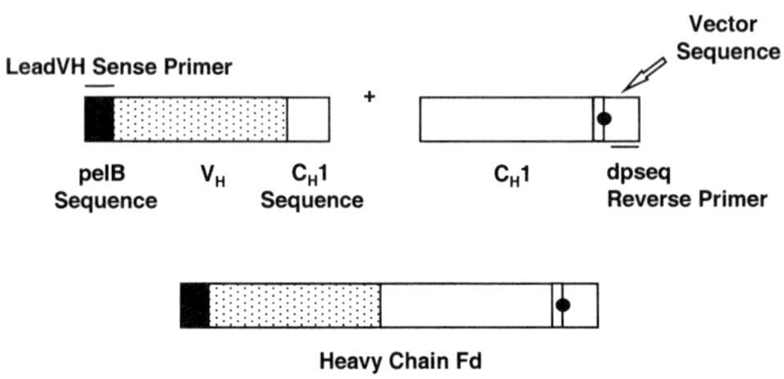

Figure 9.46. The heavy-chain (Fd) overlap extension PCR amplification for the construction of chicken chimeric Fab libraries. The chicken V_H PCR products, which have a human C_HI sequence tail, are combined with the human C_HI PCR products to create the chimeric Fd fragment. The sense extension primer (LeadVH) recognizes the pelB leader sequence tail created by the first-round PCR V_H primer. The same reverse primer (dpseq) that was used for the generation of the C_HI fragment is used in the overlap extension PCR of the heavy-chain fragment. The 3′ end of the pelB leader sequence serves as the overlap region in the final overlap extension PCR.

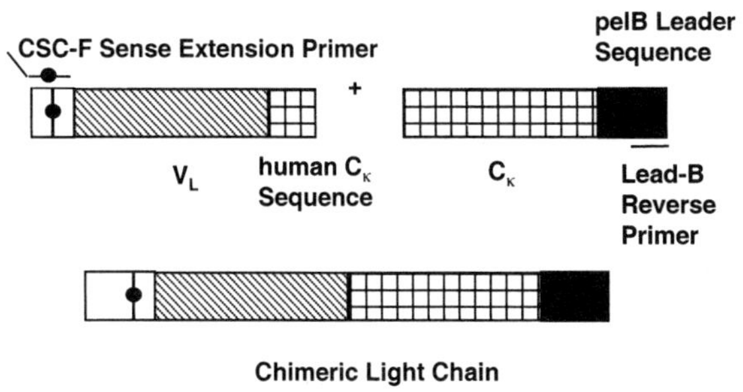

Figure 9.47. The light-chain overlap extension PCR amplification for the construction of chicken chimeric Fab libraries. The chicken light-chain variable region PCR products, which have a human C_κ sequence tail, are combined with the human C_κ–pelB products to create chimeric light-chain–pelB fragments. The sense extension primer (CSC-F) recognizes the sequence created by the first-round PCR V_L primer, and the Lead-B reverse primer recognizes the 3′ end of the pelB leader sequence that was amplified together with the human C_κ region from pComb3XTT. The 3′ end of the pelB sequence serves as the overlap region in the final overlap extension PCR.

Light-chain template–primer combination:

- 100 ng of V_λ product
- 100 ng of C_κ product
- 60 pmole of CSC-F primer
- 60 pmole of Lead-B primer

Perform the second-round PCR under the following conditions:

- 94°C for 5 minutes
- 15 cycles of
  - 94°C for 15 seconds
  - 56°C for 15 seconds
  - 72°C for 2 minutes

followed by 72°C for 10 minutes

Evaluate 5–10 µl of each reaction on a 2% agarose gel using DNA gel loading dye and appropriate MWM. For each type of reaction, expect a ~750-bp product.

Note: The long PCR extension time favors full-length product. If a strong band of the correct size is not clearly visible in the gel, the number of PCR cycles can be increased, or more overlap PCRs can be performed. To retain diversity of the library, it is better to increase the number of PCRs.

4. **Isolate the PCR products.**

Pool the products of each type of reaction, ethanol-precipitate, and wash as described in Appendix 3. Purify the DNA on a 2% agarose gel as in Step 2. Quantitate yields by reading the O.D. at 260 nm. If yields are too low, repeat the overlap PCR and combine the end products. Approximately 2–4 µg of product is required to proceed.

5. **Third round of PCR (second overlap extension).**

   In this PCR, the appropriate second-round products are mixed in equal ratios to generate the third-round overlap product. The primers in the second-round PCR create identical sequences in the downstream region of the light chain and the upstream region of the heavy-chain Fd that serve as the overlap for the extension of the full-length Fab product (see Fig. 9.48). Assemble ten 100-μl reactions for each overlap. Each reaction should contain:

100 ng	heavy-chain Fd product
100 ng	light-chain product
60 pmole	CSC-F primer
60 pmole	dp-EX primer
10 μl	10x PCR buffer
8 μl	2.5 mM dNTPs
0.75 μl	Expand HF Polymerase mix

   Add water to a final volume of 100 μl.

   Perform the PCR under the following conditions:

   10 cycles of

   94°C for 15 seconds

   56°C for 30 seconds

   72°C for 3 minutes

   followed by 72°C for 10 minutes

   Evaluate 5–10 μl on a 2% agarose gel using DNA gel loading dye and appropriate MWM. For each type of reaction expect a ~1500-bp product.

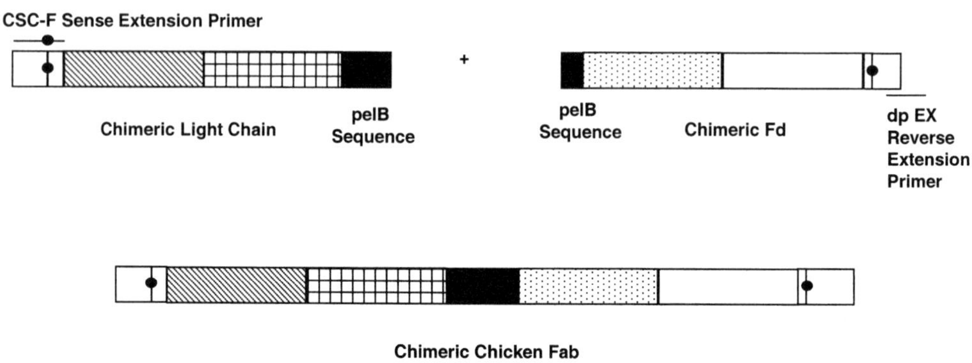

Figure 9.48. The final overlap extension PCR to combine the chimeric light chain–pelB fragment and the Fd fragment for the construction of chicken chimeric Fab libraries. The 3′ end of the pelB leader sequence serves as the overlap region for the two PCR products. The sense extension primer (CSC-F) used in this round of PCR was also used in the overlap extension amplification of the light chain. The reverse extension primer recognizes the decapeptide sequence tag downstream of the C_H1 region.

Note: Efficiency in generating the 1.5-kb overlap PCR product is often improved by using a long extension time and the Expand High Fidelity PCR System. This system has a unique enzyme mix that contains thermostable *Taq* and *Pwo* DNA polymerases. It is designed to give high yield, high fidelity, and high specificity PCR products.

The overlap products can also be generated with *AmpliTaq* polymerase. To amplify with *AmpliTaq* polymerase, use 0.5 µl of enzyme in the 100-µl amplification reaction. Perform the PCR under the following conditions:

94°C for 5 minutes

25 cycles of

- 94°C for 15 seconds
- 56°C for 15 seconds
- 72°C for 3 minutes

followed by 72°C for 10 minutes.

6. **Isolate the PCR products.**

Pool the PCR products, ethanol-precipitate, and wash as described in Appendix 3. Purify the DNA on a 2% agarose gel as in Step 2. Quantitate yields by reading the O.D. at 260 nm. If yields are too low, repeat the second overlap PCR and combine the end products. 10–15 µg is usually sufficient to continue.

7. **Restriction-digest the overlap Fab PCR products and the pComb3HSS or pComb3XSS vector.**

Prepare the PCR products and the vector for cloning by performing a restriction digest with *Sfi*I. We prefer the high-concentration (40 Units/µl) *Sfi*I and the 10x buffer M from Roche Molecular Biochemicals.

The digest of the PCR products should contain:

10 µg of purified Fab product

160 Units of *Sfi*I (16 Units per µg of DNA)

20 µl of buffer M

Add water to a volume of 200 µl.

The digest of the vector should contain:

20 µg of pComb3HSS or pComb3XSS (contains stuffer fragment between the two *Sfi*I cloning sites)

120 Units of *Sfi*I (6 Units per µg of DNA)

20 µl of 10x buffer M

Add water to a volume of 200 µl.

Note: If the loss of digested PCR product in the final purification step is high or if several large size ligations are needed to obtain a library of the desired size, digest more PCR product with *Sfi*I. The amount of added enzyme should not exceed 10% of the reaction volume. If desired, larger quantities of *Sfi*I-cut and purified vector DNA can be prepared and tested in advance.

Incubate the two digests for 5 hours at 50°C. Ethanol-precipitate and wash as described in Appendix 3. Purify the PCR products on a 2% agarose gel, and purify the vector and stuffer fragment on a 1% agarose gel. We recommend electroelution with an Elutrap for extraction of the vector from the gel (see Appendix 3). Suboptimal digestion or purification will yield vector with low transformation efficiencies and high background after ligation. Use the stuffer fragment in a test ligation to determine the quality of the vector prior to use in library ligations (see Step 8). The stuffer fragment is slightly larger than an Fab fragment, approximately 1600 bp. Quantify the purified, digested PCR products, vector, and stuffer fragment by measuring the O.D. at 260 nm.

Note: In some cases, *Sfi*I-digested DNA is partly insoluble after ethanol precipitation. To avoid this problem, the DNA can be loaded directly after digestion onto the agarose gel. When purifying the *Sfi*I-digested vector DNA, let the DNA run long enough to separate linearized vector DNA (size ~ 5000 bp; cut only once) and uncut vector DNA from the desired double-cut product (size ~ 3400 bp).

8. **Ligate the digested overlap PCR products with the vector DNA.**

Perform small-scale ligations to assess the suitability of the vector and inserts for high-efficiency ligation and transformation. The ligation efficiency of the vector DNA can be tested by ligating it with the gel-purified stuffer fragment that is generated during the *Sfi*I digestion of the vector DNA. The vector preparation should contain little uncut vector DNA or DNA that is cut only once. The amount of uncut or partly cut vector DNA can be estimated by setting up a ligation reaction that contains only vector DNA. These two types of control ligations should be done in parallel with a small-scale test ligation and should both include the same amount of vector DNA.

Test Ligations

Assemble the following small-scale ligation reactions:

small-scale test ligation

140 ng	pComb3HSS or pComb3XSS, *Sfi*I-digested and purified
140 ng	*Sfi*I-digested and purified Fab overlap PCR product
4 μl	5x ligase buffer
1 μl	ligase

Add water to a volume of 20 μl.

small-scale control ligation 1 (control insert)

140 ng	pComb3HSS or pComb3XSS, *Sfi*I-digested and purified
140 ng	*Sfi*I-digested and purified stuffer fragment
4 μl	5x ligase buffer
1 μl	ligase

Add water to a volume of 20 μl.

small-scale control ligation 2 (test for vector self-ligation)

140 ng	pComb3HSS or pComb3XSS, *Sfi*I-digested and purified
4 µl	5x ligase buffer
1 µl	ligase

Add water to a volume of 20 µl.

Incubate between 4 hours and overnight at room temperature. Transform 0.5–1 µl of each reaction by electroporation into 50 µl of XL1-Blue or ER2537 (see Support Protocol). Dilute the transformed cultures 10-fold and 100-fold with prewarmed (37°C) SOC, and plate 100 µl of each dilution on LB+carb plates. Incubate the plates overnight at 37°C. Count the colonies on the vector + Fab insert plates and calculate the number of transformants per µg of vector DNA. If this number does not exceed 1×10^7, do not proceed with the large-scale library ligation. Ideally, the final library size should be several times 10^8 but at least 5×10^7 total transformants. Determine the number of scaled-up library ligations needed to achieve this size. Count the colonies obtained from control ligation 1 for an indication of vector quality and ligation efficiency. A good vector DNA preparation should yield at least 10^8 colony-forming units (cfu) per µg of vector DNA and should have less than 10% (ideally less than 5%) background ligation (calculated cfu per µg of vector DNA in control ligation 2). If the background is greater than 10% or the ligation efficiency is low, digest new vector and repeat the ligations. If the ligations are reasonably good, but there is not enough vector or insert to make a library of the desired size, perform additional digests.

Library Ligation

Assemble enough reactions to produce at least 5×10^7 transformants. Each reaction should contain the following:

1.4 µg	pComb3HSS or pComb3XSS, *Sfi*I-digested and purified
1.4 µg	Fab overlap product, *Sfi*I-digested and purified
40 µl	5x ligase buffer
10 µl	ligase

Add water to a volume of 200 µl.

Incubate overnight at room temperature. Ethanol-precipitate and wash the pellet as described in Appendix 3. The ethanol-precipitated ligation reaction can be stored for months at –20°C. Transform the ligation reaction by resuspending in 15 µl of water and electroporating into XL1-Blue or ER2537 electrocompetent cells, as described in Chapter 10.

Proceed with the preparation and panning of the primary library (see Chapter 10).

PROTOCOL 9.9

Chicken scFv Libraries

The following is a protocol for generating chicken scFv libraries with a short linker (GGSSRSS) or a long linker (GGSSRSSSSGGGGSGGGG). For references regarding the use of alternative linker sequences, refer to Protocol 9.2. For background information regarding the use of immunized chickens for the construction of phage-display libraries, see Protocol 9.8.

Materials

cDNA (prepared from immunized female chickens, see Protocol 8.4)
oligonucleotide primers (listed below; see Appendix 1 for letter code)
2.5 mM dNTP mix (dATP/dCTP/dGTP/dTTP set, 100 mM, Amersham Pharmacia Biotech, Cat. # 27-2035-02)
AmpliTaq DNA polymerase (Perkin-Elmer, Cat. # N808-0152)
10x PCR buffer (Perkin-Elmer, supplied with polymerase)
MicroAmp PCR tubes (Perkin-Elmer, Cat. # N801-0533)
MicroAmp PCR caps (Perkin-Elmer, Cat. # N801-0534)
MicroAmp PCR trays (Perkin-Elmer, Cat. # N801-5530)
GeneAmp 9600 PCR cycler (Perkin-Elmer)
agarose (Gibco-BRL, Cat. # 15510-027)
100-bp DNA molecular weight marker (Amersham Pharmacia Biotech, Cat. # 27-4001-01) or 1-kb DNA molecular weight marker (Gibco-BRL, Cat. # 15615-024)
1x TAE electrophoresis running buffer (see Appendix 2)
DNA gel loading dye (see Appendix 2)
pComb3HSS or pComb3XSS (available from the Barbas laboratory)
*Sfi*I restriction enzyme (Roche Molecular Biochemicals, 40 Units/μl, Cat. # 1 288 059)
10x restriction enzyme buffer M (supplied with *Sfi*I restriction enzyme)
T4 DNA ligase (Gibco-BRL, 1 Unit/μl, Cat. # 15224 090)
5x T4 DNA ligase buffer (supplied with enzyme)
glycogen, 20 mg/ml (Roche Molecular Biochemicals, Cat. # 901 393)
electroporation apparatus and materials
electrocompetent *E. coli* cells (see Chapter 10): XL1-Blue (Stratagene, Cat. # 200228) or ER2537 (New England Biolabs, Cat. # 801-N)
LB + 100 μg/ml carbenicillin plates (see Appendix 2)

CAUTION: ***E. coli*** ***(see Appendix 4)***

CHICKEN scFv PRIMERS, SHORT AND LONG LINKER

V_H Primers

CSCVHo-F (sense), Short Linker
5′ GGT CAG TCC TCT AGA TCT TCC GCC GTG ACG TTG GAC GAG 3′

CSCVHo-FL (sense), Long Linker
5′ GGT CAG TCC TCT AGA TCT TCC GGC GGT GGT GGC AGC TCC GGT GGT GGC GGT TCC GCC GTG ACG TTG GAC GAG 3′

CSCG-B (reverse)
5′ CTG GCC GGC CTG GCC ACT AGT GGA GGA GAC GAT GAC TTC GGT CC 3′

V_λ Primers

CSCVK (sense)
5′ GTG GCC CAG GCG GCC CTG ACT CAG CCG TCC TCG GTG TC 3′

CKJo-B (reverse)
5′ GGA AGA TCT AGA GGA CTG ACC TAG GAC GGT CAG G 3′

Overlap Extension Primers

CSC-F (sense)
5′ GAG GAG GAG GAG GAG GAG GTG GCC CAG GCG GCC CTG ACT CAG 3′

CSC-B (reverse)
5′ GAG GAG GAG GAG GAG GAG GAG CTG GCC GGC CTG GCC ACT AGT GGA GG 3′

1. **First round of PCR.**

   Note: PCR products and other DNA samples, precipitated or in solution, can be stored for years at –20°C. The construction of antibody fragment libraries, therefore, can be interrupted at any step.

   To amplify the V gene rearrangements, one V_H amplification and one V_λ amplification are performed (see Figs. 9.49 and 9.50). Assemble four reactions for each type of amplification, using the spleen/bone marrow cDNA template.

   Each PCR consists of the following:

1 µl	cDNA (~0.5 µg)
60 pmole	5′ primer (primer combinations listed below)
60 pmole	3′ primer
10 µl	10x PCR buffer
8 µl	2.5 mM dNTPs
0.5 µl	*Taq* DNA polymerase

   Add water to a final volume of 100 µl.

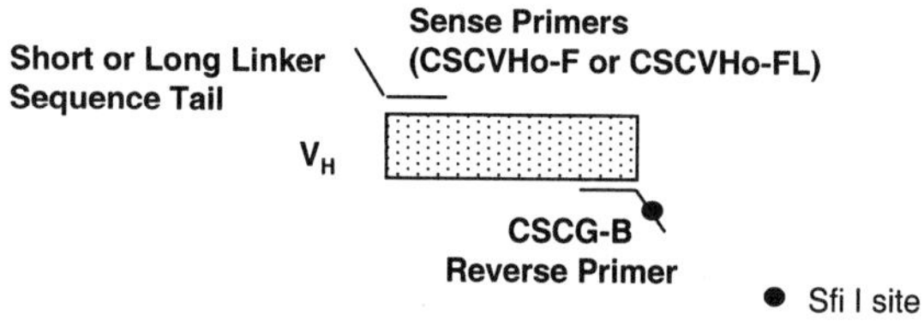

Figure 9.49. The first-round PCR amplification of chicken V_H sequences for the construction of scFv libraries (short or long linker). The primers CSCVHo-F (short linker) or CSCVHo-FL (long linker) are paired with the CSCG-B reverse primer to amplify V_H segments from chicken cDNA. The sense primers have a sequence tail that corresponds to the linker sequence that is used in the overlap extension PCR. The reverse primer has a sequence tail containing an *Sfi*I site; this tail is recognized by the reverse extension primer used in the second-round PCR.

V_H short-linker primer combination:

CSCVHo-F

CSCG-B

V_H long-linker primer combination:

CSCVHo-FL

CSCG-B

V_λ short- and long-linker primer combination:

CSCVK

CKJo-B

Perform the PCR under the following conditions:

94°C for 5 minutes
30 cycles of

94°C for 15 seconds

56°C for 15 seconds

72°C for 90 seconds

followed by 72°C for 10 minutes

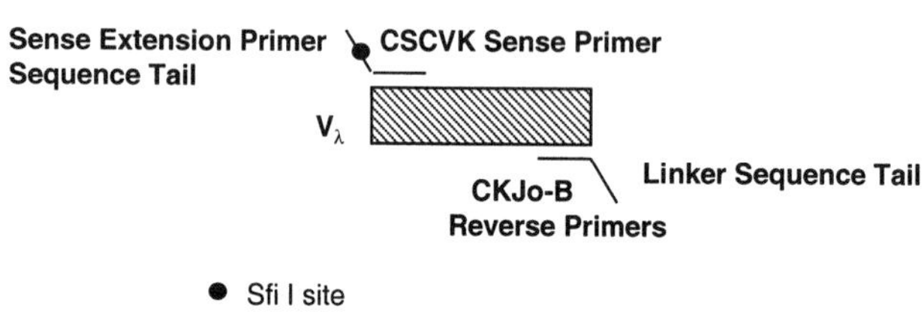

Figure 9.50. The amplification of chicken V_λ sequences for the construction of scFv libraries (short or long linker). The CSCVK sense primer is combined with the CKJo-B reverse primer to amplify V_λ gene segments from chicken cDNA. CSCVK has a 5′ sequence tail that contains an *Sfi*I site and is recognized by the sense extension primer in the second-round PCR. The reverse primer has a linker sequence tail that is used in the overlap extension.

Note: A "hot start" PCR protocol can improve specificity, sensitivity, and yield. In hot start PCR, either an essential reaction component is not added until the first denaturing step, or a reversible inhibitor of the polymerase is used. This protocol prevents low-stringency primer extension, which can generate nonspecific products.

Evaluate 5–10 µl on a 2% agarose gel using DNA gel loading dye and appropriate MWM. From these amplifications expect a ~400-bp product for the V_H short- and long-linker reactions and a ~350-bp product for the V_λ reactions.

Note: Try a test amplification with the cDNA sample. If good amplification is not obtained (if a strong amplified band is not clearly visible in the agarose gel), lithium chloride precipitation of the RNA might improve the results (see Protocol 8.3).

2. **Isolate the PCR products.**

Pool the products of each type of reaction and ethanol-precipitate. Run the products on a 2% agarose gel, cut out the correct-sized bands, and purify the DNA by freeze-squeeze, electroelution with an Elutrap (see Appendix 3), or resin bindng (e.g., QIAEX II Gel Extraction Kit, QIAGEN). Quantitate yields by reading the O.D. at 260 nm (1 O.D. unit = 50 µg/ml). If yields are too low, repeat the first round of PCR and combine the end products. 2–4 µg of each pool are needed to continue.

3. **Second round of PCR (overlap extension).**

In the second round of PCR, the appropriate first-round products are mixed in equal ratios to generate the overlap product. The primers in the first-round PCR create identical sequences that serve as the overlap for the extension of the full-length product (see Fig. 9.51). Assemble ten 100-µl reactions for each overlap.

Each reaction should include:

100 ng	V_H product (short or long linker)
100 ng	V_λ product
60 pmole	CSC-F primer
60 pmole	CSC-B primer
10 µl	10x PCR buffer
8 µl	2.5 mM dNTPs
0.5 µl	*Taq* DNA polymerase

Add water to a final volume of 100 µl.

Perform the PCR under the following conditions:

94°C for 5 minutes

25 cycles of

94°C for 15 seconds

56°C for 15 seconds

72°C for 2 minutes

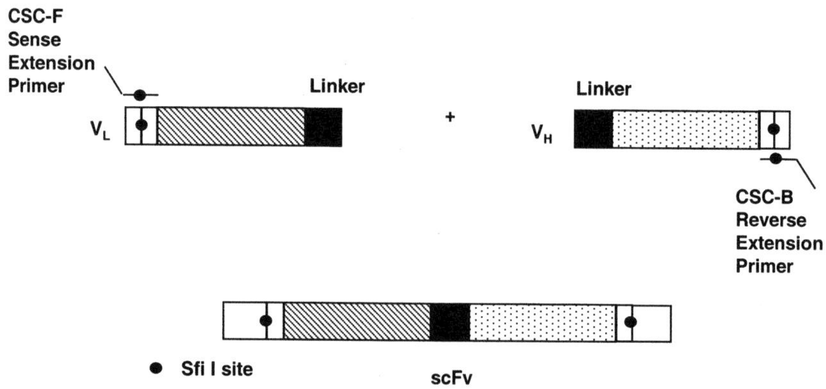

Figure 9.51. The overlap extension PCR to combine the chicken V_L and V_H fragments for the construction of scFv libraries (short or long linker). The sense and reverse extension primers used in this second round of PCR (CSC-F and CSC-B) recognize the sequence tails that were generated in the first round of PCR.

followed by 72°C for 10 minutes

Evaluate 5–10 µl on a 2% agarose gel using DNA gel loading dye and appropriate MWM for reference. For each type of reaction expect a ~750-bp product.

Note: The long PCR extension time favors full-length product. If a strong band of the correct size is not clearly visible in the gel, the number of PCR cycles can be increased, or more overlap PCRs can be performed. To retain diversity of the library, it is better to increase the number of PCRs.

4. **Isolate the PCR products.**

Pool the products of each type of reaction, ethanol-precipitate, and wash as described in Appendix 3. Purify the products on a 2% agarose gel as in Step 2. Quantitate yields by reading the O.D. at 260 nm. If yields are too low, repeat the overlap PCR and combine the end products. 10–15 µg of PCR product should be sufficient to proceed.

5. **Restriction-digest the overlap scFv PCR product and the pComb3HSS or pComb3XSS vector.**

Prepare the PCR products and the pComb3HSS or pComb3XSS vector for cloning by performing restriction digests with *Sfi*I. We prefer the high-concentration (40 Units/µl) *Sfi*I and the 10x buffer M from Roche Molecular Biochemicals.

The digest of the PCR products should contain:

10 µg	purified scFv-short or scFv-long product
360 Units	*Sfi*I (36 Units per µg of DNA)
20 µl	10x buffer M

Add water to a final volume of 200 µl.

The digest of the vector should contain:

20 μg	pComb3HSS or pComb3XSS (contains stuffer fragment between the two *Sfi*I cloning sites)
120 Units	*Sfi*I (6 Units per μg of DNA)
20 μl	10x buffer M

Add water to a final volume of 200 μl.

Note: If the loss of digested PCR product in the final purification step is high or if several large-size ligations are needed to obtain a library of the desired size, digest more PCR product with *Sfi*I. The amount of added enzyme should not exceed 10% of the reaction volume. If desired, larger quantities of *Sfi*I-cut and purified vector DNA can be prepared and tested in advance.

Incubate both reactions for 5 hours at 50°C. Ethanol-precipitate and wash as described in Appendix 3. Purify the scFv products on a 2% gel, and purify the vector and the stuffer fragment on a 1% agarose gel. We recommend electroelution with an Elutrap for extraction of the vector from the gel (see Appendix 3). Suboptimal digestion or purification will yield vector with low transformation efficiencies and high background after ligation. Use the stuffer fragment in a test ligation to determine the quality of the vector prior to use in library ligations (see Step 6). The stuffer fragment is approximately 1600 bp. Quantify the purified, digested PCR products, vector, and stuffer fragment by measuring the O.D. at 260 nm.

Note: In some cases *Sfi*I-digested DNA is partly insoluble after ethanol precipitation. To avoid this problem, the DNA can be loaded directly after digestion onto the agarose gel. When purifying the *Sfi*I-digested vector DNA, let the DNA run long enough to separate linearized vector DNA (size ~ 5000 bp; cut only once) and uncut vector DNA from the desired double-cut product (size ~ 3400 bp).

6. **Ligations.**

Perform small-scale ligations to assess the suitability of the vector and inserts for high-efficiency ligation and transformation. The ligation efficiency of the vector DNA can be tested by ligating it with the gel-purified stuffer fragment that is generated during the *Sfi*I digestion of the vector DNA. The vector preparation should contain little uncut vector DNA or DNA that is cut only once. The amount of uncut or partly cut vector DNA can be estimated by setting up a ligation reaction that contains only vector DNA. These two types of control ligations should be done in parallel with a small-scale test ligation and should both include the same amount of vector DNA.

Test Ligations

Assemble the following small-scale ligation reactions:

small-scale test ligation

140 ng	pComb3HSS or pComb3XSS, *Sfi*I-digested and purified
70 ng	scFv-short or scFv-long overlap PCR product, *Sfi*I-digested and purified

4 μl	5X ligase buffer
1 μl	ligase

Add water to a volume of 20 μl.

control ligation 1 (control insert)

140 ng	pComb3HSS or pComb3XSS, *Sfi*I-digested and purified
140 ng	stuffer fragment, *Sfi*I-digested and purified
4 μl	5X ligase buffer
1 μl	ligase

Add water to a volume of 20 μl.

control ligation 2 (test for vector self-ligation)

140 ng	pComb3HSS or pComb3XSS, *Sfi*I-digested and purified
4 μl	5X ligase buffer
1 μl	ligase

Add water to a volume of 20 μl.

Incubate between 4 hours and overnight at room temperature. Transform 0.5–1 μl of each reaction by electroporation into 50 μl of XL1-Blue or ER2537 (see Support Protocol). Dilute the transformed cultures 10-fold and 100-fold with prewarmed (37°C) SOC, and plate 100 μl of each dilution on LB+carb plates. Incubate the plates overnight at 37°C. Count the colonies on the vector + scFv insert plates and calculate the number of transformants per μg of vector DNA. If this number does not exceed 1×10^7, do not proceed with the large-scale library ligation. Ideally, the final library size should be several times 10^8 but at least 5×10^7 total transformants. Determine the number of scaled-up library ligations needed to achieve this size. Count the colonies obtained from control ligation 1 for an indication of vector quality and ligation efficiency. A good vector DNA preparation should yield at least 10^8 colony-forming units (cfu) per μg of vector DNA and should have less than 10% (ideally less than 5%) background ligation (calculated cfu per μg of vector DNA in control ligation 2). If the background is greater than 10% or the ligation efficiency is low, digest new vector and repeat the ligations. If the ligations are reasonably good, but there is not enough vector or insert to make a library of the desired size, perform additional digests.

Library Ligation

Assemble enough reactions to produce at least 5×10^7 transformants:

1.4 μg	pComb3HSS or pComb3XSS, *Sfi*I-digested and purified
700 ng	scFv (short or long linker) overlap product, *Sfi*I-digested and purified
40 μl	5X ligase buffer

10 μl ligase

Add water to a volume of 200 μl.

Incubate overnight at room temperature. Ethanol-precipitate and wash the pellet as described in Appendix 3. The ethanol-precipitated ligation reaction can be stored for months at –20°C. Transform the ligation reaction by resuspending in 15 μl of water and electroporating into XL1-Blue or ER2537 electrocompetent cells, as described in Chapter 10.

Proceed with the preparation and panning of the primary library (see Chapter 10).

SUPPORT PROTOCOL

Electroporation of Test Ligations or Other Small Ligations

Although there are a number of protocols available for the transformation of DNA into competent bacterial cells, electroporation is recommended for the preparation of phage-display libraries because it results in the highest transformation efficiency, thereby yielding larger libraries. The following protocols describe the steps involved in electroporation and preparation of primary combinatorial antibody libraries. Note that these protocols are for transformation of ampicillin-resistant plasmids/ phagemids. For other antibiotic resistance, LB agar plates should be prepared with the appropriate antibiotic.

Materials

tetracycline, 5 mg/ml stock in ethanol (Sigma, Cat. # T3258)
carbenicillin, 100 mg/ml stock (Sigma, Cat. # C1389)
LB+ 100 μg/ml carbenicillin plates (see Appendix 2)
Gene Pulser II electroporation system (BioRad, Cat. # 165-2105 and 165-2109)
electrocompetent ***E. coli*** cells (see Chapter 10)
  XL1-Blue (Stratagene, Cat. # 200228)
  ER2537 (New England Biolabs, Cat. # 801-N)
0.2-cm gap electroporation cuvettes (Invitrogen, Cat. # P450-50)
SOC Medium (see Appendix 2)
14-ml polypropylene pop top tubes (Fisher Scientific, Cat. # 14-956-1J)

CAUTIONS: ***tetracycline, E. coli (see Appendix 4)***

Procedure

1. Warm an appropriate quantity of SOC medium to room temperature (at least 3 ml per sample). Label one 14-ml polypropylene tube for each sample and aliquot 2 ml of SOC into each tube. Label and chill one electroporation cuvette for each sample.

2. Set the Gene Pulser II apparatus to 25 μF capacitance, 2.5 kV, and 200 Ω on the pulse controller unit.

3. Gently thaw the electrocompetent cells on ice. Thaw 50 μl per sample.

   Note: The competency of cells used for the transformation of small-scale ligations is crucial and should be at least 3×10^9/μg of supercoiled control plasmid DNA.

4. After the cells are thawed, dispense 50 μl of cells and 0.5–1.0 μl of a 20-μl test ligation into a chilled 1.5-ml tube. Incubate on ice for 1 minute.

5. Transfer the DNA–cell mixture to a chilled electroporation cuvette and shake the mixture to the bottom.
6. Place the cuvette into the sliding cuvette chamber and apply one pulse at the above settings. This should result in a pulse of 12.5 kV/cm with a time constant of 4–5 milliseconds. DNA containing too much salt will make the sample too conductive and cause arcing to occur at high voltage. Arcing ruins the samples, so if this happens, decrease the amount of DNA that is added to the cells and/or increase the quantity of cells to 60–70 µl.
7. Immediately add 1 ml of SOC medium to the cuvette and gently resuspend the cells. A delay of 1 minute before adding the SOC can cause a 3-fold decrease in transformation efficiency.
8. Transfer the cells and medium to a labeled polypropylene tube containing an additional 2 ml of SOC.
9. Incubate the cultures for 1 hour at 37°C while shaking at 225–250 rpm.
10. Plate 1 µl, 10 µl, and 100 µl on LB+carbenicillin plates (equivalent to diluting the cultures 10- and 100-fold, and plating 100 µl of each dilution). Incubate overnight at 37°C.
11. Count the colonies on each plate and calculate the total number of transformants. Average the numbers from the three plates.

$$\left(\text{\# of colonies}\right) \times \left(\frac{\text{culture vol. (µl)}}{\text{plating vol. (µl)}}\right) \times \left(\frac{\text{total ligation vol. (µl)}}{\text{ligation vol. transformed (µl)}}\right) = \left(\begin{array}{c}\text{total}\\\text{transformants}\end{array}\right)$$

12. Multiply the total number of transformants in the test ligation by a factor of 10 to determine the size of the library that would result from transforming a library ligation. Determine how many library ligations would be required to reach a total library size of 5×10^7.

REFERENCES

Anderson M.L., Szajnert M.F., Kaplan J.C., McColl L., and Young B.D. 1984. The isolation of a human Ig V lambda gene from a recombinant library of chromosome 22 and estimation of its copy number. *Nucleic Acids Res.* **12:** 6647–6661.

Andris J.S., Miller A.B., Abraham S.R., Cunningham S., Roubinet F., Blancher A., and Capra J.D. 1997. Variable region gene segment utilization in rhesus monkey hybridomas producing human red blood cell-specific antibodies: Predominance of the VH4 family but not VH4-21 (V4-34). *Mol. Immunol.* **34:** 237–253.

Andris-Widhopf J., Rader C., Steinberger P., Fuller R., and Barbas C.F., III. 2000. Methods for the preparation of chicken monoclonal antibody fragments by phage display. *J. Immunol. Methods* (in press).

Burton D.R., Barbas C.F., III, Persson M.A.A., Koenig S., Chanock R.M., and Lerner R.A. 1991. A large array of human monoclonal antibodies to type 1 human immunodeficiency virus from combinatorial libraries of asymptomatic seropositive individuals. *Proc. Natl. Acad. Sci.* **88:** 10134–10137.

Cary S., Lee J., Wagenknecht R., and Silverman G.J. 2000 Characterization of superantigen-induced clonal deletion with a novel clan III-restricted avian monoclonal antibody: Exploiting evolutionary distance to create antibodies specific for a conserved VH region surface. *J. Immunol.* **164:** 4730–4741.

Cook G., Tomlinson I., Walter G., Riethman H., Carter N., Buluwela L., Winter G., and Rabbitts T. 1994. A map of the human immunoglobulin VH locus compelted by analysis of the telomeric region of chromosome 14q. *Nat. Genet.* **7:** 162–168.

Cox J.P., Tomlinson I.M., and Winter G. 1994. A directory of human germline V kappa gene segments reveals a strong bias in their usage. *Eur. J. Immunol.* **24:** 827–836.

Currier J.S., Gallarda J.L., and Knight K.L. 1998. Partial molecular genetic map of the rabbit VH chromosomal region. *J. Immunol.* **140:** 1651–1656.

Davies E.L., Smith J.S., Birkett C.R., Manser J.M., Anderson-Dear D.V., and Young J.R. 1995. Selection of specific phage-display antibodies using libraries derived from chicken immunoglobulin genes. *J. Immunol. Methods* **186:** 125–135.

Foti M., Granucci F., Ricciardi-Castagnoli P., Spreafico A., Ackermann M., and Suter M. 1998. Rabbit monoclonal Fab derived from a phage display library. *J. Immunol. Methods* **213:** 201–212.

Holliger P., Prospero T., and Winter G. 1993. "Diabodies": Small bivalent and bispecific antibody fragments. *Proc. Natl. Acad. Sci.* **90:** 6444–6448.

Huse W.D., Lakshmi S., Iverson S.A., Kang A.S., Alting-Mees M., Burton D.R., Benkovic S.J., and Lerner R.A. 1989. Generation of a large combinatorial library of the immunoglobulin repertoire in phage lambda. *Science* **246:** 1275–1281.

Kang A.S., Burton D.R., and Lerner R.A. 1991. Combinatorial immunoglobulin libraries in phage λ. *Methods* **2:**111.

Kawamura S. and Ueda S. 1992. Immunoglobulin CH gene family in hominoids and its evolutionary history. *Genomics* **13:** 194–200.

Kawamura S., Omoto K., and Ueda S. 1990. Evolutionary hypervariability in the hinge region of the immunoglobulin alpha gene. *J. Mol. Biol.* **215:** 201–206.

Knight K. and Winstead C.R. 1997. Generation of antibody diversity in rabbits. *Curr. Opin. Immunol.* **9:** 228–232.

Krebber A., Bornhauser S., Burmeister J., Honegger A., Willuda J., Bosshard H.R., and Plückthun A. 1997. Reliable cloning of functional antibody variable domains from hybridomas and spleen cell repertoires employing a reengenieert phage display system. *J. Immunol. Methods.* **201:** 35–55.

Lang I., Barbas C.F., and Schleef R. 1996. Recombinant rabbit Fab with binding activity to type-1 plasminogen activator inhibitor derived from a phage-display library against human α-granules. *Gene* **172:** 295–298.

Matsuda F., Shin E., Nagaoka H., Matsumura R., Makoto H., Fukita Y., Takaishi S., Imai T., Riley J., Anand R., Soeda E., and Honjo T. 1993. Structure and physical map of 64 variable segments in the 3′0.8-megabase region of the human immunoglobulin heavy chain locus. *Nat. Genet.* **3:** 88–94.

McCormack W.T., Tjoelker L.W., and Thompson C. 1993. Immunoglobulin gene diversification by gene conversion. *Prog. Nucleic. Acid Res. Mol. Biol.* **45:** 27–45.

McGuinness B.T., Walter G., FitzGerald K., Shuler P., Mahoney W., Duncan A.R., and Hoogenboom H.R. 1996. Phage diabody repertoires for selection of large numbers of bi-specific antibody fragments. *Nat. Biotechnol.* **14:** 1149–1154.

Meek K., Eversole T., and Capra J.D. 1991. Conservation of the most JH proximal Ig VH gene segment (V_HVI) throughout primate evolution. *J. Immunol.* **146:** 2434–2438.

Meindl A., Klobeck H.-G., Ohnheiser R., and Zachau H.G. 1990. The V kappa repertoire in the human germline. *Eur. J. Immunol.* **20:** 1855–1863.

Newman R., Alberts J., Anderson D., Carner K., Heard C., Norton F., Raab R., Reff M., Shuey S., and Hanna N. 1992. "Primatization" of recombinant antibodies for immunotherapy of human disease: A macaque/human chimeric antibody against human CD4. *Bio/Technology* **10:** 1455–1460.

Persson M.A.A., Caothien R.H., and Burton D.R. 1991. Generation of diverse high-affinity human monoclonal antibodies by repertoire cloning. *Proc. Natl. Acad. Sci.* **88:** 2432–2436.

Rader C., Ritter G., Nathan S., Elia M., Gout I., Jungbluth A.A., Cohen L.S., Welt S., Old L.J., and Barbas C.F., III. 2000. The rabbit antibody repertoire as novel source for the generation of therapeutic human antibodies. *J. Biol. Chem.* **275:** 13668–13676.

Reynaud C.-A., Anquez V., Grimal H., and Weill J.-C. 1987. A hyperconversion mechanism generates the chicken light chain preimmune repertoire. *Cell.* **48:** 379–388.

Reynaud C.-A., Dahan A., Anquez V., and Weill J.-C. 1989. Somatic hyperconversion diversifies the single VH gene of the chicken with a high incidence in the D region. *Cell.* **59:** 171–183.

Ridder R., Schmitz R., Legay F., and Gram H. 1995. Generation of rabbit monoclonal antibody fragments from a combinatorial phage display library and their production in the yeast *Pichia pastoris. Bio/Technology* **13:** 255–260.

Roben P., Barbas S.M., Sandoval L., Lecerf J.-M., Stollar B.D., Solomon A., and Silverman G.J. 1996. Repertoire cloning of lupus anti-DNA autoantibodies. *J. Clin. Invest.* **98:** 2827–2837.

Samuelsson A., Chiodi F., Ohman P., Putkonen P., Norrby E., and Persson M.A.A. 1995. Chimeric macaque/human Fab molecules neutralize simian immunodeficiency virus. *Virology* **207:** 495–502.

Spieker-Polet H., Sethupathi P., Pi-Chen Y., and Knight K.L. 1995. Rabbit monoclonal antibodies: Generating a fusion partner to produce rabbit-rabbit hybridomas. *Proc. Natl. Acad. Sci.* **92:** 9348–9352.

Tang Y., Jiang N., Parakh C., and Hilvert D. 1996. Selection of linkers for a catalytic single-chain antibody using phage display technology. *J. Biol. Chem.* **271:** 15682–15686.

Thompson C.B. and Neiman P.E. 1987. Somatic diversification of the chicken immunoglobulin light chain gene is limited to the rearranged variable gene segment. *Cell* **48:** 369–378.

Tonegawa S. 1983. Somatic generation of antibody diversity. *Nature* **302:** 575–581.

Turner D.J., Ritter M.A., and George A.J.T. 1997. Importance of the linker in expression of single-chain Fv antibody fragments: Optimisation of peptide sequence using phage display technology. *J. Immunol. Methods* **205:** 43–54.

Williams S.C. and Winter G. 1993. Cloning and sequencing of human immunoglobulin V lambda gene segments. *Eur. J. Immunol.* **23:** 1456–1461.

Yamanaka H.I., Inoue T., and Ikeda-Tanaka O. 1996. Chicken monoclonal antibody isolated by a phage display system. *J. Immunol.* **157:** 1156–1162.

Zhu Z., Zapata G., Shalaby R., Snedecor B., Chen H., and Carter P. 1996. High level secretion of a humanized bispecific diabody from *Escherichia coli. Bio/Technology* **14:** 192–196.

10 Selection from Antibody Libraries

CHRISTOPH RADER, PETER STEINBERGER, AND CARLOS F. BARBAS III
Department of Molecular Biology, The Scripps Research Institute, La Jolla, California 92037

THE FOLLOWING PROTOCOLS DESCRIBE the preparation of the tools necessary for the screening of pComb3 phagemid libraries, namely electrocompetent *E. coli* (Protocol 10.1) and helper phage (Protocol 10.2). The ligation and transformation of newly generated libraries are described in Protocol 10.3, and the reamplification of existing libraries is described in Protocol 10.4. The final two protocols describe library screening based on a technique called panning. Panning consists of several rounds of binding phage to immobilized (Protocol 10.5) or cell surface (Protocol 10.6) antigens, washing, elution, and reamplification.

CONTENTS

PROTOCOL 10.1

Preparation of Electrocompetent *E. coli*

The *E. coli* that can be used for library screening must be male, i.e., they must harbor the F′ factor encoding proteins that form the pili. Through these pili, filamentous bacteriophage infect *E. coli*. The F′ factor tends to segregate from the cells, rendering them insensitive to infection by filamentous bacteriophage. It is therefore necessary to select for *E. coli* harboring the F′ factor. We have used two different strains for library screening, XL1-Blue and ER2537. The genotypes are as follows:

XL1-Blue F′ *proA*$^+$*B*$^+$ *lacI*^{q} Δ(lacZ)M15 Tn10 / *recA1 endA1 gyrA96 thi-1 hsdR17 supE44 relA1 lac*

Selection for F′factor with tetracycline

ER2537 *tonA: F′proA*$^+$*B*$^+$*, lacI*^{q}*, Δ(lacZ)M15* / *fhuA2(tonA)* Δ(*lac-proAB*) *supE thi-1* Δ*(hsdMS-mcrB)5*

Selection for F′ factor on minimal medium with thiamine and without proline

Both strains allow high-efficiency transformations by electroporation. Because of persistent contamination with environmental lytic phage of the T1 type, we usually screen our libraries with the resistant strain ER2537 (genotype *tonA*). In contrast to XL1-Blue, ER2537 are not recombination-deficient (*recA1*), which might affect the stability of the phagemids propagated in this strain. We have successfully used ER2537 in the selection of both scFv and Fab to a variety of antigens, however, and no evidence for recombination events was found. The helper phage VCSM13 (Protocol 10.2) can be used for both strains.

The size of an antibody library is restricted mainly by the efficiency of its introduction into *E. coli* (see Protocol 10.3). Electroporation is the most efficient method for transforming *E. coli* with plasmid DNA. To be electrocompetent, a suspension of *E. coli* has to have a high resistance, i.e., a low ionic strength. The ionic strength of an *E. coli* culture grown to mid-log phase is reduced by washing with low-salt buffer and eventually resuspending in 10% (v/v) glycerol.

Materials

XL1-Blue strain of ***E. coli*** (Stratagene, Cat. # 200228) or ER2537 strain of ***E. coli*** (New England Biolabs, Cat. # 801-N)
minimal medium agar plates (see Appendix 2)
LB agar+**tetracycline** (12 μg/ml) plates
SB medium (see Appendix 2)
5 mg/ml **tetracycline** in **ethanol**
20% (w/v) glucose, filter-sterilized
1 M **magnesium chloride** (**$MgCl_2$**), autoclaved

10% (v/v) glycerol
LB agar+carbenicillin (100 μg/ml) plates
LB agar+kanamycin (30 μg/ml) plates
LB top agar
plain LB agar plates

CAUTIONS: ***E. coli*, tetracycline, ethanol, magnesium chloride** ***(see Appendix 4)***

Procedure

1. Bleach the work surface, centrifuge, and rotor thoroughly. It is essential to keep the *E. coli* cultures and preparations free of phage contamination. Use shakers, pipet aids, culture flasks, and centrifuge bottles that have never been in contact with phage (see Appendix 3).
2. Inoculate 15 ml of prewarmed (37°C) SB in a 50-ml polypropylene tube with a single *E. coli* colony from a glycerol stock that has been freshly streaked onto an agar plate. Add tetracycline to 30 μg/ml for XL1-Blue (90 μl of 5 mg/ml tetracycline); add no antibiotics for ER2537. Grow overnight at 250 rpm and 37°C.

   Note: To select for the F′ factor, streak ER2537 on a minimal media agar plate with thiamine and without proline, and streak XL1-Blue on an LB agar+tetracycline plate.
3. Dilute 2.5 ml of the culture into each of six 2-liter flasks with 500 ml of SB, 10 ml of 20% (w/v) glucose, and 5 ml of 1 M $MgCl_2$. Add no antibiotics. Shake at 250 rpm and 37°C until the optical density (OD) at 600 nm is about 0.7 for XL1-Blue and between 0.8 and 0.9 for ER2537. Start checking the OD after 2.5 hours. The bacterial doubling time is approximately 20 minutes.
4. After the proper OD is reached, chill the six flask cultures as well as twelve 500-ml centrifuge bottles on ice for 15 minutes. From here on, everything should be kept on ice and done as rapidly as possible.
5. Pour the flask cultures into six prechilled 500-ml centrifuge bottles and spin at 3000*g* (e.g., 4000 rpm in a Beckman JA-10 rotor) for 20 minutes at 4°C.
6. Pour off the supernatant and resuspend each of the pellets in 25 ml of prechilled (on ice) 10% (v/v) glycerol using 25-ml prechilled plastic pipets. Combine two resuspended pellets in one prechilled 500-ml centrifuge bottle and add prechilled 10% (v/v) glycerol up to about 500 ml. Combine the other pellets similarly. Spin as before.
7. Resuspend each pellet in 500 ml of 10% prechilled glycerol. Spin as before.
8. Pour off the supernatant and resuspend each pellet in 25 ml of prechilled 10% (v/v) glycerol until complete homogeneity is reached. Transfer the suspensions into prechilled 50-ml polypropylene tubes and spin at 2500*g* (e.g., 3500 rpm in a Beckman GS-6R tabletop centrifuge) for 15 minutes at 4°C. Meanwhile, set up about 50 1.5-ml screw-cap microcentrifuge tubes in a rack in a dry ice/ethanol bath.

9. Carefully pour off the supernatant from each tube until the pellet begins to slide out. Discard the supernatant. Using a 25-ml prechilled plastic pipet, resuspend each pellet in the remaining volume (about 5 ml) and combine the three suspensions. Use a 1-ml pipet tip with a snipped-off end to immediately aliquot 300-μl volumes into the microcentrifuge tubes that were placed in the dry ice/ethanol bath. Cap the tubes and store them at –80°C.

10. The next day, test the prepared *E. coli* for competency and potential contamination with phagemids, helper phage, and lytic phage.

Competency test

1. Place 1 μl of 10 pg/μl amp^R plasmid (e.g., pUC18) in a 1.5-ml microcentrifuge tube on ice for 10 minutes. Do the same with a 0.2-cm cuvette. Thaw, on ice, an aliquot of the prepared *E. coli*.

2. Add 50 μl of the thawed *E. coli* to the plasmid, mix by pipetting up and down once, and transfer to the cuvette. Store on ice for 1 minute.

3. Electroporate at 2.5 kV, 25 μF, and 200 Ω. Expect τ to be in the range of 4.6–4.7 msec. Flush the cuvette immediately with 1 ml and then with 2 ml of SOC medium at room temperature. Combine the 3 ml in a 12-ml polypropylene tube and shake at 250 rpm for 1 hour at 37°C.

4. Add 7 ml of SB to the culture and plate 1 μl, 10 μl, and 100 μl on LB agar + carbenicillin plates. Incubate overnight at 37°C.

5. Calculate the competency in colonies per μg of plasmid as follows:

$$\frac{\text{\# colonies}}{\text{plating volume } (\mu l)} \times \frac{1 \times 10^4\ \mu l}{10\ pg} \times \frac{1 \times 10^6\ pg}{1\ \mu g}$$

The competency should be well above 1 x 10^9 colonies/μg plasmid. Preparations with lower competency can still be used for helper phage preparation and phage reamplification (Protocols 10.2, 10.4–10.6).

Contamination tests

1. Contamination with pComb3 phagemids or phage derived thereof:

   Plate 25 μl of the prepared *E. coli* directly onto an LB agar+carbenicillin plate. Incubate overnight at 37°C. No colonies should grow.

2. Contamination with helper phage:

   Plate 25 μl of the prepared *E. coli* directly onto an LB agar+kanamycin plate. Incubate overnight at 37°C. No colonies should grow.

3. Contamination with lytic phage (and secondary test for contamination with helper phage):

   Mix 25 μl of the prepared *E. coli* with 3 ml of liquefied LB top agar (cooler than 50°C) and pour onto a plain LB agar plate. Incubate overnight at 37°C. No plaques should form.

PROTOCOL 10.2

Preparation of Helper Phage

Because the pComb3 phagemids contain only the origin of replication of the filamentous bacteriophage f1, and lack all of the genes required for replication and assembly of phage particles, screening of pComb3 phagemid libraries requires a helper phage to provide the replication and assembly proteins. The helper phage VCSM13 is particularly suitable because it also contains a gene coding for kanamycin resistance, so kanamycin can be used to select for *E. coli* infected with VCSM13. A mutation in the VCSM13 origin of replication makes production of the helper phage less efficient than that of pComb3 phage. Thus, addition of helper phage VCSM13 to cells that have been transformed with the pComb3 phagemid will ultimately produce a mixed phage population that predominantly contains pComb3 phage.

VCSM13 can be produced in both XL1-Blue and ER2537 strains. Because we generally obtain higher titers with ER2537, the following protocol is based on this strain. It can be adapted to XL1-Blue by adding tetracycline to the cultures at a final concentration of 10 μg/ml.

Materials

VCSM13 helper phage (Stratagene, Cat. # 200251)
E. coli, strain ER2537 (prepared according to Protocol 10.1)
SB medium (see Appendix 2)
LB top agar (see Appendix 2)
LB agar plates (100 x 15 mm, see Appendix 2)
50 mg/ml kanamycin in water, filter-sterilized

CAUTION: ***E. coli (see Appendix 4)***

Procedure

1. Inoculate 2 ml of SB medium with 2 μl of ER2537, and shake at 250 rpm for 1 hour at 37°C. Prepare 10^{-6}, 10^{-7}, and 10^{-8} dilutions of the commercially obtained VCSM13 preparation (usually in the range of 1 x 10^{11} pfu/ml) in SB medium. Add 1 μl of each of these dilutions to 50 μl of the ER2537 culture and incubate for 15 minutes at room temperature. Add 3 ml of liquefied LB top agar (cooler than 50°C) and pour onto plain LB agar plates that have been prewarmed to 37°C. Incubate overnight at 37°C. The aim is to obtain a plate from which single VCSM13 plaques (i.e., *E. coli* colonies that are growing slower because of infection with VCSM13) can be picked easily.

2. In a 50-ml polypropylene tube, inoculate 10 ml of prewarmed (37°C) SB medium with 10 μl of ER2537, and allow growth at 250 rpm for one hour at 37°C.

3. Use a pipet tip to transfer a single VCSM13 plaque from a freshly prepared plate in Step 1 to the culture and shake at 250 rpm for 2 hours at 37°C.
4. Transfer the infected 10-ml culture to a 2-liter Erlenmeyer flask containing 500 ml of prewarmed (37°C) SB. Add 700 μl of 50 mg/ml kanamycin to a final concentration of 70 μg/ml, and continue shaking overnight at 250 rpm and 37°C.
5. Transfer the culture to ten 50-ml polypropylene tubes and spin at 2500*g* (e.g., 3500 rpm in a Beckman GS-6R tabletop centrifuge) for 15 minutes.
6. Transfer the supernatants to fresh 50-ml polypropylene tubes and incubate in a water bath at 70°C for 20 minutes.
7. Spin at 2500*g* for 15 minutes.
8. Transfer the supernatants to fresh 50-ml polypropylene tubes and store at 4°C.
9. Determine the titer of the VCSM13 preparation as follows:
   a. Inoculate 2 ml of SB medium with 2 μl of ER2537, and allow growth at 250 rpm for 1 hour at 37°C.
   b. Prepare 10^{-7},10^{-8}, and 10^{-9}dilutions of the VCSM13 preparation in SB medium. Use 1 μl of each of these dilutions to infect 50 μl of the ER2537 culture and incubate for 15 minutes at room temperature.
   c. Add 3 ml of liquefied LB top agar (< 50°C) and pour onto plain LB agar plates. Incubate overnight at 37°C.
   d. Determine the titer of the VCSM13 preparation from the number of plaques. For example, 50 plaques on the plate derived from the 10^{-8} dilution corresponds to a titer of 5×10^{12} plaque-forming units (pfu) per ml. Expect the titer to be in the range of 10^{12} to 10^{13} pfu/ml. Although this titer will decrease over time, VCSM13 preparations are stable for months at 4°C.

PROTOCOL 10.3

Library Ligation and Transformation

A major factor determining the quality of an antibody library is its complexity, i.e., the number of different antibodies in the library. The greater the complexity of the library, the more likely one is to select antibodies of the required affinity and/or specificity. There is no way to determine the absolute complexity of an antibody library, but the complexity cannot be higher than the number of independent transformants after library ligation and transformation. Thus, the number of independent transformants minus background (as estimated from the test ligation, see Chapter 9) is used to describe the complexity of an antibody library. This number is sometimes referred to as the library size. For antibody libraries derived from immune animals, reasonable library sizes are in the range of 10^7 to 10^8 independent transformants. Depending on the outcome of the test ligation (Chapter 9), one to ten library ligations of the format given in the following protocol are necessary to achieve this number.

Materials

*Sfi*I-cut pComb3 and *Sfi*I-cut PCR product (see Chapter 9)
1 Unit/µl T4 DNA ligase and 5x ligase buffer (Life Technologies, Cat. # 15224-025)
glycogen (Roche Molecular Biochemicals, Cat. # 901 393)
3 M sodium acetate, pH 5.2, autoclaved
ethanol
70% (v/v) **ethanol**
electrocompetent *E. coli,* prepared according to Protocol 10.1
Gene Pulser apparatus with Pulse Controller (Bio-Rad)
0.2-cm cuvettes (Bio-Rad, Cat. # 165-2086 or Invitrogen, Cat. # P450-50)
SOC medium (see Appendix 2)
SB medium (see Appendix 2)
100 mg/ml carbenicillin in water, filter-sterilized
5 mg/ml **tetracycline** in **ethanol**
LB agar+carbenicillin (100 µg/ml) plates
VCSM13 helper phage, prepared according to Protocol 10.2
50 mg/ml kanamycin in water, filter-sterilized
PEG-8000 (**polyethylene glycol;** Sigma, Cat. # P 2139)
sodium chloride
TBS (50 mM Tris-HCl, pH 7.5, 150 mM NaCl), autoclaved
1% (w/v) BSA in TBS, filter-sterilized (Sigma, Cat. # A 7906)
2% (w/v) **sodium azide**

CAUTIONS: ethanol, *E. coli*, tetracycline, polyethylene glycol, sodium azide ***(see Appendix 4)***

Procedure

1. For a single library ligation, combine 1.4 μg of *Sfi*I-cut pComb3 (~ 3.3 kb) with an equimolar amount of *Sfi*I-cut PCR product, e.g., 650 ng of an approximately 1.5-kb Fab-encoding cDNA. Use 10 μl of T4 DNA ligase and 40 μl of 5x ligase buffer in a total volume of 200 μl. Incubate overnight at room temperature.

   Note: Ligation efficiency may be improved by using a *Sfi*I-cut PCR product in twofold molar excess, e.g., 1300 ng of an Fab-encoding cDNA or 650 ng of an scFv-encoding cDNA.

2. Precipitate by adding 1 μl of glycogen, 20 μl (0.1 volume) of 3 M sodium acetate, pH 5.2, and 440 μl (2.2 volumes) of ethanol. Mix, and store overnight at –80°C. Precipitated library ligations are stable for months at –80°C.

3. Spin at full speed in a microcentrifuge for 15 minutes at 4°C. Remove and discard the supernatant; rinse the pellet twice with 1 ml of 70% (v/v) ethanol, drain inverted on a paper towel, and dry briefly in a SpeedVac. Dissolve the pellet in 15 μl of water by brief heating at 37°C followed by gentle vortexing.

4. Place the ligated library samples and a corresponding number of cuvettes on ice for 10 minutes. At the same time thaw, on ice, 300 μl of electrocompetent *E. coli* for each library ligation. If more than four transformations are to be performed, thaw the electrocompetent *E. coli* in series to avoid storage on ice for more than 20 minutes.

5. Use a 1-ml pipet tip with a snipped-off end to add the electrocompetent *E. coli* to each ligated library sample, mix by pipetting up and down once, and transfer to a cuvette. Store on ice for 1 minute. Electroporate at 2.5 kV, 25 μF, and 200 Ω. Expect τ to be approximately 4.0 msec. Flush the cuvette immediately with 1 ml and then twice with 2 ml of SOC medium at room temperature, and combine the 5 ml in a 50-ml polypropylene tube. Shake at 250 rpm for 1 hour at 37°C.

6. Add 10 ml of prewarmed (37°C) SB medium and 3 μl of 100 mg/ml carbenicillin (when XL1-Blue is used, also add 30 μl of 5 mg/ml tetracycline). To titer the transformed bacteria, dilute 2 μl of the culture in 200 μl of SB medium, and plate 100 μl and 10 μl of this 1:100 dilution on LB agar+carbenicillin plates. Incubate the plates overnight at 37°C. Calculate the total number of transformants by counting the number of colonies, multiplying by the culture volume, and dividing by the plating volume. Shake the 15-ml culture at 250 rpm for 1 hour at 37°C, add 4.5 μl of 100 mg/ml carbenicillin, and shake for an additional hour at 250 rpm and 37°C.

7. Add 2 ml of VCSM13 helper phage (10^{12} to 10^{13} pfu/ml; Protocol 10.2) and transfer to a 500-ml polypropylene centrifuge bottle. Add 183 ml of prewarmed (37°C) SB medium and 92.5 μl of 100 mg/ml carbenicillin (when XL1-Blue is used, add 370 μl of 5 mg/ml tetracycline). Shake the 200-ml culture at 300 rpm for 1.5–2 hours at 37 °C.

8. Add 280 μl of 50 mg/ml kanamycin and continue shaking overnight at 300 rpm and 37°C.

9. Spin at 3000*g* (e.g., 4000 rpm in a Beckman JA-10 rotor) for 15 minutes at 4°C. Save the bacterial pellet for phagemid DNA preparations using, for example, the QIAprep Spin Miniprep Kit (QIAGEN, Cat. # 27106). For phage precipitation, transfer the supernatant to a clean 500-ml centrifuge bottle and add 8 g of PEG-8000 (to 4% [w/v]) and 6 g of sodium chloride (to 3% [w/v]). Dissolve the solids by shaking at 300 rpm for 5 minutes at 37°C. Store on ice for 30 minutes.

10. Spin at 15,000*g* (e.g., 9000 rpm in a Beckman JA-10 rotor) for 15 minutes at 4°C. Discard the supernatant, drain the bottle by inverting on a paper towel for at least 10 minutes, and wipe off the remaining liquid from the upper part of the centrifuge bottle with a paper towel.

11. Resuspend the phage pellet in 2 ml of 1% (w/v) BSA in TBS by pipetting up and down along the side of the centrifuge bottle. Transfer the suspension to a 2-ml microcentrifuge tube. Resuspend further by pipetting up and down with a 1-ml pipet tip. Spin at full speed in a microcentrifuge for 5 minutes at 4°C, and pass the supernatant through a 0.2-μm filter into a 2-ml microcentrifuge tube. Sodium azide can be added to 0.02% (w/v) for storage at 4°C. The phage preparation should be used for panning only if it has been freshly prepared (prepared on the same day); stored phage preparations should be reamplified prior to panning (Protocol 10.4).

The phage library is now ready for panning (Protocols 10.5 and 10.6).

PROTOCOL 10.4

Library Reamplification

The rate of production of antibody-displaying phage is influenced by variations in antibody sequence, such that phage which display different antibodies are produced at different rates. Thus, it can be assumed that reamplification of an existing antibody library reduces its complexity and should be avoided as far as practically possible. Phage preparations generally should be used for library selection only if they have been prepared on the same day, because proteases present in trace levels cleave the displayed antibodies; therefore, library reamplification is sometimes necessary. For reamplification, we recommend the use of only original phage preparations that were directly obtained from library ligation and transformation (see Protocol 10.3).

Materials

E. coli (ER2537 or XL1-Blue, prepared according to Protocol 10.1)
SB medium (see Appendix 2)
100 mg/ml carbenicillin in water, filter-sterilized
5 mg/ml **tetracycline** in **ethanol**
phage library (see Protocol 10.3)
LB agar+carbenicillin (100 μg/ml) plates
VCSM13 helper phage, prepared according to Protocol 10.2
50 mg/ml kanamycin in water, filter-sterilized
PEG-8000 (**polyethylene glycol**; Sigma Cat. # P 2139)
sodium chloride
TBS (50 mM Tris-HCl, pH 7.5, 150 mM NaCl), autoclaved
1% (w/v) BSA in TBS (Sigma, Cat. # A 7906)
2% (w/v) **sodium azide**

CAUTIONS: ***E. coli*, tetracycline, ethanol, PEG, sodium azide (*see Appendix 4*)**

Procedure

1. Inoculate 50 ml of SB medium in a 250-ml Erlenmeyer flask with 50 μl of an *E. coli* preparation (when XL1-Blue is used, add 100 μl of 5 mg/ml tetracycline. It is convenient to start from electrocompetent *E. coli* prepared according to Protocol 10.1. Shake at 250 rpm for 1.5–2.5 hours at 37°C, until the optical density (OD) at 600 nm is about 1. As detailed in Protocol 10.1 and Appendix 3, take precautions to avoid contaminating the culture with phage.
2. Add 10 μl of the phage library preparation to the culture and incubate at room temperature for 15 minutes.

3. Add 10 µl of 100 mg/ml carbenicillin. To titer the phage-infected bacteria (the resulting number should be well above the library size), plate 1 µl and 10 µl of a 10^{-4} dilution of the infected culture on LB agar+carbenicillin plates (see Note below). Incubate the plates overnight at 37°C. Calculate the number of transformants by multiplying the number of colonies by the culture volume and dividing by the plating volume. Transfer the infected 50-ml culture to a 500-ml polypropylene centrifuge bottle and shake at 300 rpm for 1 hour at 37°C. Add 15 µl of 100 mg/ml carbenicillin, and shake at 300 rpm for an additional hour at 37°C.

   Note: For the 10^{-4} dilution, prepare serial dilutions as follows: Dilute 10 µl of the phage preparation in 1 ml of SB (a 10^{-2} dilution), mix, and dilute 10 µl of the 10^{-2} dilution in 1 ml of SB (resulting in a 10^{-4} dilution).

4. Add 2 ml of VCSM13 helper phage (10^{12} to 10^{13} pfu/ml; see Protocol 10.2). Add 148 ml of prewarmed (37°C) SB medium and 75 µl of 100 mg/ml carbenicillin (when XL1-Blue is used, also add 300 µl of 5 mg/ml tetracycline). Shake the 200-ml culture at 300 rpm for 1.5–2 hours at 37°C .

5. Add 280 µl of 50 mg/ml kanamycin and continue shaking overnight at 300 rpm and 37°C.

6. Spin at 3000*g* (e.g., 4000 rpm in a Beckman JA-10 rotor) for 15 minutes at 4°C. Save the bacterial pellet for phagemid DNA preparations using, for example, the QIAprep Spin Miniprep Kit (QIAGEN, Cat. # 27106). For phage precipitation, transfer the supernatant to a clean 500-ml centrifuge bottle and add 8 g of PEG-8000 (to 4% [w/v]) and 6 g of sodium chloride (to 3% [w/v]). Dissolve the solids by shaking at 300 rpm for 5 minutes at 37°C. Incubate on ice for 30 minutes.

7. Spin at 15,000*g* (e.g., 9000 rpm in a Beckman JA-10 rotor) for 15 minutes at 4°C. Discard the supernatant, drain the bottle by inverting on a paper towel for at least 10 minutes, and wipe off remaining liquid from the upper part of the centrifuge bottle with a paper towel.

8. Resuspend the phage pellet in 2 ml of 1% (w/v) BSA in TBS by pipetting up and down along the side of the centrifuge bottle. Transfer the suspension to a 2-ml microcentrifuge tube. Resuspend further by pipetting up and down using a 1-ml pipet tip, spin at full speed in a microcentrifuge for 5 minutes at 4°C, and pass the supernatant through a 0.2-µm filter into a sterile 2-ml microcentrifuge tube. Sodium azide can be added to 0.02% (w/v) for storage at 4°C. Only freshly prepared phage (prepared the same day) should be used for panning.

The library is now ready for panning (Protocols 10.5 and 10.6).

PROTOCOL 10.5

Library Panning on Immobilized Antigens

In this protocol, panning consists of several rounds of binding phage to an antigen immobilized to the well of an ELISA plate, a defined number of washing steps, elution by trypsinization or low pH, and reamplification (Fig. 10.1). During each round, specific binding clones are selected and amplified. These clones predominate after three to four rounds. The input of each round is usually in the range of 10^{12} phage, and the output is usually in the range of 10^5 to 10^8 phage, depending on the number of washing steps and the degree of enrichment occurring at a given round. A 10- to 100-fold increase in output after round 3 or 4 is typical.

Materials

96-well ELISA plate (Corning Costar, Cat. # 3690)
coating buffer, e.g., TBS (50 mM Tris-HCl, pH 7.5, 150 mM NaCl) or 0.1 M sodium bicarbonate, pH 8.6, autoclaved
Linbro plate sealer with adhesive back (ICN Biomedicals, Cat. # 76-401-05)
3% (w/v) BSA in TBS, filter-sterilized (Sigma, Cat. # A-7906)
freshly prepared phage library (see Protocols 10.3 and 10.4)
SB medium (see Appendix 2)
E. coli (ER2537 or XL1-Blue, prepared according to Protocol 10.1)
1% (w/v) BSA in TBS, filter-sterilized (Sigma, Cat. # A-7906)
0.5% (v/v) Tween 20 (Polyoxyethylene sorbitan monolaurate; Sigma, Cat. # P 1379) in TBS, filter-sterilized
10 mg/ml trypsin in TBS, freshly prepared and filter-sterilized (Becton Dickinson, Difco, Cat. # 0152-13-1)
acidic elution buffer (optional): 0.1 M **HCl**, pH adjusted to 2.2 with **glycine** (can be stored at room temperature for several weeks)
neutralization solution (optional; used for acidic elution): 2 M Tris base in water
100 mg/ml carbenicillin in water, filter-sterilized
5 mg/ml **tetracycline** in **ethanol**
LB agar/carbenicillin (100 µg/ml) plates
VCSM13 helper phage, prepared according to Protocol 10.2
50 mg/ml kanamycin in water, filter-sterilized
PEG-8000 (**polyethylene glycol;** Sigma, Cat. # P 2139)
sodium chloride
2% (w/v) **sodium azide**

CAUTIONS: ***E. coli,* HCl, glycine, tetracycline, ethanol, PEG, sodium azide *(see Appendix 4)***

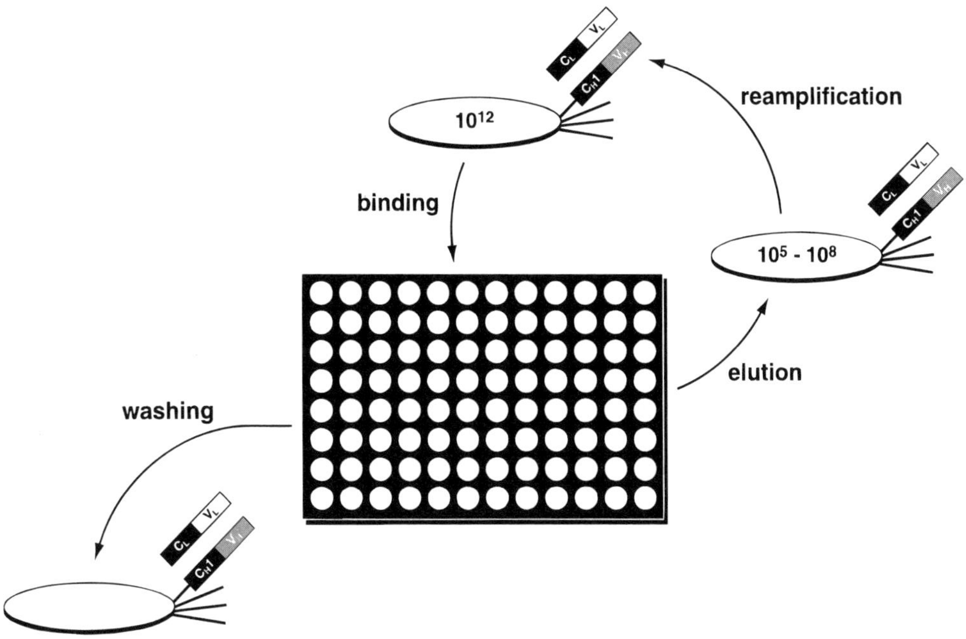

Figure 10.1. Library panning on immobilized antigens. Panning consists of several rounds of binding phage to an antigen immobilized to the well of an ELISA plate, a defined number of washing steps, elution by trypsinization or low pH, and reamplification. During each round, specific binding clones are selected and amplified. Note that elution through low pH leaves the antibody fragment displayed on the phage surface, whereas elution through trypsinization cleaves the antibody fragment off the phage surface.

Procedure

1. Coat the wells of a 96-well ELISA plate with 0.1–1 µg of antigen in 25 µl of an appropriate coating buffer, e.g., TBS or 0.1 M sodium bicarbonate, pH 8.6. Cover the plate with a plate sealer. Coating can be performed overnight at 4°C or for 1 hour at 37°C. In the first round of panning, coat two wells per library to be screened; one well is sufficient in each of the subsequent rounds. For double recognition panning (see Note after Step 5), double the number of coated wells.

2. Shake out the coating solution and block the wells by adding 150 µl of 3% (w/v) BSA in TBS. Seal and incubate for 1 hour at 37°C.

3. Shake out the blocking solution and add 50 µl of freshly prepared phage library to each well. Seal the plate and incubate for 2 hours at 37°C. Meanwhile, inoculate 2 ml of SB medium in a 12-ml polypropylene tube with 2 µl of *E. coli* (when XL1-Blue is used, add 4 µl of 5 mg/ml tetracycline). It is convenient to start from electrocompetent *E. coli* prepared according to Protocol 10.1. Incubate while shaking at 250 rpm for 1.5–2.5 hours at 37°C, until the optical density (OD) at 600 nm reaches about 1. Grow one culture for each library that is screened and an additional culture for input titering. As detailed in Protocol 10.1 and Appendix 3, take precautions to avoid contaminating the cultures with phage. These cultures will be used in Steps 5 and 6.

4. Shake the phage solution out of the ELISA plate, add 150 μl of 0.5% (v/v) Tween 20 in TBS to each well, and pipet 5 times vigorously up and down. Wait 5 minutes, shake out the wash solution, and repeat this washing step. In the first round, wash in this fashion 5 times, in the second round wash 10 times, and in the third and fourth rounds wash 10–15 times. An ELISA plate washer can be used if cross-contamination of libraries screened in parallel is not an issue.

   Note: Tween 20 in TBS can be used at 0.5% (v/v) or 0.05% (v/v) in the washing step.

5. After shaking out the final washing solution, add 50 μl of freshly prepared 10 mg/ml trypsin in TBS to each well. Seal the plate, and incubate for 30 minutes at 37°C. Pipet 10 times vigorously up and down. Transfer the eluates to the prepared 2-ml *E. coli* cultures (2 x 50 μl eluate per culture in the first round, 1 x 50 μl in the subsequent rounds) and incubate at room temperature for 15 minutes.

   Note: Instead of using trypsinization, the phage can be eluted with low pH. Add 50 μl of acidic elution buffer, e.g., 100 mM glycine-HCl, pH 2.2, instead of trypsin solution. Incubate for 10 minutes at room temperature. Pipet 10 times vigorously up and down and transfer the eluate to a microfuge tube containing 3 μl of neutralization solution (2 M Tris base) per 50 μl of acidic elution buffer used. Add the neutralized eluate to the prepared 2-ml *E. coli* culture, incubate at room temperature for 15 minutes, and proceed with Step 6.

   If low pH is used for elution of phage, they can be reincubated with coated antigen in the same round (double recognition panning). This procedure can result in a better elimination of nonspecific phage and, thus, fewer panning rounds are usually necessary to obtain sufficient enrichment of specific phage. Elute phage with low pH and neutralize the eluate as described above. Add 50 μl of 1% (w/v) BSA in TBS to 50 μl of the neutralized eluate, and incubate the solution with coated antigen as described above (100 μl per well). Seal the plate and incubate for at least 1 hour at 37°C, and wash as described above. Wash 5 times the first round and 10 times the following rounds. Elute as described above by using either trypsin or low pH. Add the eluate to 2 ml of *E. coli* culture and incubate at room temperature for 15 minutes. Proceed with Step 6. This protocol can be shortened by incubating the phage only for 1 hour with the antigen in each recognition step. Because the number of output phage obtained is expected to be lower, 100 μl and 10 μl of undiluted culture should be plated for the output titering (see Step 6). Keep in mind that for double recognition panning, twice the number of wells have to be coated with antigen in Step 1.

6. Add 6 ml of prewarmed (37°C) SB medium and 1.6 μl of 100 mg/ml carbenicillin (when XL1-Blue is used, also add 12 μl of 5 mg/ml tetracycline). Transfer the culture to a 50-ml polypropylene tube. For output titering, dilute 2 μl of the sample in 200 μl of SB medium, and plate 100 μl and 10 μl of this 1:100 dilution on LB agar+carbenicillin plates. Shake the 8-ml culture at 250 rpm for 1 hour at 37°C, add 2.4 μl of 100 mg/ml carbenicillin, and shake for an additional hour at 250 rpm and 37°C. Meanwhile, proceed with the input titering by infecting 50 μl of the prepared 2-ml *E. coli* culture (from Step 3) with 1 μl of a 10^{-8} dilution of the phage preparation (see Note below). Incubate for 15 minutes at room tempera-

ture, and plate on an LB agar/carbenicillin plate. Incubate the output and input titering plates overnight at 37°C. Calculate the output and input by multiplying the number of colonies by the culture volume and dividing by the plating volume.

Note: For the 10^{-8} dilution, prepare serial dilutions as follows: Dilute 1 µl of the phage preparation in 1 ml of SB (a 10^{-3} dilution), mix, dilute 1 µl of the 10^{-3} dilution in 1 ml of SB (for a 10^{-6} dilution), mix, and dilute 10 µl of the 10^{-6} dilution in 1 ml of SB (for a 10^{-8} dilution).

7. Add 1 ml of VCSM13 helper phage (10^{12} to 10^{13} pfu; Protocol 10.2) to the 8-ml culture and transfer to a 500-ml polypropylene centrifuge bottle. Add 91 ml of prewarmed (37°C) SB medium and 46 µl of 100 mg/ml carbenicillin (when XL1-Blue is used, add 184 µl of 5 mg/ml tetracycline). Shake the 100-ml culture at 300 rpm for 1.5–2 hours at 37°C.
8. Add 140 µl of 50 mg/ml kanamycin and continue shaking overnight at 300 rpm and 37°C. Meanwhile, for the next round, coat the well of an ELISA plate with antigen as described in Step 1.
9. (Next day) Centrifuge the culture at 3000*g* (e.g., 4000 rpm in a Beckman JA-10 rotor) for 15 minutes at 4°C. Save the bacterial pellet for phagemid DNA preparations using, for example, the QIAprep Spin Miniprep Kit (QIAGEN, Cat. # 27106). For phage precipitation, transfer the supernatant to a clean 500-ml centrifuge bottle and add 4 g of PEG-8000 (to 4% [w/v]) and 3 g of sodium chloride (to 3% [w/v]). To dissolve the solids, shake at 300 rpm and 37°C for 5 minutes. Store on ice for 30 minutes. Meanwhile, block the wells that were coated with antigen in Step 8, as described in Step 2.
10. Centrifuge the PEG precipitation at 15,000*g* (e.g., 9000 rpm in a Beckman JA-10 rotor) for 15 minutes at 4°C. Discard the supernatant, drain the centrifuge bottle by inverting on a paper towel for at least 10 minutes, and wipe off remaining liquid from the upper part of the bottle with a paper towel.
11. Resuspend the phage pellet in 2 ml of 1% (w/v) BSA in TBS by pipetting up and down along the side of the centrifuge bottle. Transfer to a 2-ml microcentrifuge tube and resuspend further by pipetting up and down using a 1-ml pipet tip. Spin at full speed in a microcentrifuge for 5 minutes at 4°C, and pass the supernatant through a 0.2-µm filter into a sterile 2-ml microcentrifuge tube. Continue from Step 1 for the next round or add sodium azide to 0.02% for storage at 4°C. Phage preparations can be stored for months at 4°C but have to be reamplified (Protocol 10.4) before panning. Only freshly prepared phage (prepared the same day) should be used for each round.

Monitor the selection progress as described in Chapter 11. For panning on immobilized antigens, 3–6 rounds of panning are required.

PROTOCOL 10.6

Library Panning on Eukaryotic Cells

Although panning on immobilized antigen is usually an easy way to select antigen binders from an antibody library, in some cases selection on cells may be required (Rader and Barbas 1997).

Eukaryotic cells express a vast number of different surface antigens. Enrichment of undesired clones that bind to epitopes unrelated to the surface antigen of interest during the selection procedure can therefore be a problem. There are several ways to minimize nonspecific selection. One method is to preincubate the phage pool with cells that do not express the antigen of interest but are otherwise closely related to the target cells. If different cell lines expressing the target surface antigen are available, switching cell lines during the panning procedure can also prevent enrichment of undesired clones. The use of cells that express the target surface antigen at high levels greatly increases the chance of a successful selection.

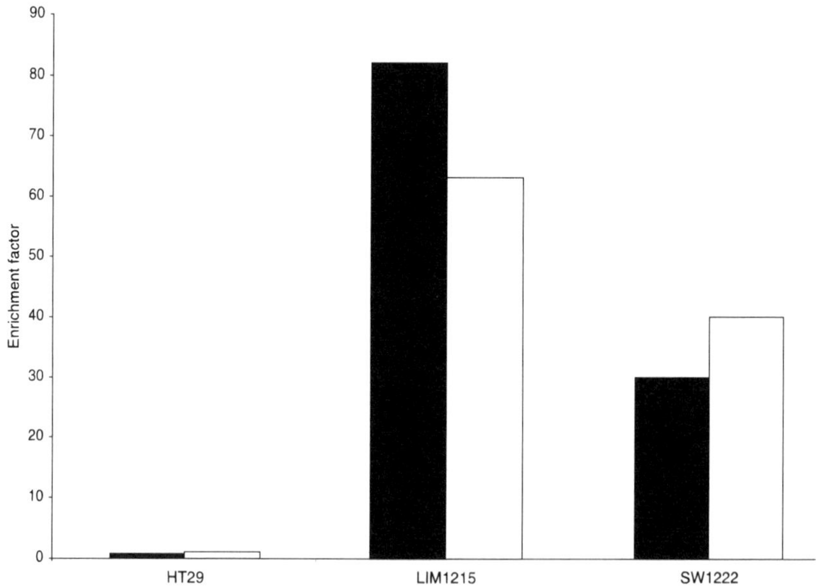

Figure 10.2. Enrichment of phage after cell-surface panning. In this experiment, two clonal phage populations were mixed in a 1:1 ratio. One population, which was generated using a pComb3 vector with a carbenicillin selection marker, displays an antibody fragment that binds to the A33 antigen. The other population, which was generated using a pComb3 vector with a chloramphenicol selection marker, displays an antibody fragment that binds to tetanus toxoid. The mixture was panned against three cell lines. HT29 is a human colon carcinoma cell line that does not express the A33 antigen; LIM1215 and SW1222 are human colon carcinoma cell lines that express the A33 antigen on their surface. After one round of panning according to Protocol 10.6, phage were titered as described using both LB agar + carbenicillin and LB agar + chloramphenicol plates. The graph shows the enrichment factor, i.e., the ratio of anti-A33 antigen phage to anti-tetanus phage. Light and dark columns represent enrichment factors obtained from two independent panning experiments.

The enrichment factor of specific clones over clones not binding to the cells (Fig. 10.2) is typically lower with cell panning than with panning on immobilized antigens; therefore, more rounds of panning are usually needed. Depending on the target antigen and the cells used, modifications of the cell panning protocol given below might be necessary to successfully select on cells. For a more sophisticated cell panning protocol that involves biotinylation and magnetically activated cell sorting, see Chapter 23.

Materials

cells expressing the target surface antigen (see Step 1)
PBS (phosphate buffered saline, see Appendix 2)
E. coli (ER2537 or XL1-Blue, prepared according to Protocol 10.1)
SB medium (see Appendix 2)
10 mg/ml trypsin in TBS, freshly prepared and filter-sterilized
(Becton Dickinson, Difco, Cat. # 0152-13-1)
flow cytometry buffer: PBS containing 20 mM HEPES pH 7.4; 1% BSA; 0.03%
sodium azide; filter-sterilized; store at 4°C up to a month
nonfat dry milk (BioRad, Cat. # 170-6404)
96-well V-bottom plates (Corning Costar, Cat. # 3896)
flow cytometry tubes (Becton Dickinson, Falcon, Cat. # 352054)
Linbro plate sealer with adhesive back (ICN Biomedicals, Cat. # 76-401-05)
100 mg/ml carbenicillin in water, filter-sterilized
5 mg/ml **tetracycline** in **ethanol**
LB agar+carbenicillin (100 μg/ml) plates
VCSM13 helper phage, prepared according to Protocol 10.2
50 mg/ml kanamycin in water, filter-sterilized
PEG-8000 (**polyethylene glycol,** Sigma, Cat. # P 2139)
sodium chloride
2% (w/v) **sodium azide**

CAUTIONS: ***E. coli*, sodium azide, tetracycline, ethanol, PEG *(see Appendix 4)***

Procedure

1. Grow a sufficient number of target cells for the panning procedure (usually 4–6 rounds). Usually cells are seeded in 60-mm dishes so that one dish of cells grown to the right density (~ 0.5 to 2 x 10^6 cells/ml or, in the case of adherent cells, a confluency of 60–80%) is available for each day of panning. Related cells that do not express the target antigen should also be grown at this point (optional, see above).

2. In the morning of every day of panning, inoculate 2 ml of SB medium in a 12-ml polypropylene tube with 2 μl of an *E. coli* preparation (when XL1-Blue is used, add 4 μl of 5 mg/ml tetracycline). It is convenient to start from electrocompetent *E. coli* prepared according to Protocol 10.1. Incubate while shaking at 250 rpm for

1.5–2.5 hours at 37°C, until the cells reach an optical density (OD) at 600 nm of about 1. Grow one culture for each library that is screened and an additional culture for input titering. As detailed in Protocol 10.1, take precautions to avoid contaminating the culture with phage. These cultures will be used in Steps 11 and 12.

3. If using adherent cells, detach either mechanically (by scraping) or enzymatically (e.g., by trypsinizing with 0.25% trypsin, 1 mM EDTA for 5–10 minutes at 37°C).

   Note: If trypsin is used to detach adherent cells, care must be taken not to destroy the antigen of interest. An additional washing step using flow cytometry buffer or cell medium containing fetal calf serum (FCS) is also necessary after trypsinization.

4. Wash the cells (add 10 ml of PBS, mix carefully, and spin at 750*g* for 5 minutes). Resuspend the cells in flow cytometry buffer at around 1×10^7 cells per ml. Follow the same procedure with the related cell line that does not express the antigen of interest.

5. Incubate 50 μl of freshly prepared phage with 150 μl of PBS containing 5% milk proteins on ice for 5 minutes to 1 hour (phage titer typically 10^{12}–10^{13}/ml).

   Note: If a related cell line that does not express the antigen of interest is used for preincubation, continue with Steps 5a–d. Otherwise, go directly to Step 6.

   a. Add the cells not expressing the antigen of interest (100 μl with ~ 1×10^6 cells) to a V-bottom plate. Use two wells of cells for each library.

   b. Add 80 μl of preincubated phage suspension to each well containing cells, mix carefully by pipetting up and down once, and incubate for 30 minutes at room temperature.

   c. Add target cells (100 μl with ~ 1×10^6 cells) to different wells of the V-bottom plate. Again, use two wells of cells for each library.

   d. Spin the plate at 750*g* (e.g., 2000 rpm in a Beckman GS-6R tabletop centrifuge with Beckman Microplus Carriers) for 2–3 minutes (room temperature, brake set to low) and transfer the supernatant from cells not expressing the antigen of interest to the target cells. Resuspend the target cells in the supernatant and incubate for 30 minutes at room temperature. Continue with Step 8.

6. Add target cells (100 μl with ~ 1×10^6 cells) to the wells of a V-bottom plate. Use two wells of cells for each library.

7. Add 80 μl of preincubated phage suspension to each well containing cells, mix carefully by pipetting up and down once, and incubate at room temperature for 30 minutes.

8. Spin the plate at 750*g* for 2–3 minutes (room temperature, brake set to low), remove and discard the supernatant, and resuspend each well of cells in 180 μl of flow cytometry buffer. Repeat this washing step 5–8 times.

9. Resuspend each well of cells in 180 µl of PBS and transfer the cell suspensions to flow cytometry tubes. Add 1 ml of PBS to each tube, and spin cells at 750*g* for 2–3 minutes (room temperature, brake set to low).

10. Resuspend cells in 150 µl of freshly prepared 10 mg/ml trypsin in TBS and incubate on a shaker (200–300 rpm) for 30 minutes at 37°C to elute phage.

11. Combine the two wells per library and transfer to a prepared 2-ml *E. coli* culture (from Step 2). Incubate for 15 minutes at room temperature.

12. Add 6 ml of prewarmed (37°C) SB medium and 1.6 µl of 100 mg/ml carbenicillin (when XL1-Blue is used, also add 12 µl of 5 mg/ml tetracycline) and transfer the culture into a 50-ml polypropylene tube. For output titering, dilute 2 µl of the sample in 200 µl of SB, and plate 100 µl and 10 µl of this 1:100 dilution on LB agar+carbenicillin plates. Shake the 8-ml culture at 250 rpm for 1 hour at 37°C, add 2.4 µl of 100 mg/ml carbenicillin, and shake for an additional hour at 250 rpm and 37°C. Meanwhile, proceed with the input titering by infecting 50 µl of another 2-ml *E. coli* culture (from Step 2) with 1 µl of a 10^{-8} dilution of the phage preparation. Incubate for 15 minutes at room temperature, and plate on an LB agar+carbenicillin plate. Incubate the output and input titering plates overnight at 37°C. Calculate the output and input by multiplying the number of colonies by the culture volume divided by the plating volume.

Note: For the 10^{-8} dilution, prepare serial dilutions as follows: Dilute 1 µl of the phage preparation in 1 ml of SB (a 10^{-3} dilution) and mix. Dilute 1 µl of the 10^{-3} dilution in 1 ml of SB (for a 10^{-6} dilution) and mix. Finally, dilute 10 µl of the 10^{-6} dilution in 1 ml of SB (for a 10^{-8} dilution) and mix.

13. Add 1 ml of VCSM13 helper phage (10^{12} to 10^{13} pfu; see Protocol 10.2) to the 8-ml culture and transfer to a 500-ml polypropylene centrifuge bottle. Add 91 ml of prewarmed (37°C) SB medium and 46 µl of 100 mg/ml carbenicillin (when XL1-Blue is used, add 184 µl of 5 mg/ml tetracycline). Shake the 100-ml culture at 300 rpm for 1.5–2 hours at 37°C.

14. Add 140 µl of 50 mg/ml kanamycin and continue shaking overnight at 300 rpm and 37°C.

15. (Next day) Centrifuge the culture at 3000*g* (e.g., 4000 rpm in a Beckman JA-10 rotor) for 15 minutes at 4°C. Save the bacterial pellet for phagemid DNA preparations using, for example, the QIAprep Spin Miniprep Kit (QIAGEN, Cat. # 27106). For phage precipitation, transfer the supernatant to a clean 500-ml centrifuge bottle and add 4 g of PEG-8000 (to 4% [w/v]) and 3 g of sodium chloride (to 3% [w/v]). To dissolve the solids, shake for 5 minutes at 300 rpm and 37°C. Store on ice for 30 minutes.

16. Spin at 15,000*g* (e.g., 9000 rpm in a Beckman JA-10 rotor) for 15 minutes at 4°C. Discard the supernatant, drain the centrifuge bottle by inverting on a paper towel for at least 10 minutes, and wipe off remaining liquid from the upper part of the bottle with a paper towel.

17. Resuspend the phage pellet in 2 ml of 1% (w/v) BSA in TBS or flow cytometry buffer by pipetting up and down along the side of the centrifuge bottle. Transfer to a 2-ml microcentrifuge tube, and resuspend further by pipetting up and down with a 1-ml pipet tip. Spin at full speed in a microcentrifuge for 5 minutes at 4°C, and pass the supernatant through a 0.2-μm filter into a sterile 2-ml microcentrifuge tube. Alternatively, flow cytometry buffer can be used to resupend the phage pellet. In this case, sterile filtering of the phage solution is optional. Continue from Step 1 for the next round of panning, or add sodium azide to 0.02% (w/v) for storage at 4°C. Phage preparations can be stored for months at 4°C but have to be reamplified (Protocol 10.4) before panning. Only freshly prepared phage (prepared the same day) should be used for each round.

Monitor the selection progress as described in Chapter 11. For panning on eukaryotic cells, 4–8 rounds of panning are required.

REFERENCES

Rader C. and Barbas C.F., III. 1997. Phage display of combinatorial antibody libraries. *Curr. Opin. Biotechnol.* **8:** 503–508.

11 Analysis of Selected Antibodies

PETER STEINBERGER, CHRISTOPH RADER, AND CARLOS F. BARBAS III
Department of Molecular Biology, The Scripps Research Institute, La Jolla, California 92037

FAST AND RELIABLE ANALYSIS of a panned antibody-fragment library at a polyclonal as well as at a clonal level is imperative to quickly assess whether or not the panning experiment was successful and to select antibody fragment clones that have the desired specificity and affinity. This chapter describes a number of procedures to assess clone pools and single clones obtained after panning (see Fig. 11.1). After several rounds of panning (described in Chapter 10), the resultant phage pools are usually tested by ELISA (Protocol 11.1). If the selected phage pools show specific binding to the antigen of interest, single clones are analyzed. This can either be done with antibody fragment-displaying phage prepared from single clones (Protocol 11.2) or with antibody fragments from IPTG-induced cultures (Protocol 11.3). Especially if a large number of positive clones are obtained, *Bst*OI fragment analysis can facilitate the identification of clones that have different DNA sequences (Protocol 11.4) and can be analyzed further. Antibody fragment-displaying phage or antibody fragments from IPTG-induced cultures can be tested for binding to cell surface antigen by flow cytometry (Protocol 11.5). The pComb3 vectors can be used to express clones without the gene III product (Protocol 11.6). Phagemid DNA is prepared from selected clones and used for DNA sequence analysis (Protocol 11.7). Selected clones can be used for large-scale expression and purification as described in Chapter 12.

CONTENTS

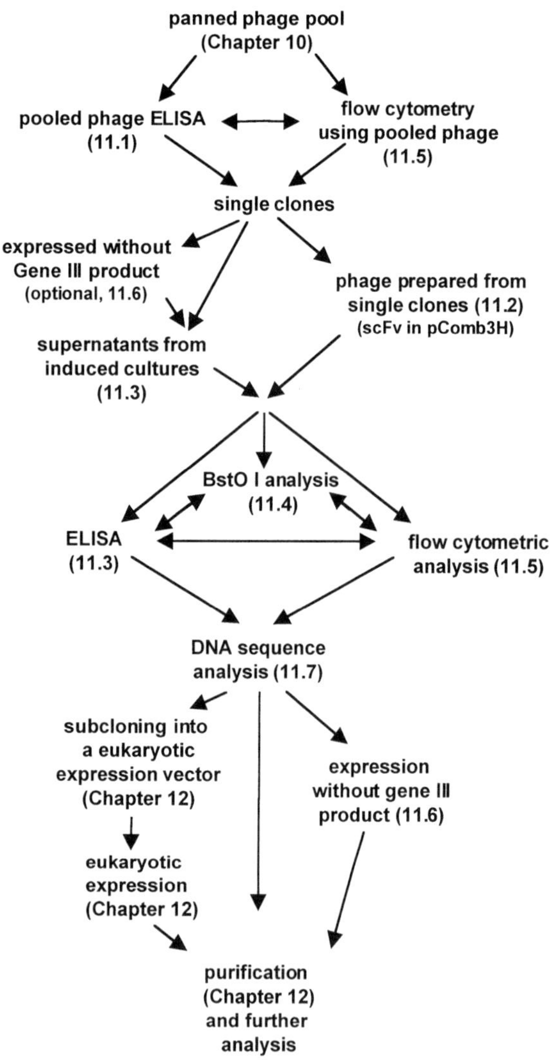

Figure 11.1. Flow chart for the analysis of phage pools and clones. After a panning experiment, the pooled phage can be tested by ELISA (Protocol 11.1) or flow cytometry (Protocol 11.5) to assess whether it is worthwhile to continue with the analysis of single clones. If the phage pools bind specifically to the antigen of interest, single clones are inoculated into culture, and the cultures are induced and tested for binding (Protocols 11.3 and 11.5). For scFv clones derived from pComb3H libraries, antibody fragment-displaying phage preparations are usually analyzed (Protocols 11.2 and 11.3). *BstOI* analysis of antibody fragment-encoding PCR products can be performed in parallel to test for diversity among the clones (Protocol 11.4). Selected clones are subjected to DNA sequence analysis (Protocol 11.7) and can be expressed without the gene III product for further analysis (Protocol 11.6 and Chapter 12). If larger amounts of antibody fragment are needed, subcloning into a eukaryotic expression vector is recommended (Chapter 12).

PROTOCOL 11.1

Analysis of Antibody Fragment-displaying Phage Pools by ELISA: Phage ELISA

After several rounds of panning, the resultant phage pool can be tested in an ELISA to evaluate the success of the panning experiment. Such a "phage ELISA" has several advantages over the often tedious analysis of single clones. Because the precipitated phage can be used right away, there is no need to grow and induce cultures. If specific binders were enriched during the panning procedure, a positive signal will usually be obtained when testing the phage pools, even in cases where the percentage of positive clones is low. The phage ELISA works regardless of the kind of antibody fragment that is expressed; this property is especially advantageous when analyzing single-chain antibody fragment libraries that do not contain a peptide tag for detection. Poor expression of antibody fragments, which might impair the analysis of induced cultures, seems to affect the phage ELISA to a lesser extent.

PEG-precipitated phage from single clones (see Protocol 11.2) can also be tested in ELISA according to the procedure described below.

To monitor the selection process, phage pools obtained in each round of panning are tested in parallel with phage from the unpanned library. Although we recommend the use of fresh phage preparations for panning, PEG-precipitated phage stored at 4°C can be used for the phage ELISA for up to several weeks and sometimes even longer. Phage preparations that have been stored longer can be reamplified as described in Protocol 10.4 (library phage pool) or Protocol 11.2 (single clones).

Materials

antigen of interest
BSA or another suitable control antigen
PBS (phosphate-buffered saline, see Appendix 2)
ELISA plates (e.g., 96-well flat-bottom assay plates with high-binding surface, half area or standard area, Corning Costar, Cat. # 3690 or # 3590)
multichannel pipettor (optional)
5% milk (w/v) in PBS (Bio-Rad, Cat. # 170-6404)
phage (**PEG** precipitated — see Chapter 9 or Protocol 11.2; phage titer is usually around 5×10^{12} cfu/ml)
conjugated secondary antibody:
- HRP-conjugated anti-M13 monoclonal antibody from mouse, use at 1:5000 (Amersham Pharmacia Biotech, Cat. # 27-9421-01)
- HRP-conjugated anti-HA, clone 12CA5 from mouse, use at 1:1000 (Roche Molecular Biochemicals, Cat. # 1 667 475) or
- HRP-conjugated anti-HA, clone 3F10 from rat, high affinity, use at 1:2000 (Roche Molecular Biochemicals, Cat. # 2 013 819)

HRP substrate (e.g., **ABTS** solution, Roche Molecular Biochemicals, Cat. # 1 684 302) or another appropriate substrate, depending on the conjugate used
ELISA reader for 96-well plates

CAUTIONS: **PEG, ABTS** ***(see Appendix 4)***

Procedure

1. Dilute the antigens (the antigen of interest, as well as BSA or another control antigen) in PBS. Make 30 µl of antigen solution containing 0.05–0.5 µg of antigen for each well to be coated. Depending on the antigens, an alternative coating buffer can be used (for example, 0.1 M $NaHCO_3$, pH 8.6).

   Note: The volume and amounts of antigen given above are for half-area assay plates. If standard size plates are used, double the volumes throughout the protocol.

2. Add 25 µl of antigen solution to the ELISA plate wells and incubate overnight at 4°C, or for at least 1 hour at 37°C.

   Note: Avoid evaporation during the incubation steps by covering the ELISA plates with plastic film.

3. Shake out the wells and wash twice. Washing is most easily accomplished by holding the plate under running deionized tap water for 20 seconds and then shaking it out.
4. Block each well by adding 150 µl of 5% milk and incubating for 1 hour at 37°C.
5. Dilute the phage threefold in 5% milk. Prepare 55 µl per well. Incubate for 5 minutes to 1 hour at room temperature.
6. Shake the blocking solution out of the wells, add 50 µl of the dilute phage preparation to each well, and incubate for 1–2 hours at 37°C.
7. Dilute HRP-conjugated anti-M13 antibodies or HRP-conjugated anti-HA antibodies with 5% milk in PBS (see Materials list for dilution factors). Prepare 55 µl per well.

   Note: Anti-HA conjugates can be used only with pComb3X libraries.

8. Wash the plate 10 times with deionized tap water as described above (Step 3).
9. Add 50 µl of diluted secondary antibody conjugate to each well, and incubate for 1 hour at 37°C.
10. Prepare the substrate (e.g., ABTS solution for HRP-conjugated antibodies) as recommended by the supplier. Prepare 55 µl per well.
11. Wash the plate 10 times with deionized tap water as described above (Step 3).
12. Add 50 µl of substrate solution to each well.

13. Incubate at room temperature and use an ELISA plate reader to read the optical density at 405 nm (OD_{405}) at appropriate time points, depending on the strength of the signal and the strength of background signals from the BSA/unrelated antigen wells (see Fig. 11. 2).

Note: Two readings, the first around 5 minutes and the second around 30–60 minutes after addition of the substrate solution, are usually sufficient to see specific binding of antibody fragment-displaying phage to the antigen.

The ELISA signal from the antigen of interest should increase during the course of panning, but the signal from the control antigen should remain low throughout the panning. An increasing signal from the control antigen might indicate nonspecific sticky clones or plastic binders. At the end of panning, the ELISA reading obtained with the antigen of interest is usually three or more times the reading with the control antigen. If this reading drops considerably from, for example, the third to the fourth round, this could indicate that the fourth-round selection should be repeated using the phage pools from the third round.

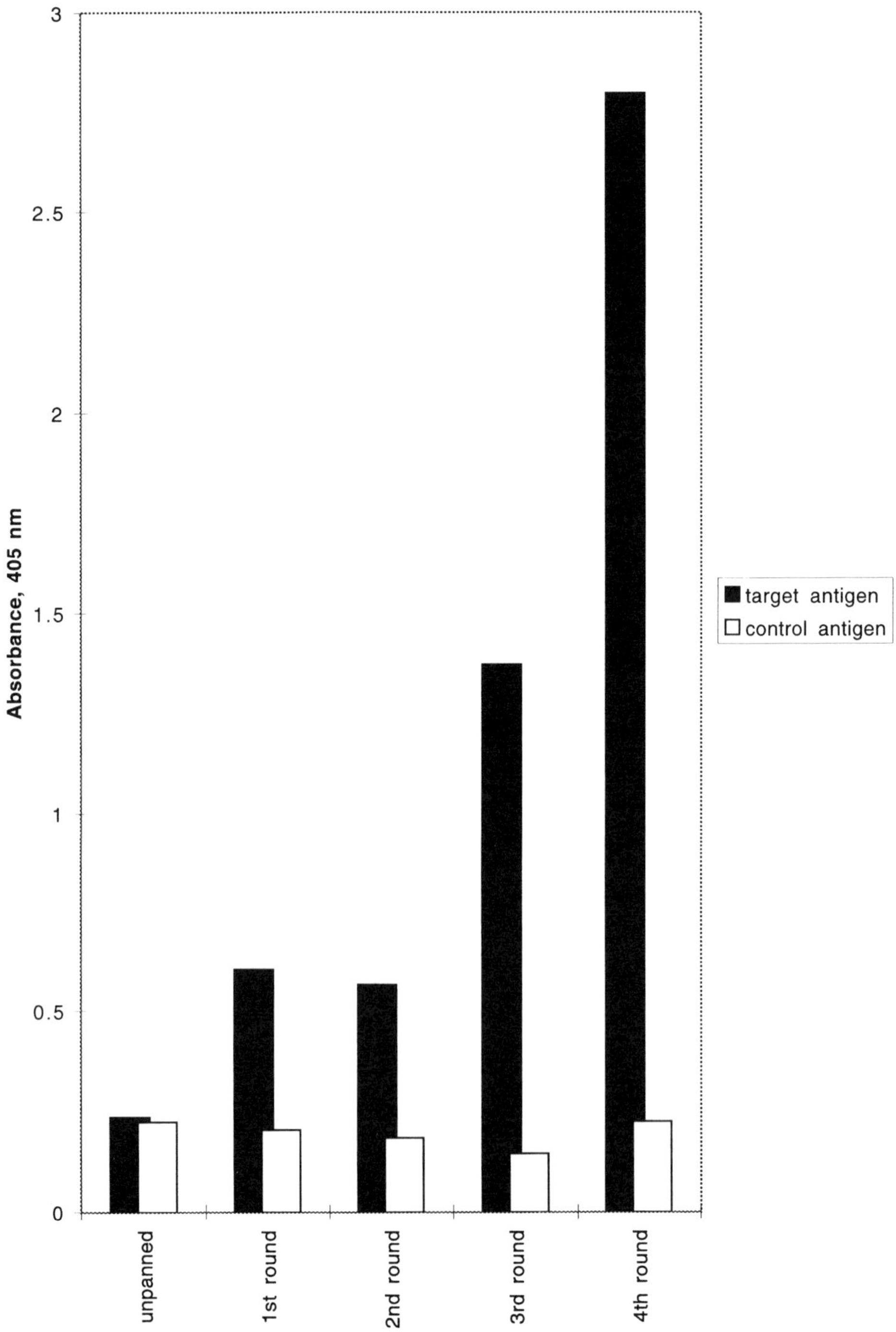

Figure 11.2. Phage ELISA. Phage pools from an unpanned chimeric rabbit/human Fab library and the phage pools obtained after different rounds of panning were tested for binding to the target antigen and a control antigen in ELISA. Binding phage were detected with an HRP-conjugated anti-M13 antibody, and ELISA plates were read 30 minutes after the addition of substrate. The ELISA was performed as described in Protocol 11.1.

PROTOCOL 11.2

Small-scale Preparation of PEG-precipitated Antibody Fragment-displaying Phage

When testing single clones for specific binding, the use of antibody-displaying phage is sometimes preferable to analysis of antibody fragments prepared from induced cultures. For example, if clones from a pComb3H single-chain library need to be analyzed for antigen binding, the phage that is linked to the single-chain antibody fragment can be used as a "tag" for detection. In this way, single clones can be analyzed without subcloning. Another advantage of using phage for single-clone analysis is that after a successful binding assay with phage pools derived from panning, single clones can be tested using the same assay conditions. Furthermore, in case of poor expression of the antibody fragment in *E. coli*, analysis of phage seems to work better than analysis of antibody fragments prepared from induced cultures.

The following protocol describes the preparation of antibody-displaying phage derived from single clones. It is a slightly modified and downscaled version of the procedure for the generation of PEG-precipitated phage for panning (Protocol 10.4). The following protocol facilitates the generation of PEG-precipitated phage from many single clones. The volume and concentration of PEG-precipitated phage obtained is usually sufficient for testing by ELISA (Protocol 11.1) and by flow cytometry (Protocol 11.5). If a more extensive analysis needs to be done, the procedure can be modified according to Protocol 10.4.

PEG-precipitated phage can be used in binding assays for several weeks. If older phage samples need to be tested, use 1 µl of phage preparation to infect 100 µl of an *E. coli* culture ($OD_{600} \sim 1$; see Protocol 10.4). Incubate for 15 minutes at room temperature, and add the infected *E. coli* to 5 ml of SB containing 50 µg/ml carbenicillin. Proceed with Step 2 of this protocol. Properly stored phage preparations can be used for reinfection for up to 1 year (see Chapter 10).

Materials

SB medium (see Appendix 2)
100 mg/ml carbenicillin (see Appendix 2)
output titering plates (see Protocols 10.5 and 10.6)
E. coli cells (optional; for reamplification of phage, see Protocol 10.4)
VCSM13 helper phage (see Protocol 10.2)
50 mg/ml kanamycin (see Appendix 2)
5x **PEG**/NaCl (see Appendix 2)
PBS + 1% BSA + 0.03% **sodium azide** or flow cytometry buffer (sterile filtered, see Appendix 2)

CAUTIONS: *E. coli*, PEG, sodium azide *(see Appendix 4)*

Procedure

1. Inoculate 5-ml volumes of SB containing 50 μg/ml carbenicillin with single colonies.

   Note: 10–30 colonies from output titer plates from the last one or two rounds of a panning experiment are typically used (see Protocols 10.5 and 10.6). If phage pools were analyzed by ELISA as described in Protocol 11.1, the results will reveal which output titer plates should be used for the analysis of single clones.

2. Shake the cultures at 300 rpm for 4–6 hours at 37°C. Increase the incubation time for larger cultures; for example, a 100-ml culture should be grown for 8 hours before the helper phage is added.

3. *Optional:* For backup, streak or dot (e.g., 2 μl) the cultures on prewarmed (37°C) LB+carbenicillin plates. Incubate overnight at 37°C, seal the plates with Parafilm, and store at 4°C for up to 3 months.

4. Add helper phage to the cultures (50 μl per 5-ml culture).

5. Shake the cultures for another 2 hours at 300 rpm and 37°C.

6. Add 7 μl of 50 mg/ml kanamycin to each culture, to a final concentration of 70 μg/ml.

7. Shake the cultures for at least 10 hours (typically overnight) at 300 rpm and 37°C.

8. Centrifuge the cultures at 2800*g* (3500 rpm in a tabletop centrifuge) for 15 minutes. Transfer 1.2 ml of each culture supernatant to each of two (or more) microfuge tubes. Add 300 μl of 5x PEG/NaCl to each tube, mix, and precipitate phage by incubating for 30 minutes on ice. The bacterial pellet can be used to prepare phagemid using a standard plasmid mini-preparation method (e.g., QIAprep Spin Miniprep Kit, QIAGEN).

9. Spin the tubes in a microcentrifuge at 14,000 rpm and 4°C. Carefully aspirate and discard the supernatants with a pasteur pipet. Spin the tubes for another 20 seconds and aspirate the remaining supernatants.

   Note: A small white pellet is usually visible after the centrifugation step. It is important to remove as much of the PEG as possible, because it can interfere with resuspension of the phage.

10. Resuspend each phage pellet in 50 μl of PBS + 1% BSA + 0.03% sodium azide or 50 μl of flow cytometry buffer. Combine the samples derived from each clone and use them for binding assays (Protocols 11.1 and 11.5) or store them at 4°C.

The PEG-precipitated phage can be used in binding assays for several weeks and are good for reinfection for up to a year.

PROTOCOL 11.3

Analysis of Antibody Fragment-expressing Clones in ELISA

Once a phage pool shows specific binding to the antigen of interest (see Protocol 11.1), single clones are screened. The following protocol describes the easiest and fastest method by which antibody-containing supernatants obtained from induced cultures can be tested in ELISA.

If pComb3X was used for library construction, antibody fragments that bind to the antigen can be detected with a conjugated anti-HA antibody, regardless of the type of antibody fragment (see Chapter 2). If chimeric Fab libraries constructed in pComb3H or pComb3X are being analyzed, antibody conjugates that recognize human Fab fragments or human light chains can be used. To analyze single-chain antibody clones derived from a library constructed in pComb3H, the DNA fragment that encodes the single chain usually has to be subcloned into a different vector for expression of a single-chain antibody fragment with a detection tag. Such vectors include pPho, pAra, and pComb3X. Alternatively, any antibody fragment derived from pComb3H or pComb3X libraries can be analyzed by phage ELISA using PEG-precipitated phage (Protocols 11.1 and 11.2).

For the preparation of antibody fragments in culture supernatants, 2 ml of SB containing 50 μg/ml carbenicillin is usually inoculated with colonies from the output titer plates of the round of interest (Protocols 10.5 and 10.6). If the output titer plates are not available, 1 μl of the PEG-precipitated phage pool is diluted 10^8-fold in SB medium. 1 μl of the diluted phage is then used to infect 50 μl of an *E. coli* culture (XL1-Blue or ER2537) and plated on carbenicillin plates as described for the titration of the input phage (Protocol 10.5). The next day, colonies from those plates can be inoculated into 2 ml of SB containing 50 μg/ml carbenicillin.

For the initial analysis of clonal binding and diversity, antibody fragments can be expressed and analyzed as fusions to the *gene III* product (pIII), because pIII does not interfere with antigen–antibody fragment interaction or with detection of the antibody fragment.

For further analysis and for purification, antibody fragments should be expressed without pIII. For clones derived from pComb3H libraries, the pIII gene fragment is eliminated by digestion with *Spe*I and *Nhe*I, followed by re-ligation. In pComb3X, antibody fragments are expressed without pIII by transforming the phagemid into a nonsuppressor strain of *E. coli*, e.g., TOP10 cells. (Protocol 11.6).

Single clones can be analyzed simultaneously for specific binding in ELISA and by *Bst*OI digest (Protocol 11.4). This parallel analysis allows fast identification of specific clones with different DNA sequences that can be analyzed further, e.g., by DNA sequencing (Protocol 11.7), or used for large-scale expression and purification (see Chapter 12). Alternatively, *Bst*OI analysis can be performed on selected clones that are positive in ELISA.

Materials

output titering plates (see introduction above, and Step 1)
SB medium (see Appendix 2)
100 mg/ml carbenicillin (see Appendix 2)
LB+carbenicillin plates (see Appendix 2)
0.5 M **IPTG** (see Appendix 2)
dry ice/**ethanol** bath (optional, used for breaking up *E. coli* cells)
ELISA plates (e.g., 96-well flat-bottom assay plates with high binding surface; half area or standard area, Corning Costar, Cat. # 3690 or Cat. # 3590)
multichannel pipettor (optional)
antigen of interest
BSA or another suitable control antigen
PBS (see Appendix 2)
5% milk (w/v) in PBS (e.g., Bio-Rad, Cat. # 170-6404)
antibody conjugate or conjugate that reacts with human Fab fragments:
  HRP-conjugated anti-HA, clone 12CA5 from mouse, use at 1:1000 (Roche Molecular Biochemicals, Cat. # 1 667 475)
  HRP-conjugated anti-HA, clone 3F10 from rat, high affinity, use at 1:1000 (Roche Molecular Biochemicals, Cat. # 2 013 819)
  HRP-conjugated donkey antihuman IgG (H+L) antibodies, use at 1:2000 (Jackson ImmunoResearch Laboratories, Cat. # 709-035-149)
HRP substrate: e.g., **ABTS** solution (Roche Molecular Biochemicals, Cat. # 1 684 302) or depending on the conjugate used, another appropriate substrate
ELISA reader for 96-well plates

CAUTIONS: IPTG, ethanol, ABTS ***(see Appendix 4)***

Procedure

1. For each clone, add 2 ml of prewarmed (37°C) SB containing 50 µg/ml carbenicillin to a 14-ml culture tube.

   Note: 10–30 colonies from output titer plates from the last one or two rounds of a panning experiment are typically used (see Protocols 10.5 and 10.6). If phage pools were analyzed by ELISA as described in Protocol 11.1, the results will reveal which output titer plates should be used for the analysis of single clones.

2. Use a pipettor to pick colonies and inoculate cells into the SB+carbenicillin.

   Note: If the clones also will be analyzed by *Bst*OI digest, first introduce the cells into PCR tubes containing 10 µl of nuclease-free water, and then inoculate the SB+carbenicillin. Continue with Protocol 11.4, Step 4 for *Bst*OI analysis.

3. Shake the 14-ml tubes containing the bacterial cultures at 300 rpm for 5–8 hours at 37°C.

4. Spot a few μl of each culture onto LB+carbenicillin plates. Incubate the plates overnight at 37°C, seal with Parafilm, and store at 4°C.

5. Induce the cultures (they should be turbid, with an OD_{600} ~ 0.5) by adding IPTG to 2 mM (8 μl of 0.5 M stock). Shake at 300 rpm for an additional 15–24 hours at 37°C. *Optional:* Divide the cultures before adding the IPTG, and grow one culture overnight without IPTG induction for phagemid DNA preparation with a standard plasmid minipreparation method (e.g., QIAprep Spin Miniprep Kit, QIAGEN).

> *Note:* The optimal time point for harvesting culture supernatants for ELISA testing varies, but it has been found that the amount of antibody fragments in culture supernatants is usually higher when induced cultures are grown for a relatively long time. Alternatively, antibody fragments can be prepared from the cell pellet. If antibody fragments are prepared from cell lysates, induce the cultures for less time, around 10 hours.
>
> To prepare antibody fragments from cell lysates, centrifuge the induced cultures at 2000*g* (3000 rpm in a tabletop centrifuge) for 10 minutes. Resuspend the cell pellets in 0.3 ml of PBS and transfer the cell suspension to a microcentrifuge tube. Lyse cells by freezing in a dry ice/ethanol bath for 3–5 minutes followed by thawing in a 37°C water bath for 3 minutes. Repeat this freeze/thaw process three times. Pellet the cell debris by centrifuging at 15,000 rpm in a microcentrifuge for 5 minutes and transfer the supernatant to a new tube. The supernatant contains the antibody fragments.

6. Dilute the antigens (the antigen of interest, as well as BSA or another control antigen) in PBS. Make 30 μl of antigen solution containing 0.05–0.5 μg of antigen for each well to be coated. Depending on the antigens used, an alternative coating buffer can be used (for example, 0.1 M $NaHCO_3$, pH 8.6).

> *Note:* The volume and amounts of antigen given above are for half-area assay plates. If standard size plates are used, double the values throughout the protocol.

7. Add 25 μl of antigen solution to the ELISA plate wells and incubate overnight at 4°C, or for at least 1 hour at 37°C.

> *Note:* Avoid evaporation during incubation steps by covering the ELISA plate with plastic film.

8. Shake out the wells and wash twice. Washing is most easily accomplished by holding the plate under running deionized tap water for 20 seconds and then shaking it out.

9. Block the wells by adding 150 μl of 5% milk in PBS and incubating for at least 1 hour at 37°C.

10. Centrifuge the induced *E. coli* cultures (from Step 5) at 2800*g* (3500 rpm in a tabletop centrifuge) for 15 minutes at room temperature or 4°C.

11. Shake the blocking solution out of the wells. Add to each well 25 μl of 5% milk in PBS and 25 μl of supernatant from induced cultures or from cell lysates. Transfer the remaining supernatants to microcentrifuge tubes. Supernatants can be stored at –20°C for years. Incubate the ELISA plate for at least 1 hour at 37°C.

12. Dilute the conjugated secondary antibody in PBS containing 5% milk (see Materials list for dilution factors). Make 55 μl per well.

13. Wash the ELISA plate 10 times, as in Step 8.

14. Add 50 μl of antibody conjugate solution to each well. Incubate the plate for 1 hour at 37°C.

15. Wash the plate 10 times as in Step 8. Add 50 μl of substrate solution to each well.

16. Incubate at room temperature and use an ELISA plate reader to measure the optical density at 405 nm (OD_{405}) at appropriate time points, depending on the strength of the signal and the strength of background signals from the BSA/unrelated antigen wells.

Note: Two readings, the first around 10 minutes and the second around 30–60 minutes after addition of the substrate solution, are usually sufficient to see specific binding of antibody fragments to the antigen of interest. Binding of antibody fragments to the antigen is usually detected, but considerable variation in the ELISA signal strength is seen frequently. This variation does not necessarily reflect binding affinity but is often due to the amount of antibody fragment expressed in *E. coli*.

PROTOCOL 11.4

*Bst*OI Fragment Analysis of Antibody Fragment-encoding PCR Product to Test Diversity among Specific Clones

Variations in *Bst*OI recognition site patterns due to DNA sequence differences can be used to quickly and inexpensively analyze a relatively large number of antibody fragment clones for diversity. This approach is useful for testing unselected libraries for diversity. More importantly, diversity can be checked in a pool of selected clones that show specificity to the antigen of interest. If contamination by antibody fragment clones obtained in earlier selections is a problem, this method can also be used to identify such contaminants before DNA sequence analysis.

Single clones can be analyzed simultaneously by *Bst*OI digest and for specific binding in ELISA (Protocol 11.3). This parallel analysis makes possible the fast identification of specific clones with different DNA sequences; these clones can then be used for further analysis (e.g., DNA sequence analysis, Protocol 11.7; large-scale expression and purification, Chapter 12). The oligonucleotide primers described here are specific for sequences upstream (ompseq) and downstream (gback) of the antibody-encoding insert. They can be used for single-chain clones and for Fab clones in pComb3X or pComb3H. Alternatively, other appropriate oligonucleotide primers can be used for PCR amplificaton.

Materials

nuclease-free water
SB medium (see Appendix 2)
100 mg/ml carbenicillin (see Appendix 2)
output titering plates, phagemid DNA, or **PEG**-precipitated phage (see Step 3)
oligonucleotide primers for PCR amplification of the antibody fragment-encoding insert:
ompseq: 5′ AAGACAGCTATCGCGATTGCAG 3′
gback: 5′ GCCCCCTTATTAGCGTTTGCCATC 3′
Taq DNA polymerase (e.g., *AmpliTaq*, 5 units/ml, Perkin-Elmer Biosystems, Cat. # N808-0160)
10x PCR buffer containing 15 mM **$MgCl_2$**(supplied with *Taq* DNA polymerase)
2.5 mM dNTP mix (dATP/dCTP/dGTP/dTTP set, 100 mM, Amersham Pharmacia Biotech Inc., Cat. # 27-2035-02)
LB+carbenicillin agar plates (see Appendix 2)
*Bst*OI restriction endonuclease, 10 units/ml, recognition sequence CC(A/T)GG (supplied with 4-CORE 10x incubation buffer C and 100x BSA solution [10 mg/ml]; Promega Corporation, Cat. # R6931)
agarose and low-melting agarose
gel loading dye for agarose gels (see Appendix 2)
100-bp DNA molecular weight marker (e.g., Amersham Pharmacia Biotech Inc., Cat. # 27-4001-01) or another appropriate DNA molecular marker

CAUTIONS: PEG, $MgCl_2$ *(see Appendix 4)*

Procedure

1. For each clone to be analyzed, add 10 μl of nuclease-free water to a PCR tube. Prepare an additional tube for a negative control PCR.

2. Add 2 ml of prewarmed SB+carbenicillin (50 μg/ml) to one 14-ml culture tube per clone.

3. Use a pipettor and sterile tips to pick colonies and introduce *E. coli* cells first into the PCR tubes and then into the SB+carbenicillin.

   Note: 10–30 colonies from output titer plates from the last one or two rounds of a panning experiment are typically used (see Protocols 10.5 and 10.6). If phage pools were analyzed by ELISA as described in Protocol 11.1, the results will reveal which output titer plates should be used for the analysis of single clones.

   *Bst*OI fragment analysis can also be performed with phagemid DNA prepared from single clones (use 10 ng to 1 μg DNA) or PEG-precipitated phage obtained from single clones (see Protocol 11.2; use 1 μl of phage preparation, around 5×10^{10} phage). Combine phagemid DNA or phage in PCR tubes with nuclease-free water to 10 μl, and continue with Step 4.

4. Assemble a PCR master mix containing 3 μl of 10x PCR buffer, 2.4 μl of dNTP mix, 0.15 μl of *Taq* polymerase, 20 pmole of each oligonucleotide primer (ompseq and gback), and nuclease-free water to a final volume of 20 μl per reaction. To avoid shortage of reagent because of pipet inaccuracies, make enough PCR master mix for about 10% more reactions than needed.

5. Add 20 μl of PCR master mix to each PCR tube.

6. Perform PCR under the following conditions:

   94°C for 5 minutes

   followed by 30 cycles of

   94°C for 15 seconds,

   56°C for 30 seconds, and

   72°C for 2 minutes.

7. Incubate the 14-ml tubes containing the bacterial cultures in a shaker at 300 rpm for several hours at 37°C.

8. Spot 2 μl of each culture onto LB+carbenicillin plates. Incubate the plates overnight at 37°C, seal with Parafilm, and store for up to 3 months at 4°C. If clones are to be analyzed for specific binding by ELISA, continue with the cultures as described in Protocol 11.3, Step 5.

9. Analyze 10 μl of each PCR reaction in a 1.5% agarose gel. A single strong band should be visible; the expected size for a single-chain clone is about 900 base pairs

and for a Fab clone is around 1600 base pairs. A good yield of PCR product is important to ensure that smaller DNA fragments are visible when analyzed after the *Bst*OI digest.

10. Make a *Bst*OI master mix containing 2.4 µl of 10x reaction buffer, 0.24 µl of 100x BSA solution (10 mg/ml); 5 units (0.5 µl) of *Bst*OI and nuclease-free water to a final volume of 16 µl per reaction. To avoid shortage of reagent because of pipet inaccuracies, make enough master mix for about 10% more reactions than needed.
11. Add 8 µl of each PCR reaction and 16 µl of *Bst*OI master mix to a microcentrifuge tube. Incubate for 1.5–2 hours at 60°C.
12. Prepare a 3% or 4% agarose gel for analysis of the *Bst*OI digest (use a 1:1 mixture of standard agarose and low-melting agarose).
13. Add loading dye to 10–15 µl of the *Bst*OI digest and load on the 3% or 4% agarose gel. Use a 100-base pair DNA ladder for reference. Separate the DNA fragments for about 1 hour at 100 V (see Fig. 11.3).

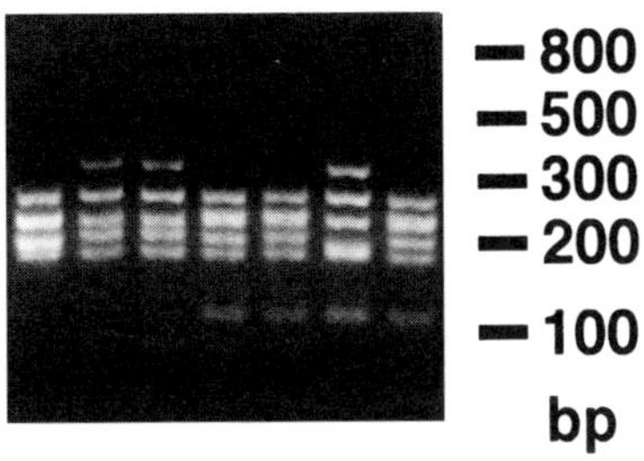

Figure 11.3. *Bst*OI fragment analysis of ELISA-positive clones analyzed after the panning of a chimeric human–rabbit Fab library constructed from animals immunized with purified integrins. It is possible, for example, that lanes *2* and *3* contain the same clone, because they have the same *Bst*OI pattern.

PROTOCOL 11.5

Analysis of Antibody Fragment-displaying Phage Pools and Clones by Flow Cytometry

The desired products of a selection are often antibody fragments that bind to properly folded cell surface antigens in the context of the membrane. Thus, it is sometimes advantageous to test the specificity of the panning output phage pools and clones using cells that express the antigen of interest on their surface rather than testing by ELISA with immobilized antigens. PEG-precipitated phage pools as well as phage or antibody fragments obtained from single clones can be tested for binding to a cell surface antigen by flow cytometric analysis (see Fig. 11.4).

Generally, the protocol for flow cytometric analysis using antibody fragment-displaying phage or antibody fragments is comparable to protocols that use purified antibodies or sera. If a protocol for staining for the cell type of interest is established in your laboratory, we recommend modifying the phage flow cytometry protocol below according to the already existing procedure. It is preferable to use a homogeneous cell population that expresses the antigen of interest at high levels.

PEG-precipitated phage (titer about 5×10^{12} cfu/ml), small-scale phage preparations (Protocol 11.2), and antibody fragments prepared from induced cultures (Protocol 11.3) can all be used in flow cytometry. Sterile phage preparations can be used for several weeks. Phage clones or phage pools that have been stored longer should be reamplified for binding analysis as described in Protocols 10.4 or 11.2. Because PEG can harm cell membranes, it is important that the PEG solution is removed carefully from precipitated phage. There are no anti-phage–fluorophore conjugates currently available commercially, so a three-step staining protocol for phage flow cytometry is described below. The reagents used for staining depend on the antibody fragments and the phagemid vectors (pComb3H or pComb3X, see also Protocol 11.3 and Fig. 11.4). For example, chimeric Fab fragments can be detected with anti-human $F(ab')_2$ fragment conjugates. Phage and antibody fragments derived from pComb3X can be stained with antibodies specific to HA and a fluorophore-conjugated antibody specific to the anti-HA antibody. We do not recommend the use of anti-HA reagents that are directly fluorophore-conjugated.

Proper controls are imperative for flow cytometric analysis. If possible, a "cellular control" should be included in the experiment. This control consists of cells that are the same as or similar to the cells to be tested, but do not express the surface antigen of interest. If cells that have been stably or transiently transfected to express the antigen of interest are being tested, the untransfected parental cell line is a good cellular control. Alternatively, the same cell type transfected to express a different surface protein could be used. In addition, control phage should be used to stain the cells. A good control for staining is a phage pool that displays antibody fragments that do not have specificity for the analyzed cells; the unpanned phage pool obtained after library construction can be used for this purpose. The control phage correspond to the isotype control in a conventional staining protocol for flow cytometry.

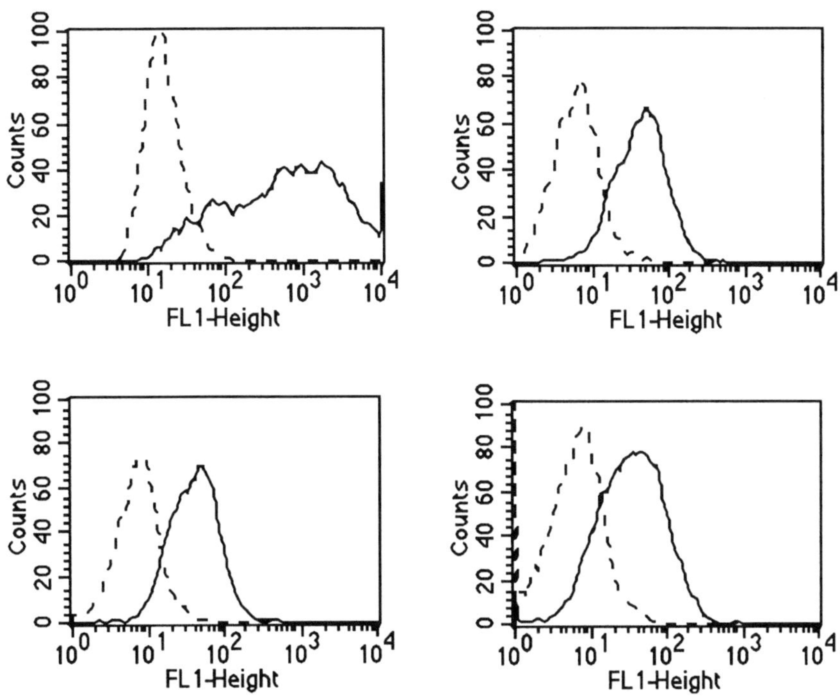

Figure 11.4. Fluorescence histograms of eukaryotic cells that have been stained for flow cytometry with antibody fragment-displaying phage pools and antibody fragments. "Counts" refers to the number of cells, and "FL1-Height" refers to fluorescence intensity. *Upper left panel:* Three-step staining using phage pools, an anti-M13 mouse monoclonal antibody, and an anti-mouse conjugate. *Upper right panel:* Three-step staining using phage pools, an anti-HA monoclonal antibody from rat, and an anti-rat conjugate. *Lower left panel:* Two-step staining using phage pools and an anti-human $F(ab')_2$ conjugate. *Lower right panel:* Two-step staining using supernatants from induced *E. coli* cultures containing chimeric rabbit–human Fab fragments and an anti-human $F(ab')_2$ conjugate. The solid lines represent the binding activity of phage pools and antibody fragments selected by panning. The dashed lines represent the binding activity of unselected control phage pools and a control antibody fragment specific for an antigen that is not expressed on the analyzed cells. The phage pools shown are derived from a chimeric rabbit–human Fab library in pComb3X. The phage pool that shows specific binding to cells was obtained after four rounds of panning against an immobilized antigen representing a domain of the surface antigen transiently expressed on the measured cells. No binding was seen to cells not expressing the target antigen (not shown). The negative control phage pool was prepared from the unpanned library.

Materials

freshly prepared phage, or antibody fragments from induced cultures (see Step 1)

cells expressing the antigen of interest (prepare around 2×10^5 to 5×10^5 cells per sample), and the same amount of negative control cells if available

cell dissociation reagents (optional, for adherent cells):

- trypsin-EDTA (Gibco BRL, Cat. # 25200-056) or Cell Dissociation Buffer, enzyme-free, PBS based (Gibco BRL, Cat. # 13151-014)

PBS (phosphate-buffered saline, see Appendix 2)

5% milk (w/v) in PBS (e.g., Bio-Rad, Cat. # 170-6404)
V-bottom 96-well plates (Corning Costar, Cat. # 3896)
96-well plate holder for centrifuge; alternatively, cells can be stained in flow cytometry tubes
flow cytometry tubes (Becton Dickinson, Falcon, Cat. # 352054)
multichannel pipettor (optional)
flow cytometry buffer (see Appendix 2)
10% **formaldehyde** in PBS, protect from light (optional, for fixing the cells)
access to a flow cytometer

Antibodies:

3-step staining for antibody fragment-displaying phage preparations:

anti-M13 mouse monoclonal antibody, use at 10 μg/ml, (Amersham Pharmacia Biotech, Cat. # 27-9420-01)

FITC-labeled goat anti-mouse $F(ab')_2$ fragment-specific antibodies, use at 1:100 (Jackson ImmunoResearch Laboratories, Cat. # 115-095-006)

3-step staining for antibody fragment-displaying phage preparations and antibody fragments obtained from pComb3X libraries:

anti-HA high-affinity rat monoclonal antibody, use at 2 μg/ml (Roche Molecular Biochemicals, Cat. # 1 867 423 or 1 867 431)

FITC-conjugated goat anti-rat Ig antibodies, use at 1:100 (Becton Dickinson, PharMingen, Cat. # 12114D)

2-step staining (chimeric Fab–antibody fragments):

FITC-conjugated donkey antihuman IgG (H+L) (Jackson ImmunoResearch Laboratories, Cat. # 709-095-149) or **FITC**-conjugated goat antihuman antibodies, $F(ab')_2$ fragment specific (Jackson ImmunoResearch Laboratories, Cat. # 109-095-006), use at 1:100.

CAUTIONS: formaldehyde, FITC *(see Appendix 4)*

Procedure

1. Preincubate 30 μl of freshly prepared phage per sample to be stained with 20 μl of 5% milk in PBS for 5 minutes to 1 hour on ice (see Protocols 10.3–10.6 or 11.2; the phage titer is typically 10^{12}–10^{13}cfu/ml). Alternatively, antibody fragments from induced cultures (Protocol 11.3) can be used: Per sample to be stained, preincubate 60 μl of antibody fragment preparation with 40 μl of 5% milk in PBS for 5 minutes to 1 hour on ice.

2. Wash the antigen-expressing cells (and negative control cells, if available) by adding 10 ml of PBS, mixing carefully, and centrifuging at 700–800*g* for 5 minutes at room temperature. Resuspend in flow cytometry buffer at around 0.5×10^7 to 1×10^7 cells per ml.

Note: A single-cell suspension is needed for flow cytometric analysis, and adherent or connected cells must be detached and individualized mechanically (by scraping) or enzymatically. When separating cells enzymatically, e.g., by using trypsin-EDTA, care must be taken not to destroy the antigen of interest. An additional washing step using tissue culture medium containing FCS is also necessary if trypsin is used to detach cells.

3. Distribute 50-μl aliquots containing about 2.5×10^5 to 5×10^5 cells into the wells of V-bottom plates.
4. Add 50 μl of phage suspension or 100 μl of antibody fragment preparation to the cells, mix carefully by pipetting up and down once, and incubate for 20–40 minutes at room temperature (seal plates by covering them with plastic film).
5. Dilute the secondary unconjugated antibodies (e.g., Anti-M13-specific antibody) or conjugated antibodies (Anti-Human $F(ab')_2$-specific FITC conjugate) in flow cytometry buffer (see Materials list for dilution factors). Prepare 60 μl per sample. Protect fluorescent dye conjugates from light throughout the staining procedure.
6. Centrifuge the cells at 1600–1800 rpm in a tabletop centrifuge (~700*g*) for 2–3 minutes with the brake set to low. Carefully shake out and discard the supernatant.
7. Wash the cells by adding 200 μl of flow cytometry buffer to each well and carefully resuspending the cells by pipetting up and down once. Centrifuge the cells at 1600–1800 rpm in a tabletop centrifuge (~700g) for 2–3 minutes with the brake set to low. Carefully shake out and discard the supernatants.
8. Add 50 μl of secondary antibody to each well and resuspend the cells by carefully pipetting up and down. Stain by incubating for 20–40 minutes at room temperature. (Seal the plates by covering them with plastic film.)
9. If secondary antibody is (already) conjugated with fluorescent dye, continue with Step 14.
10. Dilute tertiary antibody (e.g., anti-mouse, $F(ab')_2$-specific, FITC-conjugated) in flow cytometry buffer; prepare 60 μl per sample (see Materials list for dilution factors). Protect fluorescent dye conjugates from light throughout the staining procedure.
11. Centrifuge the cells at 1600–1800 rpm in a tabletop centrifuge (~700*g*) for 2–3 minutes with the brake set to low. Carefully shake out and discard the supernatant.
12. Wash the cells as described in Step 7.
13. Add 50 μl of tertiary antibody per well. Resuspend the cells by carefully pipetting up and down, and stain by incubating for 20–40 minutes at room temperature. (Seal plates by covering them with plastic film.)

14. Centrifuge the cells at 1600–1800 rpm in a tabletop centrifuge (~700*g*) for 2–3 minutes with the brake set to low. Carefully shake out and discard the supernatant.

15. Wash the cells as described in Step 7.

16. Resuspend the cells in 200 μl of flow cytometry buffer, and transfer to labeled flow cytometry tubes.

17. *Optional:* Fix the cells by adding 20 μl of 10% formaldehyde solution, and mix gently. Fixed cells can be stored for several days at 4°C when protected from light. Cells that have not been fixed can be stored for a few hours.

18. Measure the fluorescence intensity of the stained cells in a flow cytometer (see Fig. 11.4 and Radbruch 1992).

PROTOCOL 11.6

Expression of Antibody Fragments without the *Gene III* Product in pComb3H and pComb3X

Antibody fragments can be analyzed while fused to the *gene III* product (pIII), because pIII does not interfere with detection or antigen binding. For initial analysis after panning, therefore, single clones from output plates can be used directly to produce antibody fragments in induced cultures (Protocols 10.5 and 10.6).

An alternative approach is to use the *E. coli* cell pellet that is obtained after each round of panning to prepare phagemid DNA. This phagemid DNA represents a pool of clones. Phagemid DNA obtained from rounds that showed a positive pooled phage ELISA is usually used to produce clones that express antibody fragments without pIII, and single clones are then screened for specific binding in ELISA (Protocol 11.3) or by flow cytometry (Protocol 11.5). DNA obtained from promising clones is the starting point for purification or large-scale expression of antibody fragments without pIII, as described below and in Chapter 12.

With pComb3X, antibody fragments without pIII can be produced simply by introducing phagemid DNA into a nonsuppressor strain of *E. coli* (procedure A). For clones derived from pComb3H libraries, expression without pIII requires a restriction endonuclease digest and re-ligation (procedure B).

Materials

Procedure A - pComb3X:

plasmid preparation kit (e.g., QIAprep Spin Miniprep Kit, QIAGEN Inc., Cat. # 27104)
nonsuppressor ***E. coli*** cells (e.g., electrocompetent TOP10 cells, Invitrogen, Cat. # C664-55, C664-11, or C664-24)
SOC medium (see Appendix 2)
LB+carbenicillin agar plates (see Appendix 2)

Procedure B - pComb3H:

plasmid preparation kit (e.g., QIAprep Spin Miniprep Kit, QIAGEN Inc., Cat. # 27104)
*Spe*I (e.g., Roche Molecular Biochemicals, Cat. # 1 008 943)
*Nhe*I (e.g., Roche Molecular Biochemicals, Cat. # 885 843)
restriction endonuclease buffer M; 10x concentrated; supplied with *Nhe*I restriction endonuclease
QIAEX II extraction kit (QIAGEN, Cat. # 20021)
T4 DNA ligase (1 unit/ml, Gibco BRL, Cat. # 15224-090)
5x T4 DNA ligase buffer (supplied with T4 DNA ligase)
competent ***E. coli*** cells (ER2537 or XL1-Blue, see Protocol 10.1)
SOC medium (see Appendix 2)
LB+carbenicillin agar plates (see Appendix 2)

CAUTION: *E. coli (see Appendix 4)*

Procedure

Procedure A - pComb3X

1. Prepare phagemid DNA from *E. coli* cells containing pComb3X clones of interest or from *E. coli* pellets obtained during panning. Use a commercially available small-scale plasmid preparation kit according to the manufacturer's protocol.
2. Transform 20 ng of phagemid DNA into 40 μl of competent nonsuppressor *E. coli* cells (e.g., Top10 cells) by electroporation or heat shock, as described in Protocol 10.1 and Appendix 3.
3. Add 3 ml of SOC to the cells and incubate while shaking at 300 rpm for 1 hour at 37°C.
4. Dilute the cells 10-, 100-, and 1000-fold with prewarmed (37°C) SOC medium. Plate 100 μl of each dilution on prewarmed LB+carbenicillin plates.
5. Incubate plates overnight at 37°C.

Use clones from the plates to inoculate and induce cultures as described in Protocol 11.3 and Chapter 12. The amber codon that is suppressed in strains such as XL1-Blue or ER2537 leads to production of antibody fragment clones without pIII in Top10 cells and other nonsuppressor strains.

Procedure B - pComb3H

1. Prepare pComb3H phagemid DNA from *E. coli* cells containing clones of interest or from *E. coli* pellets obtained during panning. Use a commercially available small-scale plasmid preparation kit according to the manufacturer's protocol.
2. Assemble a restriction endonuclease reaction containing:

   1 μg of phagemid DNA
   3 Units of *Spe*I
   9 Units of *Nhe*I
   3 μl of 10x buffer M (supplied with *Nhe*I)
   Add water to a total reaction volume of 30 μl.

   Digest for 2–3 hours at 37°C.
3. Gel-purify the linearized phagemid DNA without the *gene III*-encoding fragment (usually around 3.6 kb for an scFv, and around 4.4 kb for a Fab fragment). Use a 0.8% gel and extract the DNA with a QIAEX extraction kit, by electroelution, or with the freeze/squeeze method (see Appendix 3 and Sambrook et al. 1989).
4. Assemble the following ligation reaction:

0.3–0.5 μg of linearized phagemid DNA

4 μl of 5x ligase buffer

1 μl of ligase (1 Unit)

Add water to a total volume of 20 μl.

5. Incubate for at least 2 hours (or overnight) at room temperature.
6. Transform 1 μl of the ligation reaction into 40 μl of competent *E. coli* cells (e.g., ER2537 or XL1-Blue) by electroporation or heat shock (see Chapter 9 Support Protocol and Appendix 3).
7. Add 3 ml of SOC to the cells and incubate while shaking at 300 rpm for 1 hour at 37°C.
8. Dilute the cells 10-, 100-, and 1000-fold with prewarmed (37°C) SOC medium, and plate 100 μl of each dilution on LB+carbenicillin plates.
9. Incubate the plates overnight at 37°C.

Use clones from the plates to inoculate and induce cultures as described in Protocol 11.3 and Chapter 12.

PROTOCOL 11.7

Sequence Analysis of Selected Clones

For DNA sequence analysis, purify phagemid DNA from selected clones by using a commercially available small-scale plasmid preparation kit. Perform sequence analysis of the antibody fragment clones with the oligonucleotide primers given below or any other suitable primers. The selected immunoglobulin sequences can then be compared with published protein or DNA sequences to determine, for example, which V genes are used (Kabat et al. 1991). Sequences can be compared by using the alignment program BLAST, which also allows access to DNA and protein sequence databases. BLAST can be found on the internet at http://www.ncbi.nlm.nih.gov/BLAST/.

A compilation of human germ-line V genes (V BASE) can be used to find the germ-line genes that are most closely related to human variable region immunoglobulin sequences. The alignment software DNAPLOT and the antibody database can be accessed at http://www.mrc-cpe.cam.ac.uk/imt-doc/public/INTRO.html.

Materials

plasmid preparation kit (e.g., QIAprep Spin Miniprep Kit, QIAGEN, Cat. # 27104)
sequencing primers:

The primers given below are "sense" primers; they prime upstream of the sequence of interest in both pComb3H and pComb3X. The primers ompseq and pelseq are specific for the omp and pel leader sequences, respectively. HRML-F, which is used to sequence the V_H region in scFv clones, is specific for the linker sequence between V_L and V_H.

single-chain antibody fragments (scFvs):

V_L: ompseq: 5′ AAGACAGCTATCGCGATTGCAG 3′

VH: HRML-F: 5′ GGTGGTTCCTCTAGATCTTCC 3′

Fab-antibody fragments:

V_L: ompseq: 5′ AAGACAGCTATCGCGATTGCAG 3′

V_H: pelseq: 5′ ACCTATTGCCTACGGCAGCCG 3′

REFERENCES

Radbruch A., ed. 1992. *Flow cytometry and cell sorting.* Springer Verlag, Berlin.

Kabat E.A., Wu T.T., Perry H.M., Gottesman K.S., and Foeller C. 1991. *Sequences of proteins of immunological interest,* 5th edition. U.S. Department of Health and Human Services, Washington, D.C.

Sambrook J., Fritsch E.F., and Maniatis T. 1989. *Molecular cloning: A laboratory manual.* 2nd edition. Cold Spring Harbor Laboratory Press, Cold Spring Harbor, New York.

12 Production and Purification of Fab and scFv

MARIKKA ELIA, JENNIFER ANDRIS-WIDHOPF, ROBERTA FULLER, AND CARLOS F. BARBAS III

Department of Molecular Biology, The Scripps Research Institute, La Jolla, California 92037

THIS CHAPTER DESCRIBES THE PURIFICATION of the Fab and scFv antibodies that have been selected using phage-display technology. An important aspect of the pComb3 phage-display system is the ability to select, express, and purify antibody fragments from a single vector. Realistically, however, all antibodies may not express equally well and may require different expression vectors. Therefore, the pComb3 system has been designed such that the genetic information for the production of selected antibodies can be readily transferred to other vectors as an *Sfi*I cassette. The protocols presented in this chapter describe the methods currently used to express and purify recombinant antibodies using pComb3H, pComb3X, and vectors that are compatible with pComb3 (see Chapter 2). These protocols include methods for both bacterial and yeast expression and purification.

CONTENTS

Protocols 12.1–12.4 describe the use of two bacterial expression vectors, pComb3H and pComb3X. Antibody proteins that are expressed in the pComb3 system are purified by using a peptide tag present on the mature protein or by using the constant region present on the antibody Fd. Single-chain antibodies do not have a constant region; therefore, they must be expressed in a vector that contains a peptide tag, such as pComb3X (see Protocol 12.1). pComb3X contains a HIS6 tag, and purification is performed with a Ni^{++}–NTA column (metal chelate chromatography, Protocol 12.2). Fab molecules can be expressed in both pComb3X and pComb3H. Because Fab molecules have a constant region domain on the heavy-chain Fd, they can be purified by protein G affinity chromatography. Alternatively, protein G can be coupled to an anti-Fab monoclonal antibody that is used to purify the recombinant Fab (we generally use a goat anti-human Fab, as we express chimeric human Fab proteins; see Protocol 12.4). The Fab can also be expressed in a vector that contains the HIS6 tag such that metal chelate chromatography can be used (Protocol 12.2).

The different methods used for purification require the use of different conditions, buffers, etc. to obtain maximum quantities of protein. We generally perform HIS6 purifications at 4°C in a nonautomated protocol, whereas protein G and anti-Fab purifications are carried out using FPLC technology.

To express protein in yeast, we use a pGAPZαA vector that has been modified to be compatible with the pComb3 system. A map of pGAPZαA is available in the pGAPZ product manual from Invitrogen. This expression vector uses the glyceraldehyde-3-phosphate dehydrogenase promoter (GAP), which has been shown to express recombinant proteins at high levels in *Pichia pastoris* (Waterham et al. 1997). In addition, recombinant proteins are fused to an amino-terminal peptide encoding the *Saccharomyces cerevisiae* α-factor secretion signal. Transformants are selected by using the dominant selectable marker for Zeocin resistance. Zeocin is functional in both yeast and bacteria.

pGAPZαA has been modified to include two *Sfi*I sites, such that antibody inserts are readily subcloned from the pComb3 vectors. The two resultant vectors are called pGAPZαA-SSFab and pGAPZαAscFvH6HA. Expression of scFv requires the direct subcloning of the *Sfi*I insert from pComb3H or pComb3X into *Sfi*I-digested pGAPZαAscFvH6HA. This vector has also been engineered to express carboxy-terminal HIS6 and HA peptide tags. The expression of Fab molecules requires two subcloning steps (Protocol 12.5). In the first step, a modified GAP promoter region is transferred into the Fab-containing pComb3 vector. In the second step, the same promoter region plus the Fab is transferred from the pComb3 vector into pGAPZαA-SSFab. This results in a Fab construct in which light-chain expression is driven by the GAP promoter already present in the pGAP vector and heavy-chain expression is driven by the inserted promoter.

The scFv- or Fab-containing pGAPZαA constructs are transformed into the *Pichia pastoris* strain X33. Transformation into X33 cells can be accomplished either through a chemical method (Protocol 12.6) or by electroporation (Protocol 12.7). Electroporation has a much higher transformation efficiency, but electrocompetent cells must be prepared and transformed on the same day. Depending on the individual requirements for protein quantity, the expression can be carried out in shake flasks (shake flasks aid aeration of the cultures while shaking, see Protocol 12.8) or by fermentation (Protocol 12.9).

PROTOCOL 12.1

Expression of Fab and scFv in *E. coli*

This protocol describes the use of two bacterial expression vectors, pComb3H and pComb3X. The use of these vectors depends on the antibody being expressed. pComb3H can be used only for purification of Fab molecules, as it does not contain any peptide tags for purification or detection. pComb3H also requires the excision of *gene III* by restriction digestion and religation before expression of soluble protein (see Protocol 11.6). pComb3X was designed for the expression of both Fab and scFv, as it contains histidine (HIS6) and hemagglutinin (HA) tags that are incorporated at the carboxyl terminus of the antibody molecule. pComb3X also contains the amber codon, and thus, when transformed into a nonsuppressor bacterial strain, can express soluble protein without the coat protein III (see Protocol 11.6). Both vectors express β-lactamase and, therefore, require ampicillin or carbenicillin for selection (we use carbenicillin because it is more stable). Both vectors contain the lacZ promoter and require IPTG for induction of protein expression.

Materials

antibody clone
electrocompetent ***E. coli*** cells (see Step 1): XL1-Blue (Stratagene, Cat. # 200228), ER2537 (New England Biolabs, Cat. # 801-N), or TOP10F′ (Invitrogen, Cat. # C3030-03)
LB+carbenicillin agar plates (see Appendix 2)
SB (super broth, see Appendix 2)
carbenicillin, 100 mg/ml stock (Sigma, Cat. # C 1389)
1 M **$MgCl_2$**
IPTG, 0.5 M stock (Fisher Scientific, Cat. # BP1620-10)
1x PBS (see Appendix 2)
100 mM **PMSF**
sonic disrupter (Tekmar-Dohrmann, TSD-300, Cat. # 10-0456-000)
Pellicon 2 filtration system (Millipore, Cat. # P17508)
PLGC 10K regenerated cellulose membrane (Millipore, Cat. # P2PL GCC05)

CAUTIONS: ***E. coli*, $MgCl_2$, IPTG, PMSF *(see Appendix 4)***

Procedure

1. To express soluble Fab in pComb3H, transform the antibody clone of choice into fresh XL1-Blue or ER2537 electrocompetent cells as described in Protocol 10.1 and in the Chapter 9 Support Protocol. To express soluble scFv or Fab in

pComb3X, transform into TOP10F′ electrocompetent cells. Plate the transformed cells onto LB+carbenicillin plates and incubate overnight at 37°C.

2. Inoculate 10 ml of SB containing 50 μg/ml of carbenicillin with a single colony from a fresh transformation plate. Incubate overnight in a 37°C shaker (255 rpm).
3. Dilute the 10-ml culture in 1 liter of SB supplemented with 50 μg/ml of carbenicillin and 20 ml of 1 M $MgCl_2$. Incubate for 8 hours in a 37°C shaker (255 rpm).
4. Induce the culture by adding 2 ml of 0.5 M IPTG and incubate overnight in a 37°C shaker.
5. Centrifuge the culture in two 500-ml centrifuge bottles for 30 minutes at ~3000*g* and 4°C. Transfer the supernatants to clean centrifuge bottles. The antibody will be purified from both supernatant and pellet.
6. Resuspend each pellet in 10 ml of 1x PBS supplemented with PMSF at 200 μM final concentration.
7. To prepare lysate, combine the resuspended pellets (total volume = 20 ml) and sonicate on ice in a Tekmar sonic disrupter for 180 seconds, pulsing at 50% duty cycle, output control set at 5.
8. Pellet the cellular debris by centrifuging at ~48,000*g* for 30 minutes at 4°C. Transfer the supernatant (the lysate) to a clean tube.
9. Concentrate the induced culture supernatant (from Step 5) to approximately 200 ml by tangential filtration using the Millipore Pellicon 2. Exchange the buffer by passing 500 ml of appropriate column buffer through the concentrated sample.
10. Store the lysate and concentrated supernatant for up to one month at –20°C.

The Fab- or scFv-containing lysate and concentrated supernatant are now ready for purification (Protocols 12.2–12.4).

PROTOCOL 12.2

Purification with Metal Chelate Chromatography, Using Ni–NTA Agarose

Proteins that are engineered to express six histidine residues in tandem can be purified using a resin that contains Ni^{++} ions that are immobilized by covalent linkage to nitrilotriacetic acid (NTA). There is a high-affinity interaction between the histidine residues and the Ni^{++} ions that enables the specific purification of tagged recombinant proteins. The proteins are eluted with imidazole, which competes for binding to the Ni^{++} on the column. This is an attractive method for purifying recombinant proteins because the purification is specific, yet the method is universal, i.e., it can be used for essentially any protein. Approximately 5–10 mg of protein can be purified from 1 ml of resin. It is important to use an appropriate quantity of resin, as an excess of resin will result in an inability to elute the protein. Although the protocol described below indicates the use of 250 mM imidazole for elution, it is possible to use higher concentrations if there is difficulty in extracting the protein from the column.

Materials

Ni–NTA agarose resin (supplied as a 50% suspension, QIAGEN, Cat. # 30230)
gravity column (Bio-Rad, Cat. # 737-1051)
6xHis antibody sample (~200 ml of induced culture supernatant and ~ 20 ml of lysate can be combined and run over column together; see Protocol 12.1)
wash buffer: 50 mM **NaH_2PO_4**, pH 8.0; 300 mM NaCl; 20 mM **imidazole**
elution buffer: 50 mM **NaH_2PO_4**, pH 8.0; 300 mM NaCl; 250 mM **imidazole**
0.22 μM bottle-top filter (Corning, Cat. # 431117)
PBS (see Appendix 2)
Centriprep 10 (Millipore, Amicon, Cat. # 4305)
Acrodisc syringe filter (VWR, Cat. # 28142-340)
SDS–PAGE materials

CAUTIONS: **NaH_2PO_4, imidazole** ***(see Appendix 4)***

Procedure

1. Prepare the column by loading 2 ml of resuspended Ni–NTA resin onto a gravity column to form a 1-ml packed column. Allow the resin to settle by opening the stopcock. Never let the column run completely dry.

2. Equilibrate the column by washing it with 10 column volumes of wash buffer. The column is ready to be loaded with the sample.

3. Before loading the sample, equilibrate it with the components of the wash buffer such that it has a pH and buffer composition similar to that of the column.

Equilibration can be accomplished by exchanging the buffer during the concentration of the culture supernatants (see Protocol 12.1), exchanging the buffer via dialysis, or adding dry buffer components to the sample. Spin the sample at ~48,000*g* for 30 minutes at 4°C to pellet precipitated debris. This ensures that the column will run properly and decreases the possibility of nonspecific binding. Filter the sample through a 0.22-µm bottle-top filter.

4. Pass the sample through the column at 4°C to optimize binding. Collect the flow-through fractions and reapply to the column 2–3 times. Collect the final flow-through and store at 4°C for SDS–PAGE analysis.
5. Wash the column with 10 column volumes (10 ml) of wash buffer. Collect the wash fraction and store at 4°C for SDS–PAGE analysis. The purpose of the 20 mM imidazole in this buffer is to minimize the binding of untagged contaminant proteins. The concentration of imidazole can be optimized for each protein by using a series of wash buffers containing increasing concentrations of imidazole between 1 and 20 mM. The higher the concentration of imidazole that is tolerated by the proteins, the cleaner the eluate.
6. Elute the proteins in one fraction with 5 column volumes (5 ml) of elution buffer. As mentioned above, the concentration of imidazole in the elution buffer can also be optimized for individual proteins. Concentrations as high as 1 M have been used.
7. Wash the elution fraction with 3 sample volumes of PBS (protease inhibitors can be added) using a Centriprep 10 filtration system. This system has a MW cutoff of 10,000 daltons. Concentrate the fraction to a volume of 500–1000 µl, and sterile-filter through a low-binding syringe filter. The concentrate can be stored for short-term periods (several months) at –4°C or –20°C. Store for longer periods at –80°C.
8. Monitor antibody purity by SDS–PAGE on a 15% gel (see Sambrook et al. 1989). Load approximately 10 µl of sample with 5 µl of loading dye, and stain with Coomassie. Compare the flowthrough and wash fractions with the eluate to determine the efficiency of the purification.
9. Monitor antibody activity by ELISA (see Protocol 11.3). For scFv molecules, use a detecting reagent that is specific for the HA tag (anti-HA antibody). For Fab molecules, the detecting reagent can be an anti-Fab antibody, or it can be anti-HA antibody if pComb3X is used for expression.

PROTOCOL 12.3

Purification of Fab, Using Protein G Affinity Chromatography and FPLC Technology

The protein G column is an affinity chromatography column in which antibody binds the protein G via a specific epitope on the constant region domain of the heavy chain (Gronenborn and Clore 1993). Protein G binds to the Fc region of IgG from a variety of mammalian species.

The protocol below is a general procedure for protein G purification using an FPLC system. The specific FPLC program will depend on the system used by individual laboratories.

Materials

FPLC System (Amersham Pharmacia Biotech)
Column:
  Protein G Sepharose 4 Fast Flow for a 10-ml packed column (binding capacity = 20 mg of human IgG/ml; Amersham Pharmacia Biotech, Cat. # 17-0618-02), or
  HiTrap Protein G 5-ml prepacked column (binding capacity = 25 mg of human IgG/ml; Amersham Pharmacia Biotech, Cat. # 17-0405-01), or
  HiTrap Protein G 1-ml prepacked column (binding capacity = 25 mg of human IgG/ml; Amersham Pharmacia Biotech, Cat. # 17-0404-03)
buffer A: PBS, pH 7.4 (see Appendix 2)
buffer B: 0.5 M **acetic acid,** pH 3.0
Fab sample (~200 ml of induced culture supernatant and ~20 ml of lysate can be combined and run over column together; see Protocol 12.1)
0.22-µm bottle-top filter (Corning, Cat. # 431117)
1 M Tris-HCl, pH 9.0
PBS (see Appendix 2)
Centriprep 10 (Amicon, Cat. # 4305)
Pellicon 2 filtration system (Millipore, Cat. # P17508)
PLGC 10K regenerated cellulose membrane (Millipore, Cat. # P2PL GCC05)

CAUTION: **acetic acid** ***(see Appendix 4)***

Procedure

1. Equilibrate the FPLC pumps with the appropriate buffers. Equilibrate the column with a minimum of 3 column volumes of buffer A.

   Note: Before carrying out any FPLC protocol, filter all buffers and samples with a 0.22-µm filter.

2. Equilibrate the sample by exchanging the buffer using the Pellicon 2 filtration system. This is accomplished by doubling the volume of the sample with buffer A and concentrating to the initial volume of the sample via tangential filtration. Repeat this process twice. The pH of the sample should be 7.0. Centrifuge the sample for 30 minutes at ~48,000*g* and filter the supernatant through a 0.22-μm filter before column loading. The sample and the column should both be at room temperature.
3. Turn the fraction collector on for collection of 4-ml fractions. Load the sample onto the column, using a flow rate of 3 ml/minute for a 5-ml or larger column and 1 ml/minute for a 1-ml column.
4. Wash the column with 10 column volumes of buffer A.
5. Elute the column with 5 column volumes of buffer B.
6. Combine the fractions containing the protein peak as visualized on a UV monitor at 280 nm. Neutralize each 4-ml fraction with 2 ml of 1 M Tris-HCl, pH 9.0.
7. Concentrate the combined peak fractions and exchange the buffer to PBS using a Centriprep 10 filtration unit (final volume ~ 1 ml). Evaluate the purity and activity of the antibody as described in Protocol 12.2. Store for short-term periods (several months) at 4°C or –20°C. Store for longer periods at –80°C.

PROTOCOL 12.4

Purification of Chimeric Human Fab, Using Antihuman Fab Affinity Chromatography and FPLC Technology

In the protocol below, the goat IgG antihuman $(Fab')_2$ is first conjugated to Protein G Sepharose to form the goat antihuman Fab/Protein G column. The goat antibody binds to the protein G in the same manner that is described above for protein G purification.

Materials

Goat antihuman $(Fab')_2$ (Pierce, Cat. # 31122ZZ)
Protein G Sepharose 4 Fast Flow (Amersham Pharmacia Biotech, Cat. # 17-0618-02)
0.2 M sodium borate, pH 9.0
DMP (**dimethyl pimelimidate**, Sigma, Cat. # D-8388)
0.2 M **ethanolamine**, pH 8.0
20% **ethanol**
Coomassie Brilliant Blue R-250 Kit (Bio-Rad, Cat. # 161-0435)
SDS–PAGE Mini-Protean II System (Bio-Rad)
FPLC system (Amersham Pharmacia Biotech)
XK 16/20 column (Amersham Pharmacia Biotech, Cat. # 18-8773-01)
Buffer A: PBS, pH 7.4
Buffer B: 0.5 M **acetic acid**, pH 3.0
chimeric human Fab sample (~200 ml of induced culture supernatant and ~20 ml of lysate can be combined and run over column together; see Protocol 12.1)

CAUTIONS: **dimethyl pimelimidate, ethanolamine, ethanol, acetic acid** ***(see Appendix 4)***

Procedure

Conjugation (see Harlow and Lane 1988, page 522)

1. Add 16 mg of goat IgG antihuman $(Fab')_2$ to 8 ml (16 ml of a 50% slurry) of Protein G Sepharose beads.
2. Mix for 1 hour on a rocker at room temperature.
3. Centrifuge at 3000*g* for 5 minutes at room temperature, and discard the supernatant. Wash the beads by adding 10 volumes of 0.2 M sodium borate, pH 9.0. Mix well by hand. Centrifuge at 3000*g* for 5 minutes. Repeat.
4. Resuspend the beads in 10 volumes of sodium borate, pH 9.0. Save a 10-μl aliquot at 4°C to test the coupling efficiency by SDS–PAGE in Step 10. Add DMP to a final concentration of 20 mM.

5. Mix for 30 minutes on a rocker at room temperature. Save a 10-μl aliquot at 4°C for SDS–PAGE analysis in Step 10.
6. Stop the reaction by washing the beads one time in 80 ml of 0.2 M ethanolamine, pH 8.0. Centrifuge at 3000*g* for 5 minutes at 4°C.
7. Discard the supernatant. Resuspend the beads in 80 ml of 0.2 M ethanolamine, pH 8.0. Incubate for 2 hours at room temperature on a rocker. Centrifuge at 3000*g* for 5 minutes at 4°C.
8. Discard the supernatant. Resuspend the beads by adding 24 ml of deionized water. Centrifuge at 3000*g* for 5 minutes.
9. Discard the supernatant. Resuspend the beads by adding 24 ml of 20% ethanol.
10. Perform SDS–PAGE on the two aliquots of beads (Steps 4 and 5) and stain the gel with Coomassie Blue protein stain (see Sambrook et al. 1989). Good coupling efficiency is indicated if the Fab band is present in the aliquot prior to DMP addition and absent in the aliquot obtained after DMP addition.

Purification

Use the resin/20% ethanol slurry from Step 9 to prepare an 8-ml FPLC column. Wash the column with three column volumes of deionized water. Then wash the column with three column volumes of buffer B to remove the non-DMP-bound IgG. The goat anti-human/protein G FPLC protocol is identical to the protein G FPLC protocol described in Protocol 12.3. As with the previous protocol, the precise FPLC program depends on the FPLC system used.

PROTOCOL 12.5

Expression of Fab in Yeast: Subcloning Fab into the pGAPZαA-SSFab Vector

The expression of Fab molecules in yeast requires two subcloning steps. In the first step, a modified GAP promoter region (modified to remove the *Avr*II linearization site) is purified from pGAPZαA-SSFab by restriction digest with *Xba*I and *Xho*I. pGAPZαA-SSFab contains the SS "dual" stuffer, with the modified GAP promoter inserted between the light-chain stuffer and the heavy-chain stuffer. It essentially serves as the vector for subcloning and provides a readily available source of the modified GAP promoter for the preparation of any Fab. The digested GAP promoter is subcloned into a pComb3-Fab plasmid that has been prepared by restriction digest with the same two enzymes (a digest that results in the excision of the pelB leader). The GAP promoter is ligated in the place of pelB and the entire *Sfi*I insert (Fab + GAP promoter) is then subcloned into the pGAPZαA-SSFab vector. This results in a Fab construct in which light-chain expression is driven by the GAP promoter already present in the GAP vector and heavy-chain expression is driven by the inserted promoter.

Materials

pComb3XFab (pComb3X containing the Fab of interest)
*Xho*I restriction enzyme (Roche Molecular Biochemicals, 40 Units/μl, Cat. # 703 770)
*Xba*I restriction enzyme (Roche Molecular Biochemicals, 40 Units/μl, Cat. #1 047 663)
10x restriction enzyme buffer H (supplied with the *Xho*I and *Xba*I restriction enzymes)
ethanol
3 M sodium acetate, pH 5.2
agarose (Gibco-BRL, Cat. # 15510-027)
100-bp DNA molecular weight marker (Amersham Pharmacia Biotech, Cat. # 27-4001-01)
1-kb DNA molecular weight marker (Gibco-BRL, Cat. # 15615-024)
1x TAE electrophoresis running buffer (see Appendix 2)
DNA gel loading dye (see Appendix 2)
pGAPZαA-SSFab (available from the Barbas laboratory)
T4 DNA ligase (Gibco-BRL, 1 Unit/μl, Cat. # 1 5224 090)
5x ligase buffer (supplied with the T4 DNA ligase)
Gene Pulser II electroporation system (Bio-Rad, Cat. # 165-2105 and 165-2109)
electrocompetent *E. coli* cells (see Protocol 10.1): XL1-Blue (Stratagene, Cat. # 200228)
0.2-cm gap electroporation cuvettes (Invitrogen, Cat. # P450-50)
SOC medium (see Appendix 2)
LB + 50 μg/ml carbenicillin plates (see Appendix 2)

SB + 50 μg/ml carbenicillin (see Appendix 2)
QIAprep Spin Miniprep Kit (QIAGEN, Cat. # 27104)
*Sfi*I restriction enzyme (Roche Molecular Biochemicals, 40 Units/μl, Cat. # 1 288 059)
10× restriction enzyme buffer M (supplied with the *Sfi*I restriction enzyme)
QIAGEN Plasmid Midi Kit (QIAGEN, Cat. # 12143)
QIAGEN Plasmid Maxi Kit (QIAGEN, Cat. # 12162)
Zeocin (Invitrogen, Cat., #R250-01)
low-salt LB + 25 μg/ml Zeocin plates (see Appendix 2)
SB + 25 μg/ml Zeocin
*Sac*I restriction enzyme (Roche Molecular Biochemicals, 40 Units/μl, Cat. # 1047 655)

CAUTIONS: ethanol, *E. coli* (*see Appendix 4*)

Procedure

1. Digest pComb3XFab (pComb3X containing Fab of interest) with *Xho*I and *Xba*I.

Assemble the following reaction:

5–10 μg of pComb3XFab (pComb3X containing Fab of interest)
10 Units of *Xho*I per μg of DNA
10 Units of *Xba*I per μg of DNA
10 μl of 10× buffer H

Add water to a final volume of 100 μl.

Incubate for 3 hours at 37°C. Ethanol-precipitate the DNA and purify on a 1% agarose gel. Extract the vector from the gel with the squeeze-freeze method or by electroelution with an Elutrap (see Appendix 3). Suboptimal digestion or purification will yield vector with low transformation efficiencies and high background after ligation. Digestion with these two enzymes will result in the excision of an approximately 100-bp fragment.

2. Digest pGAPZαA-SSFab with *Xho*I and *Xba*I.

Prepare the modified GAP promoter for subcloning into the *Xho*I/*Xba*I-digested pComb3XFab (above) by digesting with *Xho*I and *Xba*I. Assemble the following reaction:

5–10 μg of pGAPZαA-SSFab
10 Units of *Xho*I per μg of DNA
10 Units of *Xba*I per μg of DNA
10 μl of 10× buffer H

Add water to a final volume of 100 μl.

Incubate for 3 hours at 37°C. Ethanol-precipitate and purify on a 1% agarose gel. We recommend freeze-squeeze or electroelution with an Elutrap for extraction of the fragment from the gel (see Appendix 3). The resulting GAP promoter fragment is ~1150 bp.

Note: The recommended amount of enzyme would normally be considered "overkill," but it has yielded the cleanest DNA for the most efficient and consistent subcloning.

Note: "SS" in the vector names refers to the "dual" stuffer (nonsense DNA)that is part of these vectors. "SS" does not mean "single-stranded!"

3. Ligate the promoter fragment.

Ligate the purified promoter fragment into the *Xho*I/*Xba*I-digested pComb3XFab to yield a construct with the GAP promoter inserted between the light-chain and the heavy-chain Fd. The ligation reaction should contain:

50 ng of pComb3XFab, *Xho*I/*Xba*I-digested and purified
50 ng of GAP promoter, *Xho*I/*Xba*I-digested and purified
4 µl of 5x ligase buffer
1 µl of ligase
Add water to a final volume of 20 µl.

Assemble a control ligation with vector alone (no GAP promoter fragment):

50 ng of pComb3XFab, *Xho*I/*Xba*I-digested and purified
4 µl of 5*X* ligase buffer
1 µl of ligase
Add water to a final volume of 20 µl.

Incubate the reactions for 4 hours at room temperature. Transform 0.5–1 µl of each reaction into 50 µl of XL1-Blue electrocompetent cells (see Chapter 9 Support Protocol). Plate 10 µl and 50 µl on LB + carbenicillin plates. The number of colonies obtained with the control ligation should not exceed 10% of the number obtained with the promoter fragment.

4. Screen transformants for positive clones.

Inoculate 5 ml of SB medium containing 50 µg/ml carbenicillin with a single colony from the above plates. Generally 5–10 randomly chosen colonies (a total of 5–10 cultures) is enough to find several positive clones. Grow the cultures overnight at 37°C in a shaking incubator (250 rpm). Prepare plasmid DNA using the QIAprep Spin miniprep kit and the manufacturer's recommended protocol. Digest an aliquot of miniprep DNA with *Xho*I and *Xba*I to check for the presence of the modified GAP promoter in the pComb3XFab plasmid. Each digest should contain:

1–5 µg of miniprep DNA
10 Units of *Xho*I per µg of DNA
10 Units of *Xba*I per µg of DNA
3 µl of 10x buffer H
Add water to a final volume of 30 µl.

Incubate for 3 hours at 37°C. Run 15–30 μl on a 1% agarose gel. Positive clones will have an insert that is about 1150 bp in length.

Alternatively, the clones can be screened by restriction digest with *Sfi*I. Each digest should contain:

1–5 μg of miniprep DNA
10 Units of *Sfi*I per μg of DNA
3 μl of 10x buffer M
Add water to a final volume of 30 μl.

Incubate for 5 hours at 50°C. Run 15–30 μl on a 1% agarose gel. Positive clones will have an insert that is about 2950 bp in length.

5. Prepare the Fab containing the GAP promoter for subcloning into the pGAPZαA-SSFab vector.

   Select a positive clone as identified from the above screening process. Digest an aliquot of the plasmid with *Sfi*I to isolate the Fab-GAP promoter insert for subcloning. The digest should contain:

   5–10 μg of pComb3XFabGAP (pComb3X containing the Fab of interest with the GAP promoter inserted)
   10 Units of *Sfi*I per μg of DNA
   10 μl of 10x buffer M
   Add water to a final volume of 100 μl.

   Digest for 5 hours at 50°C. Ethanol-precipitate and purify on a 1% agarose gel. Electroelute with an Elutrap or use the squeeze-freeze method to extract the fragment from the gel (see Appendix 3). The excised fragement is about 2950 bp.

6. Prepare pGAPZαA-SSFab for subcloning of the Fab-GAP fragment.

   Assemble the following reaction for digestion of pGAPZαA-SSFab:

   20 μg of pGAPZαA-SSFab
   200 Units of *Sfi*I (10 Units per μg of DNA)
   20 μl of 10x buffer M
   Add water to a final volume of 200 μl.

   Incubate for 5 hours at 50°C. Ethanol-precipitate and purify on a 1% agarose gel. Electroelute with an Elutrap or use the squeeze-freeze method to extract the fragment from the gel (see Appendix 3).

7. Ligate the Fab–GAP fragment.

Assemble the following ligation reaction with the Fab–GAP fragment:

30 ng of pGAPZαA-SSFab, digested with *Sfi*I and purified

90 ng of Fab–GAP purified fragment

4 μl of 5x ligase buffer

1 μl of ligase

Add water to a final volume of 20 μl.

Assemble a control ligation with vector alone (no Fab–GAP fragment):

30 ng of pGAPZαA-SSFab, digested with *Sfi*I and purified

4 μl of 5x ligase buffer

1 μl of ligase

Add water to a final volume of 20 μl.

Incubate both reactions for 4 hours to overnight at room temperature. Transform 0.5–1 μl of each reaction into 50 μl of XL1-Blue electrocompetent cells (see Chapter 9 Support Protocol). Plate 1, 10, and 100 μl on low-salt LB + Zeocin. The number of colonies obtained with the control ligation should not exceed 10% of the number obtained with the Fab–GAP fragment.

Note: It is critical in this ligation to have a vector:insert molar ratio of approximately 1:3. Because the vector and insert are almost the same size, 3 times more insert than vector needs to be added to the ligation reaction. Low-salt LB agar is used for plating the transformed cells because Zeocin is salt-sensitive.

8. Screen transformants for expression of desired Fab.

Screen the transformants for the presence of the correct-sized insert by restriction digest of miniprep DNA. Inoculate 5 ml of SB medium containing 25 μg/ml Zeocin with a single colony from the plates in Step 7. Generally 5–10 randomly chosen colonies (a total of 5–10 cultures) are enough to find several positive clones. Grow the cultures overnight at 37°C in a shaking incubator (250 rpm). Prepare plasmid DNA using the QIAprep Spin Miniprep kit and the manufacturer's recommended protocol. Digest an aliquot of miniprep DNA with *Sfi*I to identify positive clones. Each digest should contain:

1–5 μg of miniprep DNA

10 Units of *Sfi*I per μg of DNA

3 μl of 10x buffer M

Add water to a final volume of 30 μl.

Incubate for 5 hours at 50°C. Run 15–30 μl of each reaction on a 1% agarose gel. Positive clones will have an insert that is about 2950 bp long. This process will identify clones with inserts of the correct size, but it may be difficult to distinguish between clones expressing the stuffer and the desired Fab, as the sizes may only differ by about 100 bp. An alternative is to digest an aliquot of the miniprep DNA with *Sac*I and *Xba*I. This will drop out either a 1200-bp band (if the clone contains stuffer) or an approximately 700-bp band (if the clone contains the desired Fab).

Each digest should contain:

1–5 μg of miniprep DNA

10 Units of *Sac*I per μg of DNA

10 Units of *Xba*I per μg of DNA

3 μl of 10x buffer H

Add water to a final volume of 30 μl.

Incubate for 3 hours at 37°C. Run 15–30 μl on a 1% agarose gel. Clones that contain the correct-sized insert must be further screened for expression of the Fab protein and binding to the appropriate antigen. For this purpose, use the miniprep DNA to transform X33 yeast cells, and screen by Western blot and ELISA (see Protocols 12.6 and 12.7).

PROTOCOL 12.6

Expression of Fab and scFv in Yeast: Chemical Transformation of X33 Yeast Cells

Chemically competent *Pichia* cells are prepared using the EasyComp Kit from Invitrogen. To obtain the highest efficiency transformation, the pGAPZαA plasmid containing the Fab or scFv should be linearized. The transformation of the chemically competent *Pichia* cells is also performed using the EasyComp Kit. The following information has been adapted from the Invitrogen product manual. Please refer to the manual for more detailed information.

Materials

X33 yeast strain (Invitrogen, Cat. # C180-00)
YPD media (see Appendix 2)
Pichia EasyComp transformation kit (includes Solutions I, II, and III; Invitrogen, Cat. # K1730-01)
antibody clones in pGAPZαAscFvH6HA or pGAPZαA-SSFab (see Chapter 12 introduction)
*Avr*II restriction enzyme (New England Biolabs, Cat. # 174L)
*Bsp*HI restriction enzyme (New England Biolabs, Cat. # 517S)
YPDS media (see Appendix 2)
Zeocin (100 mg/ml, Invitrogen, Cat. # R250-05)
YPDS+Zeocin plates (100 μg/ml)
1.5-ml screw-cap microcentrifuge tubes (RPI, Cat. # 144500)

Procedure

Preparation of Chemically Competent Pichia *Cells*

1. Inoculate 10 ml of YPD with a single colony of X33 and incubate overnight in a 30°C shaking incubator.
2. Dilute the cells from the overnight culture in 10 ml of fresh YPD to an optical density at 600 nm (OD_{600}) of 0.1–0.2. Incubate the cells in a 30°C shaking incubator until the OD_{600} reaches 0.6–1.0 (this generally requires 4–6 hours).
3. Pellet the cells by centrifugation at 500*g* for 5 minutes at room temperature. Discard the supernatant.
4. Resuspend the cells in 10 ml of Solution I (a sorbitol solution containing ethylene glycol and DMSO).
5. Pellet the cells immediately by centrifugation at 500*g* for 5 minutes. Discard the supernatant.

6. Resuspend the cells in 1 ml of Solution I. The cells are now competent.

7. Aliquot 50- to 200-μl volumes of the competent cells into sterile screw-cap microcentrifuge tubes. 50 μl of cells is used for a single transformation (see below). The cells can be used immediately or stored at –80°C for future use. The cells can be thawed and refrozen several times without significant loss of competency.

Transformation

1. Linearize 100 μg of scFv- or Fab-containing plasmid. pGAPZα AscFvH6HA can be linearized with either *Bsp*HI or *Avr*II, and pGAPZαA-SSFab can be linearized with *Avr*II. Use 4 Units of enzyme per μg of DNA (see instructions that accompany enzyme). After digestion, determine the extent of linearization by using agarose gel electrophoresis (load 0.1 μg of DNA).

2. Ethanol-precipitate the remainder of the digested DNA and resuspend the pellet in a small volume of sterile water (20–30 μl; see Appendix 3). The final concentration should be at least 1 μg/μl. Higher concentrations yield higher transformation efficiencies. The ethanol precipitation is only necessary to achieve a high concentration of DNA in a small volume; DNA can be used directly from the digestion without affecting the transformation efficiency.

3. For each transformation, use 50 μl of competent cells. If frozen, thaw on ice.

4. Add a minimum of 3 μg of linearized DNA to the competent cells. More than 3 μg can be used, but the volume of DNA should not exceed 5 μl.

5. Add 1 ml of Solution II to the tube and mix by vortexing (Solution II contains PEG, which is necessary for the transformation). Incubate the reactions for 1 hour at 30°C. Mix the tube every 15 minutes by vortexing to ensure the highest transformation efficiency.

6. Heat the cells for 10 minutes at 42°C.

7. Pellet the cells by centrifugation at 3000*g* for 5 minutes at room temperature. Discard the supernatant.

8. Resuspend the cells in 1 ml of Solution III (Solution III is a salt solution for washing and plating the cells).

9. Pellet the cells by centrifugation at 3000*g* for 5 minutes at room temperature. Discard the supernatant. Resuspend the cells in 100–150 μl of Solution III.

10. Plate the entire transformation on YPDS+Zeocin plates and incubate the plates for 2–4 days at 30°C.

11. Once colonies grow on the transformation plate, pick 10 clones for analysis. Inoculate 10 ml of YPD (no antibiotic) with a single colony from the transformation plate (one culture per clone). Incubate for 2 days at 30°C in a shaking incubator.

12. Centrifuge a 1-ml aliquot of each culture at 14,000 rpm in a microcentrifuge at room temperature. Use 25 μl of the supernatant to test for antibody binding by ELISA (see Protocol 11.3). Continue growing the remaining cultures for ELISA testing on Day 3 and possibly on Day 4.
13. Preserve the positive clones for later use by spotting from the cultures onto a YPDS+Zeocin plate.
14. Assess protein production of ELISA-positive clones by Western blotting (see Sambrook et al. 1989). Load 10 μl of the supernatant from Step 12 on a 15% gel, and use anti-HA or antihuman Fab as a detecting antibody.

The positive transformed clones can now be used for large-scale protein expression (Protocols 12.8 and 12.9).

PROTOCOL 12.7

Expression of Fab and scFv in Yeast: Transformation of X33 Yeast cells by Electroporation

Electrocompetent *Pichia* cells cannot be stored and, thus, must be newly prepared before each transformation. To obtain the highest efficiency transformation, the pGAP plasmid should be linearized. The following information has been adapted from the pGAP product manual available from Invitrogen. Please refer to the manual for more details.

Materials

X33 yeast strain (Invitrogen, Cat. # C180-00)
YPD media (see Appendix 2)
ice-cold sterile water
1 M sorbitol
antibody clones in pGAPZαAscFvH6HA or pGAPZαA-SSFab (see Chapter 12 introduction)
*Avr*II (New England Biolabs, Cat. # 174L)
*Bsp*HI (New England Biolabs, Cat. # 517S)
YPDS media (see Appendix 2)
YPDS+Zeocin plates (100 µg/ml, 200 µg/ml, and 400 µg/ml)
Zeocin (100 mg/ml, Invitrogen, Cat. # R250-05)
Bio-Rad Gene Pulser apparatus (Bio-Rad, Cat. # 165-2105 and 165-2109)
0.2-cm cuvettes (Bio-Rad, Cat. # 165-2086 or Invitrogen, Cat. # P450-50)
50-ml conical tubes
15-ml conical tubes

Procedure

Preparation of Electrocompetent X33 Cells

1. Inoculate 5 ml of YPD in a 50-ml conical tube with a single colony of *Pichia*. Incubate overnight in a 30°C shaker.
2. Inoculate 500 ml of fresh YPD in a 2-liter flask with 0.1–0.5 ml of the overnight culture. Incubate the cells in a 30°C shaker until an OD_{600} of 1.3–1.5 is reached (approximately 16–17 hours).
3. Pellet the cells by centrifugation at 1500*g* for 5 minutes at 4°C. Discard the supernatant, and resuspend the pellet in 500 ml of ice-cold sterile water.
4. Pellet the cells by centrifugation at 1500*g* for 5 minutes at 4°C. Discard the supernatant, and resuspend the pellet in 250 ml of ice-cold sterile water.

5. Pellet the cells by centrifugation at 1500*g* for 5 minutes at 4°C. Discard the supernatant, and resuspend the pellet in 20 ml of 1 M sorbitol.

6. Pellet the cells by centrifugation at 1500*g* for 5 minutes at 4°C. Discard the supernatant, and resuspend the pellet in 1 ml of ice-cold 1 M sorbitol for a final volume of ~1.5 ml. The cells are now electrocompetent. Incubate the cells on ice until use.

Transformation of Electrocompetent* Pichia *cells

1. Linearize 100 µg of scFv- or Fab-containing plasmid. pGAPZαAscFvH6HA can be linearized with either *Bsp*HI or *Avr*II, and pGAPZαA-SSFab can be linearized with *Avr*II. Use 4 Units of enzyme per µg of DNA (see instructions that accompany enzyme). After digestion, determine the extent of linearization using agarose gel electrophoresis (load 0.1 µg of DNA).

2. Ethanol-precipitate the remainder of the digested DNA and resuspend the pellet in a small volume of sterile water (20–30 µl; see Appendix 3). The final concentration should be at least 1 µg/µl, but higher concentrations yield higher transformation efficiencies.

3. Mix 80 µl of electrocompetent cells with 5–10 µg of linearized DNA. The volume of DNA should be 5–10 µl. Transfer the mixture to an ice-cold 0.2-cm electroporation cuvette.

4. Incubate the cuvette on ice for 5 minutes.

5. Pulse the cells according to the parameters required for yeast. For the Bio-Rad Gene Pulser, the conditions are as follows: 1500 V, 25 µF capacitance, 200 Ω resistance.

6. Immediately add 1 ml of ice-cold 1 M sorbitol to the cuvette and transfer the cuvette contents to a sterile 15-ml conical tube.

7. Incubate the tube for 1–2 hours at 30°C without shaking.

8. Plate 25 µl and 100 µl on 100 µg/ml, 200 µg/ml, and 400 µg/ml YPDS+Zeocin plates. Plating at low cell density favors Zeocin selection.

9. Incubate the plates for 2–3 days at 30°C.

10. Once colonies grow on the transformation plate, pick 10 clones for analysis. Inoculate 10 ml of YPD (no antibiotic) with a single colony from the transformation plate (one culture per clone). Incubate for 2 days at 30°C in a shaking incubator.

11. Centrifuge a 1-ml aliquot of each culture at 14,000 rpm in a microcentrifuge at room temperature. Use 25 µl of the supernatant to test for antibody binding by ELISA (see Protocol 11.3). Continue growing the remaining cultures for possible ELISA testing on Day 3 and possibly on Day 4.

12. Preserve the positive clones for later use by spotting from the cultures onto a YPDS+Zeocin plate.

13. Assess protein production of ELISA-positive clones by Western blotting (see Sambrook et al. 1989). Load 10 μl of the supernatant from Step 11 on a 15% gel, and use anti-HA or anti-human Fab as a detecting antibody.

The positive transformed clones can now be used for large-scale protein expression (Protocols 12.8 and 12.9).

PROTOCOL 12.8

Expression of Soluble scFv or Fab in X33 Yeast Cells, Using Shake Flasks

Depending on the individual requirements for protein quantity, yeast expression can be performed in shake flasks or by fermentation. If only a small quantity of protein is needed (such as 200 μg), expression is readily performed using shake flasks. However, if a larger quantity is needed, expression should be carried out using a fermentor (see Protocol 12.9).

Materials

Pichia colony transformed with clone of interest (see Protocol 12.6 or 12.7)
YPD media (see Appendix 2)
Bio-Shield Sterilization Wraps (VWR Scientific Products, Cat. # 4008)
Pellicon II filtration system (Millipore, Cat. # P17508)
PLGC 10K regenerated cellulose membrane (Millipore, Cat. # P2PL GCC05)

Procedure

1. Inoculate 5 ml of YPD in a 50-ml conical tube with a *Pichia* colony selected from the transformation in Protocols 12.6 or 12.7. Add no antibiotic. Cover the conical tube with a cotton BioShield and incubate overnight in a 30°C shaker.
2. Split 1 liter of YPD into three 2-liter shake flasks (~ 333 ml each). Divide the 5-ml overnight culture among the three shake flasks.
3. Incubate for approximately 4 days in a 30°C shaker.
4. Pellet the cells by centrifugation at approximately 3000*g* for 30 minutes at 4°C.
5. Discard the pellet. Concentrate the supernatant to approximately 200 ml via tangential filtration using the Pellicon II filtration system. If the antibodies will be purified on a protein G column, exchange the buffer to PBS. If they will be purified on a nickel column, use a nickel column wash buffer (Protocol 12.2).
6. The concentrated supernatant can be stored at –20°C until purification.

Purify the antibody molecules from the yeast supernatant using the same protocols described for expression and purification in bacteria (Protocols 12.2–12.4).

PROTOCOL 12.9

Expression of Soluble scFv or Fab in X33 Yeast Cells, Using Fermentation

Depending on individual requirements for protein quantity, yeast expression can be performed in shake flasks or by fermentation. If only a small quantity of protein is needed (such as 200 μg), expression is readily performed using shake flasks (see Protocol 12.8). However, if a larger quantity is needed, expression should be carried out using a fermentor. One advantage to using fermentation is that it can be extended until enough protein has been obtained.

The following protocol has been optimized for protein expression in *Pichia pastoris* using the New Brunswick BioFlo 3000 fermentor. The original protocol was given to us by Elizabeth Komives of UCSD.

Materials

Pichia colony transformed with clone of interest (see Protocol 12.6 or 12.7)
YPD medium (see Appendix 2)
Bio-Shield Sterilization Wraps (VWR Scientific Products, Cat. # 4008)
BioFlo 3000 fermentor (New Brunswick Scientific, Cat. # BF-3000)
50% glycerol (11 liters, autoclaved)
Basal salts medium (3 liters, see Appendix 2)
30% **NH_4OH** (2 liters, sterile-filter before use)
Antifoam (Sigma, Cat. # A5551)
0.02 M **KOH** (1 liter, sterile-filtered)
biotin solution: 25 mg of **biotin** in 100 ml of 0.02 M KOH, prepare fresh for each fermentor run
PTM salts (1 liter, sparge before use, see Appendix 2)
Pellicon II filtration system (Millipore, Cat. # P17508)
PLGC 10K regenerated cellulose membrane (Millipore, Cat. # P2PL GCC05)

CAUTIONS: **NH_4OH, KOH, biotin** ***(see Appendix 4)***

Procedure

Day 1

1. Inoculate 5 ml of YPD with the preselected colony in a 50-ml conical tube. Add 50% glycerol to a final concentration of 2% glycerol. Add no antibiotic. Cover with a Bio-Shield. Incubate overnight in a 30°C shaker at 255 rpm.

Day 2

1. Add the 5-ml overnight culture to 200 ml of YPD in a 500-ml shake flask. Add 8 ml of 50% glycerol (final concentration = 2% glycerol). Cover with a Bio-Shield. Incubate overnight in a 30°C shaker at 255 rpm.

2. Calibrate the pH probe, according to the instructions in the New Brunswick Guide to Operations. The pH electrode is calibrated with two external buffers of known pH; we use buffers of pH 7.0 and pH 4.0. Check the pH calibration after autoclaving with an external pH meter, before inoculation. The fermentor should not be turned on for more than 10 minutes without running the water.
3. Perform the sterilization procedure: Prepare 3 liters of basal salts medium and transfer to the fermentor vessel. Install the pH and dissolved oxygen (DO) probes into the vessel. Cover all openings with foil and autoclave tape. Cap the pH and DO probes. Place cover on top of the bearing house. Close the sampler but leave the inoculation port screwed on loosely so that air can vent during the autoclave procedure. Wrap the end of the condenser, two pieces of tubing and a sparge filter in foil. Change the filter. Clamp off all tubes that are open to the medium in the vessel. Autoclave the vessel, the condenser, and the tubing at 121°C, 15 PSI, 25 minutes. Slow exhaust is required.
4. Assemble the console after the fermentor vessel is relatively cool. Tighten the addition port. Place the vessel on the console. Connect the heat exchanger and exhaust condenser. Place the motor on top of the vessel. Turn on the water to a pressure of 20 PSI. Connect the pH, DO, and temperature probes. Connect the motor and sparge filter.
5. Prepare the fermentor by adjusting the following settings. After turning on the fermentor, set the agitation to 500 rpm and the mode to PID. Set the temperature to 30. "Prime" for 10 minutes, then set mode to PID. Set the DO value to 30 on master screen, then set mode to PID. Set the mode to manual on gases screen. Turn the air knob on the right side of the cabinet all the way up. Turn on the pH system. Set up 500 ml of 30% NH_4OH at feed 1. Set feed 1 to base with a value of 10. Set the pH at 5.0 and then the mode to PID. By feeding the 30% NH_4OH, the fermentor will raise the pH of the medium from approximately 1.5 to 5 overnight, and a precipitate will form.

Day 3

1. Check the pH calibration by taking a sample and measuring the pH with an external pH meter.
2. Calibrate the DO probe according to the manufacturer's Guide to Operations.
3. Inoculate the fermentor: After calibrating the DO probe, add 500 µl of antifoam, 12 ml of biotin solution, and 12 ml of PTM salts. Add 200 ml of the overnight culture (from Day 2, Step 1). The DO should start at 100 and begin dropping in about 8–10 hours.

Day 4

1. Turn the agitation up to 800.
2. Determine the wet cell weight by transferring a 1-ml aliquot of culture into a preweighed microcentrifuge tube. Centrifuge for 10 minutes at 14,000 rpm in a microcentrifuge. Discard the supernatant and weigh the tube. Record the weight of the cells in mg/ml.

3. When the wet cell weight is approximately equal to 120 mg/ml, the cells have used nearly all of the glycerol in the batch phase. Add 12 ml of PTM salts and 12 ml of biotin solution to 1 liter of 50% glycerol. Begin the glycerol feed by connecting this solution to feed 2. Set feed 2 to 15.0, manual.

Subsequent Days

1. Each day, the culture will have more than doubled in volume. Feed the glycerol solution constantly. Remove approximately 75% of the culture each morning and each evening. The wet cell weight (see Step 2) for a culture of healthy cells should reach approximately 450–500 mg/ml. The pH should remain at about 5.0.
2. Centrifuge the removed culture at ~4000*g* for 30 minutes at 4°C. Discard the pelleted cells, and store the supernatant at 4°C until concentration. Don't store for more than 2 days.
3. Concentrate the supernatant to 200 ml via tangential filtration with the Pellicon 2 filtration system.
4. Store the concentrated supernatant for up to one month at –20°C until purification.

Purify the antibody molecules from the yeast supernatant using the same protocols described for expression and purification in bacteria (Protocols 12.2–12.4).

REFERENCES

Gronenborn A.M. and Clore G.M. 1993. Identification of the contact surface of a streptococcal protein G domain complexed with a human Fc fragment. *J. Mol. Biol.* **233:** 331–335.

Harlow E. and Lane D. 1999. *Using antibodies: A laboratory manual.* Cold Spring Harbor Laboratory Press, Cold Spring Harbor, New York.

Sambrook J., Fritsch E.F., and Maniatis T. 1989. *Molecular cloning: A laboratory manual,* 2nd. edition. Cold Spring Harbor Laboratory Press, Cold Spring Harbor, New York.

Waterham H.R., Digan M.E., Koutz P.J., Lair S.V., and Cregg J.M. 1997. Isolation of the *Pichia pastoris* glyceraldehyde-3-phosphate dehydrogenase gene and regulation and use of its promoter. *Gene* **186:** 37–44.

13 Antibody Engineering

CHRISTOPH RADER AND CARLOS F. BARBAS III
Department of Molecular Biology, The Scripps Research Institute, La Jolla, California 92037

AS WE HAVE DESCRIBED IN OTHER CHAPTERS, filamentous phage display makes possible the rapid selection of antibodies from libraries prepared from immune animals, including humans. For therapeutic purposes, antibodies of human sequence are required. In this chapter, we review the application of phage display as a method for the creation of human antibodies without immunization, the conversion of nonhuman antibodies to human antibodies, and the evolution of antibody affinity.

CONTENTS

DE NOVO CREATION OF HUMAN ANTIBODIES: SYNTHETIC AND NAIVE ANTIBODY REPERTOIRES

The in vitro production of synthetic antibodies relies on our increasingly sophisticated understanding of how the structure and genetics of antibodies facilitate their rapid evolution in vivo. The three complementarity determining regions (CDRs) provided by each heavy (H) and light (L) chain protein act together by heterodimerization to form the antibody-binding site (Padlan 1994). The binding site, therefore, results from the convergence of six hypervariable peptide loops. It is primarily the variation in amino acid sequence in these regions which produces antibodies of differing specificities, i.e., antibodies that bind different antigens. In vivo, diversity in antibody-binding sites is initially encoded in the germ line by multiple variable (V), diversity (D), and joining (J) gene segments. CDRs1 and 2 of H and L chains are encoded within the V regions. Light-chain CDR3 (LCDR3) is produced by the genetic recombination of V and J regions, whereas heavy-chain CDR3 (HCDR3) is formed by the recombination of V, D, and J regions. The mechanisms for the fusions of the gene segments are complex and in the case of HCDR3 have the potential to generate more than 10^{14} peptides in this region which differ both in length and sequence (Sanz 1991). The diversity in the length of the expressed peptide in the HCDR3 produced by these rearranged segments is astounding in structural terms, as HCDR3 length may vary from 2 to more than 26 amino acid residues (Wu et al. 1993). From structural studies it has become apparent that despite diversity in sequence exhibited in the V gene segments, the structures of these segments are relatively constrained (Al-Lazikani et al. 1997). Only HCDR3 is not restricted to a small evolutionarily restricted repertoire of main-chain conformations or canonical structures (Morea et al. 1998). Following genetic rearrangement and selection, somatic mutation mechanisms then build on the encoded diversity and the diversity that is built by recombination in the CDR3 regions.

The first synthetic antibodies were prepared by Barbas et al. in 1992. In this study, a repertoire of synthetic random HCDR3 sequences was constructed on the backbone of an existing human antibody of defined specificity, an anti-tetanus toxoid antibody. The 16-amino-acid HCDR3 region of this antibody was completely randomized using the polymerase chain reaction and synthetic oligonucleotides. Typically, segmental mutagenesis of this type is performed using oligonucleotides synthesized with degeneracies at the position where randomization is desired (see Fig. 13.1). The most common oligonucleotide doping strategies use NNK or NNS codon equivalents. N represents an equal mixture of the 4 possible nucleotides A, C, G, and T, whereas K is a mixture of just G and T, and S is a mixture of only G and C. Both NNK and NNS doping strategies encode all 20 amino acids and a single amber stop codon within a total of 32 codons. Thus, each amino acid is not equivalently represented in the mixture, and the biases present in nature's 64-codon genetic code remain. The single amber codon present in NNK or NNS mixes can be suppressed when the library is selected in suppressor strains of *Escherichia coli*. These mixes currently represent the best strategy for allowing all possible amino acids to be introduced into a segment without going to codon synthesis strategies that are currently not commercially available. To facilitate DNA sequencing, the NNK doping strategy is preferred over NNS, because NNS results in sequences with higher G/C content, which sometimes complicates DNA sequencing. The number of transformants required to survey random libraries of this type with confidence in obtaining at least one copy of a rare polypeptide sequence is indicated in Table 13.1.

To generate novel binding specificities from the single anti-tetanus toxoid-binding antibody used to construct this first synthetic antibody library, randomization over the antibody's extended HCDR3 length ensured structural diversity even though such a library would be incomplete, as more than 10^{24} clones would be required for each possible amino acid sequence to be represented. The synthetic genes were cloned to yield a library of approximately 10^8 clones, a size on the order of the number of B cells in a mouse. The native light chain of this antibody was used and was of particular interest because it was a "universal light chain." Its light chain was encoded by the Humkv325 germ-line sequence, which is very much overrepresented in the expressed repertoire of human light chains (Burton and Barbas 1994; Cox et al. 1994). The Humkv325 light chain, by virtue of its fitness in the natural repertoire, seemed an ideal choice for incorporation into a synthetic library.

Selection of this synthetic repertoire for binding fluorescein resulted in antibodies with affinities in the 20-nM range for the fluorescein–BSA conjugate. Thus, an antibody was remodeled by randomization of a single CDR segment to bind a hapten for which it originally had no measurable affinity. If this change in specificity is expressed in terms of affinity improvement, then a greater than 10^4-fold improvement in affini-

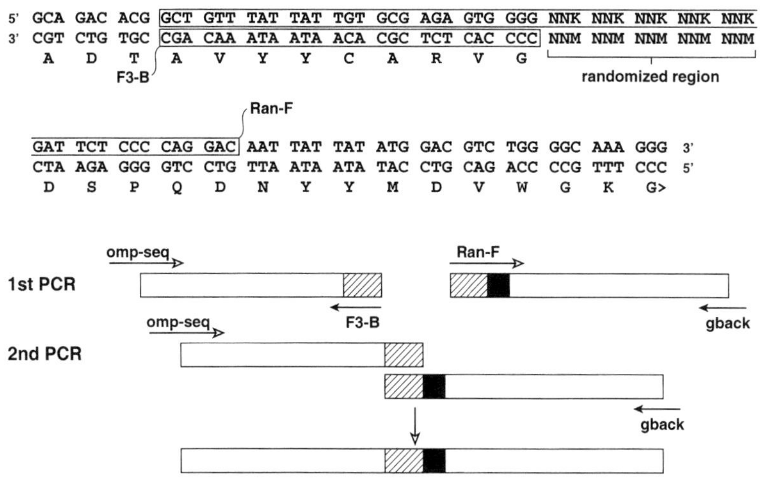

Figure 13.1. Saturation mutagenesis of an antibody CDR. This strategy involves the design of two oligonucleotide primers that are specific for the antibody under consideration. One primer (Ran-F) introduces the randomized gene segment shown here as the NNK region. The second primer is used to amplify the other half of the gene, FR3-B. Note that the two primers are complementary over 27 bp of sequence. The two other PCR primers are common to the vectors used, omp-seq and gback. Initial PCR assays combine omp-seq/FR3-B and Ran-F/gback. The products of these two PCRs are then gel-purified. A second PCR is then performed using these two purified PCR products in equimolar ratios. Approximately 1.7 pmole of each PCR product is used in this fusion step so that as many as 10^{12} different DNA molecules are represented depending on the randomization performed. During this PCR step, the two gene segments become fused to create the fragment required for *Sfi*I cloning. To optimize the success of this strategy, template contamination must be removed. This is critically important in the generation of libraries for CDR walking, where the parental clone has substantial affinity for the target antigen. We recommend the preparation of a nonbinding template antibody for use in construction of the library. A nonbinding template antibody is created by performing all the steps required for the actual library construction, but on a small scale. Following the test ligation of the library, antibody clones are expressed and studied for binding to antigen and expression of functional Fab or scFv fragments. A clone is then chosen in which the antibody fragment is well expressed but lacks binding to the antigen because key residues in the binding site have been mutated. Expression ensures that PCR amplification has not generated frameshift mutations that might compromise the library. Use of a nonbinding template ensures that the PCR template antibody that may find its way through the library construction as a contaminant will not compete for antigen binding with the mutant clones. Once the appropriate mutant template antibody is identified, it is used for the construction of the actual library by PCR as described above.

ty for the hapten was achieved. The use of a single antibody template in these studies was essential in establishing the link between the selected protein and the initial template antibody. Use of a library of antibodies on which HCDR3 diversity would be built would not have allowed the key nature of HCDR3 mutations in hapten binding to be revealed. An interesting result of these studies was the selection of an aspartate (Asp) residue at position 101 in the HCDR3. In naturally occurring antibodies, Asp-101 has been shown to participate in a structurally important salt bridge with Arg-94, the last residue in framework 3. The synthetic antibodies, therefore, recapitulated a structurally conserved feature of natural antibodies. The amino acid distribution within the HCDR3 of natural antibodies is not completely random. The most con-

Table 13.1. Number of transformants required for representation of a degenerate library mutated with NNK or NNS at η codons

No. of codons η	Peptide diversity (20^{η})	DNA diversity (32^{η})	No. of transformants required for 90% confidence	99% confidence
1	20	32	74	149
2	400	1.0×10^3	2.4×10^4	4.8×10^4
3	80×10^3	3.3×10^4	7.6×10^4	1.5×10^5
4	1.6×10^5	1.1×10^6	2.4×10^6	4.9×10^6
5	3.2×10^6	3.4×10^7	7.7×10^7	1.6×10^8
6	6.4×10^7	1.1×10^9	2.5×10^9	5.0×10^9
7	1.3×10^9	3.4×10^{10}	7.9×10^{10}	1.6×10^{11}
8	2.6×10^{10}	1.1×10^{12}	2.5×10^{12}	5.1×10^{12}
16	6.6×10^{20}	1.2×10^{24}	2.8×10^{24}	5.6×10^{24}

Confidence values and table are based on the Poisson distribution calculation of Lowman and Wells (1991).

served portion of HCDR3 is the carboxy-terminal region, which is contributed by the J-gene segment, is characteristically rich in aromatics, and in most cases encodes aspartate at position 101. This distribution is also clearly seen in selected synthetic repertoires. A recurring theme in synthetic antibody selection studies is that synthetic antibodies are not typically artificial in their sequence character and are often constrained to sequences commonly found in natural antibodies (Lerner et al. 1992; Barbas 1995). Indeed, sequence similarities between natural and synthetic antibodies can be close enough to allow the accurate prediction of cross-reactivity for some synthetic antibodies (Barbas and Wagner 1995). With the demonstration that a single antibody could be remodeled to bind many new antigens, collections of V genes in combination with synthetic H or LCDR3 provide an even more structurally diverse pool of antibodies from which to select the desired antibody. This suggests that many of the amino acids selected or encoded in natural repertoires are under considerable selective pressure as a result of demands on the protein's folding and stability and not just antigen recognition.

Another way to create synthetic antibodies involves the use of peptide leads to program a synthetic antibody library. Peptide leads can come from a variety of sources. For a number of ligand–receptor systems, the minimal ligand sequences required for receptor binding have been described and can be used. Alternatively, novel peptide ligands can be selected from random peptide phage libraries. The peptide ligand is placed in a solvent-accessible region of a CDR, most typically in HCDR3, and a library is constructed by randomizing the residues flanking the peptide motif. A library-based approach is preferred over simple transplantation of the motif, because binding selections allow the optimization of the motif's conformational display. Additional receptor–antibody contacts can also be selected. For example, within the integrin receptor family the peptide motif Arg-Gly-Asp has been shown to be the critical minimal peptide sequence for the binding of protein ligands such as fibrinogen and vitronectin. An HCDR3 library was constructed in which the Arg-Gly-Asp sequence was placed near the apex of the CDR loop and flanked on either side by three randomized amino acids.

Phage selection for integrin binding provided human antibodies against the integrin self-antigens $\alpha_V\beta_3$ and $\alpha_{IIb}\beta_3$ (Barbas et al. 1993a; Smith et al. 1994). The affinities of the selected antibodies, 10^{10} M^{-1}, are among the highest reported for Fabs. Because of the tremendous range of peptide lengths that can be accommodated in HCDR3, much longer peptide motifs (> 20 amino acids) can be optimized using this approach.

The synthetic antibody approach has been used to isolate monoclonal antibodies (mAbs) with specificity for virtually any given antigen (Barbas 1995; Pini et al. 1998; Knappik et al. 2000). The antigens that have been used for selection range from the smallest possible antigens, metal ions, to haptens, peptides, proteins, carbohydrates, DNA, and inorganic surfaces. Synthetic antibodies that catalyze chemical reactions have also been selected (Janda et al. 1994). Although synthetic libraries promise to remove the need for immunization, their large size makes maintenance and curation difficult and has prevented them from being widely accessible.

Another combinatorial approach that has shown promise in the selection of antibodies without immunization involves the use of naive libraries (Vaughan et al. 1996; de Haard et al. 1999). Naive libraries attempt to recapitulate the primary repertoire available to the animal before antigen contact. The essence of this approach is the preparation of a large combinatorial library of heavy and light chains. Random recombination of heavy- and light-chain genes and hence the native CDRs then provides a diversity of binding sites. The source of the genetic material is key to the construction of such a library. Typically, mRNA from B cells encoding surface IgM or IgD heavy chains are preferred representative sources of naive immunoglobulins. To further ensure library diversity, mRNA from a large number of healthy donors is used. As with the synthetic antibody approach, antibodies of many different specificities have been selected from large naive antibody libraries. The main limitation of this approach is in ensuring that the library is diverse and unbiased in its composition. Because mRNA is used as the genetic material for library construction, the library is susceptible to contamination by mRNA derived from plasma cells. Such contamination would limit the diversity of the library. In contrast to synthetic libraries, it is virtually impossible to accurately determine the diversity of a naive library. Regardless of this fact, both types of libraries have been used successfully. The affinity of fragments derived from synthetic and naive antibody library approaches ranges from 1 μM to 0.3 nM. For antibodies at the low end of the affinity spectrum, additional mutagenesis and selections may be required to mature the affinity of the selected antibody before its practical application.

ANTIBODY HUMANIZATION

Therapeutically useful antibodies have three characteristics: high specificity, high affinity, and lack of immunogenicity. Although rodent mAbs have long been regarded as potential therapeutic agents, a major obstacle for clinical applications has been their immunogenicity in humans. To overcome this problem, antibody engineering has been used to humanize rodent antibodies. More than 20 years of mAb generation by classic hybridoma technology have yielded a number of promising pharmaceutical candidates, and their humanization compares well with the de novo generation and characterization of human mAbs for use in clinical applications.

CDR grafting is currently the most frequently used strategy for the humanization of rodent mAbs (Riechmann et al. 1988). In this approach, the six CDR loops that make up the antigen-binding site of the rodent mAb are grafted into the corresponding human framework regions. CDR grafting takes advantage of the conserved structure of the variable immunoglobulin domains, with the four framework regions serving as a scaffold that supports the CDR loops. This method often yields humanized antibodies with much lower affinity than the original rodent antibodies, because framework residues are involved in antigen binding indirectly, by supporting the conformation of the CDR loops, and directly, by contacting the antigen (Foote and Winter 1992). Therefore, it is usually necessary to replace certain framework residues in addition to CDR grafting. The fact that about 30 framework residues potentially contribute to antigen binding (Foote and Winter 1992) makes this fine-tuning step very laborious. Another humanization strategy is the method of resurfacing (Pedersen et al. 1994). In this approach, only the surface residues of a rodent antibody are humanized. These residues are likely to be the most immunogenic and have not been seen to contribute to antigen binding (Pedersen et al. 1994).

In vitro selection and evolution of antibodies derived from phage-display libraries has become a powerful tool in antibody engineering (for reviews, see Hoogenboom 1997; Rader and Barbas 1997). Recently, we used a selective approach that combines CDR grafting with framework fine-tuning by phage display (Rosok et al. 1996; Baca et al. 1997) for the first reported humanization of a rabbit antibody (Rader et al. 2000). As outlined in Figure 13.2, this method involves selection between the original nonhuman amino acid and a human amino acid at a small number of framework positions. The residues at these positions are known to be involved in antigen binding and have been identified structurally, by crystallography and molecular modeling, as well as empirically, by antibody humanization. We applied this strategy to the humaniza-

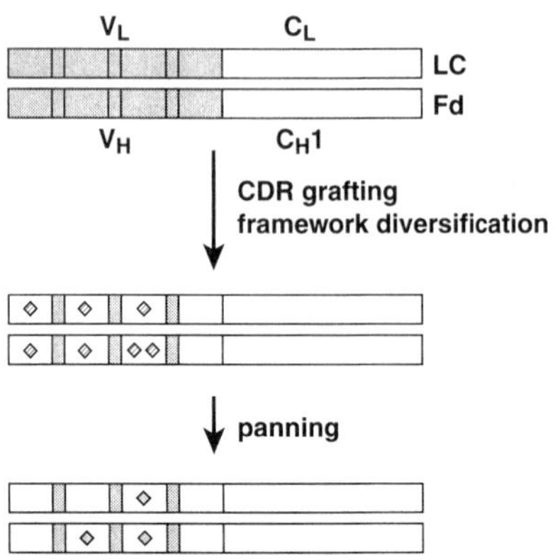

Figure 13.2. Selection strategy for antibody humanization that combines CDR grafting with framework fine-tuning. Nonhuman sequences are shown in gray, human sequences are white. The six CDRs of the nonhuman Fab are grafted into a human framework that contains a set of diversified residues (*hatched*). These residues are diversified to allow the selection of either the human or the original nonhuman residue.

tion of a rabbit antibody directed against the human A33 antigen (Rader et al. 2000), a tumor antigen that is expressed on the surface of colon cancer cells. The rabbit antibody had been selected by phage display from antibody libraries generated from immune rabbits as described in Chapters 8–11. The humanized antibodies retained both high specificity and affinity for human A33 antigen.

An entirely selective humanization strategy based on the use of phage-display libraries has been reported by Jespers et al. (1994). In two steps, each polypeptide of the rodent antibody, either light chain or heavy chain, is replaced by a corresponding human polypeptide library. The resulting hybrid antibody library is selected by panning against antigen. An advantage of this strategy over CDR grafting is the elimination of the arduous fine-tuning steps, but a selection of the entire antigen-binding site, i.e., a selection of all six CDRs, bears the risk of diminishing both affinity and specificity for the antigen. In fact, this sequential chain-shuffling procedure may lead to the production of a humanized antibody that recognizes a slightly different epitope (Kang et al. 1991; Zebedee et al. 1992; Ohlin et al. 1996). It has been observed that antibodies consisting of the same heavy chain paired with light chains that differ in LCDR3 and elsewhere in V_L can bind different epitopes on the same antigen. Although in some cases alteration of antigen specificity after antibody humanization might be interesting, it is not usually desired.

On the basis of these considerations, we designed a humanization strategy that recognizes the key roles of HCDR3 and LCDR3 in antigen recognition and combined it with a selective approach that avoids the arduous fine-tuning steps associated with CDR grafting (Fig. 13.3). In the first step, a human light-chain library with the grafted, original LCDR3 replaces the original light chain. The original Fd fragment is replaced by a chimeric Fd fragment consisting of the original V_H Ig domain fused to a human C_H1 Ig domain. The original V_H domain is maintained in the first selection step, because it typically plays the most dominant role in antigen recognition. Framework region 4 (FR4) of both the heavy and light chains is directly humanized by simple point mutations before the first selective step. Sequence changes required in this region should generally be minimal, given the substantial homology between mouse and human J genes, and should have little effect on affinity. The hybrid Fab of the first selection step is stabilized by the interaction of two matching human constant domains, C_κ and C_H1. In addition, Fab carrying human constant regions are often better expressed in *E. coli* (Carter et al. 1992; Ulrich et al. 1995), a prerequisite for phage display. In the second step, the selected human light chains from the first step are paired with a human Fd fragment library containing the original HCDR3.

By preserving the original LCDR3 and HCDR3 sequences while subjecting the remaining sequence to selection, our humanization strategy was designed to ensure antigen specificity and epitope conservation. LCDR3 and HCDR3 contain the hypervariable joints of the V/J and V/D/J gene rearrangements that participate in direct antigen contact in all studied antigen/antibody complexes (Wilson and Stanfield 1993). Unlike the other CDR regions, both LCDR3 and HCDR3 interact with all three CDRs of the other variable domain (Padlan 1994). Thus, although generalizations might be misleading (Wilson and Stanfield 1993), LCDR3 and HCDR3 can be considered to make the most significant contributions to affinity and specificity. Given the

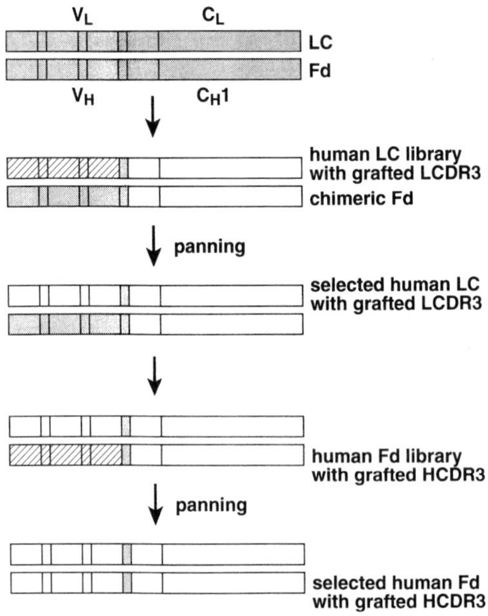

Figure 13.3. Antibody humanization using designed combinatorial V-gene libraries. Nonhuman sequences are shown in gray, human sequences are white. Libraries are shown hatched. This strategy involves two selection steps for the sequential humanization of the light chain (LC) and the Fd fragment of the heavy chain. Throughout these selections, the only preserved sequences in the variable domains of light chain (V_L) and heavy chain (V_H) are LCDR3 and HCDR3. In the first step, a chimeric nonhuman/human Fd fragment is used as a template for the selection of a human light chain that contains the grafted LCDR3 loop of the original nonhuman light chain. In the second step, a human Fd fragment that contains the grafted HCDR3 loop of the original nonhuman Fd fragment is selected.

tremendous sequence diversity displayed by human antibodies in these regions and the mechanism of its generation, it is difficult, if not impossible in most cases, to classify sequences of these regions as either mouse or human. Human HCDR3s are, on average, longer than mouse HCDR3s and encompass the full range of lengths used by mice (Wu et al. 1993). From the perspective of sequence, however, the antibody products of our humanization strategy can be considered completely human.

We applied our strategy to the humanization of mouse mAb LM609 (Rader et al. 1998). LM609 is directed against the human integrin $\alpha_v\beta_3$ and has potential applicability in cancer therapy as an anti-angiogenic agent. We demonstrated that this approach provides a rapid route for antibody humanization. It produces antibodies with affinities as high as or higher than the affinity of the original antibody, and the original antigen- and epitope-specificity are retained. Whereas CDR grafting alone generates only a single humanized antibody, this method produces several humanized versions of the original antibody. The expression level of antibodies in culture is antibody-dependent. The availability of multiple humanized variants, therefore, increases the chance that at least one will express well enough to be suitable for large-scale production. For cancer therapy, it is anticipated that an anti-angiogenic strategy would require long-term administration of antibody. Repeated therapeutic application, however, may produce an anti-idiotypic response that ablates the antibody's efficacy. Administration of an equally potent antibody that is unreactive to the anti-idiotypic response generated by the first would allow therapy to continue. Indeed, introduction of modest changes within the variable domain of an antibody can dramatically alter its reactivity to an anti-idiotypic response (Glaser et al. 1992).

We recently humanized a rabbit mAb using designed combinatorial V-gene libraries (Steinberger et al. 2000). Rabbit antibody ST6 specifically binds human chemokine receptor CCR5, which is a principal HIV-1 coreceptor and a target for immunotherapy. ST6 had been selected by phage display from antibody libraries generated from immune rabbits. Interestingly, the humanization of ST6 required the conservation of HCDR2 in addition to LCDR3 and HCDR3. HCDR2 of ST6 might have an important role in antigen binding, and the human V_H repertoire might not contain HCDR2 sequences that are similar enough to support antigen binding. Containing three grafted CDRs, humanized ST6 was found to retain both high specificity and affinity for human chemokine receptor CCR5.

DIRECTED EVOLUTION OF ANTIBODY AFFINITY

The systems described for producing new antibodies can also be used to improve the properties of existing antibodies. The first experiments involving the monovalent display of Fab fragments on phage clearly demonstrated that phage selection facilitates the sorting of antibody libraries based on their affinity (Barbas et al. 1991). Thus, phage display provides an appropriate format for the directed molecular evolution of antibody fragments. Several strategies have been applied to improve the affinity of antibody fragments. For all of these studies, it is important to consider the affinity range of the antibody to be improved. It is expected, and indeed observed, that greater relative improvements are more readily obtained when starting with low-affinity antibodies. As described earlier, a Fab with little or no binding to fluorescein can be randomized and selected to provide an antibody with a 20-nM dissociation constant for this hapten, a greater than 10^4-fold improvement in affinity (Barbas et al. 1992). Antibodies that already possess nanomolar dissociation constants for their antigen require more subtle refinements of an already tight antigen–antibody interface to improve their affinity.

Practically, when the affinity of a monovalent antibody fragment, such as a Fab, reaches a dissociation constant of about 0.1 nM, further improvements become difficult. The dissociation constant can be expressed in kinetic terms, where $K_d = k_{off}/k_{on}$. Study of natural antibodies with respect to their binding kinetics has revealed that the on-rates of antibodies typically fall within a reasonably narrow window that varies over approximately an order of magnitude. A typical k_{on} for an antibody interaction is about 10^5 M^{-1}s^{-1}, and affinity differences between antibodies are generally accounted for by differences in k_{off}. A 0.1-nM dissociation constant, therefore, can be expected to result from a k_{off} of about 10^{-5}s^{-1}. An off-rate of this magnitude results in a half-life ($t_{1/2}$) for the antibody–antigen complex of approximately 19 hours ($t_{1/2} = \ln 2/ k_{off}$). This is considered a very high affinity monovalent interaction. In comparison, a typical antibody derived from immunization, therefore a natural affinity-matured antibody, will have a monovalent affinity of approximately 10 nM, or a $t_{1/2}$ of just 11 minutes.

Phage-based selections are typically driven by off-rate. Washing time and stringency are varied such that lower-affinity phage that dissociate from the target on the solid phase are washed away and the high-affinity phage are left behind. The need to select over several half-lives to enrich higher-affinity clones presents a practical barrier that is difficult to overcome. If the antigen is on a cell surface, for example, the limitation may actually be imposed by instability of the antigen. Solution-based selections appear to overcome these problems associated with solid-phase off-rate-based selections (Schier et al. 1996).

Dispersed Mutagenesis

Natural somatic mutation provides V-gene diversification by dispersing point mutations throughout the V gene. To accomplish this diversification in vitro, error-prone PCR conditions have been used to construct libraries of point mutations (Gram et al. 1992). This approach was used to improve a low-affinity mouse scFv selected from a naive library to bind progesterone. The monovalent display system pComb3 allowed the selection of an antibody fragment with 30-fold improved affinity. The final scFv had an affinity of about 1 µM. Error-prone PCR has been used in a number of studies. Hawkins et al. (1992), for example, used error-prone PCR and the multivalent display system fd-tet-DOG1 to improve a mouse anti-hapten scFv. With fd-tet-DOG1, scFv is displayed on all copies of gIII. To select for higher-affinity variants, the antigen was biotinylated and the selection was performed in solution with streptavidin capture. Solution-phase selection was required to avoid avidity-based selection that would be the consequence of using this multivalent display vector. A fourfold improvement in hapten binding resulted. In a subsequent report, Hawkins et al. (1993) used the same strategy to optimize a high-affinity mouse anti-lysozyme scFv. In this study, a structure-based strategy could be applied, because the crystal structure of the antibody–protein complex had been solved. By using site-directed mutagenesis to combine two sets of mutations found in independent clones, an scFv with an affinity of 0.6 nM, a fivefold overall improvement in affinity, was obtained. When comparing the results of this study with other studies not guided by structure, no significant advantage was provided by the structural information.

Mutator strains of *E. coli* have also been used for dispersed mutagenesis (Low et al. 1996). An extensive amount of panning and cycling through the mutator strain was required and, as with the error-prone PCR method, only modest improvements in affinity were achieved. One of the potential problems with random mutagenesis of the entire gene using either PCR or mutator strains is that many of the mutations occur outside the CDRs and thus might more easily generate an antigenic antibody. Antigenicity, of course, is only of concern if an antibody is to be used in vivo.

Chain shuffling, originally proposed by Huse et al. (1989), can be used with great success when starting with a low-affinity clone. In one study, starting with low-affinity anti-hapten antibody and shuffling against a library of light chains led to the selection of a clone with 20-fold improved affinity (Marks et al. 1992). This antibody was diversified by combination with a V_H library and selected for an additional 15-fold improvement, providing an antibody with a nanomolar dissociation constant. This approach is of less use for the improvement of high-affinity antibodies that already possess nanomolar dissociation constants (Barbas et al. 1993b), as the sequence changes that result are too great to achieve the fine-tuning required for antibodies in this affinity range. Shuffling of V genes, as performed in the humanization of antibodies, can result in modest improvements in affinity (Rader et al. 1998). The more recently developed strategy of DNA shuffling has yet to be convincingly applied to the evolution of antibody affinity (Stemmer 1994). This method makes use of point mutations created by error-prone PCR but also involves recombination of mutations in subsequent rounds of DNA shuffling and selection. Although DNA shuffling could be used to shuffle families of V genes, the resulting antibodies would likely be chimeras derived from different germ-line V genes and would present problems with antigenicity if the antibodies were to be used in vivo.

Focused Mutagenesis

To date, a methodology known as CDR walking has proven to be the most general method for affinity optimization of antibodies. This focused mutagenesis strategy is a variant of the synthetic antibody approach (Barbas 1995) with one important difference. In this case, library completeness is stressed over structural diversity, and randomization is limited to six residues or less. Construction of libraries of 10^7–10^8 is generally routine, so that segments of 4–5 codons are readily optimized (see Table 13.1). Modifications to the parental antibody are constrained to the hypervariable regions of the antibody, because changes in the primary sequence in these regions are less likely to generate immunogenic antibodies than changes in the more sequence-constrained framework regions.

There are two strategies for CDR walking. In parallel CDR walking, several libraries are constructed that target different CDRs. After selection, the improved CDRs are then assembled together to give the best antibodies, assuming additivity. In sequential CDR walking, the improvements that result from one CDR library selection are incorporated into the antibody sequence before constructing the next CDR library. For both strategies, the free-energy changes that result from the combination of sets of mutations can be expressed as

$$\Delta\Delta G_{AB} = \Delta\Delta G_A + \Delta\Delta G_B + \Delta\Delta G_I$$

where $\Delta\Delta G_A$ and $\Delta\Delta G_B$ are changes in the free energy of CDR mutants A and B, $\Delta\Delta G_{AB}$ is the change in free energy for the protein that combines the mutations of A and B, and $\Delta\Delta G_I$ is the change in the interaction energy between the two sets of residues. The differences between parallel and sequential CDR walking result from the dependence of the combined mutants on the term $\Delta\Delta G_I$.

Parallel CDR walking assumes that $\Delta\Delta G_I$ will be negligible. Sequential CDR walking takes into account that $\Delta\Delta G_I$ may not always be negligible and that optimal binding may result from the interdependence of CDR loops. Such interdependence could result from coordinated structural changes that occur when the antibody binds antigen. Indeed, crystallographic evidence suggests that induced-fit mechanisms describe the binding observed in some antibody–antigen complexes (Stanfield and Wilson 1994; Wilson and Stanfield 1994).

With a human anti-HIV antibody as a model system, both parallel and sequential optimization strategies have been studied (Barbas et al. 1994; Yang et al. 1995). Parallel optimizations were studied by incorporating independently selected CDRs into the parental anti-HIV Fab. Six new antibodies were created by shuffling optimized CDRs. In most cases, improvement in affinity was overwhelmed by large unfavorable interaction energies, with only one combination of six yielding additivity. This result differs from those seen with other protein systems, such as human growth hormone, in which additivity of mutations was usually observed (Lowman and Wells 1993). Nonetheless, a 420-fold improvement of the parental Fab was obtained, providing an antibody with picomolar affinity (Yang et al. 1995). Interestingly, the combination of HCDR1 and HCDR3 mutants was additive. Application of this strategy to a human antibody directed against erbB-2 also resulted in production of an antibody with picomolar affinity (Schier et al. 1996).

In the anti-HIV antibody system, three sequential optimizations improved the parental HIV-I gp120-binding Fab approximately 100-fold, starting from a 6-nM affinity. Optimization of the CDR segments H3(100A–E) and H1(31–35) also provided significant affinity gains, suggesting that continued sequential selection through these two CDR regions could result in an overall improvement in affinity of approximately 3000-fold. Again, it is important to stress the role of $t_{1/2}$ of the antibody–antigen interaction in selections. The half-life experimentally limits the likelihood of isolating such ultrahigh affinity clones from libraries built on an already very high affinity parental antibody.

The potential therapeutic value of very high affinity antibodies is discussed in detail elsewhere (Barbas and Burton 1996) but includes reduction in therapeutic dose and increased therapeutic duration. Studies of several different antibodies now suggest that CDR-walking mutagenesis is the most effective method for the optimization of antibody affinity (Chowdhury and Pastan 1999). The most significant gains in affinity have been observed following optimization of the CDR3 regions in either the light or heavy chain. We suggest, therefore, that the most expeditious route to an antibody of improved affinity involves CDR-walking mutagenesis over these two regions.

REFERENCES

Al-Lazikani B., Lesk A.M., and Chothia C. 1997. Standard conformations for the canonical structures of immunoglobulins. *J. Mol. Biol.* **273:** 927–948.

Baca M., Presta L.G., O'Connor S.J., and Wells J.A. 1997. Antibody humanization using monovalent phage display. *J. Biol. Chem.* **272:** 10678–10684.

Barbas C.F., III. 1995. Synthetic human antibodies. *Nat. Med.* **1:** 837–839.

Barbas C.F., III and Burton D.R. 1996. Selection and evolution of high-affinity human antiviral antibodies. *Trends Biotechnol.* **14:** 230–234.

Barbas C.F., III and Wagner J. 1995. Synthetic human antibodies: Selecting and evolving functional proteins. *Methods* **8:** 94–103.

Barbas C.F., III, Languino L.R., and Smith J.W. 1993a. High affinity self-reactive human antibodies by design and selection: Targeting the integrin ligand binding site. *Proc. Natl. Acad. Sci.* **90:** 10003–10007.

Barbas C.F., III, Bain J.D., Hoekstra D.M., and Lerner R.A. 1992. Semi-synthetic combinatorial antibody libraries: A chemical solution to the diversity problem. *Proc. Natl. Acad. Sci.* **89:** 4457–4461.

Barbas C.F., III, Kang A.S., Lerner R.A., and Benkovic S.J. 1991. Assembly of combinatorial antibody libraries on phage surfaces: The gene III site. *Proc. Natl. Acad. Sci.* **88:** 7978–7982.

Barbas C.F., III, Hu D., Dunlop N., Sawyer L., Cababa D., Hendry R.M., Nara P.L., and Burton D.R. 1994. In vitro evolution of a neutralizing human antibody to HIV-I to enhance affinity and broaden strain cross-reactivity. *Proc. Natl. Acad. Sci.* **91:** 3809–3813.

Barbas C.F., III, Collet T.A., Roben P., Binley J., Amberg W., Hoekstra D., Cababa D., Jones T.M., Williamson R.A., Pilkington G.R., Haigwoods N.L., Satterthwait A.C., Sanz I., and Burton D.R. 1993b. Molecular profile of an antibody response to HIV-1 as probed by combinatorial libraries. *J. Mol. Biol.* **230:** 812–823.

Burton D.R. and Barbas C.F., III. 1994. Human antibodies from combinatorial libraries. *Adv. Immunol.* **57:** 191–280.

Carter P., Kelley R.F., Rodrigues M.L., Snedecor B., Covarrubias M., Velligan M.D., Wong W.L.T., Rowland A.M., Kotts C.E., Carver M.E., and Claire E. 1992. High level *Escherichia coli* expression and production of a bivalent humanized antibody fragment. *Bio/Technology* **10:** 163–167.

Chowdhury P.S. and Pastan I. 1999. Improving antibody affinity by mimicking somatic hypermutation in vitro. *Nat. Biotechnol.* **17:** 568–572.

Cox J.P., Tomlinson I.M., and Winter G. 1994. A directory of human germ-line V kappa segments reveals a strong bias in their usage. *Eur. J. Immunol.* **24:** 827–836.

de Haard H.J., van Neer N., Hufton S.E., Roovers R.C., Henderikx P., de Bruine A.P., Arends J.W., and Hoogenboom H.R. 1999. A large non-immunized human Fab fragment phage library that permits rapid isolation and kinetic analysis of high affinity antibodies. *J. Biol. Chem.* **274:** 18218–18230.

Foote J. and Winter G. 1992. Antibody framework residues affecting the conformation of the hypervariable loops. *J. Mol. Biol.* **224:** 487–499.

Glaser S.M., Yelton D.E., and Huse W.D. 1992. Antibody engineering by codon-based mutagenesis in a filamentous phage vector system. *J. Immunol.* **149:** 3903–3913.

Gram H., Marconi L.-A., Barbas C.F., III, Collet T.A., Lerner R.A., and Kang A.S. 1992. In vitro selection and affinity maturation of antibodies from a naive combinatorial immunoglobulin library. *Proc. Natl. Acad. Sci.* **89:** 3576–3580.

Hawkins R.E., Russell S.J., and Winter G. 1992. Selection of phage antibodies by binding affinity. Mimicking affinity maturation. *J. Mol. Biol.* **226:** 889–896.

Hawkins R.E., Russell S.J., Baier M., and Winter G. 1993. The contribution of contact and non-contact residues of antibody in the affinity of binding to antigen. The interaction of mutant D1.3 antibodies with lysozyme. *J. Mol. Biol.* **234:** 958–964.

Hoogenboom H.R. 1997. Designing and optimizing library selection strategies for generating high-affinity antibodies. *Trends Biotechnol.* **15:** 62–70.

Huse W.D., Sastry L., Iverson S.A., Kang A.S., Alting-Mees M., Burton D.R., Benkovic S.J., and Lerner R.A. 1989. Generation of a large combinatorial library of the immunoglobulin repertoire in phage lambda. *Science* **246:** 1275–1281.

Janda K.D., Lo C.-H.L., Li T., Barbas C.F., III, Wirsching P., and Lerner R.A. 1994. Direct selection for a catalytic mechanism from combinatorial antibody libraries. *Proc. Natl. Acad. Sci.* **91:** 2532–2536.

Jespers L.S., Roberts A., Mahler S.M., Winter G., and Hoogenboom H.R. 1994. Guiding the selection of human antibodies from phage display repertoires to a single epitope of an antigen. *Bio/Technology* **12:** 899–903.

Kabat E.A., Wu T.T., Perry H.M., Gottesman K.S., and Foeller C. 1991. *Sequences of proteins of immunological interest.* U.S. Dept. of Health and Human Services, Public Health Service, National Institutes of Health, Washington, D.C.

Kang, A. S., Jones T.M., and Burton D.R. 1991. Antibody redesign by chain shuffling from random combinatorial immunoglobulin libraries. *Proc. Natl. Acad. Sci.* **88:** 11120–11123.

Knappik A., Ge L., Honegger A., Pack P., Fischer M., Wellnhofer G., Hoess A., Wölle J., Plückthun A., and Virnekäs B. 2000. Fully synthetic human combinatorial antibody libraries (HuCAL) based on modular consensus frameworks and CDRs randomized with trinucleotides. *J. Mol. Biol.* **296:** 57–86.

Lerner R.A., Kang A.S., Bain J.D., Burton D.R., and Barbas C.F., III. 1992. Antibodies without immunization. *Science* **258:** 1313–1314.

Low N.M., Holliger P.H., and Winter G. 1996. Mimicking somatic hypermutation: Affinity maturation of antibodies displayed on bacteriophage using a bacterial mutator strain. *J. Mol. Biol.* **260:** 359–368.

Lowman H.B. and Wells J.A. 1991. Monovalent phage display: A method for selecting variant proteins from random libraries. *Methods* **3:** 205–216.

———. 1993. Affinity maturation of human growth hormone by monovalent phage display. *J. Mol. Biol.* **234:** 564–578.

Marks J.D., Griffiths A.D., Malmqvist M.S., Clackson T.P., Bye J.M., and Winter G. 1992. Bypassing immunization: Building high affinity human antibodies by chain shuffling. *Bio/Technology* **10:** 779–783.

Morea V., Tramontano A., Rustici M., Chothia C., and Lesk A.M. 1998. Conformations of the third hypervariable region in the VH domain of immunoglobulins. *J. Mol. Biol.* **275:** 269–294.

Ohlin M., Owman H., Mach M., and Borrebaeck C.A.K. 1996. Light chain shuffling of a high affinity antibody results in a drift in epitope recognition. *Mol. Immunol.* **33:** 47–56.

Padlan E.A. 1994. Anatomy of the antibody molecule. *Mol. Immunol.* **31:** 169–217.

Pedersen J.T., Henry A.H., Searle S.J., Guild B.C., Roguska M., and Rees A.R. 1994. Comparison of surface accessible residues in human and murine immunoglobulin Fv domains. Implication of humanization of murine antibodies. *J. Mol. Biol.* **235:** 959–973.

Pini A., Viti F., Santucci A., Carnemolla B., Zardi L., Neri P., and Neri D. 1998. Design and use of a phage display library. Human antibodies with subnanomolar affinity against a marker of angiogenesis eluted from a two-dimensional gel. *J. Biol. Chem.* **273:** 21769–21776.

Rader C. and Barbas C.F., III. 1997. A phage display of combinatorial antibody libraries. *Curr. Opin. Biotechnol.* **8:** 503–508.

Rader C., Cheresh D.A., and Barbas C.F., III. 1998. A phage display approach for rapid antibody humanization: Designed combinatorial V gene libraries. *Proc. Natl. Acad. Sci.* **95:** 8910–8915.

Rader C., Ritter G., Nathan S., Elia M., Gout I., Jungbluth A.A., Welt S., Old L.J., and Barbas C.F., III. 2000. The rabbit antibody repertoire as a novel source for the generation of therapeutic human antibodies. *J. Biol. Chem.* **275:** 13668–13676.

Riechmann L., Clark M., Waldmann H., and Winter G. 1988. Reshaping human antibodies for therapy. *Nature* **332:** 323–327.

Rosok M.J., Yelton D.E., Harris L.J., Bajorath J., Hellström K.-E., Hellström I., Cruz G.A., Kristensson K., Lin H., Huse W.D., and Glaser S.M. 1996. A combinatorial library strategy for the rapid humanization of anticarcinoma BR96 Fab. *J. Biol. Chem.* **271:** 22611–22618.

Sanz I. 1991. Multiple mechanisms participate in the generation of diversity of human H chain CDR3 regions. *J. Immunol.* **147:** 1720–1729.

Schier R., McCall A., Adams G.P., Marshall K.W., Merritt H., Yim M., Crawford R.S., Weiner L.M., Marks C., and Marks J.D. 1996. Isolation of picomolar affinity anti-c-erbB-2 single-chain Fv by molecular evolution of the complementarity determining regions in the center of the antibody binding site. *J. Mol. Biol.* **263:** 551–567.

Smith J.W., Hu D., Satterthwait A.C., Pinz–Sweeney S., and Barbas C.F., III. 1994. Building synthetic antibodies as adhesive ligands for integrins. *J. Biol. Chem.* **269:** 32788–32795.

Stanfield R.L. and Wilson I.A. 1994. Antigen-induced conformational changes in antibodies: A problem for structural prediction and design. *Trends Biotechnol.* **12:** 275–279.

Steinberger P., Sutton J.K., Rader C., Elia M., and Barbas C.F., III. 2000. Generation and characterization of a recombinant human CCR5-specific antibody: A phage display approach for rabbit antibody humanization. *J. Biol. Chem.* (in press).

Stemmer W.P. 1994. Rapid evolution of a protein in vitro by DNA shuffling. *Nature* **370:** 389–391.

Ulrich H.D., Patten P.A., Yang P.L., Romesberg F.E., and Schultz P.G. 1995. Expression studies of catalytic antibodies. *Proc. Natl. Acad. Sci.* **92:** 11907–11911.

Vaughan T.J., Williams A.J., Pritchard K., Osbourn J.K., Pope A.R., Earnshaw J.C., McCafferty J., Hodits R.A., Wilton J., and Johnson K.S. 1996. Human antibodies with sub-nanomolar affinities isolated from a large non-immunized phage display library. *Nat. Biotechnol.* **14:** 309–314.

Wilson I.A. and Stanfield R.L. 1993. Antibody-antigen interactions. *Curr. Opin. Struct. Biol.* **3:** 113–118.

———. 1994. Antibody-antigen interactions: New structures and new conformational changes. *Curr. Opin. Struct. Biol.* **4:** 857–867.

Wu T.T., Johnson G., and Kabat E.A. 1993. Length distribution of CDRH3 in antibodies. *Proteins* **16:** 1–7.

Yang W.-P., Green K., Pinz-Sweeney S., Briones A.T., Burton D.R., and Barbas C.F., III. 1995. CDR walking mutagenesis for the affinity maturation of a potent human anti-HIV-1 antibody into the picomolar range. *J. Mol. Biol.* **254:** 392–403.

Zebedee S.L., Barbas C.F., III, Hom Y.-L., Caothien R.H., La Polla R., Burton D.R., and Thornton G.B. 1992. Human combinatorial antibody libraries to hepatitis B surface antigen. *Proc. Natl. Acad. Sci.* **89:** 3175–3179.

14 Overview: Peptide Libraries

JAMIE K. SCOTT
Institute of Molecular Biology and Biochemistry and Department of Biological Sciences, Burnaby, British Columbia Canada V5A 1S6

THIS SECTION DESCRIBES METHODS FOR WORKING with peptide libraries of two types: those displayed by the major coat protein, pVIII, and those displayed by the minor coat protein pIII. Chapters 15–18, written by J. K. Scott and the members of her laboratory, describe the production and screening of peptide libraries displayed on the pVIII protein of the filamentous phage vector f88-4 (Zhong et al. 1994). As described in more detail in Chapter 2, this vector is a derivative of fd-tet (Zacher et al. 1980) and contains a large DNA insertion in the minus-strand origin of replication of the fd phage. The insert includes genes encoding resistance to tetracycline, and a cassette that expresses a second copy of recombinant pVIII, to which the library peptides are fused. With this vector, approximately 50–100 peptide–pVIII fusions are displayed among the approximately 3000 copies of pVIII that form the body of each phage. A restriction map of the f88-4 vector is shown in Figure 2.1C (Chapter 2), along with the nucleotide and amino acid sequences of the cassette encoding the recombinant pVIII (Fig. 2.1D).

Chapter 19, written by C. J. Noren and colleagues, covers the production and screening of pIII-displayed peptide libraries in the vector M13KE. Also described in Chapter 2, this vector is a derivative of M13mp19, to which cloning sites have been added in the regions encoding the carboxy-terminal end of the pIII signal peptide and the amino terminus of the mature pIII protein. Thus, recombinant peptide is displayed on all copies of pIII, and the physiology of the phage is more like that of wild-type Ff phages. Figure 2.1A shows the restriction map of the M13KE vector, along with the nucleotide and amino acid sequences of the signal peptide and amino terminus of mature pIII (Fig. 2.1B).

Our purpose in this section is to provide the reader with the skills for working with almost the entire range of phage-displayed peptide libraries by using examples of both pVIII and pIII display, and by using phage vectors whose physiology differs greatly. After reading this section, one should be able to apply techniques described for one phage vector to the other, and vice versa. The chapters covering pVIII-display libraries begin with general methods used in preparing and analyzing phage (Chapter 15). Most of these techniques, although designed with f88-4 phage in mind, can be applied to M13KE phage as well. In some cases, such an application would require some experimental changes, because of the differences in physiology between the two vectors. For example, the protocol for cycle sequencing phage DNA (15.9) would have to be altered to accommodate the different growth conditions required by M13KE phage; a method for preparing M13KE phage for standard dideoxy sequencing is described in Chapter 19. Whereas the general physiological differences between f88-4 and M13KE phage are described in Chapter 2, the reader should be able to glean the specific information needed for applying cycle sequencing to M13KE phage from Chapter 19. Similarly, Chapter 16 describes two methods for making libraries using single-stranded phage DNA, whereas in Chapter 19, a method for making libraries using double-stranded RF DNA is presented. Again, with a little planning by the reader, any of the three methods can be applied to either phage vector.

Protocols for panning f88-4 libraries are described in Chapter 17, and for M13KE libraries in Chapter 19. The Chapter 17 overview describes in detail how panning works and the two main panning methods. Chapter 17 describes the simultaneous panning of up to 96 samples on microplates, as well as sublibrary screening on magnetic beads. Protocols in Chapter 19 describe the screening of one or a few libraries in polystyrene dishes, in tubes, or on beads. Thus, the two chapters include most of the currently used methods for screening phage-displayed peptide libraries. The effects of using pVIII versus pIII and the behavior of f88-4 phage versus M13KE phage during panning are discussed in Chapter 2. Guided by knowledge of the differences in physiology between the two phage vectors, one should be able to modify the panning method described for one type of phage for use with the other.

Finally, methods are described in Chapters 18 and 19 for analyzing phage pools and clones for their ability to bind to a target molecule. Several of these methods, such as the phage ELISAs, appear quite similar. Also included are methods for transferring peptides from the amino terminus of pVIII (Chapter 18) or pIII (Chapter 19) to the amino terminus of the *E. coli* maltose-binding protein, which allows the production of "monovalent" peptide in an easy-to-purify form. At the risk of redundancy, we include the methods from both laboratories, because each was developed for and tested with vector systems whose physiology and means of peptide display differ greatly.

ACKNOWLEDGMENTS

I thank X. Gong, K. Brown, S. Kim, M. Rashed, and J. Mehroke for their contributions in developing the methods presented in Chapters 15–18. I also thank The Ha and Anna Day for help with figures and tables.

REFERENCES

Zacher A.N., III, Stock C.A., Golden J.W., II, and Smith G.P. 1980. A new filamentous phage cloning vector: fd-tet. *Gene* **9:** 127–140.

Zhong G., Smith G.P., Berry J., and Brunham R.C. 1994. Conformational mimicry of a chlamydial neutralization epitope on filamentous phage. *J. Biol. Chem.* **269:** 24183–24186.

15 General Phage Methods

LORI L.C. BONNYCASTLE, ALFREDO MENENDEZ, AND JAMIE K. SCOTT
Department of Molecular Biology and Biochemistry and Department of Biological Sciences, Simon Fraser University, Burnaby, B.C., Canada V5A 1S6

THIS CHAPTER COVERS GENERAL METHODS that are essential for working with filamentous phage and their DNA, and, in particular, with fd-tet phage and their derivatives. It begins with descriptions of how to grow and titer phage (Protocols 15.1 and 15.2). The two methods that follow deal with purifying phage; the first explains how to partially purify and concentrate phage with polyethylene glycol (Protocol 15.3), and the second describes how to further purify phage by CsCl equilibrium density-gradient centrifugation (Protocol 15.4). Two methods are presented for analyzing phage concentration and purity. Agarose gel electrophoresis of phage DNA (Protocol 15.5) is useful for obtaining a gross estimate of phage concentration and for confirming its molecular weight (phage readily undergo deletion mutations). The spectroscopic analysis of phage with UV light (Protocol 15.6) gives a more accurate measure of phage concentration; however, it requires purified phage, whereas the phage in culture supernatants can be analyzed directly by agarose gel electrophoresis. Protocols 15.7 and 15.8 provide methods for isolating single-stranded viral and double-stranded replicative-form (RF) DNA, respectively; the latter method can also be used to purify plasmid and phagemid DNA. Protocol 15.9 describes a very efficient means of directly sequencing the DNA from multiple phage clones that have been PEG purified. Protocol 15.10 describes how phage should be stored long-term.

CONTENTS

PROTOCOL 15.1

Preparation of Cells for Phage Infection

We make frozen stocks of each of our bacterial cell lines by picking a single colony from a minimal-medium, selective plate, growing it in minimal, selective liquid medium, and freezing the cells as a 50% (v/v) glycerol stock. Each week, we streak out fresh colonies from the frozen stock onto a minimal medium plate and use this plate as a source of cloned cells for the week.

We get the best infectivity from freshly prepared starved cells, so we use these cells for infections for which we want the lowest possible ratio of phage particles to transducing units (TUs), such as after the first round of panning. In addition, because growth in super-enriched media (e.g., Superbroth, Terrific broth) seems to induce the production of fast-growing fd-tet mutants, we avoid using such media to amplify libraries. For initial phage infection, however, growing cells to high density in Superbroth (or even LB) is an acceptable, easy alternative to making starved cells. See Figure 15.1 for growth curves of K91 cells in NZY medium.

PREPARATION OF STARVED CELLS

Materials

K91 strain of ***E. coli*** (CGSC; http://cgsc.biology.yale.edu)
sterile, capped 15 x 150 mm culture tube
sterile, capped 125-ml Erlenmeyer culture flasks
NZY medium (see Appendix 2)
sterile, 40-ml, screw-cap, round-bottom polypropylene (Oak Ridge) tubes

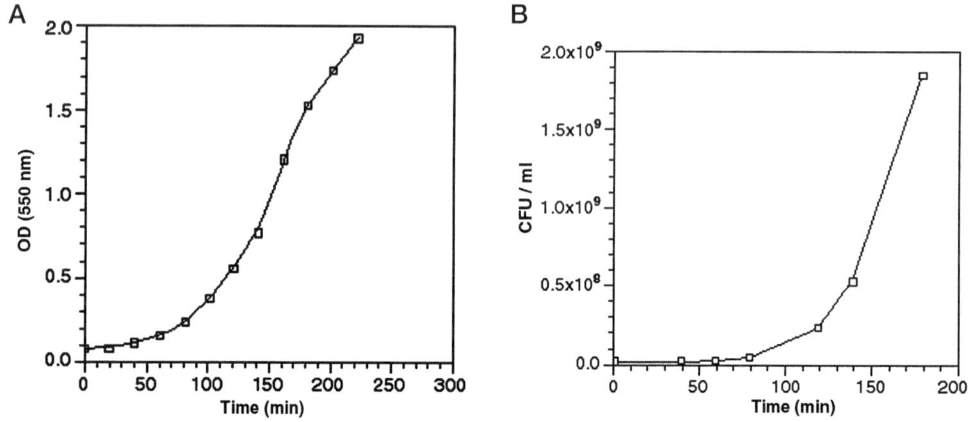

Figure 15.1. Growth curve of K91 cells in NZY medium at 37°C and shaking at 250 rpm (Protocol 15.1). The culture was inoculated with a 1/50 volume of cells from an overnight culture of cells begun from a single colony. (*A*) OD_{550} versus time in minutes. (*B*) cfu/ml versus time in minutes.

sterile, 80 mM NaCl
NAP buffer, 4°C (see Appendix 2)

CAUTION: ***E. coli* (*see Appendix 4*)**

Procedure

1. Transfer a single colony of K91 cells to a culture tube containing 2 ml of NZY. Incubate overnight at 37°C with vigorous shaking (250 rpm).
2. Inoculate a 125-ml flask containing 20 ml of NZY with 400 µl of the overnight culture. Shake vigorously (250 rpm) at 37°C until the cells have reached mid-log phase (OD_{595} = ~0.45; this takes about 1.5–2 hours).
3. Use slow shaking to 100 rpm for 10 minutes to allow the bacteria to regenerate sheared F pili. Measure the OD_{595}; it should not be over 0.65 (best at 0.55–0.65).
4. Transfer the culture to an Oak Ridge tube and centrifuge at 600*g* (2200 rpm) in a SORVALL SS34 rotor (or equivalent) for 10 minutes at room temperature or 4°C.
5. Discard the supernatant and gently resuspend the cells in 20 ml of 80 mM NaCl.
6. Transfer the mixture to a 125-ml culture flask and shake gently (100 rpm) at 37°C for 45 minutes.
7. Transfer the mixture to an Oak Ridge tube and centrifuge at 850*g* (2800 rpm) in an SS34 rotor (or equivalent) for 10 minutes at 4°C.
8. Pour off the supernatant and resuspend the cells gently in 1 ml of 4°C NAP buffer.
9. Keep the cells on ice when in use, and store on ice in the refrigerator. The cells are best used immediately, but will stay competent for phage infection for a few days. The cells are no longer competent if they remain aggregated after gentle shaking of the tube in which they are stored. The final concentration of cells should be approximately 5×10^9/ml.

HIGH-DENSITY CELLS GROWN IN LB OR SUPERBROTH

Materials

K91 strain of ***E. coli*** (CGSC; http://cgsc.biology.yale.edu)
sterile, capped 15 x 150 mm culture tube
NZY medium (see Appendix 2)
sterile, capped 125-ml Erlenmeyer culture flask
Luria broth or Superbroth (LB or SB; see Appendix 2)

CAUTION: ***E. coli* (*see Appendix 4*)**

1. Transfer a single colony of K91 cells to a culture tube containing 2 ml of NZY. Incubate overnight at 37°C with vigorous shaking (250 rpm).
2. Inoculate a 125-ml flask containing 20 ml of LB or SB with 200–250 µl of the overnight culture. Shake vigorously (250 rpm) at 37°C until the cells have reached an OD_{595} = 1.0 to 1.5.
3. Use slow shaking to 100 rpm for 10–15 minutes to allow the bacteria to regenerate sheared F pili. Measure the OD_{595}; it should not be over 2.0.
4. Place cells on ice, and use within 2–3 hours. The final concentration of cells should be approximately 5×10^9/ml.

PROTOCOL 15.2

Infecting Cells with fd-tet Phage and Titering the Phage

Infections are best performed at cell-to-phage-particle ratios of 50:1 or higher to avoid coinfection. Following infection, the tetracycline-resistance genes are induced with a low concentration of the antibiotic (0.2 µg/ml), and then the antibiotic concentration is brought up to static levels (15–20 µg/ml). For the prevention of "ghost" colonies, the agar plates contain 40 µg/ml tetracycline. All media must be cooled before adding the tetracycline. In addition, light inactivates tetracycline, so plates and media should be stored in the dark.

Materials

fd-tet-derived phage for infection (e.g., fUSE, f88-4, or fd-tet phage; G.P. Smith, University of Missouri-Columbia, Department of Biological Sciences)
TBS/gelatin (see Appendix 2)
K91 strain of *E. coli,* starved or fresh high density (see Protocol 15.1)
sterile, 1.5-ml microcentrifuge tubes and/or round-bottom, polystyrene ELISA plates
NZY medium containing 0.2 µg/ml **tetracycline** (see Appendix 2)
NZY medium containing 15 µg/ml **tetracycline**
NZY agar plates containing 40 µg/ml **tetracycline**

CAUTIONS: ***E. coli,* tetracycline (*see Appendix 4*)**

Procedure

1. Determine the concentration of the phage by agarose gel electrophoresis (Protocol 15.5) or spectrophotometry (Protocol 15.6). If phage from a cleared culture supernatant are used, be sure to kill remaining cells by heating the supernatant at 70°C for 30 minutes. Large volumes may take longer to reach 70°C, so measure the temperature, and begin timing when it reaches 70°C. Dead cells should be removed by a second clearing centrifugation step. Dilute the phage in TBS/gelatin to a maximum concentration of approximately 10^8 particles/ml. The typical yield of fd-tet phage from an overnight culture in NZY medium containing 15 µg/ml tetracycline is about 10^{12} phage particles/ml, and this can be used as a "ballpark" figure for titering (wild-type filamentous phage produce about ten times more phage).

2. Infect cells by distributing 10 µl of diluted virions ($\leq 10^6$ particles) into microcentrifuge tubes or into wells across a row of a flexible ELISA plate. Add 10 µl of starved cells or fresh, high-density cells (~5×10^7 cells). Incubate at room temperature for 10–15 minutes.

3. Induce the tetracycline-resistance genes by adding 150 µl (for microtiter plates) to 1 ml (for microcentrifuge tubes) of NZY containing 0.2 µg/ml tetracycline and mixing. Incubate at 37°C with intermittent shaking (or tapping the ELISA plate) for 30 minutes. If using 1.5-ml microcentrifuge tubes, place the tubes upside-down in a beaker and shake in an incubator. Briefly microfuge the tubes before opening, and mix the cells by pipetting.

4. Dilute the infected cells in flexible microtiter plates with NZY containing 15 µg/ml tetracycline. Mix 20 µl of cells with 180 µl of NZY+tet for 1:10 dilutions, and 2 µl in 198 µl of NZY+tet for 1:100 dilutions.

5. The concentration of tetracycline-resistant transducing units (TU) is determined by titering. Titering of an unknown sample is usually done alongside cells infected with a positive control phage whose number of particles/ml is known and that has been previously titered. A good positive control is a stock of CsCl-purified fd-tet or f88-4 phage. Cells that have been treated with phage-dilution buffer alone, in parallel with the diluted phage samples, serve as the negative control. When counting colonies, 1 colony = 1 TU.

6. For plate titering, spread 100 µl of diluted, infected cells onto NZY agar plates containing 40 µg/ml tetracycline. Spread one plate per dilution. Invert the plates and incubate overnight at 37°C.

7. For spot titering, use plates that have been dried by incubating them with their lids askew in a 37°C incubator or sterile hood for a few hours. Before spotting, mark the plates where each spot (up to 16 per plate) will go. Carefully dot 15–20 µl of diluted phage over each spot (a multichannel pipettor can be used for this). Let the agar absorb the drops (the spots should become flat), then invert the plates and incubate overnight at 30°C.

8. For colony counting (especially from spot titers), the tetracycline-resistant colonies should be small, but visible. To prevent the overgrowth of colonies, we usually (1) take the plates out of the 37°C incubator before leaving for the night, let them sit overnight at room temperature, and put them back at 37°C the following day until the colonies reach the right size; or (2) if the infections are done late at night, incubate overnight at 30°C and have someone check the colonies at the beginning of the following day. Once the colonies have reached optimal counting size, they should be stored at 4°C until they are counted. They can be stored for several weeks if the plates are sealed with Parafilm.

   Note: If colonies are to be picked and grown for phage production, the colonies should be as well-separated and fresh as possible. Phage can slowly diffuse and contaminate nearby colonies.

PROTOCOL 15.3

PEG Purification of Phage

Two phage purification procedures, adapted from Smith and Scott (1993), are described here: one for large-scale and one for small-scale phage production. The latter procedure is useful for preparing large numbers of clones for analysis by DNA sequencing and ELISAs. Because the phage can mutate, we recommend that the phage from every preparation, especially the large-scale preparations, be resequenced to confirm the displayed peptide sequence.

For fd-tet-based vectors, there are approximately 10^{12} phage particles produced per ml of culture in NZY medium, yielding approximately 10^{15} phage/liter (the number is about 10-fold higher for wild-type f1, fd, and M13 phage). At approximately 10^{14} particles/ml and higher concentrations, filamentous phage precipitate out of solution and form a gel, so keep the concentration of phage below 3×10^{13} particles/ml.

LARGE-SCALE PREPARATION OF PHAGE

Materials

transfected K91 *E. coli* colonies (see Protocol 16.6 or Appendix 3)
NZY medium containing 15 μg/ml **tetracycline** (see Appendix 2)
sterile, plugged 2-liter culture flasks
0.5-ml microcentrifuge tubes
Lysis Mix, 5x stock (see Appendix 2)
f88-4 control phage for agarose gel electrophoresis
0.8% and 1.2% agarose gels
4x GBB electrophoresis buffer (see Appendix 2)
sterile, screw-cap 250-ml polypropylene centrifuge bottles
PEG/NaCl solution (see Appendix 2)
TBS (see Appendix 2)
sterile screw-cap 40-ml polypropylene round-bottom tubes (Oak Ridge) tubes
sterile, disposable 15-ml polypropylene snap-cap tubes
sodium azide, 5% stock

CAUTIONS: ***E. coli*, tetracycline, PEG, sodium azide (*see Appendix 4*)**

Procedure

1. Pick a single transfected colony and mix it in a 1.5-ml microcentrifuge tube containing 220 μl of NZY medium. Because phage can diffuse and contaminate nearby colonies, pick well-separated colonies from a fresh plate. Inoculate each of two

2-liter flasks containing 500 ml of NZY and 15 μg/ml tetracycline with 100 μl of the diluted cells. Shake vigorously (~200 rpm) for about 20 hours at 37°C.

2. Transfer 100 μl of culture to a 0.5-ml microcentrifuge tube, and pellet the cells in a microfuge by spinning full-speed (14,000 rpm) for 2 minutes. Transfer 20 μl of the phage-containing supernatant to a fresh 0.5-ml microcentrifuge tube containing 5 μl of 5x Lysis Mix (warm the Lysis Mix to make sure the SDS is in solution); mix by pipetting. Incubate tube in a 70°C water bath for 15–20 minutes and microfuge briefly. Load a 15-μl sample onto a 0.8% agarose gel in 4x GBB, and run according to Protocol 15.5. Use a known quantity of f88.4 phage as a control (2×10^{10} phage particles = 100 ng DNA). Run this gel either before going on in the procedure (to be sure the phage yield is good; ~10^{12} particles/ml of culture supernatant), or at the end of the procedure (to determine the percent yield of PEG-purified phage).

3. Divide the culture from Step 1 among four 250-ml centrifuge bottles (~230 ml/bottle). Centrifuge at 2,400*g* for 10 minutes at 4°C. Without disturbing the cell pellet, transfer the phage-containing supernatant into clean bottles, and recentrifuge at 6,200*g* for 10 minutes at 4°C. Carefully pour the supernatant into fresh, tared, 250-ml centrifuge bottles and determine the culture volume in each bottle (1 g = 1 ml).

4. Add 0.15 volume of PEG/NaCl solution to each bottle of measured supernatant. Screw caps on tightly, and mix throughly by inverting the bottles gently approximately 100 times. Incubate the mixtures for at least 4 hours on ice or overnight at 4°C.

5. Pellet the phage by centrifuging at 6,200*g* for 40 minutes at 4°C. Discard the supernatant, being careful not to disturb the pellet. Remove the residual supernatant by briefly recentrifuging the bottles, tilting each bottle so that the pellet is opposite the remaining supernatant, and aspirating with a 1-ml pipettor.

6. Add 7.5 ml of TBS to each bottle and shake at 150 rpm in a 37°C incubator for approximately 30 minutes to resuspend the pellets. Centrifuge briefly to drive the solution to the bottom of each bottle. Transfer the solution from all four bottles to two tared Oak Ridge tubes. Rinse each bottle with another 7.5 ml of TBS and add to the Oak Ridge tubes. Each tube should now have a total volume of 30 ml. Balance the tubes with TBS and mix the phage thoroughly by inversion.

7. At this point, the phage can be heat-treated to kill remaining cells. Incubate the tubes for 30 minutes in a 70°C water bath. This step is not necessary if the phage are to be CsCl purified.

   Note: The heat-treatment step is optional and may denature some proteins displayed on phage.

8. Centrifuge the tubes at 10,100–22,700*g* for 10 minutes at 4°C to clear the supernatants. Transfer the supernatants to fresh, tared Oak Ridge tubes, and determine the volumes (1 g = 1 ml).

9. Add 0.15 volume of PEG/NaCl solution to each tube, and invert gently approximately 100 times. Allow the phage to precipitate by incubating the tubes for at least 1 hour on ice. A heavy precipitate should appear.

10. Collect the precipitated phage by centrifuging at 10,100*g* for 40 minutes at 4°C. Remove the supernatant as in Step 5 above.

11. Add 10 ml of TBS to each tube. Resuspend the phage pellet by gently vortexing, and then allowing the pellet to soften at room temperature for about 1 hour. Vortex again, and briefly centrifuge to drive the solution down. (If the phage are to be further purified on a CsCl density-gradient, add only 5 ml of TBS to each tube, resuspend the phage, and combine the two supernatants into a single Oak Ridge tube.)

12. Clear the supernatants by centrifuging the tubes at 10,100–22,700*g* for 10 minutes at room temperature or 4°C. Pour the cleared supernatant from each tube into a 15-ml polypropylene snap-cap tube and store at 4°C in the dark.

13. To determine the concentration and yield of phage particles, treat an aliquot of phage with 5x Lysis Mix (as in Step 2). Run 1-, 5-, and 10-μl samples on a 1.2% agarose gel in 4x GBB, using as a standard a known amount of phage treated in the same way (e.g., 1 x 10^{10} and 3 x 10^{10} particles; see Protocol 15.5). Include on the gel a sample from the original culture supernatant (Step 2) to calculate the percent yield. Electrophoretic analysis is also important for demonstrating that only one DNA species is present; this is especially important for fd-tet derivatives, which can delete the tetracycline genes, generating phage with approximately 6-kb genomes. The concentration of phage particles can be more accurately assessed by spectrophotometric analysis (Protocol 15.6); however, this is better done with CsCl-purified phage (Protocol 15.4). The infectious properties of the phage (TU/ml) can be analyzed by titering (Protocol 15.2); the infectivity of fd-tet-derivatives is about 20 particles/TU, whereas that of wild-type derivatives is about 1 particle/pfu.

14. The final concentration of phage (if they are not to be CsCl-purified) should not exceed approximately 3 x 10^{13}/ml, so once the phage concentration is known, it should be adjusted accordingly with TBS. To impede cell growth, the solution can be adjusted to a final concentration of 0.02% (w/v) sodium azide or 20 mM Na_2EDTA. The phage can be stored long term in 50% (v/v) sterile glycerol at –18°C (see Protocol 15.10).

SMALL-SCALE PREPARATION OF PHAGE

Materials

transfected K91 *E. coli* cells
sterile 15 x 150 mm capped, glass culture tubes
NZY medium and 15 μg/ml **tetracycline** (see Appendix 2)

sterile 1.5-ml microcentrifuge tubes
PEG/NaCl solution (see Appendix 2)
TBS (see Appendix 2)
sodium azide, 5% stock

CAUTIONS: ***E. coli*, tetracycline, PEG, sodium azide (*see Appendix 4*)**

Procedure

1. Inoculate tubes containing 2 ml of NZY and 15 μg/ml tetracycline with well-separated colonies from a freshly grown transfection. Shake vigorously at 250 rpm overnight at 37°C.
2. Transfer the cultures to 1.5-ml microcentrifuge tubes (fill the tubes), and centrifuge at full speed (14,000 rpm) in a microfuge for 5 minutes at room temperature or 4°C.
3. Transfer 1.2 ml of the culture supernatants to fresh microcentrifuge tubes containing 180 μl (0.15 volume) of PEG/NaCl solution. Mix thoroughly by vortexing, and incubate the tubes for at least 1 hour on ice.
4. Centrifuge the tubes for 40 minutes at full speed (14,000 rpm) at 4°C. A pellet should be visible. Aspirate and discard the supernatants while avoiding the pellet. Centrifuge the tubes briefly, and aspirate the remaining supernatant.
5. Resuspend each pellet in 100 μl of TBS, or 100 μl of TBS containing 0.02% sodium azide, or 100 μl of TBS containing 20 mM EDTA. Gently vortex and mix by pipet. To kill remaining cells, briefly centrifuge the tubes and place them in a 70°C water bath for 20 minutes. Centrifuge the tubes at full speed (14,000 rpm) for 10 minutes at 4°C or room temperature.

   Note: The heat treatment may denature some proteins displayed on phage.
6. Transfer the supernatants to fresh microcentrifuge tubes, and store away from light for weeks to months at 4°C. For the analysis of the phage, their infectivity and their yields, as well as their long-term storage, see Steps 13 and 14 above.

PROTOCOL 15.4

Phage Purification on CsCl Gradients

This method, adapted from Smith and Scott (1993), is particularly effective as a means of removing residual PEG, nucleases, proteases, and other contaminants from PEG-purified phage. DNA-grade CsCl is not necessary for phage purification; thus, we use the less expensive type. We especially recommend that phage libraries be CsCl-purified before long-term storage (see Protocol 15.10).

Materials

same materials as for Protocol 15.3
CsCl, optical grade (Sigma, Cat. # C 3139)
sterile 50-ml glass or plastic beakers
TBS (see Appendix 2)
12-ml polyallomer tubes for a Beckman SW40 or 70.1 Ti rotor
sterile polyethylene transfer pipets or sterile glass transfer pipets (short-tip)
26-ml, screw-cap polycarbonate centrifuge tubes for a Beckman 60Ti rotor
sterile, 40-ml, screw-cap polypropylene (Oak Ridge) tubes
sterile, disposable, 15-ml polypropylene snap-cap tubes
5% **sodium azide** or 250 mM EDTA
TBS/gelatin (see Appendix 2)

CAUTIONS: ***E. coli*, tetracycline, PEG, CsCl, sodium azide (*see Appendix 4*)**

Procedure

1. Follow steps 1–12 of Protocol 15.3 (Large-scale preparation of phage); the final volume of each phage preparation should be 10 ml.
2. For each phage preparation, weigh out 4.83 g of CsCl into a *tared* 50-ml beaker. Tare the beaker again, and add the resuspended phage from Step 12 of Protocol 15.3. Add TBS to a final weight of 10.75 g (over and above the weight of the vessel and CsCl; phage + TBS = 10.75 g). This should give 12 ml of a 31% (w/w) solution of CsCl with a density of 1.3 g/ml. The density of the solution can be checked by weighing 1 ml in a tared beaker or plastic cup, and then returning it to the beaker it came from (be sure to first check that the pipet on that setting weighs 1 ml of water at 1 g). If necessary, adjust the density to 1.3 g/ml with CsCl or buffer.
3. Transfer the volume of each beaker into a polyallomer tube. Be sure that the tubes are filled to the top (as they can collapse during centrifugation), and if necessary, add extra volume with a 31% (w/w) solution of CsCl in TBS. Balance the tubes

by transferring solution from one to another, or by adding more CsCl solution to one tube. Place the tubes in a SW40 swinging-bucket rotor and centrifuge at 37,000 rpm for 48 hours at 5°C. For the 70.1 Ti rotor, spin at 58,000 rpm for 20 hours.

4. Carefully remove the tubes from the rotor, place in a rack, and set them up, one at a time, in a clamp stand. Illuminate the clamped tube from the top with a strong, visible light source (such as a halogen desk lamp). There will be two bands toward the top of the tube. The phage band, which will be faint, bluish, and homogeneous (smoky looking), should be just visible above a narrow, stringy, flocculent, opaque white band (which is probably PEG). In a good phage preparation, the phage band is about 5 mm in width, and its density is approximately 1.33 g/ml.

5. Attach a sterile pipet tip to an aspirator pump and adjust the aspirator to a moderate speed. Hold the pipet tip at the meniscus, and aspirate the fluid overlying the phage band to within 2 mm of the upper edge, being careful not to disturb the phage band. Withdraw the phage band with a polyethylene transfer pipet (preferably sterile), or, better yet, a sterile, glass transfer pipet attached to a peristaltic pump. The phage band will be viscous; try to avoid the flocculent band that lies underneath. If using a quickseal polyallomer tube, remove the band with a syringe, as for CsCl-purified plasmid (see Sambrook et al. 1989).

6. Transfer the extracted phage bands to a 26-ml, screw-cap polycarbonate centrifuge tube for the Beckman 60 Ti rotor. 4–6 phage bands from the same clone can be pooled in a single bottle. Fill the tube to the shoulder with TBS, close the cap firmly, and invert repeatedly to mix. For centrifugation, balance against another tube filled with water.

7. Centrifuge the tubes in a Beckman 60 Ti fixed-angle rotor at 50,000 rpm for 4 hours at 4°C to pellet the phage. Pour off and discard the supernatant, recentrifuge the pellet briefly at a low speed on a tabletop centrifuge, and discard the remaining supernatant, with the pellet pointed away from the liquid.

8. Resuspend the pellet in 10 ml of TBS; vortex gently, centrifuge briefly in a tabletop centrifuge to drive the solution down, and allow the pellet to soften overnight at 4°C. Vortex to dissolve the pellet. Heat the phage for 30 minutes at 70°C to kill contaminants without losing titer.

   Note: Heat treatment is optional, and may denature some proteins displayed on the phage, especially complex ones like Fabs.

9. Top the bottle with TBS, recentrifuge to pellet the phage, and remove the supernatant as in Step 7.

   Note: This step is optional, giving somewhat purer phage.

10. Resuspend the pellet in TBS as in Step 9, using 12 ml of TBS per liter-equivalent of starting culture; this gives an anticipated concentration of 3×10^{13} virions/ml. Transfer the phage to a sterile Oak Ridge tube and centrifuge at 6,500*g* for 10 minutes.

11. Transfer the phage-containing supernatant to a 15-ml polypropylene, snap-cap tube. At this point, sodium azide can be added as a preservative to the cooled phage to a final concentration of 0.02% (w/v). Alternatively, Na_2EDTA can be used at a final concentration of 20 mM.

12. Measure the concentration of phage particles spectrophotometrically (Protocol 15.6), and/or by agarose gel electrophoresis (Protocol 15.5).

13. Dilute the phage 10^{-7}, 10^{-8}, and 10^{-9} in TBS/gelatin and titer, as in Protocol 15.2. Include the proper positive and negative controls. A good infective titer (TU/ml) is approximately 5% of the concentration of physical particles (virions/ml).

14. Store the phage at 4°C away from light, or in 50% glycerol at –18°C for long-term storage (see Protocol 15.10). Under these conditions, titers are stable for at least several years.

PROTOCOL 15.5

Agarose Gel Electrophoresis for the Analysis of Viral and RF DNA

We analyze phage by agarose gel electrophoresis for several reasons. It is the easiest way to determine the concentration of phage particles produced by a given culture. It is also a way to confirm that only a genome of the correct size is present; this is important, because fd-tet-derived vectors can delete the tetracycline gene complex and mutate into a fast-growing wild-type-like phage. Agarose gel electrophoresis is also required after the syntheses involved in phage library production and in making phage mutants (see Protocols 16.1–4).

Materials

fd-tet phage (in cleared supernatant, TBS, PBS, etc.)
5x Lysis Mix (see Appendix 2)
agarose (Gibco BRL; UltraPure, Cat. # 15510-027)
1x TBE electrophoresis buffer (see Appendix 2)
4x GBB electrophoresis buffer (see Appendix 2)
5x electrophoresis loading buffer (see Appendix 2)
0.5 mg/ml **ethidium bromide** stock solution

CAUTION: **ethidium bromide (*see Appendix 4*)**

Discriminating between the linear and covalently closed circular (ccc) forms of viral DNA

Thoroughly mix 4 volumes of suspended virions equivalent to 50–100 ng of viral DNA with 1 volume of 5x Lysis Mix (100 ng DNA = 2×10^{10} virions of fd-tet or 2.8×10^{10} virions of fd or M13). Incubate for 20 minutes in a 70°C water bath. Centrifuge briefly and load on a 1.2% agarose gel. Run samples in 1x TBE buffer (but be aware that with TBE, smearing is more likely to occur with overloading). As a marker, use viral DNA, which can be obtained from culture supernatants from virally infected cells that have been grown to mid-stationary phase or later. To run purified phage DNA on the gel, dilute in 5x loading buffer before loading. As shown in Figure 15.2, ccc viral DNA will run slowly compared to the linear form, whose mobility in 1.2% gels is about the same as that of xylene cyanol. Be aware that relative mobilities depend on the type of gel and running buffer.

Discriminating between the linear, open-circular (oc), and ccc forms of replicative form (RF) DNA

Run 50–100 ng samples of purified RF DNA diluted in 5x loading buffer (for example, see Protocol 15.8) on a 0.8% agarose gel in 4x GBB. The ccc RF DNA molecules

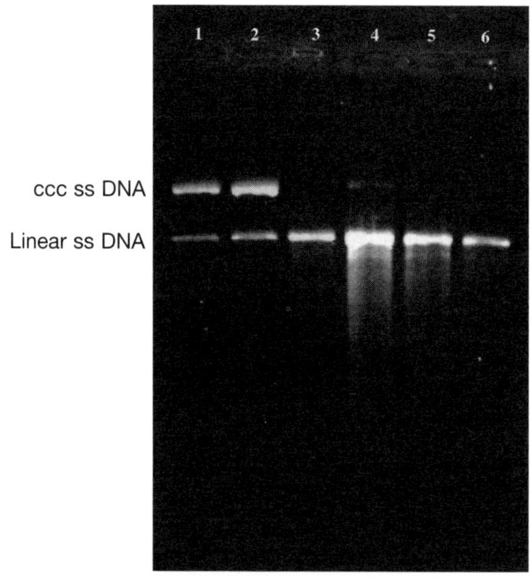

Figure 15.2. Analysis of ccc and linear viral (ss) DNA by 0.8% agarose gel electrophoresis in 1x TBE buffer (Protocol 15.5). This gel analyzes the cleavage of viral DNA as described in Protocol 16.2. Lanes: (*1*) 50 ng of viral DNA; (2) 100 ng of viral DNA hybridized to the *Hin*dIII oligonucleotide; (3) 100 ng of the lane 2 sample after cleavage by *Hin*dIII; (*4*) 1 µl of *Hin*dIII-cleaved viral DNA:*Hin*dIII-oligonucleotide duplex after oligonucleotide removal; (*5*) a 0.5-µl equivalent of the lane *4* sample; (*6*) a 0.2-µl equivalent of the lane *4* sample.

will be mixed with regard to the number of supercoils, with each number conferring a slightly different mobility on the DNA molecule. Their mobilities can be made identical by the addition of 1.0 µg/ml ethidium bromide to the gel and the loading and running buffers. By intercalating as much ethidium bromide as possible, the RF molecules supercoil to their maximal extent, and thus reach the same effective volume, which allows them to comigrate. In addition to molecular weight markers, include samples of uncleaved (ccc) and cleaved (linear) RF DNA (the uncleaved DNA prep will probably contain some linear and oc DNA, too). As shown in Figure 15.3, the oc form runs just behind the linear form (at ~11 kbp), whereas the ccc form(s) will run far ahead of both (at ~5 kbp); ccc single-stranded viral DNA runs even farther ahead (at ~2 kbp).

Most likely, both oc and ccc double-stranded DNAs form transducing units (TUs) very efficiently by electrotransformation of *E. coli* cells, so both forms should be considered when estimating the amount of "electrotransformable" DNA present in a sample.

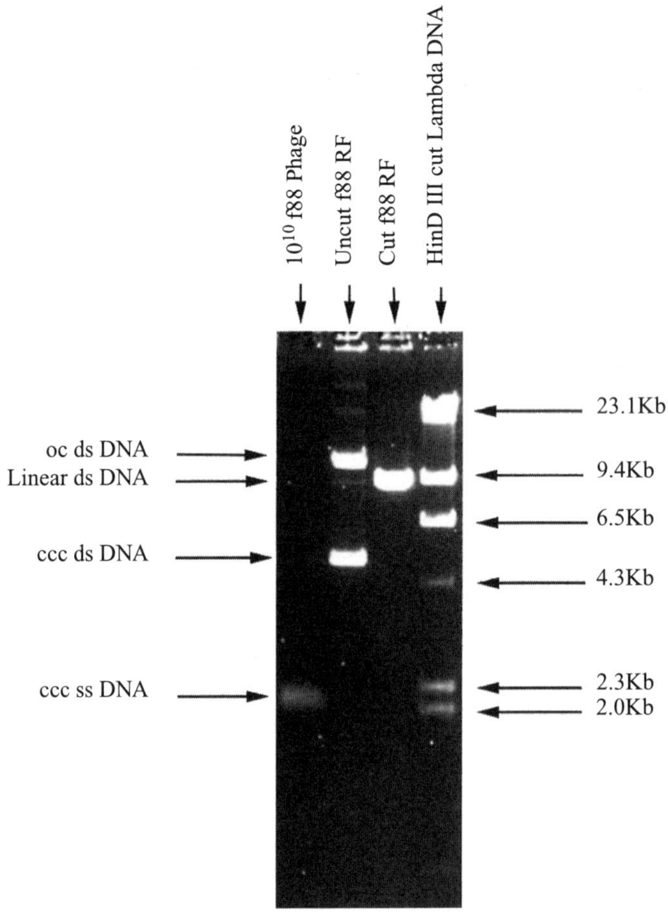

Figure 15.3. Analysis of linear, oc, and ccc RF DNA and ccc viral (ss) DNA by 0.8% agarose gel electrophoresis in 4× GBB buffer (Protocol 15.5).

PROTOCOL 15.6

Spectrophotometric Quantitation of Phage

Phage concentration (in viral particles per volume) is best measured spectroscopically with CsCl-purified phage (Day 1969; Smith and Scott 1993). Phage concentration can be determined, however, with phage that have been purified by two PEG precipitations. It is best, in this latter case, to verify the concentration by agarose gel analysis as well (see Protocol 15.5). The equation in Step 3 below is based on the nucleotide content of the phage (330 daltons per base) and a molar extinction coefficient of $1.006 \times 10^4\ \mathrm{M}^{-1}\mathrm{cm}^{-1}$. The genome sizes of the fd-tet-based phage vectors are very similar (fd-tet = 9183 bases; fUSE5 = 9206 bases; f88-4 = 9235 bases; compared with M13 = 6407 bases). Thus, for example, a 1×10^{12} particle/ml solution of fd-tet derived phage will give an absorbance at 269 nm (A_{269}) = 0.153.

Materials

CsCl- or PEG-purified phage (see Protocols 15.4 and 15.3)
TBS (see Appendix 2)
clean quartz cuvettes for samples and blank (preferably acid washed)
sterile microcentrifuge tubes

CAUTIONS: **CsCl, PEG (*see Appendix 4*)**

Procedure

1. Dilute phage in TBS to a concentration of approximately 10^{12} particles/ml. Scan samples from 240 to 320 nm with a UV spectrophotometer. As shown in Figure 15.4, there should be a broad peak from 260 nm to 280 nm, with a slight maximum at 269 nm.
2. Measure the A_{269} and A_{320}, and calculate the "adjusted" A_{269} (A_{269}-baseline):

   adjusted A_{269} = measured A_{269} – measured A_{320}
3. Calculate the virion concentration (in phage particles/ml), based on the equation below.

$$\text{phage particles per ml} = \frac{(\text{adjusted } A_{269}) \times (6 \times 10^{16})}{(\text{the number of nucleotides in the phage genome})}$$

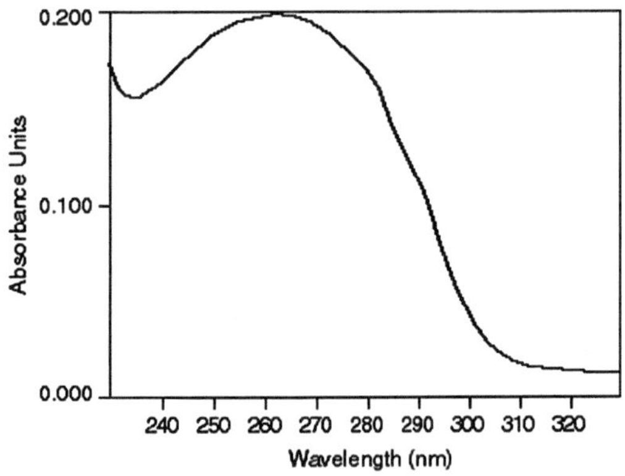

Figure 15.4. UV scan from 240 to 320 nm of f88-4 phage (Protocol 15.6). The phage were purified by two consecutive PEG precipitations (Protocol 15.3) and diluted 1:35 in 1x TE. At this dilution, the phage gave an absorbance at 269 nm of 0.198 absorbance units (AU); this equates with a phage concentration in the undiluted phage stock of 4.6×10^{13} particles/ml. This phage concentration correlated well with that estimated from an agarose gel electrophoresis experiment on the same phage stock (Protocol 15.5).

PROTOCOL 15.7

Preparation of ss-Phage DNA

Single-stranded phage DNA is used as a template for most of our cloning procedures (see Protocols 16.2–4). The amount of ccc ss DNA in 1 ml of NZY culture containing 1×10^{12} f88-4 phage particles (at 9300 bases/genome and 330 daltons per nucleotide) is approximately 5 μg. Expect a yield of about 80% from this method. Be sure to use fresh phenol and freshly grown phage (or even better, use freshly grown, CsCl-purified phage) to ensure that the resulting DNA is mostly ccc.

Materials

purified phage (see Protocols 15.3 and 15.4)
sterile, 15-ml, screw-cap, conical polypropylene tubes (Falcon)
sterile microcentrifuge tubes
TBS (see Appendix 2)
buffer-saturated **phenol**
phenol/chloroform/isoamyl alcohol 25:24:1 (v/v/v)
chloroform/isoamyl alcohol 24:1 (v/v)
3 M sodium acetate, pH 6.0
absolute **ethanol**
70% (v/v) **ethanol**, –18°C
0.5x TE (see Appendix 2)

CAUTIONS: **phenol, chloroform, isoamyl alcohol, ethanol** (*see Appendix 4*)

Procedure

1. Start with approximately 6 ml of purified phage in a 15-ml conical tube. Add one volume of phenol and vortex vigorously. Centrifuge at 1000*g* in a tabletop centrifuge for 15 minutes at room temperature. Transfer the aqueous layer, avoiding the interphase, to a fresh 15-ml tube. Back-extract the interphase layer by transferring it to a 1.5-ml microcentrifuge tube, adding TBS to expand the aqueous volume, and vortexing, then centrifuging the mixture in a microfuge for 1 minute at full speed. Add the aqueous layer from the back-extraction to that from the original extraction.

2. Add one volume of phenol/chloroform to the aqueous solution,vortex vigorously, and centrifuge as above for 15 minutes. Transfer the aqueous layer to a fresh 15-ml tube, back-extract the interphase if it is white, and combine the aqueous fractions.

3. Add one volume of chloroform to the aqueous fraction, vortex vigorously, and centrifuge as above for 15 minutes. Transfer the aqueous layer to a fresh, 15-ml tube, and if droplets of aqueous layer are still apparent, transfer this to a microcentrifuge tube, microfuge, and remove the remaining aqueous layer. Repeat the chloroform extraction, transferring the resulting aqueous layers to a tared Oak Ridge tube.

4. Determine the volume of the aqueous fraction and add 1/10 volume of 3 M sodium acetate. Mix thoroughly, add 2 volumes of absolute ethanol, and mix again. Incubate at 4°C for at least 4 hours. Centrifuge at 7150*g* at 4°C for 30 minutes.

5. Pour off the supernatant and spin the tube briefly to bring down excess ethanol. Aspirate the remainder with a pipet, avoiding the pellet. Wash the pellet by carefully pipetting approximately 1 ml of 70% ethanol down the side of the tube. Return the tube to the rotor with the pellet against the outermost wall of the rotor well, centrifuge as above for 5 minutes, and carefully aspirate the ethanol. Centrifuge briefly, and aspirate the last of the ethanol. Allow the pellet to air-dry, then resuspend it in 0.5 ml of 0.5x TE.

6. Run several dilutions of the DNA sample on a 1.2% agarose gel in 1x TBE buffer to check the quality and gauge the concentration of DNA (i.e., the fraction of the total sample that runs as ccc ssDNA, rather than linear ssDNA; see Protocol 15.5 and Fig. 15.2). Scan the sample from 240 to 310 nm to determine the concentration of DNA (1 absorbance unit at 260 nm = 33 µg/ml of single-stranded DNA). Look for a shoulder at 270 nm, a sign of residual phenol contamination. If this occurs, split the DNA between two 1.5-ml microcentrifuge tubes, bring the volume to 450 µl with 0.5x TE, and re-extract the DNA in chloroform/isoamyl alcohol twice. Precipitate the DNA in 1/10 volume of sodium acetate and 2 volumes of absolute ethanol. Wash the pellet with 70% ethanol. Dissolve the pellet in each tube in 250 µl of 0.5x TE, and combine the DNA into one tube.

7. Store the DNA at 4°C over the short term, and frozen at –18°C over the long term.

PROTOCOL 15.8

Purification of RF DNA by PEG Precipitation

This method is adapted from Sambrook et al. (1989). For an even purer grade of RF DNA, exclude Step 15 (digestion of the RNA with RNase A) and further purify the DNA on a CsCl equilibrium gradient (see Smith and Scott 1993). Alternatively, RF DNA can be made by the usual mini- and midi-prep procedures that are used for the preparation of plasmid DNA; a number of kits are also available.

Materials

transfected cells (see Appendix 3 or Protocol 16.6)
TBS (see Appendix 2)
sterile, 1.5-ml microcentrifuge tubes
sterile, plugged, 2-liter culture flasks
LB medium containing 15 μg/ml **tetracycline** (see Appendix 2)
250- and 500-ml, screw-cap, polycarbonate centrifuge bottles
50 mM Na_2EDTA, pH 8.0
buffered glucose (see Appendix 2)
0.2 N **NaOH**/1% (w/w) **SDS**, freshly made
ice-cold 5 M potassium acetate
cheesecloth
250-ml beaker
40-ml screw-cap, polypropylene, round-bottom (Oak Ridge) tubes
absolute **ethanol**
70% (v/v) **ethanol**, –18°C
1x TE (see Appendix 2)
ice-cold 5 M **LiCl**
isopropanol
RNase A (and, if desired, RNase T1)
sterile 1.6M NaCl/13% **PEG** 8000 (see Appendix 2)
buffered **phenol**
phenol/chloroform/isoamyl alcohol 25:24:1 (v/v/v)
chloroform/isoamyl alcohol 24:1 (v/v)
7.5 M **ammonium acetate**
3 M sodium acetate, pH 6.0

CAUTIONS: tetracycline, NaOH, SDS, ethanol, LiCl, isopropanol, PEG, phenol, chloroform, isoamyl alcohol, ammonium acetate (*see Appendix 4*)

Procedure

1. Pick a freshly grown, well-separated colony of transfected cells with a loop or sterile toothpick and mix the cells in a 1.5-ml microcentrifuge tube containing 150 μl of TBS. Inoculate each of two 2-liter culture flasks containing 500 ml of LB medium and 15 μg/ml tetracycline with 50 μl of the mixture. Shake vigorously (at least 200 rpm) overnight at 37°C.

2. Transfer the cells to 500-ml centrifuge bottles and spin at 2750*g* for 15 minutes at 4°C. Pour off and discard the supernatant, spin again briefly, and aspirate the remaining liquid.

3. Resuspend the cells from the two bottles in a total volume of 150 ml of 50 mM EDTA, and transfer to a single 250-ml centrifuge bottle. Pellet the cells and remove the supernatant as in Step 2 above. At this point the cell paste can be frozen at –18°C.

4. Resuspend the cells in 40 ml of buffered glucose with vigorous shaking or vortexing. Add 80 ml of freshly made 0.2 N NaOH/1% SDS, and mix by gentle inversion. Incubate the suspension for 15 minutes on ice.

5. Add 60 ml of ice-cold 5 M potassium acetate to the suspension, and mix by gentle inversion. Incubate for 10 minutes on ice.

6. Centrifuge at 6,200*g* for 15 minutes at 4°C. A gelatinous pellet will be present at the bottom of the bottle, and a stringy, viscous precipitate will pervade the supernatant.

7. Remove most of the precipitate by pouring the supernatant through 2–3 layers of cheesecloth into a beaker. The supernatant will appear cloudy. Divide the supernatant between several Oak Ridge tubes. Centrifuge the tubes at 22,700*g* for 15 minutes at 4°C. The supernatant should be clear, but it may have a slightly green tinge.

8. Pour the supernatants from all of the tubes into a single, tared 250-ml centrifuge bottle. Determine the weight (and therefore the volume, 1 g ≅ 1 ml) and divide the supernatant among three 250-ml centrifuge bottles.

9. Remove 50 μl of the supernatant for electrophoresis samples. Determine the yield of DNA by running a 15-μl sample on a 0.8% agarose gel in 4x GBB; use 25-ng and 50-ng samples of previously prepared RF DNA for comparison, as well as a molecular weight marker (see Protocol 15.5). A band should be apparent at approximately 5 kbp (representing the 9.2-kbp ccc RF molecule), with fainter bands at approximately 11 kbp (9.2-kbp oc RF) and 2 kbp (9.2-kb ccc, ssDNA). The bands will usually be superimposed on a smear of residual DNA comprising the bacterial chromosome, and there will be a heavy RNA band near the bromophenol blue dye front.

10. To each 250-ml bottle from Step 8, add 150 ml of absolute ethanol and incubate the mixtures for at least 2 hours (or overnight) at 4°C.

11. Centrifuge the bottles at 6,200*g* for 20 minutes at 4°C. Pour off the supernatant, recentrifuge briefly, and aspirate the residual supernatant. Wash each pellet by carefully pouring 20 ml of –18°C, 70% ethanol down the side of the centrifuge bottle. Centrifuge at 6,200*g* for 10 minutes at 4°C. Pour off the supernatant, recentrifuge briefly, and aspirate the residual supernatant. Dry the pellets briefly under vacuum or air-dry.

12. Resuspend the pellets from the three bottles in a total volume of 6 ml of 1x TE, and transfer to a single, tared Oak Ridge tube. Add 1 volume of ice-cold 5 M LiCl, mix well, and let stand for 5 minutes at room temperature.

13. Centrifuge the tube at 10,100*g* for 10 minutes at 4°C. Transfer the supernatant to a fresh, tared Oak Ridge tube. Add 1 volume of isopropanol and mix well by inversion. Let stand for 10 minutes at room temperature.

14. Centrifuge at 10,100*g* for 10 minutes at room temperature. Pour off and discard the supernatant, recentrifuge briefly, and aspirate the residual supernatant. Wash the pellet with 70% ethanol, recentrifuge, and discard the supernatant. Recentrifuge briefly, aspirate the residual supernatant, and let the pellet air-dry.

15. Resuspend the pellet in 500 μl of 1x TE and transfer to a 1.5-ml microcentrifuge tube. Add RNase A to 20 μg/ml (the same amount of RNAse T1 may also be added), and mix gently. Incubate the mixture for 30 minutes at room temperature or 37°C.

16. Centrifuge the tube briefly and add 500 μl of 1.6 M NaCl/13% PEG 8000. Mix well and let stand for 15 minutes at room temperature. Centrifuge in a microfuge at full speed (14,000 rpm) for 5 minutes at 4°C. Aspirate the supernatant, recentrifuge, and aspirate the residual supernatant.

17. Dissolve the pellet in 400 μl of 1x TE and extract once with phenol as described in Appendix 3. Back-extract the interphase, extract once with phenol/chloroform/isoamyl alcohol, and twice with chloroform/isoamyl alcohol.

18. Divide the aqueous phase between two 1.5-ml microcentrifuge tubes. Add 0.5 volume of 7.5 M ammonium acetate to each tube and mix thoroughly. Add 2 volumes of absolute ethanol, and let stand at 4°C for at least 1 hour. Centrifuge in a microfuge at full speed (14,000 rpm) for 15 minutes at 4°C. Aspirate the supernatant, recentrifuge briefly, and aspirate the residual supernatant. Wash the pellet with 70% ethanol, recentrifuge briefly, and aspirate the supernatant. Let the pellet air-dry for 30 minutes.

19. Resuspend the pellet in 500 μl of 1x TE, and store at –18°C. Because ammonium ions can inhibit some polymerases, if the DNA is to be used with a polymerase, consider reprecipitating the DNA in sodium acetate to remove all traces of ammonium ions: Precipitate the DNA in 1/10 volume of 3 M sodium acetate and 2 volumes of absolute ethanol, and wash the pellet with –18°C 70% ethanol, as described in Appendix 3. Dissolve the DNA in 500 μl of 1x TE.

20. Analyze a sample of the DNA on a 0.8% agarose gel to determine yield and quality, as in Step 9. Include 15 μl of the sample taken in Step 9 to assess the purity, quality, and yield of DNA. Analyze and quantitate the DNA by spectrophotometric scanning from 240 to 320 nm (1 absorbance unit at 260 nm = 50 μg of plasmid DNA per ml; see Appendix 3). Look for a shoulder at 270 nm, a sign of residual phenol contamination. If this occurs, re-extract the DNA twice with chloroform/isoamyl alcohol, and precipitate with ethanol and sodium acetate, as described in Appendix 3.

PROTOCOL 15.9

Cycle Sequencing of Phage DNA Using ^{33}P ddNTPs

Cycle DNA sequencing is an efficient and convenient method for the analysis of large numbers of selected phage clones. It is based on the polymerase chain reaction (PCR) and thus requires very small amounts of template DNA. The protocol described below follows the method of Bonnycastle (1998); it describes the direct use of polyethylene-glycol-purified phage for sequencing without additional steps for DNA purification. The phage used for sequencing usually come from panning experiments in which clones from a given phage pool are selected from well-separated colonies on NZY-agar plates containing 40 μg/ml tetracycline. The phage are then grown overnight in 2-ml cultures and PEG-purified (see Protocol 15.3, Small-scale preparation of phage). This protocol uses the Thermo Sequenase kit from Amersham Pharmacia Biotech and a 96-tube PCR format, allowing 24 phage clones to be sequenced simultaneously.

This procedure works well with 10^9–10^{10} phage per reaction, using programs of 50–60 cycles, depending on the starting amount of phage. As the volume of the PCRs is small, condensation can occur during cycling and greatly affect the reactions. A heated-lid PCR machine is recommended for the prevention of condensation, rather than oil overlays on the reactions. Because PCR is involved, extra care should be observed to avoid sample cross-contamination; hence filtered pipet tips are recommended.

As a consequence of the high glycerol content of the enzyme supplied in the kit, the reactions must be run in a glycerol-tolerant gel and buffer system rather than in 1x TBE, which is usually used for DNA sequencing gels. In addition, because the radioisotope used is ^{33}P, the gel must be fixed and dried before exposure.

PREPARING THE PCR SEQUENCING REACTIONS

Materials

PEG-precipitated phage resuspended in 1x TBS at approximately 5×10^{12} to 2×10^{13} particles/ml

sequencing primer (for f88-4, the primer for sequencing the region encoding the amino terminus of recombinant pVIII is: 5′-CTG.AAG.AGA.GTC.AAA.AGC-3′)

filtered pipet tips (10 and 20 μl)

GeNunc Modules (Nunc, Cat. # 2-32298) or similar; the volume of the tubes in the strips is 25 μl, making them excellent for work with very small samples.

Thermo Sequenase kit (Amersham Pharmacia Biotech, Cat. # US79750). It includes the four ^{33}**P**-labeled dideoxynucleotides ([^{33}**P**]-ddNTPs)

0.2-ml strip tubes and caps for PCR (VWR, Cat. # 20170-004 or similar), autoclaved (in a clean-steam autoclave, see Appendix 3) and then oven-baked (at 105°C for 4 hours)

PCR machine

CAUTION: ^{33}P *(see Appendix 4)*

1. Use filtered pipet tips to place 1 µl of each template (0.5 x 10^{10} to 2.0 x 10^{10} phage) at the bottom of the tubes of GeNunc Modules. Be sure to place the pipet tip directly on the bottom of the tube and gently dispense the template, changing tips between samples. Aliquot into 4 tubes per template.

2. In an area prepared for work with radioactive materials, prepare the Termination Master Mixes. Four different termination mixes should be prepared (one for each labeled ^{33}P-ddNTP). Handle ^{33}P-ddNTPs with filtered pipet tips. As an example, the Termination Master Mix for one reaction-equivalent of ddGTP-terminated sequencing is shown below:

	1 Reaction
dGTP labeling mix	2.0 µl
^{33}P-ddGTP	0.5 µl
	2.5 µl

Label four 0.2-ml PCR tubes (A,C,G,T) for each clone to be sequenced. Dispense 2.5 µl of each Termination Master Mix to the bottom of the appropriate tubes by placing the pipet tip directly on the bottom of the tube and gently dispensing (e.g., add ddATP-Termination Master Mix to all of the "A"-labeled tubes). Place the tubes on ice.

3. Prepare one Reaction Master Mix using the amounts indicated below. Make enough mix for four reactions per clone. Prepare a different Reaction Master Mix for each sequencing primer used. Add the Thermo Sequenase last, mix thoroughly, centrifuge briefly if necessary, and go immediately to Step 4.

	1 Reaction
10x Sequenase Reaction buffer	2 µl
primer (2 pmole/µl)	1 µl
water	14 µl
Thermo Sequenase (4 Units/µl)	2 µl
	19 µl

4. Add 19 µl of Reaction Master Mix (from Step 3) to each of the phage template samples in the GeNunc Modules (from Step 1) by pipetting the liquid down the wall of each tube. Mix by pipetting up and down.

5. Aliquot 4.5 µl of each mix from Step 4 (Reaction Master Mix plus phage template) into each of the corresponding four termination tubes (labeled A, C, G, T) from Step 2. Pipet each sample down the side of the tube, changing tips between phage samples, but not between termination tubes for a given phage sample.

6. Run the PCR according to the following thermocycler conditions:

   Step 1. Incubate at 95°C for 3 minutes to lyse the phage (only in the first cycle)
   2. 95°C for 30 seconds
   3. 53°C for 30 seconds
   4. 72°C for 60 seconds
   5. Repeat Steps 2–4 for 50–60 cycles, depending on the amount of phage template.
   6. Incubate at 4°C until ready to load the sequencing gel.

PREPARING GLYCEROL-TOLERANT SEQUENCING GELS

Materials

DNA sequencing gel electrophoresis system
397 x 330 mm (short) and 422 x 330 mm (long) glass plates
0.4-mm spacers and sharkstooth combs
gel-binding solution (see Appendix 2)
gel-repellent solution (see Appendix 2)
glycerol-tolerant gel solution (see Appendix 2)

CAUTIONS: **ethanol, acetic acid, γ-methacryloxypropyl-trimethoxysilane, taurine, acrylamide, bisacrylamide (*see Appendix 4*)**

Procedure

Many models of sequencing gel electrophoresis apparatus are available from several manufacturers. The conditions for gel casting and running should be adjusted to each specific case, depending on the brand and model. The electrophoresis system we use is the Model S2 from Gibco BRL (Cat. # 21105-010). In this apparatus, 18 sequencing reactions can be run simultaneously. We use 0.4-mm spacers and sharkstooth combs, which allow one to load up to 2.5 μl of sample per well.

Before pouring the gel, the inner surface of the long glass plate is treated with "binding solution" to keep the gel bound to it during the washes. The short plate is treated with "repellent solution" to help separate it from the gel. Both treatments are performed twice and consist of placing approximately 5 ml of the corresponding solution on the surface of a plate that has been laid flat on a bench top and spreading the solution evenly over the plate's surface with a Kimwipe until the plate is dry and the surface clear.

To cast a gel following the method of Haas and Smith (1993), set up a stand of four 100-ml plastic centrifuge bottles inside a large plastic tub (the plates' edges must be within the perimeter of the tub), and lay the long plate on top of the bottles with the inner, treated surface facing up. In this way the acrylamide spilled is isolated from the gel and also contained within the tub. Wet the spacers in purified water and align them to the edges of the plate, also making sure that the bottom end of the spacers coincides with the bottom of the glass plate. In a 250-ml flask, combine 80 ml of gel solution and 800 μl of 10% (w/v) ammonium persulfate, and mix well by swirling. Add 16 μl of TEMED, swirl the solution quickly, and pour it immediately on the long plate. Slide the short plate (with the treated surface facing down) onto the long plate, avoiding the creation of bubbles. Clamp the plates together, place the comb, and let the acrylamide polymerize.

RUNNING THE SEQUENCING GEL

Materials

0.8x glycerol-tolerant buffer (see Appendix 2)
sequencing stop solution (see Appendix 2)
gel washing solution (see Appendix 2)
autoradiographic film (Fuji Medical X Ray, Cat. # 03E010 or Kodak BioMax, Cat. # 8715187)
1 M NaOH

CAUTIONS: **taurine, formamide, bromophenol blue, methanol, acetic acid, NaOH (*see Appendix 4*)**

Procedure

1. Place the gel in the electrophoresis apparatus, fill the upper and lower buffer chambers with 0.8x glycerol-tolerant buffer, and pre-run the gel at 60 Watts for 40 minutes. Use a stick-on thermometer to follow the temperature, which should reach 40–45°C.
2. Add 4.5 μl of stop solution to each PCR sample from Step 6 above. Heat the samples for 5 minutes at 70°C and place them on ice immediately.
3. Stop the electrophoresis, and wash the wells of the gel with a syringe filled with 0.8x glycerol-tolerant buffer. Use filtered pipet tips to load 2.5 μl of each sample onto the gel. Run the gel at 50 Watts and a surface temperature of 40–45°C. Stop the gel when the xylene cyanol dye front is 19–21 cm from the top of the gel (after about 1.5–2 hours).
4. Fix the gel twice for 15 minutes in gel washing solution with occasional shaking. Let it dry overnight, using a fan to speed drying if desired.
5. Expose the dried gel in the dark for 7–24 hours by placing autoradiography film in direct contact with the gel. Make sure that the gel is completely dry before exposure.
6. Develop the film and read sequences.

7. To wash the plates, place them in a plastic tub with the gel side facing up. Soak in a solution of 1 M NaOH for one day. The gel should come off, and any remaining material can be removed with a plastic spatula. Filter the solution through a plastic mesh, and dispose of the acrylamide in radioactive solid waste. The liquid can be kept in a radioactive storage area and reused 4–6 times. When it no longer removes the gel from the plates, it should be disposed of in radioactive waste.

PROTOCOL 15.10

Long-term Storage of Phage Stocks

There are several problems that can occur with the long-term storage of phage, especially with the long-term storage of phage libraries. For example, bacteria or molds can grow in the stocks and affect the phage and their displayed proteins. To combat microbial growth, phage are usually stored at high concentrations (10^{12}–10^{13} particles/ml) in 0.02% sodium azide; they also may be heat-treated at 70°C for 20–40 minutes, depending on the volume, to kill bacteria and inactivate some proteases (be aware that heat treatment can denature some displayed proteins, especially larger ones like Fabs). CsCl purification (Protocol 15.4) of phage also removes bacteria and other contaminants.

Long-term storage also can affect the phage DNA directly. We have noticed that phage stocks stored for several months at 4°C make poorer template DNA for cloning than freshly made stocks. That is, the viral DNA, when released by heat treatment in the presence of SDS (from Lysis Mix), runs with a trailing smear in agarose gel electrophoresis, and the yield of ccc-DNA after fill-in reactions is poor, compared to reactions with freshly made phage stocks. In addition, light affects phage, so stocks should be stored in the dark.

Even with these precautions, we and other investigators have noticed drops in phage titer over years of storage. This may be due to denaturation of the phage coat proteins, leading to their aggregation and inactivation. We now store phage stocks in 50% glycerol at –18°C to impede this process, and routinely check the titers of the stocks every 6–12 months. After 6 months' storage, we have noticed no change in phage titer with storage of identical stocks at 4°C and –18°C; the benefit of storage at the lower temperature, if any, will become apparent only after a longer time has elapsed.

ACKNOWLEDGMENTS

We thank T. Ha and A. Day for making the figures. We also thank X. Gong, K. Brown, S. Kim, M. Rashed, and J. Mehroke for contributions in developing the methods presented in this chapter.

REFERENCES

Bonnycastle L.L.C. 1998. PCR sequencing of phage-encoded peptide sequences. *Anal. Biochem.* **256:** 140–142.

Day L.A. 1969. Conformations of single-stranded DNA and coat protein in fd bacteriophage as revealed by ultraviolet absorption spectroscopy. *J. Mol. Biol.* **39:** 265–277.

Haas S J. and Smith G.P. 1993. Rapid sequencing of viral DNA from filamentous bacteriophage. *BioTechniques* **15:** 422–431.

Smith G.P. and Scott J.K. 1993. Libraries of peptides and proteins displayed on filamentous phage. *Methods Enzymol.* **217:** 228–257.

Sambrook J., Fritsch E.F., and Maniatis T. 1989. *Molecular cloning: A laboratory manual,* 2nd. edition. Cold Spring Harbor Laboratory Press, Cold Spring Harbor, New York.

16 Production of Peptide Libraries

LORI L.C. BONNYCASTLE, JUQUN SHEN, ALFREDO MENENDEZ, AND JAMIE K. SCOTT

Department of Molecular Biology and Biochemistry and Department of Biological Sciences, Simon Fraser University, Burnaby, B.C., Canada V5A 1S6

TWO METHODS FOR MAKING RANDOM PEPTIDE LIBRARIES are described in this chapter. For both, library construction begins with the design and synthesis of a degenerate oligonucleotide that will encode the "random" peptide sequence to be displayed. Each variable residue (X) in the recombinant sequence of a phage-displayed library is usually encoded by a degenerate codon, either NNK or NNS. In these codons, N represents the nucleotides A, G, C, or T, whereas K represents G or T, and S represents G or C. Both NNK and NNS degenerate codons encode all 20 amino acids and one (amber) stop codon. Although NNK and NNS comprise different codon sets, they encode identical amino acid distributions (i.e., there are three codons for Arg, Leu, Ser; two codons for Pro, Val, Gly, Ala, Thr; and one codon for the remaining 12 amino acids); thus, both degenerate codons encode the same amino acid bias. Ideally, one would prefer to synthesize degenerate oligonucleotides having equal representation of codons for each amino acid and no stop codons for production of a completely randomized library. Other means of making degenerate oligonucleotides for libraries and "sublibraries" are discussed in Chapter 4.

CONTENTS

The choice of vector and cloning method depends largely on the design of the library. The preference for a simple randomized peptide or a randomized region on the surface of a stably folding protein (such as a toxin or zinc finger) is particularly important. For instance, a 3 or 3+3 display vector would likely be best for displaying a large protein scaffold, whereas 3 or 8+8 vectors work well for peptide or small protein fusions (see Chapter 2). In choosing a method for constructing the library, there are several points to consider. Two protocols for library construction are described in this chapter (16.3 and 16.4), and they differ in several aspects. In Protocol 16.3, a degenerate oligonucleotide insert encoding the peptide library is cloned using linearized single-stranded DNA from f88-4 phage as template. This method (Bonnycastle et al. 1996) results in libraries with a low frequency of vector without insert, and therefore produces fewer phage that do not display foreign sequences fused to pVIII. The main disadvantages of this procedure are (1) restriction endonuclease cleavage sites are required on either side of the insert and (2) the length of the insert is limited to about 90 bp, because of limitations in making long synthetic oligonucleotides in large enough quantities for cloning. Thus, this method is best suited for making a peptide library with inserts of up to 30 random amino acids, and for using the f88-4 phage vector, which has unique cloning sites (whose enzymes will not cleave any sequences within an NNK degenerate-codon repeat) for insertion of foreign sequences. In contrast, the latter protocol (16.4) is best for making randomized libraries in the context of a folded protein. It uses Kunkel's site-directed mutagenesis strategy (Kunkel et al. 1991) and works best if the wild-type vector DNA contains a "template" sequence that will be altered by the degenerate oligonucleotide. This method has the advantage that unique cloning sites flanking the mutagenized region are not necessary for making a library; however, a fixed sequence, which will be changed in the mutagenesis, must be present in the vector template.

There are several further considerations involved in making a random peptide library in the context of a folded protein scaffold. First, with either cloning method, a DNA fragment encoding the protein scaffold must be inserted into the vector before the library can be made. Thus, before the library is constructed, phage displaying the wild-type protein scaffold should be shown by gel electrophoresis, Western blot, ELISA, and other methods to be expressed and properly folded on the viral surface. With the oligonucleotide insert method (see Protocol 16.3 and Fig. 16.1A), but not with the site-directed mutagenesis method (see Protocol 16.4 and Fig. 16.1B), restriction endonuclease sites must be added on either side of the region to be randomized. However, with both methods (especially with Protocol 16.4) there will be a relatively high frequency of clones arising in the library that have the wild-type sequence of the template strand which encodes the protein scaffold fused to the viral coat protein. Thus, phage displaying the wild-type protein should also be tested for their level of background binding to the molecule(s) that will be used to screen the library, as well as to the other components in the screening (the plate, beads, streptavidin, etc.; for more details, see Chapter 17). If phage bearing the wild-type protein scaffold turn out to have binding activity for the screening molecule, they should be modified to produce a protein with acceptably low binding activity before going forward to make the library. For example, binding region residues that will ultimately be changed to randomized residues could be changed into glycine or alanine residues that reduce phage binding to low levels (Blancafort et al. [1999] use this approach in making a zinc-finger library.) However, one must also show that folding of the protein scaffold is not greatly affected by the changed sequence in this region. Such modifications would allow the recombinant pVIII to be produced, while ensuring that the displayed wild-type protein will not bind the screening molecule. It is not a good idea to simply place stop codons in the nucleotide sequence encoding the template strand for the library. It has been reported that stop codons are sometimes read through and have appeared in clones selected from phage libraries (Carcamo et al. 1998; our unpublished observations). Even if this modification did prevent the production of the wild-type scaffold, and hence, prevent the phage encoding it from causing high background during library screening, significant biological bias favoring phage production in cells that are not producing the scaffold protein could cause these clones to overgrow the library. Selection may thus significantly favor the growth of phage clones that do not produce a scaffold protein fused to the coat protein. It is safer to keep the level of biological selection on the library clones and the wild-type template phage about equal; however, this requires the production of the protein scaffold encoded by the cloning template.

The design of peptide libraries requires thoughtful consideration of the vector used for the library, the degenerate oligonucleotide that will ultimately encode the library, and the cloning method used. One may also wish to consider the screening method to be used before making final choices, especially regarding the vector to be used (see the overview to Chapter 17). Further considerations should be made if the library is to comprise a variegated region on a folded protein, rather than a simple peptide fusion to a viral coat protein.

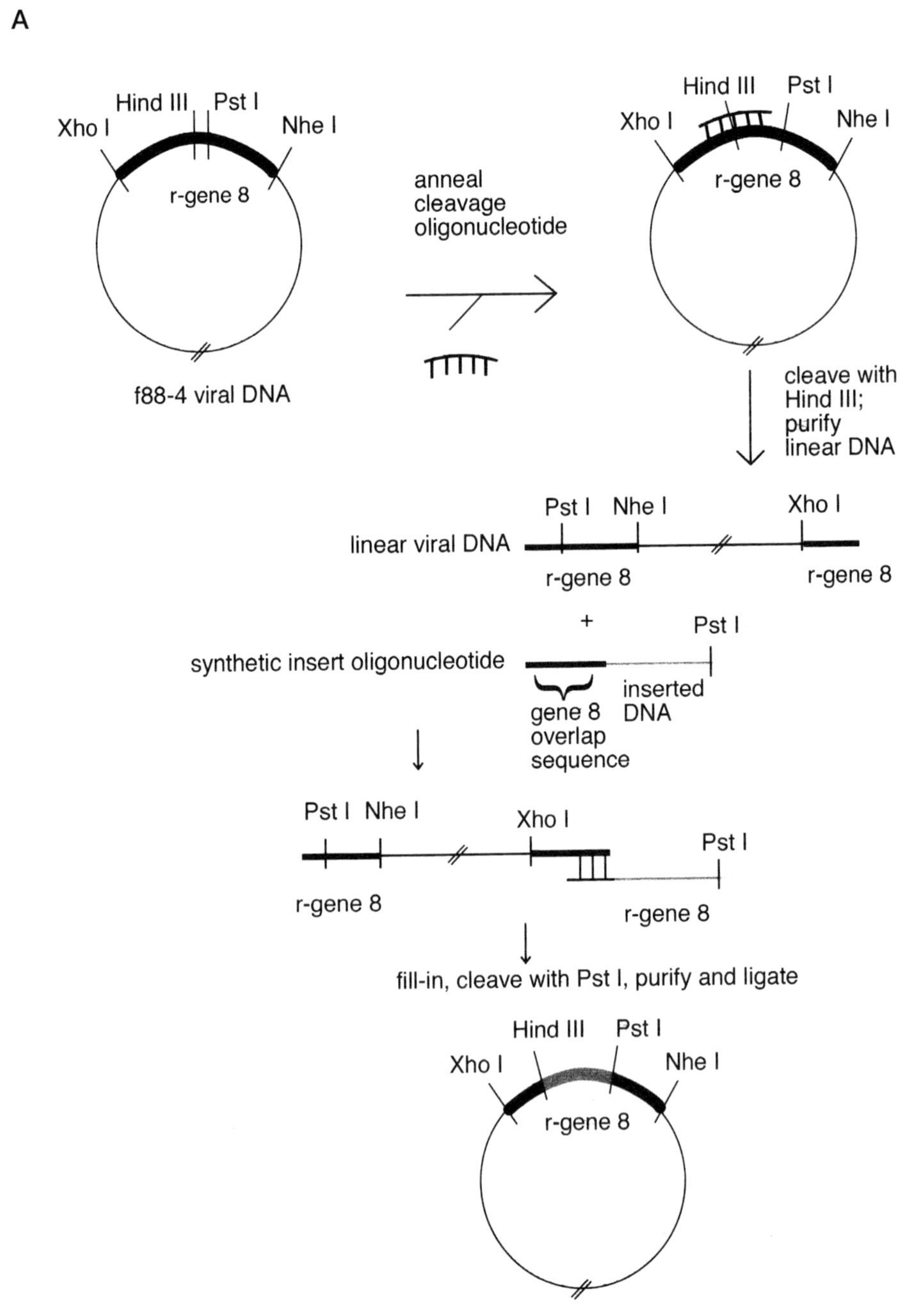

Figure 16.1. (*See facing page for legend.*)

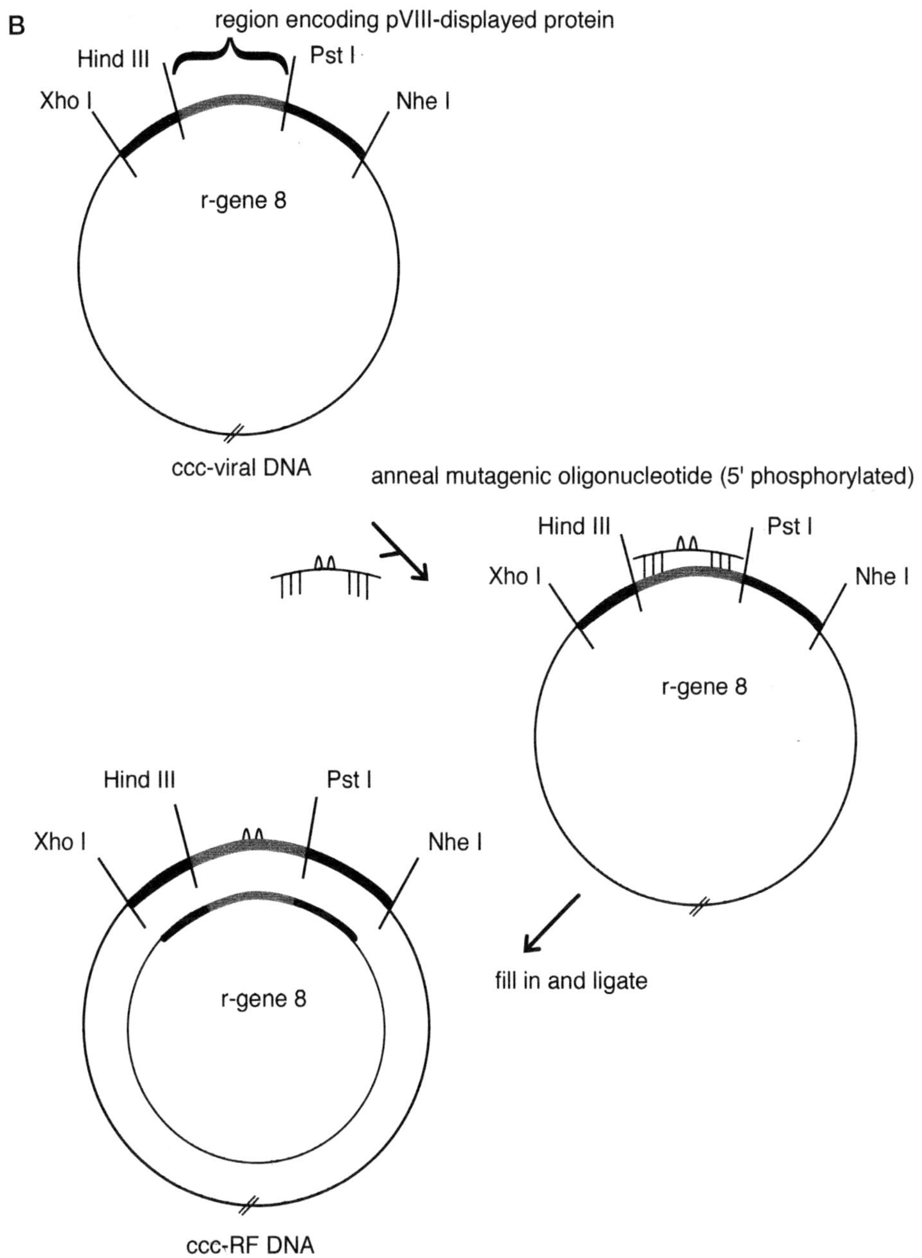

Figure 16.1. (*A*) Preparing a peptide library by the insertion of a single-stranded, degenerate oligonucleotide into single-stranded, linear f88-4 DNA, as described in Protocol 16.3. (*B*) Preparing a peptide library by Kunkel's site-directed mutagenesis method, as described in Protocol 16.4.

PROTOCOL 16.1

Purification of Oligonucleotides by Urea-PAGE

Short oligos need not be purified, but long ones of more than 40 bases should be. Run a sample of the oligonucleotide on an analytical polyacrylamide gel; if there is a smear under a high-molecular-weight band, the oligonucleotide should be purified further.

Materials

degenerate oligonucleotide in TE or water
polyacrylamide gel electrophoresis (PAGE) system, for example, model SE 400 from Hoefer Scientific Instruments
urea (DNA grade)
polyacrylamide solution (ReadySol DNA PAGE 40% **acrylamide**, 5% **bisacrylamide**, Amersham Pharmacia, Cat. # 17-1308-01)
5x TBE (see Appendix 2)
10% (w/v) **ammonium persulfate** (**APS**)
N,N,N′,N′-tetramethylethlylenediamine (TEMED)
oligo gel loading buffer (see Appendix 2)
fluorescent thin-layer chromatography plate (silica gel)
short wave, portable **UV** lamp (254 nm)
0.5x TE buffer (see Appendix 2)
Micropure separator, 0.45 μm (Amicon, Cat. # 42522)
Ultrafree-MC centrifuge filter, NMW 10000 (Millipore, Cat. # UFC3 LGC 00)

CAUTIONS: **urea, acrylamide, bisacrylamide, ammonium persulfate, TEMED, UV light** (*see Appendix 4*)

Procedure

1. Make a denaturing polyacrylamide gel. The percentage of polyacrylamide in the gel will depend on the length of the oligonucleotide. For oligonucleotides of up to 60 bases use 10% acrylamide; for longer oligos use 8%. Use 1.5- or 3-mm thick spacers and combs, depending on the amount of sample.

   Mix the required amounts of water, acrylamide, 5x TBE, and urea in a 100-ml beaker and dissolve the urea on a hot stirring plate. Allow the mixture to cool to room temperature, and add APS and TEMED. Pour the gel immediately. Prepare 50 ml of gel solution for 1.5-mm-thick gels and 75 ml for 3-mm-thick gels.

	8% gel	8% gel	10% gel	10% gel
Urea	25 g	37.5 g	25 g	37.5 g
40% acrylamide/5% bis	10 ml	15 ml	12.5 ml	18.8 ml
5x TBE	10 ml	15 ml	10 ml	15 ml
ddH_2O	9.7 ml	14 ml	7.2 ml	10.2 ml
10% APS	250 µl	375 µl	250 µl	375 µl
TEMED	50 µl	75 µl	50 µl	75 µl
Total volume	50 ml	75 ml	50 ml	75 ml

2. Assemble the electrophoresis apparatus and pre-run the gel in 1x TBE buffer at 290 V for 45 minutes to 1 hour. The temperature of the gel should be 40–45°C.

3. Add 0.66 volume of oligo gel loading buffer to the oligonucleotide solution, and mix thoroughly. Boil each sample for 5 minutes to disrupt secondary structure, and place immediately on ice.

4. Wash out the wells of the gel with 1x TBE, and immediately load the samples. Work quickly to avoid cooling the gel. Run the gel at 280 V for approximately 2 hours or until the fast-running, bromophenol blue dye is near the bottom of the gel.

5. Disassemble the apparatus and check the gel by UV shadowing: Place the gel over a fluorescent thin-layer chromatography plate coated with plastic wrap and examine it with UV illumination from above. The oligonucleotide should be visible as a dark band in a bright background. Use the UV light for as short a time as possible, as its rays break DNA strands.

6. Use a sterile blade to cut out the desired oligo band, which should be the slowest migrating and thickest band. Be sure to leave behind the oligonucleotide just at the lower edge of the thick band, as that may include DNA containing deletions of 1–2 bases, or more.

7. Transfer the gel slice to a 1.5-ml microcentrifuge tube and crush the slice with a 1-ml syringe plunger or a 1-ml pipet tip. Cover the gel slurry with 0.5x TE (in which it can be stored at 4°C for a few days). Place the slurry at 37°C for 2 hours, with gentle shaking or rocking.

8. Pass the slurry through a Micropure filter (0.45 µm). Perform the filtration according to the manufacturer's instructions, in a microfuge run at 14,000*g* for 30–60 seconds. Collect the filtrate.

9. Transfer the filtrate to an Ultrafree-MC 10,000 filter to desalt and concentrate. Make sure that the oligonucleotide molecular weight is well above the molecular weight cutoff of the filter. Wash the oligo 2–3 times with 300–400 µl of 0.5x TE according to the manufacturer's instructions, in a microcentrifuge at 2000*g* for 20 minutes or until 50–100 µl remains. Transfer the liquid on top of the membrane to a microcentrifuge tube.

10. Measure the optical density at 260 nm (OD_{260}) of a 1:50 or 1:100 dilution of the sample in 0.5x TE. Calculate the concentration of oligo as follows:

1 Absorbance unit (AU) at 260 nm ≅ 33 μg/ml ssDNA (this varies somewhat with the nucleotide content, but is a good "ballpark" estimate)

The purified degenerate oligonucleotide is now ready for library construction (Protocols 16.3 and 16.4).

PROTOCOL 16.2

Oligonucleotide-directed Cleavage of f88 Viral DNA

Filamentous phage virions contain a single-copy DNA genome that is single-stranded (ss), covalently closed circular (ccc), and plus (+) sense. This procedure describes how viral DNA, isolated from the f88-4 phage vector, is prepared for cloning. Purified viral DNA is made linear by oligonucleotide-directed cleavage using the restriction endonuclease *Hin*dIII. The *Hin*dIII-cleavage oligonucleotide has the sequence shown below; it is minus (–) sense, and the anticodons are separated by dots.

5′-AC.GTT.GGC.AAA.GCT.TAG.CAT.AGG.A.-3′

Materials

200 μg of purified f88-4 viral DNA (see Protocol 15.7)
*Hin*dIII cleavage oligonucleotide (see above)
*Hin*dIII restriction enzyme
10x *Hin*dIII reaction buffer (supplied by enzyme manufacturer)
5x Lysis Mix (see Appendix 2)
buffer-saturated **phenol**, pH 8.0
phenol/chloroform/isoamyl alcohol 25:24:1 (v/v/v)
chloroform/isoamyl alcohol 24:1 (v/v)
sterile 3 M sodium acetate, pH 5.2
absolute **ethanol**
70% (v/v) **ethanol**, –18°C
sterile purified water
100,000 MW exclusion ultrafiltration apparatus (Centricon 100; Amicon, Cat. # 4212)
sterile 5 M NaCl
freshly made 5 M **NaOH**
sterile 1x TE buffer (see Appendix 2)

CAUTIONS: **phenol, chloroform, isoamyl alcohol, ethanol, NaOH** ***(see Appendix 4)***

Procedure

Cleavage and purification of ss, ccc DNA

1. Mix in a 1.5-ml microcentrifuge tube (use a 5:1 molar ratio of oligonucleotide to f88-4 genome):

f88 viral DNA	200 μg
*Hin*dIII cleavage oligo	2.8 μg
10x *Hin*dIII reaction buffer	40 μl

Add water to a final volume of 360 µl and mix. Heat for 15 minutes at 70°C, microfuge briefly, and let stand for 10 minutes at 37°C. Save a 100-ng (0.2 µl) aliquot to serve as the undigested control DNA for analysis by agarose gel electrophoresis (see Step 13). To preserve the condition of the DNA and prepare it for gel electrophoresis, mix it with 10 µl of 1x Lysis Mix and store it at –20°C.

2. Add 40 µl (400 Units) of *Hin*dIII, mix, and incubate at 37°C for 4 hours. Save another 0.2-µl aliquot for gel analysis as the digested DNA sample; put this sample in 10 µl of 1x Lysis Mix and heat it at 70°C for 20 minutes to inactivate the enzyme, then store at –20°C (see Step 13).

   Note: One can optimize this step by first running an assay to determine the minimal digestion time for each particular ssDNA prep.

3. Phenol-extract the reaction once and back-extract, as described in Appendix 3. Extract once with phenol/chloroform/isoamyl alcohol, and back-extract if a white interphase is visible. Extract twice with chloroform/isoamyl alcohol.

   Note: The condition of phenol is crucial to the yield of linear ssDNA, because oxidized phenol will break DNA strands. Be sure the phenol used for this experiment is fresh.

4. Ethanol-precipitate as described in Appendix 3.

5. Dissolve the cleaved ssDNA in 400 µl of water. Use 100 ng to assess the DNA yield by gel analysis (see Step 13).

Removal of the Cleavage Oligonucleotide

6. Bring the volume of the DNA solution to 1 ml in 0.15 M NaCl, 0.1 M NaOH by adding 30 µl of 5 M NaCl, 20 µl of 5 M NaOH, and 550 µl of water. Incubate for 30 minutes at room temperature, occasionally flicking the tube to mix the contents.

7. Transfer the solution to a Centricon 100 ultrafiltration apparatus and centrifuge at 1000*g* (2700 rpm on a tabletop centrifuge). Reduce the volume to 30–40 µl.

8. Wash the DNA once with 1.4 ml of 0.15 M NaCl, and reduce the volume to approximately 40 µl, as in Step 7 above.

9. Transfer the solution to a fresh Centricon 100. Rinse the old Centricon with 30 µl of 0.03 M NaOH, back-spin this rinse solution into a cup, and tranfer to the fresh Centricon.

10. Wash the DNA twice with 1.5 ml of 0.03 M NaOH, reducing the volume each time to 30–40 µl.

11. Wash 4 times with 1.5 ml of 0.5x TE. Reduce the final volume to 25–30 µl, the minimum possible.

12. Transfer the washed solution to a 0.5-ml microcentrifuge tube, and wash the Centricon with 50 µl of 0.5x TE. Back-spin this solution into a cup and transfer to the microcentrifuge tube. Adjust the final volume to 200 µl. Remove a 100-ng aliquot for the gel analysis in Step 13 below.

13. Check the DNA from Steps 1, 2, and 5 of the cleavage reaction, and from Step 12 above on a 1.2% 1x TBE gel (see Protocol 15.5). Compare with a standard amount of viral DNA. Most of the cleaved DNA should run as a linear ss band (see Fig. 15.2), and the yield of DNA, compared with the starting material, should be approximately 80–90%.

The linear viral DNA is now ready for use in creating a library (Protocol 16.3).

PROTOCOL 16.3

Making a Library Using Linear Viral DNA as Template

In this procedure, linearized viral DNA is annealed to a degenerate oligonucleotide primer across a 15- to 20-base complementary region that generates very long 5′ overhanging ends. This allows the Klenow fragment of *Escherichia coli* DNA polymerase and Sequenase (T7 DNA polymerase lacking exonuclease function) to fill in the ss-DNA of the phage genome and the degenerate oligonucleotide insert. After fill-in, both ends of the linear, dsDNA are cut with the restriction endonuclease *Pst*I, and the DNA is treated with ligase to form open-circular (oc) and covalently closed-circular (ccc) RF DNA, which can subsequently be electrotransformed into *E. coli* cells to make the phage library.

We routinely obtain libraries of 10^8 to 10^9 clones using the fd-tet-based vector f88-4. Library size may be increased by roughly 20-fold if wild-type phage vectors that do not have mutations in the (–)-strand origin of replication are used. The major advantages of this library construction method are that (1) viral DNA, which is easy to prepare, is used instead of the harder-to-prepare RF DNA; (2) the chance of obtaining multiple, tandem inserts is low, compared to insertion of a purified fragment of DNA into cleaved RF DNA; and (3) because PCR is not used, there is little chance for the overrepresentation of particular DNA species, besides that coming from the oligonucleotide synthesis itself. The disadvantage is that only relatively short oligonucleotides (less than 100 bases) can be used for cloning, as the synthetic yields and quality of longer ones are usually poor. This procedure also requires that cloning sites be present on either side of the oligonucleotide insert. The resulting library is only as good as the oligonucleotide used to make it, so the degenerate oligonucleotide should be purified by denaturing gel electrophoresis and extraction from a gel fragment (see Protocol 16.1).

Materials

purified library oligonucleotide in 0.5X TE (see Protocol 16.1)
linear, ss f88 DNA (from Protocol 16.2), approximately 1 µg of DNA/µl
10X annealing buffer (see Appendix 2)
sterile purified water
10X synthesis buffer (see Appendix 2)
sterile 0.5 M NaCl
2.5 mM dNTP mix (see Appendix 2)
Klenow fragment
T7 Sequenase (13 Units/µl; Amersham Pharmacia Biotech, Cat. # E70775Y)
buffer-saturated **phenol**, pH 8.0 (see Appendix 3)
phenol/chloroform/isoamyl alcohol, 25:24:1 (v/v/v)
chloroform/isoamyl alcohol, 24:1 (v/v)
sterile 3 M sodium acetate, pH 5.2
absolute **ethanol**

70% (w/v) **ethanol**, –18°C
*Pst*I restriction endonuclease and reaction buffer
sterile 1X and 0.1X TE (see Appendix 2)
30-kD exclusion ultrafiltration apparatus with a 2-ml cup (MC 30,000; Millipore, Cat. # UFC4LTK25)
T4 DNA ligase (400 Units/μl; NEB, Cat. # 202S)
10X ligase buffer (supplied by ligase manufacturer)

CAUTIONS: phenol, chloroform, isoamyl alcohol, ethanol (*see* **Appendix 4**)

Procedure

Preparing linear dsDNA from linear, viral DNA and the library oligonucleotide

1. Prepare the linear ssDNA and purified degenerate oligonucleotide encoding the peptide library for a fill-in reaction as follows. Use a 1:3 molar ratio of ssDNA to oligonucleotide.

   Combine:

*Hin*dIII-cleaved f88 DNA	17 pmole (50 μg)
library oligonucleotide	52 pmole
10X annealing buffer	20 μl

   Bring the final volume to 200 μl with water. Heat the sample for 15 minutes at 70°C. Meanwhile, place approximately 150 ml of 70°C water in a 400-ml Pyrex beaker. Float the sample on the water in the beaker and allow it to slowly cool to room temperature. Leave the sample at room temperature for at least 5 minutes.

2. Centrifuge the sample briefly, cool it on ice, and add:

10X synthesis buffer	30 μl
0.5 M NaCl	10 μl (final: 50 mM)
2.5 mM dNTP mix	24 μl (final: 0.2 mM of each dNTP)
Klenow fragment	50 units

   Add enough water to bring the final volume (from Steps 1 and 2) to 300 ml.

   Mix. Incubate for 10 minutes at 25°C, then for 30 minutes at 37°C.

3. Inactivate the enzyme by incubating the reaction for 10 minutes in a 70°C water bath. Centrifuge briefly and cool on ice.

4. Add:

10X synthesis buffer	5 μl
0.5 M NaCl	5 μl (final: 50 mM)
2.5 mM dNTP mix	18 μl (final: 0.3 mM)
water	20.85 μl

Mix and then add 1.15 µl (15 units) of Sequenase. The final volume should be 350 µl (Steps 1+2+4). Mix and incubate for 30 minutes at 37°C. Inactivate the enzyme for 10 minutes at 70°C. Microfuge the tube briefly to bring down the condensate. After mixing, remove 0.8 µl (about 112 ng) for analysis of the yield of linear, double-stranded (ds) DNA by agarose gel electrophoresis (see Protocol 15.5 and Fig. 15.3).

5. Extract the DNA once with one volume of phenol, and back-extract the interphase, as described in Appendix 3. Extract once with phenol/chloroform/isoamyl alcohol, and back-extract the interphase if it is white. Extract twice with chloroform/isoamyl alcohol.
6. Precipitate the DNA with 3 M sodium acetate and ethanol, wash the pellet with 70% ethanol, and air-dry as described in Appendix 3.
7. Dissolve the pellet in 320 µl of water if the sample is to be digested immediately, or in 0.1X TE if the sample is to be stored (at –20°C).
8. A 0.5-µl aliquot of the sample from Step 7 should contain 125 ng of dsDNA, assuming an 80% yield. Use this volume to analyze the yield of full-length, linear dsDNA by electrophoresis on a 0.8% agarose gel (see Protocol 15.5 and Fig. 15.3). Use 125 ng of linear, f88-4 RF-DNA as a marker.

Pst*I digestion of the linear dsDNA*

9. Add 40 µl of 10X *Pst*I reaction buffer to the 320-µl sample. Mix the reaction and add 400 Units of *Pst*I and enough water to bring the final volume to 400 µl. Incubate at 37°C for 4 hours.

   Note: Do not allow the enzyme volume to exceed 10% of the reaction volume. If necessary to accommodate more enzyme, increase the reaction volume by adding more buffer and water.

10. Microfuge the sample briefly, then phenol-extract once and back-extract the interphase, as described in Appendix 3. Extract once with phenol/chloroform/isoamyl alcohol and back-extract if a white interphase is visible. Extract twice with chloroform/isoamyl alcohol.
11. Ethanol-precipitate the sample, wash the pellet with 70% ethanol, and air-dry, as described in Appendix 3.
12. Dissolve the DNA in 150 µl of 1X TE.
13. Remove the *Pst*I end fragments by transferring the sample to a 30-kD cutoff, size exclusion membrane filter, filling the 2-ml cup, and reducing the volume to approximately 30–40 µl. Wash the DNA two or three times by filling the cup with 1X TE. Reduce the final volume to the minimal possible, 20–30 µl. Transfer the concentrated solution to a fresh 1.5-ml microcentrifuge tube. For the analyses in Step 14 below, transfer 1 µl of the solution to a 0.5-ml microcentrifuge tube containing 14 µl of 1X TE and mix. Wash the filter with 25 µl of water, and add this to the 1.5-ml microcentrifuge tube containing the washed DNA. Note the final volume.

14. Use 10 µl of the sample in the 0.5-ml microcentrifuge tube from Step 13 for spectroscopic analysis (use 0.1x TE to dilute the sample and for the blank; 1 AU_{260} = ~50 µg/ml dsDNA) and analyze the remaining 5 µl by agarose gel electrophoresis (see Protocol 15.5 and Fig. 15.3). From these, determine the yield of full-length, linear dsDNA, and adjust the concentration of linear dsDNA in the sample in the 1.5-ml microcentrifuge tube to 10 µg/ml. We routinely get ~50% recovery from Steps 10 and 11, and ~80% recovery from Step 13, giving a final yield of about 30 µg of DNA.

15. Set up a dilute ligation (at 6–8 µg of DNA/ml) to favor the formation of monomeric, oc and ccc forms: Assuming an approximate 30 µg yield, aliquot the washed DNA evenly between five 1.5-ml microcentrifuge tubes (including the tube that the DNA is already in). Adjust the volume of each to 897.5 µl with water. Add 100 µl of 10x ligase buffer, mix, and cool the sample on ice. Add 2.5 µl (1000 cohesive end units, 15 Weiss units) of T4 DNA ligase, mix, and incubate for 2 days in a 15°C water bath.

16. Centrifuge the tubes briefly, add 5 µg of poly(A) RNA to each tube, and mix.

    Note: This step is optional and is meant to ensure that the DNA loss due to nonspecific binding will be minimal, and that the precipitate will be efficiently collected. A small pellet should be visible after phenol extraction and ethanol precipitation.

17. Transfer half the volume of each tube (~500 µl) to a fresh microcentrifuge tube, then phenol-extract all of the samples once, as described in Appendix 3. Combine the interphases for back-extraction. Extract once with phenol/chloroform/isoamyl alcohol, combine the interphases and back-extract, then extract twice with chloroform/isoamyl alcohol.

18. Concentrate the samples by ethanol precipitation with sodium acetate as described in Appendix 3. Wash the pellet with 70% ethanol and dry. Dissolve the DNA in 250 µl of 1x TE. Store the DNA solution at –20°C.

19. Transfer the DNA sample to a 30-kD exclusion ultrafiltration apparatus and fill the 2-ml cup with 0.1x TE. Concentrate the DNA to 30–40 µl, and wash it twice with 1 ml of 0.1x TE to desalt. Reduce the final volume to the minimum possible, 20–30 µl. Transfer the DNA solution to a 0.5-ml microcentrifuge tube (including a 30-µl, 0.1x TE rinse of the ultrafiltration apparatus), and adjust the volume to 50–100 µl with 0.1x TE.

20. Transfer 2 µl of the DNA solution to a 0.5-ml microcentrifuge tube containing 7 µl of 0.1x TE. Run 3 µl of this aliquot on an 0.8% agarose gel to determine the form of the DNA and the final yield (see Protocol 15.5 and Fig. 15.3). Dilute the remainder in 0.1x TE and determine the DNA concentration by UV scanning using a 0.1x TE blank (1 AU_{260} = ~50 µg/ml dsDNA).

The DNA is now ready for transforming cells, either by electroporation (Protocol 16.6) or heat shock in $CaCl_2$ (Appendix 3). The final concentration in Step 19 and the low concentration of TE are necessary for electrotransformation, as any remaining salt in the DNA sample can cause a short circuit. These steps are probably not necessary for $CaCl_2$ transformation; however, there is an approximate 10^4-fold larger yield of transformants per µg of DNA with electroporation, as compared with heat shock.

PROTOCOL 16.4

Making a Library Using ccc Viral DNA as Template

This protocol includes a site-directed mutagenesis procedure that uses fill-in of a uracil-containing, ccc-ssDNA template primed from a synthetic, degenerate oligonucleotide (Kunkel et al. 1991). We mainly use this procedure to change fixed residues in a protein scaffold (such as a toxin or zinc finger) into "randomized" ones (see chapter introduction). DNA encoding the protein scaffold is cloned into the phage-display vector, and then a phage clone bearing this sequence is used to produce phage in *E. coli* strain CJ236. The resulting phage DNA contains uracil instead of thymidine and is used as the template plus strand in the minus-strand synthesis reaction. A degenerate oligonucleotide encoding the library primes the reaction, and should be 5′-phosphorylated to allow ligation with the 3′ end of the minus strand after completion of the fill-in reaction. (*Note:* Depending on the homogeneity of its length after synthesis, the degenerate oligonucleotide may have to be gel purified; see Protocol 16.1). The minus-strand synthesis can be performed with either T4 or T7 DNA polymerase for the fill-in reaction (our lab mostly uses T7 DNA polymerase for this) and T4 DNA ligase for ligation of the 5′ and 3′ ends. Completed, heteroduplexed oc and ccc RF molecules are electroporated into the highly electrocompetent *E. coli* strain MC1061 to make the library (see Protocol 16.6). The uracil-containing plus strand is destroyed in these cells and replaced with the sequence complementary to the minus strand, which includes the degenerate oligonucleotide.

Because of either random priming during the synthesis, or strand-displacement in the fill-in reaction, phage bearing the wild-type (template) sequence usually comprise 10–40% of the resulting library. After completion of the reactions, and depending on the design of the template DNA and degenerate oligonucleotide, the presence of mutagenized (heteroduplexed) molecules can be shown by digesting a small sample of the filled-in DNA with a "diagnostic" restriction enzyme. This assays a restriction endonuclease site that was originally either (1) present in or (2) absent from the template DNA, and that will be lost or added, respectively, during a successful mutagenesis because of the sequence encoded by the degenerate oligonucleotide. The addition or loss of a cleavage site can also be used to assay the RF DNA from clones that have been isolated after the mutagenesis mixture has been used to transform cells; RF DNA is simply purified in a "miniprep" procedure (e.g., the UltraClean Mini Plasmid Prep Kit, Mo Bio, Cat. # 12300-100), then cleaved with the diagnostic restriction enzyme to confirm the presence/absence of the site. From this assay the frequency of clones bearing library sequences can be determined easily.

Materials

5′-phosphorylated library oligo—Enzymatic phosphorylation is inefficient, and the yield of phosphorylated molecules is difficult to quantitate. Order the synthetic oligonucleotide with a 5′-phosphorylated end.

E. coli strain CJ236 (F′ *cat* (pCJ105;M13^{s}Cm^r)/*dut ung1, thi-1, relA1, spoT1, mcrA*; New England Biolabs, Cat. # 801-M)
f88-4-derivative phage bearing a template sequence to be randomized
LB plates containing 25 µg/ml **chloramphenicol** (see Appendix 2)
NZY containing 25 µg/ml **chloramphenicol** (see Appendix 2)
NZY agar plates containing 40 µg/ml **tetracycline**
LB containing 15 µg/ml **tetracycline** and 25 µg/ml **chloramphenicol**
10x annealing buffer (see Appendix 2)
10x T4 DNA polymerase reaction buffer or 10x T7 DNA polymerase reaction buffer (supplied by enzyme manufacturer)
acetylated BSA (100x, 10 mg/ml, NEB)
10 mM ATP
5 mM dNTP mix (see Appendix 2)
T4 DNA ligase (NEB, Cat. # 202S)
T7 DNA polymerase or T4 DNA polymerase
30-kD MW cutoff, size-exclusion membrane-filter with a 2-ml cup (Millipore, Cat. # UFC4LTK25)
buffered **phenol**
phenol/chloroform/isoamyl alcohol 25:24:1 (v/v/v)
0.1x TE (see Appendix 2)

CAUTIONS: ***E. coli*, chloramphenicol, tetracycline, phenol, chloroform, isoamyl alcohol** (*see* ***Appendix 4***)

Procedure

1. Streak an LB agar plate containing 25 µg/ml chloramphenicol with *E. coli* strain CJ236 cells. Pick a colony and prepare for infection as in Protocol 15.1.

2. Infect *E. coli* cells with phage whose DNA will be used for the template strand in the library synthesis (see Protocol 15.2). Titer the infection on NZY plates containing 40 µg/ml tetracycline. Amplify a single tetracycline-resistant colony by growing it in LB medium containing 15 µg/ml tetracycline and 25 µg/ml chloramphenicol. Prepare phage DNA from this culture (Protocols 15.3 and 15.7; Protocol 15.4 is optional) and quantitate the DNA by spectrophotometric analysis (1 AU_{260} = ~33 µg/ml ssDNA). Assess the quality of the DNA by agarose gel electrophoresis (Protocol 15.5), looking for the fraction of viral DNA that is ccc rather than linear (see Fig. 15.2). If the fraction of linear DNA is greater than 20–30%, consider repeating the preparation.

3. Make an annealing mix using a 1:15 molar ratio of template to oligonucleotide as follows:

viral DNA (for f88 there are 3 µg/pmole)	0.33 pmole
5′-phosphorylated library oligonucleotide	5.09 pmole
10x annealing buffer	5.00 µl

Bring the final volume to 50 µl with sterile, purified water. Heat the sample for 10 minutes at 70°C. Meanwhile, place approximately 150 ml of 70°C water in a 400-ml Pyrex beaker. Float the sample on the water in the beaker, and allow it to slowly cool to room temperature. Place the sample on ice until use. Follow the same procedure (through Step 6) with a control reaction that is missing the oligonucleotide to assay nonspecifically primed molecules that may be present in the reaction.

4. Make a reaction mix as follows:

10x T4 or T7 DNA polymerase reaction buffer	12.0 µl
BSA	1.2 µl
ATP (10 mM)	12 µl
dNTPs (5 mM each)	6.0 µl
T4 DNA ligase	1650 cohesive end ligation units (or 25 Weiss units)
T4 or T7 DNA polymerase	9.0 units

Bring the final volume to 60 µl with sterile purified water. Mix briefly, cool the solution on ice, and transfer 50 µl to the annealing reaction. Follow the same procedure with the control reaction.

5. Mix and incubate the fill-in reactions for 5 minutes on ice, followed by 5 minutes at room temperature, and then 60 minutes at 37°C. Stop the reactions by heating for 10 minutes at 70°C.

6. Analyze 10 µl of the reactions (including the control) for the presence of ccc and oc dsDNA by electrophoresis on a 0.8% agarose gel in 4x-GBB buffer (Protocol 15.5, see Fig. 15.3).

7. Phenol-extract the reaction once and back-extract, as described in Appendix 3. Extract once with phenol/chloroform/isoamyl alcohol, and back-extract if a white interphase is visible. Extract twice with chloroform/isoamyl alcohol.

8. Desalt the sample by transferring the aqueous phase to a 30-kD MW cutoff, size exclusion membrane filter; fill the 2-ml cup with 0.1x TE, and reduce the volume to 30–40 µl. Wash the DNA two or three times with 1 ml of 0.1x TE. Reduce the final volume to 20–30 µl.

9. Transfer 1 µl of the DNA to a 0.5-ml microcentrifuge tube containing 2 µl of sterile, purified water. Mix. Use 2 µl for spectroscopic analysis in 0.1x TE (1 AU_{260} = 50 µg/ml dsDNA) and 1 µl for agarose gel analysis of the linear RF DNA (Protocol 15.5). From these analyses, determine the yield of oc and ccc RF DNA (see Fig. 15.3). If possible, assay the percent DNA that has incorporated the degenerate oligonucleotide sequence by digesting a sample of the filled-in DNA with a "diagnostic" enzyme (see introduction; use RF DNA from the template phage as a control).

The DNA is now ready for transforming cells, either by electroporation (Protocol 16.6) or heat shock in $CaCl_2$ (Appendix 3). The final concentration in Step 8 and the low concentration of TE are necessary for electrotransformation, as any remaining salt in the DNA sample causes the electroporation to short out. These steps are probably not necessary for $CaCl_2$ transformation; however, there is an approximate 10^4-fold larger yield of transformants per μg of DNA with electroporation, as compared with heat shock.

PROTOCOL 16.5

Preparation of Competent Cells for Electroporation

Electrocompetent cells are best for making libraries when used within 1–2 days after their preparation; their competence, as assessed by the number of transfectants produced by a given amount of RF DNA, appears to decrease over time during storage at –80°C. The protocol here loosely follows the method of Dower et al. (1988); we normally obtain efficiencies of >10^8 TU/µg from f88-4 RF DNA (genome size: 9235 bases) with freshly prepared cells. Cells up to 6 months old are fine for transformations for which high yields are not required (e.g., when constructing a new vector).

We use the *E. coli* strain MC1061 because it is "supercompetent"; that is, in our hands, it produces the highest transformation efficiency for fd-tet-based vectors. Because of fd-tet's tetracycline-resistance genes, electrotransfected cells will dominate in the culture if tetracycline is present, even though MC1061 is a female strain. Thus, each phage clone in the culture will derive from a single electrotransfected cell. Because the tac operon has deleted the part of the lac operon that mediates glucose suppression, f88-derived phage will synthesize recombinant pVIII constitutively; synthesis appears to be slightly increased by the addition of IPTG.

In contrast, a male cell line should be used for electrotransfections involving the DNA of wild-type filamentous phage vectors. That is because the electrotransfected cells will grow more slowly than those in the culture that were not transfected. Without antibiotic selection to limit the growth of the latter cells, they will overgrow the culture, and phage production will be low. If the cells are male, however, the phage produced by the small fraction of electrotransfected cells will go on to infect the other untransfected cells in the culture. Hence, as long as there are male cells in the culture, infected cells will dominate and phage production will be high.

Materials

E. coli strain MC1061 [F^- *araD139* Δ*(ara-leu)7696 galE15 galK16* Δ*(lac)X74 rpsL* *(Str^r)* *hsdR2* $(r_K{-}m_K{+})$ *mcrA mcrB1*] (available from ATCC; http://www.atcc.org)
sterile 125-ml capped culture flask
sterile 3-liter plugged culture flasks
NZY + 100 µg/ml streptomycin (see Appendix 2)
superbroth (SB) medium (see Appendix 2)
a large pan containing an ice-water bath
sterile 250-ml screw-cap, polypropylene or polycarbonate centrifuge bottles
sterile 40-ml screw-cap, round-bottom, polypropylene (Oak Ridge) tubes
sterile 1 mM HEPES buffer (see Appendix 2)
sterile 10% (v/v) glycerol
sterile 1.5-ml microcentrifuge tubes
pan containing powdered dry ice
labeled box for storing cells at –70°C

CAUTION: *E. coli (see Appendix 4)*

Procedure

1. Inoculate a 125-ml flask containing 20 ml of NZY and 100 μg/ml streptomycin with MC1061 from a stock culture. Incubate overnight at 37°C with brisk shaking (at least 200 rpm).
2. Inoculate each of two 3-liter flasks containing 500 ml of SB with 2 ml of the overnight culture.
3. Grow the cells with vigorous shaking to an OD_{550} = 0.50–0.60.
4. Chill the flasks for 15 minutes in a large pan of ice water. Gather prechilled centrifuge tubes and bottles, pulverized dry ice, and prechilled 1.5-ml microcentrifuge tubes. *From this point on, all steps must be performed at 4°C. This is best done in a 4°C cold room.*
5. Pour the cultures into four prechilled 250-ml bottles and centrifuge at 2000*g* (4700 rpm on a Beckman JA-14 rotor) for 15 minutes at 4°C. Gently pour and pipet off the supernatant.
6. Resuspend the cells in each bottle in 150 ml of ice-cold, 1 mM HEPES buffer. Pellet the cells and remove the supernatant as in Step 5.
7. Resuspend each pellet in 100 ml of ice-cold, 1 mM HEPES buffer. Pour the cells from two of the bottles into the other two bottles, reducing the number of bottles to two. Pellet the cells, as in Step 5.
8. Gently resuspend the cells in each bottle in 10 ml of ice-cold, sterile 10% glycerol in water. Pool the cells into a single Oak Ridge tube and centrifuge at 2500*g* (5000 rpm on a Beckman JA-17 rotor) for 15 minutes at 4°C. Gently pour and pipet off the supernatant.
9. Resuspend the cells in 1.2 ml of ice-cold, sterile 10% glycerol. Place on ice and transfer 90- or 100-μl aliquots to ice-cold, 1.5-ml microcentrifuge tubes. Close each cap securely, and push each tube into a tray containing pulverized dry ice.
10. Transfer the tubes to a labeled box that has been equilibrated to –70°C, and store at –70°C. Testing of the cells for competency and contamination is described in Protocol 16.6.

The cells are now ready for use in electroporations (Protocol 16.6).

PROTOCOL 16.6

Electroporation and the Preparation of Primary Libraries

Electroporation is the most efficient means of making libraries. By transfecting *E. coli* cells with recombinant phage DNA, 10^8 to 10^9 phage clones can be produced per µg of oc and/or ccc f88-4 RF DNA. We initially grew libraries on large agar plates to reduce bias in the library; this restricted the degree of growth competition between clones (Scott and Smith 1990). The libraries we have since made by producing the phage in liquid culture, however, appear to function just as well and are much easier to prepare. Still, to reduce clonal dominance during growth, the transfected cells for each library are split among >20 cultures for phage production.

Materials

electroporation device
electroporation cuvettes with 0.2-cm or 0.1-cm gap
frozen, electrocompetent MC1061 ***E. coli*** cells (Protocol 16.5), 90 µl/tube
SOC medium containing 0.2 µg/ml **tetracycline** (see Appendix 2)
NZY medium containing 0.2 µg/ml **tetracycline** (see Appendix 2)
NZY medium containing 15 µg/ml **tetracycline**
NZY agar plates
NZY agar plates containing 40 µg/ml **tetracycline**
disposable, capped 50-ml conical polypropylene tube
disposable, snap-cap, 14-ml polypropylene tube
22 sterile, capped 125-ml culture flasks
4 sterile, capped 250-ml polypropylene centrifuge bottles

CAUTIONS: ***E. coli*, tetracycline *(see Appendix 4)***

Procedure

1. Make up samples of library DNA at no more than 200 ng of total DNA/µl in sterile purified water or 0.1x TE. As positive controls to check for competency of the cells and efficiency of the electroporation, include samples of CsCl-purified f88-4 RF DNA at 200 ng/µl and 20 µg/µl. As a negative control to check for contamination of the competent cells, include a sample of 5 µl of the water or TE that is being used to dilute the DNA. Prechill the electroporation cuvettes and DNA samples on ice. Slightly thaw a 90-µl aliquot of frozen competent cells (by rolling it between your fingers for a few seconds), and add 5 µl of DNA sample (we use a maximum of 1 µg of DNA/electroporation). Gently mix the DNA into the cell sample with a pipet tip and place the sample on ice. Remove 1-µl samples from each mixture for "pre-electroporation" titers, and keep on ice.

2. Transfect the cells by holding an ice-cold cuvette by the plastic rim, wiping it down with a KimWipe, and immediately transferring a cell–DNA mixture to the bottom of it. Transfer 95 μl for a 0.2-cm gap cuvette and 90 μl for a 0.1-cm gap cuvette. If there are any air bubbles in the mixture, give the cuvette a hard shake to remove them. Immediately electroporate the cells at 2.5 kV, 25 μFd, 200 Ohm (for cuvettes with a 0.2-cm gap), or at 1.6 kV, 25 μFd, 200 Ohm (for cuvettes with a 0.1-cm gap).

   Note: Sometimes, especially if there is too much salt present in the DNA sample, a spark and noise will occur during electroporation. These shorted samples are usually no good because electroporation of DNA into the cells does not occur at a high frequency. Be sure that your DNA sample is salt-free before use.

3. Immediately add 1 ml of SOC containing 0.2 μg/ml tetracycline to the cuvette. Combine the library electroporations in a 50-ml conical tube; cap it loosely. Transfer the 1-ml control electroporations to labeled, 15-ml polypropylene snap-cap tubes. Incubate the samples for 1 hour at 37°C with shaking at 250 rpm to induce tetracycline resistance.

4. Return to the 1-μl pre-electroporation samples that are still on ice (from Step 1). Transfer each sample to a 1.5-ml microcentrifuge tube containing 1 ml of NZY medium and 0.2 μg/ml tetracycline, and mix. Incubate for 40 minutes at 37°C with shaking. Titer each sample on NZY agar plates and NZY agar plates containing 40 μg/ml tetracycline to determine the number of viable cells and transfected colonies in each sample; see Step 8 below.

5. After the incubation period in Step 3, remove 8 μl from the electroporated library and from the controls. Dilute with NZY to 200 μl for titering on agar plates containing NZY and plates containing NZY and 40 μg/ml tetracycline. See Steps 8 and 9 below for titering of the electroporation samples, and analysis of clones from the library, respectively.

6. Split the remainder of the cells electroporated with the library DNA among 22 125-ml flasks, with each flask containing 45 ml of NZY and 15 μg/ml tetracycline; the total volume will be 1 liter. Shake the cultures at 250 rpm overnight (~20 hours) at 37°C. Combine the library cultures in 250-ml centrifuge bottles and remove a sample for analysis of phage concentration by agarose gel electrophoresis (Protocol 15.5) and for titering (see Step 8 below).

7. Isolate the phage by PEG precipitation (Protocol 15.3), followed by purification on a CsCl density gradient (Protocol 15.4). Analyze the phage by UV spectroscopy (Protocol 15.6) and agarose gel electrophoresis (Protocol 15.5), and determine their viability by titering (Protocol 15.2). Store the phage long term as in Protocol 15.10. Store short term at 4°C.

8. Collect samples of the cell/DNA mixtures before and after electroporation for titering (Steps 4–8 and Protocol 15.2). We calculate the concentration of viable MC1061 cells (colony forming units, cfu) in pre- and post- electroporation samples (from Steps 4 and 5 above) by diluting the cells in NZY and plating on NZY

agar plates (containing no antibiotic); this gives an estimate of the percent cell death caused by electroporation (an important number to follow when optimizing this experiment). The number of transfected cells in the pre- and post-electroporation samples is calculated from their titers on NZY plates containing 40 μg/ml tetracycline. If the amount of DNA in each sample is known, the number of tetracycline-resistant electrotransfectants/μg of DNA can be calculated (transducing units, TU), and the values for the library DNA can be compared with those for the RF control DNA samples (the latter should be approximately 10^8/μg of DNA). Remember that the number of colonies will be 3 or more logs different between the total cell counts (cfu, on NZY plates) and the transfected cell counts (TU, on NZY plates containing 40 μg/ml tetracycline), so calculate the dilutions to be plated accordingly.

9. Pick phage clones from the library at two different points for analysis. First, 30–40 fresh, well-separated colonies from the electroporation titering plate (Step 5) should be picked, amplified, PEG-purified, and stored (see Small-scale preparation of phage, Protocol 15.3). Second, 30–40 fresh clones should be picked off the titering plates of the fully grown library from Step 6 above. The peptide-encoding region of each clone's genome should be sequenced (Protocol 15.9), and the sequences analyzed for the correct sequence and for bias in the amino acid distribution of the "randomized" residues (see Bonnycastle et al. 1996 for an analysis of this type). If the recombinant peptide is >6–7 residues long, the phage clones can also be analyzed for the level of expression of peptide-recombinant pVIII fusion by SDS–PAGE (Protocol 18.8). These data will give an idea of the sequence and protein expression bias in the beginning library compared to that of the fully grown library.

PROTOCOL 16.7

Amplification of a Phage Library from Phage

The number of cells used for the infection depends on the number of clones in the library to be amplified. Try to keep the ratio of cells to phage particles above 5, although this may not be possible for libraries of more than 5×10^9 clones. Maintenance of this ratio is not too crucial, because infected cells continually cure themselves of fd-tet-based phage, and then are re-infected. In other words, over the course of infection, it is possible for one cell to express more than one clone. For M13-based vectors, re-infection is not as great a factor, because unlike fd-tet-based vectors, each phage particle can produce an infected cell (see Chapter 2 for more details).

Materials

minimal medium plate containing fresh colonies of K91 ***E. coli*** cells (Protocol 15.1)
NZY medium (see Appendix 2)
sterile, capped 125-ml culture flask
2-liter culture flask
4 sterile, capped 250-ml polypropylene centrifuge bottles
80 mM NaCl
ice-cold NAP buffer (see Appendix 2)
14-ml snap-cap tubes
2 sterile, capped 250-ml culture flasks
stock of phage library of known titer (TU/ml) and phage concentration (particles/ml)
22 sterile, capped 125-ml culture flasks
NZY medium containing 0.2 µg/ml **tetracycline**
20 mg/ml **tetracycline** stock
NZY containing 15 µg/ml **tetracycline**
NZY agar plates containing 40 µg/ml **tetracycline**

CAUTIONS: ***E. coli*, tetracycline** (*see Appendix 4*)

Procedure

Preparation of starved cells

1. Inoculate a 125-ml culture flask containing 15 ml of NZY medium with a single fresh colony of K91 cells; shake vigorously (225–250 rpm) overnight at 37°C.
2. In a 2-liter culture flask, inoculate 300 ml of NZY with 6 ml of the overnight K91 culture. Shake at 225–250 rpm until the cell concentration reaches $OD_{595} \cong 0.45$. Slow the shaker to 100 rpm for 10 minutes to allow the cells to regenerate their pili.

Measure the OD_{595}; it should not be over 0.65 (aim for 0.55–0.65). At $OD_{595} = 0.6$, the cell concentration ≅ 4 x 10^8 cells/ml, yielding a total of about 1.2 x 10^{11} cells.

3. Transfer the cells to two 250-ml centrifuge bottles and spin them at 1000*g* for 10 minutes. Pour off the supernatant, and briefly spin the bottles again. Remove the remaining supernatant.

4. Gently resuspend each pellet in 150 ml of 80 mM NaCl. Incubate the mixture for 45–60 minutes at 37°C with shaking at low speed (50 rpm).

5. Transfer the cells to two 250-ml centrifuge bottles and spin them at 1100*g* for 10 minutes at 4°C.

6. Resuspend each pellet in 5 ml of ice-cold NAP buffer, then transfer the cells to a 14-ml snap-cap tube. Wash the bottles with a single ml of NAP buffer, and transfer this 1 ml to the snap-cap tube. This procedure should give a final concentration of ~10^{10} cells/ml in 11 ml. Store on ice at 4°C.

Infection and titering

1. Transfer 30 µl of cells to a 1.5-ml microcentrifuge tube. Infect them with a known number of control phage (e.g., 10^6 particles of f88-4) as described in Protocol 15.2; use the same reagents and media as the large-scale infection that follows. These cells are a positive control infection that assesses the infectivity of the starved-cell preparation. Add to the remaining cells 2 x 10^{10} phage library particles and mix by gentle inversion; this should give a multiplicity of infection (moi) of approximately 1/5 (5 cells per phage particle).

2. Let the infections stand for 10 minutes at room temperature with occasional gentle swirling. Briefly centrifuge at low speed on a tabletop centrifuge to bring down droplets on the walls of the tube. Mix the cells with a 1-ml pipet.

3. Add 0.5 ml of infected cells to each of 22 125-ml flasks, each containing 45 ml of NZY and 0.2 µg/ml tetracycline; the total volume will be 1 liter. Shake the cultures at 250 rpm and 37°C for 30–45 minutes to induce tetracycline resistance.

4. Add to each culture 333 µl of 2 mg/ml tetracycline (made by diluting the stock 1:10 in NZY) to bring the antibiotic to 15 µg/ml. Mix and remove 20-µl samples from a few cultures for titering; this will quantitate the number of infected cells that will produce the library, and hence the number of phage clones that are contained in the amplified library. Dilute the 20-µl samples with NZY containing 15 µg/ml tetracycline and titer on NZY plates containing 40 µg/ml tetracycline. Well-separated colonies from a fresh overnight plate can be picked and analyzed as in Step 9 of Protocol 16.6.

5. Shake the cultures at 250 rpm overnight (~20 hours) at 37°C. Combine the library cultures in four 250-ml centrifuge bottles, and remove a sample from each for analysis of phage concentration by agarose gel electrophoresis (Protocol 15.5) and for titering. In this case, the phage will be titered as in Protocol 15.2, and the cells should be first heat-killed, and removed by centrifugation. Well-separated

colonies from a fresh titering plate can be picked and analyzed as in Step 9 of Protocol 16.6.

6. Isolate the phage by PEG precipitation (Protocol 15.3), followed by purification on a CsCl density gradient (Protocol 15.4). Quantitate the phage and check their quality by titering (Protocol 15.2), and store as in Protocol 15.10.

ACKNOWLEDGMENTS

We thank T. Ha and A. Day for making figures and tables. We thank X. Gong, K. Brown, S. Kim, M. Rashed, and J. Mehroke for contributions in developing the methods presented in this chapter.

REFERENCES

Blancafort P., Steinberg S.V., Paquin B., Klinck R., Scott J.K., and Cedergren R. 1999. The recognition of a noncanonical RNA base pair by a zinc finger protein. *Chem. Biol.* **6:** 585–597.

Bonnycastle L.L.-C., Mehroke J.S., Rashed M., Gong X., and Scott J.K. 1996. Probing the basis of antibody reactivity with a panel of constrained peptide libraries displayed by filamentous phage. *J. Mol. Biol.* **258:** 747–762.

Carcamo J., Ravera M.W., Brissette R., Dedova O., Beasley J.R., Alam-Moghe A., Wan C., Blume A., and Mandecki W. 1998. Unexpected frameshifts from gene to expressed protein in a phage-displayed peptide library. *Proc. Natl. Acad. Sci.* **95:** 11146–11151.

Dower W.J., Miller J.F., and Ragsdale C.W. 1988. High efficiency transformation of *E. coli* by high voltage electroporation. *Nucleic Acids Res.* **16:** 6127–6145.

Kunkel T.A., Bebenek K., and McClary J. 1991. Efficient site-directed mutagenesis using uracil-containing DNA. *Methods Enzymol.* **204:** 125–139.

Scott J.K. and Smith G.P. 1990. Searching for peptide ligands with an epitope library. *Science* **249:** 386–390.

17 Screening Peptide Libraries

ALFREDO MENENDEZ, LORI L.C. BONNYCASTLE, OSCAR C.C. PAN, AND JAMIE K. SCOTT

Department of Molecular Biology and Biochemistry and Department of Biological Sciences, Simon Fraser University, Burnaby, B.C., Canada V5A 1S6

IN THIS CHAPTER WE COVER METHODS for screening multiple phage-library samples and their controls side by side on 96-well microplates and with magnetic beads, in Eppendorf tubes. The steps of panning presented here include:

- preadsorption of the libraries to remove plate and streptavidin binders
- affinity purification of phage with a biotinylated screening molecule and streptavidin (biopanning)
- washing to remove weakly bound and nonbinding phage
- acid elution to release bound phage
- infection of bacterial cells with the eluted phage to amplify the selected phage clone

CONTENTS

Yields of phage from the panning (phage enrichment) can be assessed by titering the eluted phage before their amplification, and phage production by the infected bacterial cultures can be determined by agarose gel electrophoresis. The phage pools produced by the infected cells can be assessed for binding to the screening molecule by ELISA (discussed in Chapter 18).

The Purpose of Panning a Library Determines Which Screening Experiments to Use

The primary purpose of panning a peptide library is always the same—to find peptide ligands for the screening molecule—however, the way in which the screening is performed can vary, depending on the types of data one seeks and the screening molecule(s) used. For instance, one may wish to identify the "tightest" binding peptides in a library *and* to determine the extent of sequence variation among binding clones. Achieving these two goals may require two types of panning experiments. To obtain the tightest binders, highly discriminatory screening conditions should be used, and it may be worthwhile to construct and screen "sublibraries" if only weak-binding phage are isolated after a set of thorough pannings has been performed. In contrast, less stringent panning conditions should be used to obtain the greatest sequence diversity. Reasons for assessing sequence diversity include an expectation of more than one peptide-binding site on the screening molecule and a heterogeneous population of screening molecules, e.g., polyclonal antibody. Thus, it is important to decide ahead of time the purpose of the screening so that appropriate screening conditions are designed. In designing screening experiments, it is useful to consider the different phage populations present in a given library:

- nonbinding phage (this constitutes the dominant population)
- nonspecific, plastic-binding phage and phage that bind to streptavidin or BSA (known as "plate binders")
- phage that specifically bind the screening molecule weakly
- phage that specifically bind the screening molecule tightly

Figure 17.1 shows a conjectural subdivision of a phage library of ~10^8 clones into different populations. The library is overwhelmingly dominated by clones that do not bind to either the screening molecule "X" or the other components of the screening system (i.e., the plate, blocking proteins, streptavidin). In the process of panning, these background clones are removed by repeated washing (see Table 17.1). Parmley and Smith (1988) showed that a Tris-buffered saline wash containing 0.5% (v/v) Tween 20 gave phage yields with lower backgrounds than those obtained with the same buffer containing lower concentrations of the detergent. Similarly, we have observed that high (150 mM) salt concentrations are also important for maintaining low background binding (O. Pan et al., unpubl.). We use the phage vector that was used to construct our libraries (f88-4) as a negative control phage in side-by-side pannings with the libraries to mimic the effect of washing on the reduction in background phage within the library. Thus, if the yield of negative control phage is high, one can assume that the washing conditions were not as effective as they should have been, and one should consider repeating the panning.

Also shown in Figure 17.1 is the next-smaller population of phage, the plate binders. This population comprises several subgroups of phage, each of which binds to a different component of the screening system besides the screening molecule itself. Adey et al. (1995) showed that one population of phage, displaying peptides with high tryptophan content, bound selectively to polystyrene plates, especially under conditions that included high salt and Tween 20 concentrations. Further work by Gebhardt et al. (1996) identified a WXXWXXXW motif (in which X = any amino acid) that binds polystyrene/polyurethane magnetic particles even in the presence of blocking agents and surfactants. Several groups (Kay et al. 1993; Giebel et al. 1995; Caparon et al. 1996), most notably Devlin et al. (1990), have identified streptavidin-binding phage. We routinely preadsorb the libraries in microwells containing all of the com-

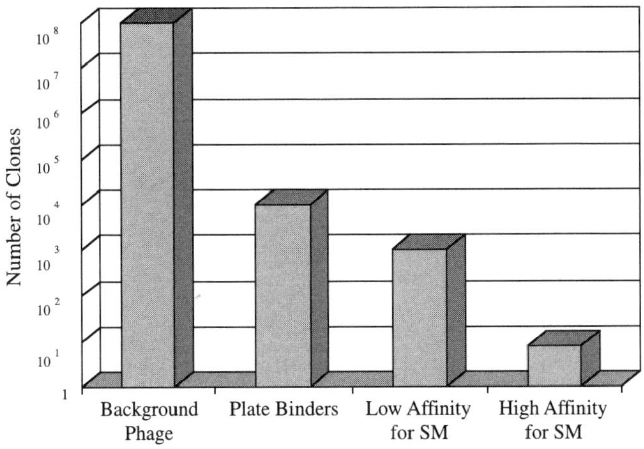

Figure 17.1. Relative sizes of different populations of phage in a hypothetical library of 100 million clones. Background phage are those that do not bind to the plate at all. Plate binders are phage that bind to some component of the screening other than the screening molecule itself (i.e., the plastic plate, immobilized BSA, or streptavidin). Phage clones that bind specifically include those with low and high affinity for the screening molecule (SM).

ponents of the screening, except the screening molecule, to remove plate-binding phage before the actual panning experiment (Table 17.1). We perform ELISA on the plates used for preadsorption to determine the extent to which plate-binding phage are present in the libraries or amplified, enriched phage pools before each round of panning. When preadsorption ELISA signals are high and yields from the panning are low or moderately high, it signals that the phage binding may be nonspecific.

The last two populations in Figure 17.1 represent phage clones that bind specifically to the screening molecule with low and high affinty. There are many more low-affinity clones than high-affinity clones. We use the following strategy to obtain a population of tighter-binding clones. In the first round of panning, the phage are incubated on a plate that has been coated first with streptavidin (Sa), and then with the biotinylated screening molecule. Because Sa (like most proteins) sticks to plates in patches, the immobilized "sandwich" consists of closely apposed streptavidin molecules to which biotinylated screening molecules are attached. Because the off-rate of

Table 17.1. Different types of "nuisance" phage clones, methods for removing them, and methods of assessing the success of the removal methods

Type of nuisance clone	Method for removing	Method of assessing removal
Nonbinding background	repeated washing	measure background phage binding with negative control phage
Plate-binders	preadsorption	ELISA of the preadsorption plate: library phage should give the same signals as negative control phage
Phage that bind the screening molecule with low affinity	phage capture in solution	higher ELISA signals are obtained from samples that are incubated with lower concentrations of screening molecule

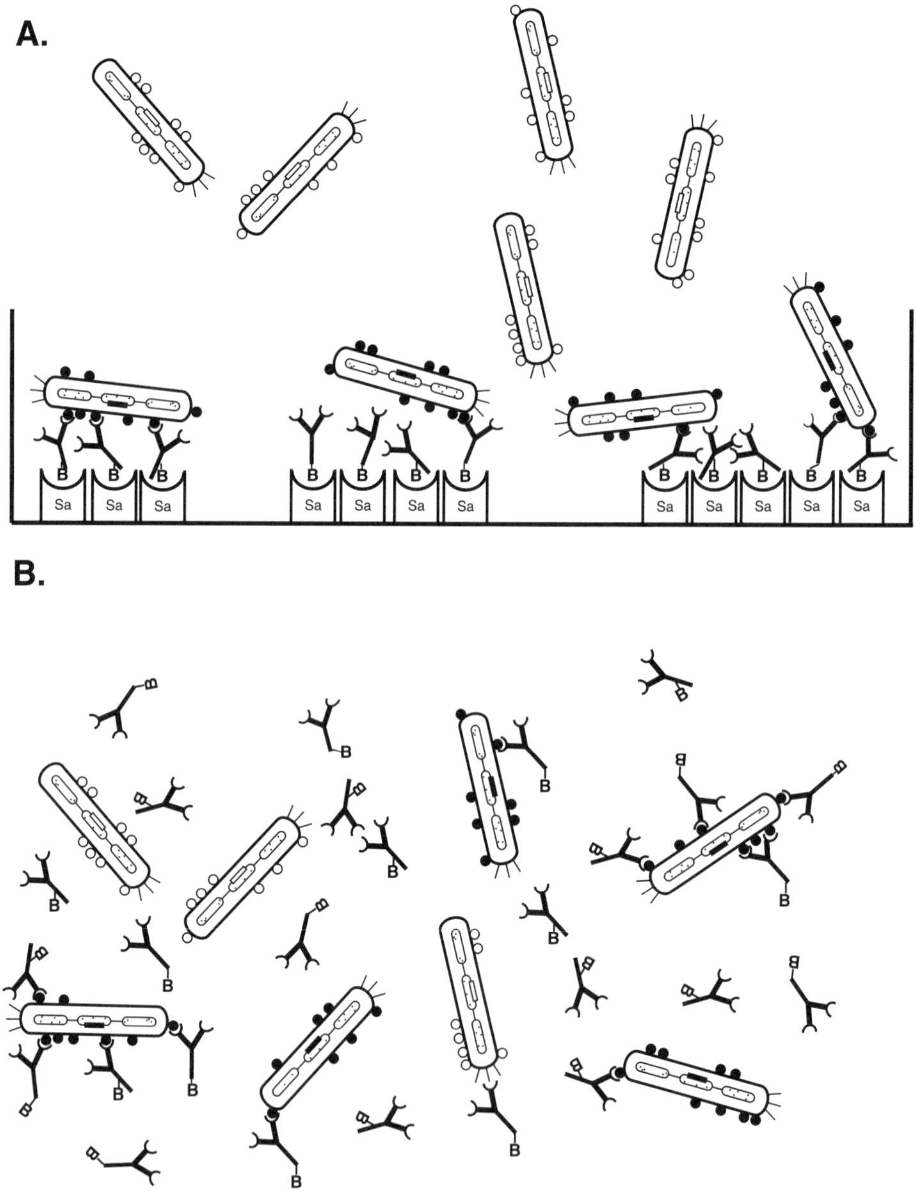

Figure 17.2. (*A*) Phage capture on solid phase. Phage bearing pVIII-displayed peptides are captured by multivalent binding to biotinylated (B) IgG that has been immobilized on streptavidin (Sa). Note that streptavidin binds plastic in small "islands" and thus is very closely apposed. This configures the bound IgG in a high density that effectively captures even phage bearing weak-binding peptides. (*B*) Binding of biotinylated IgG to phage in solution. At equilibrium, more antibody molecules are bound to phage bearing tight-binding peptides (*filled circles*) relative to phage bearing weak-binding peptides (*open circles*). Because of this difference, phage bearing tight-binding peptides have a better chance of being multivalently captured onto the streptavidin-coated plate via the biotinylated IgG. Captured phage would look something like those shown in *A*. Note that free biotinylated IgG would diffuse faster and bind the streptavidin sites before the phage could bind. Also note that if the incubation on the streptavidin lasts too long, even weak-binding phage will be captured multivalently by the immobilized IgG. The total molar amount of biotinylated IgG should be less than the capacity of the streptavidin, which we gauge at 2 pmole/well.

biotin from Sa is very slow (corresponding to $K_d \cong 10^{-15}$ M for free biotin), virtually all of the screening molecules remain immobilized throughout the incubation and washing steps. As shown in Figure 17.2A, the multivalent nature of peptides displayed by f88-4 phage allows binding to the screening molecule via several peptides. The avidity boost resulting from this multivalent capture dramatically decreases the apparent off-rate of the phage; thus, phage clones bearing tight- and weak-binding peptides are both captured effectively during the 2-hour incubation period. These conditions give a high yield per clone but low discrimination of binding affinities; thus, the goal is to capture, as effectively as possible, all binding phage. For a rigorous analysis of the dynamics of phage-library screening described above, based on a stochastic model, see Levitan (1998).

In contrast to solid-phase panning, binding the libraries to biotinylated screening molecules in solution and then capturing the complexes on Sa promotes affinity discrimination. Figure 17.2B shows that for a given concentration of screening molecule at equilibrium with the phage, phage bearing high-affinity peptides bind more of the screening molecule than phage displaying low-affinity peptides. Equilibrium at the "right" concentration of screening molecule allows the clones bearing tight-binding peptides to bind multiple screening molecules, whereas those bearing low-affinity peptides bind few or no screening molecules. The drawback of this strategy is that capture of the phage from solution onto immobilized Sa is inefficient, causing the yield per clone to be low compared to the solid-phase panning method described above. Affinity discrimination, however, is maximized, because the phage bearing high-affinity peptides are more likely to be captured multivalently and survive the washing step than are those bearing low-affinity peptides. For pVIII-displayed peptides, there is a linear relationship between peptide affinity and the number of screening molecules bound per phage; thus, as long as the screening molecule concentration during the overnight incubation to equilibrium is not so high as to bind in multiple copies to both low- and high-affinity peptides, discrimination should occur. (Of course, the concentration cannot be too low, because under such conditions, neither phage population would be captured.) When designing a panning experiment, later rounds of panning should involve in-solution binding of the screening molecule to the amplified phage pools. As the "right" concentration of screening molecule cannot be known a priori, it works best to incubate the pools at several different concentrations of screening molecule and to perform multiple rounds of panning at each concentration. ELISAs of the amplified phage pools from each different condition can be compared in side-by-side studies to identify the pools containing the tightest-binding clones (Table 17.2).

Table 17.2 shows the results from screening a sublibrary that was constructed on the basis of the consensus sequence identified from a previous set of pannings with the anti-cytochrome-*c* antibody E8 (Malvaganam et al. 1998). The library comprised approximately 200 million clones bearing a 4-residue, fixed consensus sequence interspersed with 8 randomized residues encoded by NNK codons. The library was panned with biotinylated E8 IgG captured on Sa-coated microtiter wells (Protocol 17.2) for the first round (see Figure 17.3 for the panning phylogeny). For the second round of panning, biotinylated Fab was incubated with the libraries at concentrations of 1, 10,

Table 17.2. Results of the E8 biopannings in percent yields

BP Fab concentration (nM)	R2a Input from R1, SP, IgG	R3a Input from R2a [1 nM], Sln, Fab	R3b Input from R2b, SB, IgG	R3c Input from R2c, SB, Fab	f88 control Input: fresh phage
1	1.3 e-3 (.588)	—	1.6 e-2 (.534)	1.4 e-2 (.549)	3.1 e-4
10	1.2 e-2 (.326)	—	1 e-1 (.363)	4.2 e-1 (.354)	4 e-4
50	1.4 e-1 (.065)	—	15 (.077)	15 (.083)	1.2 e-5
0.1	—	1.1 e-2 (.665)	—	—	<2.2 e-3 (.023)

(BP) Biopanning; (SP) solid capture on plate; (SB) solid capture on beads; (Sln) in-solution; (R) round.

See Fig. 17.3 for phylogeny of pannings. ELISA data for each pool are shown in parentheses. R2a, 3b, and 3c were performed side by side; each was performed using three concentrations of biotinylated Fab (1, 10, and 50 nM) and 10^{10} input phage from the rounds indicated. R3a was performed using 0.1 nM Fab and an input of 10^9 phage particles from R2a (1 nM). For the ELISA, 10^{10} phage were captured on immobilized anti-phage antibody, followed by incubation with 80 nM biotinylated-E8 Fab, then detection with ABC and ABTS developer (see Protocols 18.1 and 18.4 for ELISA details). The f88 controls are biopannings using 10^{10} particles of f88-4 vector phage at the specified Fab concentrations. This gives a measure of "background" yields and ELISA signals.

ELISA controls: The ELISA signal for 10^{10} R2c phage was 0.039; the ELISA signal for 10^{10} E8 sublibrary phage was 0.019; the ELISA signal for a BSA-blocked well was 0.027; the ELISA signals for the parental E8 clones (10^{10} particles) on which the sublibrary is based were Clone R4E2.7, 0.044; Clone R3E4.7, 0.025.

and 50 nM, and captured onto Sa-coated beads (Protocol 17.4), or the phage were captured on biotinylated IgG or Fab immobilized on Sa-coated beads. For the third round of panning, biotinylated Fab was incubated in solution at 0.1, 1, 10, and 50 nM with the amplified phage from round 2 and captured onto Sa-coated beads (Protocol 17.4). The ELISA data clearly show that, when using in-solution capture, one can select tighter-binding phage at lower Fab concentrations, even though the phage yields decreased as the Fab concentration was decreased. This decrease in phage yield occurs because at a concentration of 1 nM there are approximately 2 x 10^{10} Fab molecules in the 35-µl reaction. At that low concentration, only a small fraction of the available phage is captured by multiple Fabs. This makes it difficult to compare percent yields between rounds of panning as a measure of phage enrichment. Instead, one must rely on the ELISA data when identifying the pools having the tightest-binding phage.

Unfortunately, one never knows at the beginning of a library screening experiment whether the best-binding phage in the library will have high or low affinity for the screening molecule, or whether the frequency of binding clones will be relatively high or low. Thus, we take a conservative approach and pan the library through three rounds of solid-phase screening to ensure that binding phage can be isolated. We then return to the amplified phage from the first round of panning and repeat the second and third rounds of panning with in-solution capture at relatively high concentrations (~10 nM) of screening molecule. Finally, if the yields and ELISA signals are high for these pannings, we may repeat the screening again using a 10- to 100-fold lower concentration of screening molecule, or we may limit our focus to a few libraries and pan them simultaneously at several concentrations of screening molecule, as in Figure 17.3. When we are finally satisfied with the affinities of the enriched phage pools, we isolate clones from the most interesting ones, sequence them, and perform ELISA on them. If we discover consensus sequences between multiple independent clones, and we are not yet satisfied with the ELISA signals, we may turn to constructing sublibraries in which the consensus residues are fixed and the other residues in the sequence are randomized (see Chapter 16). Careful screening of these libraries, following the approach used to screen the E8 sublibrary (Fig. 17.3 and Table 17.2), should identify clones that bind even more tightly to the screening molecule.

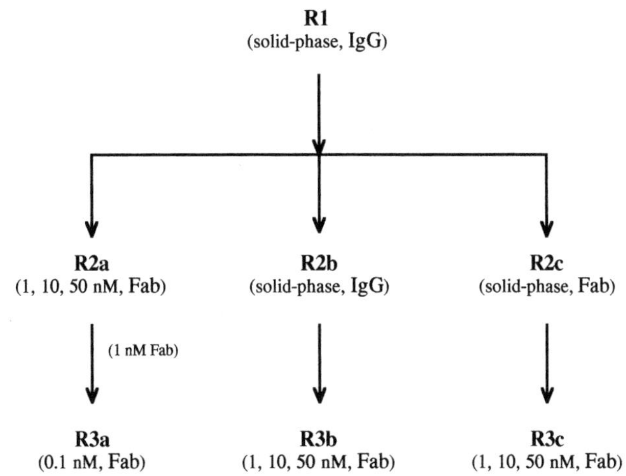

Figure 17.3. Phylogeny of pannings of the E8 sublibrary. R1 denotes Round 1, R2a denotes Round 2a, and so forth. See Table 17.2 for percent phage yields and ELISA data.

PROTOCOL 17.1

Biotinylation of Proteins

The biotinylation reaction described here conjugates a biotin molecule attached to a 22.4-Å-long carbon linker arm to primary amines on a protein (in this case, an IgG molecule) via an *N*-hydroxysuccinimide reactive group. The most reactive primary amines are the ε-amines of lysine residues, followed by amino-terminal amines. The linker places the biotin group far enough from the protein surface to make it more accessible to avidin's biotin-binding site, which is reported to be 9 Å below avidin's surface (Green et al. 1971). It is assumed that the presence of the linker does not affect the off-rate of the biotin–avidin interaction, whose K_d is on the order of 10^{-15} M; the affinity of biotin for streptavidin is about the same. After the conjugation step, Tris or ethanolamine is added to the reaction to inactivate any unreacted biotin-linker. BSA is then added to act as a carrier, and the unconjugated biotin-linker is removed by ultrafiltration, gel filtration, or dialysis. The biotinylated antibody is stored long term in 50% (v/v) glycerol.

This procedure adds about six biotins per whole IgG molecule (MW ~150K). One can roughly assume that if the same amount (in µg) of another protein is used, the same total amount of biotin will be added. Thus, for example, if Fab (MW ~50K) was biotinylated, about two biotins should be added per molecule. The number of biotins added to a protein can be checked with the colorimetric assay described below. Note that the Centricon ultrafiltration apparatus used in this protocol to remove biotin and exchange buffers was chosen for IgG purification; proteins of other molecular weights may require other types of Centricon.

Unfortunately, the reaction conditions for biotinylation destroy the activity of some proteins, either because they denature the protein or because biotin has been conjugated to crucial lysine residues. Denaturation by high pH can be circumvented by separating the protein from unconjugated biotin linker on a desalting column or by using buffered Tris at neutral pH, rather than using ethanolamine to inactivate the NHS group on the linker. Methods are available from Pierce for conjugating biotin to carbohydrates (biotin-hydrazide) or to free cysteine residues (biotin-BMCC), but these procedures also have drawbacks. Conjugating to carbohydrates uses harsh oxidizing conditions that are unsuitable for most glycoproteins, and free cysteine residues are often not available on the target protein. To circumvent the harsh conditions (100 mM sodium meta-periodate) used for the coupling of biotin-hydrazides to carbohydrate functionalities, Pierce suggests scaling back to milder oxidizing conditions (5–10 mM). For the biotinylation of monoclonal antibodies, Pierce recommends the use of sulfhydryl-reactive biotin. With this method, the reducing agent 2-mercaptoethylamine is used to split the antibody into two identical 75-kD fragments bearing free sulfhydryls from the hinge region, and biotin is conjugated to these groups, usually without affecting the antigen-binding site. This results in the IgGs becoming biotinylated Fabs, which in turn may cause decreased ELISA signals. See the Pierce catalog, their website (http://www.piercenet.com), or contact their technical support for more information.

Materials

protein to be biotinylated (see Step 1)
Centricon-50 ultrafiltration apparatus (Millipore Corp., Cat. # 4243)
refrigerated tabletop centrifuge
fixed-angle rotor (Fisher, Cat. # 04-976-694)
sterile phosphate-buffered saline (1x PBS), pH 7.4, or sterile Tris-buffered saline, pH 7.5 (1x TBS; see Appendix 2)
siliconized 1.5-ml microcentrifuge tubes
1 M sodium bicarbonate buffer ($NaHCO_3$), pH 8.6, filter-sterilized (see Appendix 2)
sulfo-NHS-LC biotin (Pierce, Cat. # 21335ZZ)
2 mM sodium acetate, pH 6.0 (see Appendix 2)
1 M **ethanolamine,** pH 9.0, or 1 M Tris-HCl, pH 7.4 (see Appendix 2)
50 mg/ml BSA in 1x TBS, filter-sterilized (Fraction V; Sigma Chemical Co., Cat. # A-3912)
sterile 100% (v/v) glycerol
6.5% (w/v) **sodium azide (NaN_3)** stock solution

CAUTIONS: ethanolamine, sodium azide (*see Appendix 4*)

Procedure

1. If amines are present in the storage buffer of the protein to be biotinylated, remove them by washing the protein on a Centricon-50 apparatus with five 2-ml volumes of 1x PBS (see Step 6). The solution should be at a final volume of approximately 50 µl and should contain 40–100 µg of protein.

   Note: For IgG purification, use a Centricon-50 or -30, because large losses of IgG will occur with a Centricon-100.

2. Transfer the protein solution to a siliconized, 1.5-ml microcentrifuge tube. Add 11 µl of 1 M bicarbonate buffer, and mix. (If the protein is unstable at higher pH, do not add the bicarbonate.)

3. Make a 0.5 mg/ml solution of Sulfo-NHS-LC biotin in 2 mM sodium acetate, pH 6.0. Make this solution right before Step 4, as the *N*-hydroxysuccinimide moiety of the biotin-linker will begin to hydrolyze as soon as the solution is made.

4. Add 50 µl of Sulfo-NHS-LC biotin solution to the protein solution from Step 2, mix, and incubate for 2 hours at room temperature.

5. Add 500 µl of 1 M ethanolamine, pH 9.0, or 1 M Tris-HCl, pH 7.4, and mix. Continue to incubate for another 30 minutes at room temperature. The ethanolamine or Tris molecules need to be in at least 20-fold excess over the biotin-linker molecules. This step inactivates the remaining free biotin-linker. Add 20 µl of 50 mg/ml BSA to the solution as carrier, followed by 1 ml of 1x TBS or 1x PBS.

6. Transfer the solution to a Centricon-50 apparatus, fill to the top (2 ml) with TBS or PBS, and centrifuge on a tabletop centrifuge at 5000*g* (7000 rpm) for 50 minutes, or until the volume has been reduced to approximately 50 µl.

   Note: If desired, the reservoir can be sealed with Parafilm. Be sure to poke a hole in it with a needle, otherwise the vacuum created by the movement of the liquid out of the reservoir during centrifugation will suck the Parafilm into the reservoir. This Centricon step can be replaced with a dialysis or gel filtration step.

7. Wash the biotinylated protein three times as above with 2 ml of TBS or PBS. Transfer the protein solution to the cup provided by Millipore by fitting the cup onto the top of the apparatus, carefully flipping it over, and centrifuging at 5000*g* (7000 rpm) for 2 minutes. Transfer the retentate from the cup to a 1.5-ml microcentrifuge tube, and gauge the volume of the solution by taking it into a micropipet tip and adjusting the volume dial on the pipettor until the meniscus of the solution reaches the tip of the pipet; the volume reading on the dial will be roughly the volume of the solution. Add an equal volume of sterile 100% glycerol and mix thoroughly. If desired, add sodium azide to a final concentration of 0.02% (w/v). Calculate the final concentration of the biotinylated protein assuming a 100% yield. Store at –18°C.

8. Sometimes biotinylation damages the functional activity of the protein sample. Thus, if possible, the protein′s activity should be checked; for example, by ELISA for binding activity or by spectrophotometry for catalytic activity if the protein is an enzyme. If functional activity is lost, consider an alternative coupling method.

DETERMINING THE AVERAGE NUMBER OF BIOTINS CONJUGATED PER PROTEIN MOLECULE

Green (1965) developed this assay based on his observation that 4′-hydroxyazobenzene-2-carboxylic acid (HABA) forms a red complex with avidin (MW 67K) or streptavidin (MW 60K) that can be spectrophotometrically monitored at 500 nm. Each avidin molecule also has four biotin-binding sites, but only about 3.72 biotins are bound at maximum; streptavidin behaves similarly. Because of a large difference in K_d between HABA (~10^{-6} M) and biotin (~10^{-15} M), one biotin molecule can displace one HABA dye molecule from its binding site on avidin or streptavidin. Thus, for a mixture of avidin with an excess of HABA, absorbance will decrease as biotin is added. Avidin is used in this assay because it is less expensive than streptavidin.

Materials

2.40 mg **4′-hydroxyazobenzene-2-carboxylic acid** (2-[4′-hydroxyazobenzene] **benzoic acid; HABA**) in 10 ml of 1x PBS (1 mM; Sigma Chemical Co., Cat. # H-5126)
1 mg/ml (15µM) avidin from egg white in 1x PBS (Sigma Chemical Co., Cat. # A-9275)
250 µM D-**biotin** stock solution in 1x PBS (from 5 mM stock, ICN Biomedical Inc., Cat. # 101023)
100-µl microcuvettes

CAUTIONS: **HABA, biotin (*see Appendix 4*)**

1. Make a standard curve for HABA displacement by biotin: Make a 200-μl solution of 100 μM HABA and 10 μM avidin in PBS. Measure the initial OD_{500} of 80 μl of the mixture in a 100-μl microcuvette. Add 1-μl aliquots of 250 μM biotin and record the change in OD_{500}. Plot the OD_{500} versus the nanomoles of biotin added.
2. Estimate the number of biotins incorporated per protein: Measure the initial OD_{500} of 80 μl of the HABA/avidin mixture from Step 1 in a 100-μl microcuvette. Add 5 μl of biotinylated protein solution at a concentration of about 1 mg/ml. Record the change in OD_{500}. Compare this reading with the standard curve to determine number of moles of biotin present in the sample. The number of moles of biotin divided by the number of moles of biotinylated protein in the 5-μl sample equals the number of biotins per protein.

PROTOCOL 17.2

Panning Phage Libraries on Solid Phase with an Automatic Plate Washer

This procedure is based loosely on that described in Bonnycastle et al. (1996), which was adapted for multiple samples of fd-tet-based phage libraries from the single-sample panning method of Parmley and Smith (1988; see also Smith and Scott 1993 for a more complete description of this single-sample panning method). There are several advantages to panning libraries on plates. The main advantage is that positive and negative control pannings can be performed side by side with the test screening molecule. The positive control pannings are performed with library reconstructions comprising the phage vector mixed with different amounts of a phage clone known to bind a given screening molecule. For example, we use phage displaying the peptide sequence YDVPDYA and screen with the monoclonal, antipeptide antibody 17/9 (Rini et al. 1992). We screen library sample sizes of 10^{11} phage particles (equivalent to 10^2–10^3 particles per clone in a library, depending on each library's size) and use as reconstructions 10^{11} f88-4 phage mixed with 10^2, 10^4, and 10^6 YDVPDYA phage (the K_d for this peptide and the 17/9 antibody is 0.67 nM; L. Craig et al., unpubl.). We expect to see the reconstruction with 10^6 YDVPDYA phage show increases in yield after the second round of panning, whereas the yields from the 10^2-phage reconstruction typically won't begin to increase until at least the third round of panning. If the yields from pannings with our test screening molecules have not increased by the round in which the 10^2-phage control yield has increased, it is likely that even phage with nanomolar affinity have not been captured. A negative control panning with 10^{11} f88-4 phage is also performed on every screening molecule tested as a means of showing the level of background binding of phage under the given panning conditions.

The results of side-by-side, identical pannings with an automatic plate washer are very consistent, so the results from the positive and negative control pannings are meaningful. Instead of a plate washer, one could use a wash bottle filled with wash solution and slap the plates dry (see Day 2, Step 2), but results from side-by-side replicates will not be as consistent, and there is a greater chance that wells will be contaminated by neighboring pannings.

Panning, as we now perform it, is preceded by a preadsorption step, in which the libraries are mixed in high-salt buffer containing 0.5% Tween 20 and incubated on a streptavidin-coated microplate that has been blocked with BSA. This removes both plate-binding (Adey et al. 1995) and streptavidin-binding (Devlin et al. 1990; Kay et al. 1993; McLafferty et al. 1993; Roberts et al. 1993; Giebel et al. 1995; Caparon et al. 1996; Gebhardt et al. 1996) phage that can contribute to high background binding in the panning reactions. We routinely preadsorb the phage before each round of panning and then perform direct ELISA on the preadsorption plates using anti-phage antibody–HRP conjugate to detect bound phage. This ELISA allows us to compare the level of nonspecific plate binding by phage in the library panning reactions with binding of the negative control panning performed with the phage vector. Typically, the amount of phage that bind nonspecifically decreases after the first round of panning. If the adsorption signal from a given well increases, compared to the f88-4 control well, a positive yield on panning may indicate that plate- or streptavidin-binding phage are being isolated. The preadsorbed libraries are panned in wells that have first been coated with streptavidin, blocked with BSA, and loaded with biotinylated screening molecule. As a further precaution against selection for streptavidin-binding phage, we block the remaining free sites on streptavidin with biotin before adding the preadsorbed libraries.

Phage elution can be performed two ways (see Day 2, Step 9). The phage can be released from the screening molecule by denaturation at pH 2.2 and transferred to wells containing a neutralizing solution. Alternatively, bacteria can be added directly to the washed wells at the end of the panning reaction, as our studies indicate that direct infection of bound phage results in, at most, 50% fewer infected cells as compared to acid elution. One caution about the direct infection procedure: Phage bound to a screening molecule via pIII may not be as infective as their pVIII-display counterparts (see Fig. 17.2A), because binding via pIII-displayed peptides may block the phage's pIII molecules, which mediate infection.

For M13-, fd-, or f1-based phage or phagemid libraries, as little as 100 equivalents of library may be used rather than the 1000 equivalents recommended for fd-tet-based libraries. This is because it takes about 20 f88-4 phage particles to produce a single transfected cell (transducing unit, TU), whereas approximately 1 particle of wild-type phage or phagemid is needed to produce a single plaque-forming unit (pfu). Thus, about the same number of infected cells will be produced at the end of a given screening of 20 times fewer library-equivalents of wild-type-like phage compared to fd-tet-derived ones. When deciding how many library equivalents to use, one should consider that although background increases with the amount of phage screened, so does the chance that a given clone will be captured by the screening molecule or even be present in the sample screened. The number of library-equivalents of phage that would be necessary to ensure the presence of each clone in the sample can be calculated statistically, but this number is also subject to biological chance, because some clones are overrepresented in the library, whereas others are not expressed at all. All of these uncertainties make it difficult to determine the optimal number of phage needed in the panning to ensure successful capture and subsequent infection.

Be aware of the procedural steps that should be changed if a buffer other than TBS is required for the screening; for example, if the screening molecule is unstable in TBS. Moreover, if the screening molecule is unstable in the library storage buffer (in our case, Tris-HCl), then the samples of phage library should be PEG-precipitated and resuspended in the buffer of choice (see Protocol 15.3) before commencing with the screening. This buffer should be used in the incubation and washing steps to maintain the stability and binding-site conformation of the screening molecule. It is wise to determine ahead of time whether the screening molecule will be stable under the screening conditions, especially when mixed with 0.5% (v/v) Tween 20. This can be checked by assaying the molecule in its biotinylated form after binding to streptavidin-coated microtiter wells and washing with an automatic plate washer containing the buffer of choice.

The buffer of choice should also be checked for its effects on the nonspecific adsorptive properties of phage and, if the buffer creates harsh conditions, on phage viability. It is important that the high concentrations of salt and Tween 20 be maintained in the washing buffer, as the absence of either will result in a high background of nonspecifically bound phage. If the buffer of choice does not contain salt, one should check the background binding of the vector phage (like f88-4) in that buffer before moving on. Incubation time, temperature, the number of washes, and time elapsing between washes can all be varied to optimize screening molecule viability and phage background binding and viability.

Materials

sterile 14-ml snap-cap polypropylene tubes
flat-bottom 96-well microtiter plates (Corning Costar, Cat. # 3369)
sterile plastic reservoirs for a multichannel pipettor (Fisher, Cat. # 21-380-3)
sterile 1x TBS, pH 7.5 (see Appendix 2)
streptavidin (Roche Molecular Biochemicals, Cat. # 973190)
sterile 1.5-ml microcentrifuge tubes
plastic lids for microtiter plates (Corning Costar, Cat. # 3930)
plastic box and lid (such as Rubbermaid makes) that will accommodate a 96-well microplate
plastic wash bottle filled with sterile 1x TBS (store at 4°C or on ice)
dialyzed BSA (Sigma Chemical Co., Cat # A-3912)
filter-sterilized 1x TBS containing 2% (w/v) dialyzed BSA (make fresh or add 0.02% [w/v] **sodium azide** [NaN_3] if the solution is to be stored)
phage library stock solutions (~10^{13} particles/ml)
filter-sterilized 1x TBS containing 2% (w/v) dialyzed BSA and 1.5 % (v/v) Tween 20 (as above, **NaN_3** is optional)
96-well Falcon flexiplates (Becton Dickinson, Cat. # 353911)
FALCON flexiplate lids (Becton Dickinson, Cat. # 353913)
K91 *E. coli* cells (see Protocol 15.1)
NZY medium (see Appendix 2)
automatic plate washer (BioTek Instruments, model EL 403)
biotinylated screening molecule (see Protocol 17.1)
5 mM **biotin** stock solution in purified water
filter-sterilized 1x TBS containing 1 mM D-biotin (ICN, Cat. # 101023) and 2% BSA (as above, **NaN_3** is optional)
1x TBS containing 0.5% (v/v) Tween 20 (TBS/Tween)
filter-sterilized 1x TBS containing 2% (w/v) dialyzed BSA and 0.5% (v/v) Tween 20 (as above, **NaN_3** is optional)
panning elution buffer (see Appendix 2)
panning neutralization buffer (see Appendix 2)
Luria broth (LB, see Appendix 2)
20 mg/ml **tetracycline** stock solution (for fd-tet-based libraries, see Appendix 2)

tabletop shaker incubator
840 mM stock solution of **IPTG** (for f88-4 libraries)
tabletop centrifuge rotor for microplates (Fisher, Cat. # 04-976-410)
sterile round-bottom 96-well microtiter plate (Corning Costar, Cat. # 3791)
materials for titering infected cells (see titering method at the end of this protocol)
refrigerated tabletop centrifuge
materials for agarose gel electrophoresis of viral DNA (see Protocol 15.5)
ELISA reagents (see Protocol 18.7)

CAUTIONS: sodium azide, *E. coli*, biotin, tetracycline, IPTG (*see Appendix 4*)

Procedure

ROUND 1

Day 1

1. *Coat the wells of two flat-bottom microtiter plates with streptavidin.* Make a fresh solution of streptavidin (1 µg/35 µl) as follows. Place 1 ml of 1x TBS in a 1.5-ml microcentrifuge tube and add 214 µg of streptavidin. Mix gently, microfuge briefly, and transfer the protein solution to a multichannel pipettor reservoir containing 6.5 ml of 1x TBS. Mix with a 1-ml pipettor. Add 35 µl of streptavidin solution to each well of two flat-bottom microtiter plates. If, in the last round of pipetting, the multichannel pipettor does not fill a few tips, use a 100-µl pipettor to transfer 35 µl of streptavidin solution from the reservoir to each well that was missed. Cover each microtiter plate with a plastic lid, and incubate one plate for 1 hour at 37°C sealed in a humidified plastic box (i.e., put a paper towel wetted with warm tap water on the bottom of the box and place the microplate on it). Place the other plate in a sealed plastic box and incubate overnight at 4°C. Alternatively, both plates can be loaded 1–3 days before Day 1 and incubated at 4°C in a sealed plastic box.

2. *Preadsorb the phage libraries on one of the two streptavidin-coated plates.* Remove the streptavidin solution from the plate that was incubated 1 hour at 37°C by filling each well with cold TBS from a wash bottle, shaking out the plate's contents in a sink, and firmly slapping the plate dry on a stack of paper towels. Immediately add 200 µl of 2% dialyzed BSA in TBS to each well and incubate for 30 minutes at 37°C. Slap out the BSA solution and wash the plate three times with 1x TBS from the wash bottle. Immediately after slapping out the plate from the last wash, add 10^3 to 10^4 equivalents of each phage library (e.g., 2×10^{11} particles for a library of 2×10^8 clones would be 10^3 equivalents) and 1x TBS containing 2% dialyzed BSA plus 1.5% Tween 20 to bring the final volume to at least 35

Materials

automatic plate washer (BioTek Instruments, model EL403)
jug containing washing buffer (usually 1x TBS/0.5% Tween 20, see Appendix 2)
jug containing **bleach** diluted 1:10 with deionized water
jug containing deionized water
a clean polystyrene microplate

CAUTION: **bleach (*see Appendix 4*)**

Procedure

1. Attach the red (washing buffer) hose to a jug containing 10% bleach and prime twice. Set the DISPENSE function at 200 μl and dispense the bleach in a plate. Aspirate and repeat two more times.
2. Run the DAY MAINTENANCE routine using 10% bleach and leave the pins soaking for 30 minutes.
3. Rinse the end of the red hose with distilled water and connect it to the water jug. Run the OVERNIGHT MAINTENANCE routine twice.
4. Connect the blue (water) hose to the water jug and run the OVERNIGHT MAINTENANCE routine once.
5. Connect the red hose to the TBS/Tween jug and prime it once. DISPENSE 200 μl and aspirate three times, then test the pH of several wells with litmus paper; it should be 7.5 before proceeding.

TITERING PHAGE ELUATES TO DETERMINE PERCENTAGE YIELDS

Before titering, first review Protocol 15.2, paying special attention to Steps 5–8. Each row of the infections from Step 14 above will be diluted three times and, thus, will require three rows of a flexiplate. Set up the flexiplate for 1/10, 1/100, and 1/1000 dilutions by adding 45 μl of NZY medium containing 15 μg/ml tetracycline to each of three rows. These dilutions will be spotted on well-dried NZY agar plates containing 40 μg/ml tetracycline.

For a first round of panning, using 10^{12} f88-4 phage particles, the dilutions above should suffice. The yields of phage should be in the range of 10^{-3} to 10^{-5}%; given 20 particles/TU, a spot size of 15 μl, and a culture size of 200 μl, one should obtain 37 colonies in the 10^{-3} dilution spot from a yield of 10^{-3}% and 37 colonies in the 10^{-1} dilution spot from a yield of 10^{-5}%.

In the rounds following the first round of panning, the titer of phage being retained in the pannings should start to increase, with yields ranging from as high as 1% to as low as 10^{-5}%. This should be reflected in the dilutions used for titering. If 10^{10} f88-4 phage are screened, the dilutions of phage should remain at 10^{-1}, 10^{-2}, and 10^{-3}. Given fd-tet phage with a particle/TU ratio of 20, the lowest detectable yield would be in the mid-10^{-4}% range, as a 10^{-4}% yield at the lowest (10^{-1}) dilution would give 4 colonies (TU). At the highest yield (1%), there would be ~375 colonies in the 10^{-3} dilution spot. This density of colonies is hard to count, but the number can be gauged by comparison to other colonies having a similar density and for which there is another dilution that is countable. Thus, dilutions of 10^{-1}, 10^{-2}, and 10^{-3} ought to include countable spots for yields from all rounds of panning if the amount of f88-4 phage used is as stated above.

If M13-type phage are used under the same conditions, about 20 times more TUs will be produced by yields in the same range. You will, however, be counting plaques on plates rather than spots, so the number of TU plated can be higher and still give readable counts. See Protocol 19.1 for more details on titering M13 and other "wild-type" phage.

A. fd-tet-Based Libraries

Materials

phage infections (see Round 1, Day 2, Step 14)
20 mg/ml **tetracycline** stock solution (see Appendix 2)
NZY + 15 µg/ml **tetracycline** (NZY/tet, see Appendix 2)
100 x 15 mm plates containing NZY-agar and 40 µg/ml **tetracycline** (dry the plates in a laminar-flow hood for ~2 hours before using them for titering)

CAUTIONS: ***E. coli*, tetracycline (*see Appendix 4*)**

Procedure

1. Use a multichannel pipettor to transfer 5-µl samples from each infection to flexiplate wells containing 45 µl of NZY and 15 µg/ml tetracycline (NZY/tet); this makes a 1/10 dilution. Mix well, change tips, and transfer 5 µl of the dilution to the corresponding well of 45 µl of NZY in the next row (a 1/100 dilution). Mix well, change tips, and transfer 5 µl to the corresponding well of 45 µl of NZY in the third row (a 1/1000 dilution).
2. Change tips again, and use the multichannel pipet to carefully spot 15 µl of each dilution onto labeled culture plates containing NZY agar and 40 µg/ml tetracycline. To conserve tips, pipet the highest dilutions first and the lowest last, using the same set of tips. Let the drops diffuse into the agar before moving the plates; otherwise the drops may migrate and fuse. For the infections with f88-4 control phage (10^6 particles, Day 2, Step 11b), spot 15 µl of the 1/10, 1/100, and 1/1000 dilutions in NZY/tet onto labeled plates.
3. Incubate the plates overnight in the dark at room temperature and transfer them to a 37°C incubator the next day. Check the size of the colonies periodically throughout the day, and when they are of optimal size for counting, either count immediately or seal the plates with Parafilm and store at 4°C for 1–2 days. Do not let the colonies become so large that they fuse.

B. M13-Based Libraries (See also Protocol 19.1)

Materials

phage infections (see Round 1, Day 2, Step 14)
overnight culture of K91 *E. coli* cells
sterile, capped 125-ml culture flask containing 20 ml of NZY medium (see Appendix 2)
60 x 15-mm plates (Fisher 08-757-13A) containing NZY agar, warmed at 37°C
sterile, capped 100 x 12-mm glass tubes containing 2 ml of top agar, heated at 50°C

CAUTION: ***E. coli*** ***(see Appendix 4)***

Procedure

1. Start a fresh culture of K91 cells by adding 100 µl of an overnight culture to a 125-ml culture flask containing 20 ml of LB. Grow cells at 37°C with shaking at 250 rpm to an OD_{550} approximately equal to 1.0 (10^9 cells/ml); place the flask on ice to slow cell growth.
2. Incubate tubes containing 2 ml of melted top agar in LB in a 50°C heating block.
3. On a flexiplate, dilute 5 µl of eluted-phage infections 1/10, 1/100, 1/1000 as above, but in NZY medium without tetracycline.
4. In glass tubes containing 50 µl of the cultured K91 cells, add 10 µl of each infection dilution; gently shake the tubes to mix. To plate the cells, flame the lips of a top-agar tube and of a sample of infected cells to be plated. Pour the contents of the top-agar tube into the sample tube and quickly pour the mixture onto a preheated agar plate. Immediately rotate the plate to spread the top agar evenly across the plate's surface. Let the agar solidify, then place the plates in a 37°C incubator overnight.

   Note: Practice this plating procedure before starting to titer. If the agar cools too quickly, try preheating the plates at 42°C and/or using a larger volume of top agar.

5. For the M13 control vector phage (10^6 input particles), plate 10 µl of the 1/100 and 1/1000 dilutions of the infections.

PROTOCOL 17.3

Solution-phase Panning

These pannings are carried out in essentially the same fashion as the solid-phase pannings. The differences between procedures are in Day 1 and in the first part of Day 2:

- After preadsorption of the libraries, transfer the phage solution to a flexiplate containing the biotinylated screening molecule. The phage/screening molecule mixture should be in TBS containing 2% BSA and 0.5% Tween 20. Incubate overnight at room temperature or 4°C (see Protocol 17.2, Day 1, Step 2).
- On Day 2, Step 4, do *not* add the biotinylated screening molecule to the blocked, streptavidin-coated wells from Step 2. Instead, wash the streptavidin-coated wells with 1x TBS, and block for 30 minutes at 37°C with 200 μl of TBS containing 2% (w/v) dialyzed BSA. Wash the wells three times with TBS containing 0.5% (v/v) Tween 20 (TBS/Tween).

  Add the phage/screening molecule reactions to each well and incubate for 5–15 minutes at room temperature. The longer the incubation continues, the greater the chance that weak-binding phage will be captured multivalently. This can be limited to some degree by limiting the amount of biotinylated screening molecule used in the panning. There are about 2 pmole of biotin-binding sites on wells coated with 1 μg of streptavidin; thus, if 100 fmole of biotinylated screening molecule are used in the overnight in-solution binding reaction (this would roughly equate to 3 nM of Fab in a 35-μl reaction volume), there will be many free biotin-binding sites remaining after capture of the biotinylated screening molecule.
- On Day 2, Step 5, do not use biotin to block the streptavidin-coated plates *before* the library/screening molecule reactions are added to the wells. Instead, for the last 5 minutes of the incubation of the library/screening molecule reaction on the plate, add 20 μl of 1 mM biotin in TBS containing 2% BSA, and mix. This blocks biotin-binding sites of the streptavidin to which phage might bind.
- Omit Day 2, Step 6, and follow the rest of the procedure, starting with Step 7. Direct bacterial infection (skipping Steps 9 and 10) is fine.
- Titer the phage, but remember that if the concentration of screening molecule is changed between pannings, the percent yields will change. Different rounds of the same library panning, therefore, will not be comparable, and neither will side-by-side pannings of the same library at different concentrations of screening molecule. This especially applies if the concentration of the screening molecule is in deficit with respect to the phage-borne peptides, because this will limit the number of phage that can be captured. In that case, ELISA will be used mainly to identify the phage pools having the best-binding phage. Titer anyway, just to make sure the infections are working well.

PROTOCOL 17.4

Solution-phase Panning: Capture with Streptavidin-coated Magnetic Beads

The rationale behind these pannings is the same as for the capture of solution-phase complexes onto streptavidin (Sa)-coated plates (Protocol 17.3). However, the use of magnetic beads in the procedure has resulted in the changes described below in the materials and methods. We find that this procedure (either performed on filter plates, as described here, or in microcentrifuge tubes, as described in Protocol 17.5) results in the lowest yields of background phage of all of the methods that we use (as measured by control screenings with f88-4 phage).

As with the solid-phase and in-solution capture methods of screening on plates, the Sa beads can also be saturated with biotinylated screening molecule for solid-phase screening, or used for in-solution capture of complexes of phage and biotinylated screening molecules. The latter method is described here, but could easily be altered to accommodate solid-phase screening.

Materials

Materials are essentially as in Protocol 17.2, with the addition of:

streptavidin-coated magnetic beads; bead concentration 10 mg/ml (Dynabeads M-280, Dynal, Cat. # 112.06)

Magnetic Particle Concentrator for 96-well plates (Dynal, Cat. # 120.05)

MultiScreen Assay System 96-well vacuum manifold (Millipore)

MultiScreen-BV, clear filtration plates (Millipore, Cat. # MABVN1210)

The automatic plate washer is not required.

Procedure

The differences from the solid-phase panning procedure (Protocol 17.2) are the following:

Day 1

- *Prewash the streptavidin (Sa)-coated magnetic beads.* Estimate the volume of beads to use, based on 6 μl per well. Mix the beads well and transfer that amount to a microcentrifuge tube. To wash the beads, add 10 volumes of TBS containing 2% BSA and 0.2% NaN_3, mix, and centrifuge at 12,000 rpm for 30 seconds.

Aspirate the supernatant and repeat the wash twice. (If you have a magnet, use it to hold the beads in place while aspirating the wash solution.) Resuspend the washed beads in TBS containing 2% BSA and 0.2% NaN_3, at the original volume (i.e., a final Sa concentration of 10 mg/ml). After washing, the beads can be kept for several weeks at 4°C.

The binding capacity of the beads is 5–10 μg (30–60 pmole) of biotinylated antibody per mg of Sa beads, according to the manufacturer. 12 μl of Sa beads, at 10 mg Sa/ml, should bind 4–8 pmole of biotinylated antibody. We use 100 nM screening molecule (that is 3.5 pmole in 35 μl) in the pannings; therefore, most, but not all, of the antibody is captured.

- *Preadsorb the libraries on the Sa beads in a BSA-coated microplate.* Coat a flat-bottom 96-well plate with 200 μl of 2% BSA in TBS and incubate it for 1–2 hours at 37°C. Remove the BSA solution from the plate by filling each well with cold TBS squirted from a wash bottle, shaking out the plate's contents in a sink, and firmly slapping the plate dry on a stack of paper towels. Before the wells dry completely, quickly add the phage libraries in TBS containing 2% BSA and 0.5% Tween 20 in a final volume of 35–40 μl (see Day 1, Step 2, of Protocol 17.2). Thoroughly mix the washed Sa beads, add 3 μl of beads to each well, and rock the plate for 4 hours at room temperature.
- *After preadsorption of the libraries.* Resuspend ALL of the Sa beads by pipetting up and down, and transfer the beads/libraries mixture to a flexiplate. Place the flexiplate on the magnetic particle separator and let it stand for 15–30 seconds or until all the beads are grouped to the side of the well. With the multichannel pipet, transfer the supernatant containing the libraries to a new flexiplate. Add the biotinylated screening molecule to each well of the new flexiplate (with the libraries), place a lid on the plate, and incubate the plate overnight at room temperature or at 4°C in a humidified box (such as a sealed plastic box that has a wet paper towel on the bottom).
- Prepare the preadsorption plate and beads for ELISA later in the week. Be sure not to let the wells of the ELISA plate or the beads dry after transferring the libraries to the flexiplate:
  1. For the plate: Use the multichannel pipet to wash each well twice with 200 μl of 1x TBS. Aspirate (make sure that any remaining beads are removed from the plate), and then add 200 μl of 1x TBS. Store the plate in a sealed plastic box at 4°C until you are ready to perform the ELISA (Day 3).
  2. For the beads: Use the multichannel pipet to resuspend the beads in the flexiplate with 200 μl of 1x TBS. Place the flexiplate on the magnetic particle separator and let it stand for 15–30 seconds or until all the beads are grouped to the side of the well. Aspirate the TBS, and repeat the wash two more times. After the last wash, add 200 μl of 1x TBS and keep the flexiplate with the beads in a sealed plastic box at 4°C until you are ready to perform the ELISA (Day 3).

Day 2

- Omit Steps 2 and 3.
- Replace Step 4 with the following: *Capture of phage/screening molecule complexes on Sa-coated beads.* Mix the 10 mg/ml Sa beads, and add 6 µl to each well containing a phage/screening molecule reaction in the flexiplate; incubate for 10–15 minutes at room temperature.

  Note: The amount of Sa beads used will be a function of the amount of the screening molecule to be captured, but it should never be less than 3 µl or it will be almost impossible to see the beads.

  The longer the incubation continues, the greater the chance that weak-binding phage will be captured multivalently. This can be reduced, to some degree, by limiting the amount of biotinylated screening molecule used in the panning.

- Replace Step 5 with the following: *Block the remaining Sa sites with biotin.* Transfer the mixture of libraries/screening molecule/Sa beads to the filtration plate; be careful not to leave any of the Sa beads behind. Add 100 µl of 1 mM biotin in TBS containing 2% BSA to the wells of the filter plate, and mix. This will compete off phage which may have bound to streptavidin.
- Omit Steps 6 and 7.
- Replace Step 8 with the following: *Wash the beads.* Apply vacuum to the manifold until all the liquid is removed, add 200 µl of TBS/Tween (0.5%), and apply vacuum again. When all the liquid is gone, remove the plate from the manifold and dab the bottom of it briefly on several layers of Whatman 3MM paper to eliminate excess liquid on the bottom of the filter plate; this is intended to minimize the chances of cross-contamination between wells. Replace the plate on the manifold, and repeat this wash-and-blot step as many times as desired. Phage and screening molecules that are not captured on the beads are washed through the filter, while the Sa beads remain in the well.

  Note: Make sure that the filter membranes don't dry during the washes (this may affect phage viability and the integrity of the membrane) and that the manifold reservoir doesn't overfill (this may cross-contaminate the samples).

- Replace Step 9 with the following: *Elute the phage.* After the last wash, add 35 µl of elution buffer to each well of the filtration plate. Mix by pipetting up and down, trying to resuspend ALL of the Sa beads in the elution buffer. Transfer the entire solution (including the beads) to a fresh flexiplate. Incubate the phage in the elution buffer for a total of 10 minutes (at pH 2.2; phage viability begins to decrease after that).
- Replace Step 10 with the following: *Neutralize the eluates.* Place the flexiplate on the magnetic particle separator and let it stand for 15–30 seconds or until all the magnetic particles have adhered to the side of the well. Use a pipet to transfer the eluate to a new flat-bottom plate containing 5–7 µl of neutralization buffer, and mix by pipetting up and down.
- Follow the rest of the procedure, starting with Step 11.

Day 3

- *Perform an ELISA on the preadsorption plate and beads.* Use anti-phage antibody conjugated to horseradish peroxidase and ABTS for detection, following the ELISA for the plates described in Protocol 17.2, Day 3, Step 1.

  For the Sa-bead ELISA, place the flexiplate on the magnetic particle separator and aspirate the TBS; wash twice with TBS containing 0.1% Tween 20. Add the anti-phage antibody solution and resuspend the beads. Follow the same basic procedure (incubation times and reagents) as for the preadsorption plate, but note that the automatic plate washer cannot be used to wash the beads. The washes must be done by hand with a multichannel pipet and the magnetic particle separator, as described above (Day 1, last step). Wash six times. Also note that to take an OD reading of the ABTS developing reaction, the solution will have to be transferred to a new plate. Make sure that no beads are carried over (or they may interfere with the readings).

- Follow the rest of the procedure, starting with Step 2.

PROTOCOL 17.5

Affinity Discrimination by Screening a Sublibrary with Streptavidin-coated Magnetic Beads

This procedure follows essentially the same rationale as Protocol 17.4, but because screening a sublibrary often involves only one or a few samples, we perform the first round(s) in microcentrifuge tubes in which the pannings are carried out on streptavidin (Sa)-coated beads, and the washes are performed by centrifugation and resuspension of the Sa beads. Because sublibraries usually contain a larger population of binding phage, the pannings are optimized, typically after the first round(s) of panning. Multiple pannings of the same phage pools are carried out under conditions in which a number of parameters can be varied, including the number of input phage, the concentration of biotinylated screening molecule, the state of the screening molecule (IgG versus Fab), screening-molecule ligands for "competitive elutions," the buffer used for screening, the pH of elutions, or eluants that denature by other means (such as reducing agents). Such optimization increases the number of samples, in which case it may be useful to return to a plate format, such as the one described in Protocol 17.4, to avoid the handling of many tubes. We have successfully used such a combination of tube and plate formats for screening several sublibraries with antibodies (see Fig. 17.2 for an example).

As with the solid-phase and in-solution capture methods of screening on plates, the Sa beads can also be saturated with biotinylated screening molecule for solid-phase screening, or used for in-solution capture of complexes of phage and biotinylated screening molecules. The latter method is described here, but could easily be altered to accommodate solid-phase screening.

For the screening in microcentrifuge tubes:

Materials

Materials are essentially as in Protocol 17.2, except that:

Streptavidin-coated magnetic beads (Dynabeads M-280, Dynal, Cat # 112.06) are required.

Flat-bottom 96-well microtiter plates are not required.

Round-bottom 96-well microtiter plates are not required.

96-well FALCON plates are not required.

Automatic plate washer is not required.

Tabletop centrifuge equipped with plate rotor is not required.

Procedure

The same principles apply to Sa-bead panning in microcentrifuge tubes as apply to the Sa-bead pannings described in Protocol 17.4. The differences in protocol are as follows:

Day 1

- *Preadsorption of the sublibrary(ies).* Preadsorb the sublibrary(ies) in 100 μl final volume in 1.5-ml microcentrifuge tube(s) by mixing the starting amount of phage (in 1X TBS, 2% BSA, and 0.5% Tween 20) with 3–6 μl of prewashed Sa beads. Usually between 10^9 and 10^{11} phage particles are used in the screening, depending on the number of clones in the library (i.e., the library size) and the round of panning. Incubate with rocking for 4 hours at room temperature (the beads will fall to the bottom of the tube if not rocked).

  Note: The ELISA procedure for assaying phage binding to the beads is identical to that described in Protocol 17.4.

- *Assaying the preadsorbed Sa beads.* Centrifuge the tubes at 12,000 rpm (15,300*g*) for 30 seconds. Transfer the supernatant (preadsorbed library) to a fresh microcentrifuge tube. Wash the beads twice by resuspending the pellets in 1 ml of TBS and centrifuging as before. If the beads are going to be tested for bound phage by ELISA, add 100 μl of TBS to each tube after the second wash and store them at 4°C until it is time to assay them.
- *Reacting the preadsorbed libraries with the biotinylated screening molecule.* Add the biotinylated screening molecule to each tube with the preadsorbed library(ies), and incubate overnight at room temperature or 4°C.

Day 2

- *Capture of phage/screening molecule complexes with Sa beads:* To the phage/screening molecule reactions in the tubes, add 6 μl or more of fresh, washed Sa beads, and incubate for 10–15 minutes at room temperature on a rocking platform. At the end of that time, block the remaining biotin-binding sites with biotin, as described in Protocol 17.4.
- Washes are carried out by centrifugation, as described for Day 1.
- Elution is performed in microcentrifuge tubes, with the same solutions and volumes as described in Protocol 17.4. The phage eluates are recovered by centrifugation of the beads, transferred to a fresh microcentrifuge tube containing neutralization solution, mixed thoroughly, and used to infect starved cells.

ACKNOWLEDGMENTS

We thank T. Ha and A. Day for making figures and tables. We thank X. Gong, K. Brown, S. Kim, M. Rashed, and J. Mehroke for contributions in developing the methods presented in this chapter.

REFERENCES

Adey N.B., Mataragnon A.H., Rider J.E., Carter J.M., and Kay B.K. 1995. Characterization of phage that bind plastic from phage-displayed random peptide libraries. *Gene* **156:** 27–31.

Bonnycastle L.L.C., Mehroke J.S., Rashed M., Gong X., and Scott J.K. 1996. Probing the basis of antibody reactivity with a panel of constrained peptide libraries displayed by filamentous phage. *J. Mol. Biol.* **258:** 747–762.

Caparon M.H., De Ciechi P.A., Devine C.S., Olins P.O., and Lee S.C. 1996. Analysis of novel streptavidin-binding peptides, identified using a phage display library, shows that amino acids external to a perfectly conserved consensus sequence and to the presented peptides contribute to binding. *Mol. Divers.* **1:** 241–246.

Devlin J.J., Panganiban L.C., and Devlin P.E. 1990. Random peptide libraries: A source of specific protein binding molecules. *Science* **249:** 404–406.

Gebhardt K., Lauvrak V., Gabaie E., Eijsink V., and Lindqvist, B.H. 1996. Adhesive peptides selected by phage display: Characterization, applications and similarities with fibrinogen. *Pept. Res.* **9:** 269–278.

Giebel L.B., Cass R.T., Milligan D.L., Young D.C., Arze R., and Johnson C.R. 1995. Screening of cyclic peptide phage libraries identifies ligands that bind streptavidin with high affinities. *Biochemistry* **34:** 15430–15435.

Green N.M. 1965. A spectrophotometric assay for avidin and biotin based on binding of dyes by avidin. *Biochem. J.* **94:** 23C–24C.

Green N.M., Konieczny L., Toms E.J., and Valentine R.C. 1971. The use of bifunctional biotinyl compounds to determine the arrangement of subunits in avidin. *Biochem. J.* **125:** 781–791.

Kay B.K., Adey N.B., He Y.S., Manfredi J.P., Mataragnon A.H., and Fowlkes D.M. 1993. An M13 phage library displaying random 38-amino-acid peptides as a source of novel sequences with affinity to selected targets. *Gene* **128:** 59–65.

Levitan B. 1998. Stochastic modeling and optimization of phage display. *J. Mol. Biol.* **277:** 893–916.

Malvaganam S.E., Paterson Y., and Getzoff E.D. 1998. Structural basis for the binding of an anti-cytochrome c antibody to its antigen: Crystal structures of FabE8-cytochrome c complex to 1.8 Å resolution and FabE8 to 2.26 Å resolution. *J. Mol. Biol.* **281:** 301–322.

McLafferty M.A., Kent R.B., Ladner R.C., and Markland W. 1993. M13 bacteriophage displaying disulfide-constrained microproteins. *Gene* **128:** 29–36.

Parmley S.F. and Smith G.P. 1988. Antibody-selectable filamentous fd phage vectors: Affinity purification of target genes. *Gene* **73:** 305–318.

Rini J.M., Schulze-Gahmen U., and Wilson I.A. 1992. Structural evidence for induced fit as a mechanism for antibody-antigen recognition. *Science* **255:** 959–965.

Roberts D., Guegler K., and Winter J. 1993. Antibody as a surrogate receptor in the screening of a phage display library. *Gene* **128:** 67–69.

Smith G.P. and Scott J.K. 1993. Libraries of peptides and proteins displayed on filamentous phage. *Methods Enzymol.* **217:** 228–257.

18 Analysis of Phage-borne Peptides

Michael B. Zwick, Alfredo Menendez, Lori L.C. Bonnycastle, and Jamie K. Scott

Department of Molecular Biology and Biochemistry and Department of Biological Sciences, Simon Fraser University, Burnaby, B.C., Canada V5A 1S6

This chapter describes methods for analyzing the binding activity of peptides that have been isolated from peptide library screenings. Two general ELISA methods are presented that assay phage-borne peptides for their ability to bind a given screening molecule. As shown in Figure 18.1A, the direct phage ELISA (Protocol 18.4) measures binding of a selecting molecule to the peptides displayed by an immobilized phage. As described in Protocol 18.1, the phage can be immobilized by direct adsorption onto microwells, or they can be captured by immobilized anti-phage antibody. Alternatively, phage can be grown in cells that supply a third coat protein bearing a streptavidin-binding or fibrinogen-binding peptide, and these phage can then be bound to the appropriate immobilized molecule (Fig. 18.2). In the reversed phage ELISA (Protocol 18.7), and as shown in Figure 18.1B, what was the detecting molecule in the direct ELISA is now used to immobilize phage, and the bound phage are detected by anti-phage antibody. This method is more sensitive to weak interactions between peptide and selecting molecule, because phage can bind multivalently to the immobilized selecting molecule. Moreover, phage binding is detected by anti-phage antibody, which, by virtue of the availability of approximately 3000 copies of pVIII, often produces a stronger signal than the selecting molecule, whose binding is limited to about 100 peptides.

CONTENTS

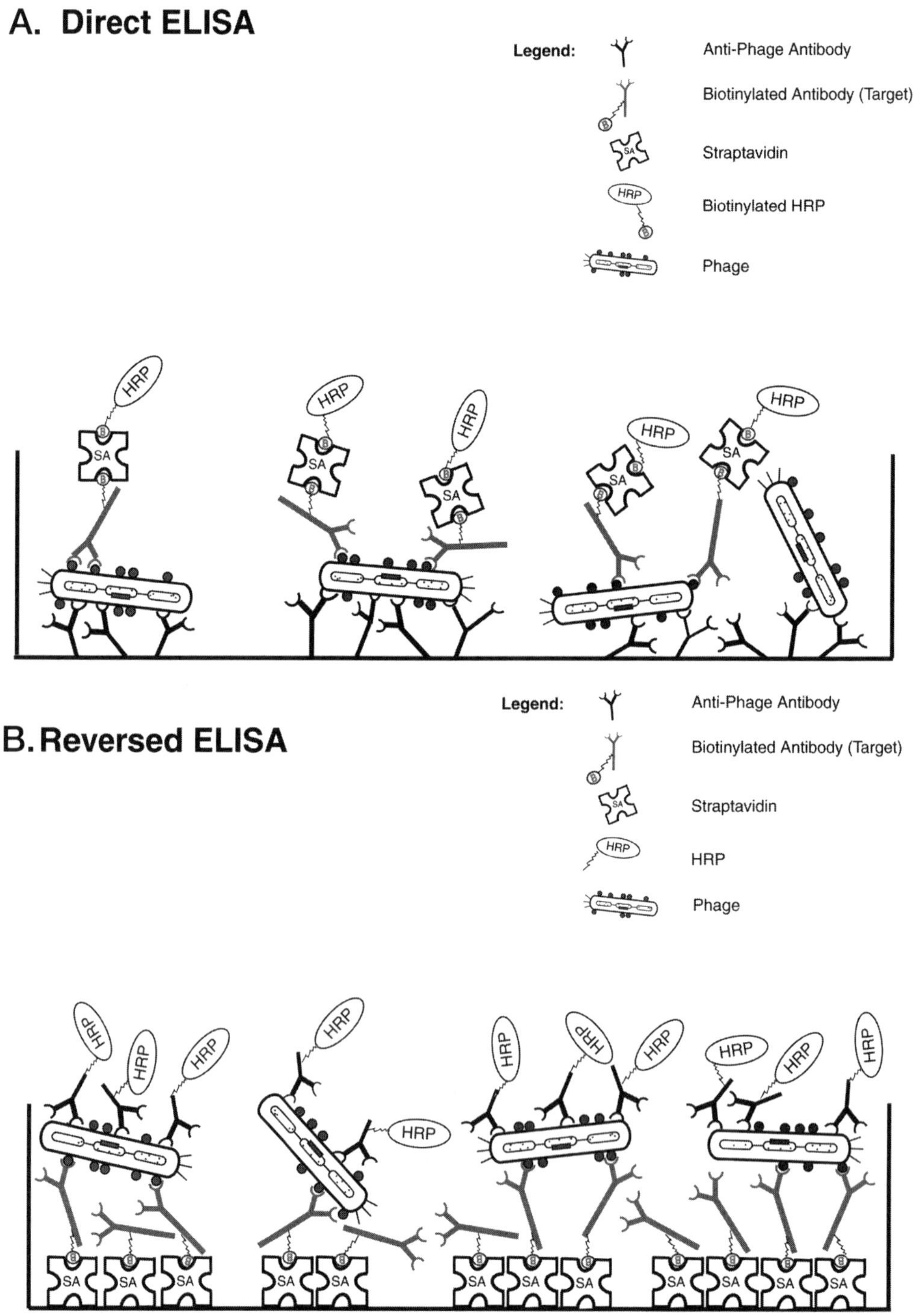

Figure 18.1. (*A*) Direct ELISA. Phage are captured on anti-phage antibody that has been non-specifically adsorbed to microwells. Phage-borne peptides are detected by the target screening molecule (Protocol 18.4). (*B*) Reversed phage ELISA. Phage are captured via their displayed peptides by immobilized target screening molecule, then detected with anti-phage antibody (Protocol 18.7).

Accompanying these protocols are methods for producing ELISA reagents that can also be found from commercial sources. These methods include the production of anti-phage antibody in rabbits (Protocol 18.2), the purification of IgG antibodies from serum (Protocol 18.3), and the conjugation of horseradish peroxidase to NeutrAvidin (Protocol 18.5) and to anti-phage antibody (Protocol 18.6).

Peptide fusion to the pVIII coat protein can also be analyzed by electrophoretic methods. This is useful when one wishes to know the level of peptide display, as reflected by the copy number of peptides per phage. Protocol 18.8 describes the electrophoretic resolution of pVIII–peptide fusions from their wild-type counterparts and the detection of disulfide bridging between pVIII molecules. The accompanying protocol, 18.9, covers the detection of antibody binding to electrophoretically resolved pVIII–peptide fusions by Western blotting.

The final protocol in this section (18.10) describes methods for transferring peptides from the phage coat proteins (either pIII or pVIII) to the amino terminus of the maltose-binding protein (MBP) of *E. coli.* MBP is a monovalent, soluble protein that contains no cysteine residues and can be purified easily on amylose. Peptide–MBP fusions serve two functions. First, they are a means of displaying peptides monovalently in their active form; this may be necessary if synthetic peptide analogs of a phage-displayed sequence are not active. Second, they can be used to determine whether the phage coat is itself required for peptide-binding activity.

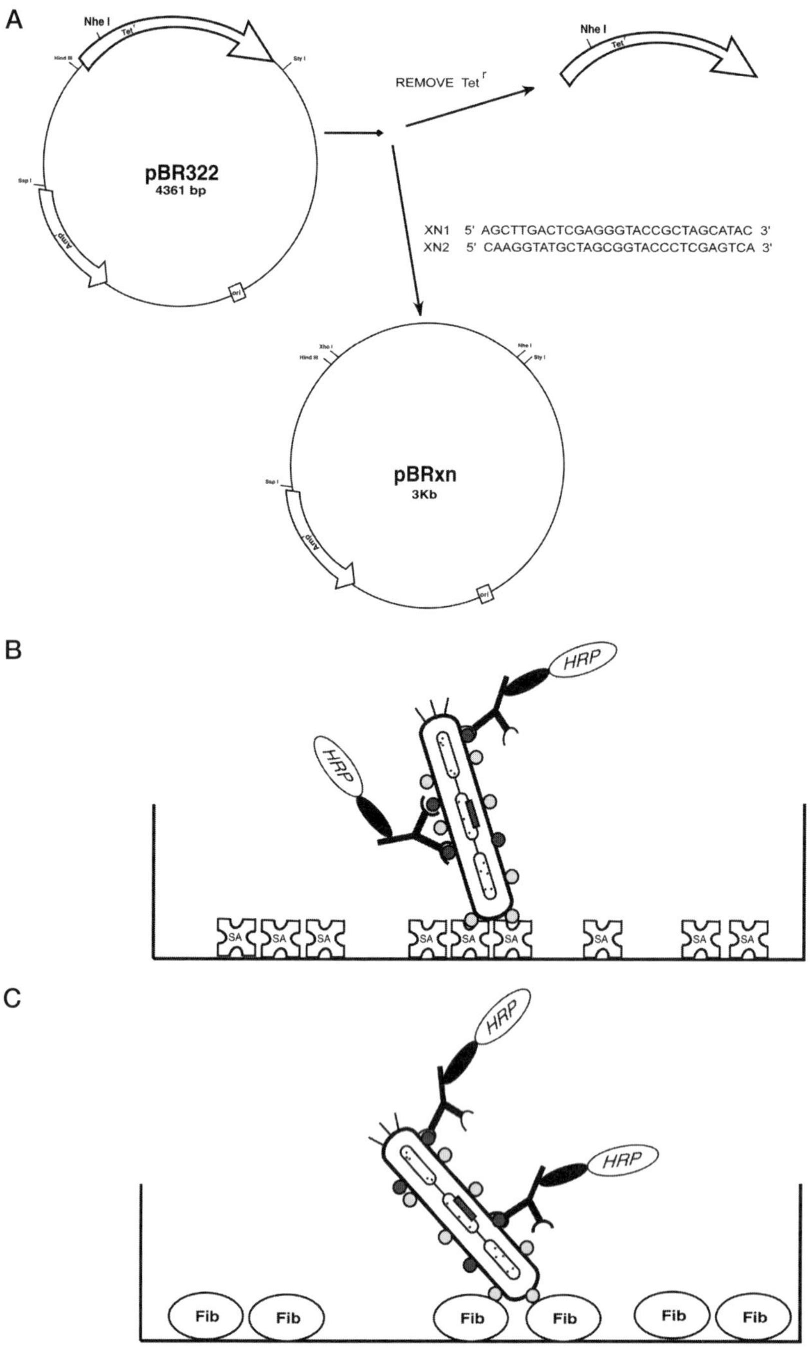

Figure 18.2. (*A*) Map of the pBRxn vector from Bonnycastle et al. (1997). Plasmid pBRxn is derived from plasmid pBR322, and bears a deletion of the tetracycline (Tn10) gene cassette. Replacing it is a *Xho*I–*Nhe*I cloning site for insertion of the recombinant *gene VIII* cassette from f88-4, which is borne on an *Xho*I–*Nhe*I fragment of approximately 400 bp (see Fig. 2.1B for this sequence). This site allows the entire gene to be PCR-amplified from phage, cleaved, and inserted into the vector. (*B*) Phage bearing a streptavidin-binding peptide fused to pVIII. (*C*) Phage bearing a fibrinogen-binding peptide fused to pVIII (Protocol 18.1).

PROTOCOL 18.1

Immobilization of Phage for Assay

There are several ways to immobilize phage in microwells for ELISA or other assay. All such assays start with a "high-protein-binding" surface, such as Immulon microplates (Dynex Technologies, Inc.), MaxiSorp plates (Nunc), or Costar plates (Corning). Simplest of all is to directly coat the wells with phage. For this method, the phage must be at least partially purified (PEG precipitated once or, preferably, twice; see Protocol 15.3) to remove proteins and debris that might interfere with phage adsorption. We normally add 1×10^{10} to 5×10^{10} phage particles in 35 μl of buffer (PBS, TBS, water, or $NaHCO_3$) to each well and allow them to adsorb overnight at 4°C. The number of phage that bind is uncertain, although usually consistent.

Alternatively, the phage can be captured on an immobilized molecule. The advantage of this approach is that the phage can be captured directly out of a cleared culture supernatant. This obviates the need for PEG precipitation and makes it relatively simple to assay phage pools from microtiter plate pannings or large numbers of clones (see Protocols 17.2–17.4). Moreover, if the saturation level of the capture molecule is known, one can be more certain of capturing nearly the same number of phage in multiple wells (that is, if the number of particles used is below the saturation level of the plate-immobilized capture molecule). Anti-phage antibody adsorbed to microwells (1 μg/well) is most often used to capture a known number of phage; we normally use 10^{10} phage particles for capture. One should note, however, that capture with an anti-phage antibody will not work if the target molecule is also an antibody and binding is to be detected with an Ig-specific reagent (like protein A, G, or A/G, anti-light-chain antibody, or an isotype-specific antibody), because such a reagent will also detect the immobilized anti-phage antibody. This problem can be avoided by using a biotinylated target antibody (see Protocol 17.1); however, in some instances, biotinylation destroys the binding activity of the target antibody.

To circumvent this problem, we developed a system for immobilizing phage via a pVIII-displayed peptide that is constitutively expressed by phage-producing bacteria. As described in Bonnycastle et al. (1997), we transferred the expression cassette for recombinant pVIII from the f88-4 phage vector to an ampicillin-selectable bacterial plasmid (a derivative of pBR322, whose tetracycline-resistance genes were deleted; see Fig. 18.2A). The recombinant *gene VIII* has *Hin*dIII and *Pst*I sites at DNA regions encoding the amino- and carboxy-terminal sides, respectively, of the pVIII signal peptidase cleavage site. Into these restriction sites, we cloned fragments encoding peptides that are specific ligands for fibrinogen and streptavidin. The resulting plasmid, bearing the recombinant genes for either the fibrinogen- or streptavidin-binding peptides fused to pVIII, was then used to transform the "supermale" K91 strain to ampicillin resistance.

These transformed cells can be infected with phage bearing pIII or pVIII fusions (for example, phage that have been affinity-selected from a library) and will produce "hybrid" phage that bear both the coat protein fusion molecule and the pVIII fusion to either the fibrinogen- or streptavidin-binding peptides. As shown in Figure 18.2B and C, hybrid phage can be captured directly out of culture supernatant by plate-immobilized fibrinogen, or out of media containing low amounts of biotin by plate-immobilized streptavidin. Because capture involves multiple binding events that result in a large "avidity boost," binding of the phage can be effective even if each individual peptide binds to the immobilized capture molecule relatively weakly (as is the case for streptavidin-binding peptides whose K_d values are μM). Expression of peptide is not limited to the f88-4 phage, as long as the phage (or phagemid) to be tested does not require ampicillin selection. Thus, one can produce hybrid phage bearing a preselected peptide fused to pVIII and another peptide or larger protein fused to pIII.

Problems can arise in the selection of the capture molecule. Fibrinogen is inexpensive, very large (~750,000), and adsorbs well to plates; in these respects it is great for coating wells. It can, however, nonspecifically adsorb to some substances. Thus, the target molecule (for example, an antibody) and the detection molecules (such as protein G–horseradish peroxidase conjugate) should both be tested for nonspecific binding to fibrinogen before doing the assay. If streptavidin is to be used for capturing phage bearing streptavidin-binding peptides, the target molecule should not be biotinylated, and detection should not involve streptavidin.

We have not been able to observe the binding of phage bearing a His_6 tag to nickel-coated wells; the reasons for this failure are unclear. Very likely, one should be able to capture phage bearing the S peptide or other S-protein-binding sequences on immobilized S protein, the product of RNase-A cleavage by subtilisin (Smith et al. 1993).

PROTOCOL 18.2

Immunizing Rabbits and Mice with Phage

Several groups have used phage as immunogens for making antibody against peptides displayed on pIII (de la Cruz et al. 1988) or pVIII (Greenwood et al. 1991; Minenkova et al. 1993; Galfre et al. 1996). This procedure explains how to produce anti-phage antibody in rabbits and mice. The rabbit IgG is then further purified using caprylic acid (Protocol 18.3) and is used to immobilize phage in direct ELISAs (Protocol 18.4) and for making anti-phage antibody–horseradish peroxidase conjugate (Protocol 18.6) for detection of phage in reversed ELISAs (Protocol 18.7). For immunization, phage are often mixed with adjuvant in PBS; we have used incomplete Freund's adjuvant and TiterMax (Sigma). The phage itself has adjuvant properties, so it is usually sufficient to immunize with phage alone (Willis et al. 1993), although T-cell help may be enhanced by the presence of immunogenic peptides (di Marzo Veronese et al. 1994). For rabbits, 1–10 ml of blood is collected from the marginal ear vein. For mice, 100–300 µl of blood is collected from the tail vein. Refer to Harlow and Lane (1988) for more details on immunization.

> *Remember:* At most institutions, procedures like the ones described here must be approved by an animal care committee or the equivalent. The animals' comfort and care should be of high priority in the planning of these experiments. Animal experimentation is an acquired skill; if this is a new procedure for you, find someone who is experienced and can teach you how to handle the animals in a way that will cause the least amount of pain, fear, and anxiety — for the animal *and* you!

RABBIT IMMUNIZATION

Materials

rabbits: New Zealand white, outbred females
phage in 1x PBS (see Day 1)
incomplete Freund's adjuvant (IFA; Sigma, Cat. # F5506) or TiterMax Gold adjuvant (Sigma, Cat. # T 2684)
1x PBS, pH 7.4 (see Appendix 2)
needles: 21-gauge for rabbit ear bleeds, 23-gauge for subcutaneous injections
syringes: 1–3 ml capacity
5-ml polypropylene, snap-cap tubes for blood
BLOTTO or 2% BSA in TBS (see Appendix 2)
Protein A/G conjugated to horseradish peroxidase (Pierce Chemical Co., Cat. # 32490), mixed 1:1 with sterile glycerol and stored at –20°C, or anti-rabbit antibody conjugated to horseradish peroxidase (Protocol 18.4)

CAUTION: **animal treatment (*see Appendix 4*)**

Procedure

Day 1

For each rabbit, prepare 100 μg of phage (equivalent to ~2.5×10^{12} phage particles, assuming a viral MW ~2.4×10^{7}) in 1x PBS to a total volume of 0.5 ml, with or without incomplete Freund's adjuvant (IFA). To use IFA, mix 250 μl of IFA with 250 μl of phage in 1x PBS, and vortex vigorously for approximately 30 minutes until a persistently thick, white emulsion develops. Alternatively, use TiterMax adjuvant at a concentration of 2 mg/ml mixed with the phage in 1x PBS. If pre-immune serum is desired, take 1–5-ml pre-bleeds from the ear. Immunize each rabbit subcutaneously on the back, at one or two sites per rabbit, for a total dose of 100 μg of phage per rabbit.

Day 14

Give the first boost (immunize as on Day 1).

Day 21

Take a primary bleed of 1–5 ml from the marginal ear vein (optional).

Day 28

Give the second boost (immunize as on Day 1).

Day 35

Take a secondary bleed (1–10 ml, as on Day 21).

Prepare serum from the blood as described in Steps 1–3, Protocol 18.3. Prepare for an ELISA by making a series of 1:5 serum dilutions in 1x PBS. As described in Protocol 18.4, immobilize 10^{10} phage particles (quantitate by spectrophotometry, Protocol 15.6) in microwells by direct adsorption, and block with BLOTTO or BSA. Detect bound antibody with horseradish peroxidase conjugated to Protein A, Protein G, Protein A/G, or anti-rabbit antibody (anti-IgG would be best). Positive ELISA signals should be detectable at dilutions of greater than 1:50,000 after the second boost. Further boosts might be required if the ELISA signals are low.

Purify the IgG from serum with caprylic acid as described in Protocol 18.3. Purified antibody can be mixed 1:1 with sterile, 100% glycerol and stored at -20°C.

MOUSE IMMUNIZATION

Materials

mice: Balb/c females
phage in 1x PBS (see Day 1)

adjuvant (optional): IFA (Sigma, Cat. # F5506) or TiterMax Gold adjuvant (Sigma, Cat. # T 2684)
1x PBS, pH 7.4 (see Appendix 2)
needles: 27-gauge for injections; 21-gauge for cardiac puncture
syringes: 1–3 ml capacity
polypropylene microfuge tubes (0.5 ml for tail bleeds, 1.5 ml for final bleeds)
BLOTTO or 2% BSA in TBS (see Appendix 2)
Protein A/G conjugated to horseradish peroxidase (Pierce Chemical Co., Cat. # 32490), mixed 1:1 with sterile glycerol and stored at –20°C, or anti-mouse antibody conjugated to horseradish peroxidase (Protocol 18.4)

CAUTION: animal treatment (*see Appendix 4*)

Procedure

Day 1

Immunize mice intraperitoneally (or subcutaneously) with 100 μg of phage in PBS, pH 7.4, with or without adjuvant, in a total volume = 100 μl (see Day 1, rabbit immunization).

Day 14

Give first boost (immunize as on Day 1).

Day 21–28

Take a primary tail-bleed (optional).

Day 28

Give second boost (immunize as on Day 1).

Day 35–42

Take a tail-bleed (optional).

Perform an ELISA as for rabbit immunization; phage binding should be detected at dilutions greater than 1:50,000 with serum taken after the second boost. Further boosts might be required if the ELISA signals are low.

If no further boosts are required

Day 42

Anesthetize the mice with CO_2, and immediately bleed them by cardiac puncture.

Prepare serum as for rabbit blood following Steps 1–3 of Protocol 18.3. The volume will not be large enough for caprylic acid purification; however, the IgG fraction can be affinity-purified by Protein A or Protein G column chromatography (see Harlow and Lane 1988).

PROTOCOL 18.3

Purification of Serum IgGs from Rabbit Blood

In this simple two-step procedure we describe the purification of rabbit IgG from whole blood. It is based on the 1988 modification by Harlow and Lane of the method of McKinney and Parkinson (1987). Caprylic acid precipitation is effective in purifying mainly IgGs from whole serum, and greatly reduces the amount of most other serum proteins. The acidic pH used in caprylic acid precipitation (pH 4.0) is significantly milder than that used for elution of IgGs from Protein A columns (typically pH 2.5). However, some antibodies may still be inactivated by this pH. Thus, we suggest that the method be tried on a small amount of serum first, and that the activity of the purified IgG be tested in ELISA side by side with the whole serum to verify that the purified IgG has retained its binding activity. Because the pH is important in this procedure, it must be measured with a pH meter. The minimum volume of blood used will depend on the minimum amount of diluted serum that can be monitored accurately by a pH meter. We usually purify no less than 2 ml of serum using this procedure.

Materials

polypropylene, screw-cap, conical 15-ml centrifuge tubes (Gordon Technologies, Cat. # 15-750)
sterile 50-ml glass beaker
sterile stirring bar
60 mM sodium acetate (NaOAc, pH 4.0)
0.5 N **NaOH**
caprylic (octanoic) acid (Sigma, Cat. # C-2875 or purest possible)
0.45-μm filter for a syringe
sterile 10–50 cc syringe
sterile 125-ml Erlenmeyer flask
sterile 10x PBS stock (see Appendix 2)
1 N and 5 N NaOH
materials for SDS–PAGE (see Sambrook et al. 1989)
saturated **ammonium sulfate** solution (Anachemia, Cat. # AC-0646)
sterile, polypropylene, screw-cap 40-ml round-bottom (Oak Ridge) tubes
18-mm, 3500 MW exclusion dialysis tubing (Spectra/Por; Spectrum Medical Industries, Cat. # 132720; or use larger pore tubing up to a MW exclusion of 50,000)
sterile 1x PBS, 3x PBS, and 2x PBS, stored at 4°C
6.5% (w/v) **sodium azide** stock solution (see Appendix 2)

CAUTIONS: NaOH, caprylic acid, ammonium sulfate, sodium azide (***see Appendix 4***)

Procedure

1. Collect blood (5–7 ml, see Protocol 18.2). Transfer to a 15-ml conical tube and let stand for 30–60 minutes at 37°C for clot formation.

2. Separate the clot from the sides of the tube with a sterile Pasteur pipet, and allow it to form completely by incubating the blood overnight at 4°C.

3. Centrifuge at 10,000*g* for 10 minutes at 4°C, and transfer the serum to a fresh 50-ml beaker. Measure the volume of the serum with a pipettor or sterile graduated cylinder. Store a small aliquot of serum on ice for SDS–PAGE analysis in Step 12 (load 1 µl of a 1:20 dilution of serum).

Perform the following steps at room temperature

4. While monitoring the pH and stirring, add 4 volumes of 60 mM NaOAc, pH 4.0, to one volume of serum. Adjust the pH to 4.5 with 0.5 N NaOH.

5. Measure a volume of caprylic acid that is 2.5% the volume of the pH 4.5 serum. Add the caprylic acid SLOWLY, drop-wise, while vigorously stirring the solution. Caprylic acid at a final concentration of 2.5% (v/v) preferentially precipitates most serum proteins, with the exception of IgG. Continue stirring for an additional 30 minutes.

6. Remove insoluble material by centrifuging the solution at 10,000*g* for 30 minutes in a warm centrifuge that is at least above 10°C. Avoid mixing the pellet and the supernatant by keeping the tubes on their sides when removing them from the rotor.

7. To remove the fines, place a 0.45-µm filter on a 10–50 cc syringe and pour the supernatant into the barrel. Replace the plunger, and filter the supernatant into a 125-ml Erlenmeyer flask (or a flask appropiate for the volume of supernatant). Mix the supernatant with 1/10 volume of 10x PBS, and adjust the pH to 7.4 with 5 N NaOH (use 1 N NaOH when the pH reaches 7.0–7.3).

8. Remove a 20-µl aliquot for SDS-PAGE analysis in Step 12. The IgG concentration at this point is likely around 1–2 mg/ml. A reasonable amount to load on a 12% polyacrylamide gel in SDS is 1 µg, but two or more dilutions may be loaded to get a better idea of the protein concentration.

Perform the following steps at 4°C

9. Cool the supernatant to 4°C and add cold, saturated ammonium sulfate to 45% (v/v) saturation (0.277 g/ml). Stir the sample for 30 minutes with a stir bar, then transfer the solution to an Oak Ridge tube. Centrifuge at 7000*g* for 15 minutes in a prechilled rotor. Discard the supernatant and dissolve the pellet in a volume of 1x PBS equal to the original volume of serum.

10. Remove a 10-μl aliquot for SDS-PAGE analysis in Step 12. The concentration of IgG at this point is probably around 5 mg/ml.

11. Transfer the sample to 3500 MW exclusion dialysis tubing and dialyze the sample against 50 volumes of 3x PBS for 1 hour, 50 volumes of 2x PBS for 1 hour, and then 50 volumes of 1x PBS overnight (all at 4°C). Transfer the dialyzed solution to a fresh Oak Ridge tube, and inactivate any remaining complement proteins by heating the sample for 20 minutes in a 56°C water bath. Clear the sample by centrifugation at 5000*g* for 20 minutes at 4°C. Transfer the supernatant to a fresh Oak Ridge tube and discard any pellet. (*Optional:* Add sodium azide to a final concentration of 0.02% [w/v].) Aliquot the IgG. Save a small aliquot for SDS-PAGE analysis in Step 12, and store the rest at –20°C.

12. Estimate the yield and purity of samples from Steps 3, 8, 10, and 11 using SDS-PAGE analysis. Run the samples next to 0.5, 1.0, and 2 μg of purified IgG as a standard (Sigma, Cat. # I 4506), along with the appropriate MW markers (see Sambrook et al. 1989 for further details). Our yields are usually 50–60% of the serum IgG; however, McKinney and Parkinson (1987) report yields of more than 80%.

PROTOCOL 18.4

Direct Phage ELISA under Standard and Reducing Conditions

In this ELISA, as shown in Figure 18.1A, phage are bound to plate-immobilized anti-phage antibodies, and binding of a biotinylated target molecule is detected with biotinylated horseradish peroxidase (HRP) in complex with NeutrAvidin. We favor this method because of the high affinity of avidin–biotin interactions. Rather than making biotinylated HRP and titering it with NeutrAvidin (Protocol 18.5), the VECTASTAIN Elite ABC kit can be purchased from Vector Laboratories.

This assay can also be performed with a nonbiotinylated antibody target molecule by directly adsorbing 1×10^{10} to 5×10^{10} PEG- or CsCl-purified phage per well, instead of anti-phage antibody (Step 1) and using HRP conjugated to Protein A, Protein G, or Protein A/G for detection instead of HRP–NeutrAvidin (Step 5). The assay can be performed with a nonbiotinylated, non-antibody target molecule, but the detecting molecule (e.g., anti-target antibody) may have to be conjugated to HRP (by modification of Protocol 18.6).

As an alternative to ABTS, OPD (*o*-Phenylenediamine; Sigma, Cat. # P-9029) can be used for detection. It is a more sensitive reagent than ABTS and gives low background signals, but in our experience, the signals for replicate samples are more apt to vary. Because OPD gives a much larger signal than ABTS, we sometimes use it to enhance the signal for very weak phage–target molecule interactions. This allows us to discriminate between background and weak target molecule binding. For the same purpose, we devised the "reversed" phage ELISA described in Protocol 18.7 and shown in Figure 18.1B; we usually use this assay, rather than OPD, for detecting weak binding interactions.

Also described here are two methods for performing phage ELISAs under reducing conditions. The purpose of these assays is to determine the role in peptide binding played by intra- and intermolecular disulfide bridging between peptidic cysteine residues. Phage-borne peptides are subjected to either reducing conditions (with dithiothreitol, DTT) or cysteine-blocking conditions (with N-ethylmaleamide, NEM, which reacts with free sulfhydryls). A combination of the two conditions allows reduced disulfides to react with NEM, which prevents the re-formation of disulfides in subsequent steps of the assay. Thus, side-by-side samples treated with DTT alone, DTT+NEM, NEM alone, and buffer alone, followed by binding of the biotinylated target molecule, should allow one to deduce whether disulfide bridging is required for peptide-binding activity. Treatment with DTT alone and buffer alone can also indicate a requirement for disulfide bridging. This information is important for understanding the structural basis of peptide recognition and for planning the production of synthetic analogs of a given phage-borne peptide.

Materials

flat-bottom, polystyrene microtiter plates (Corning Costar, Cat. # 3690)
polystyrene lids for the microtiter plates (Corning Costar, Cat. # 3930)
polypropylene plastic box and lid (such as those made by Rubbermaid)
anti-phage antibody (see Protocols 18.2 and 18.3)
0.1 M $NaHCO_3$
1x PBS or 1x TBS (see Appendix 2)
BSA (Fraction V, Sigma, Cat. # A-3912)
BLOTTO or 2% (w/v) dialyzed BSA in 1x TBS (see Appendix 2; for long-term storage, add **NaN_3** to 0.02% [w/v])
1x TBS containing 0.1% (v/v) Tween 20 (TBS/Tween), stored at 4°C or ice-cold
automatic plate washer (BioTek Instruments)
10^{10} phage/well of purified phage or phage in cleared culture supernatant
BLOTTO or 2% (w/v) BSA in TBS/Tween (see Appendix 2)
biotinylated target molecule (see Protocol 17.1)
NeutrAvidin stock solution (see Protocol 18.5)
biotinylated horseradish peroxidase stock solution (see Protocol 18.5)
0.2 M Na_2HPO_4 (no pH adjustment necessary)
0.1 M **citric acid**
ABTS (2,2′-azino-bis (3-ethylbenz-thiazoline-6-sulfonic acid); Sigma Chemical Co., Cat. # A-1888)
29–35% v/v **H_2O_2** (BDH Inc., Cat. # ACS 399-76)
microplate reader (BioTek Instruments)

For reducing ELISAs only

dithiothreitol (Cleland's Reagent; Gibco BRL, Cat. # 15508-013)
***N*-ethylmaleimide** (Sigma, Cat. # E-3876)

CAUTIONS: **sodium azide, citric acid, ABTS, hydrogen peroxide, dithiothreitol, *N*-ethylmaleimide (*see Appendix 4*)**

Procedure

Standard ELISA

1. Coat each well of an ELISA plate with 1 μg of anti-phage antibody in 35 μl of 0.1 M $NaHCO_3$, 1x PBS, or 1x TBS. Cover the plate and incubate in a humidified plastic box overnight (or several days) at 4°C, or for at least 4 hours at room temperature.

2. Aspirate the contents of the wells, then block each well by adding 300 μl of BLOTTO or 2% BSA in TBS; mix and incubate for 30 minutes at 37°C in a

humidified plastic box. Aspirate the blocking solution and wash the wells three times with TBS/Tween (preferably with a plate washer).

Note: For washes with an automatic plate washer, use these settings: six washes, 200-μl volume, shaking at low for 10 seconds, with a 5-second wait between washes. See Protocol 17.2 for details on using a BioTek automatic plate washer.

3. Add 10^{10} phage to each well. The phage can be in BLOTTO, 2% BSA in TBS/Tween, or culture supernatant. Use TBS/Tween to bring the total volume added to each well to 35 μl. Allow the phage to bind the anti-phage antibody on a rocking platform in a humid box for 1 hour at room temperature. Aspirate and wash the wells six times with TBS/Tween.

4. Add to each well 35 μl of biotinylated target molecule in BLOTTO or TBS/Tween containing 2% BSA. Transfer the plate to a humidified plastic box and incubate for 2–4 hours on a rocking platform at room temperature (or, if required for the stability of the target molecule, at 4°C). Aspirate and wash the wells six times with TBS/Tween.

5. Make HRP–NeutrAvidin complexes: In a 14-ml polypropylene snap-cap tube containing 5 ml of TBS/Tween, add 1–2 μl of biotinylated HRP and 1–2 μl of NeutrAvidin, and mix thoroughly by gentle inversion (see Protocol 18.5). To allow the complexes to form, let the solution stand at room temperature for 30 minutes. Add 35 μl of the solution to each well of the microtiter plate, and incubate for 30 minutes at room temperature (or for 45 minutes at 4°C, if required for the stability of the target molecule). Aspirate and wash the wells six times with TBS/Tween.

6. Make ABTS solution in a 14-ml polypropylene snap-cap tube containing 1.93 ml of phosphate buffer ($NaHPO_4$) and 3.07 ml of citric acid. Add 2 mg of ABTS and mix by inversion. Just before adding the solution to the ELISA wells, add 5 μl of H_2O_2 to the tube and mix. Transfer 35 μl of ABTS solution to each well, and incubate the plate in the dark at room temperature for 20–45 minutes. Read the absorbance of the wells at 405 and 495 nm with a microplate reader, and subtract them to calculate the $OD_{405\text{-}495}$.

7. Use as positive controls:
   - a ligand that binds the biotinylated target molecule; adsorb this to the well instead of anti-phage antibody (Step 1 above). Block as in Step 2.
   - 1 μl of HRP–NeutrAvidin solution mixed into a well containing 35 μl of ABTS solution (Step 6).
   - phage bearing a peptide that binds to another target, like the peptide sequence YDVPDYA, which binds mAb 17/9 (Rini et al. 1992), plus biotinylated mAb 17/9 to detect it. (Remember to use f88-4 phage as a negative control for mAb 17/9.)

8. Use as negative controls:

- wells blocked with BLOTTO only (i.e., not coated with anti-phage antibody and no phage added). Use these wells to test the reagents, HRP–NeutrAvidin followed by washing and ABTS, and ABTS alone for nonspecifically generated signals.
- wells coated with anti-phage antibody and blocked with BLOTTO, but with no phage added.
- f88-4 control phage instead of the phage to be assayed.

Reducing ELISA

The reducing ELISA is essentially the same as the standard one, but it incorporates treatment of the captured phage with dithiothreitol (DTT) and/or DTT and *N*-ethylmaleimide (NEM) before the step in which biotinylated target molecule is bound to the phage (that is, between Steps 3 and 4 of the standard protocol above). We perform the assay according to either of two different protocols, depending on how the phage are immobilized and whether both NEM and DTT are used in the assay.

Simple reducing ELISA for phage captured on anti-phage antibody: Follow the steps for the standard ELISA, but with the following changes.

3a. After the washes in Step 3, add 35 μl of 1x TBS or 1x PBS containing 15 mM DTT to the "+ DTT" wells. To the "–DTT" wells add buffer alone. Incubate for 30 minutes at 4°C, then aspirate the solution from the wells.

4. During Step 4, incubation with the biotinylated target in the "+ DTT wells" is performed in the presence of a low concentration of DTT (5 mM) to ensure that peptidic disulfides remain reduced.

7. Along with the positive controls mentioned above, use a ligand (such as a "linear" peptide) whose target molecule-binding activity is not affected by DTT, and, if available, a peptide whose binding activity depends on intact disulfide bridges. For instance, we use phage bearing the sequence YDVPDYA for the linear peptide and phage bearing the DYA motif bounded by Cys residues for the cyclic peptide; these are both recognized by the 17/9 monoclonal antibody (Rini et al. 1992) when in buffer alone, but only the linear peptide is bound if the phage are pretreated with DTT. We also incubate replicates of the "+ DTT" and "–DTT" samples with anti-phage antibody to ensure that DTT is not dissociating phage from the capture antibody.

8. Use the same negative controls as for the standard phage ELISA.

Reducing ELISA with DTT and NEM for plate-adsorbed phage (see Protocol 18.1): Follow the steps above, but with the following changes. Prepare four wells for each sample.

3a. After the washes in Step 3, add 35 μl of PBS containing 100 mM DTT to the "+ DTT/+ NEM" and "+ DTT" wells. Add PBS alone to the "+ NEM" and "–DTT" wells. Incubate for 30 minutes at 4°C.

3b. Aspirate the solution from the wells and add 35 μl of PBS containing 100 mM NEM and 15 mM DTT to the "+ DTT/+ NEM" wells. Add 35 μl of PBS containing 100 mM NEM to the "+ NEM" wells. Add 35 μl of 100 mM DTT in PBS to the "+ DTT" wells and PBS alone to the "–DTT" wells. Incubate for 30 minutes at 4°C. Aspirate the wells and proceed to Step 4.

7. Use the same sorts of positive and negative controls as for the ELISA with DTT treatment above.

PROTOCOL 18.5

Preparation of NeutrAvidin in Complex with Biotinylated Horseradish Peroxidase

This protocol (Zwick et al. 1998) describes how to make biotinylated horseradish peroxidase (HRP), which can be used in ELISA as a reagent for detecting bound, biotinylated target molecule (see Step 4, Protocol 18.4). We use NeutrAvidin instead of avidin, because it is purported to have lower nonspecific binding properties than most preparations of avidin. Rather than making your own biotinylated HRP and titering it with NeutrAvidin, the VECTASTAIN Elite ABC kit can be purchased from Vector Laboratories; its complexes are made with avidin.

Materials

sterile 1.5-ml microcentrifuge tubes
horseradish peroxidase (Sigma, St. Louis, MO, Cat. # P-6782)
1x PBS, pH 7.4 (see Appendix 2)
sulfo-NHS-LC Biotin (Pierce, Cat. # 21335ZZ)
2 mM sodium acetate, pH 6.0
1 M $NaHCO_3$
1 M **ethanolamine**, pH 9.6 (Fisher Scientific, Cat. # M251-1)
1x TBS, pH 7.4 (see Appendix 2)
BSA (Fraction V, Sigma Chemical Co., Cat. # A-3912)
Centricon-30 device (Amicon, Inc., Beverly, MA, Cat. # 4242)
6.5% (w/v) **sodium azide** stock solution
sterile glycerol
ImmunoPure NeutrAvidin (Pierce, Cat. # 31000ZZ)

CAUTIONS: **ethanolamine, sodium azide (*see Appendix 4*)**

Procedure

1. In a 1.5-ml microcentrifuge tube, dissolve 500 µg of horseradish peroxidase (HRP) in 250 µl of 1x PBS. Divide into five 50-µl aliquots in 1.5-ml microcentrifuge tubes.
2. Prepare a fresh solution of 0.5 mg/ml sulfo-NHS-LC biotin in 2 mM sodium acetate, pH 6.0. Add the following to each 50-µl aliquot of HRP, and mix:

   11 µl of 1 M $NaHCO_3$

   50 µl of sulfo-NHS-LC biotin in sodium acetate

3. Incubate the solutions for 2 hours at room temperature. Add 250 μl of 1 M ethanolamine to inactivate any remaining sulfo-NHS-LC biotin.

4. Incubate for an additional hour at room temperature, then add 20 μl of 1x TBS containing 1 mg/ml BSA.

5. Combine the mixtures in a Centricon-30 device and wash three times with 2 ml of 1x TBS, then once with 1x TBS containing 0.02% NaN_3. Adjust the concentration of the biotinylated HRP to 0.66 mg/ml in 1x TBS/0.02% NaN_3 (assuming no losses during the washing step). Gauge the volume, add an equal volume of sterile glycerol, and mix thoroughly. Divide the HRP/glycerol solution into 50- to 100-μl aliquots and store at –20°C.

6. The optimal amount of NeutrAvidin to use with biotinylated HRP in ELISAs should be determined by an ELISA in which biotinylated BSA or fibrinogen (see Protocol 17.1) is adsorbed to a microwell and detected with different amounts and ratios of NeutrAvidin and biotinylated HRP. We found 50 ng of NeutrAvidin and 12 ng of biotinylated HRP per microwell to be optimal, but this should be tested for each lot of biotinylated HRP prepared. We prepare 50- to 100-μl aliquots of a 1.4 mg/ml stock NeutrAvidin in 50% (v/v) glycerol and store them at –20°C.

7. Prepare complexes of NeutrAvidin and biotinylated HRP when ready for use (see Protocol 18.4, Step 5). In a 1.5-ml microcentrifuge tube, add 1–2 μl of each stock to 1 ml of 1x TBS containing 0.1% Tween 20, and immediately mix the solution by inversion. Let the mixture stand for 30 minutes at room temperature to allow formation of biotinylated HRP–NeutrAvidin complexes.

PROTOCOL 18.6

Preparing Conjugates of Horseradish Peroxidase and Anti-phage Antibody

We mainly use conjugates of horseradish peroxidase (HRP) and rabbit anti-phage IgG to detect bound phage in reversed ELISA (Protocol 18.7). However, this protocol, which is based on the Pierce method outlined in Hermanson (1996), can also be applied to making HRP conjugates with other molecules, such as antibody against target molecule for detection in direct ELISA (Protocol 18.4).

Materials

purified rabbit anti-phage IgG
1x PBS, pH 7.4 (see Appendix 2)
500 mM Na_2EDTA stock (pH 8.0)
sterile 1.5-ml microcentrifuge tubes
2-mercaptoethylamine (Pierce, Cat. # 20408ZZ)
Centricon-30 device (Amicon, Inc., Cat. # 4242AM)
horseradish peroxidase (HRP; Sigma, Cat. # P-6782)
sulfo-SMCC (MW 436.4; Pierce, Cat. # 22322ZZ)
1 M Tris-HCl, pH 7.5
1x TBS, pH 7.4 (see Appendix 2)

CAUTION: **2-mercaptoethylamine** (*see Appendix 4*)

Procedure

1. Prepare 4 mg of anti-phage IgG in 450 μl of PBS, 5 mM EDTA in a 1.5-ml microcentrifuge tube.
2. Add 50 μl of 0.5 M 2-mercaptoethylamine in PBS containing 5 mM EDTA. Mix. Incubate the solution for 90 minutes in a 37°C water bath, with occasional mixing.
3. Wash the reduced IgG three times in a Centricon-30 apparatus with 2 ml of PBS, 5 mM EDTA. Wash three times with PBS alone. Back-spin the washed IgG into the cup provided and transfer to a 1.5-ml microcentrifuge tube. Gauge the volume of the IgG solution.
4. In a 1.5-ml microcentrifuge tube, dissolve 4 mg of HRP in 500 μl of PBS.

5. Add 1 mg of sulfo-SMCC to the enzyme solution, and incubate for 30 minutes at 37°C. This activates the HRP for coupling. Transfer the mixture to a Centricon-30 device and wash six times with 2 ml of PBS. Use PBS to bring the volume of the washed HRP to twice that of the IgG solution.

6. Mix the reduced IgG (Step 3) with the activated HRP (Step 5). Incubate 2 hours to overnight at 4°C.

7. Add 10 µl of 0.1 M 2-mercaptoethylamine to block unreacted maleimide groups, and incubate for 20 minutes at room temperature. Check the pH, and if necessary, adjust it to 7.5–7.8 with 1 M Tris-HCl, pH 7.5.

8. Centricon-wash the conjugate three times with 2 ml of 1x TBS. Gauge the volume of the washed solution and mix with an equal volume of 100% glycerol. Bring the mixture to 0.02% (w/v) NaN_3 and store at –20°C.

The appropriate amount of this reagent for ELISA samples is determined empirically. Follow Protocol 18.4, and titer the reagent in an ELISA in which 10^{10} phage particles have been directly immobilized per well (and not captured on anti-phage antibody; see Protocol 18.1). Determine the dilution of the HRP–IgG conjugate at which the ELISA signal begins to decrease. For maximal signals in subsequent ELISAs, use the amount of HRP–IgG used in the dilution that precedes the one giving a decreased signal.

PROTOCOL 18.7

Reversed Phage ELISA

In this ELISA, as shown in Figure 18.1B, the roles of the phage and target molecule are reversed, compared to the direct phage ELISA. The biotinylated target molecule is immobilized on plate-adsorbed streptavidin. Next, phage are allowed to bind the target molecule, and their presence is detected by a horseradish peroxidase (HRP)–anti-phage antibody conjugate (Protocol 18.6). This procedure allows multiple peptides displayed by a phage or phagemid (especially those of a pVIII-display vector) to bind the target. The combination of the avidity boost involved in phage binding and the strong signal produced by the anti-phage antibody gives about a tenfold stronger signal than that produced by the same peptide–target molecule interaction in a direct ELISA. A nonbiotinylated target molecule can be used in this assay simply by directly coating it to the microplate wells in Step 1 (see Protocol 18.1), followed by blocking in Step 2. A nonbiotinylated target can also be captured on plate-immobilized anti-target antibody, which would replace streptavidin in Step 1.

Materials

flat-bottom, polystyrene microtiter plates (Corning Costar, Cat. # 3690)
polystyrene lids for the microtiter plates (Corning Costar, Cat. # 3930)
polypropylene plastic box and lid (such as Rubbermaid makes)
streptavidin (Roche Molecular Biochemicals, Cat. # 973 190)
0.1 M $NaHCO_3$
1x TBS (see Appendix 2)
dialyzed BSA (Sigma, Cat # A-3912)
2% w/v dialyzed BSA in 1x TBS (if stored long-term, add **NaN_3** to 0.02% w/v)
4°C, or ice-cold, 1x TBS containing 0.1% (v/v) Tween 20 (TBS/Tween)
2% w/v BSA in TBS/Tween (if stored long-term, add **NaN_3** to 0.02% w/v)
biotinylated target molecule (see Protocol 17.1)
automatic plate washer (BioTek Instruments)
10^{10} phage/well of purified phage or phage in cleared culture supernatant
horseradish peroxidase (HRP)–anti-phage antibody conjugate (see Protocol 18.6)
sterile 14-ml snap-cap polypropylene tube
0.2 M Na_2HPO_4 (no pH adjustment necessary)
0.1 M **citric acid**
ABTS (**2,2′-azino-bis (3-ethylbenz-thiazoline-6-sulfonic acid**), Sigma, Cat. # A-1888)
29-35% v/v **H_2O_2** (BDH Inc., Cat. # ACS 399-76)
microplate reader (BioTek Instruments)

CAUTIONS: **sodium azide, citric acid, ABTS, H_2O_2 (*see Appendix 4*)**

Procedure

1. Coat each well of an ELISA plate with 1 μg of streptavidin in 35 μl of 0.1 M $NaHCO_3$, 1x PBS, or 1x TBS. Incubate the plate overnight (or for several days) at 4°C, or for at least 4 hours at room temperature.

2. Aspirate the contents of the wells, then block each well by adding 300 μl of 2% BSA in TBS. Mix. Incubate for 30 minutes with rocking at 37°C in a humidified plastic box. Aspirate the blocking solution and wash the wells three times with TBS/Tween (preferably using a plate washer).

   Note: For washes with an automatic plate washer, use these settings: six washes, 200-μl volume, shaking at low for 10 seconds, with a 5-second wait between washes. See Protocol 17.2 for details on using a BioTek automatic plate washer.

3. Add to each well 35 μl of TBS/Tween containing 2% BSA and at least 30 pmole of biotinylated target molecule. Transfer the plate to a humidified plastic box and incubate 1 hour at room temperature (or 1–1.5 hours at 4°C, if necessary for maintaining stability of the target molecule) on a rocking platform. This procedure should load the streptavidin with biotinylated target. Aspirate and wash the wells six times with TBS/Tween to remove unbound target molecule.

4. Add to each well 10^{10} phage particles in TBS/Tween + 2% BSA or culture supernatant, with the volume made up to 35 μl with TBS/Tween. Mix and incubate on a rocking platform for 4 hours at 4°C or 1 hour at room temperature. Aspirate and wash the wells six times with TBS/Tween.

5. Add to each well 35 μl of HRP–anti-phage antibody conjugate at the optimal concentration (see Protocol 18.6), and incubate at room temperature for 1 hour. Aspirate and wash the wells six times with TBS/Tween.

6. Make ABTS solution. Mix 1.93 ml of phosphate buffer (Na_2HPO_4) and 3.07 ml of citric acid in a 14-ml snap-cap tube. Add 2 mg of ABTS and mix by inversion. Just before adding the solution to the ELISA, add 5 μl of H_2O_2 to the tube and mix again. Add 35 μl of ABTS solution to each ELISA well, and incubate the plate in the dark for 20–45 minutes at room temperature. Use an ELISA plate reader to measure the absorbance of the wells at 405 and 495 nm, and subtract the readings to calculate the $OD_{405\text{-}495}$.

7. Use as positive controls:
   - 1 μl of the working dilution of HRP–anti-phage antibody conjugate mixed into a well containing 35 μl of ABTS solution (Step 6).
   - phage bearing a peptide that binds to another target, like the peptide sequence YDVPDYA that binds mAb 17/9 (Rini et al. 1992), plus biotinylated mAb 17/9 to capture it. It will also be detected by anti-phage Ab–HRP. (Remember to use f88-4 phage as a negative control for mAb 17/9.)

8. Use as negative controls:

- wells blocked with 2% BSA only (i.e., not coated with streptavidin and no phage added). Use these wells to test the reagents, anti-phage Ab–HRP followed by washing and ABTS, and ABTS alone for nonspecifically generated signals.
- wells coated with streptavidin and blocked with 2% BSA, but with no phage added.
- f88-4 control phage instead of test phage.

PROTOCOL 18.8

Gel Electrophoresis of the Major Coat Protein of Filamentous Phage

There are many instances in which DNA sequencing indicates that a given peptide–pVIII fusion is present on the surface of a phage particle; SDS-PAGE is the method of choice for proving this. Recombinant pVIII display, by virtue of the wild-type protein's high copy number and low molecular weight (~5.5 kD), is best suited for the detection of peptide fusions. As shown in Figure 18.3, recombinant pVIII proteins often migrate more slowly than their wild-type counterpart, and hence can be detected by SDS-PAGE.

Our recent work (Zwick et al. 2000) has shown that pVIII proteins containing cysteine residues can form dimers that will assemble on the budding phage particle, as long as wild-type pVIII is also present. As an example of this, Figure 18.3 shows that a pVIII-displayed fos peptide containing cysteine residues will form a high-molecular-weight band that is ablated by pretreatment of the phage with DTT, whereas the mobility of recombinant pVIII bearing the fos sequence, but no cysteine residues, is faster and is unaffected by DTT.

The level of expression of a given peptide–pVIII fusion is also important. SDS-PAGE can also be used to determine the number of copies of recombinant pVIII per phage particle. Figure 18.4 shows an example of such an experiment; usually the copy number varies from 20 to 200 copies per f88-4-derived phage.

The procedure for SDS-PAGE of phage proteins was kindly communicated to us by R. N. Perham (University of Cambridge). It is based on the method of Schagger and von Jagow (1987), who showed that the resolution of proteins, especially those of low molecular weight, is improved by the use of tricine, rather than glycine, as the trailing ion in the running buffer. Thus, this method improves the resolution of the pVIII protein, which is small (MW ~5.5 kD), hydrophobic, and normally difficult to resolve electrophoretically.

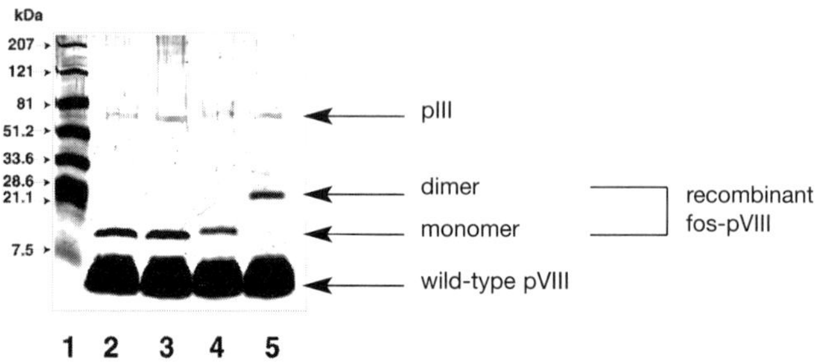

Figure 18.3. Silver stain of a gel containing phage clones that have been preincubated in the presence or absence of DTT, and separated by SDS-PAGE. (*Lane 1*) Molecular-weight size marker; (*lanes 2 and 3*) phage bearing the fos peptide on recombinant pVIII; (*lanes 4 and 5*) phage bearing a version of the fos peptide on recombinant pVIII in which the peptide is flanked on either side by a single cysteine residue. The samples run in lanes 2 and 4 were treated with DTT, whereas those run in lanes 3 and 5 were not.

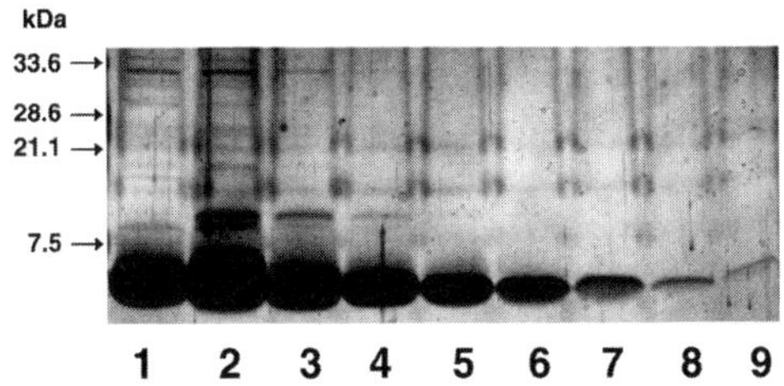

Figure 18.4. Estimating the number of recombinant pVIII molecules per phage particle. (*Lane 1*) f88-4, 6 x 10^{10} phage particles; (*lanes 2–9*) decreasing amounts of recombinant phage (threefold dilutions). (*Lane 2*) 6.0 x 10^{10} particles; (*lane 3*) 2.0 x 10^{10} particles; (*lane 4*) 6.7 x 10^{9} particles; (*lane 5*) 2.2 x 10^{9} particles; (*lane 6*) 7.4 x 10^{8} particles; (*lane 7*) 2.5 x 10^{8} particles; (*lane 8*) 8.2 x 10^{7} particles; (*lane 9*) 2.7 x 10^{7} particles. Based on the band intensities of the wild-type pVIII band in lane *8* and the recombinant pVIII band in lane *3*, one can estimate that the expression of the recombinant band is five threefold dilutions, or 243-fold, lower than that of wild-type pVIII. Given a pVIII copy number of about 4000 for f88-4 phage, the copy number of recombinant peptide–pVIII fusion per phage is approximately 16.5 (Protocol 18.8).

We modified Perham's procedure to accommodate a Hoefer SE400 electrophoresis apparatus that uses 18-cm x 16-cm plates and 1.5-mm spacers. It can also be scaled down to accommodate a smaller apparatus, such as the Bio-Rad Mini PROTEAN II system, with excellent results. In our hands, these mini-gels give more reliable results, require less reagent, and save time because they are run over a shorter period.

The silver-staining procedure that follows the SDS-PAGE method was also provided to us by R.N. Perham, who adapted the procedure of Morrissey (1981).

Materials

plates, spacers, combs, etc., for a Hoefer SE400 electrophoresis apparatus (Amersham Pharmacia Biotech)
two 125-ml vacuum flasks
glycerol
49% **acrylamide**/0.5% **bisacrylamide** stock (see Appendix 2)
3 M Tris-HCl, pH 8.3
two 50-ml plastic beakers
25% (w/v) **ammonium persulfate** (Sigma, Cat. # A-3678)
TEMED, (N,N,N′,N′-Tetramethylethylenediamine, ICN Biomedicals Inc., Cat. # 805615)
SDS-PAGE cathode buffer (upper chamber; see Appendix 2)
SDS-PAGE anode buffer (lower chamber; see Appendix 2)
2x SDS-PAGE sample buffer (see Appendix 2)
phage samples (10^{10}–10^{11} particles per lane)
dithiothreitol (DTT)

CAUTIONS: **acrylamide, bisacrylamide, ammonium persulfate, TEMED, DTT (*see Appendix 4*)**

Procedure

1. Assemble the plates and prepare the resolving and stacking gel solutions in separate 125-ml vacuum flasks as follows:

Resolving Gel	Stacking Gel
7.48 ml of (9.42 mg) glycerol	10.41 ml of water
5.90 ml of water	1.50 ml of acrylamide stock
13.25 ml of acrylamide stock	3.75 ml of 3 M Tris-HCl, pH 8.3
13.35 ml of 3 M Tris-HCl, pH 8.3	

   Mix and degas both solutions for approximately 1 minute. Transfer the resolving and stacking gel solutions to two 50-ml plastic beakers. (Be sure the spouts pour well.)

2. Do the following quickly: Add to the resolving gel solution 140 μl of 10% (w/v) ammonium persulfate and 7 μl of TEMED. Mix briefly and pour the solution between the plates, leaving enough space (~1 inch) for the stacking gel. Immediately (before the resolving gel polymerizes) add the same amount of ammonium persulfate and 10 μl of TEMED to the stacking gel solution, stir briefly, and layer it *gently* (one drop at a time at first) on top of the resolving gel, so as not to disturb the interface between the two gel solutions. Insert the comb and allow the gel to polymerize (~1 hour for large gels).

3. Assemble the gel in the electrophoresis apparatus and prepare it for electrophoresis.

4. Mix the phage samples (usually between 10^{10} and 10^{11} particles) 1:1 (v/v) with 2x sample buffer. If reducing conditions are desired (i.e., to break intra- or intermolecular disulfide bonds), add 12.3 mg of DTT per 400 μl of 2x sample buffer (giving the samples a final DTT concentration of 0.1 M). Immediately heat the samples in a 90–100°C water bath for 2 minutes. Load 20-μl samples onto the gel.

   Note: As much as 10^{11} physical particles/lane may be necessary to visualize poorly expressed recombinant proteins.

   To estimate the copy number of a recombinant coat protein, titrate 10^{11} particles of the appropriate phage clone in two- or threefold dilutions on the gel. Early in the titration, the recombinant protein band will have the same intensity as the wild-type pVIII band further on in the titration. Given a pVIII copy number of about 4000 for f88-4 phage, one can estimate the number of recombinant pVIII molecules for the number of phage in a given lane directly from the dilutions necessary to make the wild-type pVIII band the same intensity as the recombinant pVIII band. Usually we get 50–200 copies of a given recombinant pVIII per phage particle (see Fig. 18.4).

5. Run the gel overnight at 100–120 V until the bromophenol blue dye front is near the bottom of the gel.

SILVER STAINING PHAGE-PROTEIN GELS

Be sure that all glassware is very clean, and use only MilliQ water or the equivalent. Do not touch the gel with your bare fingers. Instead, use a glass stirring rod and wear gloves. Two Hoefer gels or several mini-gels can be stained by this procedure.

Materials

a glass baking dish that will accommodate an 18 cm x 16 cm gel
methanol (Caledon Laboratories Ltd., Georgetown, Ontario, Canada Cat. # 6700-1)
glacial **acetic acid**, approximately 17.4 M (Anachemia, Cat. # AC-0135)
glutaraldehyde (Fisher Scientific, Cat. # O2957-1)
dithiothreitol (DTT, Gibco/BRL, Cat. # 15508-013)
silver nitrate (Sigma, Cat. # S-0139)
anhydrous sodium carbonate (see Step 8)
formaldehyde, 37% (J.T. Baker Inc., Cat. # 2106-01)

> *Note:* Check with your campus chemical waste service to find out how to dispose of formaldehyde, glutaraldehyde, and silver nitrate.

CAUTIONS: **methanol, acetic acid, glutaraldehyde, dithiothreitol, silver nitrate, formaldehyde (*see Appendix 4*)**

Procedure

1. In a glass pan, rock the gel in 200 ml of 50% (v/v) methanol, 5% (v/v) acetic acid for 15 minutes at room temperature. The gel can be left in this solution overnight if there is insufficient time to stain the gel on the same day.
2. Change the solution to 200 ml of 7% v/v methanol, 5% v/v acetic acid, and rock for 15 minutes.
3. Fix the gel by soaking it in just enough 12.5% (v/v) glutaraldehyde to cover the gel (~60 ml) and slowly rock for 10 minutes. Wear gloves and work in a fume hood.
4. Discard the fixative and wash the gel by rocking in at least 200 ml of water for 5 minutes. Repeat four more times.
5. Incubate the gel while rocking for 10 minutes in 200 ml of water containing 1 mg of DTT.
6. Incubate the gel while rocking for 15 minutes in 200 ml of water containing 0.2 g of silver nitrate.
7. Rinse the gel by rocking for 30 seconds with 200 ml of water.

8. Prepare the following developing solution:

200 ml of water

8.16 g of anhydrous sodium carbonate

125 µl of formaldehyde (Wear gloves and work in a fume hood.)

Note: Make this buffer beforehand, because the sodium carbonate dissolves slowly.

Add about 40 ml of developing solution to the pan, rock back and forth for 10 seconds (or until the solution darkens), and pour off the used solution. Add the rest of the developer to the pan. Incubate while rocking until the protein bands become visible. This usually happens within a few minutes, so watch carefully.

9. To stop development, pour off the developer, quickly rinse the gel once with water, and add 200 ml of 10% acetic acid. The gel can be kept in this solution for several days, and can be stored temporarily between two sheets of plastic. Record the data by scanning or photography.

PROTOCOL 18.9

Western Blots of Phage Protein Gels

We use this method mainly to detect binding of a selecting molecule, like an antibody, to a recombinant pVIII protein. Figure 18.5 shows a Western blot in which the binding of a monoclonal antibody specific for a phage-displayed peptide is compared to that of a polyclonal, anti-phage antibody. In situations in which the binding activity of a dimeric or cyclic peptide is being assessed, Western blotting can provide a direct correlation between binding activity and cyclization, if one aliquot of the phage to be analyzed is first treated with dithiothreitol (see Protocol 18.8).

This procedure was modified from that of Sambrook et al. (1989; pp. 18.60–18.75). It was designed for use with the ECL kit from Amersham. The reagent volumes and sizes of blotting paper and transfer membrane are meant for a Hoefer SE400 gel electrophoresis apparatus (plates: 18 cm x 16 cm; spacers: 1.5 mm), but a similar procedure can be scaled down for a smaller apparatus. The transfer procedure is for a semi-dry electrophoretic transfer system.

Materials

Western blot transfer buffer (see Appendix 2)
Whatman 3MM Chr chromatography paper (Whatman Inc., Cat. # 3030 917), or similar blotting paper
PVDF Immobilon-P^{SQ} transfer membrane (Millipore Corp., Cat. # ISEQ 262 60), or equivalent
methanol
1x BLOTTO (see Appendix 2)

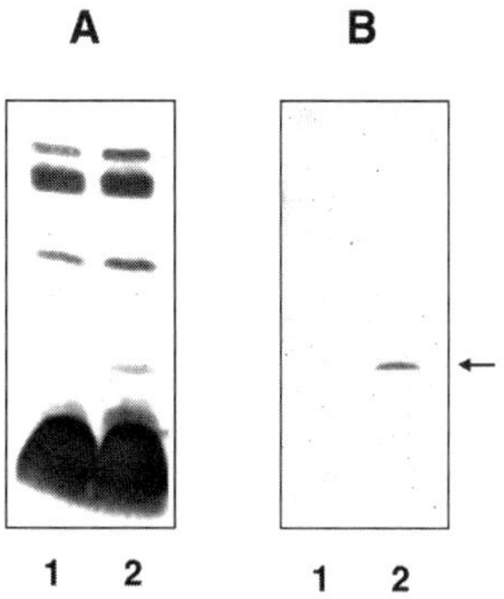

Figure 18.5. Western blots of a SDS–polyacrylamide gel run with phage bearing a variant of the malarial circumsporozoite major repeat sequence NANPNVDP(NANP)$_3$. (*A*) Western blot probing with anti-phage antibody. (*B*) Western blot probing with monoclonal antibody Pf2A.10, which is specific for the peptide displayed by the phage clone f88-NANP. (*Lane 1*) f88-4 control phage; (*lane 2*), f88-NANP phage (Protocol 18.9).

1x TBS containing 0.1% Tween 20 (TBS/Tween)
0.5x BLOTTO containing 0.05% Tween 20 (a 1:1 mixture of 1x BLOTTO and 1x TBS/0.1% Tween)
primary antibody (can be biotinylated)
secondary antibody (or streptavidin) conjugated to horseradish peroxidase
ECL Western blotting detection reagents kit from Amersham (Amersham Pharmacia Biotech, Cat. # RPN 2109)
Trans-Blot SD Semi-Dry Transfer Cell (Bio-Rad Laboratories) or equivalent

CAUTION: **methanol (*see Appendix 4*)**

Procedure

1. Soak the gel for 30 minutes (10 minutes for minigels) in 200 ml of transfer buffer. Cut four pieces of blotting paper to match the size of the gel piece that will be blotted. Also, cut one piece of PVDF transfer membrane to the same size as the blotting paper.

   Note: The gel may swell in the transfer buffer, so wait until after the incubation to cut the paper and membrane. For efficient transfer, make sure that the size of the transfer paper matches that of the gel.

2. Soak the PVDF membrane in methanol for 3 seconds. Transfer it to deionized water and soak for 1–2 minutes, and then soak in transfer buffer for a few minutes.

3. Briefly soak two sheets of blotting paper in transfer buffer and place them on the transfer apparatus (anode; see note). Be sure to remove any bubbles that may be present between the layers of paper, gel, or PVDF membrane, as they will interfere with transfer.

   Note: The order of layers in Steps 3–6 will be reversed if the bottom plate of your transfer apparatus is a cathode. In either case, the order of layers is: anode (positive electrode), blotting paper, membrane, gel, blotting paper, cathode (negative electrode).

4. Layer transfer buffer on the paper and gently place the PVDF membrane on top of the paper. Add another layer of buffer.

5. Place the gel (soaked in transfer buffer, Step 1) on top of the PVDF membrane. Place the remaining two pieces of blotting paper (also soaked in transfer buffer) on top of the membrane, layering buffer between the sheets and removing all bubbles.

6. Attach the lid of the transfer apparatus and begin the transfer according to manufacturer's directions. The following suggestion may be used as a guide: 25 V for 30 minutes for large gels; 10 V for 30 minutes for minigels.

Note: DO NOT exceed the voltage and/or current limit set by the manufacturer of the transfer apparatus.

The pVIII protein has a small molecular weight and so transfers quite readily. To detect under-transfer, silver-stain the gel after transfer to determine what amount of protein has remained in the gel (see Protocol 18.8). To detect over-transfer, include a second PVDF membrane in the transfer sandwich (immediately behind the first membrane). This extra membrane can be exposed to antibody and developed with ECL to make sure that none of the protein has transferred through the first membrane.

7. Rinse the PVDF membrane in Transfer Buffer, and rock slowly in BLOTTO (enough to cover the membrane) overnight at 4°C.

8. Wash the membrane twice by rocking for 2 minutes in TBS/Tween.

9. Incubate the PVDF membrane in primary antibody (at a concentration of ~5–50 nM in 0.5x BLOTTO containing 0.05% Tween 20); use just enough solution to cover the membrane (usually <10 ml). Rock the gel and solution slowly for 1–2 hours at room temperature, or overnight at 4°C.

Note: In a Western blot, one can usually use an antibody concentration less than that used to produce a positive signal by ELISA. If antibody concentration is a concern, test beforehand a range of antibody solutions on a small piece of PVDF that has been dotted with the target protein and then cut into pieces before incubation with antibody.

10. Rinse the PVDF membrane briefly with TBS/Tween. Wash for 20–30 minutes by rocking in BLOTTO, and twice for 5 minutes in TBS/Tween.

11. Incubate the PVDF membrane in horseradish peroxidase-conjugated secondary antibody. For our "homemade" conjugate, we use an approximate 1:2000 dilution in TBS/Tween; otherwise, refer to manufacturer's suggested dilution. Rock the membrane at room temperature for 30 minutes to 1 hour.

Note: Protein A or streptavidin conjugated to horseradish peroxidase also works fine.

12. Rinse the PVDF membrane briefly with TBS/Tween. Wash for 15 minutes by rocking in BLOTTO, then four times, each for 5 minutes, in TBS/Tween.

13. Develop the PVDF membrane using the ECL kit, and expose the membrane to X-ray film according to the manufacturer's directions. Use about 0.125 ml of each ECL reagent per cm^2 of PVDF membrane.

Note: In our hands, detection works well using roughly half the amount of ECL detection reagent suggested by the manufacturer.

PROTOCOL 18.10

Transferring Phage-displayed Peptides to the Amino Terminus of the Maltose-binding Protein

There are cases in which it would be useful to use a given phage-borne peptide as a monovalent, soluble entity. Chemical synthesis of a given peptide can achieve that; however, in some cases, synthetic peptides will have little or no binding activity with the target screening molecule, even if the peptide is perfectly soluble. This could be because of a requirement for a structural conformation of the peptide that is specifically induced by interaction with the phage coat. Alternatively, it could be that a given peptide requires a general protein milieu to fold properly. Thus, the peptide would retain binding activity if fused to the amino terminus of another soluble, monomeric protein. The maltose-binding protein (MBP) of *E. coli* is a good candidate for such a protein. It can be secreted via its amino terminus and can function perfectly with pIII or pVIII leader sequences instead of the *malE* leader (Zwick et al. 1998). It can be purified from the periplasmic space, where it accumulates, and then affinity-purified on amylose resin. This ~50-kD protein is monomeric and contains no cysteine residues, which could cause misfolding if cysteine residues were present in a given amino-terminal peptide fusion. Moreover, the binding of peptide–MBP fusions to a given target screening molecule can be detected by anti-MBP antibody, which produces an excellent signal in an ELISA.

The following protocols are derived from Zwick et al. (1998), which also contains a detailed description of the cloning vectors pMal-X and pMal-KHP. As shown in Figure 18.6, these vectors allow the 5′ end of the recombinant *gene VIII* cassette from f88-4 phage (which encodes the *tac* promoter, the ribosome-binding sites, the recombinant pVIII leader, and the peptide fusion; see Fig. 2.1) to be PCR-amplified, cleaved with *Xho*I and *Pst*I, and ligated into the region encoding the corresponding promoter, leader, and "peptide stuffer" at the 5′ end of an MBP-encoding cassette. Also shown in Figure 18.6 are the cloning sites for transfer of peptides (encoded on a *Kpn*I–*Eag*I fragment) from the amino terminus of pIII in the Ph.D. vector system to the amino terminus of MBP.

Peptide–MBP fusions are affinity-purified on amylose resin, which can be purchased from New England Biolabs, Inc., and comes with manufacturer's instructions, some of which are included below. The beads bind up to approximately 3 mg of fusion/ml bed volume. We usually pour a 15-ml bed volume and expect yields much lower than capacity. Affinity purification of peptide–MBP fusions should be followed by SDS-PAGE; a Western blot of such a gel is shown in Figure 18.7. The procedure for SDS-PAGE using 12% polyacrylamide minigels is described in detail in Appendix 3, and the Western blotting procedure is from Protocol 18.9.

The direct and reversed phage ELISAs (Fig. 18.1A and B; Protocols 18.4 and 18.7, respectively) can be used as templates for designing binding experiments for peptide–MBP fusions; the phage particle is simply replaced by MBP. In direct ELISA, fusions are either directly adsorbed to microwells or captured in wells coated with polyclonal anti-MBP antibody, and binding of biotinylated screening molecule is

```
A.      M   K   K   L   L   F   A   I   P   L   V   V   P   F   Y   S
    CAT.ATG.AAA.AAA.TTA.TTA.TTC.GCA.ATT.CCT.TTA.GTG.GTA.CCT.TTC.TAT.TCT.
     Nde I                                              Kpn I

     H   S ^ A   D   M   K   I   E   E--MBP-->
    CAC.TCG.GCC.GAT.ATG.AAA.ATC.GAA.GAA.
         Eag I

B.      M   K   K   L   L   F   A   I   P   L   V   V   P   F   Y   S
    AAC.GTG.AAA.AAA.TTA.TTA.TTC.GCA.ATT.CCT.TTA.GTG.GTA.CCT.TTC.TAT.TCT.
                                                    Kpn I

     H   S ^  Xn     G   G   G   S   A   E--pIII-->
    CAC.TCT.(NNK)n.GGT.GGA.GGT.TCG.GCC.GAA
                                  Eag I

C.   M   K   I   K   T   G   A   R   I   L   A   L   S   A   L   T   T   M   M
    ATG.AAA.ATA.AAA.ACA.GGT.GCA.CGC.ATC.CTC.GCA.TTA.TCC.GCA.TTA.ACG.ACG.ATG.ATG.

     F   S   A   S   S   F   A ^ Y   D   V   P   D   Y   A   A   A   E   E--MBP-->
    TTT.TCC.GCC.TCA.AGC.TTC.GCC.TAT.GAC.GTT.CCG.GAC.TAC.GCA.GCT.GCA.GAA.GAA.
                  Hind III                                    Pst I

D.   M   K   K   S   L   V   L   K   A   S   V   A   V   A   T
    ATG.AAA.AAG.TCT.TTA.GTT.CTT.AAA.GCA.TCT.GTT.GCT.GTT.GCG.ACT.

     L   V   P   M   L   S   F   A ^ N   V   P   A   E--pVIII-->
    CTT.GTT.CCT.ATG.CTA.AGC.TTT.GCC.AAC.GTC.CCT.GCA.GAA.
                      Hind III             Pst I

E.   M   K   K   S   L   V   L   K   A   S   V   A   V   A   T
    ATG.AAA.AAG.TCT.TTA.GTT.CTT.AAA.GCA.TCT.GTT.GCT.GTT.GCG.ACT.

     L   V   P   M   L   S   F   A ^ (X)n   A   A   E--pVIII-->
    CTT.GTT.CCT.ATG.CTA.TCG.TTT.GCC.(NNK)n.GCT.GCA.GAA.
                                             Pst I
```

Figure 18.6. Nucleotide and amino acid sequences for constructs described in this paper. *(A)* pMal-pIII vector with pIII leader sequence, including *Kpn*I(*Acc65*I) and *Eag*I sites. (*B*) Ph.D. phage display library with random peptide sequence (Xn, [NNK]n; K = G or T) followed by a GGGS linker, inserted between the pIII leader sequence and the mature protein. (*C*) pMal-X and pMal-KHP vectors with MBP:pVIII chimeric leader sequence. The two pVIII-derived residues are in bold. (*D*) Phage-display vector f88-4 with pVIII leader sequence. (*E*) pVIII-displayed peptide library (Xn, [NNK]n) in f88-4. Leader peptidase cleavage sites (^) are indicated. The nucleotide sequences of the pMal-X and pMal-pIII vectors have been deposited in GenBank (Accession numbers AF031813 and AF031088, respectively) (Protocol 18.10).

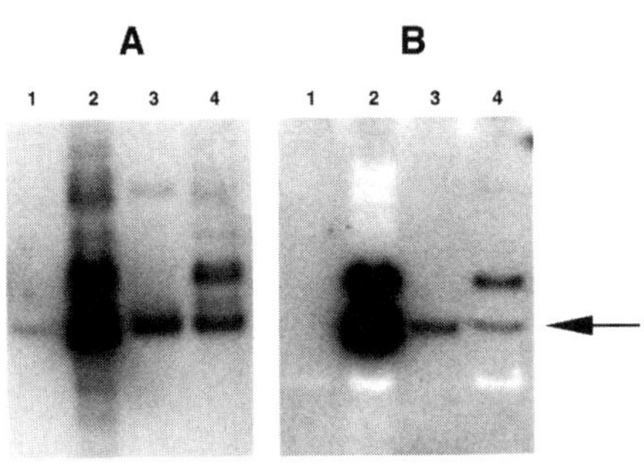

Figure 18.7. Western blot of the G5–MBP fusion expressed by the vector pMal-X. The blot was first probed with rabbit anti-MBP polyclonal antibody (*A*), then stripped and re-probed with monoclonal antibody loop2 (*B*). (*Lane 1*) Whole-cell extract harvested immediately prior to IPTG induction; (*lane 2*) cells harvested 4 hours after IPTG induction; (*lane 3*) osmotic shock supernatant containing processed G5–MBP protein (indicated by arrow); (*lane 4*) resuspended osmotic shock pellet containing processed and unprocessed protein (Protocol 18.10).

detected. In reversed ELISA, wells are coated with screening molecule and binding of the peptide–MBP fusion is detected with anti-MBP antibody. Unlike binding of the phage, binding of the peptide–MBP fusion is not polyvalent and thus does not gain an avidity boost with the reversed ELISA, but the signal from binding can be higher than in the direct ELISA because of the increased detection of the anti-MBP antibody. We have also immobilized peptide–MBP fusions on chips for K_d analysis by surface plasmon resonance (BIAcore). However, care should be taken with this approach if the target molecule used for detection is not monovalent (e.g., if it is an antibody; Zwick et al. 1998).

I. ENGINEERING pMAL-X TO EXPRESS PEPTIDE–MBP FUSIONS

Materials

pMal-X vector (Zwick et al. 1998)
*Xho*I
*Pst*I
Vent DNA polymerase (New England Biolabs, Cat. # 254)
10x ThermoPol buffer (1x buffer contains 2 mM $MgSO_4$; supplied with NEB Vent polymerase)
QIAquick gel extraction kit (QIAGEN, Cat. # 28706)
forward PCR primer oligonucleotide: 5′-TTCCCCGTCAAGCTCTAAATCG-3′
reverse PCR primer oligonucleotide: 5′-GCGGGCTGGGTATCTGAGTTC-3′
dNTPs (a mixture of 2.5 mM each of dATP, dTTP, dCTP, dGTP)
phage (see Steps 2 and 3)
QIAquick PCR purification kit (QIAGEN, Cat. # 28104)
Micropure separator (0.45 µm; Amicon Inc., Cat. # 42522)
Ultrafree-MC filters (30,000 NMWL; Millipore Corp., Cat. # UFC3 LTK 00)
T4 DNA ligase (Gibco BRL, Cat. # 15224-017)
$CaCl_2$-competent *E. coli* cells (see Appendix 3)

CAUTION: ***E. coli* (see Appendix 4)**

Procedure

1. Digest the purified pMal-X vector DNA with *Xho*I and *Pst*I according to the enzyme manufacturer's instructions. Separate the resulting DNA fragments on a 0.8% agarose gel in 1x TAE buffer. Extract and purify the 6.3-kb band (corresponding to pMal-X lacking the *tac* promoter, leader sequence, and hemagglutinin tag) with a QIAquick gel extraction kit as described in the manufacturer's instructions.

2. PEG-purify the phage clones bearing the peptides to be transferred to MBP (see Protocol 15.3). If the phage are to be amplified from an existing preparation, be sure to confirm the DNA sequence of the amplified phage.

3. The polymerase chain reaction (PCR) is used to amplify the segment of phage DNA that encodes the peptide insert. The forward PCR primer anneals immediately upstream of the *Xho*I site, which is 5′ to the *tac* promoter of f88-4. The reverse PCR primer anneals 3′ to the *Pst*I site within the synthetic *gene VIII*. The resulting PCR product encodes a unique *Xho*I site, the phage *tac* promoter, the pVIII leader sequence, the phage-displayed peptide, and a unique *Pst*I site.

Assemble the PCR as follows:

200 μM of each dNTP (8 μl of 2.5 mM stock)

300 nM each for forward and reverse primers

1 Unit of Vent DNA polymerase

10 μl of 10x ThermoPol buffer

83 pM phage (5 x 10^9 particles)

Add water to 100 μl.

4. The PCR parameters are as follows:

Pre-soak	95°C	120 seconds
Step 1	95°C	60 seconds
Step 2	56°C	30 seconds
Step 3	72°C	25 seconds

Repeat Steps 1–3 thirty times. Purify the amplified products with the QIAquick PCR purification kit.

5. Digest the purified insert DNA with *Xho*I and *Pst*I. Separate the fragments on a 6% polyacrylamide gel in 1x TBE buffer, and excise the band of correct size (usually ~240 bp).

6. Crush the gel slices and soak in sterile, deionized water at 37°C for 2 hours. Elute the DNA from the gel pieces by centrifugation through a Micropure Separator according to the manufacturer's directions. Desalt the eluates on Ultrafree-MC filters according to the manufacturer's directions.

7. Ligate approximately 20 ng (0.3 pmole) of purified insert to 30 ng (15 fmole) of the purified pMal-X fragment (Step 1) by using 0.25 Unit of T4 DNA ligase according to the manufacturer's instructions (total volume, 7 μl). Incubate the ligation reaction for 16 hours at 15°C.

8. Transform $CaCl_2$-competent DH5α, AR182, or other suitable *E. coli* strain with the ligation products from Step 7 (refer to Sambrook et al. [1989], or see Appendix 3). Pick ampicillin-resistant colonies for DNA analysis and for the production of peptide–MBP fusion protein (see below).

II. PEPTIDE–MBP FUSION PROTEIN PRODUCTION AND PURIFICATION

Materials

LB medium containing 0.2% (w/v) glucose and 100 µg/ml ampicillin (LB/G/A, see Appendix 2)
840 mM **IPTG** stock
30 mM Tris-Cl, 20% sucrose, pH 8.0, autoclaved
Bradford reagent (see Appendix 2)
MBP column buffer (see Appendix 2)
amylose resin (New England Biolabs, Cat. # 800-21L)
MBP elution buffer (see Appendix 2)
0.5 M EDTA, pH 8.0
6.5% (w/v) **sodium azide** stock solution
5 mM **$MgSO_4$**, ice cold
60-ml syringe barrel and glass wool for making a column with amylose resin (*Alternatively:* a column assembly with adapters)
materials for SDS-PAGE (see Appendix 3 and Zwick et al. 1998 for further details)

CAUTION: **IPTG, Bradford dye, sodium azide, $MgSO_4$ (*see Appendix 4*)**

Procedure

Pilot Experiment

It may be useful to carry out a small-scale pilot experiment to test for expression of a particular peptide–MBP fusion protein, and to verify that the protein is being secreted into the periplasm. This protocol may also be used to troubleshoot if no fusion protein is found in the periplasmic fraction during large-scale production of fusion protein (see below). This protocol results in six samples:

Sample 1	uninduced cells
Sample 2	induced cells
Sample 3	total cell crude extract
Sample 4	suspension of the insoluble material from the crude extract
Sample 5	fraction containing protein that binds to the amylose resin
Sample 6	periplasmic fraction prepared by the cold osmotic shock procedure

If one is testing a number of samples, a useful shortcut is to check uninduced cells, induced cells, and the periplasmic fraction on a gel, while just freezing the other samples. Check the cytoplasmic fraction (total cell extract), or insoluble material only if there is no protein in the periplasm. We have not yet encountered a peptide–MBP fusion that isn't secreted or that is insoluble. If such a situation occurs, unprocessed peptide–MBP fusion may still be recoverable from a cytoplasmic preparation (refer to NEB's instruction manual or website). If the protein is insoluble, try to produce soluble protein by modifying the conditions of cell growth. Two changes that have helped in previous cases are (1) changing to a different strain background for producing the fusions and (2) growing the cells at a lower temperature. We have found that using the *E. coli* strain AR182 (Peters et al. 1994) results in superior yields for fusions involving multiple positive charges near the signal peptidase cleavage site (see Zwick et al. 1998). When using AR182, include 25 μg/ml kanamycin in the medium. For most peptides, however, a common strain such as DH5α should work well.

1. Prepare a 2-ml overnight culture in LB/G/A medium from a single ampicillin-resistant colony containing the fusion plasmid (glucose is included to minimize expression of host MBP, 40.5 kD).
2. Inoculate 65 ml of LB/G/A medium with 0.65 ml of the overnight culture.
3. Incubate the 65-ml culture at 37°C while shaking at 200 rpm until the OD_{600} is approximately 0.5 (~2×10^8 cells/ml). Take a 1-ml sample and microcentrifuge for 2 minutes (Sample 1: uninduced cells). Discard the supernatant and resuspend the cells in 50 μl of 1x SDS-PAGE sample buffer. Vortex and freeze at –20°C.
4. Add IPTG to the remaining culture to a final concentration of 0.2 mM. Continue incubating the culture at 25°C for 2 hours. Take a 0.5-ml sample and microcentrifuge for 2 minutes (Sample 2: induced cells). Discard the supernatant and resuspend the cells in 100 μl of 1x SDS-PAGE sample buffer. Vortex to resuspend the cells and freeze the preparation at –20°C.

   Note: Additional time points at 1 and 3 hours can be helpful in deciding when to harvest the cells for a large-scale preparation.

5. Divide the remaining culture into two aliquots. Harvest the cells by centrifugation at 4000*g* for 10 minutes. Discard the supernatants. Resuspend one pellet in 4 ml of MBP column buffer, and continue with Protocol A. Resuspend the other pellet in 8 ml of 30 mM Tris-HCl, 20% sucrose, pH 8.0 (8 ml/0.1 g of cells, wet weight) and proceed with Protocol B.

Protocol A

This protocol requires some method of determining protein concentration. We use a UV monitor to measure absorbance at 280 nm. In our setup, an amylose purification column is connected to a peristaltic pump (LKB-Pump P-1, Amersham Pharmacia Biotech), which is in turn connected to a UV monitor (Single Path Monitor UV-1, Amersham Pharmacia Biotech). The elution profile is then followed by a chart recorder (LKB REC-101, Amersham Pharmacia Biotech), and the protein peak is identified. Alternatively, one can use the Bradford assay as described below and in more detail in Ausubel et al. (1987).

6A. Freeze the cells in column buffer in a dry ice–ethanol bath (freezing overnight at –20°C also works well). Thaw the suspension in cold water.

7A. Place the cells in an ice-water bath and sonicate in short pulses of 15 seconds or less. Monitor the release of protein using the Bradford assay: Add 10 μl of the sonicate to 1.5 ml of Bradford reagent in a 1.6-ml microcentrifuge tube, and quickly mix by vortexing briefly. Allow the tube to stand for 2 minutes at room temperature, pipet 1 ml of the solution into a 1-cm pathlength cuvette, and determine the absorbance at 595 nm. Continue sonicating the cells until the released protein reaches a maximum (usually about 2 minutes); a standard containing 10 μl of 10 mg/ml BSA (or 10 mg/ml BSA among a set of standards) in 1.5 ml of Bradford reagent can be helpful in gauging an approximate endpoint.

8A. Centrifuge the sample at 9000*g* and 4°C for 20 minutes. Decant the supernatant (crude extract) and save on ice. Add 5 μl of 2x SDS-PAGE sample buffer to 5 μl of the crude extract (Sample 3: total cell crude extract). Resuspend the pellet in 5 ml of column buffer (insoluble matter). Add 5 μl of 2x SDS-PAGE sample buffer to 5 μl of the insoluble matter (Sample 4: insoluble material from the crude extract).

9A. Place approximately 200 μl of amylose resin in a 1.5-ml microcentrifuge tube and microcentrifuge briefly at low speed (~3000 rpm) to settle the beads. Aspirate and discard the supernatant. Resuspend the resin in 1.5 ml of MBP column buffer, microcentrifuge briefly, and discard the supernatant. Repeat for a second wash, and resuspend the resin in 200 μl of MBP column buffer. Mix 50 μl of crude extract with 50 μl of the amylose resin slurry. Incubate for 15 minutes on ice. Microcentrifuge for 1 minute (~3000 rpm), then remove and discard the supernatant. Wash the pellet with 1 ml of MBP column buffer, microcentrifuge for 1 minute (~3000 rpm), and resuspend the resin in 50 μl of 1x SDS-PAGE sample buffer (Sample 5: protein that binds amylose). Continue protocol with Step 10, below.

Protocol B

6B. Add 16 μl of 0.5 M EDTA to the cells in Tris/sucrose (final concentration 1 mM EDTA), and incubate for 5–10 minutes at room temperature with shaking or stirring.

7B. Centrifuge the sample at 8000*g* and 4°C for 10 minutes. Remove all of the supernatant and resuspend the pellet in 8 ml of ice-cold 5 mM $MgSO_4$.

8B. Shake or stir for 10 minutes in an ice-water bath.

9B. Centrifuge as above (Step 7B). The supernatant is the cold osmotic shock fluid. Add 10 μl of 2x SDS-PAGE sample buffer to 10 μl of the cold osmotic shock fluid (Sample 6: periplasmic fraction). Continue protocol with Step 10, below.

10. Place the samples in a boiling water bath for 5 minutes. Microcentrifuge for 1 minute at low speed. Load 20 μl each of the uninduced cells, induced cells, amylose resin-purified samples (Samples 1, 2, and 5, respectively), and all of the samples 3, 4, and 6 on a 12% SDS-PAGE gel (see Appendix 3). Avoid disturbing the pellets.

11. (*Optional*) Run an identical SDS-PAGE gel(s) after diluting the samples 1:10 in 1x SDS sample buffer. Prepare a Western blot(s) and develop with anti-MBP serum and, if available, serum directed against the peptide of interest (see Protocol 18.9 for the general method, and Zwick et al. 1998, for details).

Large-scale Preparation and Amylose Affinity Chromatography of MBP Fusion Proteins

This method (Method II in the NEB instruction booklet) is designed for the purification of the secreted, or mature, form of the peptide–MBP fusion expressed from pMal-X (i.e., the leader sequence has been removed during the secretion of the protein). The method results in a periplasmic fraction that contains far fewer *E. coli* proteins than the total cell crude extract. The yields vary somewhat from one fusion protein to another. The most consistent results are obtained if the overnight culture is made from a fresh colony. From 2-liter cultures, we obtain 1–2 mg of protein. When using the AR182 strain of *E. coli* (*araD139Δ(araABC-leu)7696thrΔlacX74 galU galK hsdR mcrB rpsL(strA) thi prlA4 zhc::Tn10kan* [Peters et al. 1994; kindly provided by Peter Schatz, The Affymax Research Institute, Palo Alto, California]), include 25 μg/ml kanamycin in the medium. AR182 has given favorable results even for fusions containing peptides with multiple positive charges at or near the amino terminus. DH5α is a standard strain that grows a little slower than AR182, but should give good results for most peptides.

1. Prepare a 25-ml overnight culture in LB/G/A from a single ampicillin-resistant colony containing the fusion plasmid. (If you are using strain AR182, add kanamycin to 25 μg/ml to the agar and the culture medium.)

2. Inoculate 2 liters of LB/G/A (also containing 25 μg/ml of kanamycin, if using strain AR182) with 20 ml of the overnight cell culture.

3. Incubate the culture at 37°C while shaking at 200 rpm, until an OD_{600} approximately equal to 0.5 is reached (2×10^8 to 4×10^8 cells/ml). Add IPTG to a final concentration of 0.2 mM. Transfer the cultures to 25°C, and shake for an additional 2 hours.

   Note: The period of time and the temperature to use during expression depend on several factors (e.g., stability of the protein, host strain) and can be varied to optimize conditions (see Pilot Experiment). In addition, using less IPTG may lead to higher yields, because at higher IPTG levels, protein export in *E. coli* may not be able to keep up with full-level expression from the *tac* promoter.

4. Harvest the cells by centrifuging at 4000*g* for 20 minutes at 4°C and discard the supernatant. Resuspend the cells in 600 ml of 30 mM Tris-HCl, 20% sucrose, pH 8.0 (80 ml for each gram of cells, wet weight). Add EDTA stock to a concentration of 1 mM. Shake at about 200 rpm for 5–10 minutes at room temperature to resuspend the pellet.

5. Centrifuge the supension at 8000*g* for 20 minutes at 4°C. Discard the supernatant, and resuspend the pellet in 400 ml of ice-cold 5 mM $MgSO_4$. Shake or vigorously rock the suspension for 10 minutes at 4°C.

6. Centrifuge the suspension as in Step 5. Discard the pellet, and transfer the supernatant (the cold osmotic shock fluid) to a clean tube.

7. Add 8 ml of 1 M Tris-HCl, pH 7.4, and 400 µl of 1 M sodium azide to the osmotic shock fluid. At this stage, the sample(s) can be frozen at –20°C until you are prepared to run the affinity column.

8. Pour 15 ml of amylose resin into a 2.5 x 10 cm column. A 50-ml syringe barrel plugged with silanized glass wool can be substituted for the 2.5-cm column, but the glass wool should cover the bottom of the syringe (not just in the tip) so the column will have an acceptable flow rate. After the resin has settled evenly, wash the column with 8 column volumes of MBP column buffer.

9. Load the cold osmotic shock fluid at a flow rate of about 100–130 ml/hr (~2 ml/min). Wash the column with 12 column volumes of MBP column buffer.

   Note: It is possible to allow the column to wash overnight, if the column has a safety loop to prevent it from running dry. If the column does run dry, it is better to restart the column with elution buffer, rather than continuing the wash. Avoid loading the column overnight, because if it is allowed to run dry, or sit for a long period of time, much of the protein will come off when the flow in the column is reinitiated prior to the washing step.

10. Elute the fusion protein with MBP elution buffer (at ~2 ml/min). Collect 10–15 fractions of 3 ml each (fraction size should be 1/5 the column volume). The fusion protein usually starts to elute within the first five fractions, and can easily be detected by measuring each sample's absorbance at 280 nm.

11. Pool the protein-containing fractions. If desired, the protein can be concentrated and washed using a Centriprep-30 (30,000 MWCO, Amicon) concentrator or the equivalent. Aliquot and store the protein fusions at –70°C. We usually prepare a working aliquot (good for at least a few weeks) by adding an equal volume of 100% sterile glycerol to a sample, and storing at –20°C.

12. The column should be regenerated immediately after eluting the peptide–MBP fusion protein. Perform the following sequence of washes at room temperature: 3 column volumes of deionized water, 3 column volumes of 0.1% SDS, 1 column volume of deionized water, and 3 column volumes of column buffer.

    Note: Although the column can be washed at 4°C, 0.1% SDS will eventually precipitate at that temperature. It is therefore recommended that the SDS solution be stored at room temperature until needed, and rinsed out of the column promptly. Upon repeated use, trace amounts of amylase in the *E. coli* extract will decrease the binding capacity of the column. Thus, the column should be washed promptly after each use. The resin can be reused three to five times.

ACKNOWLEDGMENTS

We thank T. Ha and A. Day for making figures. We thank X. Gong, K. Brown, S. Kim, M. Rashed, and J. Mehroke for contributions in developing the methods presented in this chapter.

REFERENCES

Ausubel F.M., Brent R., Kingston R.E., Moore D.D., Seidman J.G., Smith J.A., and Struhl K., eds. 1987. *Current protocols in molecular biology*, vol. 2., pp. 10.1.1–10.1.3. Greene Publishing Associates and Wiley-Interscience, Toronto.

Bonnycastle L.L.C., Brown K., Tang J., and Scott J.K. 1997. Assaying phage-borne peptides by phage capture on fibrinogen or streptavidin. *Biol. Chem.* **378:** 509–515.

de la Cruz V.F., Lal A.A., and McCutchan T.F. 1988. Immunogenicity and epitope mapping of foreign sequences via genetically engineered filamentous phage. *J. Biol. Chem.* **263**: 4318–4322.

di Marzo Veronese F., Willis A.E., Boyer-Thompson C., Appella E., and Perham R.N. 1994. Structural mimicry and enhanced immunogenicity of peptide epitopes displayed on filamentous bacteriophage: The V3 loop of HIV-1 gp120. *J. Mol. Biol.* **243**: 167–172.

Galfre G., Monaci P., Nicosia A., Luzzago A., Felici F., and Cortese R. 1996. Immunization with phage-displayed mimotopes. *Methods Enzymol.* **267**: 109–115.

Greenwood J., Willis A.E., and Perham R.N. 1991. Multiple display of foreign peptides on a filamentous bacteriophage. Peptides from *Plasmodium falciparum* circumsporozoite protein as antigens. *J. Mol. Biol.* **220:** 821–827.

Harlow E. and Lane D. 1988. *Antibodies: A laboratory manual.* Cold Spring Harbor Laboratory, Cold Spring Harbor, New York.

Hermanson G.H. 1996. *Bioconjugate techniques*. Academic Press, San Diego.

McKinney M.M. and Parkinson A. 1987. A simple, non-chromatographic procedure to purify immunoglobulins from serum and ascites fluid. *J. Immunol. Methods* **96**: 271–278.

Minenkova O.O., Ilyichev A.A., Kischenko G.P., and Petrenko V.A. 1993 Design of specific immunogens using filamentous phage as the carrier. *Gene* **128**: 85–88.

Morrissey J.H. 1981. Silver stain for proteins in polyacrylamide gels: A modified procedure with enhanced uniform sensitivity. *Anal. Biochem.* **117**: 307–310.

Peters E.A., Schatz P.J., Johnson S.S., and Dower W.J. 1994. Membrane insertion defects caused by positive charges in the early mature region of protein pIII of filamentous phage fd can be corrected by *prlA* suppressors. *J. Bacteriol.* **176**: 4296–4305.

Rini J.M., Schulze-Gahmen U., and Wilson I.A. 1992. Structural evidence for induced fit as a mechanism for antibody-antigen recognition. *Science* **255:** 959–965.

Sambrook J., Fritsch E.F., and Maniatis T. 1989. *Molecular cloning: A laboratory manual,* 2nd edition, pp.18.60–18.75. Cold Spring Harbor Laboratory Press, Cold Spring Harbor, New York.

Schagger H. and von Jagow G. 1987. Tricine-sodium dodecyl sulfate-polyacrylamide gel electrophoresis for the separation of proteins in the range from 1 to 100 kDa. *Anal. Biochem.* **166**: 368–379.

Smith G.P., Schultz D.A., and Ladbury J.E. 1993. A ribonuclease S-peptide antagonist discovered with a bacteriophage display library. *Gene* **128:** 37–42.

Willis, A.E., Perham, R.N., and Wraith, D. 1993. Immunological properties of foreign peptides in multiple display on a filamentous bacteriophage. *Gene* **128**: 79–83.

Zwick M.B., Shen J., and Scott J.K. 2000. Homodimeric peptides displayed by the major coat protein of filamentous phage. *J. Mol. Biol.* **300:** 307–320.

Zwick M.B., Bonnycastle L.L.C., Noren K.A., Venturini S., Leong E., Barbas III, C.F., Noren C.J., and Scott J.K. 1998. The maltose binding protein as a scaffold for monovalent display of peptides derived from phage libraries. *Anal. Biochem.* **264**: 87–97.

19 Construction and Use of pIII-displayed Peptide Libraries

KAREN A. NOREN, LAURA H. SALTMAN, AND CHRISTOPHER J. NOREN
New England Biolabs, Beverly, Massachusetts 01915

PHAGE DISPLAY DESCRIBES A SELECTION TECHNIQUE in which a library of variants of a peptide or protein is expressed on the outside of a phage virion, while the genetic material encoding each variant resides on the inside (for review, see Wilson and Finlay 1998; Rodi and Makowski 1999). This creates a physical linkage between each variant protein sequence and the DNA encoding it, which allows rapid partitioning based on binding affinity to a given target molecule (antibodies, enzymes, cell-surface receptors, etc.) by an in vitro selection process called *panning* (Parmley and Smith 1988). In its simplest form, panning is carried out by incubating a library of phage-displayed peptides with a plate (or bead) coated with the target, washing away the unbound phage, and eluting the specifically bound phage (Fig. 19.1). The eluted phage is then amplified and taken through additional binding/amplification cycles to enrich the pool in favor of binding sequences. After 3 or 4 rounds, individual clones are characterized by DNA sequencing and ELISA.

CONTENTS

Random peptide libraries displayed on phage have been successfully used for numerous applications, including epitope mapping/vaccine development (Cwirla et al. 1990; Scott and Smith 1990; for review, see Cortese et al. 1995), identification of protein kinase substrates/SH2 ligands (Dente et al. 1997), and identification of peptide mimics of nonpeptide ligands (Devlin et al. 1990; Oldenburg et al. 1992; Scott et al. 1992; Hoess et al. 1993; Katz 1995). One particularly notable application is

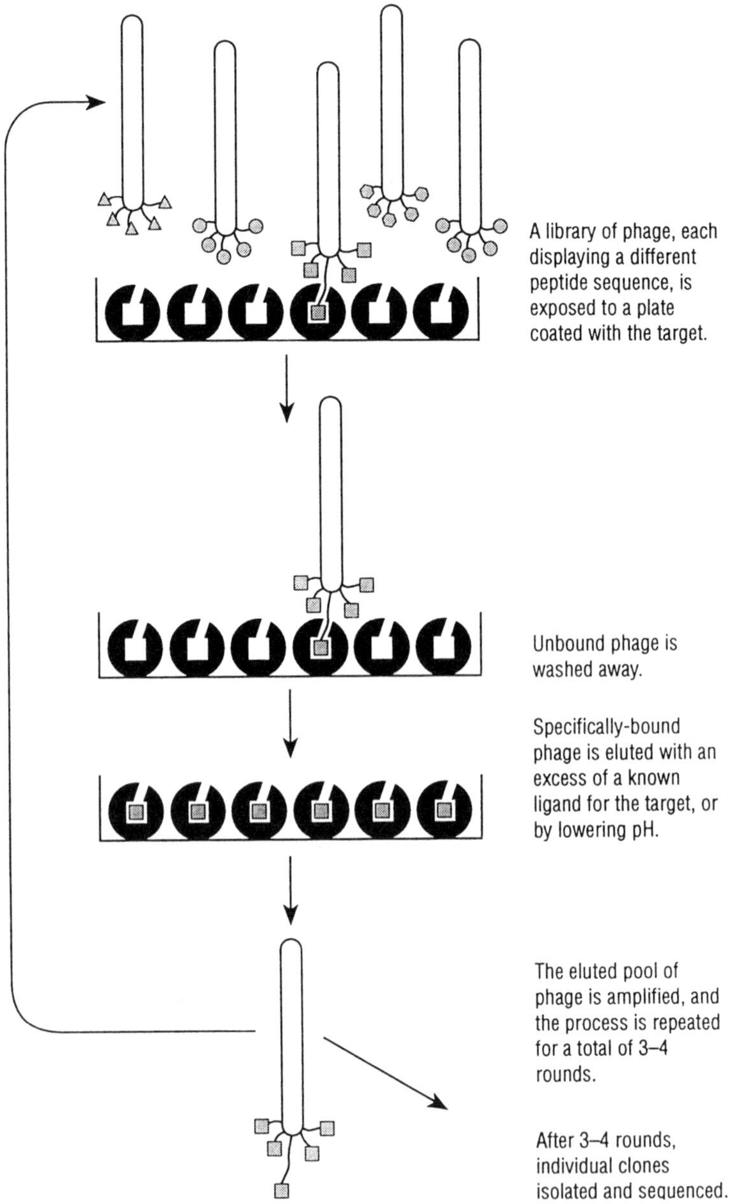

Figure 19.1. Panning with a pentavalent peptide library displayed on pIII. (Reprinted, with permission, © 1998/99 New England Biolabs Catalog.)

the identification of novel bioactive peptides by panning against immobilized cell-surface receptors (O'Neil et al. 1992; Wrighton et al. 1996; Cwirla et al. 1997) or intact cells (Doorbar and Winter 1994; Goodson et al. 1994; Barry et al. 1996; Siegel et al. 1997; Szardenings et al. 1997; see also Chapter 23). An exciting recent development is the use of phage display to select organ-specific peptides in vivo (see Chapter 22). Following injection of a peptide library into mice, the organs of interest are harvested and washed, and the eluted phage are used in subsequent rounds of injection and selection (Pasqualini and Ruoslahti 1996). Peptides selected in this manner have been successfully used to specifically deliver drugs to tumor cells (Arap et al. 1998).

The Ph.D. (for *Ph*age *D*isplay) system described in this chapter is based on a simple M13 phage vector for display of peptides as amino-terminal fusions to the minor coat protein pIII (Cwirla et al. 1990; Devlin at al. 1990; Scott and Smith 1990). This protein modulates phage infectivity by binding to the F-pilus of the recipient bacterial cell and is present in five copies clustered at one end of the mature M13 virion (Russel et al. 1997; Rodi and Makowski 1999). In contrast to pVIII libraries, which typically have on the order of 100 displayed peptides per virion, pIII libraries have a maximum valency of only 5 peptides per virion. As a result of this greatly reduced valency, pIII libraries are best suited to the discovery of higher affinity ligands (K_d of 10 μM or better). If the displayed peptide is sufficiently short, the infectivity function of pIII is not affected, and all five copies can carry displayed peptides without measurable attenuation of phage infectivity. As a result, library vector systems need only a single copy (the fused copy) of *gene III* (*gIII*). In contrast, pVIII libraries almost always involve two copies of *gene VIII* (*gVIII*): the fused copy typically present on a phagemid vector, and an unfused copy supplied by a helper phage. pIII libraries in which the displayed protein is large enough to adversely affect pIII function are also constructed in phagemid/helper systems, which can result in an average valency of 1 displayed protein per virion (Kang et al. 1991; Lowman et al. 1991).

The Ph.D. system was designed for pentavalent display of short (<30 residues) peptide libraries. The system uses a phage vector, which can be simply and rapidly propagated without the need for antibiotic selection or helper phage superinfection (Devlin et al. 1990). The cloning vector M13KE is derived from M13mp19, allowing construction and propagation of phage-display libraries using standard M13 techniques. If the peptide insert is sufficiently short, the resulting library is fully infectious and can be amplified repeatedly with little apparent loss of sequence diversity (Rodi et al. 1999). Extensive sequencing of naive libraries prepared in this vector system has revealed little sequence bias apart from selection against unpaired cysteine residues (C. Noren et al., unpubl.) and the expected reduced levels of arginine (but not lysine) residues. The reduced arginine levels are likely caused by the *secY*-dependent secretion of pIII and can be overcome if desired by the use of a *prlA* suppressor strain for library amplification (Peters et al. 1994). Experiments at New England Biolabs have identified consensus peptide-binding sequences against a variety of proteins, including enzymes, cell-surface receptors, and monoclonal antibodies (Fig. 19.2). In all cases the prepared libraries have been demonstrated to be of sufficient complexity to produce multiple DNA sequences encoding the same consensus peptide motifs. Epitope mapping experiments have been successfully carried out elsewhere using this system (Birkenmeier et al. 1997; Dore et al. 1998; Gevorkian et al. 1998; Osman et al. 1998). In a particularly dramatic application, a commercially available dodecapeptide library (Ph.D.-12) constructed in M13KE was panned against Taxol, and the selected sequences were compared to a protein database to identify the natural target for the drug as Bcl-2 (Rodi et al. 1999). This quite unexpectedly demonstrated that short peptides from an unstructured peptide library can mimic a three-dimensional ligand-binding site, greatly increasing the potential utility of these libraries.

β-endorphin:

```
Y G G F M T S E K Q T P...
```

1st round sequences:

```
Y G W I S P P L H L P T
Y Q P D N P S R Q I A N
Y W P A H I R A V P M I
R L D D I K N T L A F S
S S D V Y S L Y P F I M
E F F P H P M L H N S R
D N W P Y R P S F S L S
S H N T Y S A P R P S A
S L L H Y A S S L S L M
F N Q N A E P F S S R P
H P R Q L L H H P L S P
```

2nd round sequences:

```
Y G G F L I G L Q D A S
Y G G F H Y K E T G A L
Y Q P D N P S R Q I A N
V Y C Y I N Q S M I G N
H H D T E Y R T T Q L S
N L K F P T N P K A M W
L P N L T W A L M P R A
D N W P Y R P S F S L S
S H N T Y S A P R P S A
S L L H Y A S S L S L M
V T M N T K T P G P M P
```

3rd round sequences:

```
Y G G F M T T P S H V P
Y G G F M T T P S H V P
Y G G F I S Q T Q H Y S
Y G G F I S Q T Q H Y S
Y G G F G N S L V M P V
Y G G F S M P F L P A L
Y G A F D V T T G V T S
Y G V F N P H Y L P S L
A P S T D K Q A T M P L
A S V A V S S R Q D A A
```

Figure 19.2. Epitope mapping with the Ph.D.-12 library. The Ph.D.-12 library was panned in solution (10 nM antibody) against a monoclonal antibody (3E-7) raised against the opioid neuropeptide β-endorphin, followed by affinity capture of the antibody–phage complexes onto Protein A-agarose (rounds 1 and 3) or protein G-agarose (round 2) beads. Selected sequences are shown aligned with the first 12 residues of β-endorphin. The results clearly show that the epitope for this antibody spans the first 7 residues of β-endorphin (YGGFMTS), and the conserved position of the selected sequences within the 12-mer indicates that the free α-amino group of the amino-terminal tyrosine of β-endorphin is part of the epitope. The results also suggest that most of the binding energy for the antigen–antibody interaction is contributed by the first four residues of the epitope (YGGF), with some flexibility allowed in the third position. (Reprinted, with permission, © 1998/99 New England Biolabs Catalog.)

PROTOCOL 19.1

General M13 Methods

It is important to note that unlike phage lambda, M13 is not a lytic phage. Plaques are caused by diminished cell growth rather than cell lysis, and are turbid rather than clear. Plating on X-gal/IPTG media is strongly recommended to facilitate visualization of plaques.

STRAIN MAINTENANCE

Materials

E. coli ER2738 host strain, glycerol culture (not competent), (New England Biolabs, Cat. # E4104S)
LB medium or LB+ 20 μg/ml **tetracycline** (LB+tet, see Appendix 2)
LB+tet plates (see Appendix 2)

CAUTIONS: ***E. coli*, tetracycline (*see Appendix 4*)**

Procedure

1. The recommended *E. coli* host strain ER2738 (F′ *proA*$^+$*B*$^+$ *lacI*^{q} *Δ(lacZ)M15 zzf::Tn10 (Tet*^{R}*)/fhuA2 glnVthi Δ(lac-proAB) Δ(hsdMS_mcrB)5* [r_k^- m_k^- McrBC$^-$]) is a robust F$^+$ strain with a rapid growth rate and is particularly well-suited for M13 propagation. ER2738 is a recA$^+$ strain, but we have never observed spontaneous in vivo recombination events with M13 or phagemid vectors. Commercially available F$^+$ strains such as DH5αF′and XL1-Blue can probably be substituted for ER2738, but have not been tested with our vector system. Any strain used should be *glnV* in order to suppress amber (UAG) stop codons within the library with glutamine.
2. Because M13 is a male-specific coliphage, it is recommended that all cultures for M13 propagation be inoculated from colonies grown on media selective for presence of the F-factor, rather than directly from the glycerol culture. The F-factor of ER2738 contains a mini-transposon that confers tetracycline resistance, so cells harboring the F-factor can be selected by plating and propagating in tetracycline-containing medium.
3. Streak out ER2738 from the supplied glycerol culture onto an LB+tet plate. Invert and incubate at 37°C overnight. Store wrapped with Parafilm at 4°C in the dark for a maximum of 1 month.
4. ER2738 cultures for infection can be grown either in LB or LB+tet media. Loss of F-factor in nonselective media is insignificant as long as cultures are not serially diluted repeatedly.

AVOIDING PHAGE CONTAMINATION

The library cloning vector M13KE differs from wild-type filamentous phages in that the lacZα-peptide cloning sequence (which permits blue/white screening) has been inserted in the vicinity of the (+)-strand origin of replication, resulting in a longer replication cycle. In addition, display of foreign peptides as amino-terminal fusions to pIII (which mediates infectivity by binding to the F-pilus of the recipient bacterium) may attenuate infectivity of the library phage relative to wild-type M13. As a result, there is a reasonably strong in vivo selection for any contaminating wild-type phage during the amplification steps between rounds of panning. In the absence of a correspondingly strong in vitro binding selection, even vanishingly small levels of contamination can result in a majority of the phage pool being wild-type phage after three rounds of panning.

1. The potential for contamination with environmental bacteriophage can be minimized by using aerosol-resistant pipet tips for all protocols and wearing latex gloves for all steps.
2. Because the library cloning vector M13KE is derived from the common cloning vector M13mp19, which carries the *lacZα* gene, phage plaques appear blue when plated on media containing X-gal and IPTG. Environmental filamentous phage typically yield colorless plaques when plated on the same media. These plaques are also larger and "fuzzier" than the library phage plaques. We strongly recommend plating on LB/IPTG/X-gal plates for all titering steps and, if white plaques are evident, picking ONLY blue plaques for sequencing.
3. Severe contamination (white plaques present in large numbers) can lead to contamination of subsequent panning experiments. To prevent this, all solutions should be re-autoclaved if possible; any solutions containing heat-labile components should be remade. The work area should be wiped down with ethanol. Pipettors should be disassembled and the parts soaked in detergent, rinsed carefully with sterile water, and reassembled.

PHAGE TITERING

The number of plaques increases linearly with added phage only when the multiplicity of infection (moi) is much less than 1 (i.e., cells are in considerable excess). For this reason, it is recommended that phage stocks be titered by diluting prior to infection, rather than by diluting cells infected at a high moi. Plating at low moi will also ensure that each plaque contains only one DNA sequence.

Materials

M13KE phage: amplified culture supernatants or unamplified panning eluates
E. coli ER2738 host strain (New England Biolabs, Cat. # E4104S)
LB (see Appendix 2)
Top Agar (see Appendix 2)
LB/**IPTG/X-gal** plates (see Appendix 2)

CAUTIONS: ***E. coli*, IPTG, X-gal (*see Appendix 4*)**

Procedure

1. Inoculate 5–10 ml of LB with a single colony of ER2738 and incubate with shaking until mid-log phase (OD_{600} ~0.5, approximately 4 hours).
2. While cells are growing, melt Top Agar in microwave and dispense 3 ml into sterile culture tubes, one per expected phage dilution. Equilibrate tubes at 45°C until ready for use.
3. Prewarm one LB/IPTG/X-gal plate per expected dilution at 37°C until ready for use.
4. Prepare 10-fold serial dilutions of phage in LB. Suggested dilution ranges: for amplified phage culture supernatants, 10^8–10^{11}; for unamplified panning eluates, 10^1–10^4. Use aerosol-resistant pipet tips to prevent cross-contamination, and use a fresh pipet tip for each dilution.
5. When the culture in Step 1 reaches mid-log phase, dispense 200 μl into microfuge tubes, one for each phage dilution.
6. Add 10 μl of each dilution to each tube, vortex quickly, and incubate at room temperature for 1–5 minutes.
7. Transfer the infected cells one infection at a time to culture tubes containing 45°C Top Agar. Vortex quickly and IMMEDIATELY pour each culture onto a prewarmed LB/IPTG/X-gal plate. Spread Top Agar evenly by tilting the plate.
8. Allow the plates to cool for 5 minutes, invert, and incubate overnight at 37°C.
9. Count plaques on plates that have ~100 plaques. Multiply each number by the dilution factor for that plate to get phage titer in plaque-forming units (pfu) per 10 μl.

PROTOCOL 19.2

Construction of pIII-displayed Peptide Libraries

M13KE (Fig. 19.3) is a simple M13mp19 derivative into which cloning sites have been introduced at the 5′ end of *gene III* for display of short peptide sequences as amino-terminal pIII fusions. The sequence of M13KE is available at

http://www.neb.com/neb/tech/nucleotide_seq_maps/sequences/m13ke.seq

Because this is a phage, rather than a phagemid vector, all 5 copies of pIII on the surface of each virion will be fused to the cloned peptide. Because displayed proteins longer than 50 amino acids may have a deleterious effect on the infectivity function of pIII, this vector is suitable only for the display of peptides and small proteins (unpublished observations). The small insert size does not appreciably attenuate phage replication, allowing the vector to be propagated as phage, rather than as a plasmid (i.e., titer for plaques, not colonies). Thus, the vector carries neither a plasmid replicon nor antibiotic resistance. This simplifies the intermediate amplification steps during panning considerably, as it is not necessary to express antibiotic genes before plating or to use helper phage during amplification. The steps necessary to clone a peptide library into M13KE are outlined below. To clone a single peptide sequence, reactions can be scaled down appropriately.

PREPARATION OF ELECTROCOMPETENT CELLS

This procedure will generate a sufficient quantity of electrocompetent cells for test ligations and large-scale library production. The use of commercially available competent cells on this scale is financially prohibitive.

Materials

LB (see Appendix 2)
E. coli ER2738 host strain, overnight culture
9 liters of ice-cold autoclaved water, doubly distilled or purified with a Milli-Q or equivalent system
500 ml ice-cold 10% (v/v) glycerol in water
M13 RF DNA: M13KE (New England Biolabs, Cat. # E8101S) or M13mp19 (New England Biolabs, Cat. # N4040S), or see Sambrook et al., p. 4.31 (1989)
SOC medium (see Appendix 2)
LB/**IPTG**/**X-gal** plates (for testing electrocompetence)

CAUTIONS: ***E. coli*, IPTG, X-gal (*see Appendix 4*)**

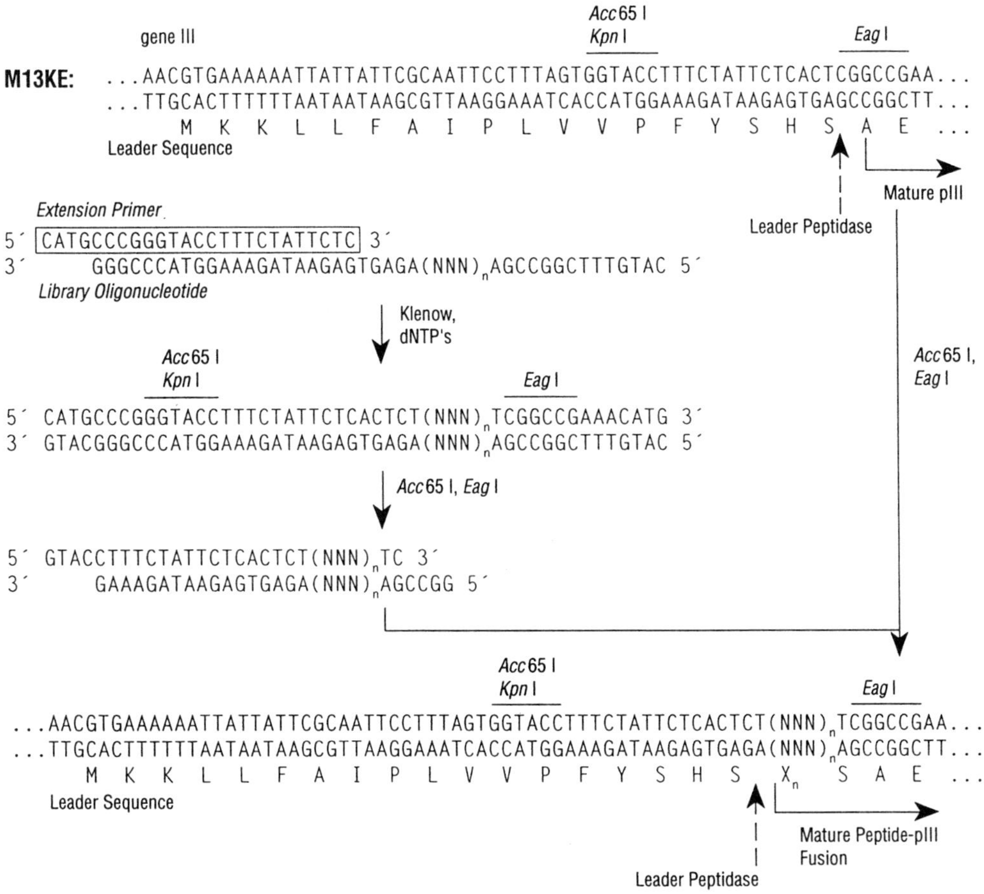

Figure 19.3. Cloning a synthetic library insert into M13KE. The amino terminus of gIII is shown, with introduced *Acc*65I and *Eag*I restriction sites. A library oligonucleotide is designed as shown, leaving constant the bases not indicated by Ns. The randomized segment, along with any spacer sequence, is introduced at the region indicated by $(NNN)_n$, taking care not to interrupt the reading frame. Randomized codons of the form 5´-NNK-3´ (K=G or T) should be encoded as 5´-MNN-3´ (M = A or C), as this sequence corresponds to the noncoding strand. The universal extension primer (boxed) is annealed and extended, and the resulting duplex is digested with *Acc*65I and *Eag*I. The insert is then cloned in M13KE that has been digested with the same enzymes, resulting in expression of the peptide insert directly at the amino terminus of mature pIII.

Procedure

1. Inoculate 6 liters of LB medium (in six 4-liter Erlenmeyer or 2.8-liter Fernbach flasks to maximize aeration) with 1/100 volume (10 ml per liter) of an overnight culture of ER2738 that has been grown at 37°C with shaking.
2. Incubate the cultures at 37°C with vigorous shaking (>250 rpm) until cultures reach an OD_{600} of 0.5–1.0.
3. Chill flasks on ice for 30 minutes, and harvest the cells by centrifugation at 5000*g* for 20 minutes at 4°C. Discard the supernatant.

4. Suspend each pellet in 1 liter of ice-cold autoclaved water. Centrifuge as before and carefully discard the supernatant.
5. Suspend each pellet in 0.5 liter of ice-cold autoclaved water. Combine in three bottles and centrifuge as before. Discard supernatant.
6. Suspend the pellets in 120 ml total of ice-cold 10% (v/v) glycerol in water. Combine in a single bottle and centrifuge at 8000*g* for 10 minutes at 4°C. Discard as much of the supernatant as possible without disturbing the pellet.
7. Suspend the pellet in 12 ml of ice-cold 10% glycerol. Dispense into 100-μl aliquots and immediately freeze in liquid nitrogen. Store at –80°C.
8. Check electrocompetence by electroporating 1 ng of M13 RF DNA (e.g., M13KE or M13mp19, diluted in a low ionic strength buffer such as TE) into a freshly thawed aliquot of electrocompetent cells, according to the manufacturer's instructions. Suggested parameters (Bio-Rad Gene Pulser): 25 μF, 200 Ω, 2.5 kV. The resulting time constant (τ) should be on the order of 5.8–6.1 milliseconds.
9. Immediately add 1 ml of SOC medium and incubate at 37°C for 30–40 minutes. This gives the cells sufficient time to recover without allowing phage production and infection of untransformed cells, which could result in inflated electroporation efficiencies.
10. Prepare 10^2-, 10^3-, and 10^4-fold dilutions of the outgrowth in LB. Transfer 10 μl of each dilution to a test tube containing 3 ml of Top Agar and 200 μl of a mid-log culture of ER2738, equilibrated at 45°C. Vortex briefly and spread on LB/IPTG/X-gal plates. Incubate overnight at 37°C and count blue plaques the next day. Electroporation efficiencies should be at least 1×10^9 transformants/μg.

DESIGN AND CLONING OF SYNTHETIC OLIGONUCLEOTIDE INSERTS

The following procedure (adapted from Devlin et al. 1990 and Scott and Smith 1990) is specific for the M13 cloning vector M13KE, but could easily be adapted for other phage (but NOT phagemid) vectors.

Materials

library oligonucleotide (see Step 1)
M13KE cloning vector and universal extension primer (available in the Ph.D. Peptide Display Cloning System, New England Biolabs, Cat. # E8101S)
TE (see Appendix 2)
100 mM NaCl in TE (see Appendix 2)
water, doubly distilled or purified with a Milli-Q or equivalent system
Klenow fragment of DNA polymerase I (New England Biolabs, Cat. # M0210S)
10x Klenow buffer (as supplied with Klenow fragment)
10 mM dNTPs (10 mM each of dATP, dTTP, dGTP, dCTP in water, pH 7)
*Eag*I (New England Biolabs, Cat. # R0505S)

*Acc*65 I (New England Biolabs, Cat. # R0599S)
10x NEBuffer 3 (supplied with *Eag*I)
phenol/chloroform
chloroform
3 M sodium acetate, pH 5.2
ethanol
8% nondenaturing **polyacrylamide** gel (see Step 5)
pBR322–*Msp*I digest (New England Biolabs, Cat. # N 3032S)
DNA elution buffer: 100 mM sodium acetate, pH 4.5, 1 mM EDTA, 0.1% SDS
Lambda *Hin*dIII digest (New England Biolabs, Cat. # N3012S)
T4 DNA ligase (New England Biolabs, Cat. # M0202L)
T4 DNA ligase buffer, as supplied with ligase
electrocompetent ER2738 ***E. coli*** (see Protocol, above)
SOC medium (see Appendix 2)
70% **ethanol**
LB medium (see Appendix 2)
20% **PEG**/2.5 M NaCl (see Appendix 2)
TBS (see Appendix 2)

CAUTIONS: **phenol, chloroform, ethanol, polyacrylamide, *E. coli*, PEG (*see Appendix 4*)**

Procedure

1. Design a library oligonucleotide following the convention in Figure 19.3. Bear in mind that the sequence VPFYSHS preceding the leader peptidase cleavage site is part of the pIII signal sequence and should not be altered. The first residue of the displayed peptide will immediately follow this sequence. For randomized positions, relative representations of each amino acid can be improved by limiting the third position of each codon to G or T (= A or C on the synthetic library oligo). We recommend including a short spacer sequence between the randomized segment and the first native pIII residue to improve target accessibility to the displayed peptide. The Ph.D. libraries from NEB contain the spacer Gly-Gly-Gly between the random peptide and the Ser-Ala-Glu (SAE) shown in Figure 19.3. This sequence can also include a protease cleavage site to allow elution of bound phage by protease digestion (Matthews and Wells 1993; Smith et al. 1995). Pentavalency of the displayed peptide does not prevent protease release of bound phage (unpublished observations). The oligonucleotide should be synthesized on a minimum of 0.2-µmole scale, gel-purified (see Protocol 16.1), and accurately quantitated by measuring the OD_{260} in a spectrophotometer (1 absorbance unit at 260 nm = 20–30 µg/ml oligonucleotide).
2. Anneal 5 µg of the library oligo with 3 molar equivalents of the universal extension primer 5′-^{HO}CATGCCCGGGTACCTTTCTATTCTC-3′ (~4 µg for a 90-nucleotide library oligonucleotide) in a total volume of 50 µl of TE containing

100 mM NaCl. Heat to 95°C and cool slowly (15–30 minutes) to less than 37°C in a thermal cycler or water bath.

3. Extend the annealed duplex as follows (mix in the given order):

water	119 µl
10x Klenow buffer	20
annealing reaction	50
10 mM dNTPs	8
Klenow fragment (10 Units/µl)	3
	200 µl

Incubate at 37°C for 10 minutes, then 65°C for 15 minutes. Save 4 µl for later analysis (Step 5).

4. Digest the extended duplex as follows:

extension reaction	196 µl
water	154
10x NEBuffer 3	40
*Eag*I (10 units/µl)	5
*Acc*65I (10 units/µl)	5
	400 µl

Incubate at 37°C for 3 hours. Purify the DNA by phenol/chloroform extraction, chloroform extraction, and ethanol precipitation (see Appendix 3). The use of *Acc65*I in place of *Kpn*I permits double digestion in the same buffer as *Eag*I.

5. Gel-purify the digested duplex on an 8% nondenaturing polyacrylamide gel (Sambrook et al. 1989). Include pBR322 DNA–*Msp*I digest and 4 µl of the undigested complex as molecular weight markers. Visualize by ethidium bromide staining, and excise the digested duplex from the gel. Mince the excised band and elute the DNA by shaking overnight in several volumes of 100 mM sodium acetate, pH 4.5, 1mM EDTA, 0.1% SDS at 37°C.

6. Briefly microfuge to separate the gel fragments from the elution buffer, and transfer the supernatant to a clean tube. Purify the DNA duplex from the supernatant by phenol/chloroform extraction, chloroform extraction, and ethanol precipitation (see Appendix 3). Resuspend the pellet in 50 µl of TE and quantitate a small amount by PAGE, comparing to a known amount of pBR322–*Msp*I digest. 1 µg of purified insert is more than sufficient for a library of complexity 10^9.

7. For a high-complexity library, digest 10–20 µg of M13KE vector with 10 units/µg of *Eag*I and *Acc65*I in a volume of 40 µl of 1x NEBuffer 3 per µg of DNA (total volume = 400–800 µl). Gel-purify using standard methods (β-Agarase, QIAGEN, etc). Quantitate a small amount of purified cut vector on an agarose gel, comparing to a known amount of λ DNA–*Hin*dIII digest.

8. Optimize the ligation conditions. Suggested starting parameters per 20-µl ligation: 40 and 100 ng of cut vector; 3:1, 5:1, and 10:1 molar excess of cut duplex; 2

µl of 10x ligase buffer; and 200 Units (= 3 Weiss units) of T4 DNA ligase. Incubate overnight at 16°C.

9. Heat-kill the test ligations at 65°C for 15 minutes, then electroporate 1 µl of each into 100 µl of electrocompetent ER2738 or other F^+ strain. Suggested electroporation parameters (Bio-Rad Gene Pulser): 25 µF, 200 Ω, 2.5 kV. Immediately add 1 ml of SOC medium to each electroporation cuvette and transfer to a culture tube. Incubate for 30–45 minutes at 37°C with shaking. Titer each culture by combining 10 µl each of 1:10, 1:100, and 1:1000 dilutions of each outgrowth with 200 µl of mid-log ER2738 and 3 µl of Top Agar, vortexing briefly, and pouring onto prewarmed (37°C) LB/IPTG/X-gal plates. Incubate the titering plates overnight at 37°C.

10. Count plaques. Scale up the ligation using the highest plaque/µg ratio to desired library complexity. For example, a library with a complexity of 1×10^9 clones would require a 5-µg ligation if the test ligations yield a ratio of 2×10^8 plaques/µg of vector. Use no more than 500 µl of ligation reaction per microfuge tube; use multiple tubes if necessary.

11. Purify the large-scale ligation by phenol/chloroform extraction, chloroform extraction, and ethanol precipitation (see Appendix 3). Wash with 70% ethanol to desalt, and resuspend the DNA in 300 µl of TE. Electroporate. To reduce the likelihood of cells picking up more than one DNA sequence, the ligation should be divided and electroporated using as many cuvettes as convenient. For a 10–20 µg scale ligation we typically carry out 100 electroporations, using 3 µl of resuspended ligated DNA per 100 µl of electrocompetent cells.

12. Add 1 ml of SOC to each cuvette immediately after electroporation. For high-complexity libraries it may be convenient to pool the SOC outgrowths in groups of 5. Each outgrowth (or pool of 5) should be incubated for 30–45 minutes (no longer) before amplification. Titer several outgrowths or pools as in Step 9 prior to amplification in order to quantitate library complexity.

13. Amplify the electroporated cells by adding each 5 ml of pooled SOC outgrowth to 1 liter of early-log cells in LB medium. Incubate with vigorous aeration (250 rpm) at 37°C for 4.5–5 hours. Centrifuge at 5000*g* for 20 minutes at 4°C. Transfer the supernatant to a clean bottle and discard the cells.

14. Recover the phage from the supernatant by adding 1/6 volume of 20% PEG/2.5 M NaCl and incubating overnight at 4°C. Pellet the phage by centrifugation at 5000*g* for 20 minutes at 4°C. Discard the supernatant.

15. Thoroughly resuspend the phage pellet in 100 ml of TBS. Remove residual cells by centrifugation at 8000 rpm (Beckman JA-17, SORVALL SS-34 or equivalent) for 10 minutes at 4°C.

16. Transfer the supernatant to a new tube and discard the pellet. Reprecipitate the phage by adding 1/6 volume of 20% PEG/2.5 M NaCl and incubating for 1 hour at 4°C. Centrifuge at 5000*g* for 20 minutes and discard the supernatant.

17. Resuspend the final library in 10–40 ml of TBS by gentle rocking for 24–48 hours at 4°C. Store at 4°C. For long-term storage, add an equal volume of sterile glycerol and store at –20°C. The titer of the library should remain constant for several years at this temperature. Further amplification of the library is not recommended, as sequence biases may occur upon reamplification.

PROTOCOL 19.3

Solid-phase Panning with Direct Target Coating

The most straightforward method of panning involves directly coating a plastic surface with the target of interest (by nonspecific hydrophobic and electrostatic interaction), washing away the excess, and passing the pool of phage over the target-coated surface (Fig. 19.1). Depending on the target, direct coating can result in an inaccessible ligand-binding site, due to either steric blocking or partial denaturation of the target along the surface. In these cases it is necessary to pre-bind the target with the phage in solution, followed by affinity capture of the phage-target complexes either onto a streptavidin-coated surface or an affinity matrix (Protocols 19.4 and 19.5) or to pre-bind biotinylated target to a streptavidin-coated surface before addition of phage (Protocol 17.2). Because of its relative simplicity, we recommend trying the direct coating method described below first. If no clear consensus binding sequence emerges, then proceed with either of the solution binding procedures.

Materials

sterile polystyrene petri dishes, 12- or 24-well plates, or 96-well microtiter plates
100 μg/ml target in 0.1 M $NaHCO_3$, pH 8.6
LB (see Appendix 2)
LB+ 20 μg/ml **tetracycline** (LB+tet, see Appendix 2)
E. coli ER2738 host strain (New England Biolabs, Cat. # E4104S)
blocking buffer (see Appendix 2)
TBS (see Appendix 2)
TBST (TBS + 0.1% [v/v] Tween 20), (TBS + 0.5% Tween 20)
M13KE phage library
elution buffer (see Step 10)
20% **PEG**/2.5 M NaCl
TBS + 0.02% **sodium azide (NaN_3)**
LB/**IPTG/X-gal** plates (see Appendix 2)
streptavidin (Prozyme, Cat. # SA10 001)
biotin (Sigma, Cat. # B-4501)

CAUTIONS: **tetracycline, *E. coli*, PEG, sodium azide, IPTG, X-gal, biotin (*see Appendix 4*)**

Procedure

Day 1

Depending on the available quantity of target molecule and the number of different targets being panned against simultaneously, panning can be carried out in individual sterile polystyrene petri dishes, 12- or 24-well plates, or 96-well microtiter plates. Coat a minimum of 1 plate (or individual well) per target. Volumes given in the following procedure are for 60 × 15 mm petri dishes, with volumes for microtiter wells given in parentheses. For wells of intermediate size adjust volumes accordingly, but in all cases the number of input phage should remain the same.

1. Prepare a solution of 100 µg/ml of the target in 0.1 M $NaHCO_3$, pH 8.6. Alternative buffers (containing metal ions, etc.) of similar ionic strength (e.g., TBS) can be used if necessary for stabilizing the target molecule.
2. Add 1.5 ml of this solution (150 µl if using microtiter wells) to each plate (or well) and swirl repeatedly until the surface is completely wet (this may take some effort as the solution may bead up).
3. Incubate overnight at 4°C with gentle agitation in a humidified container (e.g., a sealable plastic box lined with damp paper towels). Store plates at 4°C in humidified container until needed. Plates can be stored for several weeks; discard if mold is evident on the paper towels.

Day 2

4. Inoculate 10 ml of LB+tet medium with a single colony of ER2738. This culture will be used for titering in Step 11. If amplifying the eluted phage on the same day, also inoculate 20 ml of LB medium in a 250-ml Erlenmeyer flask (do not use a 50-ml conical tube) with ER2738. Incubate both cultures at 37°C with vigorous shaking. Incubate the titering culture until needed; incubate the 20-ml culture to early-log phase, for use in Step 12.
5. Pour off the coating solution from each plate and firmly slap it face-down onto a clean paper towel to remove residual solution. Fill each plate or well completely with blocking buffer. Incubate for at least 1 hour at 4°C.
6. Discard the blocking solution as in Step 5. Wash each plate rapidly 6 times with TBST (TBS + 0.1% [v/v] Tween-20). Coat the bottom and sides of the plate or well by swirling, pour off the solution, and slap the plate face-down on a clean paper towel each time. (If using a 96-well microtiter plate, an automatic plate washer may be used.) Work quickly to avoid drying out the plates.
7. Dilute a 100-fold representation of the library (e.g., 2×10^{11} phage for a library with 2×10^9 clones) with 1 ml of TBST (100 µl if using microtiter wells). Pipet onto the coated plate and rock gently for 10–60 minutes at room temperature (see Helpful Hints).
8. Discard nonbinding phage by pouring off and slapping plate face-down onto a clean paper towel.
9. Wash plates 10 times with TBST as in Step 6. Use a clean section of paper towel each time to prevent cross-contamination.
10. Elute bound phage with 1 ml (100 µl if using microtiter wells) of an appropriate elution buffer for the interaction being studied. Typically this will be a solution

of a known ligand for the target (0.1–1 mM) in TBS, or a solution of the free target (~100 μg/ml in TBS) to compete the bound phage away from the immobilized target on the plate. Rock gently for 10–60 minutes at room temperature. Pipet eluate into a microcentrifuge tube. Alternatively, a general buffer for nonspecific disruption of binding interactions is 0.2 M glycine-HCl (pH 2.2), 1 mg/ml BSA (see Appendix 2). If using this buffer, rock gently for no more than 10–20 minutes, pipet eluate into a microcentrifuge tube, and neutralize with 150 μl (15 μl for microtiter wells) of 1 M Tris-HCl, pH 9.1.

11. Titer a small amount (~1 μl) of the eluate as described in General M13 Methods (Protocol 19.1). Plaques from the first or second round eluate titering can be sequenced if desired (see Protocol 19.6).

Note: The remaining eluate can be stored overnight at 4°C at this point if necessary and amplified the next day. In this case, inoculate 10 ml of LB+tet with ER2738 and incubate with shaking overnight at 37°C. The next day, dilute the overnight culture 1:100 in 20 ml of LB in a 250-ml Erlenmeyer flask and add the unamplified eluate. Incubate with vigorous shaking for 4.5 hours at 37°C and proceed to Step 13.

12. Amplify the rest of the eluate by adding the eluate to the 20-ml ER2738 culture from Step 4 (should be early-log at this point) and incubating with vigorous shaking for 4.5 hours at 37°C.

13. Transfer the culture to a centrifuge tube and spin for 10 minutes at 10,000 rpm (Sorvall SS-34, Beckman JA-17, or equivalent) at 4°C. Transfer the supernatant to a fresh tube and re-spin (discard the pellet).

14. Transfer the upper 80% of the supernatant to a fresh tube and add to it 1/6 volume of 20% PEG/2.5 M NaCl. Allow the phage to precipitate at 4°C for at least 2 hours, preferably overnight.

Day 3

15. Spin the PEG precipitation at 10,000 rpm (Sorvall SS-34, Beckman JA-17, or equivalent) for 15 minutes at 4°C. Decant and discard the supernatant, re-spin the tube briefly, and remove residual supernatant with a pipet.

16. Suspend the pellet in 1 ml of TBS. Transfer the suspension to a microcentrifuge tube and spin at 14,000 rpm for 5 minutes at 4°C to pellet residual cells.

17. Transfer the supernatant to a fresh microcentrifuge tube and reprecipitate by adding 1/6 volume of 20% PEG/2.5 M NaCl. Incubate on ice for 15–60 minutes. Microcentrifuge at 14,000 rpm for 10 minutes at 4°C. Discard the supernatant, re-spin briefly, and remove residual supernatant with a micropipet.

18. Suspend the pellet in 200 μl of TBS, 0.02% NaN_3. Microcentrifuge for 1 minute to pellet any remaining insoluble material. Transfer the supernatant to a fresh tube. This is the amplified eluate.

19. Titer the amplified eluate as described in General M13 Methods (Protocol 19.1) on LB/IPTG/X-gal plates. Incubate the plates overnight at 37°C. The eluate can be stored for several weeks at 4°C. For long-term storage, add an equal volume of sterile glycerol and store at –20°C.

20. Coat a plate or well for the second round of panning as in Steps 1–3 above.

Days 4 and 5

21. Count blue plaques from the titering plates in Step 19 and determine the phage titer, which should be on the order of 10^{14} pfu/ml. Use this value to calculate an input volume corresponding to the input titer in Step 7. If the phage titer of the amplified eluate is too low, succeeding rounds of panning can be carried out with as little as 10^9 pfu of input phage.

22. Carry out a second round of panning: Repeat Steps 4–18 using the calculated amount of the first-round amplified eluate as input phage, and raising the Tween concentration in the wash steps to 0.5% (v/v).

23. Titer the resulting second-round amplified eluate on LB/IPTG/X-gal plates.

24. Coat a plate or well for the third round of panning as in Steps 1–3 above.

Day 6

25. Carry out a third round of panning: Repeat Steps 4–10, using the second-round amplified eluate at an input titer equivalent to what was used in the first round (Step 7), again using 0.5% Tween in the wash steps.

26. Titer the unamplified third-round eluate as in Step 11 on LB/IPTG/X-gal plates. It is not necessary to amplify the third-round eluate. Plaques from this titering can be used for sequencing: Time the procedure so that plates are incubated at 37°C for no longer than 18 hours, as deletions may occur if plates are grown longer. The remaining eluate can be stored at 4°C for at least one week.

27. For preparation of individual clones for sequencing or ELISA, set up a 10-ml overnight culture of ER2738 from a colony, not by diluting the titering culture (see Protocol 19.1). Proceed with plaque amplification, Protocol 19.6. Do *not* amplify the third-round eluate and carry out a fourth round of panning unless the characterization shows no clear consensus sequence or ELISA signal.

CONTROL PANNING EXPERIMENT

Follow the above procedure using streptavidin as the target, including 0.1 μg/ml streptavidin in the blocking solution to complex any biotin or biotinylated protein in the BSA. Elute bound phage with 0.1 mM biotin in TBS for at least 30 minutes. After 3 rounds of enrichment/amplification, the consensus sequence for streptavidin-binding peptides should include the motif His-Pro-Gln (HPQ) (Devlin et al. 1990).

PROTOCOL 19.4

Solution-phase Panning with a Biotinylated Target and Streptavidin Plate Capture

As an alternative to directly coating a plate with the target molecule, the target can be reacted with the phage in solution, followed by affinity capture of the phage–target complexes (Cwirla et al. 1990). Depending on the target, binding in solution can result in improved kinetics compared to surface binding and can bypass problems associated with partial denaturation of the target on the plastic surface. Affinity capture requires some sort of affinity tag on the target; one way this can be accomplished is by biotinylating the target and capturing the complexes with a streptavidin-coated polystyrene plate. Alternatively, the biotinylated target–phage complexes can be captured with streptavidin-agarose beads, using the general procedure described in Protocol 19.5. Capture of the target can be accomplished either before or after addition of phage. The following protocols (19.4 and 19.5) describe affinity capture of phage–target complexes formed in solution, whereas Protocol 17.2 describes pre-binding the target to the affinity surface, followed by panning phage against the resulting "sandwich."

Materials

target protein (see Step 1)
50 mM $NaHCO_3$, pH 8.5
Sulfo-NHS-LC-Biotin (Pierce, Cat. # 21335ZZ)
water, doubly distilled or purified with a Milli-Q or equivalent system
100 µg/ml streptavidin in 0.1 M $NaHCO_3$, pH 8.5 (Prozyme, Cat. # SA10 001)
LB (see Appendix 2)
LB + 20 µg/ml **tetracycline** (LB+tet, see Appendix 2)
E. coli ER2738 host strain (New England Biolabs, Cat. # E4104S)
Blocking buffer + 0.1 µg/ml streptavidin
TBS (see Appendix 2)
TBST (TBS + 0.1% [v/v] Tween 20, TBS + 0.5% [v/v] Tween 20)
M13KE phage library
biotin, 10 mM in water (Sigma, Cat. # B-4501)
elution buffer (see Step 10)
20% **PEG**/2.5 M NaCl (see Appendix 2)
TBS + 0.02% **NaN_3 (sodium azide)**
LB/**IPTG**/**X-gal** plates (see Appendix 2)

CAUTIONS: **tetracycline, *E. coli*, biotin, PEG, sodium azide, IPTG, X-gal (*see Appendix 4*)**

Procedure

Biotinylation of target

1. Dissolve or dilute 2 mg of target protein in 1 ml of 50 mM $NaHCO_3$, pH 8.5. Other buffers may be used if necessary to maintain stability of target, but do NOT use buffers with free primary or secondary amine groups (e.g., Tris, ethanolamine).
2. Immediately prior to use, dissolve 1 mg of Sulfo-NHS-LC-Biotin in 1 ml of water. Vortex vigorously. Add 74 μl of this solution to the target solution. This reagent is a water-soluble activated ester of biotin that specifically targets the ε-amine of solvent-accessible (surface) lysine residues. The remaining solution should be discarded, as the ester hydrolyzes rapidly upon storage.
3. Incubate the biotin–target mixture for 2 hours on ice.
4. Remove unreacted biotin by dialysis against a minimum of two 1-liter changes of TBS, gel-filtration, or ultrafiltration through a Centricon apparatus (Amicon) or equivalent. Alternate buffers of similar ionic strength can be used at this point.
5. Quantitate the biotinylated protein by Bradford or Lowry assay. Biotinylation can be confirmed by Western blot detection or ELISA using commercially available anti-biotin antibodies. The degree of biotinylation can be quantitated by a colorimetric assay based on displacement of the dye HABA (2-[4′-hydroxyazobenzene]-benzoic acid) (Protocol 17.1). Based on results obtained at NEB, the described reaction conditions result in an average of two biotinylated lysines per ~25-kD protein molecule.

Panning with surface streptavidin capture

Follow the panning procedure described above for the solid-phase panning, direct target coating method (Protocol 19.3), but coat the plates with streptavidin (100 μg/ml of streptavidin in 0.1 M $NaHCO_3$, pH 8.6) instead of target. The blocking buffer should contain 0.1 μg/ml streptavidin in order to complex any biotin or biotinylated protein in the BSA. Replace Step 7 in the direct target coating method (Protocol 19.3) with the following:

7a. While plates are blocking (Step 5), pre-complex the phage with the biotinylated target: Combine in a microfuge tube 0.1 μg of biotinylated target (~10 nM final for a 25-kD protein) and a 100-fold representation of the library (e.g., 2×10^{11} pfu for a library with complexity 2×10^{9}) in 400 μl of TBST. To isolate low-affinity binders it may be necessary to increase the target concentration as high as 1–2 μM (Scott et al. 1992). Alternate buffers (containing metal ions, etc.) of similarly high ionic strength (e.g., TBS) can be used if necessary for stabilizing the target molecule. Incubate for 10–60 minutes at room temperature.

7b. Add the phage–target solution to the washed, blocked plate. Incubate for 10 minutes at room temperature.

7c. Add biotin to a final concentration of 0.1 mM and incubate an additional 5 minutes. This will displace any streptavidin-binding phage (displaying the HPQ sequence) from the plate. The off-rate for the biotinylated target is sufficiently slow so that the target will not be displaced by the biotin. Continue with Step 8 in Protocol 19.3.

PROTOCOL 19.5

Solution-phase Panning with Affinity Bead Capture

As a general alternative to panning against a target that has been immobilized on a surface, the library can be reacted with the target in solution, followed by affinity capture of the target–phage complexes onto an affinity matrix (bead) specific for the target protein. For example, if the target protein has a GST, MBP, or polyhistidine affinity tag, the target–phage complexes can be captured on glutathione, amylose, or chelated nickel beads, respectively. If the target is an antibody, Protein A and/or Protein G beads can be used for capture. In addition to requiring substantially less target per experiment than surface panning, solution panning can result in improved accessibility of the putative ligand-binding site to phage-displayed peptides, as well as avoiding partial denaturation of the target on a plastic surface. Fortuitous selection of peptide sequences that specifically bind the bead can be avoided by employing a negative selection beginning with Round 2, in which the amplified phage is preincubated with the bead in the absence of target. The supernatant from this step is then reacted with the target in a positive selection. For antibodies or other target proteins that bind to more than one matrix type, bead-specific peptides can be avoided by alternating rounds between the matrix types. For example, for antibody targets, peptides specific for Protein A or Protein G are avoided by alternating rounds of panning between Protein A-agarose and Protein G-agarose (magnetic beads can also be used). For antibodies that do not bind well to Protein A (sheep, goat, chicken, and rat polyclonals, as well as some human IgG_3 and mouse IgG_1 monoclonal antibodies), Protein G-agarose can be used in all rounds, employing a negative selection strategy as described above. Alternatively, Protein A can be used for the mouse IgG_1 subclass if the pH of the binding and wash buffer (TBS) is raised to 8.6.

Materials

LB (see Appendix 2)
LB + 20 µg/ml **tetracycline** (LB+tet, see Appendix 2)
E. coli ER2738 host strain (New England Biolabs, Cat. # E4101S)
affinity beads (see introduction, above)
TBS (see Appendix 2)
TBST (TBS + 0.1% [v/v] Tween 20, TBS + 0.5% Tween 20)
Blocking buffer (see Appendix 2)
M13KE phage library
Elution buffer (see Steps 9–11)
LB/**IPTG**/**X-gal** plates (see Appendix 2)
20% **PEG**/2.5 M NaCl (see Appendix 2)
TBS + 0.02% **NaN_3 (sodium azide)**

CAUTIONS: **tetracycline, *E. coli*, IPTG, X-gal, PEG, sodium azide (*see Appendix 4*)**

Procedure

1. Inoculate 10 ml of LB+tet medium with a single colony of ER2738, for use in titering. If amplifying the eluted phage on the same day, also inoculate 20 ml of LB medium in a 250-ml Erlenmeyer flask (do not use a 50-ml conical tube) with ER2738. Incubate both cultures at 37°C with vigorous shaking. Incubate the titering culture until needed; incubate the 20-ml culture until early-log phase, for use in Step 13.

2. Transfer 50 µl of a 50% aqueous suspension of affinity beads appropriate for capture of the target to a microfuge tube. For antibody targets (see above), use Protein A-agarose (or magnetic beads) for the first round if possible. Add 1 ml of TBS + 0.1% Tween (TBST). Suspend the resin by tapping the tube or GENTLY vortexing.

3. Pellet the resin by centrifugation in a low-speed benchtop microcentrifuge (Capsulefuge [TOMY TECH USA, Inc.] or equivalent) for 30 seconds, or by magnetic capture if using magnetic beads. Carefully pipet away and discard the supernatant, taking care not to disturb the resin pellet.

4. Suspend the resin in 1 ml of blocking buffer. Incubate for 60 minutes at 4°C, mixing occasionally.

5. In the meantime, dilute a 100-fold representation of the library (e.g., 2×10^{11} pfu for a library of complexity 2×10^{9}) and 2 pmole of target (= 300 ng for an antibody) to a final volume of 200 µl with TBST. Other buffers (with metal ions, etc.) of similar ionic strength can be used if necessary for stabilizing the target. The final concentration of target is 10 nM. For low-affinity ligands it may be necessary to use target concentrations as high as 1–2 µM (Scott et al. 1992). Incubate for 20 minutes at room temperature.

6. Following the blocking reaction in Step 4, pellet the resin as in Step 3 and wash 4 times with 1 ml of TBST, pelleting the resin each time.

7. Transfer the phage–target mixture to the tube containing the washed resin. Mix gently and incubate for 15 minutes at room temperature, mixing occasionally.

8. Pellet the resin as in Step 3, discard the supernatant, and wash 10 times with 1 ml of TBST, pelleting the resin each time.

9. Elute the bound phage by suspending the pelleted resin in 1 ml of Glycine Elution Buffer (0.2 M Glycine-HCl, pH 2.2, 1 mg/ml BSA; see Appendix 2) or a solution of a known ligand in TBS (see Protocol 19.3, Step 10). Incubate for 10 minutes at room temperature.

10. Centrifuge the elution mixture for 1 minute in a low-speed benchtop microcentrifuge. Carefully transfer the supernatant to a new microfuge tube, taking care not to disturb the pelleted resin.

11. If eluting with Glycine Elution Buffer, immediately neutralize the eluate with 150 μl of 1 M Tris-HCl, pH 9.1.

12. Titer a small aliquot of the eluate on LB/IPTG/X-gal plates as described in General M13 Methods (Protocol 19.1).

13. Amplify the remaining eluate by adding it to the 20-ml ER2738 culture from Step 1 (should be early-log at this point) and incubating at 37°C with vigorous shaking for 4.5 hours.

    Note: Alternatively, the eluate can be stored overnight at 4°C and amplified the next day. In this case, inoculate 10 ml of LB+tet with ER2738 and incubate overnight at 37°C with shaking. The next day, dilute the overnight culture 1:100 in 20 ml of LB in a 250-ml Erlenmeyer flask and add the unamplified eluate. Incubate at 37°C with vigorous shaking for 4.5 hours.

14. Transfer the culture to a centrifuge tube and spin for 10 minutes at 10,000 rpm (Sorvall SS-34, Beckman JA-17, or equivalent) at 4°C. Transfer the supernatant to a fresh tube and re-spin (discard the pellet).

15. Pipet the upper 80% of the supernatant to a fresh tube and add to it 1/6 volume of 20% PEG/2.5 M NaCl. Allow the phage to precipitate at 4°C for 2 hours or (preferably) overnight.

16. Spin the PEG precipitation at 10,000 rpm (Sorvall SS-34, Beckman JA-17, or equivalent) for 15 minutes at 4°C. Decant and discard the supernatant, re-spin briefly, and remove the residual supernatant with a pipet.

17. Suspend the pellet in 1 ml of TBS. Transfer the suspension to a microcentrifuge tube and spin for 5 minutes at 4°C to pellet residual cells.

18. Transfer the supernatant to a fresh microcentrifuge tube and reprecipitate with 1/6 volume of 20% PEG/2.5 M NaCl. Incubate for 15–60 minutes on ice. Microcentrifuge at 14,000 rpm for 10 minutes at 4°C. Discard the supernatant, re-spin briefly, and remove residual supernatant with a micropipet.

19. Suspend the pellet in 200 μl of TBS, 0.02% NaN_3. Microcentrifuge at 14,000 rpm for 1 minute to pellet any remaining insoluble matter. Transfer the supernatant to a fresh tube. This is the amplified eluate.

20. Titer the amplified eluate on LB/IPTG/X-gal plates as described in General M13 Methods, Protocol 19.1. Incubate the plates overnight at 37°C. The eluate can be stored for several weeks at 4°C. For long-term storage, add an equal volume of sterile glycerol and store at –20°C.

21. The next day, count blue plaques and determine phage titer. Use this value to calculate an input volume corresponding to the input titer used in Step 5. If the titer is too low, succeeding rounds of panning can be carried out with as little as 10^9 pfu of input phage.

22. Perform a second round of panning: Repeat Steps 1–21 using the calculated amount of the first-round amplified eluate as input phage. Raise the Tween con-

centration in the binding and wash steps to 0.5% (v/v). For antibody targets, use Protein G-agarose (or magnetic beads) for this round. If using the same resin for all three rounds, a negative selection can be added at this stage if desired (do not carry out a negative selection in the first round): Prepare an additional 50 μl of washed, blocked resin as in Steps 2–4. Pellet the resin and wash 4 times with TBST. Dilute the library in 200 μl of TBST as in Step 5, but leave out the target. Add the diluted phage to the washed, blocked resin and incubate for 15 minutes at room temperature with occasional mixing. Spin out the resin (or capture magnetically) and transfer the supernatant to a fresh microfuge tube. Add 2 pmole of target to the supernatant, incubate for 20 minutes at room temperature, and continue with Step 6, using another 50 μl of washed, blocked resin.

23. Perform a third round of panning: Repeat Steps 1–12 using the calculated amount of the second-round amplified eluate as input phage. Keep the Tween concentration at 0.5% (v/v) in the binding and wash steps. For antibody targets, use Protein A-agarose (or magnetic beads) if you used Protein A for the first round. If using the same resin in all three rounds, a negative selection can be performed as described in Step 22. The eluate can be stored for up to a week at 4°C.

24. For preparation of individual clones for sequencing or ELISA, set up a 10-ml overnight culture of ER2738 from a colony, not by diluting the titering culture (see Protocol 19.1). Time the titering step so the plates are incubated no longer than 18 hours, as deletions may occur if the plates are incubated longer. Proceed with plaque amplification, Protocol 19.6. Do not amplify the third-round eluate and carry out a fourth round of panning unless the characterization shows no clear consensus sequence or ELISA signal.

PROTOCOL 19.6

Plaque Amplification for ELISA or Sequencing

Materials

ER2738 *E. coli* overnight culture
LB (see Appendix 2)
titering plate with plaques

CAUTION: ***E. coli* (*see Appendix 4*)**

Procedure

1. Dilute an ER2738 overnight culture (grown at 37°C with shaking) 1:100 in LB. Dispense 1 ml of diluted culture into culture tubes, one for each clone to be characterized. 10 clones from the third round are often sufficient to detect a consensus binding sequence.
2. Use a sterile wooden stick or pipet tip to stab a blue plaque from a titering plate (*important:* phage should have been plated no longer than 18 hours previously) and transfer to a tube containing diluted culture. Pick well-separated plaques from plates having no more than ~ 100 plaques. This will ensure that each plaque contains a single DNA sequence.
3. Incubate the tubes at 37°C with shaking for 4.5–5 hours (no longer).
4. *Optional:* For ELISA characterization, the entire pool of selected phage can be examined for binding activity, rather than individual clones. In order to detect signal, it is generally necessary to amplify the third round of eluted phage, as unamplified titers are typically no more than 10^7 pfu/ml. Add 10 μl of the unamplified eluate to 1 ml of diluted overnight culture and incubate for 4.5–5 hours with shaking at 37°C.
5. Transfer the cultures to microcentrifuge tubes, and microfuge at 14,000 rpm for 30 seconds. Transfer the supernatant to a fresh tube and re-spin. Using a pipet, transfer the upper 80% of the supernatant to a fresh tube. This is the amplified phage stock and can be stored at 4°C for several weeks with little loss of titer. For long-term storage (up to several years), dilute 1:1 with sterile glycerol and store at –20°C.

PROTOCOL 19.7

Rapid Purification of Sequencing Templates

This extremely rapid procedure (Wilson 1993) produces template of sufficient purity for manual or automated dideoxy sequencing, without the use of phenol or chromatography.

Materials

ER2738 *E. coli* overnight culture
LB (see Appendix 2)
titering plate with plaques
20% **PEG**/2.5 M NaCl (see Appendix 2)
Iodide buffer (see Appendix 2)
70% **ethanol**
TE (see Appendix 2)
purified single-stranded M13mp18 DNA (New England Biolabs, Cat. # N4040S)

CAUTIONS: ***E. coli*, PEG, sodium iodide, ethanol (*see Appendix 4*)**

Procedure

1. Perform the plaque amplification procedure described above (Protocol 19.6). After the first centrifugation in Step 5, transfer 500 μl of the phage-containing supernatant to a fresh microfuge tube.
2. Add 200 μl of 20% PEG/2.5 M NaCl. Invert several times to mix, and let stand for 10–20 minutes at room temperature.
3. Microfuge at 14,000 rpm for 10 minutes at 4°C. Discard the supernatant.
4. Re-spin briefly. Carefully pipet away and discard any remaining supernatant.
5. Suspend the pellet thoroughly in 100 μl of iodide buffer by vigorously tapping the tube. Add 250 μl of ethanol and incubate 10–20 minutes at room temperature. Short incubation at room temperature will preferentially precipitate single-stranded phage DNA, leaving most phage protein in solution.
6. Spin in a microfuge at 14,000 rpm for 10 minutes at 4°C and discard the supernatant. Wash the pellet with 0.5 ml of 70% ethanol, re-spin, discard the supernatant, and briefly dry the pellet under vacuum.
7. Suspend the pellet in 30 μl of TE buffer. The template can be suspended in water instead of TE if desired, but this is not recommended for long-term storage. In TE buffer the phage DNA should be stable indefinitely at –20°C.

8. Quantitate by agarose gel electrophoresis. 5 µl of resuspended template should give a band of comparable intensity to 0.5 µg of purified single-stranded M13mp18 DNA.

9. 5 µl (~0.5 µg) of the resuspended template should be sufficient for manual dideoxy sequencing with ^{35}S or ^{33}P, or automated cycle sequencing with dye-labeled dideoxynucleotides. More or less template may be required depending on the sequencing method used.

SEQUENCING GUIDELINES

Materials

sequencing materials, manual or automated
purified sequencing template (see protocol, above)
–28 gIII sequencing primer, 5′- ^{HO}GTA TGG GAT TTT GCT AAA CAA C –3′ (NEB, Cat. # S1258S)
–96 gIII sequencing primer, 5′- ^{HO}CCC TCA TAG TTA GCG TAA CG –3′ (NEB, Cat. # S1259S)

1. Use the –28 primer for manual dideoxy sequencing. The –96 primer should be used for automated sequencing.

2. The sequence being read corresponds to the anticodon strand of the template. Write out the complementary strand and check against the top strand of the insert sequence shown in Figure 19.3 (the primers hybridize downstream from the insert).

3. TAG stop codons are suppressed by glutamine in ER2738 (*glnV*). If the library was amplified in this strain or any *glnV* strain, TAG should thus be considered a glutamine codon when translating.

4. Libraries often contain a small percentage (<2%) of clones containing multiple inserts of the randomized region. Preferential selection and amplification of these clones may occur when panning against targets whose ligand specificity spans a length greater than that specified by the insert. When interpreting sequence data, be sure the sequence outside the restriction sites used for inserting the randomized sequence (*Acc*65I and *Eag*I) matches the sequence shown in Figure 19.3. If multiple inserts are evident in selected clones, translate the entire region downstream from the first occurrence of the leader peptidase cleavage site shown in Figure 19.3.

PROTOCOL 19.8

ELISA Binding Assay with Direct Target Coating

The following ELISA protocol is sufficient for rapidly determining whether a selected phage clone binds the target, without the need for an antibody specific for the target. In this procedure a microtiter plate is coated with the target at high density, and each purified phage clone is applied to the plate at various dilutions. Bound phage is then detected with an anti-M13 antibody. Because the amount of target coated on the plate is not quantifiable, and is present at sufficiently high density to allow multivalent binding to the phage, this method will not determine whether the selected phage binds with high or low affinity. This method is useful for qualitative determination of relative binding affinities for a number of selected clones in parallel, and will distinguish true target binding from binding to the plastic support. The latter is particularly useful if the direct coating method of panning (Protocol 19.3) is used. For much better discrimination between high and low affinity binders, a sandwich ELISA can be carried out in which the selected phage is immobilized with anti-M13 antibody and the target applied in the liquid phase (Protocol 18.4). This procedure requires an antibody against the target protein, or some other means of detecting bound target protein.

Materials

ER2738 ***E. coli*** overnight culture
LB (see Appendix 2)
titering plate with plaques
20% **PEG**/2.5 M NaCl (see Appendix 2)
TBS (see Appendix 2)
LB/**IPTG**/**X-gal** plates (see Appendix 2)
ELISA plates
100 μg/ml target protein in 0.1 M $NaHCO_3$, pH 8.6
Blocking buffer (see Appendix 2)
TBST (TBS + 0.1% [v/v] Tween 20, TBS + 0.5% Tween 20)
HRP/Anti-M13 monoclonal antibody conjugate (Amersham Pharmacia Biotech, Cat. # 27-9421-01)
HRP substrate solution (see Step 13)
ABTS (**2,2′-Azino-bis(3-ethylbenzthiazoline-6-sulfonic acid**); Sigma, Cat. # A-1888)
50 mM sodium citrate, pH 4.0
30% **hydrogen peroxide** (H_2O_2)

CAUTIONS: ***E. coli*, PEG, IPTG, X-gal, ABTS, hydrogen peroxide (*see Appendix 4*)**

Procedure

1. Perform the plaque amplification procedure described in Protocol 19.6. After the first centrifugation in Step 5, save the phage-containing supernatants at 4°C.
2. In addition to individual clones, the amplified first-, second-, and third-round eluted pools should also be assayed. For each clone or pool to be characterized, inoculate 20 ml of LB medium with a single colony of ER2738 and incubate at 37°C until slightly turbid. Alternatively, dilute an overnight culture of ER2738 1:100 in 20 ml of LB.
3. Add 5 μl of phage stock for each clone or pool to be characterized to each culture and incubate with vigorous aeration for 4.5–5 hours at 37°C.
4. Transfer the culture to a centrifuge tube and spin at 10,000 rpm (Sorvall SS-34, Beckman JA-17 or equivalent) for 10 minutes at 4°C. Transfer the supernatant to a fresh tube and re-spin (discard the pellet).
5. Pipet the upper 80% of the supernatant to a fresh tube and add to it 1/6 volume of 20% PEG/2.5 M NaCl. Allow the phage to precipitate at 4°C for at least 2 hours or overnight.
6. Spin the PEG precipitation at 10,000 rpm (Sorvall SS-34, Beckman JA-17 or equivalent) for 15 minutes at 4°C. Decant and discard the supernatant, re-spin briefly, and remove residual supernatant with a pipet.
7. Suspend the pellet in 1 ml of TBS. Transfer the suspension to a microcentrifuge tube and spin at 14,000 rpm for 5 minutes at 4°C to pellet residual cells.
8. Transfer the supernatant to a fresh microcentrifuge tube and re-precipitate with 1/6 volume of 20% PEG/2.5 M NaCl. Incubate 15–60 minutes on ice. Microcentrifuge at 14,000 rpm for 10 minutes at 4°C. Discard the supernatant, re-spin briefly, and remove residual supernatant with a micropipet.
9. Suspend the pellet in 50 μl of TBS. Titer as described in General M13 Methods, Protocol 19.1. The titer should be approximately 10^{14} pfu/ml. The eluate can be stored for several weeks at 4°C. For long-term storage, add an equal volume of sterile glycerol and store at –20°C.
10. Coat one row of ELISA plate wells for each clone or pool to be characterized with 100–200 μl of 100 μg/ml of target in 0.1 M $NaHCO_3$, pH 8.6. Incubate overnight at 4°C in an airtight humidified box (e.g., a sealable plastic box lined with damp paper towels).
11. Shake out excess target solution and slap plate face-down onto a paper towel. Fill each well completely with blocking buffer. Additionally, one row of uncoated wells per clone to be characterized should also be blocked to test for binding of each selected sequence to BSA-coated plastic (this test is extremely important, especially if panning was carried out on a polystyrene surface). A second, fully uncoated microtiter plate should be blocked for use in serial dilutions of phage

before addition to the target-coated plate. Dilutions are done in a separate blocked plate to ensure that phage are not absorbed onto the target during the course of performing dilutions, which would result in a sudden "falling-off" of signal as the phage is diluted. Incubate the plates filled with blocking buffer for 1–2 hours at 4°C.

12. Shake out the blocking buffer and wash each plate 6 times with TBS/Tween, slapping the plate face-down onto a clean section of paper towel each time. The percentage of Tween should be the same as the concentration used in the panning wash steps.
13. In the separate blocked plate, carry out fourfold serial dilutions of the phage in 200 μl of TBS/Tween per well, starting with 10^{12} virions in the first well of a row and ending with 2×10^5 virions in the 12th well.
14. Using a multichannel pipettor, transfer 100 μl from each row of diluted phage to a row of target-coated wells, and transfer 100 μl to a row without target. Incubate at room temperature for 1–2 hours with agitation.
15. Wash plate 6 times with TBST as in Step 12.
16. Dilute HRP-conjugated anti-M13 antibody in blocking buffer to the final dilution recommended by the manufacturer. Add 200 μl of diluted conjugate to each well and incubate at room temperature for 1 hour with agitation.
17. Wash 6 times with TBST as in Step 12.
18. Prepare the HRP substrate solution as follows: A stock solution of ABTS can be prepared in advance by dissolving 22 mg of ABTS (Sigma) in 100 ml of 50 mM sodium citrate, pH 4.0. Filter-sterilize and store at 4°C. Immediately prior to the detection step, add 36 μl of 30% H_2O_2 to 21 ml of ABTS stock solution per plate to be analyzed.
19. Add 200 μl of substrate solution to each well, and incubate for 10–60 minutes at room temperature with gentle agitation.
20. Read the plates using a microplate reader set at 405–415 nm. For each phage concentration, compare the signals obtained with and without target protein.

PROTOCOL 19.9

Use of Synthetic Peptides in Specificity Analysis

For more detailed binding or inhibition studies, it may be necessary to synthesize selected sequences as free, soluble peptides. This allows precise control of peptide concentration, without avidity artifacts associated with pentavalent display. Additionally, without the phage attached, the peptide can be used at much higher concentrations and can be used in vivo. When designing a synthetic peptide corresponding to a selected binding sequence, it is important to realize that, while the amino terminus of the displayed sequence was free during panning, the carboxyl terminus was fused to the phage. Furthermore, the carboxy-terminal residue of the selected sequence did NOT have a free negatively charged carboxylate during panning, so a simple synthetic peptide with a free carboxyl terminus will introduce a negatively charged group at a position occupied by a neutral peptide bond during panning. Depending on the nature of the target–ligand interaction, this negative charge can completely abolish binding. It is, therefore, recommended that the carboxy-terminal carboxylate of the synthetic peptide be amidated to block the negative charge. Additionally, if the library insert was designed to include a peptide spacer between the random sequence and pIII (as are the commercially available Ph.D. libraries, which include a Gly-Gly-Gly-Ser spacer), this spacer sequence should be added to the carboxyl terminus of the peptide. For chemical conjugation of the peptide to a reporter enzyme or a solid support, the peptide can be designed with a carboxy-terminal cysteine (if there are no other cysteines present in the sequence). The resulting peptide thiol can be easily coupled to maleimide-activated HRP, alkaline phosphatase, or agarose beads (Pierce).

PROTOCOL 19.10

Expression of Selected Sequences as Monovalent MBP Fusions

As an alternative to chemical synthesis for more detailed binding/inhibition studies, selected sequences can be expressed as monovalent, soluble fusions to the *E. coli* maltose binding protein (MBP), using the shuttle vector pMal-pIII (Fig. 19.4, GenBank accession number AF031088). This vector was designed for rapid subcloning of sequences selected from libraries constructed in the M13KE system (Zwick et al. 1998). The same enzymes used for cloning the library insert (*Acc*65I and *Eag*I) are used to transfer the selected sequence from M13KE into pMal-pIII (Fig. 19.4). Because the upstream *Acc*65I restriction site lies within the pIII leader sequence of M13KE, the transferred insert fragment contains a portion of this leader sequence. When subcloned in pMal-pIII, the complete leader sequence is regenerated, resulting in secretion of the peptide-MBP fusion into the periplasm, with concomitant removal of the pIII leader. The amino terminus of the fusion is therefore identical to the amino terminus of the peptide-pIII fusion expressed from M13KE, precisely duplicating the sequence context of the displayed peptide, but fused to a soluble, monovalent protein. The purified fusion protein can be used in a variety of analytical techniques, including solution-phase inhibition assays and surface plasmon resonance (Zwick et al. 1998). It is important to note that the selected sequence expressed in pMal–pIII is fused to the amino terminus of MBP, rather than the carboxyl terminus as in the commercially available MBP fusion vectors pMal-c2X and pMal-p2X from NEB (Riggs 1992).

Materials

single-stranded sequencing template (see above, Rapid purification of sequencing templates)
M13KE extension primer (available in the Ph.D. Peptide Display Cloning System, New England Biolabs, Cat. # E8101S)
–96 gIII sequencing primer, 5′-^{HO}CCC TCA TAG TTA GCG TAA CG –3′ (New England Biolabs, Cat. # S1259S)
10 mM dNTPs (10 mM each of dATP, dTTP, dGTP, dCTP)
thermostable polymerase for PCR
PCR buffer (supplied with polymerase)
*Acc*65I (New England Biolabs, Cat. # RO599S)
*Eag*I (New England Biolabs, Cat. # RO505S)
NEBuffer 3 (supplied with *Eag*I)
8% nondenaturing **polyacrylamide** gel
pBR322–*Msp*I digest (New England Biolabs, Cat. # N3032S)
pMal–pIII (available on request from author)
T4 DNA ligase (New England Biolabs, Cat. # MO202L)

(a)

```
1401  AGCTGTTGAC AATTAATCAT CGGCTCGTAT AATGTGTGGA ATTGTGAGCG
             -35        tac promoter   -10

1451  GATAACAATT TCACACAGGA AACAGCCAGT CCGTTTAGGT GTTTTCACGA
      NEB #1233 seq primer→

                         RBS        Nde I   pIII leader
1501  GCACTTCACC AACAAGGACC ATAGCATATG AAAAAATTAT TATTCGCAAT
                                     M   K  K  L  L   F  A  I

                Acc65 I                   Eag I
1551  TCCTTTAGTG GTACCTTTCT ATTCTCACTC GGCCGATATG AAAATCGAAG
       P  L  V   V  P  F  Y   S  H  S  A  D  M   K... malE
                                      ↑
                                  leader peptidase
```

(b)

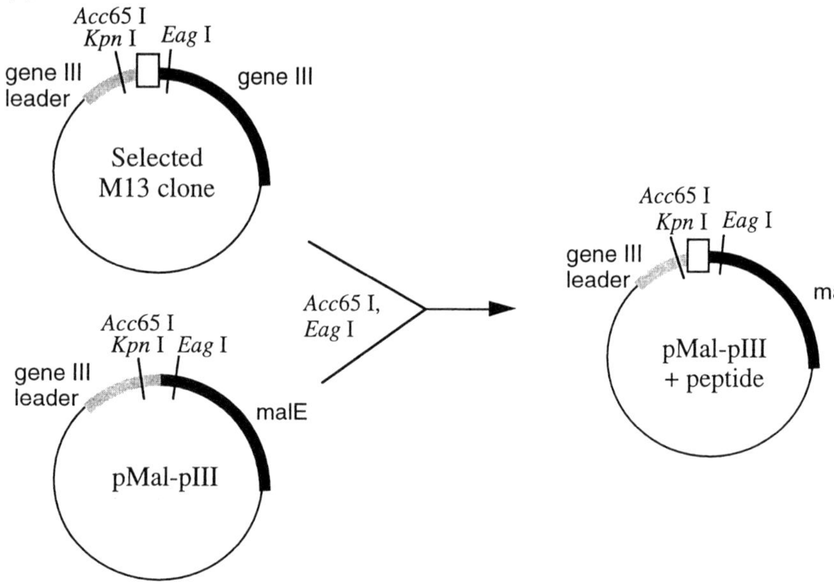

Figure 19.4. Subcloning a selected peptide sequence into the pMal–pIII shuttle vector. (*a*) The cloning region of pMal–pIII, showing restriction sites and pIII leader sequence introduced at the amino terminus of the native MBP gene (*malE*). The gene is under control of the IPTG-inducible *tac* promoter. Locations of the ribosome binding site (RBS) and the priming site for the universal pUC/M13 primer #1233 are shown. (*b*) Subcloning schematic. The small *Acc*65I–*Eag*I fragment, containing a selected peptide sequence (*open box*), is transferred into pMal–pIII along with a portion of the pIII leader sequence (*gray segment*). The pIII leader is regenerated, resulting in a precise fusion of the selected peptide to MBP, with the first residue of the peptide expressed directly at the amino terminus of the fusion. The expressed fusion can be rapidly purified by affinity chromatography on amylose resin for further binding analysis (Zwick et al. 1998).

T4 DNA ligase buffer (supplied with ligase)
ER2738 or other standard cloning strain of ***E. coli***
LB + 100 μg/ml ampicillin plates (see Appendix 2)
miniprep materials
LB + 100 μg/ml ampicillin + 0.2% (w/v) glucose
10 mM **IPTG**
30 mM Tris-HCl, pH 8 + 20% (w/v) sucrose
0.5 M EDTA
5 mM **$MgSO_4$**
1 M Tris-HCl, pH 7.4
Amylose resin (New England Biolabs, Cat. # E8021S)
Amylose column buffer (20 mM Tris-HCl, pH 7.4, 200 mM NaCl, 1 mM EDTA, 1 mM **DTT**)
Amylose column buffer + 10 mM maltose
Bradford assay materials (Bio-Rad)
SDS-PAGE materials

CAUTIONS: polyacrylamide, *E. coli*, IPTG, $MgSO_4$, DTT (*see Appendix 4*)

Procedure

1. Amplify the selected sequence by PCR from 1 μl of the single-stranded sequencing template corresponding to the desired clone (from Protocol 19.7). Perform 25 PCR cycles, each including 30 seconds at 95°C, 30 seconds at 55°C, and 30 seconds at 72°C. For primers, use the M13KE insert extension primer 5′- ^{HO}CAT-GCCCGGGTACCTTTCTATTCTC -3′ and the –96 gIII sequencing primer 5′-^{HO}CCCTCATAGTTAGCGTAACG -3′. Purify the amplified product away from excess primers, dNTPs, and polymerase by gel purification (see Sambrook et al. 1989), QIAquick (QIAGEN), or equivalent. Resuspend in 50 μl of TE and quantitate by measuring the OD_{260} (one absorbance unit = 50 μg/ml of DNA).
2. Digest 1 μg of the purified PCR product with *Acc*65I and *Eag*I in a total volume of 50 μl as in Protocol 19.2, Design and cloning of synthetic inserts, Step 4. Use 10 units of each enzyme and incubate for 2 hours at 37°C. There is no need for phenol extraction. Gel-purify the digested insert by 8% nondenaturing PAGE (Sambrook et al. 1989), using pBR322–*Msp*I digest as markers. Excise and elute the DNA as in Protocol 19.2, Design and cloning of synthetic inserts, Steps 5–6.
3. Digest 1 μg of pMal-pIII with *Acc*65I and *Eag*I, as in Step 2. Purify the digested vector by agarose gel electrophoresis or treat with alkaline phosphatase. Quantitate by gel electrophoresis (Sambrook et al. 1989).
4. Assemble a ligation reaction containing approximately 25–50 ng of digested pMal–pIII, 3–5 molar equivalents of insert, 200 NEB units (3 Weiss units) of T4 DNA ligase, and T4 DNA ligase buffer in a volume of 20 μl. As a control for back-

ground self-ligation of the vector, assemble a similar ligation reaction with no insert. Incubate overnight at 16°C.

5. Heat-kill the ligations for 20 minutes at 65°C, then transform into a suitable *E. coli* strain (ER2738, TB1, or any standard cloning strain), either by electroporation or heat shock. Plate on LB medium containing 100 μg/ml ampicillin (no IPTG), and incubate overnight at 37°C.
6. Prepare miniprep DNA from the colonies (Sambrook et al. 1989) and analyze clones by restriction mapping. The insert will increase in size by 3 times the number of amino acids displayed on pIII (e.g., 21 nucleotides for a 7-residue peptide).
7. Continue with the expression/purification steps described in Protocol 18.10.

Helpful Hints

OPTIMIZING PEPTIDE BINDING INTERACTIONS

There are several variables affecting the stringency of selection during panning. Depending on the interaction being studied, adjustment of the stringency of selection or elution may be necessary to obtain a consensus binding sequence.

1. *Detergent:* The presence of detergent (typically Tween 20) in the binding and wash buffers reduces nonspecific hydrophobic interactions between the phage and the target and/or blocking agent (BSA), which lead to higher levels of background binding. Lower Tween concentrations in early rounds will result in higher eluate titers, and the stringency can be gradually increased with each round by raising the Tween concentration stepwise to a maximum of 0.5%. In side-by-side experiments, however, we have obtained identical consensus sequences when Tween concentrations were held constant (0.5%) or increased stepwise (0.1, 0.3, 0.5%) in 3 rounds of panning. The use of lower Tween concentrations in earlier rounds is recommended when the interaction under study is so specific that the eluate titer (i.e., the number of bound sequences) in early rounds is expected to be very low.
2. *Salt:* Hydrophobic interactions are favored at high ionic strength, wheras ionic interactions are favored at low ionic strength. Nonspecific ionic interactions between surface charged groups on the target and phage, which would lead to high background binding, are thus avoided by the use of a high-salt buffer such as TBS in the binding steps. Because it is impossible to predict in advance, however, whether the peptide–target interaction will be largely ionic or hydrophobic in nature, it may be necessary to adjust the salt concentration of the binding buffer.
3. *Target concentration:* If panning against the target in solution, the stringency can be increased by lowering the concentration of target (Barrett et al. 1992). An initial target concentration of 10 nM is recommended; this can be lowered to 1 nM in later rounds for selection of ligands with high binding affinity. If no consensus sequence is evident, it is possible that the best sequences in the library bind with

very low affinity. Even with the avidity effects associated with multivalent display, target concentrations in the nanomolar range may simply be too low to see any binding. In this case it may be necessary to raise the target concentration to 1–2 μM in all rounds of panning (Scott et al. 1992). Also, low-affinity binders present in low numbers in the initial unpanned library might be lost in the first round of solution-phase panning, regardless of target concentration. In this case it may be necessary to carry out the first round by panning against target that has been immobilized by direct coating (Protocol 19.3) or pre-bound to an affinity support (Protocol 17.2), and then carrying out subsequent rounds in the solution phase.

4. *Number of rounds:* With each round of panning and amplification, the pool of phage becomes enriched in favor of sequences that bind to the target. Maintaining a constant input phage concentration in each round results in a stepwise increase in the number of particles displaying a given sequence until a point is reached where most or all of the eluted particles display a consensus binding sequence. Depending on the interaction being studied and the applied stringency, this usually happens after two or three rounds. If no clear consensus sequence emerges after three rounds, the third-round eluate can be amplified and a fourth round of panning carried out. Reports in the literature of five or more rounds being carried out (e.g., Wrighton et al. 1996) generally refer to an iterative selection technique where a low-affinity ligand is selected after three rounds and then subjected to further mutagenesis followed by additional rounds of selection.

5. *Choice of library:* If, following three or four rounds of panning, there is no clear consensus ligand sequence (or if all of the plaques are white), it is possible that the library simply does not contain any clones that bind tightly to the target. One explanation is that the ideal ligand sequence is not statistically represented in the library. With a linear library, it is also possible that the most "fit" peptide in the library does not bind with high enough affinity to be selected. In this case a structurally constrained library, e.g., in which the random residues are flanked by a pair of cysteine residues that spontaneously form a disulfide bond (such as the Ph.D.-C7C library, New England Biolabs, Cat. # E8120S), might work better. Such libraries are especially useful for targets whose native ligands are in the context of a surface loop, such as antibodies with structural epitopes. Additionally, imposing structural constraints on the unbound ligand may result in a less unfavorable binding entropy, improving the overall free energy of binding compared to unconstrained ligands (O'Neil et al. 1992). A major disadvantage of structurally constrained libraries is that the constraint may "freeze out" a conformation required for target binding, preventing binding outright rather than improving affinity (McConnell et al. 1994). Finally, if the ligand sequence needs to adopt a specific structural conformation, or requires a longer stretch of amino acids to bind, it may be necessary to increase the length of the randomized segment. Unfortunately, in the absence of detailed structural information about the target–ligand interaction, it is impossible to predict in advance which type of library will yield the most productive ligands (see, e.g., Bonnycastle et al. 1996).

SPECIFIC PROBLEMS

Extensive white plaques after four or more rounds

In a typical round of panning, approximately 10^{11} input phage are reacted with the target, and between 10^3 and 10^7 total phage are eluted after washing. This corresponds to an enrichment of 10^4- to 10^8-fold per round. Since a typical high-complexity library contains ~10^9 different clones, the eluted pool of phage should in theory be fully enriched in favor of binding sequences after only two or three rounds. Once this point is reached, further rounds of amplification and panning will result only in selection of phage that have a growth advantage over the library phage (Rodi et al. 1999). For example, vanishingly small levels of contaminating environmental wild-type phage (less than one part per billion) will completely overtake the pool if too many rounds of amplification are carried out, regardless of the strength of the in vitro selection.

ELISA indicates that background binding to the plate is as high as binding to the target

If panning against a polystyrene plate coated with the target (Protocol 19.3), it is possible to inadvertantly select peptides that specifically bind the polystyrene surface (Adey et al. 1995). These peptides will yield identical ELISA signals in the presence and absence of target, because the ELISA plate is also made of polystyrene. Such "plastic binders" are typically rich in aromatic residues (Phe, Tyr, Trp, His), which often alternate (the sequence FHWTWYW is a plastic binder discovered and characterized at NEB). Selection of plastic binders often occurs in the absence of a strong target preference for peptide sequences present in the library: Other libraries may yield the desired target-specific sequences. Selection of polystyrene-specific peptides can be avoided by using the bead capture protocol described in Protocol 19.5. The phage are reacted with the target in solution, and the phage–target complexes are then captured onto beads that specifically bind the target (Protein A-agarose for antibody targets, glutathione-agarose for GST fusions, etc.). Unbound phage are removed by extensively washing the beads in a microfuge tube. Unlike polystyrene, neither the beads (typically crosslinked agarose) nor the microfuge tube (polypropylene) is likely to select specific peptide sequences from the library, although the species conjugated to the beads (Protein A, glutathione, etc.) might. To avoid selection of bead-specific ligands, we suggest either alternating rounds between different beads specific for the target (e.g., for antibody targets, use Protein A beads for rounds one and three, and Protein G beads for round two), or adding a subtractive panning step, beginning with round two, in which the phage pool is first reacted with the beads alone (no target), the beads are discarded, and the supernatant from this step is reacted with the target.

ACKNOWLEDGMENTS

The authors are indebted to the many individuals who worked with us in developing and testing our phage-display system, in particular Christina Marchetti, Karin Andersson, Corinna Tuckey, Karen Sandman, Don Comb, Laurie Mazzola, Jennifer Ware, Fiona Stewart, Marion Sibley, Sandra Schieferl, and Roberto Polakiewicz. We also thank Devon Byrd, Ira Schildkraut, Michael Comb, and especially Jamie Scott for advice and encouragement.

REFERENCES

Adey N.B., Mataragnon A.H., Rider J.E., Carter J.M., and Kay B.K. 1995. Characterization of phage that bind plastic from phage-displayed random peptide libraries. *Gene* **156:** 27–31.

Arap W., Pasqualini R., and Ruoslahti E. 1998. Cancer treatment by targeted drug delivery to tumor vasculature in a mouse model. *Science* **279:** 377–380.

Barrett R.W., Cwirla S.E., Ackerman M.S., Olson A.M., Peters E.A., and Dower W.J. 1992. Selective enrichment and characterization of high affinity ligands from collections of random peptides on filamentous phage. *Anal. Biochem.* **204:** 357–364.

Barry M.A., Dower W.J., and Johnston S.A. 1996. Toward cell-targeting gene therapy vectors: Selection of cell-binding peptides from random peptide-presenting phage libraries. *Nat. Med.* **2:** 299–305.

Birkenmeier G., Osman A.A., Kopperschlager G., and Mothes T. 1997. Epitope mapping by screening of phage display libraries of a monoclonal antibody directed against the receptor binding domain of human α2-macroglobulin. *FEBS Lett.* **416:** 193–196.

Bonnycastle L.L., Mehroke J.S., Rashed M., Gong X., and Scott J.K. 1996. Probing the basis of antibody reactivity with a panel of constrained peptide libraries displayed by filamentous phage. *J. Mol. Biol.* **258:** 747–762.

Cortese R., Monaci P., Nicosia A., Luzzago A., Felici F., Galfre G., Pessi A., Tramontano A., and Sollazzo M. 1995. Identification of biologically active peptides using random libraries displayed on phage. *Curr. Opin. Biotechnol.* **6:** 73–80.

Cwirla S.E., Peters E.A., Barrett R.W., and Dower W.J.. 1990. Peptides on phage: A vast library of peptides for identifying ligands. *Proc. Natl. Acad. Sci.* **87:** 6378–6382.

Cwirla S.E., Balasubramanian P., Duffin D.J., Wagstrom C.R., Gates C.M., Singer S.C., Davis A.M., Tansik R.L., Mattheakis L.C., Boytos C.M., Schatz P.J., Baccanari D.P., Wrighton N.C., Barrett R.W., and Dower W.J. 1997. Peptide agonist of the thrombopoietin receptor as potent as the natural cytokine. *Science* **276:** 1696–1699.

Dente L., Vetriani C., Zucconi A., Pelicci G., Lanfrancone L., Pelicci P.G., and Cesareni G. 1997. Modified phage peptide libraries as a tool to study specificity of phosphorylation and recognition of tyrosine containing peptides. *J. Mol. Biol.* **269:** 694–703.

Devlin J.J., Panganiban L.C., and Devlin P.E. 1990. Random peptide libraries: A source of specific protein binding molecules. *Science* **249:** 404–406.

Doorbar J. and Winter G. 1994. Isolation of a peptide antagonist to the thrombin receptor using phage display. *J. Mol. Biol.* **244:** 361–369.

Dore J.M., Morard F., Vita N., and Wijdenes J. 1998. Identification and location on syndecan-1 core protein of the epitopes of B-B2 and B-B4 monoclonal antibodies. *FEBS Lett.* **426:** 67–70.

Gevorkian G., Manoutcharian K., Almagro J.C., Govezensky T., and Dominguez V. 1998. Identification of autoimmune thrombocytopenic purpura-related epitopes using phage-display peptide library. *Clin. Immunol. Immunopathol.* **86:** 305–309.

Goodson R.J., Doyle M.V., Kaufman S.E., and Rosenberg S. 1994. High-affinity urokinase receptor antagonists identified with bacteriophage peptide display. *Proc. Natl. Aad. Sci.* **91:** 7129–7133.

Hoess R., Brinkmann U., Handel T., and Pastan I. 1993. Identification of a peptide which binds to the carbohydrate-specific monoclonal antibody B3. *Gene* **128:** 43–49.

Kang A.S., Barbas C.F., Janda K.D., Benkovic S.J., and Lerner R.A. 1991. Linkage of recognition and replication functions by assembling combinatorial antibody Fab libraries along phage surfaces. *Proc. Natl. Acad. Sci.* **88:** 4363–4366.

Katz B.A. 1995. Binding to protein targets of peptidic leads discovered by phage display: Crystal structures of streptavidin-bound linear and cyclic peptide ligands containing the HPQ sequence. *Biochemistry* **34:** 15421–15429.

Lowman H.B., Bass S.H., Simpson N., and Wells J.A. 1991. Selecting high-affinity binding proteins by monovalent phage display. *Biochemistry* **30:** 10832–10838.

Matthews D.J. and Wells J.A. 1993. Substrate phage: Selection of protease substrates by monovalent phage display. *Science* **260:** 1113–1116.

McConnell S.J., Kendall M.L., Reilly T.M., and Hoess R.H. 1994. Constrained peptide libraries as a tool for finding mimotopes. *Gene* **151:** 115–118.

O'Neil K.T., Hoess R.H., Jackson S.A., Ramachandran N.S., Mousa S.A., and DeGrado W.F. 1992. Identification of novel peptide antagonists for GPIIb/IIIa from a conformationally constrained peptide library. *Proteins* **14:** 509–515.

Oldenburg K.R., Loganathan D., Goldstein I.J., Schultz P.G., and Gallop M.A. 1992. Peptide ligands for a sugar-binding protein isolated from a random peptide library. *Proc. Natl. Acad. Sci.* **89:** 5393–5397.

Osman A.A., Uhlig H., Thamm B., Schneider-Mergener J., and Mothes T. 1998. Use of the phage display technique for detection of epitopes recognized by polyclonal rabbit gliadin antibodies. *FEBS Lett.* **433:** 103–107.

Parmley S.F. and Smith G.P. 1988. Antibody-selectable filamentous fd phage vectors: Affinity purification of target genes. *Gene* **73:** 305–318.

Pasqualini R. and Ruoslahti E. 1996. Organ targeting in vivo using phage display peptide libraries. *Nature* **380:** 364–366.

Peters E.A., Schatz P.J., Johnson S.S., and Dower W.J. 1994. Membrane insertion defects caused by positive charges in the early mature region of protein pIII of filamentous phage fd can be corrected by prlA suppressors. *J. Bacteriol.* **176:** 4296–4305.

Riggs P. 1992. Expression and purification of maltose-binding protein fusions. In *Current protocols in molecular biology* (ed. Ausubel F.M. et al.), pp. 16.6.1–16.6.14. Greene Publishing and John Wiley & Sons, New York.

Rodi D.J. and Makowski L. 1999. Phage-display technology—Finding a needle in a vast molecular haystack. *Curr. Opin. Biotechnol.* **10:** 87–93.

Rodi D.J., Janes R.W., Sanganee H.J., Holton R.A., Wallace B.A., and Makowski L. 1999. Screening of a library of phage-displayed peptides identifies human bcl-2 as a taxol-binding protein. *J. Mol. Biol.* **285:** 197–203.

Russel M., Linderoth N.A., and Sali A. 1997. Filamentous phage assembly: Variation on a protein export theme. *Gene* **192:** 23–32.

Sambrook J., Fritsch E.F., and Maniatis T. 1989. *Molecular cloning: A laboratory manual,* 2nd Edition. Cold Spring Harbor Laboratory Press, Cold Spring Harbor, New York.

Scott J.K. and Smith G.P. 1990. Searching for peptide ligands with an epitope library. *Science* **249:** 386–390.

Scott J.K., Loganathan D., Easley R.B., Gong X., and Goldstein I.J. 1992. A family of concanavalin A-binding peptides from a hexapeptide epitope library. *Proc. Natl. Acad. Sci.* **89:** 5398–5402.

Siegel D.L., Chang T.Y., Russell S.L., and Bunya V.Y. 1997. Isolation of cell surface-specific human monoclonal antibodies using phage display and magnetically-activated cell sorting: Applications in immunohematology. *J. Immunol. Methods* **206:** 73–85.

Smith M.M., Shi L., and Navre M. 1995. Rapid identification of highly active and selective substrates for stomelysin and matrilysis using bacteriophage peptide display libraries. *J. Biol. Chem.* **270:** 6440–6449.

Szardenings M., Tornroth S., Mutulis F., Muceniece R., Keinanen K., Kuusinen A., and Wikberg J.E. 1997. Phage display selection on whole cells yields a peptide specific for melanocortin receptor 1. *J. Biol. Chem.* **272:** 27943–27948.

Wilson D.R. and Finlay B.B. 1998. Phage display: Applications, innovations, and issues in phage and host biology. *Can. J. Microbiol.* **44:** 313–329.

Wilson R.K. 1993. High-throughput purification of M13 templates for sequencing. *Bio/Techniques* **15:** 414–422.

Wrighton N.C., Farrell F.X., Chang R., Kashyap A.K., Barbone F.P., Mulcahy L.S., Johnson D.L., Barrett R.W., Jollife L.K., and Dower W.J. 1996. Small peptides as potent mimetics of the protein hormone erythropoietin. *Science* **273:** 458–463.

Zwick M.B., Bonnycastle L.L., Noren K.A., Venturini S., Leong E., Barbas C.F. 3rd, Noren C.J., and Scott J.K. 1998. The maltose-binding protein as a scaffold for monovalent display of peptides derived from phage libraries. *Anal. Biochem.* **264:** 87-97.

20 Construction and Selection from Gene Fragment Phage-display Expression Libraries

Gregg J. Silverman

Department of Medicine, University of California at San Diego, La Jolla, California 92093-0663

Over the past decade a wide range of structurally diverse proteins has been successfully displayed on the surface of filamentous phage with the retention of full functional properties (see Chapter 5). This extensive experience suggests that if a functionally active protein can be displayed, then phage-display technology can be used to elucidate the minimum molecular requirements for a defined functional activity or binding site. This approach has been exploited by using monoclonal or polyclonal antibodies for the localization of dominant B-cell epitopes (i.e., antibody contact sites) on hormones, cell cycle regulatory proteins, cytoskeletal proteins, enzymes, and microbial antigens (Balass et al. 1993; Bianchi et al. 1995; Bottger et al. 1995; Du Plessis et al. 1995; Stephen et al. 1995; Petersen et al. 1995; Pitard et al. 1997; Williamson et al. 1998). In particular, this approach appears to be especially well suited for the identification of linear recognition determinants.

CONTENTS

In one proven strategy, a library of gene fragments is made from a defined encoding gene, and phage-display selection methods are used to map the relevant active sites. Commonly, these libraries are created by digestion of the gene of interest into random fragments (Williamson et al. 1998), or a more defined peptide library can be created by digestion with a restriction enzyme (Tsunetsugu-Yokota et al. 1991). In the following protocols, we detail the expression cloning of DNA fragments from a gene of interest by adapting previously published methods (see Fig. 20.1) (Stanley and Herz 1987; Williamson et al. 1998).

In most cases, the first step in using phage-display approaches to define the sites responsible for dominant epitopes, or a defined functional activity, will be the demonstration that this protein of interest can be displayed in a functional form on the surface of filamentous phage (see Chapter 5). To be rigorous, these demonstrations

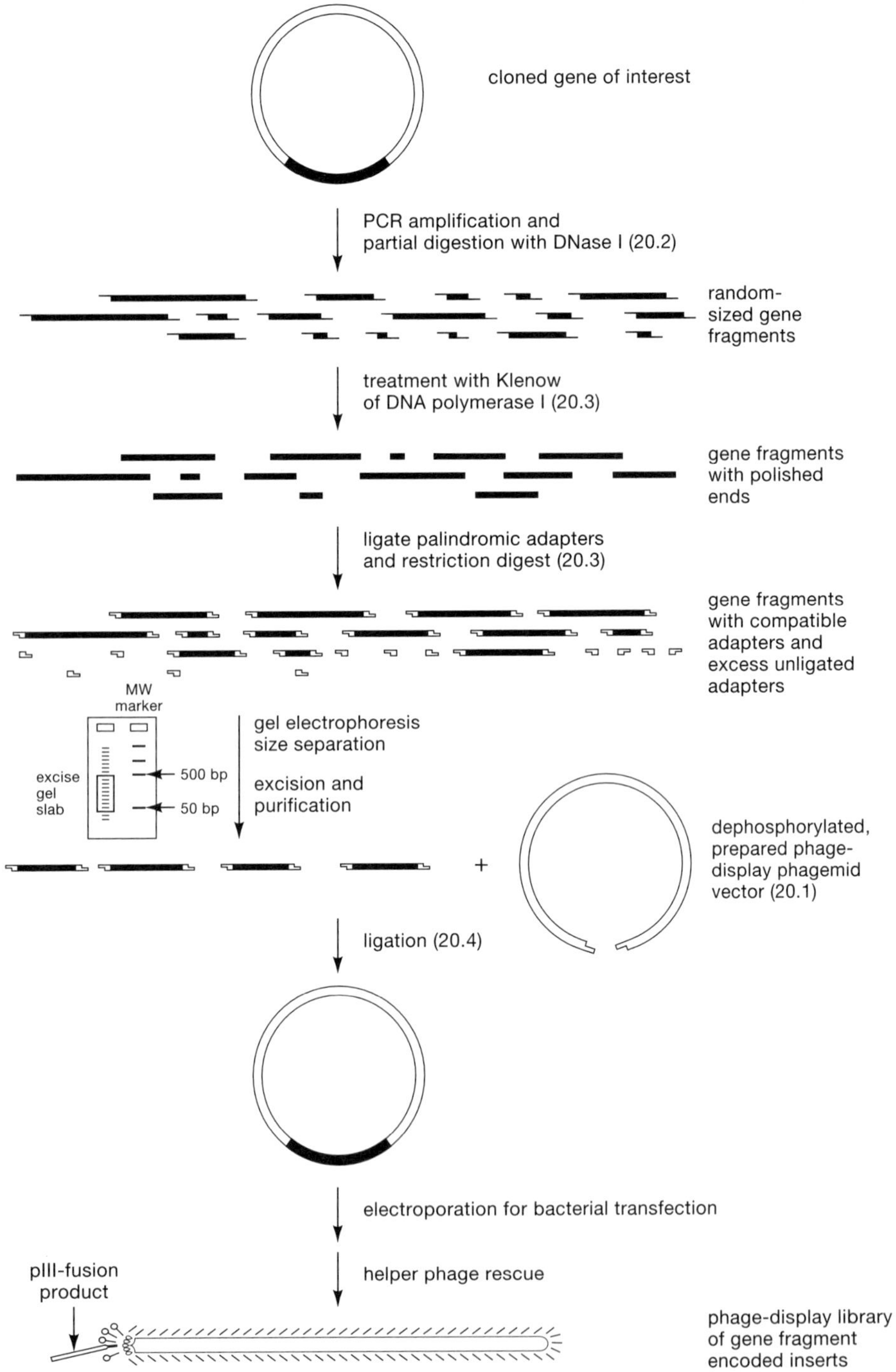

Figure 20.1. Overview of the procedures involved in creation and screening of gene fragment libraries. Each of the protocols involved in library generation is shown. The protocols described in this chapter involve cloning into the *Xho*I site of the pComb3H phagemid vector for display as a pIII fusion, but with minor adaptation this approach can be applied to library generation in other related vectors (adapted from Du Plessis and Jordaan 1996).

should incorporate several different types of assays of function and/or binding activity of the displayed protein that include side-by-side comparisons with the native protein. In addition, these studies should demonstrate that, beginning from libraries with infrequent representation of the phage encoding the gene of interest (i.e., aliquots of helper phage spiked with a functional clone at $<1{:}10^7$), the specific clones can also be efficiently isolated after only a few rounds of selection using standard selection methods. These verified selection methods should later be employed for the selection of relevant phage clones from the derivative gene fragment libraries.

In these studies, it may be reasonable to anticipate that even optimal clones from gene fragment libraries will display wild-type activities that are relatively attenuated compared to the native full-length parental clone. Hence, it is very important to control the quality of each step in construction of a gene fragment library to ensure that full-length clones are not included. Otherwise, full-length clones may be preferentially selected from the library, disallowing the acquisition of useful insights into the minimal structural determinants of the functional activity.

If possible, the design of these studies should include considerations obtained from sequence and structural analysis of the protein of interest, as these structural insights may aid in deciding the desired average size and range of sizes of the gene fragment libraries to be generated. For example, if the relevant site is likely to be primary-sequence-dependent, smaller libraries (e.g., including inserts of less than 50 codons) may be desirable. Alternatively, if the relevant site is highly conformation-dependent, libraries with larger inserts may be more likely to convey the functional activity. On the basis of these considerations, and the particulars of the system to be investigated, the most complete studies of the minimal structural requirements of activity may need to incorporate independent evaluations of libraries with sequentially smaller DNA fragment insert sizes.

To create a library of gene fragments, the parental DNA species is partially digested with deoxyribonuclease (DNase) I, leaving staggered ends that are then enzymatically filled in (i.e., "blunt-ended" or "polished"). To facilitate efficient ligation of the fragments into the library vector, special oligonucleotide adapters are blunt-end ligated to the fragments. These adapters carry complementary protruding termini and are thus more efficiently ligated into the vector than DNA fragments with blunt ends. The fragments with the adapters are then purified and size-fractionated.

The controlled cloning of the DNA inserts into phage-display libraries requires the co-ligation of two types of DNA molecules (linearized vector and gene fragment insert), which potentially poses a special problem. Unless precautions are taken, the two compatible DNA ends of the prepared vector can preferentially be ligated to one another, creating a "library" that is devoid of DNA inserts of interest. A common strategy to avoid this outcome is to enzymatically remove the phosphate groups on the 5′ termini of the digested vector with alkaline phosphatase (Sambrook et al. 1989c). This will minimize the frequency of undesirable transformants created with vector that has re-ligated without insert, and the prevalence of empty plasmids in the resulting library will be greatly decreased.

For the efficient cloning of the gene fragment inserts, in vitro ligation into the dephosphorylated vector is performed with bacteriophage T4 ligase, to form the phosphodiester bonds between the 5′-phosphates of the ends of the inserts with the 3′-hydroxyl groups of the compatible ends of the prepared dephosphorylated vector (Sambrook et al. 1989b). This results in the efficient formation of an open circular DNA molecule containing two single-stranded nicks. Bacteria are more efficiently transformed with circular DNA than with linear plasmid DNA, so after electroporation with the products of the ligation reaction, most of the transformants will contain recombinant plasmids. The nicks in the plasmids are subsequently healed by the host bacteria.

The following methods are designed especially for nondirectional cloning into the *Xho*I site of the original version of the pComb3H vector, which has a compatible DNA sequence downstream of this cloning site for creation of an in-frame fusion protein with pIII (see Fig. 20.2). With minor changes, these same protocols can be adapted for gene fragment cloning into any phage or phagemid display vector. For other papers on gene fragment phage display, see Petersen et al. (1995), Wang and Yu (1996), Sparks et al. (1996), and Fack et al. (1997).

*Xho*I *Spe*I

ATG GCC GAG GTG CAG CTG CTC GAG GGA TCC ACT AGT GGC CAG GCC GGC CAG GAG GGT GGT . . .

Met Ala Glu Val Gln Leu Leu Glu Gly Ser Thr Ser Gly Gln Ala Gly Gln Glu Gly Gly . . .

Figure 20.2. The pel B-associated cloning site of the original version of pCOMB3H. For cloning of antibody heavy-chain genes, the unique *Xho*I and *Spe*I restriction sites are used to introduce an in-frame DNA insert that has a pel B leader sequence for targeted secretion to the periplasmic space, and that also includes the phage gene III at the carboxyl terminus (for which only the 5′ end is shown). The original form of the vector includes a minimal six-nucleotide insert between these restriction sites. This form of the vector is suitable for the cloning of gene fragments into the *Xho*I site, because in-frame cloning will be associated with a carboxy-terminal sequence compatible with phage assembly and surface display.

PROTOCOL 20.1

Preparation of the pCombH3H Vector for Gene Fragment Cloning

In this protocol, bacteria carrying the plasmid vector are grown up in culture. The purified plasmid is then prepared by restriction digest with *Xho*I for nondirectional cloning of the DNA inserts. To prevent the two ends of the linearized vector from ligating together without an insert during ligation of the gene fragment libraries (Protocol 20.4), the vector must be treated with alkaline phosphatase.

Materials

pComb3H plasmid (The Scripps Research Institute)
XL1-Blue strain of *E. coli* (Stratagene, Cat. # 200228)

> *Note:* This catalog number is for commercially made electrocompetent cells; however, chemically competent and electrocompetent cells are less expensive when prepared in the lab (see Appendix 3 and Protocol 10.1).

carbenicillin (carb) 100 mg/ml, (see Appendix 2)
Luria Broth (LB) + carb (50 μg/ml) plates (see Appendix 2)
Superbroth (SB) + carb (50 μg/ml)
centrifuge with Beckman JA-10 or SORVALL GS-3 rotor
Tris/EDTA (TE) buffer (see Appendix 2)
QIAfilter Maxikit (QIAGEN, Cat. # 12263)
*Xho*I and *Xho*I buffer (supplied with enzyme)
sterile nuclease-free water
1 kb DNA marker
0.7% agarose gel
glycogen, 20 mg/ml (Roche Molecular Biochemicals, Cat. # 106 089)
3 M sodium acetate, pH 5.2
ethanol, stored at –20°C
dry ice/**isopropanol** slurry
70% **ethanol,** stored at –20°C
phenol/chloroform/isoamyl alcohol 25:24:1 (v:v:v)
chloroform
calf or shrimp alkaline phosphatase
10X alkaline phosphatase buffer
200 mM EGTA, pH 8.0 (see Appendix 2 and Note below)
3 M sodium acetate, pH 7.0

CAUTIONS: ***E. coli*, phenol, chloroform, isoamyl alcohol, ethanol, ethidium bromide, isopropanol** ***(see Appendix 4)***

Procedure

1. Transform XL1-Blue bacteria with purified pComb3H plasmid by electroporation (or by heat shock, see Protocol 10.1 and Appendix 3), and plate on LB+carb. Incubate the plates overnight at 37°C. In the morning, inoculate a single colony into a 10-ml culture of SB+carb and incubate at 37°C on a shaker at 300 rpm for approximately 8 hours.

   Note: Carbenicillin is preferred over ampicillin due to its greater thermostability. In addition, carbenicillin remains active in plates left at room temperature for many days, whereas ampicillin may degrade over time, resulting in greater problems with outgrowth of satellite colonies.

2. Inoculate the 10-ml culture into 100 ml of SB+carb, and incubate overnight (10–16 hours) at 37°C on a shaker at 300 rpm.

3. Centrifuge this 100-ml culture at 4000 rpm (2700*g*) for 15 minutes in a JA-10 or GS-3 rotor at 4°C. To recover 500 μg of plasmid, use the QIAfilter Maxikit or equivalent, and follow the manufacturer's protocol.

4. Resuspend each product in 500 μl of TE. Check the concentration of pComb3H by using an ethidium bromide agarose plate or by spectrophotometrically measuring the OD_{260} (see Appendix 3). Aliquot the DNA (e.g., 25 μl/tube) and store at –80°C. The vector stored at this temperature is stable for years.

5. Digest the purified pComb3H vector at the *Xho*I site to prepare the DNA for nondirectional cloning. The digestion reaction should contain: 30 μg of pComb3H plasmid, 40 μl of 10x *Xho*I buffer, 90 units of *Xho*I enzyme, and enough water to bring the total reaction volume to 400 μl. Incubate at 37°C for 3 hours.

   Note: Do not allow the enzyme volume to exceed 1/10 of total volume. If necessary to accommodate a larger enzyme volume, increase the total reaction volume by adding more water and reaction buffer.

6. Extract once with phenol/chloroform/isoamyl alcohol and once with chloroform as described in Appendix 3.

7. Transfer the supernatant to a fresh microcentrifuge tube. Ethanol-precipitate and wash once as described in Appendix 3.

8. Resuspend the pellet in 91 μl of water. Remove an aliquot (~500 ng, or 2 μl) for later comparisons in ligation reactions (see Protocol 20.4). To prevent self-ligation of the vector without an insert, treat the remainder of the cut vector as follows: Add 10 μl of 10x calf alkaline phosphatase buffer. Add 1 μl of calf alkaline phosphatase, mix briefly, and incubate for 2 hours at 37°C.

   Note: Shrimp alkaline phosphatase may be preferable due to greater ease of inactivation.

9. Inactivate the calf alkaline phosphatase by adding 10 μl of 200 mM EGTA, pH 8.0. Incubate at 75°C for 10 minutes, and then cool to room temperature on the bench top. Afterward, spin the tube for 10 seconds in the microfuge; the volume is now 110 μl.

   Note: In this protocol EGTA is included instead of the more commonly used EDTA, because EGTA is better than EDTA at chelating zinc, which is essential for alkaline phosphatase activity. *Do not substitute EDTA.*

10. Extract once with phenol/chloroform/isoamyl alcohol and once with chloroform as described in Appendix 3.

    Note: To create a library of the largest size, it is required to both heat-inactivate AND phenol-extract to destroy alkaline phosphatase activity.

11. Transfer the supernatant to a fresh tube, and add 1 μl of glycogen. Add 1/10 volume (11 μl) of 3 M sodium acetate, **pH 7.0** (it is important that this pH be used; see Note below). Add 3 volumes (330 μl) of 100% ethanol, and place on a dry ice/isopropanol slush for 20 minutes to precipitate. Spin at maximum speed for 30 minutes in a 4°C microfuge. Discard the supernatant and wash the pellet with 200 μl of 70% ethanol. Vortex, then microfuge for 10 minutes. Discard the supernatant and dry the pellet in a SpeedVac for 10 minutes. Dissolve the pellet in 100 μl of TE.

    Note: These ethanol precipitations employ a pH 7.0 sodium acetate buffer, because the EGTA used to inactivate the phosphatase will precipitate in acid pH, resulting in loss of the ligated DNA product.

12. Evaluate the prepared plasmid on a 0.7% agarose gel. Use an appropriate sample loading buffer to load 1 μg (~5 μl) of the prepared plasmid in a lane adjacent to 1 μg of the original plasmid preparation. The prepared vector should display a single distinct band of 3700 bp.

    Note: If you visualize your samples and you see a nonhomogeneous band or tear drops, it may indicate that the DNA did not get into solution and that good electrophoretic separation will not occur. Additional bands may represent a poor digestion or DNA degradation. Poorly separated DNA will result in suboptimal cloning efficiencies. Gel purification of the band may later improve transformation efficiency and decrease background, or another preparation of DNA may be required. If you decide to gel-purify the band, be aware that extraction with glass bead products like QIAEX II gel extraction beads (QIAGEN) or equivalents may yield a lower quality vector due to DNA shearing.

13. Determine the concentration of the prepared pComb3H vector by using the ethidium bromide agarose plate quantitation method or by measuring the OD_{260} (see Appendix 3). The vector is stable for at least 2–4 weeks when stored at –20°C. Prolonged storage results in a loss of cloning efficiency.

The prepared pComb3H vector is now ready for ligation of gene fragment libraries (Protocol 20.4).

PROTOCOL 20.2

Preparation of the Gene Fragments with DNase I for Epitope Mapping Surveys

To create a gene fragment library, the gene of interest is digested under selected conditions to provide inserts likely to include a functional binding site or B-cell epitope (adapted from Du Plessis and Jordaan 1996). For most applications, inserts of 50–400 bp may be desirable, although display of discontinuous epitopes will likely require bigger inserts (see the introduction to this chapter). To create the DNA fragment inserts, it is important to perform digestions with DNase I in the presence of manganese to create double-stranded cuts (Campbell and Jackson 1980). Otherwise, the use of magnesium in the buffer will primarily create only single-stranded breaks. After digestion and size selection, the ends of the DNA fragments must be "polished" with the Klenow fragment of *Escherichia coli* DNA polymerase I to fill in any overhangs. To enable site-specific cloning, oligonucleotide linkers are added that introduce an *Xho*I site to each DNA terminus. These sites are then cleaved and the fragments purified by agarose gel electrophoresis.

Because losses can occur during many of the steps, it is important to begin the procedure with sufficient DNA. A quantity of 20–25 μg is usually sufficient to allow several attempts at library construction. If the gene of interest has been cloned into a plasmid, it is usually acceptable to fragment the total plasmid DNA. Alternatively, the target insert can be excised and then isolated using preparative gel electrophoresis, or flanking oligonucleotide primers can be designed to enable efficient PCR amplification of only the DNA insert of interest. Optimally, pilot scale digestions of 2 μg of DNA and various concentrations of DNase I are first performed. To increase the chance that the large-scale digestions will yield fragments with the same size distribution, use the same DNase I preparation and do the scale-up soon after the test reaction. Alternatively, do several pilot scale reactions and pool the digests. The manganese-containing DNase I buffer should be made just before use, because it may degrade during storage. The following protocol describes the cloning of small pieces of DNA which often display linear B-cell epitopes, but gene fragment display libraries with larger inserts have also been used to successfully map continuous epitopes on viral antigens.

Materials

target DNA containing gene of interest, at least 36 μg
DNase I buffer (50 mM Tris buffer, pH 7.6, containing 10 mM $MnCl_2$ and 0.1 mg/ml BSA)
DNase I
16°C water bath (usually kept in the cold room)
stop solution (70% glycerol, 75 mM EDTA, 0.3% bromophenol blue)
2.5% agarose gel
DNA molecular weight markers for 50 bp to >1 kb

ethidium bromide
UV transilluminator
250 mM EDTA, pH 8.0
Tris/EDTA (TE) buffer (see Appendix 2)
phenol:chloroform, 50:50 (v:v)
3 M sodium acetate, pH 5.2
ethanol (store at –20°C)

CAUTIONS: **ethidium bromide, UV light, phenol, chloroform, ethanol** ***(see Appendix 4)***

Procedure

1. Perform pilot digestions to determine the conditions that will provide appropriately sized DNA fragments. For the mapping of linear epitopes, fragments of 50–400 bp are usually desirable. For these pilot studies, divide 16 µg of the target DNA dissolved in DNase I buffer into eight 20-µl aliquots. Keep on ice (see Fig. 20.3).
2. Dilute the DNase I to 10 units/ml in ice-cold DNase buffer, and make six additional serial 1:1 dilutions until 0.1563 units/ml. From each dilution of DNase I, take 5 µl and add to an aliquot of the DNA prepared in Step 1. One aliquot of DNA will receive no DNase I. To this tube, add 5 µl of DNase I buffer instead.
3. Place the tubes into a 16°C water bath and incubate for 10 minutes. Stop the digestion by adding 2.0 µl of stop solution to each tube (this stop solution is suitable for the loading of samples onto gels).

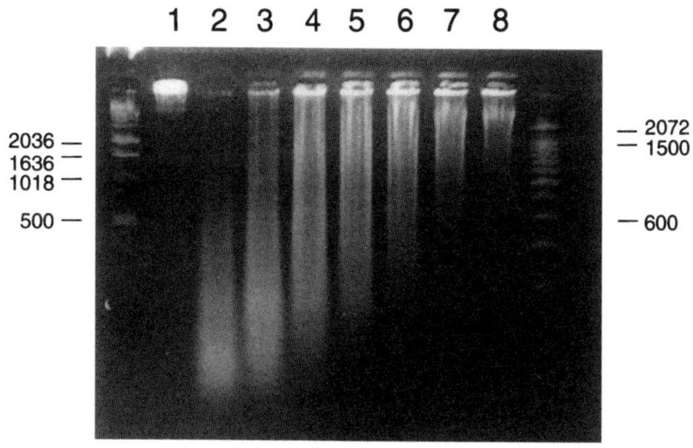

Figure 20.3. Fragmentation of DNA of interest. In this figure, DNA fragments from test DNase I digestions of a plasmid were electrophoretically separated on a 2.5% agarose gel in TBE buffer and stained with ethidium bromide. Lane *1* includes an aliquot of plasmid digested in the absence of DNase I. In lane 2, the sample was incubated with the highest concentration of DNase I, and in each subsequent lane, the amount of DNase I was decreased twofold. Each reaction was performed in a final volume of 25 µl, with an aliquot of 4 µg of plasmid mixed with 10 µl of DNase I buffer and 5 µl of a DNase I dilution, starting at 10 units/ml. The reactions were incubated for 10 minutes at 16°C and stopped with 2 µl of stop solution. 10 µl of each reaction was loaded onto the gel. Sample lanes are flanked by molecular weight markers.

4. Load 10 μl of the DNA sample from each tube onto a 2.5% agarose gel for electrophoresis. Include a lane with 1 μg of the starting material as well as a lane with appropriate size markers of 50 bp to > 1 kb. After electrophoresis, stain the gel with ethidium bromide and evaluate the size distribution of the fragmented DNA under transmitted UV light. The sample that was not treated with DNase I should comigrate with the starting target DNA material.

5. Consider the optimal size distribution that will be most valuable for your planned studies. Compare the size distributions in each of the lanes to select the optimal DNase I concentration for creating the inserts and scale-up the digestion proportionally. For the scaled-up reactions, approximately 10–20 μg of DNA is required in a total reaction volume of 100 μl.

6. In a large-scale digestion containing 20 μg of target DNA, add DNase I to the required concentration and incubate for 10 minutes at 16°C (100 μl total reaction volume). Stop the reaction with 10 μl of 250 mM EDTA, pH 8.0. Remove a 5-μl sample and confirm the size distribution by agarose gel electrophoresis, as described in Step 4.

7. If the size distribution is acceptable, dilute the remaining DNA fragments to a total volume of 500 μl in TE buffer and extract once with 500 μl of a 50:50 mixture of phenol and chloroform. Precipitate the DNA by adding 1/10th volume (50 μl) of 3 M sodium acetate, pH 5.2, and 2 volumes (1 ml) of 100% ethanol, and place at –20°C overnight or on a dry ice/isopropanol slurry for 20 minutes. The fragments can be stored for a month or longer at 80°C. If the size distribution of the fragments is not acceptable, repeat the large-scale digestion with an adjustment of DNase I concentration to attain the desired size distribution.

The DNase-generated fragments are now ready for terminus modification (Protocol 20.3).

PROTOCOL 20.3

Modifying the DNase-generated Gene Fragments Prior to Cloning

Several cloning strategies can be used to ligate the DNase-generated fragments into the library vector. These strategies include blunt-end ligation, the addition of adenosine to the fragments with *Taq* DNA polymerase for ligation into a vector with a 3′ thymidine overhang (Marchuk et al. 1991), and the addition to the fragments and vector of non-phosphorylated adapters (Haymerle et al. 1986). In the procedure described here, the fragments are blunt-ended with T4 polymerase and the Klenow fragment of *E. coli* DNA polymerase I, and a phosphorylated palindromic oligonucleotide is used as an adapter. The self-complementary oligonucleotide anneals to itself and is ligated to the blunt ends of the DNA fragments. After digestion of the internal *Xho*I site contained within the adapters, the fragments are ready for directed cloning. As described here, the use of glass bead extractions (e.g., GENECLEAN [BIO 101], QIAEX Beads [QIAGEN], or StrataClean resin [Stratagene]) in place of phenol extraction for fragment recovery may help reduce losses of DNA inserts. This method, however, is not optimal for vector recovery, as it results in shearing of larger DNA and thus adversely affects cloning efficiency.

Materials

precipitated DNase-generated fragments (see Protocol 20.2)
70% **ethanol** (store at –20°C)
polishing buffer (40 mM Tris-HCl, pH 8.0, 10 mM ammonium sulfate, 10 mM **β-mercaptoethanol,** 5 mM **magnesium chloride,** 0.5 mM EDTA)
dNTPs, in a mixture of 10 mM each of dATP, dTTP, dCTP, and dGTP
T4 DNA polymerase
Klenow fragment of *E. coli* DNA polymerase I
250 mM EDTA, pH 8.0
QIAEX II Gel extraction kit or equivalent (QIAGEN, Cat. # 20021)
3 M sodium acetate, pH 5.2
ethanol (store at –20°C)
dry ice/**isopropanol** slurry
Tris/EDTA buffer (TE, see Appendix 2)
phenol/chloroform/isoamyl alcohol, 25:24:1 (v:v:v)
chloroform
glycogen, 20 mg/ml (Roche Molecular Biochemicals, Cat. # 106 089)
5′ phosphorylated linker (5′pdCCCTCGAGGG, New England Biolabs, Cat. # 1073)
sterile nuclease-free water
T4 DNA ligase (Gibco BRL, Cat # 15224-017)
10x T4 DNA ligase buffer (supplied with enzyme)

16°C water bath (usually kept in the cold room)
*Xho*I
10x *Xho*I buffer (supplied with enzyme)
2.5% agarose gel
ethidium bromide
UV transilluminator
DNA molecular weight markers for 100 bp to 2000 bp

CAUTIONS: **ethanol, isopropanol, phenol, chloroform, isoamyl alcohol, ethidium bromide, magnesium chloride, β-mercaptoethanol, UV light** ***(see Appendix 4)***

Procedure

1. Place the microcentrifuge tubes used for precipitation of the DNase-generated gene fragments (see Protocol 20.2) into a 4°C microfuge, and centrifuge at maximum speed for 30 minutes. Remove the supernatant without dislodging the pellet. Afterward, wash the pellet with 500 μl of 70% ethanol. Allow the pellet to dry (use a SpeedVac or air-dry at 37°C), and redissolve in 100 μl of polishing buffer.
2. Add 1 μl of dNTPs to a final concentration of 100 μM each, and mix gently. Add 3 μl (30 units) of T4 DNA polymerase and incubate for 1 hour at 16°C. To complete the process, add 2 μl (8–10 units) of the Klenow fragment of *E. coli* DNA polymerase I and incubate for 30 minutes at 37°C.
3. Add 5 μl of 250 mM EDTA to stop the reaction. The polished DNA can be efficiently recovered with 10 μl of QIAEX II resin or the equivalent, by following the manufacturer's instructions. To ensure that any residual protein is removed, perform the extraction twice. Afterward, ethanol-precipitate and wash the DNA once as described in Appendix 3. Redissolve the pellet in 400 μl of TE. Mix briefly, spin out residual resin, and transfer supernatant to a clean microcentrifuge tube.
4. Extract once with phenol/chloroform/isoamyl alcohol and once with chloroform as described in Appendix 3.
5. Ethanol-precipitate the supernatant and wash the DNA once as described in Appendix 3. Redissolve the pellet (containing the blunt-ended DNA fragments) in 50 μl of water.
6. To ligate the adapters to the fragments, add 2 μl (1 μg/ μl) of 5′ phosphorylated linker (5′pdCCCTCGAGGG) to 4 μl of water and 3 μl of 10x T4 DNA ligase buffer. Denature for 5 minutes at 94°C, and anneal by cooling to 25°C over a period of 5 minutes. Add a 20-μl sample of the DNA fragments (keeping the rest of the DNA as a backup). Mix briefly, add 1 μl of T4 DNA ligase, and microfuge briefly to bring down the solution. Allow the ligation to proceed overnight at 16°C.

7. Inactivate the ligase by heating for 10 minutes at 65°C. After cooling on ice, spin down briefly in a microfuge and add 110 μl of water and 18 μl of 10x *Xho*I buffer. Next, add a total of 20 μl of *Xho*I (200 units) and incubate for 4 hours at 37°C. Add another 6 μl of *Xho*I (60 units), spin briefly in a microfuge and incubate for a further 60 minutes at 37°C.

8. Inactivate the restriction enzyme by heating for 10 minutes at 65°C. Extract the DNA twice with QIAEX resin according to the manufacturer's instructions (or once with phenol/chloroform/isoamyl alcohol). Ethanol-precipitate as described in Appendix 3. Resuspend the pellet in 60 μl of water, add 15 μl of sample loading buffer, and load on a 10% Tris-borate EDTA polyacrylamide gel for final size selection. (A 2.5% agarose gel can be used with poorer resolution. A 2% agarose gel is appropriate for fragments smaller than 500 bp.) Dye may be omitted from the sample buffer to avoid interference with visualizing the fragments. Use molecular weight markers to elute the proper size distribution of fragments from the gel and leave behind the excess linkers. QIAEX resin may be used to recover fragments from the gel, but recovery by electroelution (see Appendix 3) may provide inserts with higher cloning efficiency. At least 2 μg of the purified gene fragment DNA is required to proceed to the fragment–vector ligation step. Quantitate the product on ethidium bromide agarose plates side by side with prepared vector (see Appendix 3). These materials will be used in test ligations of defined insert to vector molar DNA ratios that evaluate cloning efficiency of these preparations (see Appendix 3). The fragments should be stored at 4°C and used as soon as possible.

The gene fragment inserts are now ready for cloning into the library vector (see Protocol 20.4).

PROTOCOL 20.4

Ligation of Gene Fragment Libraries

The successful creation of a library of useful size requires efficient ligation reactions that introduce the gene fragment inserts into the prepared vector. Appropriately designed test ligations will include an evaluation of whether dephosphorylation of the vector was attained (Sambrook et al. 1989a). Efficient insert ligation requires a molar ratio of plasmid DNA to foreign DNA ≤ 1. If the concentration of foreign DNA is substantially lower than that of vector, the number of useful ligation products will be very low. Hence, for 400 ng of prepared pComb3H vector (3700 bp), ligation reactions including about 15 ng of insert of 150 bp would be predicted to have a desirable frequency of resulting cloned inserts.

In practice, accurate determinations of small amounts of DNA are often difficult using available methods. Hence, concentration determinations are best performed for insert and vector at the same time, with comparisons to a known standard. Test ligations varying the ratios of insert to vector over a two- or threefold range may be required to determine the optimal effective molar DNA ratios. The kinetics of the ligation reaction are favored by relatively concentrated DNA, and thus, ligation reactions are most efficient with vector DNA concentrations from 20 to 60 μg/ml (or 20 to 60 ng/μl), but practical considerations may necessitate the use of lower concentrations.

It should also be appreciated that the preparation of vector and insert DNA may result in undesirable damage to termini that will adversely affect cloning. If adequate ligation efficiency is not attained, it may be necessary to independently evaluate the ligation and transformation efficiencies of the insert and plasmid DNA, each paired with other preparations of test DNAs. If a single correctly ligated clone can be isolated, this can be used as the source of high-quality inserts to test the maximum transformation efficiencies associated with a batch of prepared vector. Transformation efficiencies of $>5 \times 10^6$ colony forming units (cfu) per μg of prepared vector DNA are desirable. These frequencies are easily attainable by electroporation but are not generally feasible with chemically competent cells.

Materials

prepared vector (see Protocol 20.1)
prepared insert (see Protocols 20.2 and 20.3)
T4 DNA ligase (Gibco BRL, Cat. # 15224-017)
5x T4 DNA ligase buffer (supplied with ligase)
electrocompetent XL1-Blue (Stratagene, Cat. # 200228)

> *Note:* This catalog number is for commercially made electrocompetent cells; however, chemically competent and electrocompetent cells are less expensive when prepared in the lab (see Appendix 3 and Protocol 10.1).

glycogen, 20 mg/ml (Roche Molecular Biochemicals, Cat. # 106 089)
3 M sodium acetate, pH 5.2

ethanol (store at –20°C)
70% **ethanol** (store at –20°C)
sterile nuclease-free water
electroporation apparatus and cuvettes for bacteria
SOC liquid medium (see Appendix 2)
16°C water bath (usually kept in the cold room)
carbenicillin (carb, 100 mg/ml in water, see Appendix 2)
tetracycline (tet, 5 mg/ml in ethanol, see Appendix 2)
Superbroth (SB) + 20 μg/ml carb + 10 μg/ml tet (see Appendix 2)
Luria Broth (LB) + 100 μg/ml carb plates
LB medium (see Appendix 2)
QIAfilter Maxikit (QIAGEN, Cat. # 12263)
VCSM13 helper phage (Stratagene, Cat. # 200251)
kanamycin (10 mg/ml in water, see Appendix 2)
20%PEG 8000/15%NaCl
2% BSA/TBS (see Appendix 2)
0.20-micron push filter

CAUTIONS: ***E. coli*, ethanol, tetracycline *(see Appendix 4)***

Procedure

1. Perform test ligations on a small scale to determine ligation efficiency and background prior to large-scale library construction. Ligate 100 ng of vector with 10 ng of insert in a 20 μl final volume overnight at room temperature using 2 units of T4 DNA ligase in ligase buffer. Perform a parallel ligation with 100 ng of prepared vector alone to test for the sufficient dephosphorylation of the vector (see Protocol 20.1; also see Sambrook et al. 1989a, b for a description of controls).

   Note: This ligase buffer is formulated with polyethylene glycol (PEG), which increases viscosity and favors intermolecular interactions between different pieces of DNA. Reaction conditions may differ with other sources of T4 DNA ligase, although comparable results can be attained.

2. Mix 1 μl of each ligation with 40 μl of electrocompetent XL1-Blue (or equivalent). Incubate the mixtures on ice for 1 minute. Transform by electroporation and titer the transformants as described below in Steps 4–6. Calculate the transformation efficiency based on the use of 1 μg of prepared vector.

   Note: We have also had good results with ER2537 cells (NEB), which are comparable to XL1-Blue in most respects but have the advantage of being resistant to infection by environmental lytic phage.

   In the test ligations described above, electroporations are performed with 1/20th of the 100 ng of vector in the ligation reaction. Hence, transformation efficiencies will be multiplied by 200 to calculate the anticipated library size that can be created from 1 μg of this prepared vector.

Control ligations performed with the prepared vector alone should have a very low transformation efficiency (<10%) compared to ligations including the treated vector and inserts. Transformation efficiencies of > 10^5 cfu/µg for prepared insert-ligated vector DNA can be expected. See Sambrook et al. 1989 for troubleshooting methods.

3. For the ligation of the library, ligate 1000 ng of vector alone with 100 ng of the prepared insert, with 40 µl of 5x ligase buffer and 10 units of T4 DNA ligase in a total volume of 200 µl. Incubate overnight at room temperature. Heat-kill the ligation mix for 10 minutes at 70°C. Ethanol-precipitate and wash as described in Appendix 3. Resuspend the pellet in 15 µl of water. If resuspension is incomplete, brief heating to 50°C followed by vortexing may be required. Place the tube on ice for 10 minutes. Add 300 µl of electrocompetent cells, mix, and incubate for 1 minute on ice.

4. Add the mixture to an ice-cold electroporation cuvette, and shake the cells to the bottom. To electroporate, pulse at 2.5 kV, 0.2 cm gap cuvette, 25 µF, and 200 Ohms, or according to the cuvette manufacturer's instructions.

5. Flush the electroporation cuvette *immediately* with 1 ml of room-temperature SOC, and empty into a 15-ml culture tube with loosened top. Flush the cuvette again with an additional 2 ml of SOC and add to the culture tube. Immediately place in a shaker at 250 rpm for 1 hour at 37°C. To recover all of the cells in the electroporation cuvette, up to 5 ml of SOC can be used.

   Note: If possible, perform these steps using a shaking incubator that is phage "free" (see Appendix 3). Any phage contamination at this step can be disastrous!

6. Transfer the contents of the culture tube to a covered, sterile 50-ml Erlenmeyer flask containing 10 ml of prewarmed (37°C) SB+carb+tet and immediately titer transformants by plating 100 µl, 10 µl, and 1 µl for the test ligations and 10 µl, 1 µl, and 0.1 µl for the library ligations (diluted in each case with SB+carb+tet to a volume of 100 µl) on LB+carb plates. Incubate the plates overnight at 37°C, and when colonies are visible, quantitate titer and calculate transformation efficiency (see Step 2 and Protocol 10.3).

7. Incubate the liquid culture for 1 hour at 37°C on a shaker at 300 rpm.

8. Add carb to a final concentration of 50 µg/ml and incubate for one additional hour at 37°C on a shaker at 300 rpm.

Option 1: If panning is not to be performed the next day

9. Bring the liquid culture up to 100 ml in LB medium. Add tet to a final concentration of 10 µg/ml, carb to 50 µg/ml, and glucose to 20 mM. Grow overnight at 37°C on a shaker at 300 rpm, and harvest for plasmid preparation using the QIAfilter Maxikit or equivalent. Do not grow the culture for more than approximately 16 hours, or overgrowth of defective clones may become more prominent. Protocol is completed at this point. Store the DNA at –20°C. If stored as a precipitated pellet, it is stable for years.

Option 2: If panning will be performed the next day

(skip Step 9)

10. Add VCSM13 helper phage (at least 10^{12} plaque-forming units [pfu]) to the liquid culture and bring the total volume up to 100 ml in LB medium. Add tet to a final concentration of 10 μg/ml and carb to 50 μg/ml. Grow for 2 hours at 37°C on a shaker at 300 rpm.
11. Add kanamycin to 70 μg/ml, and grow overnight at 37°C on a shaker at 300 rpm.
12. Pellet the bacteria by centrifugation at 4000 rpm (2700*g*) in a GS3 or JA10 rotor for 15 minutes at 4°C. Transfer the supernatant to a clean tube. Store the cell pellet at –70°C for up to 2–4 weeks as a source of plasmids that can be recovered and used as unselected library. Precipitate the phage in the supernatant by adding 25 ml of a solution containing 20% PEG 8000 and 15% NaCl (w/v) to give final concentrations of 4% PEG-8000 and 3% NaCl (w/v). Incubate for 30 minutes on ice.
13. Pellet the phage by centrifugation at 9000 rpm in a Beckman JA10 rotor at room temperature. Remove the supernatant carefully, and invert the bottle to drain residual liquid on a paper towel for 10 minutes.
14. Resuspend the pellet in 1 ml of 2% BSA/TBS. Transfer to a sterile 1.5-ml microcentrifuge tube, and spin at maximum speed in a microfuge for 5 minutes at 4°C to remove debris.
15. Remove the supernatant (containing the phage library) with a 3-ml Luer-Lok syringe, and sterilize the supernatant with a 0.20-micron push filter. Store at 4°C.
16. Titer the resulting phage library (see Protocol 10.5). Acceptable libraries should have more than 10^7 independent transformants.

    Note: Use the phage form of the library within 24–48 hours of preparation for panning. It is important to use very fresh phage preparations for panning, as bacterial proteases in the phage preps may remove surface displayed recombinant proteins over time, decreasing the utility of the library. Phage can be stored for a year or more at 4°C without losing viability, but from these stored libraries fresh phage stocks must be created for panning.

PANNING OF GENE FRAGMENT LIBRARIES

Any selection strategy applicable to antibody or peptide display libraries can be utilized for gene fragment display studies. Common choices include biopanning of microtiter wells directly coated with the ligand or receptor of interest, or fixation of biotinylated receptor/ligand to streptavidin-coated wells. Alternatively, strategies employing cell surface selection can also be exploited (see Chapter 23 and Protocol 10.6). Whenever possible, it is wise to first optimize precoating conditions for ligand/receptor concentration and buffer. The possible effect of the planned blocking solution on the interaction should also be considered, using defined amounts of the target of interest in native form. Confirmation that selection conditions can provide an adequate signal for the binding interaction in an enzyme-linked immunoassay can represent a large step toward the ultimate success of the phage-display selection experiments.

SCREENING OF SELECTED CLONES FROM GENE FRAGMENT PHAGE-DISPLAY LIBRARIES

Due to the limited complexity of the gene fragment libraries, only 2–4 rounds of panning may be required to isolate specific binders. Akin to the application of other types of phage-display libraries, documentation of an increasing titer of phage eluted with each sequential round of panning will provide evidence of efficient selection of phage binders. A number of clones (e.g., 10–40 individual isolates) from the original (unselected) and each of the selected libraries should be evaluated for a decreasing heterogeneity of size distribution of the inserts. This can be evaluated from individual plasmid preps by restriction analysis (using flanking *Xho*I sites or other appropriate unique sites), but discrimination of insert sizes of less than 50 bp may be difficult. Alternatively, clonal diversity may be assessed by DNA sequence determination or by examination of the size distribution of PCR products from plasmid amplifications using oligonucleotides with DNA sequences from the vector that flank the cloning site.

Confirmation of specific binding interactions will require immunoassays such as direct phage binding ELISA. For these studies, a number of phage preparations of individual colonies are required, with confirmation by phage binding inhibition studies using the native protein encoded by the gene of interest (and irrelevant proteins as negative controls). In successful studies, DNA sequence analysis should demonstrate the recurrent isolation of a limited number of different clones. These clones would be predicted to express DNA sequences overlapping for a single minimal site for a linear B-cell epitope, or a single or small number of minimal ligand-binding sites, in the native protein, if selection employs a monoclonal antibody or receptor–ligand binding interaction, respectively.

REFERENCES

Balass M., Heldman Y., Cabilly S., Givol D., Katchalski-Katzir E., and Fuchs S. 1993. Identification of a hexapeptide that mimics a conformation-dependent binding site of acetylcholine receptor by use of a phage-epitope library. *Proc. Natl. Acad. Sci.* **90:** 10638–10642.

Bianchi E., Folgori A., Wallace A., Nicotra M., Acali S., Phalipon A., Barbato G., Bazzo R., Cortese R., and Felici F. 1995. A conformationally homogeneous combinatorial peptide library. *J. Mol. Biol.* **247:** 154–160.

Bottger V., Bottger A., Lane E.B., and Spruce B.A. 1995. Comprehensive epitope analysis of monoclonal anti-proenkephalin antibodies using phage display libraries and synthetic peptides: Revelation of antibody fine specificities caused by somatic mutations in the variable region genes. *J. Mol. Biol.* **247:** 932–946.

Campbell V.W. and Jackson D.A. 1980. The effect of divalent cations on the mode of action of DNase I. The initial reaction products produced from covalently closed circular DNA. *J. Biol. Chem.* **255:** 3726–3735.

Du Plessis D.H. and Jordaan F. 1996. Phage libraries displaying random peptides derived from a target sequence. In *Phage display of peptides and proteins: A laboratory manual* (ed. Kay B.K. et al.), pp. 141–149. Academic Press, San Diego.

Du Plessis D.H., Romito M., and Jordaan F. 1995. Identification of an antigenic peptide specific for bluetongue virus using phage display expression of NS1 sequences. *Immunotechnology* **1:** 221–230.

Fack F., Hugle-Dorr B., Song D., Queitsch I., Petersen G., and Bautz E.K. 1997. Epitope mapping by phage display: Random versus gene-fragment libraries. *J. Immunol. Methods* **206:** 43–52.

Haymerle H., Herz J., Bressan G.M., Frank R., and Stanley K.K. 1986. Efficient construction of cDNA libraries in plasmid expression vectors using an adaptor strategy. *Nucleic Acids Res.* **14:** 8615–8624.

Marchuk D., Drumm M., Saulino A., and Collins F.S. 1991. Construction of T-vectors, a rapid and general system for direct cloning of unmodified PCR products. *Nucleic Acids Res.* **19:** 1154.

Petersen G., Song D., Hugle-Dorr B., Oldenburg I., and Bautz E.K. 1995. Mapping of linear epitopes recognized by monoclonal antibodies with gene-fragment phage display libraries. *Mol. Gen. Genet.* **249:** 425–431.

Pitard B., Aguerre O., Airiau M., Lachages A.M., Boukhnikachvili T., Byk G., Dubertret C., Herviou C., Scherman D., Mayaux J.F., and Crouzet J. 1997. Virus-sized self-assembling lamellar complexes between plasmid DNA and cationic micelles promote gene transfer. *Proc. Natl. Acad. Sci.* **94:** 14412–14417.

Sambrook J., Fritsch E.F., and Maniatis T. 1989a. Test ligations and transformations. In *Molecular cloning: A laboratory manual,* 2nd edition, pp. 62. Cold Spring Harbor Laboratory Press, Cold Spring Harbor, New York.

———. 1989b. Ligation reactions. In *Molecular cloning: A laboratory manual,* 2nd edition, pp. 63–67. Cold Spring Harbor Laboratory Press, Cold Spring Harbor, New York.

———. 1989c. Dephosphorylation of linearized plasmid DNA. In *Molecular cloning: A laboratory manual,* 2nd edition, p. 60. Cold Spring Harbor Laboratory Press, Cold Spring Harbor, New York.

Sparks A.B., Rider J.E., Hoffman N.G., Fowlkes D.M., Quillam L.A., and Kay B.K. 1996. Distinct ligand preferences of Src homology 3 domains from Src, Yes, Abl, Cortactin, p53bp2, PLCgamma, Crk, and Grb2. *Proc. Natl. Acad. Sci.* **93:** 1540–1544.

Stanley K.K. and Herz J. 1987. Topological mapping of complement component C9 by recombinant DNA techniques suggests a novel mechanism for its insertion into target membranes. *EMBO J.* **6:** 1951–1957.

Stephen C.W., Helminen P., and Lane D.P. 1995. Characterisation of epitopes on human p53 using phage-displayed peptide libraries: Insights into antibody-peptide interactions. *J. Mol. Biol.* **248:** 58–78.

Tsunetsugu-Yokota Y., Tatsumi M., Robert V., Devaux C., Spire B., Chermann J.C., and Hirsch I. 1991. Expression of an immunogenic region of HIV by a filamentous bacteriophage vector. *Gene* **99:** 261–265.

Wang L.F. and Yu M. 1996. Random fragment libraries displayed on filamentous phage. *Methods Mol. Biol.* **66:** 269–285.

Williamson R.A., Peretz D., Pinella C., Ball H., Bastidas R.B., Rozenshteyn R., Houghten R.A., Prusiner S.B., and Burton D.R. 1998. Mapping of the prion protein using recombinant proteins. *J. Virol.* **72:** 9413–9418.

21 Construction and Selection from cDNA Phage-display Expression Libraries

GREGG J. SILVERMAN
University of California, San Diego, La Jolla, California 92093

PHAGE-DISPLAY cDNA EXPRESSION CLONING can be a powerful tool for the isolation of unknown genes, but to date experience in the scientific community has been relatively limited (see Chapter 5). The great strength of the system is that the functional activity of a protein structure can be used to select the biologic package that carries along the gene encoding this functionally important activity. With the most commonly used methods, libraries with 10^7 to 10^9 individual members can be readily created, and, using more recently developed methods, even larger libraries are possible (Fisch et al. 1996; Nissim et al. 1994). The practical aspects of phage selection enable the efficient isolation of species even if they are underrepresented in the original library.

There are potentially serious limitations to the utility of phage display methods. Like any bacterial expression system, this approach may not be suitable for selection of proteins that require posttranslation modification (e.g., glycosylation) or heteromeric assembly for functional activity. Due to the absence of mammalian chaperonins, conformationally dependent structures may not be efficiently expressed in these bacterial expression systems. Nonetheless, phage-display methods have significant advantages over other screening methods. Compared to conventional bacterial expression or lambda phage-based methods, which can be more laborious and time-consuming, phage-display methods inherently enable highly rapid and efficient selection in small volumes. Conventional screening methods also generally necessitate high-specific-activity label-

CONTENTS

ing of the ligand of interest. In contrast, phage-display methods do not require direct labeling of the ligand, and purification of the ligand may not be required.

During the creation of a cDNA phage-display expression library, every possible step should be taken to create the highest frequency of potentially expressible full-length cDNA inserts. Optimization of the methodological approach and continual monitoring of efficiency and quality are needed to ensure the absolute minimum frequency of defective inserts. Transfection with plasmids containing inserts that are too short to express functional binding domains and plasmids ligated without inserts may result in the overgrowth of bacterial clones with a shorter cell cycle. These undesirable clones can rapidly become overrepresented during library amplification. Although the nature of the inserts can also influence relative clonal fitness in peptide and antibody display libraries, the relative homogeneity of these libraries might prevent overrepresentation of defective clones from being a concern. In contrast, avoiding overgrowth of defective clones is especially important to the creation of large and highly heterogeneous cDNA expression libraries. In fact, it is possible that this factor may limit the maximum number and heterogeneity of cells and molecules from which useful phage-display cDNA libraries can be made.

Defective clones can have a competitive growth advantage, perhaps because they waste little energy expressing metabolically noncontributory protein, or because they do not express a protein that conveys some relative toxicity to the bacterial host (but which the researcher finds highly desirable). In practice, the library eluted from a panning well often includes a very favorable frequency of specific phage binders, while the ensuing step of thousand-fold (or greater) library amplification can be associated with differential growth rates resulting in a much less desirable clonal distribution. Common strategies to limit overgrowth of defective clones include decreasing colony growth duration. The addition of sterile filtered 1% or 2% glucose to the media may also greatly reduce the level of toxic proteins constitutively produced from vectors using the lac promoter. An overrepresentation of defective clones can occur during both liquid phase (liquid media) and solid phase (bacteriology plates) growth and expansion of a library. Although the issue is controversial (McConnell et al. 1995), many investigators still prefer library amplification on plates as a means of limiting defective clones. Finally, reported successes in in vivo panning efforts (Pasqualini et al. 1995; Pasqualini and Ruoslahti 1996) have in part been attributed to the isolation after each round of selection of only 300–500 nondefective colonies. These clones are then expanded individually and combined in equal parts to the library used for the subsequent round of panning.

Success in the selection of specific clones of interest also requires optimized incubation conditions for ligand/receptor-binding interactions. Large surface areas for binding interactions or use of liquid phase interactions (e.g., using beads, cell surfaces, or biotinylated capture ligands) can impart more favorable kinetics for selection. It is strongly advised to perform binding studies that define the optimal antigen concentrations, blocking buffer, and incubation conditions. Conditions that incorporate positive selection of desirable clones and negative selection to remove nonspecific binders may also provide the greatest efficiency of selection.

EXPRESSION CLONING AND THE JUN/FOS DISPLAY SYSTEM

Phage-display strategies employing a cloning site to create a fusion protein with the *gene III* product (pIII) are potentially problematic for cDNA expression cloning (see Chapter 5). By necessity, a cDNA library includes many genes of unknown sequence. As a direct consequence, defined upstream oligonucleotide primers cannot be used during library construction. The use of random priming (e.g., random hexamer oligonucleotides) for first-strand cDNA synthesis results in nondirectional cloning, without preference for either of the two possible cloning orientations. For expression in the most commonly used *gene III* display system, the gene of interest must be fused in-frame with the preceding bacterial leader sequence. Because the carboxyl terminus of the pIII inserts into the viral capsid, only the 5′ end of the *gene III* is available for the cloning site. Therefore, the gene of interest must also be fused in-frame at its carboxy-terminal fusion site with the *gene III* product required for display. Hence, for any gene encoding a desired activity, at most only about 1 in 18 possible cloning orientations enable appropriate display, greatly reducing the effective level of expressed genes that can be accessed in the library.

Due to these inherent limitations, there is a great advantage associated with the use of oligo dT-containing primers for first-strand synthesis, as priming from the 3′ terminal polyadenylation site of mammalian mRNA can greatly improve recovery of full-length cDNA. By defining the 3′ end of the gene, cDNA primed with oligo-dT-containing oligonucleotides can facilitate directional cloning, a strategy that can at least double the frequency in a library of potentially expressible open reading frames (ORFs). However, the otherwise highly desirable goal of cloning full-length cDNA inserts can introduce stop codons upstream of the *gene III*, preventing the creation of a fusion protein. To circumvent this problem, Crameri and Suter (1993) developed the pJuFo phagemid vector (see Chapter 5 and Figs. 21.1 and 21.2). Cloning into this vector creates a fusion product at the carboxyl terminus of the Fos protein. In transformed clones, helper phage infection induces expression of the Fos–cDNA product and the Jun–pIII product, which are both directed to the periplasmic space by their leader sequences to take part in the assembly of the recombinant phage. The Fos–cDNA product and the Jun-pIII product spontaneously form a heterodimer, which is stabilized by four engineered cysteine residues that form two disulfide bonds. The fundamental advantage of this vector design is that stop codons in the cDNA inserts do not interfere with phage display. However, the relative display efficiency of the Fos–Jun system compared to simple pIII array has not been investigated experimentally.

The phagemid vector pJuFo3H (or pJF3H) has features especially useful for cDNA cloning, and it incorporates several improvements over the original pJuFo vector of Crameri and Suter (Figs. 21.3 and 21.4) (Crameri et al. 1994; Crameri and Suter 1993, 1995; Crameri and Blaser 1995). pJF3H incorporates the backbone of pComb3H (C. Barbas, The Scripps Research Institute), and the original *Not*I and *Bgl*II sites in the pComb3H vector have been destroyed. The Fos gene and downstream cloning site have been inserted between the *Sac*I and *Xba*I sites (Figs. 21.3 and 21.4). Hence,

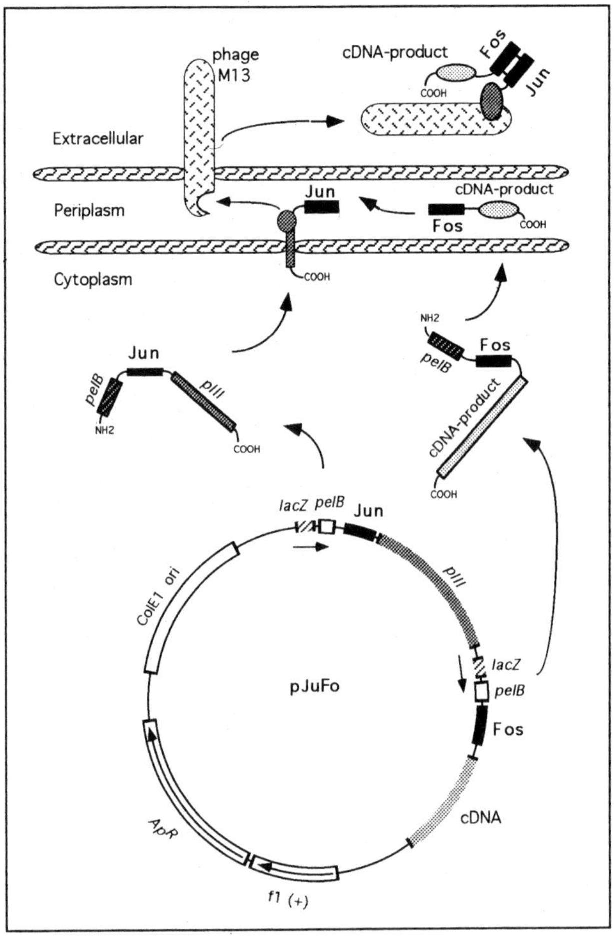

Figure 21.1. Vector map of the pJuFo phagemid vector. Upon infection of pJuFo-transfected bacteria with helper phage, filamentous phage-encoded proteins and the Jun–pIII and Fos–cDNA fusion proteins are produced. Under control of the pelB leader sequences these fusion products are directed to the periplasmic space, where they take part in assembly of phage containing a single-stranded copy of the phagemid vector. (Reprinted, with permission, from Crameri and Suter 1993, © Elsevier Science.)

cloning into the prepared vector results in the creation of a carboxy-terminal fusion product with the Fos protein. Between the unique *Xho*I and *Spe*I sites, the Jun gene fusion with the truncated phage *gene III* has been inserted. Like other phagemid display vectors, helper phage rescue of transformed bacteria results in phage packaging that incorporates both wild-type gene III products and the fusion-truncated *gene III* product. As a result, oligovalent or monovalent display of pIII occurs. In addition, in contrast with pComb3-related vectors that employ two promoters, the backbone of pComb3H provides a single promoter for monocistronic expression, which is designed to improve insert stability. Also adopted in the pJF3H vector, the Jun and Fos-associated transcripts employ different bacterial leader sequences, Pel B and

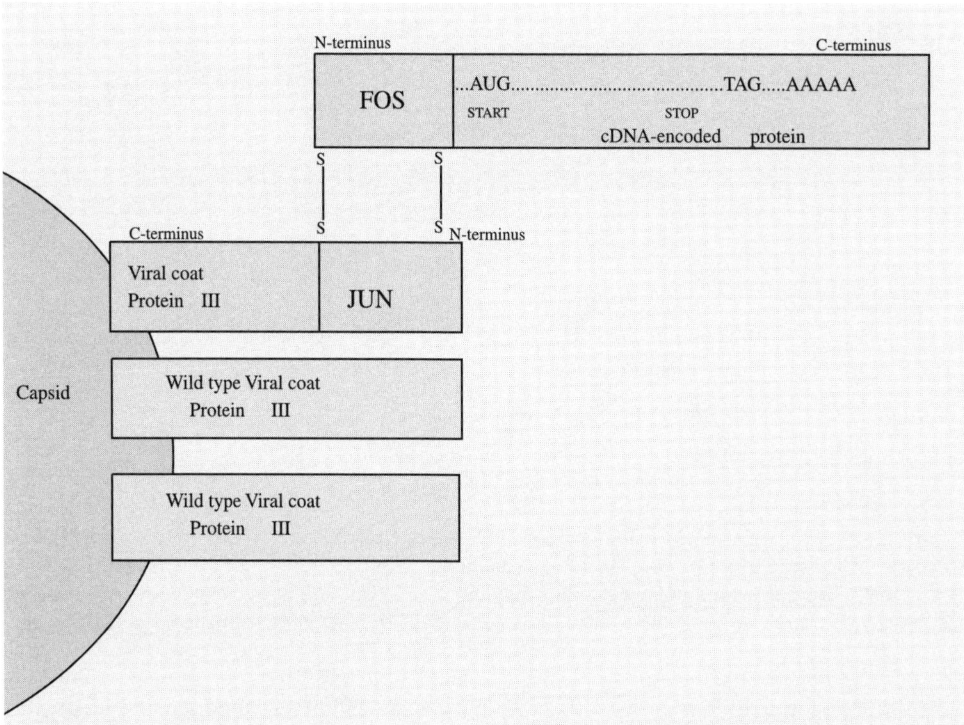

Figure 21.2. Phage display of expression products of the pJuFo and pJF3H phagemid vectors. As described, library products of vectors based on the pJuFo vector are expressed as fusions to the carboxyl terminus of the Fos protein. Induction results in accumulation of these products in the periplasmic space, under the direction of the leader sequence. There, during phage assembly, dimerization spontaneously occurs with the Jun protein fused to the amino terminus of a truncated pIII. Codons for cysteines engineered into the Jun and Fos genes create two covalent disulfide bonds after heterodimerization.

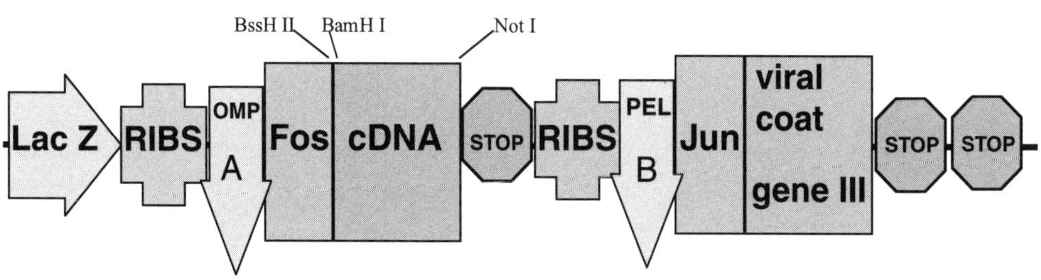

Figure 21.3. Vector map of the pJF3H phagemid vector. This vector is the result of transfer of the Fos–cDNA cloning site insert and the Jun–gene III insert from the pJuFo vector into the compatible sites in the pComb3H vector. In addition, a new multicloning site was generated, and a *Not*I site in the backbone was destroyed. The functional components of the vector include the lacZ promoter, the ribosomal binding site (RIBS), the OmpA and pelB leader sequences, the genes encoding Fos and Jun, the cDNA cloning site flanked by the unique *Bss*HII and *Not*I restriction sites, and the phage coat gene III.

```
   140             150               160             170             180
CTG CAA ACC GAA ATC GCG AAC CTG CTG AAA GAA AAA GAA AAG CTG
Leu Gln Thr Glu Ile Ala Asn Leu Leu Lys Glu Lys Glu Lys Leu

           190             200             210               220
GAG TTC ATC CTG GCG GAA CAC GGT GGT TGC AGA TCT TCT AGT GAT
Glu Phe Ile Leu Ala Glu His Gly Gly Cys Arg Ser Ser Ser Asp

   230             240               250             260             270
   BssH II                 Not I                           XbaI
CAG CGC GCT GGT ACC TCG CGG CCG CAT TAG AAG CTT CTA GAT AAT
Gln Arg Ala Gly Thr Ser Arg Pro His *** Lys Leu Leu Asp Asn

           280
TAA TTA GGA GGA
*** Leu Gly Gly
```

Figure 21.4. Cloning site of pJF3H. The DNA insert of the pComb3H vector from the *SacI* to *XbaI* sites has been altered to now include a sequence encoding a portion of the Fos gene followed by a cloning site containing unique *BssHII*, *NotI*, and *XbaI* sites. The above sequence includes the upstream bacterial leader sequence, OmpA, that precedes the Fos gene sequence and cloning site.

Omp A, respectively. Just as for the parental vectors, this vector is selectable on the basis of ampicillin resistance, and expression is inducible by the lac promoter.

In the following sections, protocols for expression cloning of cDNA gene products are detailed. The specialized methods for the de novo creation of cDNA libraries are beyond the scope of the current chapter, and the reader is directed to other references (see Gubler and Hoffman 1983) and to several high-quality kits available from commercial sources. Here, we present methods for the PCR-mediated transfer of a commercial cDNA plasmid library, an approach likely to be acceptable for most projects, given the wide diversity of high-quality libraries now available. These methods employ the pJF3H vector, but with minor modification of these methods, other phage-display vectors can be substituted.

PROTOCOL 21.1

Preparation of the pJF3H Vector for cDNA Cloning

In this approach, the vector is prepared for cloning of the appropriate double-stranded cDNA inserts into a site bounded by *Bss*HII and *Not*I restriction digestion. To create the largest library of diverse inserts, the goal is to prepare a plasmid vector with the highest competency (greatest number of transformants/µg of prepared plasmid) and the lowest background of re-ligated plasmid without insert.

It is best to start the plasmid preparation with a clone of the vector that has an irrelevant insert in the cloning site, which makes it easier to monitor efficiency of digestion and the recovery of appropriately prepared vector. After restriction digestion, the spacer in the cloning site is removed by agarose gel electrophoresis. Extra care and patience must be exercised to ensure that complete double digestion and separation are attained. The separation is best done in a preparative electrophoretic apparatus with a run of 4 hours or longer to ensure good separation and no carryover of the insert with the prepared vector.

Materials

pJF3H plasmid (The Scripps Research Institute)
XL1-Blue strain of ***E. coli*** (Stratagene, Cat. # 200228) or ER2354 (NEB)

> *Note:* The Stratagene catalog number is for commercially made electrocompetent cells; however, chemically competent and electrocompetent cells are less expensive when prepared in the lab (see Appendix 3 and Protocol 10.1).

carbenicillin (carb) 50 mg/ml in water (see Appendix 2)
Luria Broth (LB) + carb (50 µg/ml) plates (see Appendix 2)
Superbroth (SB) + carb (50 µg/ml) (see Appendix 2)
centrifuge with JA-10 (Beckman) or GS-3 (SORVALL) rotors
QIAfilter Maxikit (QIAGEN, Cat. # 12263)
*Not*I (10 Units/µl)
10x *Not*I buffer (supplied with enzyme)
sterile, nuclease-free water
phenol/chloroform/isoamyl alcohol
chloroform
ethanol
3 M sodium acetate, pH 5.2
glycogen (20 mg/ml, Roche Molecular Biochemicals, Cat. # 106 089)
*Xho*I
*Bss*HII (10 Units/µl)
10x *Bss*HII buffer (supplied with enzyme)
0.7 % agarose gel
Tris/EDTA buffer (TE, see Appendix 2)

CAUTIONS: ***E. coli*, ethidium bromide, ethanol, phenol, chloroform, isoamyl alcohol** ***(see Appendix 4)***

Procedure

1. Transform XL1-Blue or ER2354 bacteria with purified pJF3H plasmid. Plate on LB+carb and incubate overnight at 37°C. Inoculate a single colony into a 10-ml culture of SB+carb and incubate for approximately 8 hours at 37°C on a shaker at 300 rpm.

   Note: Carbenicillin is preferred over ampicillin due to its greater thermostability. In addition, carbenicillin remains active in plates left at room temperature for many days, whereas ampicillin may degrade over time, resulting in greater problems with outgrowth of satellite colonies.

2. Inoculate the 10-ml culture into 100 ml of SB+carb and incubate overnight (10–16 hours) at 37°C on a shaker at 300 rpm.

3. Centrifuge this 100-ml culture at 4000 rpm (2700*g*) in a JA-10 or GS-3 rotor for 15 minutes at 4°C. To recover approximately 500 μg of plasmid, use the QIAfilter Maxikit or equivalent. Follow the manufacturer's protocol. Ethanol-precipitate the eluant and wash as described in Appendix 3. Resuspend in 500 μl of TE.

4. Determine the concentration of pJF3H plasmid by using an ethidium bromide agarose plate or by spectrophotometrically measuring the OD_{260} (see Appendix 3). Aliquot for future use (e.g., 25 μl/tube) and store at –20°C. After precipitation, the DNA pellet is stable at –80°C for years.

5. Prepare the pJF3H vector DNA for directional cloning by digesting at the *Not*I and *Bss*HII sites. The buffer incompatibility of these enzymes requires sequential digestion. The *Not*I digestion reaction should contain 30 μg of pJF3H plasmid, 40 μl of 10x *Not*I buffer, 90 units of *Not*I enzyme, and enough water to bring the total volume up to 400 μl. Incubate for 3 hours at 37°C.

   Note: Do not allow the enzyme volume to exceed 1/10 of the total volume. If necessary to accommodate a larger enzyme volume, increase the total reaction volume by adding more water and reaction buffer.

6. Extract once with phenol/chloroform/isoamyl alcohol and once with chloroform as described in Appendix 3.

7. Ethanol-precipitate and wash once as described in Appendix 3.

8. Resuspend the pellet in 300 μl of water. Add 40 μl of 10x *Bss*HII buffer, 90 units of *Bss*HII, and enough water to bring the total volume up to 400 μl. Incubate for 3 hours at 50°C.

9. Repeat Steps 6 and 7, and resuspend the pellet in 90 μl of water.

10. Purify the vector on a 0.7% agarose gel. For this procedure, it is very important not to overload the gel. Use a comb with wide teeth for preparative separation. If necessary, tape together the teeth of a narrow-tooth comb. Load about 2 μg of plasmid per usual (~5 mm) small lane. Run the apparatus very slowly. Allow the cut vector to migrate at least three inches; more than 4 hours are required for a good separation.

Note: If you visualize your samples and see a nonhomogeneous band or tear drops, it may indicate that the DNA did not get into solution and that good electrophoretic separation will not occur. This will result in suboptimal cloning efficiencies.

11. Excise the 4000-bp band and recover the DNA by electroelution (see Appendix 3). Resuspend band in 50 μl of TE. The DNA can be stored for 2–4 weeks in solution at –20°C, but should be used as soon as possible.

Note: Extraction with glass bead products like GENECLEAN (BIO 101) or equivalents may yield a lower quality vector due to DNA shearing.

12. Check the concentration of the prepared pJF3H vector by using the ethidium bromide agarose plate method or by measuring the OD_{260} (see Appendix 3).

The pJF3H vector is now ready for cDNA cloning (Protocol 21.3).

PROTOCOL 21.2

Amplification and Preparation of cDNA Inserts for Library Construction

High-quality cDNA libraries from diverse species, organs, and cell types are available from other investigators and commercial sources. Hence, in many cases de novo library generation may not be required. Whatever the source of the library, it is wise to consider carefully which will be the best source of mRNA for the planned studies. Experimental decisions need to be made as to whether to use homogeneous cell lines, selectively stimulated cell lines, or even whole organ samples. To create cDNA expression libraries, we have used commercial kits (Invitrogen) for RNA preparation and for cDNA synthesis. To increase the representation of full-length cDNA, the method of Gubler and Hoffman (1983) has been adapted for the use of an oligo dT primer that creates a downstream *Not*I site, which enables directional cloning and improves the chances of recovery of carboxy-terminal sequence of a functional domain. Depending on the system, however, priming with random hexamers may be more desirable.

In the approach detailed below, phage-display libraries are created beginning from PCR amplification of inserts from a preexisting cDNA library. The primers are designed to introduce flanking restriction sites that facilitate the subcloning of the library into the pJF3H vector at a cloning site between the unique *Bss*HII and *Not*I sites (see Figs. 21.3 and 21.4). In our pilot studies, we have used commercially available libraries (Invitrogen) that are compatible with the cloning and expression strategy.

In the strategy outlined below, we amplify cDNA inserts from a preexisting library of plasmids in which the insert is flanked upstream by a *Bst*XI site and downstream by a *Not*I site. The *Not*I site was originally at the 5′ end of an oligo-dT primer used to prime the polyadenylation site of the mRNA (Invitrogen). Oligonucleotide-based amplification of the inserts employs specially designed primers. For specific PCR amplification, the Alb1 oligonucleotide (5′-ctg cgc gcg gat cca cta gta acg g-3′) introduces a *Bss*HII site upstream of the cDNA insert while the antisense BGH oligonucleotide (5′-tag aag gca cag tcg agg-3′) primes from downstream of the cloning site. For PCR-mediated recovery of these inserts, a relatively large amount of template is used to ensure representation of all individual DNA species. Conditions are designed with a small number of amplification cycles (20) to limit any possible skewing of the library composition. Following restriction digestion with *Bss*HII and *Not*I, directional cloning into the prepared pJF3H vector can be performed.

In any phage-display approach, there is a concern for outgrowth of species with defective, irrelevant, or nonexpressed inserts. To counter this problem, we have included a size fractionation step to remove smaller size inserts (i.e., to select for inserts of more than 500 bp that are more likely to encode a functional domain), but this may not be desirable in all applications. In addition, we have included steps to plate the selected library after each panning, which is intended to lessen the outgrowth of nonexpressing clones that might occur more frequently if solution culture was used instead for amplification of the library. Although we prefer this approach, there is experimental evidence that altered representation may occur in a library amplified by either approach (McConnell et al. 1995).

For purification of the cDNA inserts after restriction digestion, size separation with an agarose gel could be employed, but due to the diverse insert sizes, the large agarose plugs would result in poor DNA recovery efficiency. Therefore, we have opted for an alternative approach using centrifugation through a chromatography column. This easy and efficient method provides relatively high yields and potentially has fewer problems with environmental nuclease contamination.

Materials

plasmids of cDNA library (Invitrogen)
Alb1 primer (5′-ctg cgc gcg gat cca cta gta acg g-3′, custom synthesized)
BGH primer (5′-tag aag gca cag tcg agg-3′, Invitrogen, N575-02)
Pfu DNA polymerase (high-fidelity proofreading capacity, Stratagene, Cat. # 600153)
10x *Pfu* buffer (supplied with enzyme)
dNTPs, a mixture of 10 mM each of dATP, dTTP, dCTP, and dGTP
CHROMA SPIN+TE-1000 column (Clontech, Cat. # K1324-1)
CHROMA SPIN+TE-100 column (Clontech, Cat. # K1322-1)
*Not*I (10 Units/μl)
*Bss*HII (10 Units/μl)
ethanol
agarose
TE (see Appendix 2)
3 M sodium acetate (pH 5.2)
sterile nuclease-free water
phenol/chloroform/isoamyl alcohol 24:23:1 (v:v:v)
chloroform

CAUTIONS: ethanol, phenol, chloroform, isoamyl alcohol *(see Appendix 4)*

Procedure

1. PCR amplify cDNA inserts from a commercially prepared library according to the following protocol. Include a replicate that is devoid of the DNA (plasmid) template. For each reaction prepare the following:

cDNA plasmid	200 ng
Alb1 primer	50 pmole
BGH primer	50 pmole
dNTPs (10 mM)	1 μl (final concentration= 200 μM)
10x *Pfu* polymerase buffer	5 μl (final $MgCl_2$ concentration = 2.0 mM)

water to bring the final volume to 49 μl

2. "Hot Start" the PCR by raising the temperature to 96°C for 2 minutes and holding the reactions at this temperature until you have added 2.5 units (1 µl) of *Pfu* DNA polymerase to each tube.

    Perform 20 cycles of the following:

    96°C for 30 seconds,

    50°C for 90 seconds, and

    72°C for 150 seconds.

    Afterward, incubate at 72°C for 10 minutes to complete the extension of all products. Store the reactions at 4°C until analysis, which should be performed within one week.

3. Evaluate the size distribution of the PCR products by electrophoretic separation on a 1% agarose gel with appropriate m.w. markers. Load 5 µl per reaction. Products up to 3 kb or larger are anticipated, with an average size of 1.5 kb. In comparison, no DNA product (only small-sized oligonucleotides) should be detected in the lane from the tube in which DNA template was omitted.

4. Combine the PCR steps. Based on yield, 1–5 individual reactions may be required to produce at least 30 µg of DNA inserts. Because losses can occur during many of the following steps, adequate product should be obtained before continuing.

5. Extract once with phenol/chloroform/isoamyl alcohol and once with chloroform as described in Appendix 3.

6. Ethanol-precipitate and wash once as described in Appendix 3. Resuspend in 100 µl of TE.

7. Filter the DNA in a CHROMA SPIN+TE-100 column to remove residual primers. Elute the DNA on top of the filter with 1 ml of water, according to the manufacturer's directions.

8. Quantitate a 50-µl aliquot of the DNA by measuring the absorbance at 260 nm. Assess the purity by OD 260/280 ratio (see Appendix 3).

9. Digest 20 µg of library PCR product DNA with *Not*I and *Bss*HII. The different incubation temperatures of these enzymes require sequential digestion.

    The *Not*I digestion reaction should contain 20 µg of DNA, 40 µl of 10x *Not*I buffer, 120 Units of *Not*I enzyme, and enough water to bring the total volume up to 400 µl. Incubate for 3 hours at 37°C .

    Note: Do not allow the enzyme volume to exceed 1/10 of total volume. If necessary to accommodate a larger enzyme volume, increase the total reaction volume by adding more water and reaction buffer.

10. Extract once with phenol/chloroform/isoamyl alcohol and once with chloroform as described in Appendix 3.

11. Ethanol-precipitate and wash once as described in Appendix 3.
12. Resuspend the pellet in 300 μl of water. Add 40 μl of 10x *Bss*HII buffer, 120 Units of *Bss*HII, and enough water to bring the total volume up to 400 μl. Incubate for 3 hours at 50°C.
13. Repeat Steps 10 and 11, and resuspend the pellet in 100 μl of water.
14. Size-select the DNA for sizes greater than 1 kb by use of a CHROMA SPIN+TE-1000 column according to the manufacturer's instructions.

    Note: Appropriate sizing can be confirmed on a 2% agarose gel to document recovery of fragments of more than approximately 1000 bp.

15. Ethanol-precipitate as described in Appendix 3. Wash with 70% ethanol and resuspend in 20 μl of water. Quantitate the insert DNA side by side with the prepared vector. Use the ethidium bromide agarose plate method (see Appendix 3), as the diversity of cDNA insert sizes does not permit quantitation by mass ladder and agarose gel electrophoresis. Store at 4°C and use as soon as possible.

The cDNA inserts are now ready for ligation into the pJF3H vector (Protocol 21.3).

PROTOCOL 21.3

Ligation of cDNA Libraries into the pJF3H Vector

The successful creation of a library of useful size requires efficient ligation reactions for optimal transformation efficiencies of plasmids containing gene fragment inserts. Appropriately designed test ligations should be performed using about one-tenth of the scale of the final ligations (Sambrook et al. 1989a). These pilot studies should include a ligation reaction containing the prepared vector alone, to determine how often the vector will ligate to itself without insert. This represents the background of the ligation reaction, which should not represent more than 5–10% of the frequency for ligated inserts. Efficient insert ligation requires sufficient vector DNA to achieve a molar ratio of plasmid DNA:foreign DNA ≤ 1 (Sambrook et al. 1989b). If the concentration of insert DNA is substantially lower than that of vector, the number of useful ligation products will be very low. Hence, for each 1-μg quantity of prepared pJF3H vector (3700 bp), ligation reactions should include approximately 500 ng of insert of average size of about 1800 bp to provide the highest ligation and transformation efficiency (i.e., to create the largest library).

In practice, accurate determinations of small amounts of DNA are often difficult using available methods. Hence, concentration determinations are best performed for insert and vector at the same time, with comparisons to a known standard. Test ligations varying the amount of insert to vector over 2- or 3-fold ranges may be required to determine the optimal effective molar DNA ratios. The kinetics of the ligation reaction are favored by relatively concentrated DNA, and thus ligation reactions are most efficient with vector DNA concentrations between 20 and 60 μg/ml (or 20 and 60 ng/μl), but practical considerations may necessitate the use of lower concentrations.

It should also be appreciated that the preparation of vector and insert DNA may result in undesirable damage to termini that adversely affects cloning efficiency. Therefore, if an adequate ligation efficiency is not attained, it may be necessary to independently evaluate the ligation and transformation efficiencies for reactions of the insert and plasmid DNA, each paired with other preparations of test DNAs. Transformation efficiencies of $>5 \times 10^6$ colony forming units (cfu) per μg of vector DNA are desirable. These frequencies are easily attainable by electroporation but are not generally feasible with chemically competent cells.

Materials

prepared vector (see Protocol 21.1)
prepared insert (see Protocol 21.2)
T4 DNA ligase (Gibco BRL, Cat. # 15224-017)
5x T4 DNA ligase buffer (supplied with ligase)
electrocompetent XL1-Blue (Stratagene, Cat. # 200228, or prepare according to Protocol 10.1)
glycogen (20 mg/ml, Roche Molecular Biochemicals, Cat. # 106 089)

3 M sodium acetate, pH 5.2
ethanol (store at –20°C)
70% **ethanol** (store at –20°C)
sterile nuclease-free water
electroporation apparatus and cuvettes for **bacteria**
SOC liquid medium (see Appendix 2)
16°C water bath (usually kept in the cold room)
carbenicillin (carb) 50 mg/ml in water (see Appendix 2)
tetracycline (tet) 5 mg/ml in ethanol (see Appendix 2)
Superbroth (SB) + 20 μg/ml carb + 10 μg/ml tet (see Appendix 2)
Luria Broth (LB) + 100 μg/ml carb plates
LB media (see Appendix 2)
QIAfilter Maxikit (QIAGEN, # 12263)
VCSM13 helper phage (Stratagene, # 200251)
kanamycin, 10 mg/ml in water (see Appendix 2)
20% PEG/15% NaCl
2% BSA/TBS (w/v) (make fresh daily, see Appendix 2)
3-ml Luer-Lok syringe
0.20-micron push filter
UV transilluminator
DNA m.w. markers for 100 bp to 2000 bp
ethidium bromide
250 mM EDTA

CAUTIONS: ***E. coli*, ethanol, tetracycline, UV light, ethidium bromide *(see Appendix 4)***

Procedure

1. Perform test ligations on a small scale to determine ligation efficiency and background prior to use in library construction. Ligate 100 ng of vector with 50 ng of insert in a 20 μl final volume overnight at room temperature using 2 Units of T4 DNA ligase in ligase buffer. Perform a parallel ligation with 100 ng of prepared vector alone to determine the background of the ligation.

   Note: This ligase buffer is formulated with polyethylene glycol (PEG), which increases viscosity and favors intermolecular interactions between different pieces of DNA. Reaction conditions may differ with other sources of T4 DNA ligase, although comparable results can be attained.

2. Separately electroporate 1 μl of each of the two ligations into 40 μl of electrocompetent XL1-Blue (or equivalent). Incubate on ice for 1 minute. Transform by electroporation and titer transformants as described below in Steps 4 to 6. Calculate the transformation efficiency based on the use of 1 μg of vector. The background for the ligation (i.e., ligated vector alone) should be less than 10% of the ligation with insert.

Note: We have also had good results with ER2354 cells (NEB) which are comparable to XL1-Blue in most respects but have the advantage of being resistant to infection by environmental lytic phage.

In the test ligations described above, electroporations are performed with 1/20th of the 100 ng of vector in the ligation reaction. Hence, transformation efficiencies will be multiplied by 200 to calculate the anticipated library size that can be created from 1 μg of this prepared vector.

If a high colony count is produced for the vector alone, the vector preparation is inadequate and must be repeated. More efficient electrophoretic separation to remove inserts from the cut vector may improve transformation efficiency and decrease background. Alternatively, treatment with calf or shrimp alkaline phosphatase may greatly reduce this background. If the background colony count is low, but the transformation efficiency for ligations with insert is not substantially higher, the preparation of the DNA inserts may be inadequate and should be repeated. Alternatively, the ratio of vector to insert may need to be varied.

3. For the ligation of the library, ligate 1000 ng of vector alone with 500 ng of the prepared PCR insert products, with 40 μl of 5x ligase buffer and 10 Units of T4 DNA ligase in a total volume of 200 μl. Incubate overnight at room temperature. Heat-kill the ligation mix for 10 minutes at 70°C. Ethanol-precipitate and wash as described in Appendix 3. Resuspend the pellet in 15 μl of water. If resuspension is incomplete, brief heating to 50°C followed by vortexing may be required. Place tube on ice for 10 minutes. Add 300 μl of electrocompetent cells, mix, and incubate for 1 minute on ice.

4. Add the mixture to an ice-cold electroporation cuvette and shake the cells to the bottom. To electroporate, pulse at 2.5 kV, 0.2-cm gap cuvette, 25 μF, and 200 Ohms, according to the manufacturer's instructions.

5. Flush the electroporation cuvette *immediately* with 1 ml of room-temperature SOC and empty into a 15-ml culture tube with loosened top. Flush the cuvette again with another 2 ml of SOC and add to the culture tube. Immediately place in a shaker at 250 rpm for 1 hour at 37°C. To recover all of the cells in the electroporation cuvette, up to 5 ml of SOC can be used.

   Note: If possible, perform these steps in a shaking incubator that is phage "free" (see Appendix 3).

6. Transfer the contents of the culture tube to a covered, sterile 50-ml Erlenmeyer flask containing 10 ml of prewarmed (37°C) SB+carb+tet and immediately titer transformants by plating 100 μl, 10 μl, and 1 μl for the test ligations and 10 μl, 1 μl, and 0.1 μl for the library ligations (diluted in each case with SB+carb+tet to a volume of 100 μl) on LB+carb. Incubate the plates overnight at 37°C, and when colonies are visible, quantitate titer and calculate transformation efficiency (see Step 2 and Protocol 10.3).

7. Incubate the liquid culture for one hour at 37°C on a shaker at 300 rpm.

8. Add carb to a final concentration of 50 μg/ml and incubate for 1 additional hour at 37°C on a shaker at 300 rpm.

Option 1: If panning will not be performed the next day

9. Bring the liquid culture up to 100 ml in LB medium. Add glucose to 20 mM, tet to a final concentration of 10 μg/ml, and carb to 50 μg/ml. Grow overnight at 37°C on a shaker at 300 rpm, and harvest for plasmid preparation using the QIAfilter Maxikit or equivalent. Do not grow for more than approximately 16 hours, or overgrowth of defective clones may become more prominent. Protocol is completed at this point. If precipitated as a pellet, the DNA is stable for years at –20°C.

Option 2: If panning will be performed the next day

(skip Step 9)

10. Add VCSM13 helper phage (at least 10^{12} plaque-forming units [pfu]) to the liquid culture and bring the total volume up to 100 ml in LB medium. Add tet to a final concentration of 10 μg/ml and carb to 50 μg/ml. Grow for 2 hours at 37°C on a shaker at 300 rpm.
11. Add kanamycin to 70 μg/ml, and grow overnight at 30°C on a shaker at 300 rpm.
12. Pellet the bacteria by centrifugation at 4000 rpm (2700*g*) in a GS3 or JA10 rotor for 15 minutes at 4°C. Store the cell pellet at –70°C for up to 2–4 weeks as a source of plasmids that can be recovered and used as unselected library. Transfer the supernatant (containing the phage) to a clean tube. Precipitate the phage from the supernatant by adding 25 ml of a solution containing 20% PEG 8000 and 15% NaCl (w/v) to give final concentrations of 4% PEG-8000 and 3% NaCl (w/v). Incubate for 30 minutes on ice.
13. Pellet the phage by centrifugation at 9000 rpm in a Beckman JA10 centrifuge at 4°C. Remove the supernatant carefully, and invert the bottle to drain residual liquid on a paper towel for 10 minutes.
14. Resuspend the pellet in 1 ml of 2% BSA/TBS. Transfer to a sterile 1.5-ml microcentrifuge tube, and spin at maximum speed in a microfuge for 5 minutes at 4°C to remove debris.
15. Remove supernatant (containing the phage library) with a 3-ml Luer-Lok syringe, and sterilize the supernatant with a 0.20-micron push filter. Store at 4°C.
16. Titer the resulting phage library (see Protocol 10.5). Acceptable libraries should have more than 10^7 independent transformants.

Use the library within 24–48 hours for panning (see Protocol 21.5). It is important to use very fresh phage preparations for panning, as bacterial proteases in the phage preps may remove surface-displayed recombinant proteins over time, decreasing the utility of the library. Phage can be stored for a year or more at 4°C without losing viability, but from these stored libraries fresh phage stocks must be created for panning (see Protocol 21.4).

PROTOCOL 21.4

Preparation of a cDNA-pJF3H Phage Library for Panning

This protocol assumes that you have previously prepared and stored a phage library, which is now used for infection to create fresh phage stocks for panning. Alternatively, a plasmid prep can be used to electroporate competent bacterial cells for subsequent expansion and phage rescue. The protocol below represents a standard method for amplification and phage rescue.

Materials

unselected phage library
electrocompetent XL1-Blue (Stratagene, Cat. # 200228, or prepare according to Protocol 10.1)
tetracycline (tet) 5 mg/ml in ethanol (see Appendix 2)
LB + tet 10 µg/ml (see Appendix 2)
carbenicillin (carb) 50 mg/ml in water (see Appendix 2)
VCSM13 helper phage (Stratagene, Cat. # 200251)
kanamycin 10 mg/ml in water (see Appendix 2)
20% PEG/15% NaCl
2%BSA/TBS (w/v, make fresh daily, see Appendix 2)
3-ml Luer-Lok syringe
0.20-micron push filter

CAUTION: **tetracycline** ***(see Appendix 4)***

Procedure

1. Use 20 µl of the unselected phage library to infect 10 ml of fresh log-phase XL1-Blue in LB containing 10 µg/ml tet. Shake for 1 hour at 37°C at 300 rpm.
2. Add carb to 20 µg/ml. Shake for 1 hour at 37°C at 300 rpm.
3. Increase carb to 50 µg/ml. Shake for 1 hour at 37°C at 300 rpm.
4. Add helper phage (at least 10^{12} pfu) and bring up to 100 ml in LB medium. Add tet to a final concentration of 10 µg/ml and carb to 50 µg/ml. Grow for 2 hours at 37°C on a shaker at 300 rpm.
5. Add kanamycin to 70 µg/ml, and grow for 4–6 hours at 30°C at 300 rpm.
6. Pellet the bacteria by centrifugation at 4000 rpm in a GS3 or JA10 rotor for 15 minutes at 4°C. Transfer the supernatant to a clean tube. Store the cell pellet at

–70°C for up to 2–4 weeks as a source of plasmids that can be recovered and used as unselected library. Precipitate the phage in the supernatant by adding 25 ml of a solution containing 20% PEG 8000 and 15% NaCl (w/v) to give final concentrations of 4% PEG-8000 and 3% NaCl (w/v). Incubate for 30 minutes on ice.

7. Pellet the phage by centrifugation at 9000 rpm in a Beckman JA10 rotor or equivalent at 4°C. Remove the supernatant carefully, and invert the bottle to drain residual liquid on a paper towel for 10 minutes.
8. Resuspend the pellet in 1 ml of 2% BSA/TBS. Transfer to a sterile 1.5-ml microcentrifuge tube, and spin at maximum speed in a microfuge for 5 minutes at 4°C to remove debris.
9. Remove supernatant with a 3-ml Luer-Lok syringe, and sterilize the supernatant with a 0.20-micron push filter. Store at 4°C.
10. Titer the resulting phage library (see Protocol 10.5).

Use the fresh phage library within 24–48 hours for panning (Protocol 21.5). Alternatively, phage can be stored for a year or more at 4°C without losing viability, but surface-displayed proteins may be lost. From these stored libraries, fresh phage stocks must be created for panning.

PROTOCOL 21.5

Panning of a cDNA Phage-display Library

There are many options for in vitro selection of a cDNA expression library. However, it is important to appreciate that the first round of panning represents the bottleneck in selection, as most of the phage clonal diversity is discarded during the washing steps in that round. As one approach to selection of a library, we provide a protocol that uses the relatively large surface area of an immunosorbent tube for interactions with the immobilized ligand. The ligand has been biotinylated to enable interactions with a solid-phase surface precoated with streptavidin. In an alternative approach, a library can be screened after incubation with a soluble biotinylated protein, and the phage-biotinylated protein complex is later captured on wells with immobilized streptavidin (Scott and Smith 1990). The outcome of different biopanning methods can be greatly affected by the precoat concentration. Although it may not be true for every ligand–receptor interaction, in certain cases, selection using limiting concentration of soluble antigen can greatly improve the capacity to isolate higher affinity binders (Hawkins et al. 1992). Direct ligand precoats can also be useful, but in certain cases, denaturation of important ligand determinants or sites may occur following interactions with the polystyrene surface (Butler et al. 1993). As a direct consequence, panning of directly precoated antigens can occasionally select for binders to structures displaying less useful conformations.

Binding incubations of the library are commonly followed by washing with distilled water or 0.05% Tween-20/BBS to remove nonbinders. Subsequent efforts to recover the binders can employ several strategies. Most commonly, after panning, the binding phage are eluted with 0.1 M glycine-HCl, pH 2. Alternatively, phage can be released by base (e.g., ethanolamine, pH 12), 6 M guanidine-HCl, or 2 M urea without adversely affecting infectivity. Phage can also be released from the solid phase by protease treatment (e.g., trypsin or subtilisin) of the pIII-fusion protein without adversely affecting phage infectivity (Schwind et al. 1992). Direct infection of the phage in the well can also be employed.

Materials

immunotubes (Nunc, Cat.# 444202)
streptavidin (10 µg/ml in PBS)
2% and 5% BSA/TBS (w/v, make fresh daily; see Appendix 2)
biotinylated ligand (see Note, Step 3)
0.05% Tween/BBS (v/v, see Appendix 2)
fresh phage-display library (Protocol 21.3 or 21.4)
XL1-Blue strain of ***E. coli*** (Stratagene, Cat. # 200228)
carbenicillin (carb) 50 mg/ml in water (see Appendix 2)
tetracycline (tet) 5 mg/ml in ethanol (see Appendix 2)
SB + 2% glucose + tet 10 µg/ml + carb 50 µg/ml

125-mm plastic petri dishes with covers
VCSM13 helper phage (Stratagene, Cat. # 200251)
sterile nuclease-free water
T4 DNA ligase (Gibco BRL, Cat. # 15224-017)
kanamycin 10 mg/ml in water (see Appendix 2)
3-ml Luer-Lok syringe
0.20-micron push filter
20% PEG/15% NaCl (w/v)
0.1 M glycine, pH 2.2
2 M Tris-HCl, pH 7.4
LB (see Appendix 2)
SOC (see Appendix 2)
LB plates + carb 100 μg/ml

CAUTIONS: *E. coli*, tetracycline (see Appendix 4)

Procedure

1. Precoat a Nunc tube by incubating with 4.0 ml of streptavidin at 10 μg/ml in PBS overnight at 4°C. Rinse 3 times with distilled water.

   Note: It may be desirable to substitute NeutrAvidin (Pierce, Cat. # 31000), which is a recombinant form of avidin that is devoid of the glycosylation modifications and the RYD cell recognition sequence that can be responsible for undesirable binding interactions.

2. Block the Nunc tube by incubating with 4.5 ml of 5% BSA/TBS for 2 hours at room temperature.

3. Incubate the tube with 1 ml of the biotinylated ligand at 5 μg/ml in 2% BSA/TBS for 2 hours at room temperature.

   Note: Protein antigens are biotinylated using biotin-LC NHS (Pierce), and glycoconjugates can be labeled with biotin-LC-hydrazide (Pierce), using manufacturer's protocols.

4. Wash the tube three times with 0.05% Tween/BBS.

5. Incubate the tube with 4 ml of 2% BSA/PBS containing 500 μl of the phage-display library (at least 10^9 phage) for 2 hours at room temperature.

6. Discard the solution (containing the nonbinding phage), and wash the tube 10–20 times with 0.05% Tween/BBS.

7. Elute the phage by incubating the tube with 1 ml of 0.1 M glycine, pH 2.2, for 10 minutes. Neutralize the eluted phage by adding to 0.5 ml of 2 M Tris-HCl buffer, pH 7.4. Use about 0.5 ml of the phage to infect 2 ml of fresh XL1-Blue grown in

LB+10 µg/ml tetracycline to an OD_{600} about 0.5. Incubate for 15 minutes at room temperature without shaking. Store the rest of the phage at –80°C for up to several weeks.

8. In addition, add 2 ml of XL1-Blue (OD_{600} ~0.2) directly to the coated immunotube for 15 minutes at room temperature with inversion.

9. Add about 200 µl of the pooled infected bacterial solution to each of 10 large (125 mm) plates containing SB and 2% glucose, with tet at 10 µg/ml and carb at 50 µg/ml. Dry and incubate overnight at 37°C.

   Note: To determine background of nonspecific binding phage, panning can be performed in parallel against tubes that contain only streptavidin and have been blocked without adding biotinylated ligand. Titer both output phage for comparison, but subsequently discard the libraries from the nonspecific binding phage. All input phage and output phage at each round of panning should be titered (see Protocol 10.5), as these values will be useful for following progress.

10. Remove colonies from the overnight 125-mm plates by adding 10 ml of SB (total volume = 100 ml) to each plate and scraping the bacteria with a glass spreader. Collect into a 1-liter baffled Erlenmeyer flask.

11. Adjust the culture to an OD_{600} of 0.1 by diluting approximately 1:20 into SB and 2% glucose with tet at 10 µg/ml and carb at 50 µg/ml. The final volume should be 100–300 ml; the excess can be discarded. Grow on a shaker at 37°C at 300 rpm until OD_{600} is about 0.2.

12. Infect the culture with VCSM13 helper phage (at least 10^{12} pfu) and shake for 2 hours at 37°C at 300 rpm.

13. Add kanamycin to 70 µg/ml and grow for an additional 4 hours.

14. Pellet the bacteria and precipitate the phage as described in Steps 6–9 of Protocol 21.4.

15. Titer the phage at 10^{-3}, 10^{-6}, and 10^{-8} dilutions (see Protocol 10.5).

16. Repeat the panning procedure for a total of 3–6 rounds, using a newly coated immunotube for each round.

PROTOCOL 21.6

PCR Analysis of cDNA Inserts from Selected Display Libraries

With the successful selection of ligand-specific clones, there will be a change in the clonal distribution within the selected library. To survey for these changes in insert size heterogeneity, representative clones from sequential libraries (e.g., 20–40 per library) can be evaluated by modification of the approach used to transfer this library into the phage-display vector. In these reactions, the PCR primers flank the cDNA cloning site in the vector, enabling amplification of the insert. This method is simpler and more efficient than preparing individual plasmid preps prior to restriction analysis. If binders have been selected, with each sequential library the frequency of clones with small or undetectable inserts should decrease. Concurrently, the frequency of isolation of clones with a limited range of insert sizes should become increasingly apparent.

In libraries that contain a high frequency of ligand-selected clones, repeated copies of the same clonal isolate are often present. PCR amplification and agarose gel electrophoresis will enable identification of these clones with the same size insert. To further characterize the heterogeneity of these clones, it may be desirable to perform "fingerprinting" by restriction analysis of the plasmid with a frequent cutter (e.g., *Bst*NI). By this approach, identical or independent clones with inserts of the same size can be discriminated readily (Marks et al. 1992). These studies should be performed in coordination with single-colony phage-display binding immunoassays (see Protocol 21.7) to identify distinct sets of binders. By these methods, it may be possible to identify the greatest diversity of binders in the selected library prior to their further characterization by DNA sequence analysis.

Materials

colonies from selected display libraries (pJF3H-based) or minipreps from those colonies
Taq DNA polymerase
Taq buffer (supplied with enzyme)
oligonucleotides
  Alb1 (5′-ctg cgc gcg gat cca cta gta acg g-3′, custom synthesized)
  JFCB1 (5′-caa tcc agc ggc tgc cgt agg-3′, custom synthesized)
dNTPs, mix of 10 mM each of dATP, dTTP, dCTP, and dGTP
sterile nuclease-free water
thermal cycler
1% agarose gel

Procedure

1. PCR-amplify cDNA inserts from individual colonies from a library according to the following protocol. Include a replicate that is devoid of the DNA (plasmid) template. For each reaction prepare the following:

   Bacterial streak or 20–200 ng plasmid

Alb1 primer	50 pmole	
JFCB1 primer	50 pmole	
dNTPs (10 mM)	1 μl	(200 μM final concentration)
10x *Taq* polymerase buffer	5 μl	(final $MgCl_2$ concentration of 1.5 mM)

   Bring to a final volume of 49 μl with water

   In one approach, individual colonies (20–40 per library) are picked with a toothpick from a plate, separately dipped into a tube prepared for PCR amplification, and then substreaked onto a fresh plate (SB+2% glucose+tet 10 μg/ml+carb 50 μg/ml) with the position labeled and numbered. Alternatively, colonies can be picked and grown overnight in 2-ml liquid cultures (SB+glucose+tet+carb), and 20–200 ng of the plasmid miniprep used as a template for PCR.

2. "Hot Start" the PCR by raising the temperature to 96°C for 2 minutes and holding the reactions at this temperature until you have added 1.5 units (1 μl) of *Taq* DNA polymerase to each tube.

   Perform 30 cycles of the following:

   96°C for 30 seconds,

   50°C for 90 seconds, and

   72°C for 150 seconds.

   Afterward, incubate at 72°C for 10 minutes to complete the extension of all products. Store the reactions at 4°C until analysis, and use within 1 week.

3. Evaluate the size distribution of the PCR products by electrophoretic separation on a 1% agarose gel with appropriate molecular weight markers. Load 5 μl per reaction. Products up to 3 kb or larger are anticipated, with an average size of 1.5 kb. In comparison, no DNA product (only small-sized oligonucleotides) should be detected in the lane from the tube in which DNA template was omitted.

PROTOCOL 21.7

Screening of cDNA Libraries by Phage-binding ELISA

Phage display enables direct selection based on expression of functionally active protein on the surface of the filamentous phage. One of the most direct and efficient means for demonstration of successful selection of a desired specificity is the use of a phage-binding immunoassay. The advantage of this approach is that the same format used for selection of ligand binders is also directly employed to evaluate the functional capacity of the clones represented in an entire selected library. In fact, as long as they have not been stored for more than a few days, the same phage preparations used for the sequential rounds of biopanning can also be used in the phage-binding enzyme linked immunosorbent assay (ELISA). The demonstration of binders in these ELISA studies is facilitated by the use of enzyme-tagged anti-M13 phage antibodies that predominantly recognize the high-copy gene VIII product. Similar high-quality reagents of this type are available from several commercial sources (for example, Pharmacia, 5′→3′, Invitrogen).

In practice, ELISA studies are performed by evaluation of binding activities in replicate wells containing serial 10-fold dilutions of a phage preparation. Side-by-side comparisons of the original and the sequentially selected libraries are most helpful for evaluating for enrichment for a functional activity, and meaningful comparisons require the addition of equal phage titers of each library to the plates. In many cases, enrichment for functionally active phage will be detected in the same library for which there was an increased titer of phage eluted following incubation and washing on coated wells. Together, these findings provide strong evidence of efficient selection of specific ligand binders.

Notably, phage-binding ELISAs are prone to false positives in undiluted samples, especially if polyethylene glycol (PEG) has been used in phage preparation. In this case, the background signal can be diminished by evaluating phage stock at dilutions that begin at 1:10 or greater. In most assays, optimal signals can be obtained with titers of 10^6 to 10^7 cfu/ well, and a rapid dropoff of signal is often seen with less concentrated samples. By this method, one can readily identify the selected library that is most likely to yield high-affinity, ligand-specific binders.

Once a library displaying enhanced ligand-binding activity has been identified, small-scale phage preparations from individual representative colonies should be performed to identify specific binders. 10–20 clones from the original (unselected) library and 20–50 clones from each of the selected libraries demonstrating increased binding activity should be evaluated. Phage-binding inhibition studies using the ligand of interest in solution as a positive control and irrelevant proteins as negative controls can also be devised to confirm binding specificity.

It should be anticipated that even in libraries that demonstrate an enrichment for ligand-specific binders, a significant fraction of colonies without binding activity will be identified. These nonbinders will include clones without inserts or with nonexpressed inserts. Colony growth rates and morphology on plates can vary greatly, particularly in libraries created with highly heterogeneous inserts. Although not entirely predictable, defective clones often grow the most rapidly, suggesting that screening efforts should include the greatest range of colony types. It is best to evaluate fresh colonies. If possible, avoid clones that were plated after amplifications without selective pressures. It is also common to observe a dropoff of colony viability after only a few days, and undesirable plasmid rearrangements and deletions can occur even on plates stored under refrigeration.

For the specific protocols for application of phage-binding ELISA, see Chapters 10 and 11.

PROTOCOL 21.8

Screening for Expression of cDNA Inserts by Immunoblot Analysis

To evaluate the frequency and heterogeneity of inserts in pJF3H clones that are expressed at a protein level, an entire library can be subcloned into a compatible bacterial expression vector. For this purpose, an expression system, pRSET.JF (Invitrogen), has been created with restriction sites compatible with the pJF3H system (Figs. 21.5 and 21.6). Expression in the pRSET (Invitrogen)-based vectors requires the use of BL21(DE3)pLysS, an *E. coli* strain that contains the lambda lysogen expressing the T7 polymerase under the control of the lacUV5 promoter, and also a plasmid that constitutively expresses T7 lysozyme. Expression of T7 polymerase is necessary for high-level transcription of the recombinant proteins cloned into these vectors.

Transfer into the pRSET.JF vector enables in-frame cloning of the DNA inserts with a sequence coding for an amino-terminal fusion peptide. This sequence codes for

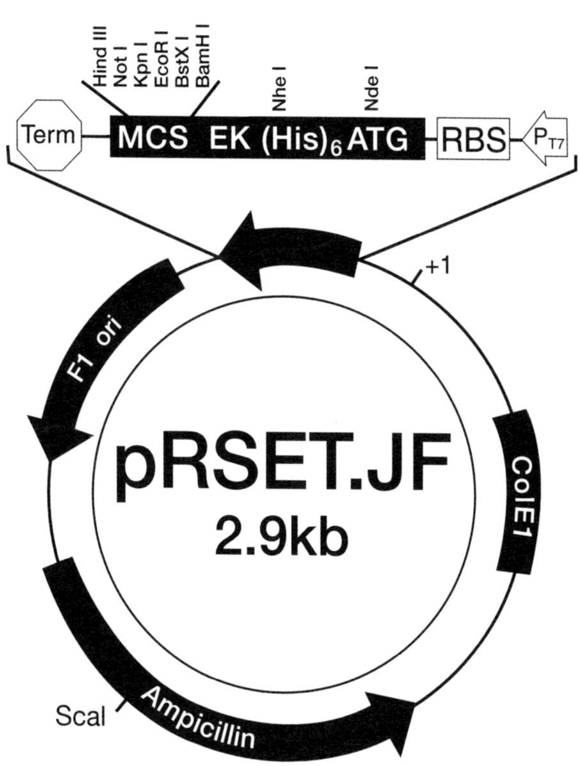

Figure 21.5. Vector map of the pRSET.JF expression vector. This vector, based on the pRSET bacterial expression vector (Invitrogen, San Diego, CA), includes a multicloning site that is compatible with pJF3H. This vector includes downstream sequences encoding a six-histidine $(His)_6$ purification tag and an enterokinase site (EK) for removal of the tag.

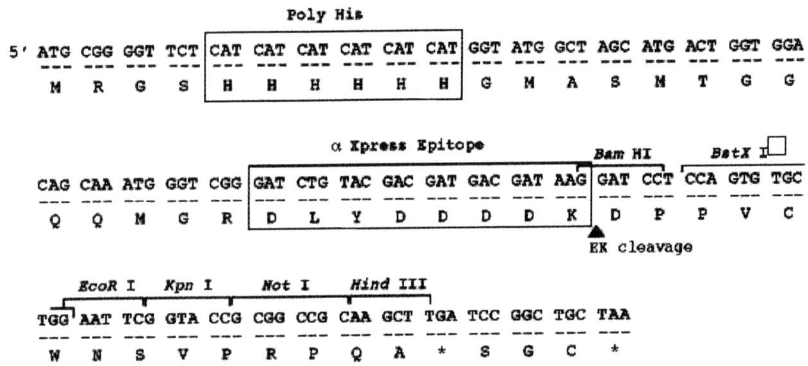

Figure 21.6. Multicloning site of the pRSET.JF expression vector.

an ATG translation initiation codon, six histidine residues that function as a metal-binding domain, a transcript stabilizing sequence from gene 10 of phage T7, a linear peptide epitope, and the enterokinase cleavage recognition sequence. The six-histidine (hexahis) sequence permits purification of these recombinant proteins by immobilized metal affinity chromatography (IMAC). The upstream epitope tag (Asp-Leu-Tyr-Asp-Asp-Asp-Asp-Lys) or Xpress Epitope (Invitrogen) enables detection of the fusion protein with a specific detection antibody. The enterokinase cleavage site in the fusion peptide, located between the hexahistidine sequence and the recombinant protein encoded by the DNA insert, enables subsequent removal of the amino-terminal fusion peptide from the purified recombinant protein.

To evaluate for expression of recombinant proteins from different clonal isolates, the anti-epitope tag antibodies can be used in Western blots. By comparison of representative clones from the original and selected library, the frequency of in-frame clonal protein producers can be assessed, and surveys for changes in the size distribution of the products can be efficiently performed. Binding reactivity of these recombinant proteins can also be evaluated by ELISA using plates precoated with the selecting ligand, and binding detected with the anti-epitope antibody.

To use this expression vector, the library of DNA inserts are amplified by PCR, and after suitable restriction digestion, are transferred into the pRSET.JF vector. After transformation into a compatible bacterial strain, individual colonies are picked and expanded in liquid culture, prior to induction of protein expression under the control of the T7 promoter system. Aliquots of these bacterial supernatants are then run on denaturing acrylamide gels prior to transfer to a membrane and Western blotting with the anti-XPress epitope antibody. Predictably, surveys of the original library generally demonstrate an overwhelming majority of clones that do not encode functional proteins of substantial (>10 kD) molecular weight. In contrast, successful panning should yield an increasing frequency of clones that encode expressed proteins of larger size with functional capacity. Successful selection of clones with a desired functional activity would also be expected to yield a library containing examples of multiple copies of the same functional clone. Hence, a detectable frequency of epitope tag-associated protein products of the same molecular weight could also be good evidence for selection of clones based on their ability to produce a functional protein product. Subsequent studies using ligand-binding Western blots, ELISA, and DNA sequence analysis would then be well warranted.

Materials

phage-display libraries, original and selected (pJF3H-based)
pRSET.JF (Invitrogen)
oligonucleotides
  Alb1 (5′-ctg cgc gcg gat cca cta gta acg g-3′, custom synthesized)
  JFCB1 (5′-caa tcc agc ggc tgc cgt agg-3′, custom synthesized)
BL21(DE3)pLysS strain of *E. coli* (Invitrogen, Cat. # C6060-10)
QIAfilter Maxikit (QIAGEN, Cat. # 12263)
glycogen, 20 mg/ml (Roche Molecular Biochemicals, Cat. # 106 089)
Taq polymerase
10x *Taq* buffer (supplied with enzyme)
dNTPs, mix of 10 mM each of dATP, dTTP, dCTP, and dGTP
sterile nuclease-free water
1% agarose gel
phenol/chloroform/isoamyl alcohol 25:24:1 (v/v/v)
3 M sodium acetate, pH 5.2
ethanol
*Not*I (10 Units/μl)
10x *Not*I buffer (supplied with enzyme)
*Bam*HI (10 Units/μl)
10x *Bam*HI buffer (supplied with enzyme)
CHROMA SPIN+TE-200 column (Clontech, Cat. # K1325-1)
carbenicillin (carb) 50 mg/ml in water (see Appendix 2)
chloramphenicol
T4 DNA ligase (Gibco BRL, Cat. # 15224-017)
Immobilon membrane (Millipore)
anti-Xpress epitope antibody (monoclonal, Invitrogen, Cat. # R910-25)
Luria Broth (LB) plates + carb (50 μg/ml) (see Appendix 2)
Tris/EDTA buffer (TE, see Appendix 2)
SOB (see Appendix 2)
PAGE sample and running buffers (see Appendix 2)
Western transfer and blocking buffers (see Appendix 2)

CAUTIONS: **chloramphenicol, phenol, chloroform, isoamyl alcohol, ethanol *(see Appendix 3)***

Procedure

1. Prepare for the subcloning of the inserts by PCR-amplifying the DNA inserts from the original and selected libraries. For each library, perform replicate reactions in 3–5 tubes and include a negative control that is devoid of DNA template.

   Include the following in each reaction:

Library plasmid prep (in pJF3H)	200 ng
10x *Taq* buffer	5 µl
dNTPs (10 mM)	1 µl
Alb1 primer	25 pmole
JFCB1 primer	25 pmole

   Bring to a final volume of 49 µl with water.

   "Hot Start" the PCR by raising the temperature to 96°C for 2 minutes and holding the reactions at this temperature until you have added 1.5 Units (1 µl) of *Taq* DNA polymerase to each tube.

   Perform 30 cycles of the following:

   96°C for 30 seconds,
   55°C for 60 seconds, and
   72°C for 90 seconds.

   Afterward, incubate at 72°C for 7 minutes to complete the extension of all products. Store reactions at 4°C until analysis. Analyze as soon as possible.

2. Pool the reactions for each library separately. Extract with phenol/chloroform/isoamyl alcohol, extract with chloroform, ethanol-precipitate, and wash as described in Appendix 3. Resuspend in 50 µl of TE, and quantitate by determination of OD_{260} (see Appendix 3).

3. Run the individual PCR products on a 1.5% agarose gel with suitable molecular weight markers. Compare clones from the unselected and selected libraries. A minimum of 20 colonies from each library is required for an adequate analysis.

4. Digest 10 µg of library PCR DNA with *Not*I and *Bam*HI, according to the following protocol. The buffer incompatibilities of these enzymes require sequential digestions.

   Combine 40 µl of 10x *Not*I buffer, 60 Units of *Not*I enzyme, and enough water to bring the total volume to 400 µl. Incubate at 37°C for 3 hours.

   Note: Do not allow the enzyme volume to exceed 1/10 of the total volume. If necessary to accommodate a larger enzyme volume, increase the total reaction volume by adding more water and reaction buffer.

5. Extract with phenol/chloroform/isoamyl alcohol, extract with chloroform, ethanol-precipitate, and wash as described in Appendix 3.

6. Resuspend the pellet in 300 μl of water. Add 40 μl of 10x *Bam*HI buffer, 60 Units of *Bam*HI, and enough water to bring the total volume to 400 μl. Incubate for 3 hours at 37°C.
7. Repeat Step 6 and resuspend the pellet in 100 μl of water.
8. Remove the cut ends by using a CHROMA SPIN+TE-200 column to select for DNA longer than 200 bp. Ethanol-precipitate the prepared inserts and resuspend in 100 μl of TE.
9. Prepare the pRSET.JF vector for directional cloning by digesting with the same enzymes. Combine 30 μg of pRSET.JF plasmid with 40 μl of 10x *Not*I buffer, 90 Units of *Not*I enzyme, and enough water to bring the total volume up to 400 μl. Incubate at 37°C for 3 hours.

   Note: Do not allow the enzyme volume to exceed 1/10 of the total volume. If necessary to accommodate a larger enzyme volume, increase the total reaction volume by adding more water and reaction buffer.
10. Extract the digested plasmid with phenol/chloroform/isoamyl alcohol and chloroform/isoamyl alcohol as described in Appendix 3. Ethanol-precipitate and wash as described in Appendix 3.
11. Resuspend the pellet in 50 μl of water. Add 40 μl of 10x *Bam*HI buffer, 90 Units of *Bam*HI, and enough water to bring the final volume up to 400 μl. Incubate for 3 hours at 37°C.
12. Extract the digested plasmid with phenol/chloroform/isoamyl alcohol, extract with chloroform/isoamyl alcohol, ethanol-precipitate and wash as described in Appendix 3. Resuspend in 100 μl of TE. To prepare the vector for cloning, remove the DNA spacer by running on a 0.6% agarose gel. Good separation will require a run of 4–6 hours. The prepared vector will appear as a 2.9-kb homogeneous band. Cut out the desired band and electroelute the DNA. Alternatively, the "freeze-squeeze" method (see Appendix 3) can provide good yields of high-quality vector.
13. Ethanol-precipitate the prepared vector as described in Appendix 3, and resuspend in 50 μl of water. Quantitate the prepared vector side by side with the insert DNA. Use the ethidium bromide agarose plate method (see Appendix 3), as the diversity of cDNA insert sizes does not permit quantitation by mass ladder and agarose gel electrophoresis.
14. For the ligation of the library, ligate 500 ng of vector alone with 250 ng of the prepared PCR insert product, with 20 μl of 5x ligase buffer and 5 Units of T4 DNA ligase in a total volume of 100 μl. Perform equivalent control ligation reactions including only the vector DNA alone. Incubate overnight at room temperature. Heat-kill the ligation mix for 10 minutes at 70°C. Ethanol-precipitate and wash as described in Appendix 3, and resuspend the pellet in 10 μl of water. If resuspension is not complete, brief heating to 50°C followed by vortexing may be required. Incubate for 10 minutes on ice.

15. Transform chemically competent or electrocompetent BL21(DE3)pLysS bacteria with 2 µl of the ligation product. High-efficiency transformation is not required, as only a limited number of colonies are needed.

Note: If maintaining viable BL21(DE3)pLysS colonies in the lab, it is important to use LB plates with 35 µg/ml of chloramphenicol, which selects for maintenance of the pLysS plasmid required for T7 lysozyme expression. In the above described experiments, the carbenicillin (or ampicillin, if you choose) selects for the pRSET.JF plasmid.

16. Titrate the transformation reaction, using 50-µl aliquots containing the transformation reaction undiluted and at 1:20, 1:200, and 1:2000. Spread onto LB+carb plates and incubate for at least 16 hours at 37°C.

17. Count the colonies from each of the ligation reactions. The ligation of vector alone should yield less than 10% of the number of colonies from the ligation of the DNA inserts from a library. If a high colony count is produced for the vector alone, the vector preparation is inadequate and must be repeated. More efficient electrophoretic separation to remove inserts from the cut vector may improve transformation efficiency and decrease background. Alternatively, treatment with calf or shrimp alkaline phosphatase may greatly reduce this background. If the background colony count is low, but the transformation efficiency for ligations with insert is not substantially higher, the preparation of the DNA inserts may be inadequate and should be repeated. It might also be necessary to vary the vector-to-insert ratio.

18. If appropriate transformation efficiencies are obtained, individual colonies are picked from the titration plates representing the original library or the selected library. For each library, 10–40 colonies are picked, substreaked, and coded on master LB+carb+chloramphenicol plates. The streak is then placed into 2 ml of fresh SOB medium with 35 µg/ml of chloramphenicol and 50 µg/ml of carb in individual 12-ml capped tubes. Tubes are placed in a phage-free incubator (see Appendix 3) at 37°C, shaking at 300 rpm overnight.

19. Inoculate 10 ml of prewarmed SOB to an OD_{600} of 0.1 with the overnight culture of each individual clone. Grow the cultures at 37°C with vigorous shaking to an OD_{600} of 0.4–0.6.

20. Add IPTG to the cultures to a final concentration of 1 mM (0.10 ml of 100 mM IPTG stock to the 10-ml culture), and continue to grow the cells. In most cases, near-optimal expression is attained at about 4 hours of induction. It is advisable not to induce overnight, as in most cases substantially lower (or undetectable) yields are then attained.

Note: Optimal kinetics of induction of recombinant protein may vary for each clone. To determine the optimal kinetics for a bacterial isolate, an initial 1-ml aliquot can be removed right before addition of IPTG, followed by removal of serial 1-ml aliquots every hour for 6–10 hours. Induction productivity is then evaluated by direct comparisons in SDS-PAGE and Western blots.

21. Pellet the cells by centrifugation, and aspirate the supernatant. Resuspend the cells in 1 ml of 20 mM phosphate buffer, pH 7.2, for every 10 ml of induced culture. Place the tube immediately in liquid nitrogen or a methanol/dry ice bath for 60 seconds. Thaw the frozen lysate at 42°C. Repeat this freeze/thaw three times or more. Centrifuge at maximum speed in a microfuge for 10 minutes at 4°C. Collect and store the supernatants and pellets at –70°C until analysis.

22. Resuspend pellets of induced cultures from individual colonies in SDS-PAGE sample buffer. Run these samples in parallel on several replicate PAGE gels alongside appropriate protein molecular weight markers. Stain one of the gels for Coomassie Blue detection of total protein. Transfer the other replicate gels to an appropriate membrane for Western blot analysis (e.g., Immobilon, Millipore). Follow manufacturer's instructions. Replicate blots can be stained using the selecting ligand, labeled with biotin or enzyme. Alternatively, recombinant proteins can be identified by reactivity with the antibody to the Xpress epitope that is upstream of the DNA inserts. If products are not detected in the cell pellets, perform the same analysis on the supernatants. For details on SDS-PAGE and Western blot analysis, see Sambrook et al. (1989).

Note: In recombinant proteins from the pRSET.JF vector, the amino-terminal fusion peptide makes an additional contribution of 3 kD to the overall molecular weight of the recombinant protein. Comparisons of induced and noninduced cultures of the same clone will demonstrate detection of a bona fide recombinant product from nonspecific staining.

REFERENCES

Butler J.E., Ni L., Brown W.R., Joshi K.S., Chang J., Rosenberg B., and Voss E.W.J. 1993. The immunochemistry of sandwich ELISAs–VI. Greater than 90% of monoclonal and 75% of polyclonal anti-fluorescyl capture antibodies (CAbs) are denatured by passive adsorption. *Mol. Immunol.* **30:** 1165–1175.

Crameri R. and Blaser K. 1995. Cloning allergens from *Aspergillus fumigatus:* The filamentous phage approach. *Int. Arch. Allergy Immunol.* **107:** 460–461.

———. 1996. Cloning *Aspergillus fumigatus* allergens by the pJuFo filamentous phage display system. *Int. Arch. Allergy Immunol.* **110:** 41–45.

Crameri R. and Suter M. 1993. Display of biologically active proteins on the surface of filamentous phages: A cDNA cloning system for selection of functional gene products linked to the genetic information responsible for their production. *Gene* **137:** 69–75.

———. 1995. Display of biologically active proteins on the surface of filamentous phages: A cDNA cloning system for the selection of functional gene products linked to the genetic information responsible for their production. *Gene* **160:** 139.

Crameri R., Jaussi R., Menz G., and Blaser K. 1994. Display of expression products of cDNA libraries on phage surfaces. *Eur. J. Biochem.* **226:** 53–58.

Fisch I., Kontermann R.E., Finnern R., Hartley O., Soler-Gonzalez A.S., Griffiths A.D., and Winter G. 1996. A strategy of exon shuffling for making large peptide repertoires displayed on filamentous bacteriophage. *Proc. Natl. Acad. Sci.* **93:** 7761–7766.

Gubler U. and Hoffman B.J. 1983. A simple and very efficient method for generating cDNA libraries. *Gene* **25:** 263–269.

Hawkins R.E., Russell S.J., and Winter G. 1992. Selection of phage antibodies by binding affinity. Mimicking affinity maturation. *J. Mol. Biol.* **226:** 889–896.

Marks J.D., Hoogenboom H.R., and Griffiths A.D. 1992. Molecular evolution of proteins on filamentous phage. Mimicking the strategy of the immune system. *J. Biol. Chem.* **267:** 16007–16010.

McConnell S.J., Uveges A.J., and Spinella D.G. 1995. Comparison of plate versus liquid amplification of M13 phage display libraries. *BioTechniques* **18:** 803–804, 806.

Nissim A., Hoogenboom H.R., Tomlinson I.M., Flynn G., Midgley C., Lane D., and Winter G. 1994. Antibody fragments from a 'single pot' phage display library as immunochemical reagents. *EMBO J.* **13:** 692–698.

Pasqualini R. and Ruoslahti E. 1996. Organ targeting in vivo using phage display peptide libraries. *Nature* **380:** 364–366.

Pasqualini R., Koivunen E., and Ruoslahti E. 1995. A peptide isolated from phage display libraries is a structural and functional mimic of an RGD-binding site on integrins. *J. Cell Biol.* **130:** 1189–1196.

Sambrook J., Fritsch E.F., and Maniatis T. 1989a. Test ligations and transformations. In *Molecular cloning: A laboratory manual*, 2nd. edition, p. 62. Cold Spring Harbor Laboratory Press, Cold Spring Harbor, New York.

———.1989b. Ligation reactions. In *Molecular cloning: A laboratory manual*, 2nd edition, pp. 63–67. Cold Spring Harbor Laboratory Press, Cold Spring Harbor, New York.

Schwind P., Kramer H., Kremser A., Ramsberger U., and Rasched I. 1992. Subtilisin removes the surface layer of the phage fd coat. *Eur. J. Biochem.* **210:** 431–436.

Scott J.K. and Smith G.P. 1990. Searching for peptide ligands with an epitope library. *Science* **249:** 386–390.

22 In Vivo Selection of Phage-display Libraries

RENATA PASQUALINI,[1] WADIH ARAP,[1] DANIEL RAJOTTE, AND ERKKI RUOSLAHTI

Cancer Research Center, The Burnham Institute, La Jolla, California 92037;
[1]Departments of Genitourinary Oncology and Cancer Biology, M.D. Anderson Cancer Center, Houston, Texas 77030

PHAGE DISPLAY IS COMMONLY USED TO SELECT for peptides, antibodies, and recombinant proteins capable of binding to a target molecule in an in vitro setting. We have devised a new approach to the use of phage-display peptide libraries: in vivo selection of phage that home to a preselected target when injected into the circulation of a mouse. Figure 22.1 illustrates the principle of the method.

By selecting from libraries of phage-displayed peptides in vivo, we have been able to isolate peptides capable of homing specifically to normal tissues (Pasqualini and Ruoslahti 1996; Rajotte et al. 1998) and to tumors (Arap et al. 1998a,b). We have also shown that the target molecules recognized by the peptides isolated in these selections are tissue-specific molecules in the vascular endothelium; in the vasculature of tumors, targets have been identified as markers of angiogenic vessels (Brooks et al. 1994; Hammes et al. 1996; Pasqualini et al. 1997; Arap et al. 1998a). The results indicate that the vasculature of most if not all tissues displays markers selective for that tissue (Rajotte et al. 1998). The peptides that home to tumors are useful in targeting anticancer drugs, protease inhibitors and pro-apoptotic peptides (Arap et al. 1998a; Ellerby et al. 1999; Koivunen et al. 1999c).

In vivo selection of peptide phage libraries is a new technique with tremendous potential. Here, we provide technical details of vascular targeting and discuss in vivo applications outside the vasculature.

CONTENTS

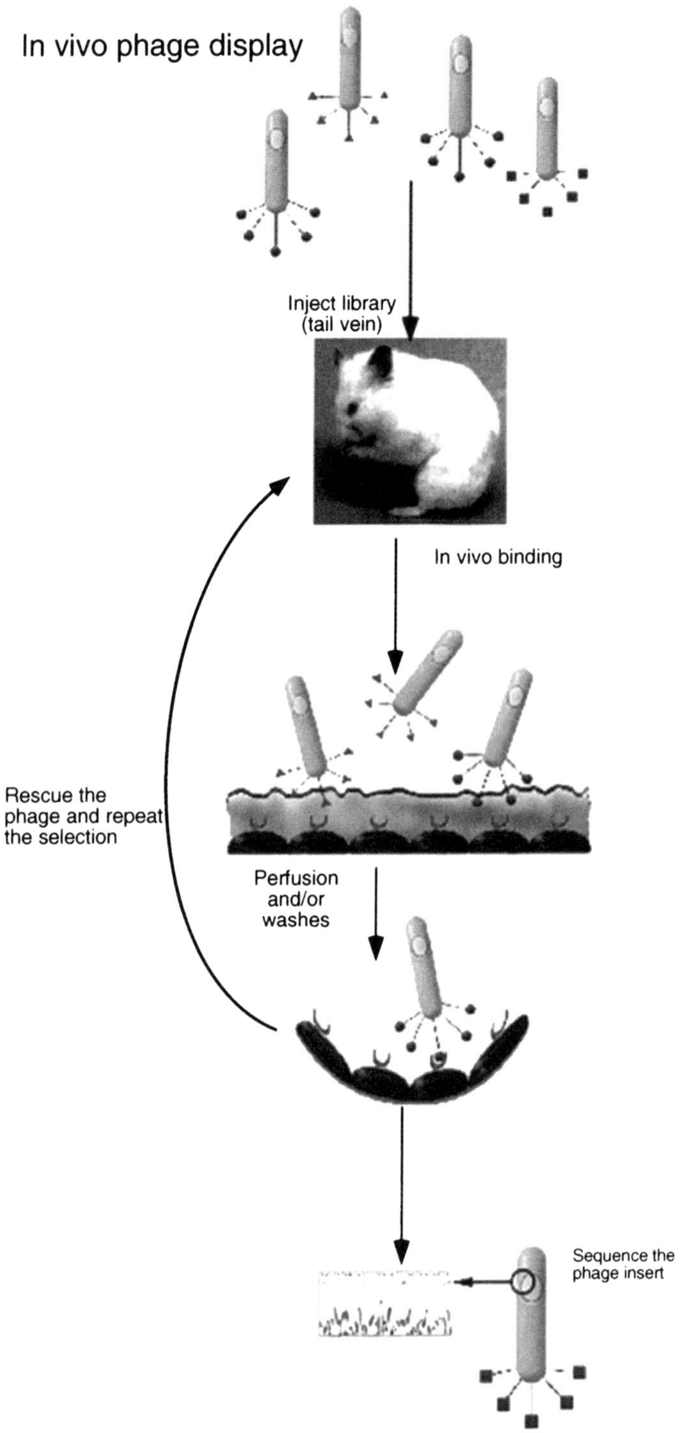

Figure 22.1. (*See facing page for legend.*)

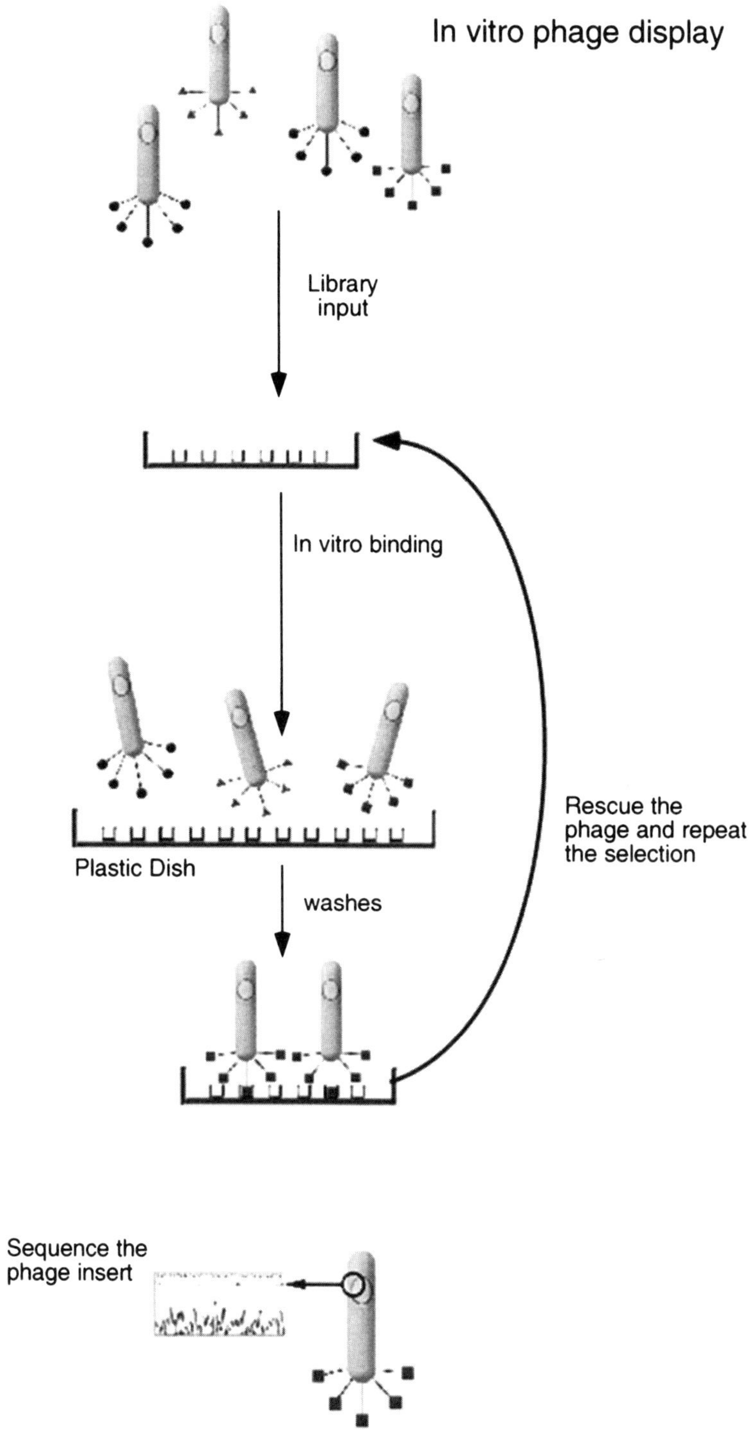

Figure 22.1. The principle of in vivo panning of phage-display peptide libraries. A comparison of the in vivo and in vitro selection methods is presented. Intravenous administration of the phage library is equivalent to the phage library input into a plastic well or tissue culture dish, and whole-body perfusion is analogous to the washing steps in vitro.

Our experience suggests that the quality of the peptide library is more critical in in vivo selection than it is in in vitro strategies, apparently because the selection has lower efficiency in vivo than in vitro. Thus, if the library contains mutant phage with a replicative advantage, such as phage that have lost the peptide-encoding insert, those phage will tend to overgrow desirable ones during successive amplification and selection steps. For this reason, we have amplified selected phage separately and then pooled individually amplified phage for the subsequent in vivo selection step (Pasqualini and Ruoslahti 1996). This procedure sets a practical limit of a few hundred phage to the number that can be processed from one selection to the next. Although our success in identifying homing peptides shows that this has not been a serious limitation, we have recently devised ways to circumvent this problem. Approaches that are still incompletely explored include the amplification of all phage in bulk between successive rounds of selection or the elution of phage from the target organ and reinjection without amplification (W. Arap et al., unpubl.).

In most studies we have used libraries that express cyclic peptides in our in vivo selections (Pasqualini and Ruoslahti 1996; Arap et al. 1998a; Rajotte et al. 1998), but nonconstrained libraries have also yielded homing peptides (Rajotte et al. 1998). In most reports of in vitro panning strategies, libraries have been engineered to express cyclic peptides that include the insertion of an even number of cysteine residues, as these libraries tend to yield peptides with higher affinities for their receptors than do linear libraries (Koivunen et al. 1993, 1994). Therefore, when the receptor is unknown, as is the case with in vivo selections, cyclic libraries of various designs are a good first choice, although linear ones should also be tried. For example, we have demonstrated that certain interactions actually favor linear peptides (Koivunen et al. 1995). In addition, linear libraries sometimes yield cyclic peptides formed from the cysteines provided by randomized codons in the library. In our experience, recovery of a cyclic peptide from a linear library often indicates a selection for a more defined secondary structure that confers preferential binding activity (Koivunen et al. 1993, 1994).

The length of the insert in pIII libraries (see Section 3, Peptide libraries) should optimally be kept under 10 codons. Experience suggests that larger inserts are likely to confer a selective disadvantage relative to defective phage clones that have deleted the insert, which can greatly decrease the efficiency of in vivo selection studies. Preliminary results suggest that it may also be possible to isolate antibodies from phage libraries in vivo.

Another potentially important variable that has not been thoroughly investigated is the length of time phage are allowed to circulate before harvesting the target tissues. Initially, we reasoned that the circulation interval should be short, because over time the target cells might internalize binding phage, resulting in their subsequent inactivation and clonal loss. As a result, we have always limited the circulation time following phage injection to only a few minutes. However, we now know from immunohistochemistry studies that following injection of specific phage that home to tumor vasculature, phage antigen can be found outside the blood vessels 24 hours later (Pasqualini et al. 1997; Arap et al. 1998a). We have not yet determined whether these deposits include infectious phage particles. Future studies are required to determine whether circulation times in the range of several hours can be used to isolate phage capable of binding to extravascular targets, provided that the recovery strategy is appropriate.

Characterization of phage clones isolated by in vivo selection methods indicates that our homing peptides specifically bind to the vasculature in the target tissues. Because endothelial cells in various locations share many surface markers, it may seem surprising that our method consistently yields tissue-specific vascular homing peptides, rather than peptides that bind nonselectively to the blood vessels within any tissue. A possible explanation is based on the fact that phage-display libraries contain only a few phage displaying any given sequence. Injecting 10^{10} phage particles into a mouse means that fewer than 100 copies of any given phage clone are included. As a direct consequence, we believe that any phage clone that binds to a ubiquitous endothelial marker will interact with binding sites elsewhere than in the tissue of interest, and is less likely to appear in the selected pool. This situation is represented schematically in Figure 22.2.

It is a common finding that the binding motif in the targeting peptides is often a tripeptide isolated in different sequence contexts. For example, the motif SRL is found within several brain-homing peptides (Pasqualini and Ruoslahti 1996), whereas lung-homing peptides contain a GFE motif, and peptides containing a RDV motif home to retina (Rajotte et al. 1998). In other experiments, peptides with NGR and GSL motifs displayed tumor-homing properties (Arap et al. 1998a). The prototype motif for these types of interactions is the RGD motif, which is known to be important for integrin binding in distinct molecular contexts, resulting in different binding affinities and specificities toward selective integrin heterodimers (Koivunen et al. 1994; Ruoslahti 1996). Thus, the specificity of many adhesive interactions seems to reside in a tripeptide recognition motif.

To date our group has identified a number of targeting peptides for diverse normal and tumor tissues (Table 22.1), and we have not yet failed to target a chosen tissue. Given that fact, one could speculate that all tissues may label their vasculature

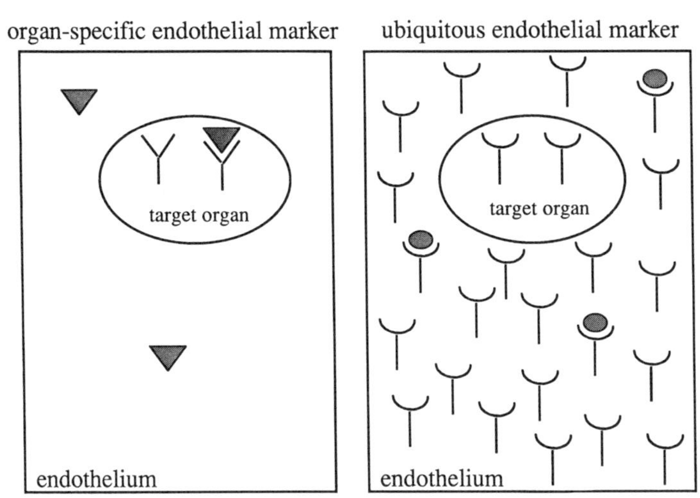

Figure 22.2. Specificity of the targeting in vivo: A hypothetical model.The behavior of motifs homing to a ubiquitous endothelial marker (*right* panel) and to a tissue-specific endothelial marker (*left* panel) is illustrated.

with distinct markers that can be detected by our in vivo phage-display method. In several instances, we have identified peptides with unrelated sequences that can home to the same vascular bed, a finding which may indicate that a given vascular bed can have multiple specific markers. The number of these markers is, however, likely to be finite, because we have often targeted the same receptor in independent experiments.

In one application, we have coupled peptides that home specifically to tumor blood vessels to the anticancer drug doxorubicin; the resulting compounds had enhanced efficacy and reduced toxicity against human breast cancer xenografts in nude mice (Arap et al. 1998a). Thus, our homing peptides may not only contribute to a basic understanding of endothelial cell biology, but may also provide molecular tools for targeting diagnostic and therapeutic substances to selected vascular beds.

The potential of in vivo phage display for identification of endothelial targeting sequences is far from fully explored. We have tested relatively few types of libraries, time frames, and experimental strategies, and many additional normal and pathological tissues remain to be targeted. Rheumatoid arthritis, diabetes mellitus, and atherosclerosis are examples of pathological conditions that have an angiogenesis component (Folkman 1995) and whose vasculature could be probed in vivo using phage-display libraries. Moreover, whereas the in vivo selection method has so far been used to isolate phage that home to vascular targets, other applications are possible. For example, phage that are transported across the blood-brain barrier would be of great interest. For the development of therapy for bladder diseases, phage targeting of urothelium may yield tools for intravesical therapy. It may also be possible to isolate phage that are taken up by the mucous membranes, the airways, or the digestive tract and are subsequently translocated into tissue or even into the circulation. It is certain that as more laboratories exploit this technology, additional applications will emerge.

Table 22.1. Tissue-targeting peptides isolated by in vivo phage display

Organ/tissue	Homing peptide
Brain	CSSRLDAC
Kidney	CLPVASC
Lung	CGFECVRQCPERC
Lung	CGFELETC
Skin	CVALCREACGEGC
Pancreas	SWCEPGWCR
Intestine	YSGKWGW
Uterus	GLSGGRS
Adrenal gland	LMLPRAD
Retina	CSCFRDVCC
Retina	CRDVVSVIC
Human tumor xenografts	CDCRGDCFC
Human tumor xenografts	CNGRCVSGCAGRC
Human tumor xenografts	CGSLVRC

From Pasqualini and Ruoslahti (1996); Rajotte et al. (1998); Arap et al. (1998).

PROTOCOL 22.1

Phage-display Peptide Libraries

Our studies are based on methods for phage display, which were first developed in the mid 1980s (Smith 1985). In phage display-based strategies, libraries of random peptides are expressed on the surface of the phage. In one common approach, the peptide is fused to the phage capsid protein pIII by insertion of its encoding DNA sequence into *gene III* of the phage genome. A similar approach using *gene VIII* has also been used. The pIII libraries display 3–5 copies of each individual peptide, whereas pVIII libraries display approximately 3000 copies (Smith and Scott 1993).

Our in vivo selection experiments have exclusively employed libraries of random peptides displayed in the context of the *gene III* fusion product. In these studies we have primarily utilized the fUSE5 vector (Smith and Scott 1993) to generate our peptide libraries (Koivunen et al. 1994; 1999a,b). The number and diversity of individual clones present in a given library is a critical factor for the success of the in vivo selection. Therefore, we routinely use primary, unamplified libraries, which are less likely to have an overrepresentation of defective phage clones. Moreover, the preparation of a library must be optimized to create the highest possible diversity—between 10^8 and 10^9 transducing units (TU)/ml. To establish that the transformation background representing self-ligated defective clones is minimal, small-scale ligations should be performed using prepared vector plus randomized oligonucleotide inserts in comparison with vector alone (Koivunen et al. 1999a).

In practice, there is no difference between primary libraries used for in vitro and in vivo applications. The basic protocols for construction of peptide libraries used for phage-display applications are described in this volume (see Chapter 16) and elsewhere (Smith and Scott 1993; Cortese et al. 1996; Koivunen et al. 1999a).

PROTOCOL 22.2

Animal Models

In vivo phage-display studies have been developed primarily in mice (Harlan Sprague Dawley, Charles River, or equivalent). We have used 2-month-old Balb/c or C57BL/6J strains for targeting the vasculature of normal tissues (Pasqualini and Ruoslahti 1996; Rajotte et al. 1998; and R. Pasqualini et al., unpubl.) and immunodeficient nude mice (athymic Balb/c Nu/Nu) for targeting the vasculature of experimental human tumors (Arap et al. 1998a). Thus far, we have found that organ-specific phage will display similar homing patterns regardless of the mouse strain (Rajotte et al. 1998). The mouse offers an excellent model to perform vascular targeting with phage libraries due to its relatively low cost, small size, and the availability of various disease models. However, the technology can be adapted easily to other suitable animal models depending on the tissue or pathology studied. The limiting factor is usually the amount of primary library needed for the selection, a concern that can at times be addressed experimentally. For example, to minimize library input when targeting retinal blood vessels, we administered phage to the carotid artery of 2-month old Simonson Albino rats (Rajotte et al. 1998).

All of our studies in mice employed intraperitoneal administration of tribromoethanol (prepared in-house, 0.017 ml/g), which is nontoxic and induces long-lasting (10–20 minutes) anesthesia (Papaioannou and Fox 1993). For rats, sodium-phenobarbital (50 mg/kg) was administered intraperitoneally after a pre-sedation with isofluorene gas.

PROTOCOL 22.3

Titration of Phage Preparations

It is critical that the titers of the phage sample—whether of a phage library or of an individual phage preparation—be determined accurately before intravenous injection. Discrepancy in the phage input due to inaccurate titration will introduce a major error into the calculation of homing efficiency. The samples and controls used in a given experiment should always be titered simultaneously, and the same titer of phage should be used for each injection.

Materials

host bacteria (e.g., ***E. coli*** K91kan)
phage library (10^9–10^{10} TU/mouse)
tetracycline (tet) stock, 5 mg/ml in **ethanol** (see Appendix 2)
kanamycin stock, 10 mg/ml in water (see Appendix 2)
LB liquid medium (see Appendix 2)
LB + 20 to 40 μg/ml tet + 100 μg/ml kanamycin, in 10-cm plates

CAUTIONS: ***E. coli*, tetracycline, ethanol *(see Appendix 4)***

Procedure

1. Infect host bacteria with serial dilutions of the original phage stock. Include a wide range of dilutions (between $1{:}10^2$ and $1{:}10^7$) to ensure that the amount of bacteria is not limiting during infection.
2. Incubate the diluted phage preparations with 1 ml of host bacteria for 30 minutes at room temperature.
3. Transfer to 50-ml FALCON tubes containing 10 ml of room-temperature LB medium supplemented with 0.2 μg/ml of tetracycline and 100 μg/ml of kanamycin. Incubate at room temperature for 30 minutes.

   Note: It has been suggested that this low concentration of tetracycline induces the promoter of the tet^R gene of fUSE5.
4. Plate 100-μl triplicates of each serial dilution on 10-cm LB plates containing 20–40 μg/ml of tetracycline and 100 μg/ml of kanamycin. Incubate the plates overnight at 37°C. Plates containing 200–500 colonies are suitable for counting and yield reproducible results.

PROTOCOL 22.4

The First Round of Selection

Several decisions need to be made before undertaking in vivo targeting. One choice to be made is whether perfusion, freezing in liquid nitrogen, or both will be performed (see Step 2). Freezing the whole animal or organs in liquid nitrogen may prevent internalization of the phage. We have used this procedure previously (Pasqualini and Ruoslahti 1996), but it does not appear to be essential (Rajotte et al. 1998). Whole-body perfusion can be used to reduce phage background in highly vascularized tissues (Rajotte et al. 1998). It is also important to decide whether single colonies of the selected phage will be grown individually for successive rounds of selection, or whether bulk recovery of phage will be used (see below, Protocol 22.6). A basic protocol with some variations is described below:

Materials

mice or another suitable animal model
phage library (10^9–10^{10} TU/mouse)
DMEM (Dulbecco's modified Eagle's medium)
1/2 cc U-100 insulin syringe with 28G1/2 needle (Becton Dickinson, Cat. # BD329461)
tribromoethanol (made in-house, according to Pappaioannou and Fox, 1993)
Vacutainer blood collection set with 25G3/4 needle, if perfusing (Becton Dickinson, Cat. # BD366255)
liquid nitrogen and **ethanol,** if snap-freezing
DMEM supplemented with 1% BSA, 1 mM **PMSF**, 1 μg/ml **leupeptin,** and 20 μg/ml **aprotinin**
Dounce tissue homogenizer
host bacteria (e.g., *E. coli* K91kan)
tetracycline (tet) stock, 5 mg/ml in **ethanol** (see Appendix 2)
kanamycin stock, 10 mg/ml in water (see Appendix 2)
LB (see Appendix 2)
LB + 20 to 40 μg/ml tet + 100 μg/ml kanamycin plates

CAUTIONS: **animal treatment, tribromoethanol, liquid nitrogen, ethanol, PMSF, leupeptin, aprotinin, *E. coli*, tetracycline *(see Appendix 4)***

Procedure

Phage administration

1. Inject anesthetized mice intravenously (right or left lateral tail vein) with approximately 10^{10} TU of the phage peptide library in 200 μl of DMEM. Injection vol-

umes larger than 500 μl should be avoided, because the large volume may have effects on blood flow, disrupt blood vessels, and affect phage distribution. Keep a small aliquot of the actual input phage sample to confirm its titer.

2. Two to five minutes after administration of the phage, perfuse the animal through the heart with 5–10 ml of DMEM (to reduce phage background, Steps 3–6) or snap-freeze in liquid nitrogen (to prevent phage internalization, Steps 7,8) or both, depending on the tissue being screened.

If performing whole-body perfusion

3. Before perfusing, it is important to ensure that the mice remain under deep anesthesia (another half-dose of tribromoethanol can be injected at this point).

4. Incise the skin at the level of the diaphragm to expose the entire thoracic and peritoneal walls. Incise the sternum, taking care to avoid damage to the heart or large vessels.

5. Expose the heart, and insert the Vacutainer catheter within the left ventricle chamber. The catheter should be attached to a 5- or 10-ml syringe filled with DMEM that has been prewarmed to 37°C.

   Note: We have used DMEM for perfusion because it appears to offer less mechanical resistance than PBS or 10% mannitol (R. Pasqualini and E. Ruoslahti, unpubl.).

6. After arterial blood starts flowing in the catheter (due to cardiac contractions), make a small incision in the right atrial chamber to provide an outlet for the blood during perfusion. Up to 50 ml of medium can be perfused this way. Larger volumes are recommended for tumor targeting, particularly when testing individual phage, but are not required during the selection rounds.

If snap-freezing with liquid nitrogen

7. Dip the deeply anesthetized mice in ethanol (first 70%, then 100%) to avoid bacterial contamination from fur before snap-freezing and processing of tissues.

8. Snap-freeze in liquid nitrogen for 5 minutes. When freezing multiple animals at the same time, use an individual container for each mouse, because the carcasses are often damaged during freezing.

Phage recovery

9. Recover the bound phage by removing the selected organs, weighing and grinding with a glass Dounce homogenizer in 1 ml of DMEM-PI (DMEM plus protease inhibitors: 1 mM PMSF, 20 μg/ml aprotinin, and 1 μg/ml leupeptin). If the animals were snap-frozen, partially thaw the carcasses at room temperature before removing the organs. For certain samples (e.g., skeletal muscle, skin, or tumor xenografts), disrupt the tissues with a razor blade or a mechanical grinder prior to Dounce homogenization.

10. Wash the tissue samples three times with 1 ml of ice-cold washing medium (DMEM-PI containing 1% BSA), according to the following: Mix the tissue with washing medium and vortex for a few seconds. Centrifuge the samples in a microfuge for 4–5 minutes at 3000–4000 rpm, and carefully discard the supernatant without disturbing the tissue pellet. Vortex the samples again for 10 seconds to ensure that the tissue is well homogenized before adding more washing medium. When processing multiple samples, keep the tissues on ice.

11. Combine the tissues with 1 ml of starved *E. coli* K91kan (OD_{600}=1–2). The bacteria prepared for this step should be used immediately for infection and not kept at room temperature after they have reached the proper density, because infectability by the phage may be reduced or lost. Mix the bacteria gently with the tissue homogenate and allow the tubes to stand for at least 15 minutes (but no more than 1 hour) at room temperature. Warm LB medium containing kanamycin (100 µg/ml) to 37°C.

12. Add 10 ml of the prewarmed LB medium plus kanamycin, and add tetracycline to 0.2 µg/ml (it has been suggested that this low concentration of tetracycline induces the promoter of the tet^R gene of fUSE5). Incubate at room temperature for 30 minutes.

13. Spread aliquots of the bacterial culture onto LB agar plates containing 20–40 µg/ml tetracycline and 100 µg/ml kanamycin to recover phage clones.

14. Incubate the plates overnight at 37°C.

Note: Triplicate platings are recommended in Step 13. Plating 2, 20, and 200 µl will usually yield plates with a suitable number of colonies to evaluate the total amount of phage recovered from the tissue and pick individual clones for the successive rounds of selection. The number of phage recovered from different organs varies considerably. However, the numbers observed in one particular type of experiment are often remarkably consistent in subsequent experiments, provided that the input phage and circulation time are kept constant.

PROTOCOL 22.5

Phage Amplification and Purification

The amplification and purification of phage are performed as described in this volume (see Chapter 17) and elsewhere (Koivunen et al. 1999b). There is essentially no difference between in vitro and in vivo procedures, although shortening the duration of steps such as PEG precipitation and the growth of the bacterial culture in the shaker appears to decrease the incidence of insertless phage (E. Koivunen and E. Ruoslahti, unpubl.).

PROTOCOL 22.6

Additional Rounds of In Vivo Selection

In general, three rounds of in vivo selection are sufficient to select for organ- and tumor-specific homing phage. In our experience, there is rarely any increase in the specificity or affinity of the motifs selected after the third round.

Materials

plated phage library from first round of selection
LB or NZY + 20 to 40 μg/ml **tetracycline** (see Appendix 2)
phage purification materials (see Protocol 22.5)
phage administration materials (see Protocol 22.4)

CAUTION: tetracycline *(see Appendix 4)*

Procedure

1. For successive rounds of selection, obtain 250–500 individual colonies from the first round and grow separately in 5 ml of LB or NZY medium containing 20–40 μg/ml tetracycline for no longer than 12–14 hours in a shaker at 37°C. Longer incubation times appear to increase the possibility of recovering insertless phage or phage with unreadable sequences.
2. Pool the bacterial cultures.
3. Purify the phage as described above (see Protocol 22.5).
4. Reinject 10^9–10^{10} TU from this phage pool into mice.

Note: An alternative to growing 250–500 colonies individually for successive rounds of selection is to directly grow the bulk phage recovered from a given organ or tissue for a few hours with bacteria. We have not yet performed this procedure successfully, possibly because incubating phage together with tissue debris for a long period is harmful. Such a procedure appears to favor degradation by tissue proteases. Another alternative is to plate all of the phage that is recovered. To do so, divide the 10 ml of LB/tissue into multiple plating aliquots of 250 μl and plate on 150-mm LB agar plates containing tetracycline. The following day, recover colonies from the plates with a scraper, pool in 200 ml of LB+tetracycline (20–40 μg/ml), and incubate in the shaker for 4 hours at 37°C. Purify the phage and use for another round of selection.

PROTOCOL 22.7

Sequencing of the DNA Phage Insert

Materials

StrataClean resin (Stratagene, Cat. # 400714)
Oligonucleotide primer. For fUSE5 vector: 5′-CCCTCATAGTTAGCGTAACG-3′
DNA sequencing kit (e.g., Dye Terminator Cycle Sequencing Ready Reaction Kit, Perkin Elmer, Cat. # 402079)

Procedure

Inserts can be sequenced once the selection for a given organ or tissue shows at least a 3-fold enrichment for the phage pool relative to the first round. This usually occurs after the second or third round of selection. We generally sequence about 50 clones after the second and third rounds of selection. To sequence the fUSE5 ssDNA directly, prepare phage ssDNA from individual clones using the StrataClean resin, and use the primer 5′-CCCTCATAGTTAGCGTAACG-3′ (antisense) to sequence by manual or automatic methods. We have noticed a difficulty in sequencing phage ssDNA when we used the commercial kit for M13 phage DNA purification from QIAGEN. It may be that contaminating genomic material containing phage-related DNA sequences coprecipitates with the phage ssDNA and produces sequencing artifacts. Commercially available computer programs (DNA Strider [Christian Mark, Service de Biochimie, Centre d'Etudes Nucleaires de Saclay, Gif-Sur-Ivette, 91191 Cedex, France], MacVector [Oxford Molecular], or equivalent) can be helpful in analysis of the sequences.

PROTOCOL 22.8:

Evaluation of Homing: Testing a Phage Clone for Specific Tissue Targeting

A successful round of selection yields enriched peptide or peptide motifs from the phage pool. A peptide motif is defined as a sequence that appears several times within different structural contexts among the phage sequenced, and as noted above, a tripeptide motif is often detected.

Phage that display motifs and/or peptides appearing repeatedly in successive rounds should be retained for further analysis. When tested individually, phage clones within a pool will often have a better homing ability than that observed for the pool. Evaluate the selected phage for specificity by comparing their homing individually to negative controls such as an unselected library, fd-tet phage, or phage with a different selectable marker. Calculate the selectivity of a given phage from two parameters. First, compare the overall number of the selected phage recovered from the target organ to the number of selected phage recovered from control organs (normalized by mass). Unless brain and kidney homing is being evaluated, those two organs are suitable controls. Second, compare these numbers to the numbers of unselected individual phage or phage mixtures recovered from both the target organ and the control organs.

Use of Phage with a Different Selectable Marker as an Internal Control

fdAMPLAY88 phage (R. Perham, University of Cambridge, England; Malik and Perham 1997) contains an ampicillin resistance marker instead of tetracycline resistance. Coinjection of these phage with the homing phage allows one to assess the targeting specificity of a given phage in the same animal rather than in a separate control. The organ distribution of fd-tet (Zacher et al. 1980) and fdAMPLAY88 phage after intravenous administration is similar (J. A. Hoffman et al., unpubl.), and as a result, the homing ratio of fd-tet phage to fdAMPLAY88 phage is close to 1. Therefore, an animal can be co-injected with an equal amount of the fd-tet-derived phage (either a library or a selected individual phage clone) and of fdAMPLAY88 phage. Homing to a specific tissue is then evaluated by comparing the number of TU rescued from the organ or tumor xenograft on both tetracycline plates (the selected phage of interest) and ampicillin plates (fdAMPLAY88 phage). Control organs are still necessary to eliminate the possibility that the selected phage homes to the vascular beds of other tissues, even when the above protocol is used.

fdAMPLAY88 co-injection is particularly useful in targeting tissues and organs with a very small mass. For example, we did not use comparisons to control organs to assess homing of individual phage to the retina because of the large error in determining retinal tissue mass. Instead, the rats were injected with an equal amount of both selected and fdAMPLAY88 phage, and both phage were individually counted from retinal tissue (Rajotte et al. 1998).

PROTOCOL 22.9

Reproduction of the Phage-displayed Peptide as Recombinant Fusion Proteins or Synthetic Peptides

Phage Peptides as Recombinant GST-fusion Proteins

The coding sequence of a given phage-displayed peptide can be fused easily in-frame to the carboxy-terminal portion of the glutathione-S-transferase (GST) gene. We rely on the pGEX series of GST fusion cloning vectors (Amersham Pharmacia), but other systems such as the His-Tag (QIAGEN), pRSET (Invitrogen), or perhaps even baculoviral vectors (Pharmingen) may also be used. The following steps summarize the cloning of a peptide insert from a fUSE5-based library into the pGEX2TK vector. This vector has a built-in kinase site that allows one to prepare a radiolabeled insert, when desired.

The major advantages of the recombinant system are convenience, speed, and low cost. However, the molar concentration of the peptide to be used for testing is limited by the large size of the GST fusion partners, and S–S bridges of cyclic peptides may not form properly. On occasion, we have seen examples of recombinant fusion peptides that do not inhibit the homing of the phage in vivo in a competition assay (Rajotte et al. 1998), suggesting that certain peptides, when fused to GST, may not bind the target with as high an affinity as the phage-displayed peptide. On the other hand, in one instance we have generated a recombinant GST-peptide that inhibited phage homing in vivo more efficiently than the corresponding peptide (D. Rajotte and E. Ruoslahti, unpubl.). It is possible that in this case the GST protein scaffold provided a structure required for that particular peptide-receptor interaction.

Materials

selected phage with fUSE5-derived inserts
primers 5′-AGGCTCGAGGATCCTCGGCCGACGGGGCT-3′ (sense) and
5′-AGGTCTAGAATTCGCCCCAGCGGCCCC-3′ (antisense)
3 M sodium acetate, pH 5.2
ethanol
glycogen, 20 mg/ml stock solution
*Bam*HI
*Eco*RI
pGEX2TK vector (Amersham Pharmacia Biotech; Cat. # 27-4587-01)
standard protein expression materials
standard SDS-PAGE materials

CAUTION: **ethanol** ***(see Appendix 4)***

Procedure

1. Purify the template from phage with StrataClean resin or equivalent, according to manufacturer's instructions.
2. PCR-amplify the peptide-coding DNA insert from phage ssDNA using the primers 5′-AGGCTCGAGGATCCTCGGCCGACGGGGCT-3′ (sense) and 5′-AGGTCTAGAATTCGCCCCAGCGGCCCC-3′ (antisense). Use standard PCR conditions (1 minute at 95°C for denaturing, 1 minute at 53°C for annealing, 1 minute at 72°C for extension; for a total of 35 cycles).
3. Ethanol-precipitate the PCR products in the presence of glycogen (see Appendix 3).
4. Digest the DNA fragment with the appropriate restriction enzymes (*Bam*HI and *Eco*RI for fUSE5-derived inserts).
5. Ligate the digested fragment into the same sites of the pGEX2TK vector, creating an in-frame insertion.
6. Use the ligated vector to produce large-scale preparations of GST-fusion proteins according to standard protein expression protocols.
7. Examine the molecular weight and purity of the GST-fusion proteins by SDS-PAGE and Coomassie Blue staining.

Synthetic Peptides

An appropriate standard to confirm that the displayed peptide mediates in vivo targeting of a specific phage is to generate a synthetic cognate peptide and test its ability to inhibit phage homing. However, competition of phage homing with the cognate peptide in vivo requires a large molar excess of peptide compared to phage (Pasqualini and Ruoslahti 1996; Arap et al. 1998a; D. Rajotte and E. Ruoslahti, unpubl.). The homing peptide can also be labeled and coupled to cells (Pasqualini and Ruoslahti 1996), or linked to a cytotoxic agent (Arap et al. 1998a) and tested directly in vivo. Tumor vasculature-homing peptides coupled to chemotherapeutic agents can be highly potent anticancer agents in experimental metastasis models (Arap et al. 1998a), and we have also demonstrated that a synthetic brain-homing peptide directs selective localization of red blood cells to the brain (Pasqualini and Ruoslahti 1996).

To perform the competition assay, dissolve the peptide from a stock solution to the appropriate concentration (0.1–1 mM) in DMEM. If the peptide is not water soluble, prepare a peptide stock (500–1,000 mM) in DMSO and dilute from this stock into the phage sample. If DMSO is used, you must include a DMSO control at the same final concentration as in the peptide sample. When diluting to a working concentration, it is important to check the pH of the peptide solution and adjust it to a physiological range (7.0–8.0). Co-inject a fixed input of phage (10^8–10^9 TU) with increasing amounts of the synthetic peptides. 100 μg and 1 mg per mouse are suitable initial amounts to try. Pilot and control experiments should be done to ensure that the peptides used do not affect the ability of the phage to infect bacteria.

PROTOCOL 22.10

Analysis of Phage Distribution In Vivo by Immunohistochemistry

Phage proteins can be detected in tissues by immunostaining (Pasqualini and Ruoslahti 1996; Pasqualini et al. 1997; Arap et al. 1998a; Rajotte et al. 1998). This technique can confirm the specificity of the targeting and also reveals localization of the phage in the target tissue. Immunostaining is generally performed as described below.

Examination of the tissue localization of phage staining at different time points following injection can be informative. Liver and spleen capture phage nonspecifically from the circulation (Geter et al. 1973; Pasqualini and Ruoslahti 1996; Pasqualini et al. 1997; Arap et al. 1998a; Rajotte et al. 1998) and exhibit phage immunostaining that increases with time (Pasqualini and Ruoslahti 1996; Pasqualini et al. 1997; Arap et al. 1998a; Rajotte et al. 1998). Some tumor-homing phage accumulate in the tumor whereas others appear to decrease with time (Arap et al. 1998a). We believe that persistence of immunostaining is correlated with the ability of the receptor to which the phage bind to translocate bound phage into cells or across the endothelium. We have shown that an integrin-binding phage accumulates in tumors, and others have found that integrin-bound phage is internalized by cells (Hart et al. 1994). Thus, it appears that if the receptor mediates internalization of the phage particle, staining is likely to persist for long periods of time. The increased permeability of tumor blood vessels (Dvorak et al. 1991; Shockley et al. 1991) may account for the spreading of phage proteins into the parenchyma of tumors.

Materials

mice or another suitable animal model
phage library (10^9–10^{10} TU/mouse)
DMEM and perfusion apparatus (see Protocol 22.4)
Bouin's fixing solution (contains **picric acid, formalin,** and **acetic acid;** Sigma, Cat. # HT10-1-32)
all the common reagents required for paraffin embedding and tissue sectioning
PBS
xylene
ethanol
ddH_2O
quenching solution (20 mM Tris/HCl, 300 mM NaCl, 1% Tween 20, 1% chicken egg albumin)
horse anti-mouse (Vector Laboratories, Elite ABC; anti-mouse, Cat. # PK6102) or goat anti-rabbit (Vector Laboratories, Elite ABC; anti-rabbit, Cat. # PK6101) serum, for quenching
antibody against M13 (Amersham Pharmacia, Cat. # 27-9410-01) or fd phage antiserum (Sigma, Cat. # B7786)

VECTASTAIN Elite ABC kit (Vector Laboratories, Cat. # PK-6200)
Metal Enhanced **DAB** Substrate kit (Pierce, Cat. # 34065)
Mayer's hematoxylin
***n*-butyl acetate** (FLUKA, Cat. # 45870)

CAUTIONS: animal treatment, picric acid, formalin, acetic acid, microtome blades, xylene, ethanol, DAB, *n*-butyl acetate *(see Appendix 4)*

Procedure

1. Inject anesthetized mice with 10^9 phage TU. Allow the phage to circulate for the desired time period. In short-term experiments (2–5 minutes), perfuse the animal through the heart with 5–10 ml of DMEM as described above (Protocol 22.4, Steps 3–6).
2. Prepare standard paraffin sections for histology by surgically removing the organs or tumor xenografts and fixing in Bouin's solution overnight at room temperature. Transfer to PBS, and follow standard procedures for paraffin embedding and tissue sectioning.
3. Deparaffinize slides by soaking in xylene for 10 minutes with gentle agitation. Wash in decreasing percentages of ethanol (100% ethanol, three times; followed by single washes of 96%, 70%, 50% ethanol, and ddH_2O). All washes are 10 minutes each, with gentle agitation.
4. Microwave the slides for 10 minutes at 600 Watts in 100 mM citrate buffer, pH 6. Use enough buffer to cover the slide or tissue section (~ 500 μl to 1 ml). Wash for 20 minutes in ddH_2O.
5. Incubate for 20 minutes in quenching solution plus three drops of horse serum (use anti-mouse for mouse tissue sections), or three drops of goat serum (use anti-rabbit for rabbit tissue sections).

   Note: The use of a quenching solution is critical to decrease the high phage background immunostaining.
6. Incubate the slide for 1 hour in PBS plus the primary antibody. Use either a monoclonal antibody against M13 phage diluted 1:20 or a rabbit polyclonal anti-fd bacteriophage diluted 1:200 in PBS/0.1% BSA. Wash three times with PBS for 5 minutes per wash.
7. Incubate the slide for 30 minutes in PBS/0.1% BSA plus the biotinylated secondary antibody from the VECTASTAIN Elite ABC kit. Wash three times with PBS for 5 minutes per wash with gentle agitation.
8. Apply the VECTASTAIN Elite ABC kit detection complex following the manufacturer's protocol. Wash three times with PBS for 5 minutes per wash, with gentle agitation.

9. React with DAB (diaminobenzidine tetrahydrochloride) for 3 minutes using the Metal Enhanced DAB Substrate kit. Wash in ddH_2O for 5 minutes with gentle agitation.
10. Soak for 2 minutes in Mayer's hematoxylin (available commercially). Hold under running tap water for 5 minutes.
11. Wash in increasing percentages of ethanol, including single washes of 50%, 70%, and 96% ethanol, followed by three washes of 100% ethanol and one wash with *n*-butyl acetate. All washes are 10 minutes each.
12. Mount.

Other Considerations

Issues in Targeting Specific Organs

For in vivo targeting of the skin, use Balb/c nude mice to avoid bacterial contamination of the harvested tissue from fur. Inject the mice intravenously with phage and perfuse as described above (Protocol 22.4). Remove the skin in large sections and place on an ice-cold plate with the hypodermis facing up. Dissect with a surgical scalpel, yielding primarily hypodermis, and leaving behind epidermis and part of the dermis. Avoid any contact of the skin with perfused blood.

When targeting the kidney blood vessels, do not use whole-body perfusion because the increased hydrostatic pressure appears to damage the kidney blood vessels, producing artificially increased phage counts in the kidney. On the other hand, post-injection perfusion with buffer is particularly useful in the targeting of organs that have a high phage background due to nonspecific trapping, as we have found to be the case in the lungs (Pasqualini and Ruoslahti 1996; Rajotte et al. 1998). In addition, perfusion may increase the stringency of the screen, selecting for phage with higher affinities for the target and washing away background levels of phage.

Targeting of Experimental Tumors/ Establishing Tumor Xenografts

We have used malignant melanoma (C8161), Kaposi's sarcoma (KS1767), and breast carcinoma (MDA-MB-435) in our in vivo screenings (Arap et al. 1998a), although in theory, any tumor type can be used to establish tumor xenografts.

Materials

tumor cells
standard tissue culture equipment
DMEM + 10% fetal calf serum
PBS + 2.5 mM EDTA (see Appendix 2)
mice or another suitable animal model

CAUTION: animal treatment *(see Appendix 4)*

Procedure

1. Culture the tumor cells in Dulbecco's modified Eagle's medium (DMEM) with 10% fetal calf serum.
2. Change the culture medium 12 hours before the cells are injected into mice.
3. At 80% confluence, detach the cells from monolayers with PBS containing 2.5 mM EDTA.
4. Wash three times with serum-free DMEM, count, and resuspend in serum-free DMEM.
5. Inject tumor cells (10^6 cells/site) subcutaneously into the mammary fat pad, intraperitoneally or intravenously in nude mice.
6. Monitor tumor growth daily. Thirty to sixty days after injection, select animals bearing tumors of similar sizes and use for targeting experiments and histological analysis. Mammary fat pad tumors reach a volume of about 1 cm^3 during this period. We have found that this size is an optimal window for recovery of tumor-homing peptides. The MDA-MB-435, C8161, and KS1767 Kaposi's sarcoma induced tumors are amply vascularized when prepared in this manner (unpublished results).

    Note: If necrosis is observed externally or internally upon dissection, such tumors should not be used for testing of phage homing or immunostaining. Avoid using fractions of larger tumors because of interexperiment variability, except if the targeting ability is being evaluated in the same animal by co-injection of a selected targeting phage with tetracycline resistance and a control phage with ampicillin resistance (see Protocol 22.8).

Blocking Nonspecific Phage Interactions

We have developed an in vivo selection approach that includes a noninfective phage as a competitor to prevent phage trapping in organs containing reticulo-endothelial tissue, such as liver and spleen. Fd-tet phage (Zacher et al. 1980) that has been heat-inactivated for 6 hours at 85°C can be used for this purpose. Thus, selection for peptides that home specifically to these tissues may also be possible (R. Pasqualini et al., unpubl.). Alternatively, fdAMPLAY88 phage (see Protocol 22.8) also can be used to block nonspecific accumulation of phage (R. Pasqualini et al., unpubl.). In addition to providing a suitable internal control, fdAMPLAY88 decreases nonspecific phage binding to tissues. The fdAMPLAY88 phage is easier to produce in large quantities than noninfective phage because the yields are higher and heat inactivation is not required (W. Arap et al., unpubl.).

ACKNOWLEDGMENTS

The authors thank Drs. Renate Kain for the phage immunohistochemistry protocol, Bradley Restel for technical assistance, and Erkki Koivunen and Jason A. Hoffman for sharing unpublished data. This work is supported by the National Institutes of Health grants CA-28896, CA-74238 to E.R., and the Cancer Center Support Grant CA-30199. W.A. is the recipient of a CapCURE Award. D.R. is a Research Fellow of the National Cancer Institute of Canada supported with funds provided by the Terry Fox Run.

REFERENCES

Arap W., Pasqualini R., and Ruoslahti E. 1998a. Cancer treatment by targeted drug delivery to tumor vasculature in a mouse model. *Science* **279:** 377–380.

———. 1998b. Chemotherapy targeted to tumor vasculature. *Curr. Opin. Oncol.* **10:** 560–565.

Brooks P.C., Clark R.A., and Cheresh D.A. 1994. Requirement of vascular integrin avb3 for angiogenesis. *Science* **264:** 569–571.

Cortese R., Monaci P., Luzzago A., Santini C., Bartoli F., Cortese I., Fortugno P., Galfre G., Nicosia A., and Felici F. 1996. Selection of biologically active peptides by phage display of random peptide libraries. *Curr. Opin. Biotechnol.* **7:** 616–621.

Dvorak H.F., Nagy J.A., and Dvorak A.M. 1991. Structure of solid tumors and their vasculature: Implications for therapy with monoclonal antibodies. *Cancer Cells* **3:** 77–85.

Ellerby H.M., Arap W., Ellerby L.M., Kain R., Andrusiak R., Del Rio G., et al. 1999. Anti-cancer activity of targeted pro-apoptotic peptides. *Nat. Med.* **5:** 1032–1038.

Folkman J. 1995. Angiogenesis in cancer, vascular, rheumatoid and other disease. *Nat. Med.* **1:** 27–31.

Geter M.R., Trigg M.E., and Merril C.R., 1973. Fate of bacteriophage lambda in non-immune germ-free mice. *Nature* **246:** 221–223.

Hammes H.P., Brownlee M., Joonczyk A., Sutter A., and Preissner K.T. 1996. Subcutaneous injection of a cyclic peptide antagonist of vitronectin receptor-type integrins inhibits retinal neovascularization. *Nat. Med.* **2:** 529–533.

Hart S.L., Knight A.M., Harbottle R.P., Mistry A., Hunger H.D., Cutler D.F., Williamson R., and Coutelle C. 1994. Cell binding and internalization by filamentous phage displaying a cyclic Arg-Gly-Asp-containing peptide. *J. Biol. Chem.* **269:** 12468–12474.

Koivunen E., Gay D.A., and Ruoslahti E. 1993. Selection of peptides binding to the α5b1 integrin from phage display library. *J. Biol. Chem.* **268:** 20205–20210.

Koivunen E., Wang B., and Ruoslahti E. 1995. Phage libraries displaying cyclic peptides with different ring sizes: Ligand specificities of the RGD-directed integrins. *Bio/Technology* **13:** 265–270.

Koivunen E., Wang B., Dickinson C.D., and Ruoslahti E. 1994. Peptides in cell adhesion research. *Methods Enzymol.* **245:** 346–369.

Koivunen E., Arap W., Rajotte D., Lahdenranta J., and Pasqualini R. 1999a. Identification of receptor ligands by using phage display peptide libraries. *J. Nucl. Med.* **40:** 883–888.

Koivunen E., Restel B.H., Rajotte D., Lahdenranta J., Hagedorn M., Arap W., and Pasqualini R. 1999b. Integrin-binding peptides derived from phage display libraries. *Methods Mol. Biol.* **129:** 3–17.

Koivunen E., Arap W., Valtanen H., Rainisalo A., Medina O.P., Heikkila P., et al. 1999c. Tumor targeting with a selective gelatinase inhibitor. *Nat. Biotechnol.* **8:** 768–774.

Malik P. and Perham R.N. 1997. Simultaneous display of different peptides on the surface of filamentous bacteriophage. *Nucleic Acids Res.* **25:** 915–916.

Papaioannou V.E., and Fox J.G. 1993. Efficacy of tribromoethanol anesthesia in mice. *Lab. Anim.* **43:** 189–192.

Pasqualini R. and Ruoslahti E. 1996. Organ targeting in vivo using phage display peptide libraries. *Nature* **380:** 364–366.

Pasqualini R., Koivunen E., and Ruoslahti E. 1997. Alpha V integrins as receptors for tumor targeting by circulating ligands. *Nat. Biotechnol.* **15:** 542–546.

Rajotte D., Arap W., Hagedorn M., Koivunen E., Pasqualini R., and Ruoslahti E. 1998. Molecular heterogeneity of the vascular endothelium revealed by in vivo phage display. *J. Clin. Invest.* **102:** 430–437.

Ruoslahti E. 1996. RGD and other recognition sequences for integrins. *Annu. Rev. Cell Dev. Biol.* **12:** 697–715.

Sambrook J., Fritsch E.F., and Maniatis T. 1989. *Molecular cloning: A laboratory manual,* 2nd edition. Cold Spring Harbor Laboratory Press, Cold Spring Harbor, New York.

Shockley T.R., Lin K., Nagy J.A., Tompkins R.G., Dvorak H.F., and Yarmush M.L. 1991. Penetration of tumor tissue by antibodies and other immunoproteins. *Ann. N.Y. Acad. Sci.* **618:** 367–382.

Smith G.P. 1985. Filamentous fusion phage: Novel expression vectors that display cloned antigens on the virion surface. *Science* **228:** 1315–1317.

Smith G.P. and Scott J.K. 1993. Libraries of peptides and proteins displayed in filamentous phage. *Methods Enzymol.* **217:** 228–257.

Zacher A.N., Stock C.A., Golden J.W., and Smith G.P. 1980. A new filamentous phage cloning vector: fd-tet. *Gene* **9:** 127–140.

23 Cell-surface Selection and Analysis of Monoclonal Antibodies from Phage Libraries

DON L. SIEGEL

University of Pennsylvania School of Medicine, Department of Pathology & Laboratory Medicine, University of Pennsylvania Medical Center, Philadelphia, Pennsylvania 19104

AFTER THE CONSTRUCTION OF PEPTIDE or protein display libraries, phage particles bearing the desired ligands are captured with antigen (or receptor), propagated in bacteria, and further enriched through successive rounds of selection referred to as "panning." In most cases, panning methods use purified antigen immobilized to a solid substrate. Sometimes, however, purification of the target antigen is not possible because the target is unknown, as with putative stem-cell markers or tumor-specific antigens. In other cases, purification procedures affect an antigen's binding properties, as with the purification of some integral membrane proteins. In these cases, it is useful to select ligand-bearing phage particles based on their ability to bind to antigen on intact cell surfaces.

CONTENTS

In all types of panning, phage tend to bind to targets nonspecifically. In other words, both specific and nonspecific phage will bind to the antigen of interest, especially in early rounds of selection. Panning with intact cell surfaces presents an additional challenge, because cell surfaces contain many nonspecific antigens in the form of proteins, carbohydrates, and lipids (Siegel and Silberstein 1994; Siegel 1995). To enrich for specific target cell binders, a negative selection strategy can be used in which the phage libraries are incubated with cells that lack the antigen of interest (antigen-negative cells) before incubation and selection with cells that express the target antigen (antigen-positive cells) (Marks et al. 1993; Portolano et al. 1993; Cai and Garen 1995). This strategy can be relatively inefficient, however, because each negative selection cycle removes only a small percentage of nonspecific phage. In addition, even negative selections can remove some specific phage, and multiple cycles of negative selection can quickly deplete a library containing only low levels of the desired phage. These problems might explain why many cell-surface panning methods have not produced as wide a diversity of positive clones as methods that use purified antigens.

Alternative methods for cell-surface panning have been described that involve simultaneous positive and negative selection. In these methods, phage are incubated with a mixture of antigen-positive "target" cells and, in excess, antigen-negative "absorber" cells. The absorber cells act as a sink for nonspecific phage, and the target cells capture the specific phage. Mixing the two populations of cells requires, of course, that they again be separated after incubation. One method of separation involves using flow cytometry and fluorescently labeled antibodies that recognize an unrelated antigen present only on the target cells (de Kruif et al. 1995). This method requires target-cell-specific antibodies and may be problematic for targets such as tumor cells. In addition, flow cytometry may be impractical for sorting large numbers of cells in a reasonable amount of time. There is also the possibility that shear forces exerted in the flow cytometry process will strip some phage from their target cells, leaving behind only the highest-affinity binders and reducing the diversity of binders recovered.

In this chapter, protocols that employ another method of competitive cell-surface panning are presented (Siegel et al. 1997). In this approach (Fig. 23.1), target cells are "labeled" with magnetic beads and then mixed with an excess of unlabeled absorber cells. The mixture of cells is incubated with the phage-display library and loaded onto a magnetic column. The column is washed to isolate the magnetically labeled target cells and remove nonspecific phage. Finally, antigen-specific phage-displayed antibodies are eluted from the retrieved target cells and propagated in bacterial culture. This approach has proven useful for the isolation and analysis of both human auto- and alloantibodies (Chang and Siegel 1997), particularly for the selection of large arrays of anti-red blood cell (RBC) Rhesus (Rh) antibodies from immune libraries (Siegel and Chang 1997; Chang and Siegel 1998; Siegel 1998). Relevant sample data obtained from these studies are provided for reference, and methods for adapting this cell-surface panning approach for use with other cell types are described below.

THEORETICAL AND PRACTICAL ISSUES IN THE DESIGN OF CELL-SURFACE PANNING EXPERIMENTS

The goal of panning in any phage-display experiment is to increase the signal-to-noise ratio; that is, to increase the number of specifically binding phage relative to the number of nonspecifically binding phage. In this respect, competitive cell-surface panning methods are superior to methods that include separate positive and negative selection cycles. As mentioned previously, simultaneous incubation of phage with target and absorber cells decreases the background binding of nonspecific phage to irrelevant antigens on target cells while minimizing the loss of specific phage that bind to antigen-negative cells. The use of a magnetic column to isolate the target cells avoids the high volumes of wash buffer and strong shear forces that are associated with other methods of separation such as batch magnetic separation or flow cytometry. The con-

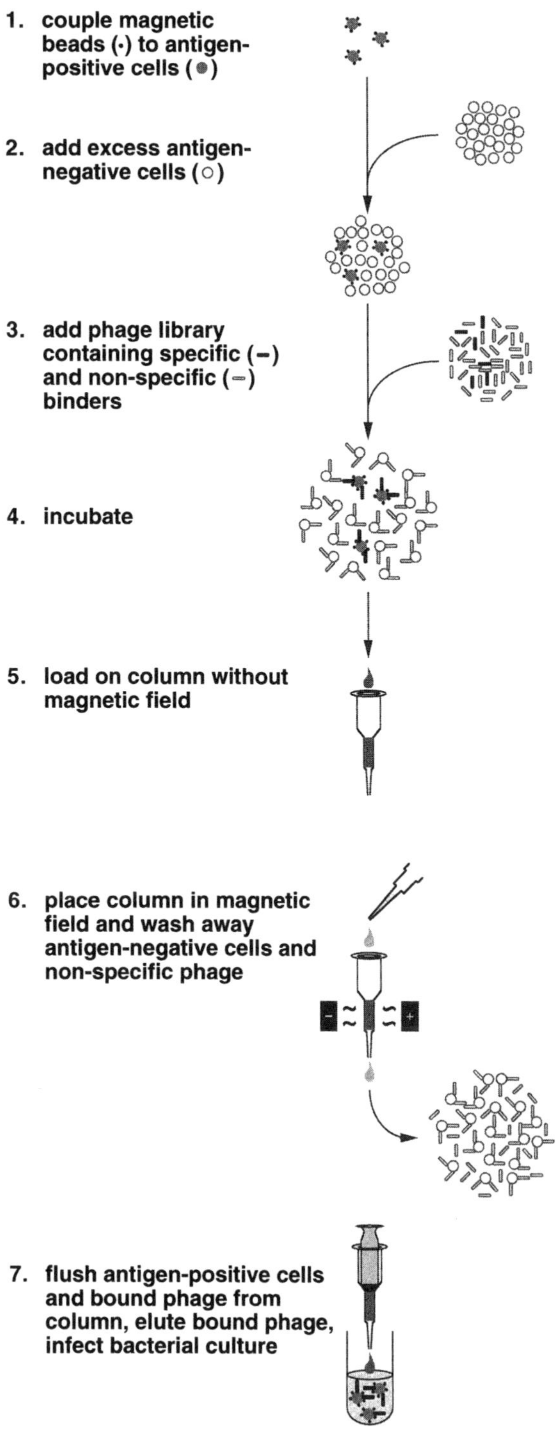

Figure 23.1. Strategy for cell-surface panning of phage-display libraries using magnetically activated cell sorting. (Reprinted, with permission, from Siegel et al. 1997.)

ditions associated with these latter methods select against ligands with rapid off-rates and, by allowing ligands of even moderate affinity to be lost, could limit the diversity of selected phage. In addition, solution-based separation that uses a magnet to pull target cells to the side of the incubation tube can lead to trapping of absorber cells that bear nonspecific phage. Time is also a concern when choosing a panning method. For example, processing 10^8 cells by flow cytometry at a standard sort rate of 2000 cells/sec requires over 12 hours. With a large magnetic column, more than 10^{10} cells can be processed in less than 1 hour.

One of the primary ways to optimize the signal-to-noise ratio is to change the concentration of antigen that is incubated with the phage library (Kretzschmar et al. 1995). For cell-surface-expressed antigens, achieving the desired concentration of antigen might not be straightforward. By definition, to capture more than 50% of the antigen-specific phage present in an aliquot of library, the concentration of antigen must be greater than the K_D of the desired binders. For example, capturing phage antibodies with K_D values less than 10^{-7} M is relatively easy to achieve with a purified antigen (e.g., 500 ng of a 50-kD protein in 100 μl is 10^{-7} M). However, achieving an antigen concentration of 10^{-7} M when the antigen is expressed on cells at 10,000 copies per cell would require approximately 6×10^8 cells per 100 μl of incubation volume. Depending on the volume of the particular cell type used, it might be impossible to fit that many cells in only 100 μl, much less have room for the aliquot of phage library. With these numbers in mind, it is best to incubate the cell–phage mixture in as small a volume as possible, with target cells that express the highest possible number of target antigens per cell. In the examples that follow, Rh(D)-positive RBCs of the genotype "-D-" (Mollison et al. 1993) were used for panning, because they express 10–20 times more Rh(D) antigens per cell (~100,000) than the more common Rh(D)-positive RBCs.

The choice of blocking agent in the incubation and wash buffers can have a significant effect on background binding. Nonfat dry milk appears to be one of the gold standards in the field; however, centrifugation of milk-containing buffers with cell–phage mixtures can actually increase background through the sedimentation of milk solids with adherent phage. This problem is avoided with magnetic columns because centrifugation of such mixtures is never required. When choosing a blocking agent, it is important to verify that it does not contain compounds that may, in fact, block the specific binding of phage-displayed antibodies to their target. In particular, milk contains numerous carbohydrate substances (including certain human blood group antigens) that may be a cause of concern depending on the particular application. In the examples shown below, it was known that the Rh(D) protein is not glycosylated and that the binding of anti-Rh(D) antisera is not inhibited by milk.

Efficient elution of specifically captured phage is also important for effective panning. Nonspecific elution methods include the use of acid, base, or chaotropic agents such as urea (Smith and Scott 1993), whereas more specific elution methods include competitive elution with monoclonal antibodies (Meulemans et al. 1994) and the use of protease-sensitive sites that are either present in *gene III* or have been engineered into the *gene III* fusion protein (Dziegel et al. 1995; Ward et al. 1996). In the case of the RBC, which is anucleated, contains no organelles, and is easily lysed, and the Rh(D) protein, which is readily denatured, simple acid elution was found to be sufficient.

OVERVIEW OF PROTOCOLS

Protocol 23.1 provides a detailed method for the panning of phage-display libraries for cell-surface binders using magnetically activated cell sorting. Protocol 23.2 shows how one can monitor enrichment for binders using either agglutination assays (suitable for RBCs) or flow cytometry (suitable for non-RBC target cells). Once enrichment is observed after one or more rounds of cell-surface panning, individual clones can be expressed, either as phage-displayed molecules (Protocol 23.3) or as a soluble form (Protocol 23.4), and assayed for binding using Protocol 23.2. After unique binders are identified through nucleotide sequencing, their fine specificities can be studied through competition assays using Protocol 23.5. Because most of the procedures described in this chapter have been optimized for the use of RBCs as targets, Protocols 23.6 and 23.7 provide approaches to optimize cell-surface panning for non-erythroid cells.

PROTOCOL 23.1

Affinity Purification of Phage-displayed Antibodies by Magnetically Activated Cell Sorting

As noted above, the procedure that follows has been optimized for the isolation of RBC-binding phage-displayed antibodies. Depending on the particular application, it is recommended that a series of pilot experiments be performed ahead of time to determine the optimal target cell biotinylation conditions, absolute cell numbers, and target cell/absorber cell ratio as described in Protocols 23.6 and 23.7. Having determined these parameters, it is further recommended that individual preparations of magnetically labeled target cells be prepared and "quality-controlled" a day prior to the actual phage library panning procedure so that proper attachment and elution of the cells from the magnetic column may be verified. This is accomplished through the use of mock panning procedures (Protocol 23.7).

To illustrate the types of results one can obtain using the following method, data from a previously published study (Siegel et al. 1997) have been summarized in Table 23.1. Briefly, separate $\gamma_1\kappa$ and $\gamma_1\lambda$ phage libraries were constructed from 2 x 10^7 mononuclear cells isolated from 25 ml of peripheral blood from an Rh(D)-negative individual previously hyperimmunized with Rh(D)-positive RBCs. The libraries,

Table 23.1. Results obtained by panning an immune human Fab/phage library on intact RBCs

Panning[a]	Phage input (cfu)[b]	Phage output (cfu)[c]	% Bound[d] (x 10^{-4})	Enrichment[e]	Agglut titer[f]	Binders/ total (%)[g]
		(*a*) $\gamma_1\kappa$ *Fab/phage library panning results*				
0					0	0/24 (0)
1	2.94 x 10^{11}	6.04 x 10^5	2.1		1/16	0/16 (0)
2	2.15 x 10^{11}	1.68 x 10^7	78.3	38.0x	1/2048	15/15 (100)
3	1.72 x 10^{11}	1.44 x 10^8	840.0	10.7x	1/2048	12/12 (100)
		(*b*) $\gamma_1\lambda$ *Fab/phage library panning results*				
0					0	0/20 (0)
1	2.28 x 10^{11}	3.48 x 10^5	1.5		0	
2	5.51 x 10^{11}	1.34 x 10^6	2.4	1.6x	1/128	32/36 (89)
3	3.93 x 10^{11}	3.86 x 10^8	980.0	404.0x	1/512	24/24(100)
4	2.87 x 10^{11}	3.08 x 10^8	1100.0	1.1x	1/1024	

[a] Panning round, where "0" represents the initial, unpanned Fab/phage library.

[b] Number of colony-forming units (cfus) of phage incubated with Rh(D)-positive/-negative RBC admixture.

[c] Total number of cfus of phage contained in eluate.

[d] Phage output/phage input) x 100.

[e] Fold increase in % bound compared to previous round of panning.

[f] Agglutination titer; see Protocol 23.2 for details.

[g] Number of individual Rh(D)-binding Fab/phage clones per total number of individual clones screened from panning round (see Protocol 23.3 for details).

cloned in the pComb3H phagemid vector, were panned independently on magnetically labeled Rh(D)-positive RBCs in the presence of a tenfold excess of unlabeled Rh(D)-negative RBCs. For the $\gamma_1\kappa$ library (approximately 7 x 10^7 independent transformants), significant enrichment for binders occurred after only one round of panning, whereas significant enrichment for the $\gamma_1\lambda$ library (approximately 3 x 10^8 independent transformants) occurred during the second round. This was reflected by both the sharp increase in percent phage bound during a given round as well as the ability of the polyclonal $\gamma_1\kappa$ and $\gamma_1\lambda$ phage libraries to agglutinate Rh(D)-positive RBCs after one and two rounds, respectively. Details on the agglutination experiments and analyses of individual clones are presented in Protocols 23.2 and 23.5.

Materials

antigen-positive and antigen-negative cells, ~10^8 each (e.g., **red blood cells** [RBCs], 3–4% [v/v] suspensions available from Gamma Biologicals, Houston, TX)
phosphate-buffered saline (PBS, see Appendix 2)
sulfo-NHS-LC-biotin reagent (Pierce Chemical Co., Cat. # 21335)
MACS streptavidin microbeads (Miltenyi Biotec, Cat. # 481-02)
1% and 3% BSA/PBS (see Appendix 2)
XL1-Blue strain of ***E. coli,*** glycerol stock (Stratagene, Cat. # 200268)
superbroth (SB, see Appendix 2)
tetracycline (tet) antibiotic stock: 10 mg/ml in 70% **ethanol**
SB+tet: SB with 10 μg/ml tetracycline
Luria broth (LB, see Appendix 2)
LB+tet plates: LB with 12.5 μg/ml tetracycline and 15 g/liter agar
Phage display library, ~10^{12} cfu at a concentration of ~10^{13} cfu/ml, e.g., a phage-displayed Fab library produced using the pComb3H phagemid (The Scripps Research Institute, La Jolla, CA) as described in Chapter 9, Antibody Library Construction
ddH$_2$O
ethanol
5x MPBS: 10 g of nonfat dry milk to 100 ml with PBS, titrated to pH 7.4
MPBS: 5x MPBS diluted to 1x with PBS
MiniMACS type MS columns (Miltenyi Biotec, Cat. # 422-01)
5-cc syringe
MiniMACS magnetic separation unit (Miltenyi Biotec, Cat. # 421-02)
acid elution buffer: 76 mM **citric acid**, pH 2.4 (need <<100 ml)
eluate neutralization buffer: 2 M Tris base, untitrated (pH ~ 10.5, need << 100 ml)
carbenicillin (carb) antibiotic stock: 100 mg/ml in sterile ddH$_2$O
LB+carb plates: LB with 100 μg/ml carb and 15 g/liter agar
kanamycin antibiotic stock: 25 mg/ml in sterile ddH$_2$O
VCSM13 helper phage, 10^{12} pfu per panning, amplified from commercial stock (Protocol 10.2 and Stratagene, Cat. # 200251)
5x **PEG:** 200 g of **polyethylene glycol-8000,** 150 g of NaCl to 1 liter with ddH$_2$O, and sterilize by autoclaving 20 minutes at 15 psi on liquid cycle

CAUTIONS: **blood products, citric acid, *E. coli*, tetracycline, ethanol, PEG** ***(see Appendix 4)***

Procedure

Cell-surface biotinylation and magnetic labeling of target cells

1. Prepare a stock of cell-surface biotinylated cells sufficient for at least 10 selection procedures by placing 265 µl of a 3.5% suspension of antigen-positive RBCs in a microcentrifuge tube. Pellet cells by centrifuging for about 4 seconds during the acceleration phase of a microcentrifuge set at full speed (~ 13,000*g*). Discard the supernatant and wash the cells 5 times with 400 µl of room-temperature PBS per wash. Resuspend the pellet with PBS to a final volume of 46 µl. This should yield an RBC suspension of approximately 20% (v/v).
2. Prepare at least 50 µl of a 1 mg/ml solution of sulfo-NHS-LC-biotin in room-temperature PBS. To facilitate the accurate measurement of reagent, weigh 1–10 mg of sulfo-NHS-LC-biotin, place in a 15-ml centrifuge tube, and dissolve in the appropriate volume of PBS by vortex mixing. *Immediately* add 46 µl of the biotin solution to the RBC suspension and thoroughly mix by drawing up and down with a micropipettor. It is important to work quickly once the biotin reagent is in solution, because the amino-reactive groups readily hydrolyze.
3. Incubate the cell/biotin reagent mixture for 40 minutes at room temperature on a rotator to maintain the cells in suspension. Centrifuge and wash the cells 5 times with 400 µl of room-temperature PBS per wash to remove unreacted NHS-biotin reagent.
4. Resuspend the RBCs to their original volume (265 µl) with PBS and determine their concentration using a hemacytometer or an automated cell counter. The cell count should be approximately 2.5×10^8 to 3.5×10^8 cells/ml.
5. Place 8×10^6 biotinylated RBCs (typically 30 µl of suspension) in a fresh microcentrifuge tube and bring to 90 µl with room-temperature PBS. Add 10 µl of streptavidin-coated paramagnetic microbeads. Incubate for 1 hour at room temperature on a rotator as above, centrifuge briefly (as in Step 1), discard the supernatant, and resuspend the magnetic bead-coated RBCs with 100 µl of room-temperature 3% BSA/PBS.

Preparation of bacteria

6. Dilute 250 µl of an overnight culture of XL1-Blue bacteria into 25 ml of SB+tet medium. Shake diluted culture at 300 rpm and 37°C in a 125-ml Erlenmeyer flask until they reach an OD_{600} of 1.0 (~ 4 hours) by which time they will be ready for Step 17 below.

   Note: Prepare the overnight culture by inoculating 10 ml of SB+tet with a single colony of bacteria picked from a freshly streaked LB+tet plate (< 2 weeks old) and shaking overnight (~ 15 hours) at 300 rpm and 30°C in a 50-ml centrifuge tube. The overnight incubation at 30°C avoids overgrowth of bacteria, which may render the bulk of the cells nonviable in the morning.

Preparation of cell/phage incubation mix

7. While the bacterial culture is incubating, mix an excess amount of phage library (e.g., 10^{12} cfu in 100 μl of 1% BSA/PBS) with 1/4 volume of 5x MPBS (i.e., 25 μl) in a microcentrifuge tube to achieve a final concentration of 2% milk in PBS. Incubate (block) the phage for 30–60 minutes at room temperature.

8. While the phage are in the blocking solution, place 8 x 10^7 absorber (antigen-negative) RBCs in a microcentrifuge tube (typically 200–300 μl of a 3.5% [v/v] RBC suspension). Pellet the cells as above and resuspend with 100 μl of 3% BSA/PBS. Pellet the magnetically labeled target cells (Step 5), discard the supernatants, and resuspend the labeled cells with the suspension of absorber cells. Use a second 100-μl aliquot of 3% BSA/PBS to rinse and transfer any residual absorber cells from their original tube to the tube containing the target cell/absorber cell admixture. Incubate the cells at room temperature for 15 minutes in the BSA-containing solution as a "blocking" step.

   Note: BSA is used as a blocking reagent with cells (versus milk protein) to avoid the sedimentation of milk solids that can occur when milk-containing buffers are centrifuged.

9. Pellet the cell mixture, discard the supernatant, and resuspend with 33 μl of the phage library suspension. Incubate for 2 hours at 37°C on a rotator to keep the cells in suspension.

Preparation of MiniMACS column

10. During the cell/phage library incubation, equilibrate the magnetic column with ice-cold MPBS as follows. Run several milliliters of sterile ddH_2O through the column to remove its cloudy preservative/storage material. Push a plunger from a 5-cc disposable syringe into the reservoir of the column. Place the tip of the column in a small beaker containing several ml of 100% ethanol and vigorously push and pull the plunger in and out using the ethanol to purge all of the trapped air from within the column matrix. End this procedure with a final aspiration of ethanol into the column as the plunger is pulled out. Clamp the column to a stand and allow the ethanol remaining in the reservoir to flow through. *Before the last 100–200 μl enters the column*, fill the reservoir with room-temperature PBS and let it wash out the ethanol. *Before the PBS completely flows through*, fill the reservoir with room-temperature MPBS (previously degassed for about 30 minutes) and allow approximately 5 ml to flow through. Attach a 30-gauge x 1/2-inch needle to the column outlet, fill the reservoir with MPBS, place the column in a cold room, and allow ice-cold MPBS to flow through until the column is needed. With the needle attached, the flow rate of the column should be about 10 μl/minute.

Column loading, washing, and cell elution

11. After cell/phage library incubation (Step 9), carefully remove the column running buffer from atop the column matrix and immediately pipet the incubation mixture (~ 40 μl) directly on top of the matrix. This loading step should be performed *without* a magnetic field around the column to prevent the magnetically labeled target cells from instantly adhering to the very top of the column, clogging it, and causing the trapping of antigen-negative unlabeled absorber cells. Loading the RBC/phage incubation mixture in the absence of a magnetic field causes the antigen-negative and antigen-positive cells to distribute evenly throughout the column yet not run off, since the excluded volume of the column is slightly greater than 40 μl.
12. Once the material has seeped into the column, place the column within the MiniMACS magnet unit. Leave the column undisturbed for 2 minutes to allow the labeled cells to adhere to the magnetically charged matrix.
13. Perform a series of 500-μl washes with ice-cold MPBS followed by a final wash with cold (4°C) PBS. Perform a total of three washes for the first two rounds of panning and a total of six washes for all subsequent pannings. For each panning, leave the column outlet needle in place for the first wash only. Carefully remove the needle for all subsequent washes. The flow rate should increase to about 200 μl/minute with the needle removed.
14. Remove the column from the magnet and flush the target cells with bound phage off the column into a microcentrifuge tube using the plunger and two 500-μl aliquots of ice-cold PBS. Immediately centrifuge the cells (4 seconds at 13,000*g*, as in Step 1) and discard the supernatant.
15. Resuspend the cell pellet with 200 μl of acid elution buffer and incubate at room temperature for 10 minutes with vortex mixing every 2 minutes for 15 seconds.
16. Neutralize the pH of the eluted phage and cellular debris with 18 μl of eluate neutralization buffer and transfer to a sterile 125-ml Erlenmeyer flask.

Amplification of captured phage antibodies

17. Add 10 ml of the XL1-Blue culture (Step 6) that has reached its target OD_{600} of 1.0. Incubate at room temperature without shaking for 15 minutes.
18. Add 10 ml of prewarmed (37° C) SB+tet medium containing 40 μg/ml carb (20 μl of carb stock) to achieve a final carb concentration of 20 μg/ml. Immediately remove a small aliquot of culture (~ 100 μl) for titering (see next step) and shake the balance of the 20-ml culture at 300 rpm for 1 hour at 37°C.
19. Perform serial 10-fold dilutions of the culture aliquot (100 μl each; 10^{-1} through 10^{-4} dilutions) and plate 50 μl of each dilution onto LB+carb. Incubate plates at 37°C overnight and count the number of bacterial colonies per plate the following day. These counts will be used to monitor the enrichment of antigen-specific phage (Step 31).

20. Add additional carb to the incubating culture (Step 18) to bring the final concentration of antibiotic to 50 μg/ml (i.e., add 6 μl of carb stock) and return the flask to the shaking incubator for another hour.

21. Add 10^{12} pfu of VCSM13 helper phage (typically 1–3 ml) to the culture and incubate for 15 minutes at 37°C without shaking.

22. Pour the culture into a sterile 500-ml Erlenmeyer flask containing 100 ml of prewarmed (37°C) SB+tet with 50 μg/ml of carb (50 μl of carb stock). Shake at 300 rpm at 37°C for 2 hours.

23. Add kanamycin to 70 μg/ml (~ 350 μl of antibiotic stock), lower the temperature of the incubator to 30°C, and shake overnight (~ 15 hours) at 300 rpm.

Purification of phagemid particles

24. Divide the approximately 120-ml culture into three 50-ml centrifuge tubes and centrifuge at 5300*g* for 10 minutes at 4°C. Transfer the phage-containing supernatant into four 40-ml high-speed polycarbonate centrifuge tubes and store the bacterial pellets at –20°C for the future preparation of plasmid DNA (if desired).

25. Centrifuge the phage suspension at higher speed (28,000*g* for 10 minutes at 4°C) to clarify the supernatant of any residual bacterial debris. Collect the supernatants by aspiration and pool them into a 250-ml polycarbonate spherical-bottom centrifuge bucket.

26. Add 1/4-volume of 5x PEG solution to the phage suspension (~ 30 ml). Mix well by inversion, place the bucket on its side in an ice bath, and rock gently for 30 minutes.

27. Centrifuge the precipitated phage at 14,000 rpm (30,000g) in a Beckman JA-14 rotor (or equivalent) with round-bottom bucket adapters in place for 30 minutes at 4°C.

28. Aspirate and discard the supernatant. Drain the pellet of residual PEG solution by inverting the tube for approximately 10 minutes.

29. Resuspend the phage pellet in 2 ml of 1% BSA/PBS. Transfer phage to microcentrifuge tubes and spin at 13,000*g* for 5 minutes at room temperature to remove any unsuspended debris. Transfer the supernatants to fresh microcentrifuge tubes and store at 4°C.

30. Determine the titer of the amplified library prior to the next round of panning. Perform serial 10-fold dilutions of phage in SB (100 μl each) out to a dilution of 10^{-9}. In three separate microcentrifuge tubes labeled 10^{-7}, 10^{-8}, and 10^{-9}, place 100-μl aliquots of a fresh XL1-Blue culture prepared as in Step 6 except grown to an OD_{600} of 0.5. To each tube add 2 μl of the appropriate phage dilution. Incubate for 15 minutes at room temperature and plate 51 μl (i.e., 50 μl of bacteria and 1 μl of diluted phage) on LB+carb plates. Incubate plates overnight at 37°C and count the number of colonies per plate the following day. These counts are used to monitor enrichment for antigen-specific phage (Step 31).

Data Analysis

31. For each round of panning, monitor enrichment for antigen-specific phage by calculating phage input, phage output, and percentage of phage bound as exemplified in Table 23.1.

Useful formulas:

Phage titer/μl [Step 30] = (# colonies per plate)(dilution factor)

e.g., 200 colonies on the 10^{-8} plate yields a calculated phage titer of 2×10^{10} cfu/μl

phage input [Step 9] = (phage titer/μl)(100/125 dilution from blocking solution)(33 μl)

e.g., $(2 \times 10^{10})(100/125)(33) = 5.28 \times 10^{11}$ phage input to round of panning

phage output [Step 19] = (colonies per plate)(1/50 μl)(dilution factor)(20,000 total μl culture volume)

e.g., 200 colonies on the 10^{-2} plate yields a calculated phage output of $(200/50)(10^2)(20{,}000) = 8 \times 10^6$ phage

% phage bound = (phage output)/(phage input) x 100

PROTOCOL 23.2

Cell-binding Assays Using Phage-displayed Fab and Soluble Fab Molecules

The presence of anti-RBC phage-displayed Fab binders is easily assayed by indirect agglutination using anti-M13 antibody as a secondary antibody to bridge between RBCs coated with phage-displayed Fab (Siegel et al. 1997). This assay can be used to measure anti-Rh(D) phage-displayed Fab activity in panned polyclonal libraries (Protocol 23.1) or in individual randomly picked phage-displayed Fab clones (Protocol 23.3). Figure 23.2 presents a representative example of this assay showing negative reactivity to Rh(D)-negative RBCs and strongly positive reactivity to Rh(D)-positive RBCs from the second panning of the $\gamma_1\kappa$ library (see Protocol 23.1 and Table 23.1). Negative reactions show sharp, approximately 2-mm diameter RBC spots, whereas the RBCs in positive wells form a thin carpet coating the entire floor of the well. This reactivity begins to titer out at a phage-displayed Fab dilution of 1/2048 (where "neat" is defined as 5×10^{12} cfu/ml) or when a total of about 2.3×10^{8} cfu is present in the reaction mix. As shown in Figure 23.2, the assay is quite sensitive, as a strong positive reaction required only 10–20 phage-displayed Fab particles per RBC (versus ~150 IgG for a conventional indirect antiglobulin test [Coombs and Mourant 1945; Mollison 1945]). To confirm that these phage-displayed Fab had specificity against the target antigen (in this case Rh(D)) rather than some other unrelated RBC antigen, it was necessary to screen the polyclonal phage libraries as well as individual clones (Protocol 23.3) against several dozen RBCs of varying blood group phenotype to verify target antigen specificity as per standard blood-banking serological practice (Siegel 1997). Agglutination assays can also be performed with soluble Fab (Protocol 23.4) by using antihuman $F(ab')_2$ fragment-specific antibody as the bridging reagent (Siegel and Silberstein 1994).

For cells other than RBCs for which agglutination assays cannot be employed, flow cytometry is a useful tool to monitor the enrichment or analyze individual clones. Flow cytograms of the unpanned phage-displayed Fab library and clones obtained after each round of panning can be run in parallel along with individually picked clones. Unlike the case with RBCs in which panels of phenotypically defined cells are readily available, determining the identity of the cell-surface expressed molecule(s) to which the recombinant antibodies are binding may require the use of other techniques such as immunoprecipitation or immunoblotting.

AGGLUTINATION ASSAYS

Materials

phage-displayed antibody preparations or soluble Fab extracts
96-well round-bottom microplates
PBS (see Appendix 2)

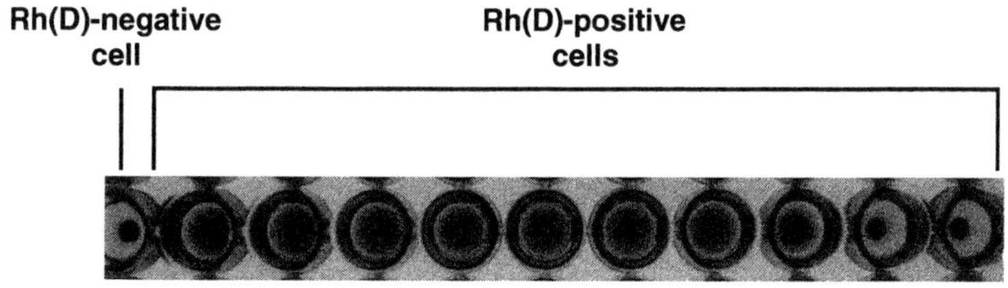

# Fab/phage added (x 10^7 cfu's):	3000	3000	1500	750	375	188	94	47	23	12	6
# RBCs added (x 10^7):	2	2	2	2	2	2	2	2	2	2	2
RATIO Fab/phage per RBC:		1500	750	375	188	94	47	24	13	6	3

Figure 23.2. Detecting the binding of anti-RBC phage-displayed Fab by agglutination. (Reprinted, with permission, from Siegel 1998, © Elsevier Science.)

1% BSA/PBS
cells of desired phenotypes as for Protocol 23.1
rotor and centrifuge for centrifuging 96-well plates
sheep anti-M13 phage antibody (5-Prime→ 3-Prime, Cat. # 916192) or goat antihuman IgG, $F(ab')_2$ fragment specific (Pierce, Cat. # 31132)

CAUTION: **blood products** ***(see Appendix 4)***

Procedure

1. Prepare 100-µl serial twofold dilutions of phage-displayed antibody preparations or soluble Fab extracts in the round-bottom wells of a 96-well microplate. Use 1% BSA/PBS as diluent. Include one well of undiluted phage to be incubated with antigen-negative cells (e.g., absorber cells) as a negative control.
2. Add 50 µl of a 3% (v/v) suspension of antigen-positive RBCs to each well (except the negative control) and mix with a micropipettor. Add 50 µl of antigen-negative RBCs to the negative control well.
3. Incubate the microplate for desired time and temperature, typically 60 minutes at 37°C. Periodically, draw samples up and down with a micropipettor to keep the RBCs in suspension.
4. Centrifuge the plate for 1 minute at 1100*g*. Aspirate and discard the supernatants. Resuspend cell pellets with 200 µl of ice-cold PBS. Spin again and wash cells two more times.
5. Resuspend the washed RBC pellets with 100 µl of either a 10 µg/ml solution of sheep anti-M13 antibody (for phage-displayed Fabs) or a 36 µg/ml solution of goat antihuman $F(ab')_2$-specific antibody (for soluble Fab preparations).
6. Place the microplate on a light box and allow the cells to settle undisturbed. Positive or negative reactions begin to appear within about 15 minutes and are complete by 1 hour.

FLOW CYTOMETRIC ASSAYS

Materials

phage-displayed antibody preparations
cells of desired phenotypes as for Protocol 23.1
1% BSA/PBS
biotinylated sheep anti-M13 phage antibody (5-Prime→3-Prime, Cat. # 7-187156)
R-phycoerythrin-conjugated streptavidin (Jackson ImmunoResearch, Cat. # 016-110-084) or R-phycoerythrin-conjugated $F(ab')_2$ fragment goat anti-human IgG, $F(ab')_2$ fragment specific (Jackson ImmunoResearch, Cat. # 109-116-097)
access to flow cytometer

CAUTION: blood products *(see Appendix 4)*

Procedure

1. Incubate approximately 10^6 cells with 100 µl of phage-displayed Fabs (~10^{11} cfu in 1% BSA/PBS) or crude soluble Fab periplasmic extract. The duration and temperature of the incubation are somewhat dependent on the cell type and particular target antigen and should mimic the conditions normally used for immunocytometry, immunoprecipitation, immunofluorescence, etc.
2. Wash the cells to remove any unbound phage-displayed Fab or soluble Fab primary antibody. Typically, three 400-µl washes of ice-cold 1% BSA/PBS are sufficient.
3. For phage-displayed Fab samples, resuspend cell pellets in 100 µl of a 1:200 dilution of biotinylated sheep anti-M13 phage antibody and incubate for an additional hour on ice. Wash cells twice in 400 µl of ice-cold 1% BSA/PBS, resuspend in 100 µl of a 1:50 dilution of R-phycoerythrin-conjugated streptavidin, incubate for 30 minutes, wash again, and analyze by flow cytometry.
4. For soluble Fab samples, resuspend cell pellets in 100 µl of a 1:50 dilution of R-phycoerythrin-conjugated goat antihuman $F(ab')_2$ antibody, incubate for 30 minutes, wash cells twice in 400 µl of ice-cold 1% BSA/PBS, and analyze by flow cytometry.

PROTOCOL 23.3

Preparation of Monoclonal Phage-displayed Fab Stocks

Once enrichment has been observed following one or more rounds of cell-surface panning, individual clones can be expressed as phage-displayed Fabs and assayed for binding using Protocol 23.2. Plasmid DNA can be isolated from clones of interest and Fab heavy-chain and light-chain nucleotide sequences can be determined.

For illustrative purposes, Table 23.2 and Figure 23.3 present results for 83 positive clones randomly picked from the anti-Rh(D) phage-displayed Fab libraries described in Table 23.1. Nucleotide sequencing showed a total of 28 unique heavy and 41 unique light chains. Because of the combinatorial effects during phage-display library construction, heavy- and light-chain gene segments paired to produce 53 unique Fab antibodies (Siegel et al. 1997). Of note, all 28 heavy chains and nearly all of the 18 κ light chains were derived from only four V_HIII or three V_κI germ-line genes, whereas the λ light chains were derived from a more diverse set of germ-line genes. A more detailed

Table 23.2. Summary of anti-Rh(D) Fab/phage clonal analysis

	Gene family	# Clones	# Unique nucleic acid sequences
V_H Germ line			
VH3-21	III	7	2
VH3-30	III	21	6
VH3-33	III	51	19
VH3-30.3	III	4	1
		(total = 83)	(total = 28)
V_κ Germ line			
A30	I	1	1
DPK8	I	1	1
DPK9	I	24	15
DPK15	II	1	1
		(total = 27)	(total = 18)
V_λ Germ line			
DPL2	I	8	4
DPL3	I	7	3
DPL5	I	2	2
DPL7	I	4	3
DPL13	II	1	1
DPL16	III	21	4
DPL18	VII	6	2
IGLV3S2	III	1	1
IGLV8A1	IV	3	1
lv2046	II	2	1
VL2.1	II	1	1
		(total = 56)	(total = 23)

Reprinted, with permission, from Siegel (1998).

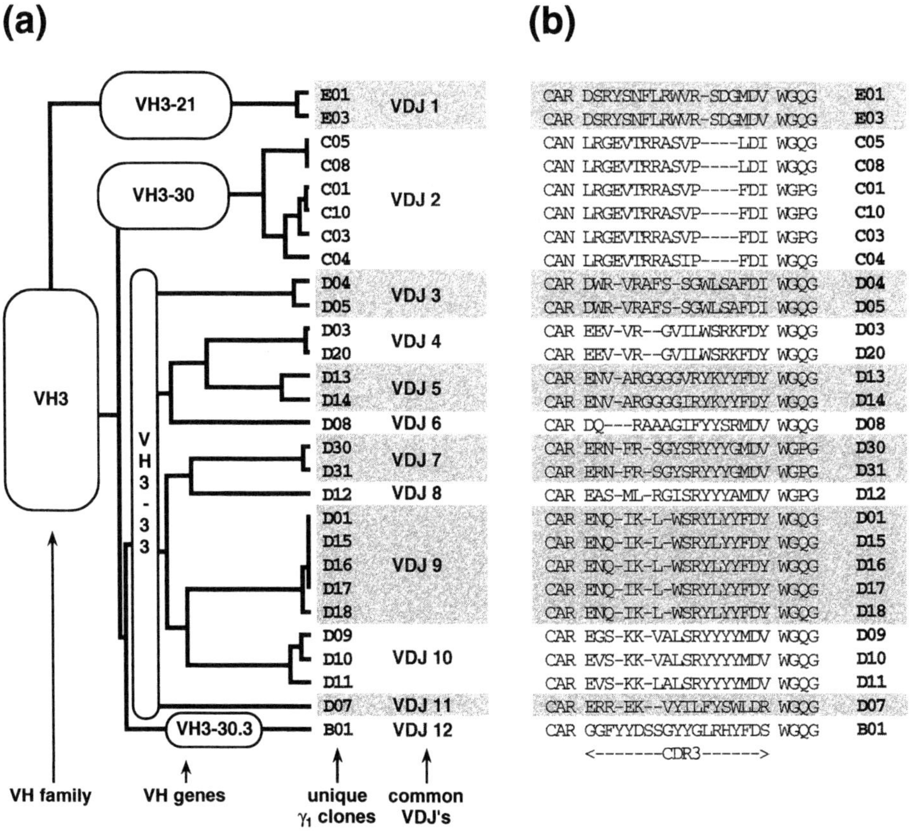

Figure 23.3. (*a*) Dendrogram and (*b*) CDR3 alignment of anti-Rh(D) antibody heavy chains. The 28 unique heavy-chain clones are organized by V_H family, V_H germ-line gene, and VDJ rearrangement. Each heavy-chain clone is identified by a numeral preceded by a letter (B through E) that denotes its germ-line gene. The 28 heavy chains comprise 12 distinct VDJ regions, designated VDJ 1 – VDJ 12. (Reprinted, with permission, from Chang and Siegel 1998.)

analysis of the heavy-chain sequences (Figure 23.3) showed that several sequences shared identical VDJ joining regions, and 12 unique VDJ rearrangements were identified. These data suggested that the 28 heavy chains resulted from the intraclonal diversity of 12 original B lymphocytes in the donor. The restriction in heavy-chain germline use to four highly homologous germ-line V_H genes (three of which share >98% homology), yet their ability to encode antibodies specific for a large number of Rh(D) epitopes, enabled the investigators to construct a model depicting Rh(D) epitope topology (Chang and Siegel 1998).

In the case of anti-Rh(D) antibodies, rare Rh(D)-positive RBCs were available that lacked one or more of the Rh(D) epitopes. These cells could be used to determine the epitope specificity for each of the 53 clones. In general, however, such mutant forms of cell-surface-expressed antigen may not be available. In these cases, competitive binding experiments between pairs of monoclonal antibodies produced in this protocol and in that which follows can be performed as described in Protocol 23.5.

Materials

panning eluate plates (Protocol 23.1, Step 19)
SB + **tetracycline** (tet, 10 µg/ml)
carbenicillin (carb): 100 mg/ml in sterile ddH_2O
LB/glycerol freezing solution stock: 70 ml of LB medium (see Appendix 2) and 30 ml of glycerol; sterilize by autoclaving 20 minutes at 15 psi on liquid cycle.
plasmid miniprep kit (commercially available)
VCSM13 helper phage as in Protocol 23.1
kanamycin: 25 mg/ml in sterile ddH_2O
5x **PEG** as in Protocol 23.1
1% BSA/PBS

CAUTIONS: tetracycline, PEG *(see Appendix 4)*

Procedure

1. Pick the desired number of single bacterial colonies from the appropriate panning eluate plates (Protocol 23.1, Step 19) and inoculate each into 10 ml of SB+tet containing 50 µg/ml carb (5 µl of carb stock) in a 50-ml tube.
2. Shake cultures overnight (~ 15 hours) at 300 rpm and 30°C. The overnight incubation at 30°C avoids the overgrowth of bacteria which would render the bulk of cells nonviable in the morning.
3. Prepare 1:100 dilutions of the overnight cultures in 30 ml of SB+tet+carb medium and shake at 300 rpm and 37°C in 125-ml Erlenmeyer flasks until the cultures reach an OD_{600} of 0.3 (~ 3–4 hours).
4. While the diluted cultures are incubating, make glycerol stocks of bacterial clones from the overnight cultures. Place 0.5 ml of culture in a microcentrifuge tube and add 0.5 ml of LB medium containing 30% glycerol. Mix well and store at –80°C. Centrifuge the remaining volumes of overnight cultures (3800*g* for 10 minutes at 4°C), discard the supernatants, and store the pellets at –20°C for the future preparation of plasmid DNA for Fab heavy chain/light chain nucleotide sequencing (if desired). The use of plasmid prep methods that employ DNA-binding resins and provide high-quality DNA suitable for all downstream applications are recommended.
5. When diluted cultures have reached an OD_{600} of 0.3 (Step 3), add 2.5×10^{11} pfu of VCSM13 helper phage to each culture (typically 250–750 µl of phage stock). Incubate the cultures without shaking for 15 minutes at 37°C.
6. Shake the cultures at 300 rpm for an additional 2 hours at 37°C.
7. Add kanamycin to 70 µg/ml (~85 µl of antibiotic stock), lower the temperature to 30°C, and shake overnight (~15 hours).

8. Purify the monoclonal phage particles as described in Protocol 23.1, Steps 24–29, except scale down the volumes appropriately. Resuspend the final phage pellet in approximately 0.5 ml of 1% BSA/PBS.
9. Determine the titer of the monoclonal phage preps (see Protocol 23.1, Step 30) and assay the individual clones for binding using an appropriate method (Protocol 23.2).

PROTOCOL 23.4

Preparation of Soluble Monoclonal Fab Antibody Stocks

An important feature of phage-display technology is the ability to convert to the production of soluble antibody molecules which may be more suitable for downstream applications than phage-displayed Fabs. In the pComb3 phagemid system, this is accomplished by isolating plasmid DNA from a clone of interest, performing a *Nhe*I/*Spe*I restriction digestion (which removes the *gene III* coat protein sequence), and re-ligating the vector to itself (*Nhe*I and *Spe*I produce compatible nucleotide overhangs). In practice, performing this procedure on a number of clones is fairly labor-intensive using the originally described methods (Barbas and Lerner 1991), because agarose gel purification is required to remove the *gene III* digestion product. The following protocol describes a method that obviates the need for gel purification by carrying out the ligation reaction in an excessively large volume which essentially reduces the chances of *gene III* reincorporation to zero. After reintroducing the modified plasmid DNA into bacteria by electroporation, cultures are induced with IPTG, and soluble Fab-containing extracts are obtained by osmotic shock of the bacterial periplasmic space.

Materials

DNA: pComb3H plasmids containing Fabs of interest
*Spe*I (40 Units/µl, Roche Molecular Biochemicals, Cat. # 1 207 644)
*Nhe*I (40 Units/µl, Roche Molecular Biochemicals, Cat. # 885 860)
10x restriction digestion buffers as provided with enzymes
Sterile ddH_2O
1% agarose
TE (see Appendix 2)
phenol/chloroform 1:1 (v:v)
chloroform
glycogen, 20 mg/ml
3 M sodium acetate, pH 5.0
ethanol
70% **ethanol**
T4 DNA ligase (1 Unit/µl, Gibco BRL, Cat. # 15224-017)
5x ligase reaction buffer (supplied with enzyme)
XL1-Blue ***E. coli*** (Stratagene)
electroporation device with disposable cuvettes (e.g., Bio-Rad Gene Pulser with 0.2-cm cuvettes, Cat. # 1652086)
SOC medium (see Appendix 2)
carbenicillin (carb) stock: 100 mg/ml in sterile ddH_2O
LB+carb plates and LB+glycerol freezing solution stock as in Protocols 23.1 and 23.3, respectively

SB+ **tetracycline** (tet), 10 µg/ml (see Appendix 2)
1 M **$MgCl_2$**
plasmid miniprep kit as in Protocol 23.3
*Eco*RI (10 Units/µl, Roche Molecular Biochemicals, Cat. # 703 737)
*Not*I (10 Units/µl, Roche Molecular Biochemicals, Cat. # 1 014 706)
molecular biology grade agarose
1 M **IPTG** stock (**isopropyl-β-D-thiogalactopyranoside**), 0.238 g/ml in ddH_2O (filter-sterilize)
osmotic shock buffer: 500 mM sucrose, 1 mM EDTA, 100 mM Tris base, pH 8.0, filter-sterilized
Slide-a-Lyzer dialysis cassettes, 10,000 MWCO (Pierce, Cat. # 66425)
PBS (see Appendix 2)

CAUTIONS: phenol, chloroform, ethanol, *E. coli*, tetracycline, $MgCl_2$, IPTG *(see Appendix 4)*

Procedure

1. Perform the following restriction enzyme digestion in microcentrifuge tubes to remove the *gene III* coat protein sequence from pComb3H plasmids that contain Fabs of interest:

DNA (1 µg)	X	µl
*Spe*I (7.5 Units)	0.19	µl
*Nhe*I (30 Units)	0.76	µl
10x digest buffer	2.50	µl
sterile ddH_2O	21.55 - X	µl

   Incubate at 37°C for 3 hours to overnight.

2. Remove 5 µl (200 ng of DNA) and electrophorese on a 1% agarose gel to verify removal of the *gene III* fragment. One should obtain DNA bands of approximately 4200 bp and 550 bp (*gene III*).

3. Add 340 µl of TE to the remaining 20-µl digest. Extract with 360 µl of phenol/chloroform followed by a second extraction with an equal volume of chloroform. For each extraction, vortex samples for 1 minute, centrifuge at 13,000*g* for 2 minutes, and transfer the DNA-containing supernatants to fresh tubes.

4. Add 20 µg of glycogen (2 µl of stock) to each sample and precipitate the DNA by the addition of 40 µl of 3 M sodium acetate, pH 5.0, and 1 ml of 100% ethanol. Place the samples on a dry ice/ethanol bath until frozen.

5. Thaw the frozen, precipitated DNA. Pellet the DNA (13,000*g* for 10 minutes) and wash one time with 600 µl of 70% ethanol.

6. Dry the DNA pellets (air-dry or by vacuum) and resuspend with 270 µl of sterile ddH_2O. Add 72 µl of 5x ligase buffer and 18 µl of T4 DNA ligase (18 Units).

7. Allow DNA to ligate for at least 5 hours at 20°C. Heat-kill the ligase enzyme (70°C for 10 minutes), precipitate and wash the DNA pellets as above (Steps 4–5), and resuspend the pellets in 30 µl of sterile ddH_2O.

8. Electroporate 5 µl of the ligated samples (~130 ng) into 40 µl of XL1-Blue electrocompetent cells. For a Bio-Rad Gene Pulser, use 0.2-cm cuvettes and the following settings: 2500 volts, 25 µFd, and 200 ohms.

9. Immediately rinse the cuvettes with three aliquots of 1 ml each of SOC medium, pooling into a sterile 50-ml tube.

10. Shake the bacterial cultures at 250 rpm for 1 hour at 37°C .

11. Plate 50-µl aliquots of serial 10-fold dilutions of cultures on LB+carb plates (see Protocol 23.1, Step 19). Following this protocol, one typically gets about 300 colonies on the 10^{-3} plate. This corresponds to about 1.8×10^7 total transformants for an efficiency of approximately 1.4×10^8 transformants per µg of electroporated, ligated DNA.

12. The following day, set up several overnight 10-ml cultures of randomly picked colonies for each Fab clone and incubate overnight at 30°C as described above (Protocol 23.3, Steps 1–2).

13. Prepare a 1:100 dilution of each culture in 30 ml of SB containing 50 µg/ml of carb (15 µl of antibiotic stock) and 20 mM $MgCl_2$ (600 µl of 1 M stock). Shake at 300 rpm and 37°C in 125-ml Erlenmeyer flasks until the cultures reach an OD_{600} = 0.4. (see Protocol 23.3, Step 4).

14. While the diluted cultures are growing, prepare glycerol stocks from the overnight cultures and perform plasmid minipreps on the bacterial pellets derived from 5 ml of the same cultures (see Protocol 23.3, Step 4). Set up *Eco*RI/*Not*I restriction digestions with the miniprep DNA to verify the absence of *gene III* in the randomly selected clones.

DNA (300 ng)	X	µl
*Eco*RI (2 Units)	0.2	µl
*Not*I (1 Unit)	0.1	µl
10x digestion buffer	1.50	µl
sterile ddH_2O	13.2 - X	µl

Incubate for 3 hours to overnight at 37°C.

15. Electrophorese 12 µl (200 ng) of digested DNA on a 0.6% agarose gel. DNA bands of approximately 2600 bp and 1600 bp should be obtained. If the lower band is approximately 2150 bp, the *gene III* fragment is still present. Following this protocol, typically 16 of 16 (i.e., 100%) randomly selected clones will restrict appropriately.

16. When the diluted cultures (Step 13) reach an OD_{600} of about 0.4, add IPTG to 1 mM (30 μl of 1 M stock), lower the temperature to 30°C, and shake at 300 rpm overnight (~15 hours).

17. In the morning, centrifuge the 30-ml cultures (3800*g* for 15 minutes at 4°C) and resuspend the bacterial pellets in 1 ml of osmotic shock buffer.

18. Transfer the suspensions to microcentrifuge tubes and incubate for 30 minutes on ice.

19. Spin the tubes at 13,000*g* for 15 minutes at 4°C. Transfer the Fab-containing supernatants to fresh tubes. Store at 4°C.

20. Depending on the particular application, it may be desirable to remove the high-sucrose-containing (high osmotic strength) extraction buffer by dialyzing the Fab extracts into PBS. Use three changes of PBS at 4°C, 500 ml each, for at least 8 hours per change.

PROTOCOL 23.5

Competition Assays between Cell-binding Clones

Having isolated a number of unique cell-surface binding clones (i.e., different amino acid sequences), it may be desirable to perform competition binding assays between pairs of antibodies to compare their fine specificity. The ability to express antibody clones as either soluble Fabs or as phage-displayed particles provides a unique opportunity to design such experiments without the need to tag one of the antibodies (e.g., by radioiodination), which can be time-consuming and can affect the antibody's binding properties. By using anti-M13 antibody to detect the binding of a phage-displayed antibody, one can ask whether its binding is inhibited by co-incubation with a soluble Fab derived from a different clone. This approach was used to demonstrate the overlapping nature of Rh(D) epitopes (Chang and Siegel 1998) as illustrated in Figure 23.4. Mutual inhibition between antibodies directed at two different Rh(D) epitopes (epD3 and epD6/7) was shown, but not between an Rh(D) antibody and an antibody to an unrelated blood group antigen (an anti-blood group B antibody (Chang and Siegel 1997). In Figure 23.4a, Rh(D)-positive RBCs were incubated with soluble Fabs only, phage-displayed Fabs only, or a combination of the two, as indicated. In Figure 23.4b, Rh(D)-positive RBCs that also expressed the blood group B antigen were used. After washing, RBCs were resuspended in anti-M13 antibody and assessed for agglutination induced by the phage-displayed Fabs.

In designing such an experiment, soluble Fabs should be used full-strength, whereas phage-displayed Fab preparations should be present in limiting amounts to

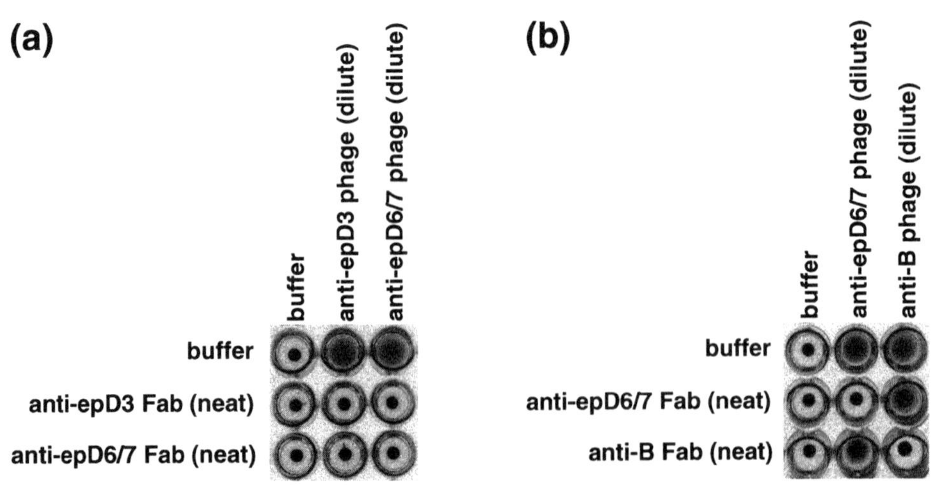

Figure 23.4. Inhibition studies with recombinant anti-Rh(D) antibodies. Panels show results of representative experiments demonstrating the mutual inhibition between antibodies directed at two different Rh(D) epitopes (in this example, epD3 and epD6/7, panel *a*), but not between an Rh(D) antibody and an unrelated recombinant anti-RBC antibody (an anti-blood group B antibody, panel *b*). (Reprinted, with permission, from Chang and Siegel 1998.)

increase the sensitivity of the inhibition assay. It is necessary to include controls that demonstrate that sufficient inhibitory amounts of soluble Fabs are present by verifying that each soluble Fab can inhibit its own phage-displayed form (e.g., Fig. 23.4a and 23.4b; samples on the diagonal).

The following protocol for RBC antigen studies uses agglutination as the method for detecting binding of the phage-displayed antibody. However, with few modifications, the protocol can be adapted for assays that use flow cytometric analyses.

Materials

96-well round-bottom microplates, rotor and centrifuge for centrifuging plates
monoclonal phage-displayed Fab (Protocol 23.3) and soluble Fab periplasmic extracts (Protocol 23.4)
cold (4°C) PBS (see Appendix 2)
1% BSA/PBS
cells of desired phenotypes (as in Protocol 23.1)
sheep anti-M13 antibody (as in Protocol 23.2)

CAUTION: blood products *(see Appendix 4)*

Procedure

1. Prepare nine incubation mixes in the wells of a round-bottom 96-well microplate as listed in Table 23.3.
2. Incubate the plate 1 hour at 37°C, wash samples three times with cold PBS, and resuspend the RBC pellets with anti-M13 antibody as in Protocol 23.2, Steps 3–5.
3. Expected results and interpretation:
    - **a.** Sample 1 should be negative; otherwise, it suggests that the source of anti-M13 had heterologous reactivity with human RBCs.
    - **b.** Samples 4 and 7 should also be negative, as monovalent soluble Fabs should not directly agglutinate cells, nor should anti-M13 antibody crosslink Fab-coated RBCs.
    - **c.** Samples 2 and 3 should be positive, indicating that the phage-displayed Fab dilutions were made correctly and were sufficient to cause the agglutination in the absence of any competitor.
    - **d.** Samples 5 and 9 should be negative, thus verifying that a given Fab is present in sufficient amount to inhibit itself when expressed on phage. If either one is positive, then the question of competition between antibodies #1 and #2 cannot be reliably interpreted.
    - **e.** Samples 6 and 8, if negative, suggest competition between clones. Additional experiments of this type should be performed between antibodies #1 and #2 and a third antibody known to bind to an unrelated antigen

Table 23.3. Worksheet for Protocol 23.5

Sample	Soluble Fab[a] (or 1% BSA/PBS)	Fab/Phage[b] (or 1% BSA/PBS)	RBCs (3% suspension)
1	100 µl BSA/PBS	100 µl BSA/PBS	50 µl
2	100 µl BSA/PBS	100 µl Fab/Phage #1	50 µl
3	100 µl BSA/PBS	100 µl Fab/Phage #2	50 µl
4	100 µl FAB #1	100 µl BSA/PBS	50 µl
5	100 µl FAB #1	100 µl Fab/Phage #1	50 µl
6	100 µl FAB #1	100 µl Fab/Phage #2	50 µl
7	100 µl FAB #2	100 µl BSA/PBS	50 µl
8	100 µl FAB #2	100 µl Fab/Phage #1	50 µl
9	100 µl FAB #2	100 µl Fab/Phage #2	50 µl

[a]Use undiluted soluble Fab periplasmic extract prepared as in Protocol 23.4 and dialyzed into PBS. It is assumed that Fab preparations #1 and #2 agglutinate cells when assayed with anti-Fab reagent as in Protocol 23.2.

[b]Use Fab/phage pre-titered with 1% BSA/PBS to its last positive dilution. This dilution, previously determined using Protocol 23.2, should be corrected for the difference in final incubation volumes; i.e., in Protocol 23.2, titers are determined in 150-µl incubation volumes (100 µl Fab/phage + 50 µl RBCs), whereas in this competition assay, the Fab/phage are in a final volume of 250 µl (100 Fab/phage + 100 µl Fab + 50 µl RBCs). For example, if the titer of Fab/phage #1 determined in Protocol 23.2 was 1:512, Fab/phage for competition assays should be 1:512(150/250) ~ 1:300. Therefore, one would prepare ~400 µl of diluted Fab/phage #1 (1.33 µl stock up to 400 µl with BSA/PBS) to have 3 100-µl aliquots for incubation samples 2, 5, and 8. Preparation of diluted Fab/phage #2 would be calculated similarly.

on the same cell (as in Fig. 23.4b). In these control experiments, Samples 6 and 8 should be positive, thus ruling out the presence of nonspecific inhibitors of agglutination in the soluble Fab preparations.

PROTOCOL 23.6

Optimizing Cell-surface Biotinylation

As noted in Protocol 23.1, target cells should be sufficiently biotinylated so that an adequate number of streptavidin-coated paramagnetic beads will bind to the cells and cause them to be retained by the magnetic column. However, overly biotinylating the cell surface would be undesirable, as it could theoretically destroy the antigenicity of the target antigen or possibly make the cells nonspecifically adsorb irrelevant phage-displayed Fab. It is recommended that a series of pilot experiments be performed before any panning procedure in which these issues are addressed.

Materials

target cells of desired phenotypes (as in Protocol 23.1)
PBS (see Appendix 2)
sulfo-NHS-biotin reagent (Pierce Chemical Co., Cat. # 21335)
Access to flow cytometer and materials for flow cytometry assay (e.g., FITC-streptavidin)
antisera to target antigen and nonspecific targets, if available
MACS streptavidin microbeads (Miltenyi Biotec, Cat. # 481-02)
MiniMACS type MS columns (Miltenyi Biotec, Cat. # 422-01)
MiniMACS magnetic separation unit (Miltenyi Biotec, Cat. # 421-02)
ddH_2O
ethanol
5x MPBS: 10 g of nonfat dry milk to 100 ml with PBS, titrated to pH 7.4
MPBS: 5x MPBS diluted to 1x with PBS

CAUTIONS: **blood products, ethanol** ***(see Appendix 4)***

Procedure

1. Using Protocol 23.1, Steps 1–4, as a guide, biotinylate aliquots comprising a fixed number of target cells over a range of final sulfo-NHS-LC-biotin concentrations (~100 μg/ml to 1 mg/ml). Semi-quantify the extent of cell-surface biotinylation through an appropriate assay with streptavidin, such as flow cytometry. Figure 23.5 shows that in the case of RBCs, a linear nonsaturating response was observed over the range of biotin reagent suggested above. If the mean fluorescence level had been a plateau over this range, it would have suggested that a lower range of biotin reagent concentrations should be tested to avoid over-biotinylation.
2. Determine that the target antigen has been unaffected by the derivatization procedure (to a reasonable level of certainty) by performing an appropriate binding

assay between each sample of biotinylated cells and antibody to that antigen. In the case of the RBC Rh(D) antigen, specific polyclonal antisera were used to show that Rh(D) antigenicity was (grossly) unaffected by cell-surface biotinylation at the concentrations used (Fig. 23.5). Furthermore, the biotinylation procedure did not cause the nonspecific uptake of antisera specific for a different RBC antigen not present on the biotinylated cells. Depending on the particular application, however, antisera to the target antigen may not be available (which may be why phage-display technology is being employed in the first place) or the target antigen may be unknown. In the case of an uncharacterized autoantigen, for example, it may be possible to use patient sera to verify that the autoantigen's ability to bind the relevant antibody(ies) has not been compromised by biotinylation. In the case of an unknown, putative tumor marker, for example, the level of sulfo-NHS-LC-biotin to use may have to be determined empirically.

3. Select an appropriate set of candidate conditions for cell-surface biotinylation, incubate each biotinylated cell sample with an excess of streptavidin-coated paramagnetic microbeads, and apply the coated cells to an appropriately prepared

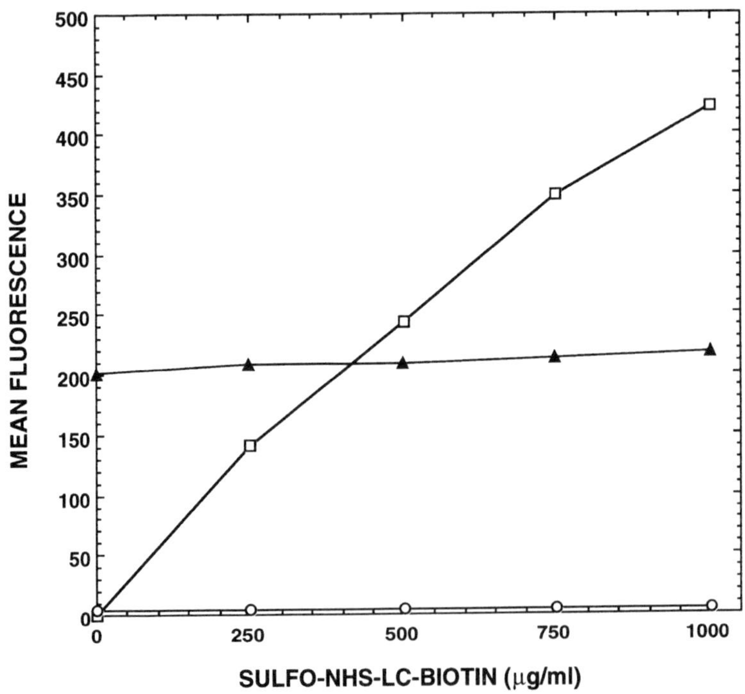

Figure 23.5. Cell-surface biotinylation of human RBCs. Rh(D)-positive/KEL-1-negative cells (KEL-1 is another human RBC antigen unrelated to Rh [Mollison et al. 1993]) were biotinylated at varying biotin concentrations, incubated with FITC-streptavidin, and analyzed by flow cytometry (curve with open boxes). Aliquots of cells were also analyzed for retention of Rh(D) antigenicity (i.e., specific staining, curve with closed triangles) or for the lack of nonspecific staining (curve with open circles) by incubating with either anti-Rh(D) or anti-KEL1 typing sera, and FITC-anti-human IgG. (Reprinted, with permission, from Siegel et al. 1997, © Elsevier Science.)

MiniMACS column (Protocol 23.1, Steps 5 and 10–14, without inclusion of phage library incubation). Assess for binding and elution by microscopic examination of column flowthrough and eluate, flow cytometry, etc. As a starting point, add 10 µl of streptavidin microbead stock, and determine the optimal amount of microbeads by titering. The number of target cells to apply to the column should mimic that which will be used in the actual panning experiment (see Protocol 23.7). As an upper limit, it appears that approximately 10^8 microbead-coated RBCs is the maximum number of cells that can be retained by a MiniMACS column under the conditions described in Protocol 23.1. The binding capacity of these columns for other cell types should be determined empirically but may be roughly proportional to the total cell volume. For example, approximately 20 times as many magnetically labeled platelets (~5 pL/cell) can bind to a column as RBCs (~100 pL/cell [Roark et al. 1998]).

PROTOCOL 23.7

Determining Target Cell Numbers and Target Cell/Absorber Cell Ratio

As noted above (Protocol 23.1, Step 11), during a panning procedure it is desirable to load the magnetic column in the absence of a magnetic field. This necessitates an incubation volume of less than or equal to 40 μl so that none of the load runs off the column. As noted in the chapter introduction, one can roughly calculate the appropriate concentration of cells required in a 40-μl volume to capture more than 50% of phage-displayed Fab specific for a given cell-surface antigen. Such a calculation is a function of the number of antigen sites per cell and the K_D of the bound phage-displayed Fab. In the case of the Rh(D) antigen, using a value of about 100,000 sites per RBC (genotype "-D-/-D-" [Mollison et al. 1993]) and a desired phage-displayed Fab K_D less than or equal to 10^{-8} M, approximately 8×10^6 Rh(D)-positive target cells in the 40-μl incubation mixture were required.

Approximate the number of target cells required per incubation, and mix aliquots of cells with increasing amounts of antigen-negative cells at ratios of 1:5, 1:10, 1:20, etc. The maximum number of absorber cells that can be effectively separated from the antigen-positive target cells should be determined by a mock panning procedure (Protocol 23.1, Steps 10–14, without inclusion of phage library incubation). Depending on the particular application, various methods can be used to assess the quality of separation of the two cell populations as a function of cell ratio. This may involve direct visualization of column washes and eluates by light microscopy (with or without histological staining) or use of flow cytometry as illustrated for the case with RBCs in Figure 23.6. In this example, biotinylated, streptavidin-microbead-coated

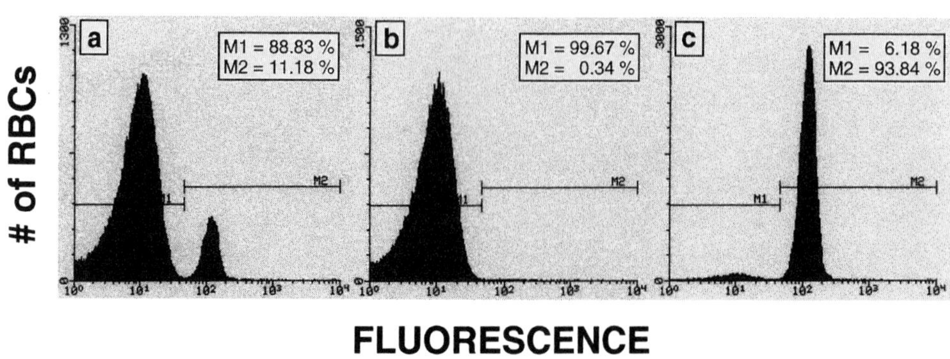

Figure 23.6. Validation of antigen-positive/antigen-negative cell separation procedure for RBCs using magnetically activated cell sorting. (*a*) Column load, (*b*) column wash, (*c*) column eluate. M1= Rh(D)-negative cells, M2= Rh(D)-positive cells. (Reprinted, with permission, from Siegel et al. 1997, © Elsevier Science.)

Rh(D)-positive RBCs (8 x 10^6 cells) were mixed with a tenfold excess of Rh(D)-negative cells (8 x 10^7 cells) in a 40-µl volume of MPBS (refer to Protocol 23.1). The cell admixture was applied to the MiniMACS column, washed, and eluted. Aliquots of RBCs contained in the original admixture (Fig. 23.6a), the pooled column washes (Fig. 23.6b), and the column eluate (Fig. 23.6c) were stained with anti-Rh(D) typing serum and FITC-goat antihuman IgG. As shown in the cytograms, although approximately 90% of the cells applied to the column were Rh(D)-negative, nearly all of them washed off of the column, yielding a column eluate that contained almost entirely Rh(D)-positive cells. Because only about 6% of the final eluate comprised Rh(D)-negative cells, and Rh(D)-negative cells were initially present in a tenfold excess to Rh(D)-positive cells, only approximately 0.6% of the initial antigen-negative absorber cells contaminated the final antigen-positive preparation. The acceptable limits of contamination of the column eluate with absorber cells must be determined by the investigator empirically for each particular application.

REFERENCES

Barbas C. F. and Lerner R. A. 1991. Combinatorial immunoglobulin libraries on the surface of phage (Phabs): Rapid selection of antigen-specific Fabs. *Methods* **2:** 119–124.

Cai X. and Garen A. 1995. Anti-melanoma antibodies from melanoma patients immunized with genetically modified autologous tumor cells: Selection of specific antibodies from single-chain Fv fusion phage libraries. *Proc. Natl. Acad. Sci.* **92:** 6537–6541.

Chang T. and Siegel D. 1997. Isolation of a highly mutated human anti-B antibody from a phage-display library. *Transfusion* **37:** 89S (abstr.).

———. 1998. Genetic and immunological properties of phage-displayed human anti-Rh(D) antibodies: Implications for Rh(D) epitope topology. *Blood* **91:** 3066–3078.

Coombs R.R.A. and Mourant A.E. 1945. Detection of weak and "incomplete" Rh agglutinins: A new test. *Lancet* **249:** 15–16.

de Kruif J., Terstappen L., Boel E., and Logtenberg T. 1995. Rapid selection of cell subpopulation-specific human monoclonal antibodies from a synthetic phage antibody library. *Proc. Natl. Acad. Sci.* **92:** 3938–3942.

Dziegiel M., Nielsen L.K., Andersen P.S., Blancher A., Dickmeiss E., and Engberg J. 1995. Phage display used for gene cloning of human recombinant antibody against the erythrocyte surface antigen, rhesus D. *J. Immunol. Methods* **182:** 7–19.

Kretzschmar T., Zimmermann C., and Geiser M. 1995. Selection procedures for nonmatured phage antibodies: A quantitative comparison and optimization strategies. *Anal. Biochem.* **224:** 413–419.

Marks J.D., Ouwehand W.H., Bye J.M., Finnem R., Gorick B.D., Voak D., Thorpe S.J., Hughes-Jones N.C., and Winter G. 1993. Human antibody fragments specific for human blood group antigens from a phage display library. *Bio/Technology* **11:** 1145–1149.

Meulemans E. V., Slobbe R., Wasterval P., Ramaekers F.C., and van Eys G. J. 1994. Selection of phage-displayed antibodies specific for a cytoskeletal antigen by competitive elution with a monoclonal antibody. *J. Mol. Biol.* **244:** 353–360.

Mollison P.L., Engelfreit C.P., and Contreras M. 1993. *Blood transfusion in clinical medicine.* Blackwell Scientific Publications, Oxford.

Portolano S., McLachlan S.M., and Rapoport B. 1993. High affinity, thyroid-specific human autoantibodies displayed on the surface of filamentous phage use V genes similar to other autoantibodies. *J. Immunol.* **151:** 2839–2851.

Roark J., Chang T., Caton M., and Siegel D. 1998. Isolation of human IgG platelet autoantibodies using phage display. *Transfusion* **38:** 1S (abstr.).

Siegel D.L. 1995. Isolation of human anti-red blood cell antibodies by repertoire cloning. *Ann. N.Y. Acad. Sci.* **764:** 547–558.

———. 1997. New approaches to monoclonal antibody production. In *Applications of molecular biology to blood transfusion medicine* (ed. Garratty G.), pp. 73–102. AABB, Bethesda.

———.1998. The human immune response to red blood cell antigens as revealed by repertoire cloning. *Immunol. Res.* **17:** 239–251.

Siegel D. and Chang T. 1997. Isolation of cell surface-specific human monoclonal antibodies using phage display and magnetically-activated cell sorting. In *Antibody engineering: New technologies, applications, & commercialization* (ed. Hori W.), pp. 263–272. Biomedical Library Series, IBC, Boston.

Siegel D.L. and Silberstein L.E. 1994. Expression and characterization of recombinant anti-Rh(D) antibodies on filamentous phage: A model system for isolating human red blood cell antibodies by repertoire cloning. *Blood* **83:** 2334–2344.

Siegel D.L., Chang T.Y., Russell S.L., and Bunya V.Y. 1997. Isolation of cell surface-specific human monoclonal antibodies using phage display and magnetically-activated cell sorting: Applications in immunohematology. *J. Immunol. Methods* **206:** 73–85.

Smith G.P. and Scott J.K. 1993. Libraries of peptides and proteins displayed on filamentous phage. *Methods Enzymol.* **217:** 228–257.

Ward R.L., Clark M.A., Lees J., and Hawkins N.J. 1996. Retrieval of human antibodies from phage-display libraries using enzymatic cleavage. *J. Immunol. Methods* **189:** 73–82.

APPENDIX

1 Useful Information

CONTENTS

The DNA Degenerate Alphabet

A Adenine	R = A or G puRine	B = C, G, or T not A
C Cytosine	Y = C or T pYrimidine	D = A, G, or T not C
G Guanine		H = A, C, or T not G
T Thymine	N = A,C,G, or T aNy	V = A, C, or G not T

K = G or T	Keto	(in large groove)	S = G or C Strong	(3 H bonds)
M = A or C	aMino	(in large groove)	W = A or T Weak	(2 H bonds)

complement of:	A C G T	R Y	K M	S W	B D H V	N
is:	T G C A	Y R	M K	S W	V H D B	N

The Genetic Code

First position (5′ end)	Second position U		C		A		G		Third position (3′ end)
U	UUU	Phe	UCU	Ser	UAU	Tyr	UGU	Cys	U
	UUC	Phe	UCC	Ser	UAC	Tyr	UGC	Cys	C
	UUA	Leu	UCA	Ser	UAA	Stop	UGA	Stop	A
	UUG	Leu	UCG	Ser	UAG	Stop	UGG	Trp	G
C	CUU	Leu	CCU	Pro	CAU	His	CGU	Arg	U
	CUC	Leu	CCC	Pro	CAC	His	CGC	Arg	C
	CUA	Leu	CCA	Pro	CAA	Gln	CGA	Arg	A
	CUG	Leu	CCG	Pro	CAG	Gln	CGG	Arg	G
A	AUU	Ile	ACU	Thr	AAU	Asn	AGU	Ser	U
	AUC	Ile	ACC	Thr	AAC	Asn	AGC	Ser	C
	AUA	Ile	ACA	Thr	AAA	Lys	AGA	Arg	A
	AUG	Met	ACG	Thr	AAG	Lys	AGG	Arg	G
G	GUU	Val	GCU	Ala	GAU	Asp	GGU	Gly	U
	GUC	Val	GCC	Ala	GAC	Asp	GGC	Gly	C
	GUA	Val	GCA	Ala	GAA	Glu	GGA	Gly	A
	GUG	Val	GCG	Ala	GAG	Glu	GGG	Gly	G

The NNK Code

First position (5′ end)	Second position U		C		A		G		Third position (3′ end)
U	UUU	Phe	UCU	Ser	UAU	Tyr	UGU	Cys	U
	UUG	Leu	UCG	Ser	UAG	Stop	UGG	Trp	G
C	CUU	Leu	CCU	Pro	CAU	His	CGU	Arg	U
	CUG	Leu	CCG	Pro	CAG	Gln	CGG	Arg	G
A	AUU	Ile	ACU	Thr	AAU	Asn	AGU	Ser	U
	AUG	Met	ACG	Thr	AAG	Lys	AGG	Arg	G
G	GUU	Val	GCU	Ala	GAU	Asp	GGU	Gly	U
	GUG	Val	GCG	Ala	GAG	Glu	GGG	Gly	G

This example of a degenerate nucleotide triplet is central to a proven molecular biology strategy in which synthetic oligonucleotides are used to randomize (or variegate) codons at a defined position in the gene encoding for a protein of interest. By introduction of this degenerate triplet, one can generate libraries of variant genes that have all possible amino acids encoded at a defined position. Other analogous strategies, in particular, VNS, have also been employed (see Chapter 6).

Amino Acids Data

Name	Code	Side group	MW	HW[a]	KD[b]
Alanine	A ala	$-CH_3$	71.0	–0.5	1.8
Cysteine	C cys	$-CH_2$-SH	103.0	–1.0	2.5
Aspartic acid	D asp	$-CH_2$-COOH	115.0	3.0	–3.5
Glutamic acid	E glu	$-CH_2$-CH_2-COOH	129.0	3.0	–3.5
Phenylalanine	F phe	$-CH_2$-ϕ	147.1	–2.5	2.8
Glycine	G gly	–H	57.0	0.0	–0.4
Histidine	H his	$-CH_2$-imidazole	137.1	–0.5	–3.2
Isoleucine	I ile	$-CH(CH_3)$-CH_2-CH_3	113.1	–1.8	4.5
Lysine	K lys	$-(CH_2)_4$-NH_2	128.1	3.0	–3.9
Leucine	L leu	$-CH_2$-$CH(CH_3)_2$	113.1	–1.8	3.8
Methionine	M met	$-CH_2$-CH_2-S-CH_3	131.0	–1.3	1.9
Asparagine	N asn	$-CH_2$-$CONH_2$	114.0	0.2	–3.5
Proline	P pro	[N]-$(CH_2)_3$-[CH]	97.1	0.0	–1.6
Glutamine	Q gln	$-CH_2$-CH_2-$CONH_2$	128.1	0.2	–3.5
Arginine	R arg	$-(CH_2)_3$-NH-CNH-NH_2	156.1	3.0	–4.5
Serine	S ser	$-CH_2$-OH	87.0	0.3	–0.8
Threonine	T thr	$-CH(CH_3)$-OH	101.0	–0.4	–0.7
Valine	V val	–CH-$(CH_3)_2$	99.1	–1.5	4.2
Tryptophan	W trp	$-CH_2$-indole	186.1	–3.4	–0.9
Randomized	X Xaa	—	—	—	—
Tyrosine	Y tyr	$-CH_2$-ϕ-OH	163.1	–2.3	–1.3

The values in this table come from DNA Strider, version 1.0 (Department of Fundamental Research of the CEA, France).

[a] Hopp and Woods hydropathy.

[b] Kyte and Doolittle hydropathy.

Nucleic Acid Data

For information about nucleic acids, restriction enzymes, bacterial strains, and common plasmid elements, we recommend the following references:

- Sambrook J., Fritsch E.F., and Maniatis T. 1989. *Molecular cloning: A laboratory manual.* 2nd Edition. Cold Spring Harbor Laboratory Press, Cold Spring Harbor, New York.
- The regularly updated *Current protocols in molecular biology* (ed. Ausubel et al.), John Wiley & Sons, New York; also available on CD-ROM
- Catalogs from commercial suppliers, e.g., New England Biolabs (http://www.neb.com)
- American Type Culture Collection (ATCC): http://www.atcc.org
- *E. coli* Genetic Stock Center (CGSC): http://cgsc.biology.yale.edu

For searching and aligning antibody sequences, we recommend the following databases:

- GenBank (http://www.ncbi.nlm.nih.gov/Genbank/GenbankSearch.html)
- V BASE (human antibody sequences; http://www.mrc-cpe.cam.ac.uk/imt-doc/)
- The Kabat database (http://immuno.bme.nwu.edu).
- Igblast (human and mouse antibody gene sequences; http://www.ncbi.nlm.nih.gov/igblast/)

Useful sources for phage display include:

- http://www.biosci.missouri.edu/smithgp

  The website of George P. Smith's lab. Includes recipes, the sequences of fd-tet and f88-4, information on phage libraries, cloning, etc.
- http://web.wi.mit.edu/kim/pub/computing.html

  A combinatorial codons program from Peter S. Kim's laboratory. Generates amino acid probability distributions.
- http://www.scripps.edu/mb/barbas/

  The Barbas laboratory website. Includes information about the pComb system.
- http://www.neb.com/neb/tech/nucleotide_seq_maps/sequences/m13ke.seq

  Includes the sequence of M13KE.
- The sequences of pComb3H and pComb3X are in the GenBank database with accession numbers AF268280 and AF268281.

Traditional Fab Primers

Over the past several years, a large number of papers have demonstrated the utility of the pComb system for combinatorial cloning and phage-display selection of antibodies with desired binding specificity. In general, these papers have used an approach in which the heavy (H) and light (L) chain genes encoding the Fab are cloned in separate, sequential steps. For human libraries, we have generally cloned the H-chain gene first, as the L-chain genes more frequently are cleaved by the *Xho*I restriction enzyme used for H-chain gene cloning. This two-step cloning approach has proven utility but is relatively challenging, as two steps of vector preparation and library cloning are required. The well-tested oligonucleotide primers used for these studies are included in the following pages. It should be noted that the original pComb3 vector has many design differences from the more recent pComb3H vector in leader sequences and other expression-related elements. In addition, the pComb3H vector includes nucleotide sequences for the 5′ end of the antibody genes that are consensus sequences for the human system, whereas the earlier pComb3 vector has sequences taken from the murine system. In the latest protocols, including those outlined in this manual, an overlap PCR approach is employed to link the V_H and V_L genes, enabling the use of a single cloning step at *Sfi*I sites.

REFERENCES

Huse W.D., Sastry L., Iverson S.A., Kang A.S., Alting-Mees M., Burton D.R., Benkovic S.J., and Lerner R.A. 1989. Generation of a large combinatorial library of the immunoglobulin repertoire in phage lambda. *Science* **246:** 1275–1281.

Kang A.S., Burton D.R., and Lerner R.A. 1991. Combinatorial immunoglobulin libraries in phage. *Methods* **2:** 111–118.

Peretz D., Williamson R.A., Matsunaga Y., Serban H., Pinilla C., Bastidas R.B., Rozenshteyn R., James T.L., Houghten R.A., Cohen F.E., Prusiner S.B., and Burton D.R. 1997. A conformational transition at the N-terminus of the prion protein features in formation of the scrapie isoform. *J. Mol. Biol.* **273:** 614–622.

Persson M.A.A., Caothien R.H., and Burton D.R. 1991. Generation of diverse high-affinity human monoclonal antibodies by repertoire cloning. *Proc. Natl. Acad. Sci.* **88:** 2432–2436.

Roben P., Barbas S.M, Sandoval L., Lecerf J.-M., Stollar B.D., Solomon A., and Silverman G.J. 1996. Repertoire cloning of lupus anti-DNA autoantibodies. *J. Clin. Invest.* **98:** 2827–2837.

Sasano M., Burton D.R., and Silverman G.J. 1993. Molecular selection of human antibodies with an unconventional bacterial B-cell antigen. *J. Immunol.* **151:** 5822–5839.

Silverman G.J., Roben P., Bouvet J.P., and Sasano M. 1995. Superantigen properties of a human sialoprotein involved in gut-associated immunity. *J. Clin. Invest.* **96:** 417–426.

Williamson R.A., Persson M.A.A., and Burton D.R. 1991. Expression of a human monoclonal anti-rhesus D Fab fragment in *E. coli* using phage lambda vectors. *Biochem. J.* **277:** 561–563.

Williamson R.A., Peretz D., Smorodinsky N., Bastidas R., Blochberger T., Serban A., DeArmond S., Prusiner S.B., and Burton D.R. 1996. Circumventing tolerance in order to generate autologous monoclonal antibodies to the prion protein. *Proc. Natl. Acad. Sci.* **93:** 7279–7282.

Human Primers for Fab Amplification: Original Set

Heavy Chain

V_H 5′ primers

V_{H1a}	5′- CAG GTG CAG CTC GAG CAG TCT GGG -3′
V_{H1f}	5′- CAG GTG CAG CTG CTC GAG TCT GGG -3′
V_{H2f}	5′- CAG GTG CAG CTA CTC GAG TCG GG -3′
V_{H3a}	5′- GAG GTG CAG CTC GAG GAG TCT GGG -3′
V_{H3f}	5′- GAG GTG CAG CTG CTC GAG TCT GGG -3′
V_{H4f}	5′- CAG GTG CAG CTG CTC GAG TCG GG -3′
V_{H4g}	5′- CAG GTG CAG CTA CTC GAG TGG GG -3′
V_{H6a}	5′- CAG GTA CAG CTC GAG CAG TCA GG -3′

C_H 3′ primers

CG1z	5′- GCA TGT ACT AGT TTT GTC ACA AGA TTT GGG -3′	IgG1
CG2a	5′- CTC GAC ACT AGT TTT GCG CTC AAC TGT CTT -3′	IgG2
CG3a	5′- TGT GTG ACT AGT GTC ACC AAG TGG GGT TTT -3′	IgG3
CG4a	5′- GCA TGA ACT AGT TGG GGG ACC ATA TTT GGA -3′	IgG4
CM1	5′- GCT CAC ACT AGT AGG CAG CTC AGC AAT CAC -3′	IgM
CD1	5′- TGC CTT ACT AGT CTC TGG CCA GCG GAA GAT -3′	IgD
CE1	5′- GCT GAA ACT AGT GTT GTC GAC CCA GTC TGT GGA -3′	IgE
CA1	5′- AGT TGA ACT AGT TGG GCA GGG CAC AGT CAC -3′	IgA

Control C_H 5′ primers

CONGa	5′- TCC ACC AAG GGC CCA TCG -3′	IgG1/IgG2/IgG3
CONG4	5′- GCT TCC ACC AAG GGC CCA TC -3′	IgG4
CONM1	5′- GCA TCC GCC CCA ACC CTT -3′	IgM
COND1	5′- CCC ACC AAG GCT CCG GAT -3′	IgD
CONE	5′- CCA CAC AGA GCC CAC CCG TCT TCC CC -3′	IgE
CONA1	5′- TCC CCG ACC AGC CCC AAG -3′	IgA

Light Chain

V_κ 5′-primers

V_{K1a}	5′- GAC ATC GAG CTC ACC CAG TCT CCA -3′
V_{K1s}	5′- GAC ATC GAG CTC ACC CAG TCT CC -3′
V_{K2a}	5′- GAT ATT GAG CTC ACT CAG TCT CCA -3′
V_{K3a}	5′- GAA ATT GAG CTC ACG CAG TCT CCA -3′
V_{K3b}	5′- GAA ATT GAG CTC AC(G/A) CAG TCT CCA -3′

V_λ 5′-primers

V_{L1}	5′- AAT TTT GAG CTC ACT CAG CCC CAC -3′
V_{L2}	5′- TCT GCC GAG CTC CAG CCT GCC TCC GTG -3′
V_{L3}	5′- TCT GTG GAG CTC CAG CCG CCC TCA GTG -3′
V_{L4}	5′- TCT GAA GAG CTC CAG GAC CCT GTT GTG TCT GTG -3′
V_{L5}	5′- CAG TCT GAG CTC ACG CAG CCG CCC -3′
V_{L6}	5′- CAG ACT GAG CTC ACT CAG GAG CCC -3′
V_{L7}	5′- CAG GTT GAG CTC ACT CAA CCG CCC -3′
V_{L8}#	5′- CAG GCT GAG CTC ACT CAG CCG TCT TCC -3′

C_L 3′ primers

κ 3′ primer

CK1d 5′- GCG CCG TCT AGA ATT AAC ACT CTC CCC TGT TGA AGC TCT TTG TGA CGG GCG AAC TCA G -3′

λ 3′ primer

CL2 5′- CGC CGT CTA GAA TTA TGA ACA TTC TGT AGG -3′

Control C_L 5′ primers

KAPPA	CONKa	5′- ACT GTG GCT GCA CCA TCT G -3′
LAMBDA	CONL1	5′- AAG GCT GCC CCC ACG GTC ACT CTG -3′

Human Primers for Fab Amplification: Modified Set

The human primer set was updated somewhat in 1994 for the following reasons:

1. The pComb3H vector had been modified to now contain consensus human (as opposed to mouse) V_H and V_κ sequences at the amino terminus prior to the restriction sites. The new primer set was better adapted to take advantage of this and produce on average a more satisfactory amino terminus.
2. Many more sequences were available than when the original primer set was designed, making some changes desirable.
3. It was felt that some reduction could be made in the total number of primers.

Heavy Chain

V_H 5′ primers

VH135 AG GTG CAG CTG CTC GAG TCT GG
VH2 CAG (A/G)TC ACC TTG CTC GAG TCT GG
VH4 CAG GTG CAG CTG CTC GAG TCG GG
VH4b CAG GTG CAG CTA CTC GAG TGG GG
VH4gs CAG GTG CAG CTA CTC GAG TGG GGC
VH6 CAG GTA CAG CTG CTC GAG TCA GG

IgG1 3′ primers

CG1d GCA TGT ACT AGT TTT GTC ACA AGA TTT GG

Extension primers: These can be used to re-amplify PCR material obtained using the above primers and provide larger overhangs for more efficient restriction enzyme digestion.

VHext $(GTC)_7$ G TAG GTG CAG CTG CTC GAG TC
VH4ext $(GTC)_7$ G TAG GTG CAG CTA CTC GAG TG
VH2ext $(GTC)_7$ G TAG CAG (A/G)TC ACC TTG CTC GAG TC

V_H2 is generally very poorly represented in human response.
CG1ext $(GA)_{10}$G GCA TGT ACT AGT TTT GTC A

κ Chain

V_κ 5′ primers

VK1 GAG CCG CAC GAG CCC GAG CTC CAG ATG ACC CAG TCT CC
VK3 GAG CCG CAC GAG CCC GAG CTC GTG (A/T)TG AC(A/G) CAG TCT CC
VK2/4 GAG CCG CAC GAG CCC GAG CTC GTG ATG AC(C/T) CAG TCT CC
VK5 GAG CCG CAC GAG CCC GAG CTC ACA CTC ACG.CAG TCT CC

C_κ 3′ primer

CK1d GCG CCG TCT AGA ATT AAC ACT CTC CCC TGT TGA AGC TCT TTG TGA CGG GCG AAC TCA G

Extension primers

CKext (GAC)$_6$ GCG CCG TCT AGA ATT AAC ACT CTC
VKext (GAG)$_6$ CCG CAC GAG CCC GA

λ Chain

V_λ 5′ primers

VL1a GAG CCG CAC GAG CCC GAG CTC GTG TTG ACG CAG CCG CCC TC
VL1b GAG CCG CAC GAG CCC GAG CTC GTG CTG ACT CAG CCA CCC TC
VL2 GAG CCG CAC GAG CCC GAG CTC GCC CTG ACT CAG CCT (C/G)CC TCC GT
VL3/9 GAG CCG CAC GAG CCC GAG CTC G(A/T)G CTG ACT CAG CCA CC(C/T) TC
VL7/8 GAG CCG CAC GAG CCC GAG CTC GTG GTG AC(T/C) CAG GAG CC(C/A) TC

C_λ 3′ primer

CL2d G CGC CGT CTA GAA TTA TGA ACA TTC TGT AGG

Extension primers

CLext A(GA)$_9$G CGC CGT CTA GAA TTA TGA ACA TTC TGT AGG
VKext = VLext

Mouse Primers for Fab Amplification

Heavy Chain

V_H 5′ primers

MoVH1	5′- SAK GTG CAG CTC GAG SAG TCA GGA CCT -3′
MoVH2	5′- GAG GTY CAG CTC GAG CAR TCT GGA CCT -3′
MoVH3	5′- CAG GTC CAA CTC GAG CAG YCT GGG KCT -3′
MoVH4	5′- GAG GTT CAG CTC GAG CAG TCT GGR GCW G-3′
MoVH5	5′- GAR GTG AAG CTC GAG GAG WCT GGA SGA -3′
MoVH6	5′- GAG GTG AAG CTT CTC GAG TCT GGA GGT -3′
MoVH7	5′- GAA GTG MAG CTC GAG GAG TCT GGG GGA -3′

C_H 3′ primers

MoIgG1	5′- AGG CTT ACT AGT ACA ATC CCT GGG CAC AAT -3′
MoIgG2a	5′- GTT CTG ACT AGT GGG CAC TCT GGG CTC -3′
MoIgG2b	5′- CTC CTT ACT AGT AGG ACA GGG GTT GAT TGT -3′
MoIgG3	5′- GGG GGT ACT AGT CTT GGG TAT TCT AGG CTC -3′

Light κ Chain

V_L 5′ primers

MoVK1	5′- CCA GTT CCG AGC TCG TTG TGA CTC AGG AAT CT -3′
MoVK2	5′- CCA GTT CCG AGC TCG TGT TGA CGC AGC CGC CC -3′
MoVK3	5′- CCA GTT CCG AGC TCG TGC TCA CCC AGT CTC CA -3′
MoVK4	5′- CCA GTT CCG AGC TCC AGA TGA CCC AGT CTC CA -3′
MoVK5	5′- CCA GAT GTG AGC TCG TGA TGA CCC AGA CTC CA -3′
MoVK6	5′- CCA GAT GTG AGC TCG TCA TGA CCC AGT CTC CA -3′
MoVK7	5′- CCA GTT CCG AGC TCG TGA TGA CAC AGT CTC CA -3′

C_L 3′ primer

MoK3′	5′- GCG CCG TCT AGA ATT AAC ACT CAT TCC TGT TGA A -3′

APPENDIX

2 Recipes

Note: Some recipes are specific for particular protocols. In these cases, the relevant protocols or chapters are indicated in parentheses after the recipe name.

See Appendix 4 for cautions for materials in bold type.

49% Acrylamide/0.5% Bis Stock Solution

49% (w/w) **acrylamide** (EM Science, Cat. # 1150)

0.5% (w/w) **N,N′-methylene-*bis*-acrylamide** (ICN Biomedicals, Cat. # 800175)

sintered-glass apparatus (Millipore Corporation, Cat. # XX1004700)

In a 250-ml beaker containing 200 ml of deionized water and a stir bar, measure 98 g of **acrylamide** and 1 g of ***bis*-acrylamide.** Stir the reagents until dissolved. Using a sintered-glass apparatus, pass the solution through a 0.45-μm filter to remove the fines. Store the solution at 4°C in a clean, brown glass bottle.

Note: Unpolymerized **acrylamide** is a neurotoxin. Take precautions, especially during weighing, not to breathe the solid reagent. Wear a mask and/or use a hood. Wear gloves at all times while working with unpolymerized liquid, and remember that spilled liquid does not polymerize, so treat spills with caution.

10% (w/v) Ammonium Persulfate (APS)

In a 15-ml snap-cap polypropylene tube, dissolve 1 g of **ammonium persulfate** (Sigma, Cat. # A 3678) in 10 ml of purified water. Store at 4°C.

Ammonium Sulfate, Saturated

Dissolve 1 kg of ammonium sulfate in 1 liter of water; boil to dissolve completely. Immediately filter through a coffee filter. Cool while stirring. Adjust pH to 7.2 with **ammonium hydroxide**, and store at 4°C.

Ampicillin: 20 mg/ml ***Ampicillin*** *Stock*

Add 2 g of ampicillin to a beaker containing 45 ml of deionized water, while stirring. Allow the solid to dissolve; then bring the volume to 50 ml. Continue stirring, and add 50 ml of sterile glycerol until thoroughly mixed. Alternatively, dissolve in water without adding glycerol. Store at –20°C. Working concentration is usually 20–60 µg/ml.

10x Annealing Buffer (see Protocol 16.3)

100 mM Tris-HCl, pH 7.5

20 mM **$MgCl_2$**

500 mM NaCl

Basal Salts Medium

39.9 ml of 85% **phosphoric acid**

6.9 g of **calcium sulfate**•$2H_2O$

42.9 g of potassium sulfate

35.1 g of **magnesium sulfate**•$7H_2O$

11.7 g of **potassium hydroxide**

120 g of glycerol

Add water to final volume of 3 liters.

Blocking Buffer (see Chapter 19)

0.1 M $NaHCO_3$, pH 8.6

5 mg/ml BSA

0.02% (w/v) **sodium azide (NaN_3)**

Filter sterilize, store at 4°C.

1x BLOTTO

Mix 2.5 g of skim-milk powder with 25 ml of 10x TBS, and bring to 250 ml with deionized water.

Bradford Reagent

Combine the following:

100 mg of Coomassie Brilliant Blue G-250 (Sigma, Cat # B 0770)

50 ml of 95% **ethanol**

100 ml of 85% (w/v) **phosphoric acid**

When the dye has dissolved, bring the volume to 1 liter with water. Filter through Whatman #1 paper before use.

Note: For more information on the Bradford assay, refer to Ausubel F.M. et al., eds. 1987. *Current protocols in molecular biology*, Vol. 2, pp. 10.1.1–10.1.3. Greene Publishing and Wiley-Interscience, Toronto.

Buffered Glucose

50 mM glucose

25 mM Tris-HCl, pH 8.0

10 mM EDTA, pH 8.0 (adjusted with **NaOH**).

Autoclave or filter-sterilize. Store at 4°C.

Buffer M (10x; supplied with SfiI restriction enzyme)

10 mM Tris-HCl

50 mM NaCl

10 mM **$MgCl_2$**

1 mM **DTT**

pH 7.5

Carbenicillin Stock (100 mg/ml)

Dissolve carbenicillin (α-carboxybenzylpenicillin; Sigma, Cat. # C 1389) at 100 mg/ml in distilled water. Filter sterilize. Divide into 1-ml aliquots and store in the dark at –20°C. Working concentration is usually 20–60 μg/ml.

Chloramphenicol Stock Solution

Dissolve **chloramphenicol** at 35 mg/ml in 100% **ethanol**. It is not necessary to filter-sterilize. Store at –20°C. To prepare selective media, cool the medium to ~50°C after autoclaving, and add 1 ml of the stock solution per liter of medium for a final concentration of 35 μg/ml. Working concentration is usually 25–170 μg/ml.

10X DNA Gel Loading Dye (see Chapters 8–13)

3.9 ml of glycerol

500 µl of 10% SDS

200 µl of 0.5 M EDTA

0.025 g of **bromophenol blue**

0.025 g of xylene cyanol

Bring to 10 ml total volume with water.

2.5 mM dNTP Mix

10 µl of each dNTP (100 mM stocks of dATP, dTTP, dCTP, dGTP)

360 µl of sterile, purified water

This makes a total of 400 µl. Split into aliquots and freeze at –20°C.

5 mM dNTP Mix

10 µl of each dNTP (100 mM stocks of dATP, dTTP, dCTP, dGTP)

160 µl of sterile, purified water

This makes a total of 200 µl. Split into aliquots and freeze at –20°C.

500 mM EDTA stock ($Na_2EDTA \cdot H_2O$; FW 372.2; add 186 g/liter)

The EDTA will not dissolve until it is at ~pH 8.0. Autoclave and store at room temperature.

200 mM EGTA, pH 8.0 (see Chapter 20)

To make this buffer, add the EGTA to nuclease-free water, then adjust pH with 50% NaOH to aid dissolution. Under pH meter and stirring, bring the pH to 8.0, adjust to the final volume, and autoclave 20 minutes, 15 psi, liquid cycle.

5X Electrophoresis Loading Buffer (see Chapter 15)

50% (v/v) glycerol

25 mM Na_2EDTA, pH 8.0

0.02% (w/v) **bromophenol blue**

*1 M **Ethanolamine**, pH 9.0*

For 500 ml of solution, add 30 ml of **ethanolamine** (FW 61.08, d= 1.02 g/ml) to 470 ml of water. Adjust pH to 9.0.

Flow Cytometry Buffer

1% BSA

0.03% **sodium azide** (**NaN_3**)

25 mM HEPES, pH 7.4

in PBS

Sterile filter (0.45 µm); store at 4°C.

40x GBB

142.4 g of Tris base (1.68 M)

45.94 g of anhydrous sodium acetate (or 76.16 g of trihydrate) (0.8 M)

18.83 g of $Na_2EDTA \cdot 2H_2O$ (72 mM)

Dissolve in ~500 ml of water. Adjust pH to 8.3 with glacial **acetic acid**, adjust the final volume to 700 ml, and store at room temperature.

Gel-binding Solution (see Protocol 15.9)

In a plastic, 250-ml Erlenmeyer flask, mix:

100 ml of 95% **ethanol**

3 ml of purified water

300 µl glacial **acetic acid**

100 µl of γ-methacryloxypropyl-trimethoxysilane (Sigma, Cat. # M 6514)

Mix and transfer the solution to a plastic bottle. Store tightly sealed at room temperature.

Gel-repellent Solution (see Protocol 15.9)

RAIN-X ("The Invisible Windshield Wiper"), Blue Coral-Slick 50, Cleveland, Ohio. Available in supermarkets.

Gel-washing Solution (see Protocol 15.9)

To a 4-liter beaker containing 3.2 liters of purified water, add:

400 ml of **methanol** (10%)

400 ml of glacial **acetic acid** (10%)

Mix and pour into 2-liter polypropylene bottles; store at room temperature. The solution can be re-used once.

Glycerol-tolerant Buffer, 20x Stock (see Protocol 15.9)

1.78 M Tris base (216 g/liter)

0.58 M taurine (72 g/liter; Sigma, Cat. # T 0625)

10 mM EDTA (4 g of $Na_2EDTA \cdot 2H_2O$ per liter)

Mix the ingredients in 750 ml of purified water, bring to 1 liter, and autoclave.

Glycerol-tolerant Gel Solution, Stock (see Protocol 15.9)

For 1 liter of gel solution, mix in a 2-liter beaker:

350 ml of purified water

40 ml of 20x glycerol-tolerant buffer

500 g of **urea**, ultra pure (ICN, Cat. # 821527)

200 ml of 40% **acrylamide** stock solution (ReadySol DNA PAGE 40% **acrylamide**, 5% ***bis*-acrylamide**, Amersham Pharmacia Biotech, Cat. # 17-1308-01)

Dissolve the urea by stirring the solution on a hot plate for about 1 hour. The volume needs to be very close to one liter; otherwise it will take longer to dissolve. Vacuum-filter the solution through a 0.45-μm membrane in a Millipore apparatus, transfer to a brown glass bottle, and store at room temperature. The final concentrations are: 8% acrylamide, 8 M urea, and 0.8x glycerol-tolerant buffer.

0.8x Glycerol-tolerant Running Buffer (see Protocol 15.9)

71 mM Tris base

23 mM taurine

0.4 mM EDTA

Dilute the 20x buffer stock to 0.8x by mixing 40 ml of stock in 960 ml of purified water. The pH should be ~8.4. Store at room temperature.

Glycine *Elution Buffer (0.2 M* ***glycine-HCl****, pH 2.2, 1 mg/ml BSA)*

This is properly made by dissolving 1.5 g of **glycine** in ~90 ml of water, adjusting the pH to 2.2 with concentrated **HCl**, adding 100 mg of BSA (bovine serum albumin, molecular biology grade, Sigma, Cat. # B 4287), and raising the final volume to 100 ml. Sterilize by filtration and store at 4°C. This buffer is identical to what is often erroneously referred to as "HCl buffer" and prepared by adjusting the pH of 0.1 M HCl with concentrated glycine. In fact, glycine ($pK_a = 2.35$) is the buffering species at pH 2.2, not HCl ($pK_a = -7$).

1 M HEPES, pH 7 (238.3 g/liter)

Adjust pH with **NaOH**, autoclave, and store at room temperature.

Iodide Buffer

10 mM Tris-HCl, pH 8.0

1 mM EDTA

4 M **NaI** (Sigma, Cat. # S 9538)

Store at room temperature in the dark. Do not use if darker than pale yellow.

IPTG

For 0.5 M stock, dissolve 6.0 g of **IPTG (isopropyl-β-D-thiogalactoside**; Fisher, Cat. # BP 1620-10) in 50 ml of distilled water. Filter sterilize (0.2 μm). Divide into 1-ml aliquots and store at –20°C. For 1 M stock, dissolve 0.238 g/ml in water. For 840 mM stock (see Protocol 18.10), dissolve 1 g in water and bring to a final volume of 5 ml.

Kanamycin Stock

Dissolve kanamycin (kanamycin monosulfate; Sigma, Cat. # K 4000) at 50 mg/ml in distilled water. Filter-sterilize. Divide into 1-ml aliquots and store in the dark at –20°C. Working concentration is usually 10–50 μg/ml.

LB (Luria broth; see Chapters 15–18 and 23)

Dissolve 10 g of tryptone, 5 g of yeast extract, and 10 g of NaCl in 1 liter of ddH_2O. Adjust to pH 7.00 with **NaOH**, and sterilize by autoclaving 20 minutes at 15 psi on liquid cycle. For plates, make 2x stock, autoclave, and store at room temperature.

LB (see Chapter 19)

Dissolve 10 g of tryptone, 5 g of yeast extract, and 5 g of NaCl in 1 liter of ddH_2O. Autoclave; store at room temperature.

LB (Low Salt) + 25 μg/ml Zeocin (see Chapter 12)

Dissolve 10 g of tryptone peptone, 5 g of yeast extract, and 5 g of NaCl in 1 liter-sterile deionized water. Adjust pH to 7.5 with 1 N NaOH, and autoclave. When cooled, add antibiotic. Store at 4°C. For plates, add 15 g of Bacto agar before autoclaving.

*LB Medium Containing 0.2% (w/v) Glucose and 100 μg/ml **Ampicillin** (see Protocol 18.10)*

Measure 10 g of tryptone, 5 g of yeast extract, 5 g of NaCl, and 2 g of glucose into a beaker containing ~600 ml of deionized water, stirring. Dissolve, bring the volume to 1 liter, and autoclave. Cool the solution and add 5 ml of a 20 mg/ml **ampicillin** stock to achieve a final concentration of 100 μg/ml.

LB Agar Plates

Combine 32 g of LB agar (Gibco-BRL, Cat. # 22700-041) with 1 liter of water. Stir. Autoclave for 15 minutes at 121°C. When cooled to 42–45°C, add appropriate antibiotics. Pour into petri dishes and allow to solidify. Store at 4°C. Alternatively, autoclave 500 ml of water and 11 g of Bacto agar. While the agar mixture is still hot, add 500 ml of 2x LB medium. Mix gently, add antibiotics, and pour into petri dishes.

LB Top Agar

Add 0.35 g of Bacto agar and 1.25 g of LB medium (Gibco-BRL) to 50 ml of water. Autoclave and store at 4°C. Melt in a microwave before use.

LB + Mg^{++} Top Agar (see Chapter 19)

For 1 liter, in water:

10 g of Bacto-Tryptone

5 g of yeast extract

5 g of NaCl

1 g of **$MgCl_2$•$6H_2O$**

7 g of Bacto agar

Autoclave, dispense into 50-ml aliquots. Store solid at room temperature, melt in microwave (4 minutes, 50% power) as needed.

*LB/**IPTG/X-gal** Plates*

LB medium + 15 g/liter agar. Autoclave, cool to below 70°C, add **IPTG/X-gal** stock (1 ml/liter), and pour. Store plates at 4°C.

***IPTG/X-gal** stock:* Mix 1.25 g of **IPTG** (**isopropyl β-D-thiogalactoside**; Sigma, Cat. # I 6758) and 1 g of **X-gal** (**5-bromo-4-chloro-3-indolyl-β-D-galactoside**) (Sigma, Cat. # B 4252) in 25 ml of **dimethylformamide**. Store at –20°C in the dark.

5x Lysis Mix (see Chapters 15 and 16)

Combine

2 g of **SDS**

18 ml of water

2 ml of 40x GBB

40 mg of **bromophenol blue**

20 ml of glycerol

For use: add 1 volume to 4 volumes of electrophoresis sample.

MBP Column Buffer

PBS + 1 mM EDTA, pH 7.4

Note: 1 mM or 0.0065% (w/v) sodium azide may be added as a preservative. Nonionic detergents such as Triton X-100 and Tween 20 have been seen to interfere with the affinity of some fusion proteins for the amylose resin.

MBP Elution Buffer

MBP column buffer with 10 mM maltose. Filter-sterilize.

Minimal Media Agar Plates

15 g of Bacto agar

10.5 g of $\mathbf{K_2HPO_4}$

4.5 g of $\mathbf{KH_2PO_4}$

1 g of $\mathbf{(NH_4)_2SO_4}$

0.5 g of sodium citrate dihydrate

Total volume 985 ml in water.

Autoclave.

When cooled to 42–45°C add the following:

1 ml of 1 M $\mathbf{MgSO_4}$ heptahydrate (autoclaved)

0.5 ml of 1% B1 (thiamine HCl) (filter-sterilized)

4 ml of 10 mg/ml amino acids as required (filter-sterilized)

10 ml of 20% glucose (filter-sterilized)

NAP Buffer

80 mM NaCl

50 mM $\mathbf{NH_4H_2PO_4}$, pH 7.0 with $\mathbf{NH_4OH}$.

Autoclave with cap on tight; store at room temperature.

Note: Make the ammonium phosphate buffer as a 0.5 M stock, and autoclave with the cap on tight to prevent evaporation of ammonia.

NZY Medium

10 g of N-Z-amine A (Sheffield Products, Quest International)

5 g of yeast extract

5 g of NaCl

Dissolve in 1 liter of water, adjust pH to 7.5 with NaOH, autoclave. Store at room temperature.

NZY Agar Medium

For 1 liter (about 40 100-mm petri dishes), autoclave 11 g of Bacto agar (Difco) in 500 ml of water in a 2-liter plastic flask; while still hot, add 500 ml of sterile 2x NZY at room temperature (and antibiotics and other heat-labile components as appropriate), mix by gentle swirling, and pour ~25 ml per 100-mm petri dish. Allow plates to cool to room temperature. Dry overnight at room temperature or for several hours at 37°C.

Oligonucleotide Gel Loading Buffer (see Chapter 16)

95% **formamide**

20 mM Na_2EDTA (from a 250 mM stock)

0.05% **bromophenol blue** dye

Make a 10-ml stock and aliquot into 1.5-ml Eppendorf tubes. Store at room temperature.

Panning Elution Buffer (see Chapter 17)

0.1 M **HCl** (adjusted to pH 2.2 with **glycine**)

1 mg/ml BSA

Panning Neutralization Buffer (see Chapter 17)

1 M Tris base, pH 9.1

1x PBS (Phosphate-buffered Saline)

137 mM NaCl	10x stock:	80 g of NaCl
2.7 mM KCl		2.0 g of KCl
12 mM **Na_2HPO_4**		17.0 g of **Na_2HPO_4**
1.2 mM **KH_2PO_4**		1.63 g of **KH_2PO_4**
pH to 7.4 with **HCl**		Bring to 1 liter with water and autoclave.

1% (3%) BSA/PBS (see Chapter 23)

1 g (3 g) of bovine serum albumin (Sigma, Cat. # 7638) to 100 ml with PBS and pH retitrated to pH 7.4.

10x PCR Buffer (Perkin-Elmer, Supplied with Taq DNA Polymerase)

100 mM Tris-HCl, pH 8.3

500 mM KCl

15 mM **$MgCl_2$**

0.01% gelatin

5x **PEG**/NaCl solution (see Chapters 8–13)

Dissolve in water: 200 g of **polyethylene glycol**-8000 (Sigma, Cat. # P 2139), 150 g of NaCl (2.5 M). Bring volume to 1 liter; stir until dissolved. Filter through a 0.8-μm filter. Can be stored at room temperature for several weeks.

PEG/NaCl (16.7%/3.3 M stock; see Protocol 15.3)

100 g of **polyethylene glycol**-8000

116.9 g of NaCl

475 ml of water.

Total volume = 600 ml.

Brief heating to 65°C may be necessary to dissolve solids. Can be autoclaved with no apparent ill effects. Store at 4°C.

13% **PEG**/1.6 M NaCl (see Protocol 15.8)

Dissolve 65 g of **polyethylene glycol**-8000 and 46.8 of g NaCl in water. Bring the volume to 500 ml. Brief heating to 65°C may be necessary to dissolve solids. Can be autoclaved with no apparent ill effects. Store at 4°C.

5x **PEG, PEG**/NaCl (see Chapters 19 and 23)

Dissolve in ddH_2O: 200 g of **polyethylene glycol**-8000 (20% w/v; Sigma, Cat. # P 5413), 150 g of NaCl (2.5 M). Bring to 1 liter. Sterilize by autoclaving 20 minutes at 15 psi on liquid cycle.

PMSF

100 mM = 0.85 g/50 ml of **isopropanol**.

Potassium Acetate (KOAc; "5 M" stock: 3 M KOAc, 2 M **HOAc**; see Chapter 15)

For 100 ml:

- 29.4 g of potassium acetate (anhydrous)
- 71.7 g of water
- 12.1 g of glacial **acetic acid**
- store at 4°C; don't autoclave.

PTM Salts

2 g of **cupric sulfate**•5 H_2O

0.08 g of **sodium iodide**

3 g of **manganese sulfate**

0.2 g of **sodium molybdate**•2 H_2O

0.02 g of **boric acid**

0.5 g of **cobalt chloride**

7 g of **zinc chloride**

22 g of ferrous sulfate•$7H_2O$

5 ml of **sulfuric acid**

Add water to final volume of 1 liter.

Sparge with nitrogen and sterile filter.

SB (Super Broth)

10 g of **MOPS (3 (N-Morpholino) propanesulfonic acid**; Sigma, Cat. # M 8899)

30 g of tryptone (BD Biosciences, Difco, Cat. # 0123-17-3)

20 g of yeast extract (BD Biosciences, Difco, Cat. # 0127-17-9)

1 liter total volume with ddH_2O

Stir to dissolve, titrate to pH 7.0. Sterilize in 500-ml aliquots in 2-liter flasks by autoclaving 20 minutes at 15 psi on liquid cycle.

SB Agar Plates

Add 15 g of Bacto agar to 1 liter of SB, and autoclave. Allow to cool to 42–45°C, and add appropriate antibiotics. Pour into plates, allow to dry, and store at 4°C.

SDS-PAGE Cathode Buffer (Upper Chamber; see Protocol 18.8)

0.1 M Tricine

0.1 M Tris base

0.1% (w/v) **SDS**

No pH adjustment necessary.

SDS-PAGE Anode Buffer (Lower Chamber; see Protocol 18.8)

0.2 M Tris-HCl, pH 8.9

2x SDS-PAGE Sample Buffer (see Protocol 18.8)

12 ml of 10% w/v **SDS**

6 ml of glycerol

1 ml of 1 M Tris-HCl, pH 6.8

10 ml of water

1.2 mg of **bromophenol blue** dye

Sequencing Reaction Stop Solution

95% **formamide**, DNA-grade

20 mM Na_2EDTA

0.05% **bromophenol blue** dye

0.05% xylene cyanol FF dye

SOB Medium

To 950 ml of deionized water, add:

20 g of tryptone

5.0 g of yeast extract

0.5 g of NaCl

186 mg of KCl (or 10 ml of 250 mM KCl)

Mix the solution until dissolved, and adjust the pH to 7.0 with **NaOH**. Adjust the volume to 1 liter. Sterilize by autoclaving for 20 minutes at 15 psi on liquid cycle. When cooled, add 10 ml of sterile 1 M **$MgCl_2$** (see Chapters 8–13 and 19). Alternatively, autoclave in 100-ml portions, then add 1 ml of 2 M Mg^{++} (1 M **$MgCl_2$**, 1 M **$MgSO_4$**, filter sterilized) to each aliquot (see Chapters 15–18 and 23). Store at room temperature.

SOC Medium

Add 1 ml of 1 M glucose (filter-sterilized) to 50 ml of SOB (final = 20 mM). Store at room temperature.

2 M Sodium Acetate (NaOAc), pH 6

2 M sodium acetate (FW 82.03) adjusted to pH 6 with acetic acid. Autoclave with cap tight to prevent evaporation of HOAc. Store at room temperature.

3 M Sodium Acetate, pH 5.2

3 M sodium acetate adjusted to pH 5.2 with acetic acid. Autoclave and store at room temperature.

5% (6.5%) ***Sodium Azide (NaN_3)***

Dissolve 50 mg (65 mg) of solid in 950 µl (935 µl) of water. Store in 1.5-ml microfuge tube at 4°C, with appropriate warnings on tube. Toxic!!

1 M Sodium Bicarbonate ($NaHCO_3$), pH 8.6

Dissolve 84.0 g in water; bring to a total volume of 1 liter. Adjust to pH 8.6 and filter-sterilize.

Streptavidin Stock Solution (see Chapter 19)

Dissolve 1.5 mg of lyophilized streptavidin in 1 ml of 10 mM **sodium phosphate**, pH 7.2, 100 mM NaCl, 0.02% **NaN_3**. Store at 4°C or –20°C (avoid repeated freezing/thawing).

Streptomycin Stock

Dissolve in water for a 50 mg/ml stock. Filter-sterilize. Store at 4°C. Working concentration is usually 10–100 µg/ml.

10x Synthesis Buffer (see Protocol 16.3)

400 mM Tris-HCl, pH 7.5

100 mM **$MgCl_2$**

10 mM ATP

50 mM **dithiothreitol**

Aliquot and store frozen at –20°C.

5x T4 DNA Ligase Buffer (Supplied with T4 DNA Ligase, Gibco BRL)

250 mM Tris-HCl, pH 7.6

50 mM **$MgCl_2$**

5 mM ATP

5 mM **DTT**

25%(w/v) **polyethylene glycol**-8000

100x TAE (Stock for Making 1x TAE Electrophoresis Running Buffer)

484 g of Tris base

114.2 ml of **acetic acid**

200 ml of 0.5 M EDTA

1 liter total volume

TBE (Tris-borate-EDTA) Electrophoresis Buffer, 5x Stock

60.5 g of Tris-HCl

31 g of **boric acid** (**H_3BO_3**)

3.7 g of $Na_2EDTA \cdot 2H_2O$

Dissolve the ingredients in 900 ml of purified water, mix, and bring the volume to 1 liter. Autoclave and store at room temperature.

TBS

50 mM Tris-HCl, pH 7.5, 150 mM NaCl. Autoclave, store at room temperature. Can also be made as a 10x stock.

TBS/Gelatin

0.1% (w/v) gelatin in TBS (0.1 g in 100 ml). Autoclave. Swirl to mix in melted gelatin. Store at room temperature.

TE (Tris/EDTA) Buffer

10 mM Tris base or Tris HCl, pH 8.0

1 mM Na_2EDTA, pH 8.0

Sterilize by autoclaving 20 minutes at 15 psi on liquid cycle. Can also be prepared as a 10x or 100x stock. Store at room temperature.

Tetracycline (tet) Stock

20 mg/ml in **ethanol**. Store at –20°C in the dark. Alternatively, dissolve solid in water at 40 mg/ml, and filter-sterilize into an equal volume of autoclaved, cooled glycerol. Mix, and store at –20°C in the dark. Working concentration is usually 10–50 µg/ml.

Note: Mg^{++} is an antagonist of tet.

TfB I

30 mM KOAc (2.94 g/liter)

100 mM KCl (7.46 g/liter)

10 mM **$CaCl_2$**$\cdot 2H_2O$ (1.5 g/liter)

15% v/v glycerol (150 ml)

50 mM **$MnCl_2$**$\cdot 4H_2O$

Add all of the ingredients, except the **$MnCl_2$**, to 700 ml of purified water, and mix. Bring the solution to a final volume of 900 ml, and autoclave in a pan containing water, with the cap sealed tightly. Crack the seal of the bottle, and allow the solution to cool to room temperature. Dissolve 9.90 g of **$MnCl_2$** in purified water and bring to a final volume of 100 ml. Filter-sterilize the **$MnCl_2$** solution and add it to the cooled, 900-ml solution. Mix and store, tightly sealed, at 4°C.

TfB II

10 mM **NaMOPS** (0.23 g/100 ml)

75 mM **$CaCl_2$**$\cdot 2H_2O$ (1.1 g/100 ml)

10 mM KCl (75 mg/100 ml)

15% (v/v) glycerol (15 ml)

Make the solution in 80 ml of purified water, and adjust the pH to 7.0 with HCl. Bring the volume to 100 ml, then filter-sterilize and store at 4°C.

Top Agar: see LB

1 M Tris-HCl Stocks

Adjust pH with concentrated **HCl** at room temperature. Autoclave; store at room temperature. Note that pH of Tris buffers varies with temperature and concentration.

YPD Medium

10 g of yeast extract

20 g of peptone

in 900 ml of water

Autoclave; add 100 ml of 10x dextrose (20 g of glucose/100 ml water, sterile filter).

YPDS Agar Plates

10 g of yeast extract

182.2 g of sorbitol

20 g of peptone

20 g of Bacto agar

in 900 ml of water

Autoclave. When cooled, add 100 ml of 20% glucose (sterile-filtered) and antibiotics.

Western Blot Transfer Buffer

Combine 1.46 g of **glycine** (39 mM), 2.9 g of Tris base (48 mM), and 0.19 g of **SDS** (0.04%); add 100 ml of **methanol**. Bring to 500 ml with deionized water. No pH adjustment is necessary.

APPENDIX

3 General Procedures

STERILE TECHNIQUE, AND HEALTH AND SAFETY ISSUES

STERILE MICROBIOLOGICAL TECHNIQUE IS FOLLOWED for several reasons. First, it avoids contamination of phage and bacterial stocks and related solutions. Second, and very importantly, it minimizes contamination of DNA samples with outside nucleases. When pipetting, avoid touching the mouth of the bottle or tube. Do not pipet from solutions such as 500-ml bottles of media. Try to use small bottles of media for small-volume work and large bottles for large-volume work, to minimize the number of times the medium is exposed to potential contamination. Place bottle caps and tip-box covers face-up on the bench to avoid contaminating the lip. Microcentrifuge tubes are routinely centrifuged for a few seconds before opening, to avoid sample contamination of the cap or lip of the tube. These are some of the many ways in which phage, bacterial, and nuclease contamination is minimized.

CONTENTS

Be sure you know the biosafety and radiation-safety procedures specified by your institution BEFORE getting started in the lab; most institutions require attendance in special courses on these procedures. If you are exposed to a biohazard, follow the biosafety procedures of your institution, and notify your biosafety officer as well as a physician or other appropriate healthcare worker if necessary.

WORKING WITH DNA

All plasticware (e.g., pipet tips), glassware (glass bottles), and "simple" buffer solutions (that do not contain complex additives like PEG, gelatin, or medium components like

N-Z-amine or yeast extract) to be used for DNA-engineering work should be sterilized in a "clean-steam" autoclave that uses distilled water. No complex solutions, like media, should be allowed in this autoclave. This is to ensure that the trace heavy metals found in tap water, and other contaminants that may be present in complex media or agar, cannot contaminate substances which will contact the DNA that will be worked on. Bottles used for storing solutions for DNA work should be rinsed thoroughly, dried, and stored, as dishwashing probably introduces more contaminants than are left behind by simple rinsing. Keep a separate set of glassware (with differently colored tops) for complex solutions and media. Use "DNA-grade" reagents, which are specifically prepared to have low levels of nucleases and heavy metals that would cleave DNA strands, as well as those that would inhibit the activity of nucleic-acid specific enzymes.

WORKING WITH PHAGE

Sterilize all plasticware, glassware, complex solutions (such as PEG/NaCl and TBS/gelatin), and media in an autoclave that is never used to autoclave biological waste, such as spent medium (which is decontaminated with bleach), spent agar plates, or contaminated plasticware. This is meant to keep the phage load in the autoclave to a minimum, as phage are known to survive standard autoclaving conditions. Phage are killed by heat-treating dry, autoclaved materials in an oven for 4 hours at 105°C or by simply drying autoclaved materials overnight at the same temperature. (If you do the latter, be sure to watch the temperature, so your plasticware doesn't melt!) The latter procedure also kills mold and is an excellent way of sterilizing plastics (like polystyrene) that can't take autoclaving; we do this for the re-use of large 23 x 23 cm plates (Nunc), which can be used for plating up to 2 million clones apiece (see Scott and Smith 1990).

PREPARATION OF STERILE, PURIFIED WATER

To prepare sterile, purified water for the autoclave and for making media, pass distilled tap water through two Ultrapure mixed-bed cartridges (Fisher, Cat. # 09-034-14), followed by a 0.2-µm filter. To sterilize this water, autoclave in closed bottles in a pan of purified water in a clean-steam autoclave.

PHENOL, PHENOL/CHLOROFORM/ISOAMYL ALCOHOL, PHENOL/CHLOROFORM, AND CHLOROFORM EXTRACTIONS OF DNA

Phenol extraction is an excellent way to purify nucleic acids and inactivate DNA-modifying enzymes; as such, it is a good way to prepare DNA for storage. These extractions are meant to rid a solution containing DNA of unwanted proteins, lipids, and some carbohydrates. Depending on the level of contaminants in the DNA sample, a single phenol extraction may suffice. For heavier contamination, phenol/chloroform extractions can be added. Phenol extraction probably removes and inactivates different substances than does phenol/chloroform extraction; thus, the combination of the two is particularly effective at inactivating and removing DNA-modifying enzymes. Chloroform extraction is performed to remove the bulk of the remaining phenol from the aqueous layer. Ethanol precipitation is usually performed after this to ensure the removal of all traces of phenol and chloroform, as well as to change the solution in which the DNA is suspended and to concentrate the DNA. If there is any remaining phenol, it can affect enzymatic reactions and oxidatively damage the nucleic acid itself. Phenol contamination can be detected spectroscopically, as it gives the 260–280-nm absorption curve for DNA a distinctive "shoulder" at 270 nm.

Be sure to wear protective eyewear, a lab coat, and gloves whenever working with phenol, as it can burn. If it gets on your skin, rinse the majority off with lots of water, then remove the remainder with soap (do not use alcohol, as it may help the phenol penetrate the skin). If it gets into your eyes, rinse immediately with lots of water. Phenol, like chloroform, should be treated as hazardous waste. Be sure you know the biosafety and biohazard protocols of your institution.

Phenol

Buy phenol saturated in water (but not pH-adjusted) and store it at –20°C. For use, thaw a bottle containing 100 ml of phenol in water (BRL), and place the contents into a separatory funnel. Add 100 ml of 0.5 M Tris-HCl, pH 8.0. Hold the stopper and shake vigorously for about 10 seconds. Invert the funnel and turn the stopcock to release built-up gas; shake and vent several times. Vent the funnel in a hood or away from other workers, as sometimes the pressure may shoot an aspirate out of the funnel. Place the funnel on a stand, and allow the phases to separate. Pour the phenol layer into a beaker, and discard the aqueous layer in the sink with plenty of running water. Repeat the addition of Tris three times, then remove the aqueous layer and check the pH of the phenol with pH paper; it should be approximately 8.0. Transfer the phenol to a clean, screw-cap, brown bottle containing 10–15 ml of 0.1 M Tris-HCl, pH 8.0. Store it in the dark at 4°C; it should be good for 1–2 months (it goes bad by becoming oxidized; this turns it yellow). The condition of the phenol is essential to the success of procedures using single-stranded (ss) DNA; oxidized phenol can cleave ssDNA, making it useless. Thus, the use of fresh phenol is recommended for these experiments.

Chloroform/Isoamyl Alcohol

Remember that chloroform is poisonous and should be treated as hazardous waste. Be sure you know the biosafety and biohazard protocols of your institution. Isoamyl alcohol helps reduce foaming. The density of chloroform is 1.5 g/ml and the density of isoamyl alcohol is 0.81 g/ml. Sterilize a 500-ml, wide-mouth glass jar and the sintered glass stopper that goes with it; wrap the mouth of the jar and the entire cap in foil. Tare the jar and pour a volume of chloroform into it (preferably from a jar or can whose contents have only been poured, not pipetted); weigh and calculate the volume. Add 1/24 volume of isoamyl alcohol to make a 24:1 (v/v) mixture, and store at room temperature.

Phenol/Chloroform/Isoamyl Alcohol (25:24:1)

It is best to make small volumes of this that will be used fairly quickly. Make a 1:1 (v/v) mixture of phenol and chloroform/isoamyl alcohol, mix well, and centrifuge briefly. Remove or avoid the water that forms as a top layer. Store in the dark at 4°C.

Phenol, Phenol/Chloroform, and Chloroform Extraction

Estimate the volume of the solution to be extracted, and add an equal volume of phenol. Vortex thoroughly and centrifuge (centrifuge microcentrifuge tubes in a microfuge at ≥12,000 rpm for 1–2 minutes). Transfer the aqueous layer to a fresh tube; avoid the white interphase. Remove and discard most of the phenol, leaving the interphase. Back-extract the interphase by adding approximately 40 μl of purified water (for a 0.5-ml tube, add ~10 μl), vortexing, and re-centrifuging. Remove the aqueous layer (avoiding the interphase if it is opaque) and combine with the other extracted aqueous layer.

Extract the combined aqueous layers as above with an equal volume of phenol/chloroform, and transfer the aqueous layer to a fresh tube. If a white precipitate is visible at the interphase, back-extract it. Repeat the phenol/chloroform extractions (and back extractions) until the interface no longer is opaque.

Remove phenol from the aqueous solution by chloroform extraction. Add an equal volume of chloroform/isoamyl alcohol to the aqueous layer, vortex thoroughly, and centrifuge at ≥12,000 rpm for 1–2 minutes. Transfer the aqueous layer to a fresh microcentrifuge tube and repeat the chloroform extraction. Repeat the process once. Usually phenol extraction is followed by ethanol precipitation. If residual phenol is present in the DNA preparation, spectrophotometric scanning will reveal a shoulder at 270 nm. In this case, extract the DNA solution twice with choloroform/isoamyl alcohol, then precipitate it with ethanol.

ETHANOL PRECIPITATION OF DNA

This procedure is used to change the solution in which a DNA sample is suspended, to remove contaminants (such as phenol and chloroform), and to concentrate the nucleic acid. Alternatively, the sample buffer can be changed and the sample can be concentrated by filtration through size-exclusion membranes (Amicon, Millipore).

100% (Absolute) Ethanol

Sterilize a 500-ml, wide-mouth glass jar and the sintered glass stopper that goes with it; wrap the mouth of the jar and the entire cap in foil. Pour the contents of a new bottle of absolute ethanol directly into the sterile jar; cap and store at room temperature.

Note: Absolute ethanol is a controlled substance.

70% Ethanol

The density of ethanol is 0.81 g/ml. Place a sterile, 100-ml, wide-mouth, screw-cap, glass jar on a balance and tare it. Pour a volume of absolute ethanol into the jar (preferably from a bottle that has not been pipetted from); weigh, and calculate the volume. Add sterile water to make the final solution 70% (v/v) ethanol, and store at –18°C. At cold temperatures, the ethanol is viscous and thus less likely to disrupt a DNA pellet during the wash.

Precipitation of DNA with Sodium Acetate

Add a 1/10 volume of 3 M sodium acetate (NaOAc, pH ~5.2–6) to a solution of DNA. Mix. Add 2 volumes of absolute ethanol, mix, and let stand at 4°C for 2 hours. Recommended temperatures and times for this incubation vary; temperatures range from –70°C to room temperature, and times range from 0–15 minutes to overnight. More concentrated DNA solutions (>0.1 µg/ml) generally require less incubation time, whereas dilute solutions (<10 ng/ml) require much longer incubations (see TechOnLine, http://www.lifetech.com). If SDS is present above about 0.5% (w/v), it will precipitate, as might contaminants (such as carbohydrates). If there is only a small amount of DNA, or if it is very dilute, the addition of a "carrier" such as poly(A) RNA or glycogen (Roche Molecular Biochemicals) to the precipitation may help coprecipitate the DNA. Add ~5 µg of carrier; enough to see a pellet after centrifugation. Microfuge at full speed (≥12,000 rpm) for 15–30 minutes at 4°C. If the DNA concentration is low (especially if the volume is large), a longer centrifugation period is required (up to ~45 minutes). After centrifugation, gently aspirate the solution with a pipet; avoid upsetting the pellet. Centrifuge the tube briefly and remove the last bit of liquid. Air-dry the pellet, or dry with a Speed-Vac; however, avoid drying the pellet too much, especially with heat, as an over-dried pellet is sometimes difficult to dissolve.

Washing a DNA Pellet

Washes are performed to remove excess contaminants, especially salts and chelaters, that may inhibit the next step in the procedure. Immediately after aspirating the last of the ethanol/NaOAc, add enough 70% (v/v) ethanol to fill the 0.5- or 1.5-ml microcentrifuge tube containing the pellet. If the pellet is small or invisible, place the tip of the pipet on the side of the tilted microcentrifuge tube and allow the ethanol to slowly cover the pellet and fill the tube. If the pellet is large, this is not necessary, and the pellet can even be vortexed. Microfuge at ≥12000 rpm for 1–2 minutes, and aspirate the solution. Centrifuge briefly to bring the last bit of ethanol to the bottom of the tube, and aspirate. Air-dry the pellet, or briefly dry in a Speed-Vac.

Extraction of DNA from Agarose Gels by Electroelution with an Elutrap

One of the key issues in achieving success with phage-display technology is the quality and quantity of the DNA fragments that are used to construct combinatorial antibody libraries. In this manual, three methods are recommended by which DNA can be extracted successfully from agarose gel slices—electroelution, freeze/squeeze, and resin binding (e.g., QIAEX II Gel Extraction Kit, QIAGEN). The following protocol summarizes the use of the Elutrap electroelution system from Schleicher and Schuell (see Fig. A3). A detailed protocol is available upon purchase of the electroelution system.

Materials

1x TAE electrophoresis running buffer (see Appendix 2)
Elutrap Electro-Separation System (Schleicher & Schuell, Cat. # 46178)
BT1 membranes (Schleicher & Schuell, Cat. # 46180)
BT2 membranes (Schleicher & Schuell, Cat. # 46190)
forceps

Procedure

1. Clean the Elutrap chamber with water. Apply vacuum grease as a sealant to the inner threads at each end and to the outer sides and edges of the U inserts. Do not apply grease to the arched inserts or clamping plates.
2. Beginning at the end marked with a (–), place one arched insert into the Elutrap device with the flat side facing the threads. Open a BT1 membrane (handle only with forceps and gloves) and place it against the arched insert with the taller corner of the sloping edge pointing toward the arrow. Tighten the clamping plate with the supplied lever device. If the BT1 membrane is punctured or torn, replace it with a new membrane before proceeding. If the integrity of the membrane is compromised in any way, it will result in loss of sample.
3. Working at the end marked with a (+), insert the four U inserts into the notched region of the device. Open a BT2 membrane (handle only with forceps and gloves), and place it against the fourth U insert facing the elution chamber. Carefully apply the third arched insert against the BT2 membrane with the flat side facing the threads. It is important that the BT2 membrane is touching the bottom of the device and completely covers the U opening. If the BT2 membrane is punctured or torn, replace it with a new membrane. If the integrity of the membrane is compromised in any way, it will result in loss of sample.
4. Open a second BT1 membrane and place it against the fourth arched insert facing the end marked with a (+), with the taller corner of the sloping edge point-

ing toward the arrow on the Elutrap. Tighten the clamping plate with the supplied lever device. Again, take care not to puncture or tear the membrane.

5. Place the Elutrap device into the electrophoresis chamber with the (+) pointing toward the positive electrode. Position the tray slider so that the aperture is open to allow the current to flow through the device. The unused tray sliders should be positioned so that the apertures are closed. Fill the electrophoresis chamber with 1X TAE buffer so that the level of buffer on the outside of the Elutrap device covers half of the BT1 membranes.

6. Fill the device with 7–10 ml of 1X TAE. Cut the agarose gel slice into small pieces with a clean razor blade and place them next to the BT2 membrane inside the device. The buffer should cover the gel slices. Pipet 200 μl of 1X TAE buffer into the reservoir between the BT1 and BT2 membranes.

7. Place the lid on the electrophoresis chamber and connect the leads to the appropriate electrodes. Apply 150 V for 1–4 hours. Reverse the current for 30 seconds, stop the current, disconnect the leads, remove the lid, and collect the liquid in the reservoir between the BT1 and BT2 membranes. Be careful not to puncture either membrane. Replace sample with 200 μl of fresh 1X TAE buffer. Replace the lid, reconnect the leads, and apply 100 V overnight. In the morning, collect the liquid as described.

8. Ethanol-precipitate the DNA that is collected from the reservoir, as described in this Appendix.

9. Upon completion of electroelution, remove and discard all membranes. Do not reuse any membranes, as this may result in contamination from one sample to the next. Wash the device with water.

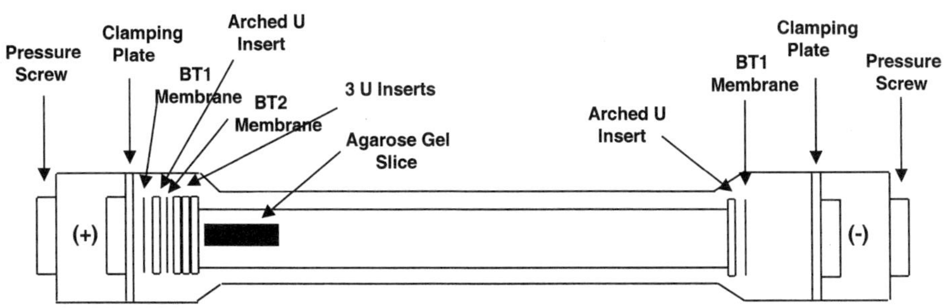

Figure A3. The Elutrap device. This figure represents a top view of the Elutrap electroelution device. The entire electroelution system can be purchased from Schleicher & Schuell. (Adapted, with permission, from the Schleicher & Schuell manual.).

Extraction of DNA from Agarose Gels Using Freeze/Squeeze

The following protocol summarizes the use of freeze/squeeze for isolation of DNA from agarose gel slices. A detailed protocol on the use and handling of the DNA filter units is available with purchase of the units from Millipore.

Materials

Ultrafree-MC Centrifugal Filter Units, 0.45 μm (Millipore, Cat. # UFC3OHVNB)
Parafilm (VWR, Cat #52858-000)

Procedure

1. Cut out the appropriate DNA band from a 1% (w/v) agarose gel, taking care to remove excess agarose. Excess agarose can result in lower yields of DNA by blocking the filter unit.

   Note: Freeze/squeeze yields from >1% agarose gels can be improved by using gels made from a mixture of half agarose and half low-melt agarose.

2. Form a pocket for the gel slice with a piece of Parafilm. Fold and seal two edges of the pocket by compressing with the conical end of a microcentrifuge tube. Insert the gel slice and fold the unsealed edge. Freeze the gel slice at –20°C for a minimum of 30 minutes. Alternatively, the gel slice can be placed at –80°C for faster freezing.

3. After the gel slice is completely frozen, thaw partially, then use the flat end of a closed microcentrifuge tube to thoroughly squeeze the gel. Thawing can be done for a few minutes at room temperature or briefly at 37°C. In general, the best yields are obtained when the gel slice is crushed as much as possible.

4. Transfer both liquid and gel to pre-labeled filter units. Spin in a microcentrifuge at 6000–7000 rpm for 5 minutes.

5. Remove and discard the top of the filter unit. Ethanol-precipitate the liquid containing the DNA, as described in this Appendix. Use glycogen (Roche Molecular Biochemicals) as a carrier.

Quantitation of DNA and RNA

QUANTITATION WITH A SPECTROPHOTOMETER

There are several ways to quantitate solutions of nucleic acid. If the solution is pure, one can use a spectrophotometer to measure the amount of ultraviolet radiation absorbed by the bases. Use a quartz cuvette, and use water or TE as a solvent. Zero the spectrophotometer with a sample of solvent. For more accurate readings, dilute the nucleic acid to give readings between 0.1 and 1.0.

For a 1-cm pathlength, the optical density at 260 nm (OD_{260}) = 1

for a 50 μg/ml solution of dsDNA,

for a 33 μg/ml solution of ssDNA,

for a 20–30 μg/ml solution of oligonucleotide,

and for a 40 μg/ml solution of RNA.

Sample calculation for a solution of dsDNA:

$$\text{DNA concentration} = 50\ \mu g/ml \times OD_{260} \times \text{dilution factor}$$
$$= 50\ \mu g/ml \times 0.65 \times 50$$
$$= 1.63\ mg/ml$$

Contamination of nucleic acid solutions makes spectrophotometric quantitation inaccurate. Calculate the OD_{260}/OD_{280} ratio for an indication of nucleic acid purity. Pure DNA has a ratio of about 1.8; pure RNA has a ratio of about 2.0. Low ratios could be caused by protein or phenol contamination.

QUANTITATION BY ETHIDIUM BROMIDE FLUORESCENCE EMISSION

(adapted from Sambrook et al. 1989)

DNA can be quantified by measuring the UV-induced emission of fluorescence from intercalated ethidium bromide. This method is useful if there is not enough DNA to quantify with a spectrophotometer, or if the DNA solution is contaminated. Run DNA samples (include several amounts ranging from 25 to 200 ng) on a 0.8% agarose minigel containing 0.5 μg/ml ethidium bromide. To maintain constant background staining of the gel, include 0.5 μg/ml ethidium bromide in the running buffer. Run the samples next to DNA standards of known concentration or use molecular mass markers (DNA MASS Ladder, GIBCO-BRL, Cat. # 10068-013 and 10496-016). Use a UV light to photograph the gel. Compare fluorescence intensities and estimate DNA concentrations.

To quantitate mixtures of DNA fragments or DNA of multiple sizes, spot samples on a 1% agarose slab gel containing 0.5 μg/ml ethidium bromide. Also spot several DNA standards of known concentration. Let the gel stand for a few hours at room temperature to allow small contaminating molecules to diffuse away. Use a UV light to photograph the gel. Compare fluorescence intensities and estimate nucleic acid concentrations.

Cellular Transformation by Heat Shock in $CaCl_2$

Transformation by heat shock in the presence of $CaCl_2$ is a widely used method for the introduction of a plasmid or phage genome into bacterial cells. Compared to electroporation, it is a simpler method, but the yield of transformants is 10^3- to 10^4-fold lower. Consequently, electroporation is recommended as the transformation method for the production of phage libraries, whereas either method can be used for producing a single phage or phagemid clone, such as a site-directed mutant. The DNA used in $CaCl_2$ transformation does not have to be salt-free, as it does with electroporation, which allows one to transform directly after DNA ligation. The protocol below describes how competent cells are made, stored, and used in heat-shock transfection.

PREPARATION OF COMPETENT CELLS

Materials

E. coli, untransfected colonies (K91, XL1-Blue, ER2537, MC1061, etc.)
sterile, capped, 15 x 150 mm glass culture tubes
LB medium (see Appendix 2)
sterile, capped 250-ml culture flask
2 M glucose (filter-sterilized)
sterile, screw-capped, 40-ml, polypropylene, round-bottom (Oak Ridge) tubes
ice-cold TfB I buffer (see Appendix 2)
ice-cold TfB II buffer (see Appendix 2)
sterile 0.5-ml microcentrifuge tubes

CAUTION: ***E. coli*** **(see *Appendix 4*)**

Procedure

1. Inoculate a 15 x 150 mm glass culture tube containing 3 ml of LB medium with a single *E. coli* colony and shake vigorously overnight at 37°C.
2. Inoculate a 250-ml culture flask containing 35 ml of LB medium and 2.1 ml of 2 M glucose with 80 µl of the overnight culture. Shake at 37°C until the cell density reaches $OD_{595} = 0.6–0.8$.
3. Place the flask on ice, swirl gently to cool rapidly, and transfer the cultures to cold Oak Ridge tubes. Pellet the cells by centrifugation at 900*g* for 10 minutes at 4°C. *At this step, and beyond, everything must be done in a 4°C cold room or on ice.*
4. Resuspend the pellet in 7 ml of ice-cold TfB I buffer by gently shaking on ice, then pellet the cells again, as above.
5. Gently resuspend the pellet in 1.4 ml of ice-cold TfB II buffer, keeping the mixture on ice.

6. Dispense the cells in 100-μl aliquots into prechilled 0.5-ml microcentrifuge tubes. Quick-freeze the cells on crushed dry ice and store at –70°C. The cells retain competence for up to six months at –70°C.

SMALL-SCALE HEAT SHOCK TRANSFORMATION OF COMPETENT CELLS

Materials

competent *E. coli* cells (K91, XL1-Blue, ER2537, MC1061, etc.)
sterile ice-cold 1.5-ml microcentrifuge tubes
purified RF phage DNA, ligated oc- or ccc-RF DNA, or phagemid DNA
ccc f88-4 vector RF DNA or pComb3 DNA (positive control)
SOC, NZY, or LB medium (containing 0.2 μg/ml tetracycline for fd-tet-derived vectors) (see Appendix 2)
NZY agar plates containing 40 μg/ml **tetracycline** (for fd-tet-derived vectors)
LB agar plates containing 50 μg/ml carbenicillin (for pComb3 vectors)

CAUTIONS: ***E. coli*, tetracycline (*see Appendix 4*)**

Procedure

1. Thaw cells briefly at room temperature, place on ice, and transfer 20-μl aliquots into ice-cold 1.5-ml microcentrifuge tubes. Keep the aliquots on ice.
2. Add DNA (1–50 ng) to each tube and mix gently with a pipet tip. As a positive control, use a known amount of ccc vector RF DNA (for example, 1 ng of f88-4 or pComb3). As a negative control, use only the buffer in which the experimental and positive control DNA is suspended (such as 0.1x TE).

   Note: Ligations (from Protocol 16.3) should be diluted 5-fold before transformation, because toxic substances in the ligation reactions can decrease transformation efficiency.

3. Incubate the mixtures on ice for 5 minutes.
4. Heat-shock the cells by incubating for 90 seconds in a 42°C water bath. Incubate on ice for 1–2 minutes.
5. Add 1 ml of SOC, NZY, or LB medium (containing 0.2 μg/ml tetracycline, if transforming with fd-tet-derived vector) to each tube. Incubate the tube for 1 hour at 37°C with shaking at 225 rpm.
6. Dilute the cells in medium and spread on plates containing NZY agar and 40 μg/ml tetracycline (for fd-tet-derived vectors) or LB agar and 50 μg/ml carbenicillin (for pComb vectors). Dilution and spreading volumes depend on the experiment and the quality and quantity of DNA. Incubate overnight at 37°C.

7. Subclone transformed colonies, grow cloned cells, and check minipreps (see Sambrook et al. 1989) for the presence of plasmid. Alternatively, check supernatants for the production of fd-tet-derived phage by agarose gel electrophoresis (Protocol 15.5).

SDS-PAGE on 12% Polyacrylamide Minigels

We follow the procedure of Sambrook et al. (1989) with some minor modifications. Gels are run on the Bio-Rad Mini PROTEAN II system.

Materials

two 15-ml polypropylene snap-cap tubes (Falcon)
1.5 M Tris-HCl, pH 8.8
20% (w/v) **SDS**
acrylamide (EM Science, Cat. # 1150)
N,N´-methylene-bis-acrylamide (ICN Biomedicals Inc., Cat. # 800175)
acrylamide stock: Prepare 29.2% (w/v) **acrylamide**, and 0.8% (w/v) **N,N´-methylene-bis-acrylamide** in deionized water, following in general the procedure for acrylamide stock solution in Appendix 2.
10% (w/v) **ammonium persulfate** in deionized water (Sigma Chemical Co., Cat. #A-3678)
1.0 M Tris-HCl, pH 6.8
TEMED (N,N,N´,N´-Tetramethylethylenediamine)
anhydrous **ethanol**
Tris-glycine running buffer: 25 mM Tris base, 0.1% **SDS**, 0.192 M **glycine**, pH 8.3 (when amounts are measured properly, there is no need to adjust the pH). Make up a stock as follows: Combine 6.06 g of Tris base, 2.0 g of **SDS**, and 28.83 g of **glycine** in a large beaker. Make up to 2 liters with ddH_2O. Store in a large plastic bottle at room temperature.
5x gel loading buffer: 2.5 ml of 1.0 M Tris-HCl, pH 6.8, 0.2 g of **SDS**, 5 mg of **bromophenol blue**, 5 ml of glycerol, 2.5 ml of deionized water. Make 0.2-ml aliquots and store frozen. If it is desired to break disulfide bonds in the protein sample, add **dithiothreitol** prior to loading (15.4 mg of **DTT** per 0.2-ml aliquot).
methanol
glacial **acetic acid,** reagent grade
Coomassie Brilliant Blue R250 staining solution (0.25 g of dye in 100 ml in 50:40:10 methanol:deionized water:glacial acetic acid; ICN Biomedicals, Inc., Cat. # 821616)
Destaining solution (50:40:10 **methanol:**deionized water:glacial **acetic acid**)

CAUTIONS: SDS, acrylamide, bisacrylamide, ammonium persulfate, TEMED, ethanol, glycine, bromophenol blue, DTT, methanol (*see Appendix 4*)

Procedure

1. Prepare the resolving gel solution in a 15-ml Falcon tube: 3.3 ml of deionized water, 2.5 ml of 1.5 M Tris-HCl, pH 8.8, 4.0 ml of acrylamide stock, 50 µl of 20% SDS. Mix by inversion and add 100 µl of 10% ammonium persulfate. Mix again and add 10 µl of TEMED. Mix and immediately pour the resolving gel, leaving approximately 1 cm of space from the top of the plates for the stacking gel. Gently layer approximately 10 drops of 50% ethanol (made with absolute ethanol) on top of the resolving gel, and allow the gel to polymerize (~15–20 minutes) at room temperature.
2. Prepare the stacking gel solution in a separate 15-ml Falcon tube: 3.4 ml of deionized water, 0.63 ml of 1.0 M Tris-HCl, pH 6.8, 0.83 ml of acrylamide stock, 25 µl of 20% SDS. When the gel has polymerized, pour the ethanol off. Mix the stacking gel solution by inversion, and add 100 µl of 10% ammonium persulfate, mix again, and add 10 µl of TEMED. Mix and immediately pour the stacking gel on top of the resolving gel. Insert the comb, and allow the stacking gel to polymerize (~30 minutes).
3. Prepare the polymerized gel for electrophoresis by removing the comb, assembling the gel apparatus, and filling the buffer reservoirs with Tris-glycine running buffer. Flush out the wells with a syringe filled with Tris-glycine running buffer, using a 23-gauge needle.
4. Prepare the samples in 0.6-ml microcentrifuge tubes. Dilute the samples in deionized water to a total volume of 16 µl. Add 4 µl of 5x loading buffer to each sample, and incubate in a boiling water bath for about 2 minutes. During this time, flush out the wells again. Briefly microfuge the samples and load them onto the gel. Run the gel at 100 V until the dye front reaches the bottom of the gel.
5. Dissemble the gel and stain it with Coomassie Brilliant Blue dye by immersing the gel in staining solution and rocking for about 1 hour. Return the staining solution to a glass bottle for reuse.
6. Destain the gel by incubating it in 100 ml of destaining solution for 15 minutes, with gentle rocking. Pour off the destain and repeat, then wash in 100 ml of 10% glacial acetic acid two or three times until the protein bands are visible and the gel matrix is nearly colorless.
7. Sandwich the destained gel between two pieces of overhead transparencies (or in plastic wrap with the wrinkles smoothed out). Photograph the gel or scan it on a desktop scanner.

REFERENCE

Sambrook J., Fritsch E.F., and Maniatis T. 1989. *Molecular cloning: A laboratory manual.* 2nd Edition. Cold Spring Harbor Laboratory Press, Cold Spring Harbor, New York.

APPENDIX

4 Cautions

The following general cautions should always be observed.

- You are absolutely required to **inform yourself** on the properties of the substances you will work with **before** beginning the procedure.
- **The absence of a warning** does not necessarily mean that the material is safe, since information may not always be complete or available.
- If you are **exposed** to toxic substances or think you have been exposed to them, contact your local safety office immediately for instructions.
- **Proper disposal procedures** must be used for all chemical, biological, and radioactive waste.
- For specific guidelines on **appropriate gloves**, consult your local safety office.
- **Acids and bases** that are concentrated should be handled with great care. Wear goggles and appropriate gloves. A face shield should be worn when handling large quantities.

  Strong acids should not be mixed with organic solvents because they may react. Especially, sulfuric acid and nitric acid may react highly exothermically and cause fires and explosions.

  Strong bases should not be mixed with halogenated solvent because they may form reactive carbenes which can lead to explosions.
- Never **pipet** solutions using mouth suction. This method is not sterile and can be dangerous. Always use a pipet aid or bulb.
- **Halogenated and nonhalogenated** solvents should be kept separately (e.g., mixing chloroform and acetone can cause unexpected reactions in the presence of bases). Halogenated solvents are organic solvents such as chloroform, dichloromethane, trichlorotrifluoroethane, and dichloroethane. Some nonhalogenated solvents are pentane, heptane, ethanol, methanol, benzene, toluene, *N,N*-dimethylformamide (DMF), dimethyl sulfoxide (DMSO), and acetonitrile.
- **Laser radiation**, visible or invisible, can cause severe damage to the eyes and skin.

Take proper precautions to prevent exposure to direct and reflected beams. Always follow manufacturer's safety guidelines and consult your local safety office. See caution below for more detailed information.

- **Flash lamps**, due to their light intensity, can be harmful to the eyes. They also may explode on occasion. Wear appropriate eye protection and follow the manufacturer's guidelines.
- **Photographic fixatives and developers** also contain chemicals that can be harmful. Handle them with care and follow manufacturer's directions.
- **Power supplies and electrophoresis equipment** pose serious fire hazards and electrical shock hazards if not used properly.
- The use of **microwave ovens and autoclaves** in the lab requires certain precautions. Accidents have occurred involving their use (e.g., to melt agar or Bacto agar stored in bottles or to sterilize). Often the screw top is not completely removed and there is not enough space for the steam to vent. When the containers are removed from the microwave or autoclave, they can explode and cause severe injury. Always completely remove bottle caps before microwaving or autoclaving. An alternative method for routine agarose gels that do not require sterile agar is to weigh out the agar and place the solution in a flask.
- Injury may be caused by the incautious use of **cutting devices** such as microtome blades, scalpels, razor blades, or needles. Microtome blades are extremely sharp! Use care when sectioning. If unfamiliar with their use, have someone demonstrate proper procedures. For proper disposal, use the "sharps" disposal container in your lab. It is recommended that used needles be discarded unshielded, with the syringe still attached. This prevents injuries (and possible infections; see Biological Safety) while manipulating used needles since many accidents occur while trying to replace the needle shield. Injuries may also be caused by broken pasteur pipets, coverslips, or slides.

GENERAL PROPERTIES OF COMMON CHEMICALS

The hazardous materials list can be summarized in the following categories:

- Inorganic acids, such as hydrochloric, sulfuric, nitric, or phosphoric are colorless liquids with stinging vapors. Avoid spills on skin or clothing. Spills should be diluted with large amounts of water. The concentrated forms of these acids can destroy paper, textiles, and skin as well as cause serious injury to the eyes.
- Inorganic bases such as sodium hydroxide are white solids that dissolve in water and under heat development. Concentrated solutions will slowly dissolve skin and even fingernails.
- Salts of heavy metals are usually colored powdered solids that dissolve in water. Many of them are potent enzyme inhibitors and therefore toxic to humans and to the environment (e.g., fish and algae).

- Most organic solvents are flammable volatile liquids. Avoid breathing the vapors, which can cause nausea or dizziness. Also avoid skin contact.
- Other organic compounds, including organosulfur compounds such as mercaptoethanol and organic amines, can have very unpleasant odors. Others are highly reactive and should be handled with appropriate care.
- If improperly handled, dyes and their solutions can stain not only your sample, but also your skin and clothing. Some of them are also mutagenic (e.g., ethidium bromide), carcinogenic, and toxic.
- All names ending with "ase" (e.g., catalase, β-glucuronidase, or zymolase) refer to enzymes. There are also other enzymes with nonsystematic names like pepsin. Many of them are provided by manufacturers in preparations containing buffering substances, etc. Be aware of the individual properties of materials contained in these substances.
- Toxic compounds are often used to manipulate cells. They can be dangerous and should be handled appropriately.

HAZARDOUS MATERIALS

ABTS, *see* **2,2′-Azino-bis(3-ethylbenzthiazoline)-6-sulfonic acid**

Acetic acid (concentrated) must be handled with great care. It may be harmful by inhalation, ingestion, or skin absorption. Wear appropriate gloves and goggles and use in a chemical fume hood.

Acrylamide (unpolymerized) is a potent neurotoxin and is absorbed through the skin (the effects are cumulative). Avoid breathing the dust. Wear appropriate gloves and a face mask when weighing powdered acrylamide and methylene-bisacrylamide. Use in a chemical fume hood. Polyacrylamide is considered to be nontoxic, but it should be handled with care because it might contain small quantities of unpolymerized acrylamide.

$AgNO_3$, *see* **Silver nitrate**

Ammonium acetate, $H_3CCOONH_4$, may be harmful by inhalation, ingestion, or skin absorption. Wear appropriate gloves and safety glasses. Use in a chemical fume hood.

Ammonium hydroxide, NH_4OH, is a solution of ammonia in water. It is caustic and should be handled with great care. As ammonia vapors escape from the solution, they are corrosive, toxic, and can be explosive. Use only with mechanical exhaust. Wear appropriate gloves and use only in a chemical fume hood.

Ammonium persulfate, $(NH_4)_2S_2O_8$, is extremely destructive to tissue of the mucous membranes and upper respiratory tract, eyes, and skin. Inhalation may be fatal. Wear appropriate gloves, safety glasses, and protective clothing. Use only in a chemical fume hood. Wash thoroughly after handling.

Ammonium phosphate, *see* **Phosphoric acid**

Ammonium sulfate, $(NH_4)_2SO_4$, may be harmful by inhalation, ingestion, or skin absorption. Wear appropriate gloves and safety glasses.

Ampicillin may be harmful by inhalation, ingestion, or skin absorption. Wear appropriate gloves and safety glasses. Use in a chemical fume hood.

Animal treatment: Procedures for the humane treatment of animals must be observed at all times. Consult your local animal facility for guidelines.

Aprotinin may be harmful by inhalation, ingestion, or skin absorption. It may also cause allergic reactions. Exposure may cause gastrointestinal effects, muscle pain, blood pressure changes, or bronchospasm. Wear appropriate gloves and safety glasses. Do not breathe the dust. Use only in a chemical fume hood.

2,2′-Azino-bis(3-ethylbenzthiazoline)-6-sulfonic acid (ABTS) causes irritation to the skin, eyes, and respiratory system. It may be harmful by inhalation, ingestion, or skin absorption. Wear appropriate gloves and safety glasses. Do not breathe the dust.

Bacterial strains (shipping of): The Department of Health, Education, and Welfare (HEW) has classified various bacteria into different categories with regard to shipping requirements (see Sanderson and Zeigler, *Methods Enzymol.* 204: 248–264 [1991]). Nonpathogenic strains of *E. coli* (such as K12) and *B. subtilis* are in Class 1 and are considered to present no or minimal hazard under normal shipping conditions. However, *Salmonella*, *Haemophilus*, and certain strains of *Streptomyces* and *Pseudomonas* are in Class 2. Class 2 bacteria are "Agents of ordinary potential hazard: agents which produce disease of varying degrees of severity...but which are contained by ordinary laboratory techniques." For detailed regulations regarding the packaging and shipping of Class 2 strains, see Sanderson and Ziegler (*Methods Enzymol.* 204: 248-264 [1991]) or the instruction brochure by Alexander and Brandon (*Packaging and Shipping of Biological Materials at ATCC* [1986]) available from the American Type Culture Collection (ATCC), Rockville, Maryland.

BCP, *see* **1-Bromo-3-chloropropane**

Benzoic acid is an irritant and may be harmful by inhalation, ingestion, or skin absorption. Wear appropriate gloves and safety glasses. Do not breathe the dust.

Biotin may be harmful by inhalation, ingestion, or skin absorption. Wear appropriate gloves and safety glasses. Use in a chemical fume hood.

Bisacrylamide is a potent neurotoxin and is absorbed through the skin (the effects are cumulative). Avoid breathing the dust. Wear appropriate gloves and a face mask when weighing powdered acrylamide and methylene-bisacrylamide.

Bleach (Sodium hypochlorite), NaOCl, is poisonous, can be explosive, and may react with organic solvents. It may be fatal by inhalation and is also harmful by ingestion and destructive to the skin. Wear appropriate gloves and safety glasses. If possible, use in a chemical fume hood to minimize exposure and odor.

Blood (human) and blood products and Epstein-Barr virus. Human blood, blood products, and tissues may contain occult infectious materials such as hepatitis B virus and HIV that may result in laboratory-acquired infections. Investigators working with EBV-transformed lymphoblast cell lines are also at risk of EBV infection. Any human blood, blood products, or tissues should be considered a biohazard and should be handled accordingly until proved otherwise. Wear disposable appropriate gloves, use mechanical pipetting devices, work in a biological safety cabinet, protect against the possibility of aerosol generation, and disinfect all waste materials before disposal. Autoclave contaminated plasticware before disposal; autoclave contaminated liquids or treat with bleach (10% [v/v] final concentration) for at least 30 minutes before disposal. Consult the local institutional safety officer for specific handling and disposal procedures.

Boric acid, H_3BO_3, may be harmful by inhalation, ingestion, or skin absorption. Wear appropriate gloves and goggles.

Bradford dye contains phosphoric acid and methanol. It is corrosive and toxic. Wear appropriate gloves and safety glasses.

1-Bromo-3-chloropropane (BCP) has a narcotic effect and may be harmful by inhalation, ingestion, or skin absorption. Wear appropriate gloves and safety glasses. Do not breathe the vapor.

Bromophenol blue may be harmful by inhalation, ingestion, or skin absorption. Wear appropriate gloves and safety glasses. Use in a chemical fume hood.

***n*-Butyl acetate** is an irritant and may be harmful by inhalation, ingestion, or skin absorption. It poses a risk of serious damage to the eyes. Wear appropriate gloves and safety goggles and use only in a chemical fume hood. Keep away from heat, sparks, and open flame.

$CaCl_2$, *see* **Calcium chloride**

Calcium chloride, $CaCl_2$, may be harmful by inhalation, ingestion, or skin absorption. Wear appropriate gloves and safety glasses. Use in a chemical fume hood.

Calcium sulfate, *see* **Sulfuric acid**

Cesium chloride, CsCl, may be harmful by inhalation, ingestion, or skin absorption. Wear appropriate gloves and safety glasses.

$CHCl_3$, *see* **Chloroform**

CH_3CH_2OH, *see* **Ethanol**

$C_6H_5CH_2SO_2F$, *see* **Phenylmethylsulfonyl fluoride**

$C_7H_7FO_2S$, *see* **Phenylmethylsulfonyl fluoride**

Chloramphenicol may be harmful by inhalation, ingestion, or skin absorption and is a carcinogen. Wear appropriate gloves and safety glasses. Use in a chemical fume hood.

Chloroform, $CHCl_3$, is irritating to the skin, eyes, mucous membranes, and respiratory tract. It is a carcinogen and may damage the liver and kidneys. It is also volatile. Avoid breathing the vapors. Wear appropriate gloves and safety glasses and always use in a chemical fume hood.

Citric acid is an irritant and may be harmful by inhalation, ingestion, or skin absorption. It poses a risk of serious damage to the eyes. Wear appropriate gloves and safety goggles. Do not breathe the dust.

Cobalt chloride, $CoCl_2$, may be harmful by inhalation, ingestion, or skin absorption. Wear appropriate gloves and safety glasses.

$CoCl_2$, *see* **Cobalt chloride**

Copper sulfate, $CuSO_4$, may be harmful by inhalation or ingestion. Wear appropriate gloves and safety glasses.

CsCl, *see* **Cesium chloride**

$CuSO_4$, *see* **Copper sulfate**

DAB, *see* **3,3´-Diaminobenzidine tetrahydrochloride**

3,3´-Diaminobenzidine tetrahydrochloride (**DAB**) is a carcinogen. Handle with extreme care. Avoid breathing vapors. Wear appropriate gloves and safety glasses and use in a chemical fume hood.

Diethanolamine may be harmful by inhalation, ingestion, or skin absorption. Wear appropriate gloves and safety glasses.

***N,N*-Dimethylformamide (DMF), HCON(CH_3)$_2$,** is irritating to the eyes, skin, and mucous membranes. It can exert its toxic effects through inhalation, ingestion, or skin absorption. Chronic inhalation can cause liver and kidney damage. Wear appropriate gloves and safety glasses. Use in a chemical fume hood.

Dimethyl pimelimidate (DMP) is irritating to the eyes, skin, mucous membranes, and upper respiratory tract. It can exert harmful effects by inhalation, ingestion, or skin absorption. Avoid breathing the vapors. Wear appropriate gloves, face mask, and safety glasses and do not inhale.

Dimethyl sulfoxide (DMSO) may be harmful by inhalation or skin absorption. Wear appropriate gloves and safety glasses. Use in a chemical fume hood. DMSO is also combustible. Store in a tightly closed container. Keep away from heat, sparks, and open flame.

Dithiothreitol (DTT) is a strong reducing agent that emits a foul odor. It may be harmful by inhalation, ingestion, or skin absorption. When working with the solid form or highly concentrated stocks, wear appropriate gloves and safety glasses and use in a chemical fume hood

DMF, *see* **N,N-dimethylformamide**

DMP, *see* **Dimethyl pimelimidate**

DMSO, *see* **Dimethyl sulfoxide**

DTT, *see* **Dithiothreitol**

Escherichia coli (E. coli), *see* **Bacterial strains**

Ethanol, CH_3CH_2OH, may be harmful by inhalation, ingestion, or skin absorption. Wear appropriate gloves and safety glasses.

Ethanolamine, $HOCH_2CH_2NH_2$, is toxic and harmful by inhalation, ingestion, or skin absorption. Handle with care and avoid any contact with the skin. Wear appropriate gloves and goggles and use in a chemical fume hood. Ethanolamine is highly corrosive and reacts violently with acids.

Ethidium bromide is a powerful mutagen and is moderately toxic. Consult the local institutional safety officer for specific handling and disposal procedures. Avoid breathing the dust. Wear appropriate gloves when working with solutions that contain this dye.

***N*-Ethylmaleimide (NEM)** may be harmful by inhalation, ingestion, or skin absorption. Wear appropriate gloves and safety glasses. Always use in a chemical fume hood.

FITC, *see* **Fluorescein isothiocyanate**

Fluorescein isothiocyanate, FITC, may be harmful by inhalation, ingestion, or skin absorption. Wear appropriate gloves and safety glasses.

Formaldehyde, HCOH, is highly toxic and volatile. It is also a carcinogen. It is readily absorbed through the skin and is irritating or destructive to the skin, eyes, mucous membranes, and upper respiratory tract. Avoid breathing the vapors. Wear appropriate gloves and safety glasses. Always use in a chemical fume hood. Keep away from heat, sparks, and open flame.

Formalin is a solution of formaldehyde in water. *See* **Formaldehyde**

Formamide is teratogenic. The vapor is irritating to the eyes, skin, mucous membranes, and upper respiratory tract. It may be harmful by inhalation, ingestion, or skin absorption. Wear appropriate gloves and safety glasses. Always use in a chemical fume hood when working with concentrated solutions of formamide. Keep working solutions covered as much as possible.

Glassware, pressurized, must be used with extreme caution. Autoclave and cool sealed bottles in metal containers, pressurize bottles behind Plexiglas shields, and encase 20-liter bottles in wire mesh. Handle glassware under vacuum, such as in desiccators, vacuum traps, drying equipment, or a reactor for working under argon atmosphere, with appropriate caution. Always wear safety glasses.

Glutaraldehyde is toxic. It is readily absorbed through the skin and is irritating or destructive to the skin, eyes, mucous membranes, and upper respiratory tract. Wear appropriate gloves and safety glasses. Always use in a chemical fume hood.

Glycine may be harmful by inhalation, ingestion, or skin absorption. Wear gloves and safety glasses. Avoid breathing the dust.

Guanidine thiocyanate may be harmful by inhalation, ingestion, or skin absorption. Wear appropriate gloves and safety glasses.

H_3BO_3, *see* **Boric acid**

$H_3CCOONH_4$, *see* **Ammonium acetate**

HCl, *see* **Hydrochloric acid**

HCOH, *see* **Formaldehyde**

H_3COH, *see* **Methanol**

H_2O_2, *see* **Hydrogen peroxide**

$HOCH_2CH_2NH_2$, *see* **Ethanolamine**

H_3PO_4, *see* **Phosphoric acid**

H_2SO_4, *see* **Sulfuric acid**

Hydrochloric acid, HCl, is volatile and may be fatal if inhaled, ingested, or absorbed through the skin. It is extremely destructive to mucous membranes, upper respiratory tract, eyes, and skin. Wear appropriate gloves and safety glasses and use with great care in a chemical fume hood. Wear goggles when handling large quantities.

Hydrogen peroxide, H_2O_2, is corrosive, toxic, and extremely damaging to the skin. It may be harmful by inhalation, ingestion, or skin absorption. Wear appropriate gloves and safety glasses and use only in a chemical fume hood.

Imidazole is corrosive and may be harmful by inhalation, ingestion, or skin absorption. Wear appropriate gloves and safety glasses. Use in a chemical fume hood.

IPTG, *see* **Isopropyl-β-D-thiogalactopyranoside**

Isoamyl alcohol may be harmful by inhalation, ingestion, or skin absorption and presents a risk of serious damage to the eyes. Wear appropriate gloves and safety goggles. Keep away from heat, sparks, and open flame.

Isopropanol is irritating and may be harmful by inhalation, ingestion, or skin absorption. Wear appropriate gloves and safety glasses. Do not breathe the vapor. Keep away from heat, sparks, and open flame.

Isopropyl-β-D-thiogalactopyranoside (IPTG) may be harmful by inhalation, ingestion, or skin absorption. Wear appropriate gloves and safety glasses.

$KH_2PO_4/K_2HPO_4/K_3PO_4$, *see* **Potassium phosphate**

KOH, *see* **Potassium hydroxide**

Leupeptin (or its **hemisulfate**) may be harmful by inhalation, ingestion, or skin absorption. Wear appropriate gloves and safety glasses. Use in a chemical fume hood.

Liquid nitrogen can cause severe damage due to extreme temperature. Handle frozen samples with extreme caution. Do not breathe the vapors. Seepage of liquid nitrogen into frozen vials can result in an exploding tube upon removal from liquid nitrogen. Use vials with O-rings when possible. Wear cryo-mitts and a face mask.

LiCl, *see* **Lithium chloride**

Lithium chloride, LiCl, is an irritant to the eyes, skin, mucous membranes, and upper respiratory tract. It may be harmful by inhalation, ingestion, or skin absorption. Wear appropriate gloves, safety goggles, and use in a chemical fume hood. Do not breathe the dust.

Magnesium chloride, $MgCl_2$, may be harmful by inhalation, ingestion, or skin absorption. Wear appropriate gloves and safety glasses, and use in a chemical fume hood.

Magnesium sulfate, $MgSO_4$, may be harmful by inhalation, ingestion, or skin absorption. Wear appropriate gloves and safety glasses.

Manganese chloride, $MnCl_2$, may be harmful by inhalation, ingestion, or skin absorption. Wear appropriate gloves and safety glasses. Use in a chemical fume hood.

Manganese sulfate may be harmful by inhalation, ingestion, or skin absorbtion. Wear appropriate gloves and safety glasses.

2-Mercaptoethylamine may be harmful by inhalation, ingestion, or skin absorption. Wear appropriate gloves and safety glasses.

Methanol, H_3COH, is poisonous and can cause blindness. It may be harmful by inhalation, ingestion, or skin absorption. Adequate ventilation is necessary to limit exposure to vapors. Avoid inhaling these vapors. Wear appropriate gloves and goggles. Use only in a chemical fume hood.

$MgCl_2$, *see* **Magnesium chloride**

$MgSO_4$, *see* **Magnesium sulfate**

Microtome blades are extremely sharp! Use care when sectioning. If unfamiliar with the microtome, have someone demonstrate its use.

3-(*N*-Morpholino)-propanesulfonic acid (MOPS) may be harmful by inhalation, ingestion, or skin absorption. It is irritating to mucous membranes and upper respiratory tract. Wear appropriate gloves and safety glasses and use in a chemical fume hood.

$MnCl_2$, *see* **Manganese chloride**

MOPS, *see* **3-(*N*-Morpholino)-propanesulfonic acid**

NaH_2PO_4/Na_2HPO_4/Na_3PO_4, *see* **Sodium phosphate**

NaI, *see* **Sodium iodide**

NaN_3, *see* **Sodium azide**

NaOCl, *see* **Bleach**

NaOH, *see* **Sodium hydroxide**

NEM, *see* ***N*-Ethylmaleimide**

NH_4OH, *see* **Ammonium hydroxide**

$(NH_4)_2S_2O_8$, *see* **Ammonium persulfate**

$(NH_4)_2SO_4$, *see* **Ammonium sulfate**

p-Nitrophenyl phosphate (**PNPP**) is toxic and may be harmful by inhalation, ingestion, or skin absorption. Wear appropriate gloves and safety glasses and use only in a chemical fume hood.

^{33}P, *see* **Radioactive substances**

PEG, *see* **Polyethylene glycol**

Phenol is extremely toxic, highly corrosive, and can cause severe burns. It may be harmful by inhalation, ingestion, or skin absorption. Wear appropriate gloves, goggles, and protective clothing. Always use in a chemical fume hood. Rinse any areas of skin that come in contact with phenol with a large volume of water and wash with soap and water; do not use ethanol!

Phenylmethylsulfonyl fluoride (**PMSF**), $C_7H_7FO_2S$ or $C_6H_5CH_2SO_2F$, is a highly toxic cholinesterase inhibitor. It is extremely destructive to the mucous membranes of the respiratory tract, eyes, and skin. It may be fatal by inhalation, ingestion, or skin absorption. Wear appropriate gloves and safety glasses and always use in a chemical fume hood. In case of contact, immediately flush eyes or skin with copious amounts of water and discard contaminated clothing.

Phosphoric acid, H_3PO_4, is highly corrosive and may be harmful by inhalation, ingestion, or skin absorption. Wear appropriate gloves and safety glasses.

Picric acid powder is caustic and potentially explosive if it is dissolved and then allowed to dry out. Care must be taken to ensure that stored solutions do not dry out. Handle all concentrated acids with great care. It is also highly toxic and may be harmful by inhalation, ingestion, or skin absorption. Wear appropriate gloves and goggles.

PMSF, *see* **Phenylmethylsulfonyl fluoride**

PNPP, *see* **p-Nitrophenyl phosphate**

Polyacrylamide is considered to be nontoxic, but it should be treated with care because it may contain small quantities of unpolymerized material (see **Acrylamide**).

Polyethylene glycol (PEG) may be harmful by inhalation, ingestion, or skin absorption. Avoid inhalation of powder. Wear appropriate gloves and safety glasses.

Potassium hydroxide, KOH and KOH/methanol, can be highly toxic. It may be harmful by inhalation, ingestion, or skin absorption. Solutions are caustic and should be handled with great care. Wear appropriate gloves.

Potassium phosphate, $KH_2PO_4/K_2HPO_4/K_3PO_4$, may be harmful by inhalation, ingestion, or skin absorption. Wear appropriate gloves and safety glasses. Do not breathe the dust. *$K_2HPO_4 \cdot 3H_2O$ is dibasic and KH_2PO_4 is monobasic.*

Radioactive substances: While planning an experiment that involves the use of radioactivity, include the physico-chemical properties of the isotope (half-life, emission type, and energy), the chemical form of the radioactivity, its radioactive concentration (specific activity), total amount, and its chemical concentration. Order and use only as much as really needed. Always wear appropriate gloves, lab coat, and safety goggles when handling radioactive material. **X-rays** and **gamma rays** are electromagnetic waves of very short wavelengths either generated by technical devices or emitted by radioactive materials. They may be emitted isotropic from the source or may be focused into a beam. Their potential dangers depend on the time period of exposure, the intensity experienced, and the wavelengths used. Be aware that appropriate shielding is usually of lead or other similar material. The thickness of the shielding is determined by the energy(s) of the X-rays or gamma rays. Consult the local safety office for further guidance in the appropriate use and disposal of radioactive materials. Always monitor thoroughly after using radioisotopes. A convenient calculator to perform routine radioactivity calculations can be found at:
http://www.graphpad.com/calculators/radcalc.cfm

SDS, *see* **Sodium dodecyl sulfate**

Silver nitrate, $AgNO_3$, is a strong oxidizing agent and should be handled with care. It may be harmful by inhalation, ingestion, or skin absorption. Avoid contact with skin. Wear appropriate gloves and safety glasses. It can cause explosions upon contact with other materials.

Sodium azide, NaN_3, is highly poisonous. It blocks the cytochrome electron transport system. Solutions containing sodium azide should be clearly marked. It may be harmful by inhalation, ingestion, or skin absorption. Wear appropriate gloves and safety goggles and handle with great care.

Sodium dodecyl sulfate (**SDS**) is toxic, an irritant, and poses a risk of severe damage to the eyes. It may be harmful by inhalation, ingestion, or skin absorption. Wear appropriate gloves and safety goggles. Do not breathe the dust.

Sodium hydroxide, NaOH, and solutions containing NaOH are highly toxic and caustic and should be handled with great care. Wear appropriate gloves and a face mask. All other concentrated bases should be handled in a similar manner.

Sodium iodide, NaI, may be harmful by inhalation, ingestion, or skin absorption. Wear appropriate gloves and safety glasses and use in a chemical fume hood.

Sodium molybdate dihydrate may be harmful by inhalation, ingestion, or skin absorption. Wear gloves and safety glasses and use in a chemical fume hood.

Sodium phosphate, $NaH_2PO_4/Na_2HPO_4/Na_3PO_4$, is an irritant to the eyes and skin. It may be harmful by inhalation, ingestion, or skin absorption. Wear appropriate gloves and safety goggles. Do not breathe the dust.

Sulfuric acid, H_2SO_4, is highly toxic and extremely destructive to tissue of the mucous membranes and upper respiratory tract, eyes, and skin. It causes burns, and contact with other materials (e.g., paper) may cause fire. Wear appropriate gloves, safety glasses, and lab coat and use in a chemical fume hood.

TEMED, *see* ***N,N,N′,N′*-Tetramethylethylenediamine**

Tetracycline may be harmful by inhalation, ingestion, or skin absorption. Wear appropriate gloves and safety glasses and use in a chemical fume hood.

***N,N,N′,N′*-Tetramethylethylenediamine** (**TEMED**) is extremely destructive to tissue of the mucous membranes and upper respiratory tract, eyes, and skin. Inhalation may be fatal. Prolonged contact can cause severe irritation or burns. Wear appropriate gloves, safety glasses, and other protective clothing and use only in a chemical fume hood. Wash thoroughly after handling. *Flammable:* Vapor may travel a considerable distance to source of ignition and flash back. Keep away from heat, sparks, and open flame.

2,2,2-Tribromoethanol may be harmful by inhalation, ingestion, or skin absorption. The vapor is also irritating to the eyes, mucous membranes, and upper respiratory tract. Wear appropriate gloves and safety glasses.

Urea may be harmful by inhalation, ingestion, or skin absorption. Wear appropriate gloves and safety glasses.

UV light and/or **UV radiation** is dangerous and can damage the retina of the eyes. Never look at an unshielded UV light source with naked eyes. View only through a filter or safety glasses that absorb harmful wavelengths. UV radiation is also mutagenic and carcinogenic. To minimize exposure, make sure that the UV light source is adequately shielded. Wear protective appropriate gloves when holding materials under the UV light source.

X-gal, *see* **5-Bromo-4-chloro-3-indolyl-β-D-galactopyranoside** (**BCIG**)

Xylene is flammable and may be narcotic at high concentrations. It may be harmful by inhalation, ingestion, or skin absorption. Wear appropriate gloves and safety glasses and use only in a chemical fume hood. Keep away from heat, sparks, and open flame.

Xylene cyanol, *see* **Xylene**

Zinc chloride is corrosive and poses a possible risk to the unborn child. It may be harmful by inhalation, ingestion, or skin absorption. Wear appropriate gloves and safety glasses. Do not breathe the dust.

APPENDIX

5 Suppliers

With the exception of those suppliers listed in the text with their addresses, all suppliers mentioned in this manual can be found in the BioSupplyNet Source Book and on the web site at:

http://www.biosupplynet.com

If a copy of the BioSupplyNet Source Book was not included with this manual, a free copy can be ordered by any of the following methods:

- Complete the Free Source Book Request Form found at the web site at:
  http://www.biosupplynet.com
- E-mail a request to info@biosupplynet.com
- Fax a request to 1-516-349-5598

APPENDIX

6 Trademarks

The following trademarks and registered trademarks are accurate to the best of our knowledge at the time of printing. Please consult individual manufacturers and other resources for specific information.

ABTS	Boehringer Mannheim GmbH
Acrodisc	Pall Gelman
AmpliTaq	Roche Molecular Systems, Inc.
ART	Molecular Bio-Products, Inc.
BioFlo	New Brunswick Scientific Co., Inc.
Biopur	Eppendorf-Netheler-Hinz GmbH
Bio-Shield	Baxter International Inc.
Centricon	Millipore Corp.
Centriprep	Millipore Corp.
CHROMA SPIN	CLONTECH Laboratories, Inc.
Coomassie Brilliant Blue	Imperial Chemical Industries, Ltd.
DH5α	Life Technologies, Inc.
DNA Strider	Commissariat a L'Energie Atomique (France)
Dynabeads	DYNAL, Inc.
EasyComp	Invitrogen Corp.
ECL	Amersham Pharmacia Biotech
EGGstract	Promega Corp.
Elutrap	Schleicher & Schuell, Inc.
Expand	Boehringer Mannheim Corp.
FALCON	Becton Dickinson and Co.
Ficoll-paque	Amersham Pharmacia Biotech
FPLC	Pharmacia LKB Biotechnology AB
GenBank	National Institutes of Health
GeneAmp	Roche Molecular Systems, Inc.
GENECLEAN	Q•BIOgene
Gene Pulser	Bio-Rad Laboratories, Inc.
His·Tag	Novagen Inc.
HiTrap	Amersham Pharmacia Biotech
Immobilon-P	Millipore Corporation

Immulon	Dynatech Laboratories, Inc.
ImmunoPure	Pierce Chemical Co.
Kimwipe	Kimberly-Clark Corp.
Linbro	ICN Biomedicals, Inc.
LKB	Amersham Pharmacia Biotech
Luer-Lok	Becton Dickinson and Co.
MACS	Miltenyi Biotec
MacVector	Oxford Molecular
MASS	Life Technologies, Inc.
MaxiSorp	Nalge Nunc International
Metofane	Mallinckrodt-Baker Laboratory Chemicals
MicroAmp	The Perkin-Elmer Corp.
Micropure	Millipore Corp.
Milli-Q	Millipore Corp.
MPC	DYNAL, Inc.
NALGENE	Nalge Co.
NeutrAvidin	Pierce Chemical Co.
Ni-NTA	QIAGEN, Inc.
N-Z-amine	Quest International
Parafilm	American National Can Co.
PCR	Hoffman-LaRoche
Pellicon	Millipore Corp.
Ph.D.	New England Biolabs, Inc.
PROTEAN	Bio-Rad Laboratories, Inc.
Pyrex	Corning, Inc.
QIAEX	QIAGEN, Inc.
QIAfilter	QIAGEN, Inc.
QIAprep	QIAGEN, Inc.
QIAquick	QIAGEN, Inc.
RAIN-X	UNELKO Corp.
Sepharose	Amersham Pharmacia Biotech
Sequenase	Amersham Pharmacia Biotech
Slide-a-Lyzer	Pierce Chemical Co.
Spectra/Por	Spectrum Laboratories, Inc.
SpeedVac	Savant Instruments, Inc.
StrataClean	Stratagene
Stripette	Corning Costar Corp.
SUPERSCRIPT	Life Technologies, Inc.
Taxol	Bristol-Meyers Squibb Co.
Thermo Sequenase	Amersham Pharmacia Biotech
Tissumizer	Tekmar Company
TiterMax	CytRx Corp.
Trans-Blot	Bio-Rad Laboratories, Inc.
TRI Reagent	Molecular Research Center Inc.
Triton X-100	Union Carbide Chemicals and Plastics Technology Corp.

Tween	ICI Americas Inc.
Ultrafree	Millipore Corp.
Vacutainer	Becton Dickinson and Co.
V BASE	MRC Centre for Protein Engineering
VECTASTAIN	Vector Laboratories Inc.
Xpress	Invitrogen Corp.
Zeocin	CAYLA

Index